THERMODYNAMICS OF CRYSTALS

THERMODYNAMICS OF CRYSTALS

Duane C. Wallace

École Polytechnique Fédérale de Lausanne
and
CSIRO National Standards Laboratory, Sydney

John Wiley & Sons, Inc.

New York · London · Sydney · Toronto

Library of Congress Catalog Card Number: 71-161495

ISBN 0-471-91855-5

Printed in the United States of America.

10 9 8 7 6 5 4 3 2 1

To Dan, Diane, and Amy

PREFACE

This book is intended to provide the basic theory for a definite area of solid state physics, namely the equilibrium thermodynamic properties of perfect crystals. In outline, the theory begins by describing the crystal as a collection of interacting ions and electrons, then finds the quantum energy levels for such a model crystal, and finally proceeds via statistical mechanics to describe the observable properties of the crystal. Even within such a limited scope, there are still a number of topics which are not treated here. In particular the theory is formulated for any type of crystal in the presence of any externally applied homogeneous stress; interactions with externally applied electric and magnetic fields are not treated, however, and the theory does not cover piezoelectric and magnetostrictive effects, magnetic contributions to the thermodynamic functions, and so on. The material which is covered is self-contained; i.e., the necessary derivations are carried out from the level of the basic definitions, and only a modest background in quantum mechanics and solid state physics is required.

In my opinion this book should prove useful and timely both for theoretical developments and for applications to material problems. Concerning the theory, it is true that the subject of lattice dynamics and thermodynamics of crystals contains many old and well-worn paths; nevertheless, these paths are liberally strewn with stumbling blocks. For example, it is only in the past few years that the thermoelastic theory of stressed crystals has been formulated completely enough to clarify the differences between elastic coefficients which govern stress–strain relations and those which govern elastic wave propagation. This clarification removes a long-standing difficulty in the lattice-dynamical definition of elastic constants of stressed crystals. In addition, there seems to be a periodic recurrence of the question of whether or not the lattice-dynamical methods of long waves and homogeneous deformation are equivalent; in the present formulation these two methods are equivalent, subject only to the approximation that nonadiabatic effects are put into a perturbation. It has also been argued in the past that the infinite lattice model requires a special equilibrium condition, namely that the externally applied stresses must vanish. This condition, which severely restricts lattice-dynamics theory, is eliminated here. Another problem area has been in the methods by which theory and experiment are compared. When the theoretical expression for a given function is too complicated, comparison of calculated and measured values of that function yields little information about the accuracy of the theory. The approach here is to use thermodynamic relations to separate experimental quantities into their theoretically simple contributions; then comparison of theory and experiment for these contributions provides detailed information about the accuracy of the theory. Many more theoretical stumbling blocks are described, and hopefully removed, in the text.

An obvious application of the present theory is in calculating or estimating the thermodynamic properties of materials at stresses and temperatures beyond where they have been measured. The heat capacity, thermal expansion, and compressibility are important properties for engineering design work, especially for applications at high temperatures and pressures. In many cases it is desirable to know the anisotropic thermal stresses which develop in composite materials at high temperatures. While I have not tried to develop approximate or empirical formulas for such effects, I have tried to develop the theory with sufficient accuracy to identify the major contributions to thermodynamic functions, and with sufficient generality to serve as a basis for estimates of the important material properties.

By definition, the audience for which this book is intended includes anyone who is interested in the theory of lattice dynamics and thermodynamics of crystals. For a graduate student doing thesis work in this area, the book should provide a useful reference. In addition, the book could be used as a text for a short special topics course in solid state physics; for this purpose Chapters 2–6 would perhaps be most useful. The materials research scientist who is interested in thermodynamic properties may find Chapters 1, 4, and 7 most useful. In any case, the prospective reader should understand that my theoretical procedures are somewhat pedestrian, and are designed to produce practical, usable results, rather than fancy equations suitable for framing.

Starting with the atomic description of a crystal, there are three main problems in the process of deriving and calculating thermodynamic functions. These three problems constitute the central part of my book plan, and are as follows.

1. Solution of the lattice-dynamics problem. This problem is formulated in Chapter 3, in terms of a set of weakly interacting phonons. The interactions arise from cubic and quartic terms in the expansion of the total crystal

potential in powers of the displacements of the ions from equilibrium; these interactions are treated as a perturbation, correct to the leading order in which they contribute to the total crystal energy levels. This perturbation theory is quite accurate for the thermodynamic properties of nearly all crystals, for temperatures up to, say, three times the Debye temperature. In order to go beyond the leading-order perturbation theory, it would seem appropriate to use a self-consistent lattice-dynamics procedure, and an introduction to such theory is also provided here.

2. Calculation of the total crystal potential. It is now time to proceed beyond the level of force constant models and simple central potential approximations. The band-structure theory of lattice dynamics is formulated in Chapter 5, in terms of the adiabatic approximation and the one-electron approximation. Nonadiabatic effects are neglected for nonmetals, and are described as electron–phonon interactions and treated as a perturbation for metals. The dielectric function plays a central role in the band-structure theory, and in order to formulate simple but physically meaningful models, my recommendation is to make approximations in the dielectric function rather than in the direct interatomic potentials.

3. Evaluation of thermodynamic functions. Procedures for numerical evaluation of the Brillouin zone sums involved in the thermodynamic functions are described in Appendix 1. Such sums should be evaluated by direct summation over the phonon wave vectors, with the use of weighting factors to avoid multiple counting of wave vectors which are shared by more than one zone, and the transformation to integrals involving the phonon distribution function should be avoided.

With the completion of this volume, I should like to suggest several research problems which will provide valuable extensions of our present knowledge. Experimental observation of the isotope effect in the low-temperature elastic constants and Debye temperature will allow the separation of harmonic and anharmonic lattice-dynamical contributions (see Section 18). It would also be useful to measure directly the stress- or pressure-dependence of the phonon frequencies, by means of inelastic neutron-scattering experiments on stressed crystals (see the model calculations of Section 34). A higher-order anharmonic contribution to the entropy has been identified in the analysis of Section 31, and more extensive analyses will establish more clearly the limitations of the leading-order anharmonic perturbation theory. For noncubic crystals, more accurate experimental data are needed, as well as the corresponding data analyses. Lattice-dynamics calculations based on the band-structure theory (Chapter 5) will be most useful. It would also be of interest to study the electronic and electron–phonon interaction contributions to the free energy for different electronic models (see Section 25). The description of transport properties in terms of renormalized phonons and electrons (Sections 13 and 25) may prove useful, and there are good possibilities for formulating self-consistent lattice dynamics in terms of the statistical perturbation theory (see Section 14).

It is a pleasure to express my appreciation to two skilled and dedicated typists, Mrs. Karen Shane and Mme. Inès Devrient. I am happy to thank Mr. Stanley Sleeter for assistance in the tabulation and treatment of experimental data, and M. Bernard Fournier for preparing the ink drawings. My colleagues have generously shared their insights and understandings regarding many details; I particularly wish to thank Nelson S. Gillis and Alfred C. Switendick. Finally, I am deeply grateful to the Président of the École Polytechnique Fédérale de Lausanne, Professeur M. Cosandey, and also to Professeurs Philippe Choquard and Antonio Quattropani of the Departement de Physique, for supporting my work on this manuscript.

Lausanne, Switzerland DUANE C. WALLACE
December 1970

CONTENTS

NOTATION NOTES

VECTORS, MATRICES, AND OPERATORS

$\hat{\mathbf{k}} = \mathbf{k}/|\mathbf{k}|$, a dimensionless unit vector
$[M_{\alpha\beta}]$ = matrix **M** whose elements are $M_{\alpha\beta}$
$[A,B] = AB - BA$, the commutator
$\{A,B\} = AB + BA$, the anticommutator
Hermitian: $A^\dagger = A$
Unitary: $A^\dagger = A^{-1}$
Orthogonal: $\tilde{A} = A^{-1}$

SYMBOLS

$\sim$ *means* is of order
>0 *means* greater than zero for a number
>0 *means* positive definite for a quadratic form
$+ \cdots$ *means* plus terms which are of higher order (in some sense) than those terms which are written explicitly
$o(x)$ *means* of order x
$\langle f \rangle$ *means* an average of f (e.g., a Brillouin zone average, a statistical average, etc.)
symmetric in i, j *means* invariant with respect to interchange of i and j

SUMMATION CONVENTIONS

When summation indices (e.g., λ) are functions of vectors (e.g., **p**):
$\lambda = 0$ *means* $\lambda(\mathbf{p} = 0)$
Σ'_λ *means* sum over all λ except $\lambda = 0$
$\Sigma'_{\lambda\lambda'}$ *means* sum over all λ and λ' except $\lambda = \lambda'$

LATTICE STRUCTURE ABBREVIATIONS

fcc = face-centered cubic
bcc = body-centered cubic
hcp = hexagonal close-packed (with arbitrary c/a in general)

CONSTANTS AND UNITS*

N_0 = number of unit cells in a crystal (not necessarily the same as Avogadro's number N_0)
K = Boltzmann constant
a_0 = Bohr radius
Ry = Rydberg
$\beta(10^{-5}\,^{\circ}\mathrm{K}^{-1})$ *means* β is in units of $10^{-5}\,^{\circ}\mathrm{K}^{-1}$

* See also Table 40

THERMODYNAMICS OF CRYSTALS

1

THERMODYNAMICS

The first chapter provides an outline of those aspects of equilibrium thermodynamics which are pertinent to the study of anisotropic solids. The requirements of thermodynamic equilibrium, invariance, and stability play a fundamental role in the theory; the corresponding mechanical requirements are again of importance in the lattice dynamics theory of the following chapters.

1. PRESSURE–VOLUME VARIABLES

Thermodynamic Functions

We begin with the first and second laws of thermodynamics, written in differential form.

$$đQ = dU + đW, \quad \text{first law.} \tag{1.1}$$

$$đQ = TdS, \quad \text{second law.} \tag{1.2}$$

Here $đQ$ is the quantity of heat which goes into the crystal, $đW$ is the work done by the crystal, T is the absolute temperature, and U and S are, respectively, the internal energy and the entropy of the crystal; $đQ$ and $đW$ are inexact differentials, i.e., they depend on the path taken by the thermodynamic process. For isotropic pressure P, the work done by the crystal is

$$đW = PdV, \tag{1.3}$$

where V is the crystal volume. In this case the combined first and second laws may be written

$$dU = TdS - PdV. \tag{1.4}$$

The Helmholtz free energy F is defined by

$$F = U - TS. \tag{1.5}$$

Differentiating F gives

$$dF = dU - TdS - SdT,$$

or in view of (1.4) for dU,

$$dF = -SdT - PdV. \tag{1.6}$$

Since the differential of U is given in terms of dS and dV, it is convenient (but not necessary) to consider U as a function of S and V; in this case the dependent variables are P and T. Similarly, it is convenient to consider F as a function of T and V, with dependent variables P and S:

$$\begin{aligned} U &= U(V,S); \\ F &= F(V,T). \end{aligned} \tag{1.7}$$

With these dependences, the state functions U and F are obviously invariant with respect to translation and rotation of the crystal.

The dependent variables may be written as partial derivatives of the state functions. For example, from (1.4) for dU, it follows that

$$P = -(\partial U/\partial V)_S, \tag{1.8}$$

$$T = (\partial U/\partial S)_V. \tag{1.9}$$

In terms of the free energy, (1.6) for dF implies

$$P = -(\partial F/\partial V)_T, \tag{1.10}$$

$$S = -(\partial F/\partial T)_V. \tag{1.11}$$

The two equations (1.8) and (1.10) for P are often referred to as equilibrium conditions, since they determine the forces on the crystal, in the form of the pressure, at equilibrium. The quantity $-(\partial U/\partial V)_S = -(\partial F/\partial V)_T$ may be considered the thermal pressure of the crystal, exerted outward and just balanced by the externally applied pressure P. On account of the equilibrium condition, the variables P, V, and T are not all independent; they are related by the equation of state, which may be written formally as

$$f(P,V,T) = 0, \quad \text{at equilibrium.} \tag{1.12}$$

If we are able to calculate the free energy as a function of volume and temperature, then the equation of state may be calculated from the equilibrium condition $P = -(\partial F/\partial V)_T$; this is illustrated in Figure 1.

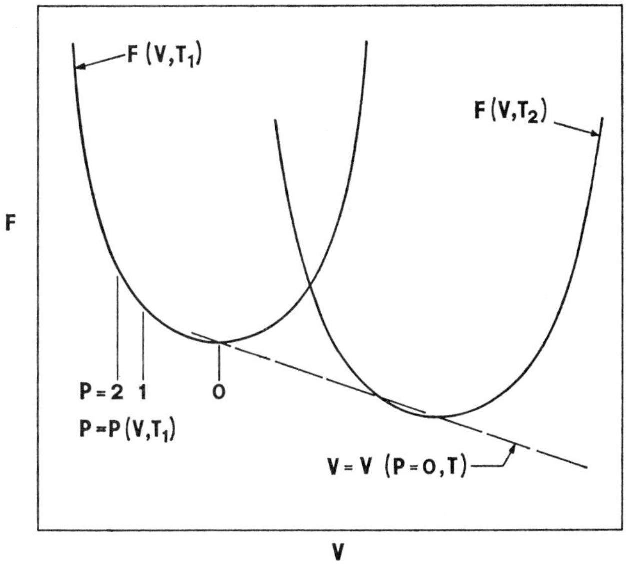

Figure 1. Schematic illustration of the determination of the equation of state from the equilibrium condition. At temperature T_1 for example, $P(V,T_1)$ is obtained from the slopes of $F(V,T_1)$ plotted as a function of V. On the other hand, $V(P = 0,T)$ is the curve passing through the minima of the $F(V)$ *vs* V curves at each temperature, as shown by the dashed line; $V(P \neq 0,T)$ can be constructed in a similar way.

In the case where X, Y, and Z are three thermodynamic variables, only two of which are independent, the manipulation of partial derivatives is aided by the following well-known identities:

$$(\partial X/\partial Y)_Z = 1/(\partial Y/\partial X)_Z; \tag{1.13}$$

$$(\partial X/\partial Y)_Z = -(\partial X/\partial Z)_Y(\partial Z/\partial Y)_X; \tag{1.14}$$

$$(\partial X/\partial Y)_Z = (\partial X/\partial W)_Z(\partial W/\partial Y)_Z, \tag{1.15}$$

where in the last equation X, Y, and Z are considered functions of the arbitrary variable W. The Maxwell equations are also quite useful; these are obtained by requiring second partial derivatives to be independent of the order of differentiation. For example, we must have $(\partial^2U/\partial V\partial S)_{SV} = (\partial^2U/\partial S\partial V)_{VS}$, and comparing this with the equations (1.8) and (1.9) for P and T gives the Maxwell equation

$$(\partial P/\partial S)_V = -(\partial T/\partial V)_S. \tag{1.16}$$

Similarly, from the free energy derivatives it follows that

$$(\partial P/\partial T)_V = (\partial S/\partial V)_T. \tag{1.17}$$

We can derive two more Maxwell equations from these, by making use of the identities (1.13)–(1.15). For example,

$$(\partial T/\partial V)_S = (\partial T/\partial P)_S(\partial P/\partial V)_S = -(\partial P/\partial S)_V,$$

from which

$$(\partial T/\partial P)_S = -(\partial P/\partial S)_V(\partial V/\partial P)_S,$$

and finally

$$(\partial T/\partial P)_S = (\partial V/\partial S)_P. \tag{1.18}$$

In a similar way, the fourth Maxwell equation is found to be

$$(\partial S/\partial P)_T = -(\partial V/\partial T)_P. \tag{1.19}$$

As an example of the use of the Maxwell equations, we show a simple derivation of the energy equation:

$$P = -(\partial F/\partial V)_T = -(\partial U/\partial V)_T + T(\partial S/\partial V)_T,$$

where the second equality follows from the definition (1.5) of F; then with (1.17) this becomes the energy equation

$$(\partial U/\partial V)_T = T(\partial P/\partial T)_V - P. \tag{1.20}$$

We are now ready to define the thermodynamic functions which are of major concern to this book. The heat capacity C measures the variation of the crystal temperature due to heat input:

$$C = đQ/dT = TdS/dT. \tag{1.21}$$

Since $đQ$ is inexact, the heat capacity depends on the path of the process; for example, the heat capacity at constant pressure is

$$C_P = T(\partial S/\partial T)_P, \tag{1.22}$$

and at constant volume is

$$C_V = T(\partial S/\partial T)_V = (\partial U/\partial T)_V, \tag{1.23}$$

where the second equality follows since $T = (\partial U/\partial S)_V$. The entropy may be obtained by integrating (1.22) along a line of constant pressure:

$$S(P,T) = \int_0^T \frac{C_P(P,T')}{T'}\,dT'; \tag{1.24}$$

this equation provides a practical way of determining $S(P = 0,T)$.

The compressibility k measures the variation of the crystal volume with pressure. The adiabatic compressibility is

$$k_S = -V^{-1}(\partial V/\partial P)_S, \tag{1.25}$$

and the isothermal compressibility is

$$k_T = -V^{-1}(\partial V/\partial P)_T. \tag{1.26}$$

The bulk modulus B is just the inverse of the compressibility:

$$B_S = 1/k_S; \qquad B_T = 1/k_T. \tag{1.27}$$

The thermal expansion coefficient β (expansivity) measures the variation of the crystal volume with temperature at constant pressure:

$$\beta = V^{-1}(\partial V/\partial T)_P. \tag{1.28}$$

The quantity βB_T is a particularly simple function:

$$\beta B_T = -\left(\frac{\partial V}{\partial T}\right)_P\left(\frac{\partial P}{\partial V}\right)_T = \left(\frac{\partial P}{\partial T}\right)_V = \left(\frac{\partial S}{\partial V}\right)_T, \tag{1.29}$$

where the last equality is a Maxwell equation. The magnitude and temperature-dependence of β at $P = 0$ is shown for some common crystals in Figure 2.

The Grüneisen parameter γ is defined by

$$\gamma = \frac{V\beta B_T}{C_V} = \frac{V\beta B_S}{C_P}, \tag{1.30}$$

where the second equality follows from (1.43) derived below. The Grüneisen parameter measures the variation of pressure with internal energy at constant volume, since from (1.29) for βB_T and (1.23) for C_V,

$$\gamma = \frac{V}{C_V}\left(\frac{\partial P}{\partial T}\right)_V = V\left(\frac{\partial T}{\partial U}\right)_V\left(\frac{\partial P}{\partial T}\right)_V = V\left(\frac{\partial P}{\partial U}\right)_V. \tag{1.31}$$

This equation provides an interesting possible way to measure γ directly. If, in a region of the crystal, the internal energy can be increased at constant volume, say by the absorption of a pulse of laser light, a pressure wave will be generated and can be detected at a crystal surface.

Since theoretical calculations are most conveniently based on the state functions, it is useful to write the thermodynamic quantities as derivatives of U and F. From the definitions, it is easy to show that

$$C_V = -T(\partial^2 F/\partial T^2)_V, \tag{1.32}$$

$$B_T = V(\partial^2 F/\partial V^2)_T, \tag{1.33}$$

$$B_S = V(\partial^2 U/\partial V^2)_S, \tag{1.34}$$

$$\beta B_T = -(\partial^2 F/\partial V \partial T)_{TV}. \tag{1.35}$$

These quantities are theoretically simple, being just second derivatives of state functions. The quantities β and γ are more complicated, since they

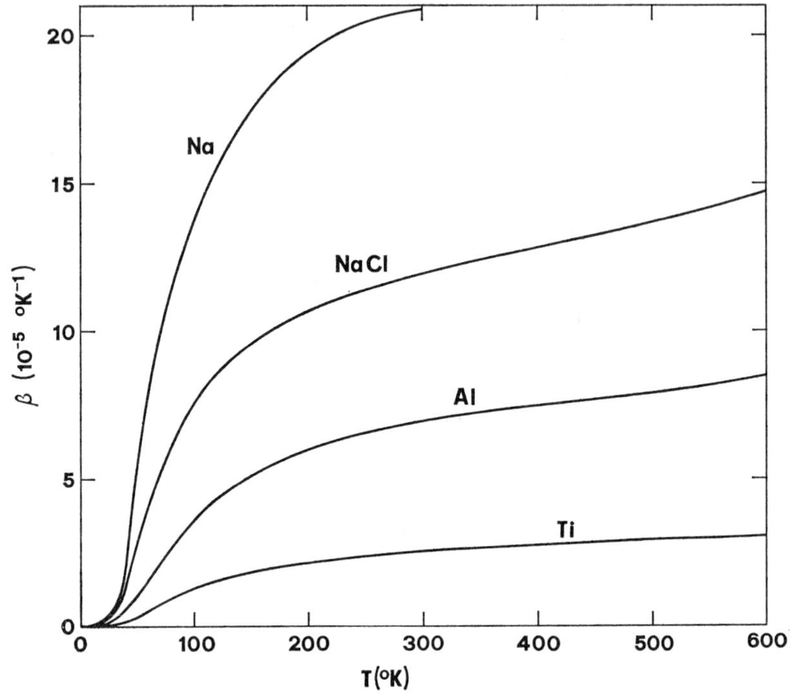

Figure 2. Experimental curves of the thermal expansion coefficient β as a function of T at $P = 0$; β is not necessarily positive, but $\beta \to 0$ as $T \to 0$. Notable examples of negative thermal expansion are Ge and Si, where β is negative at low T, then goes positive and continues to increase as T increases. In addition, for anisotropic crystals the thermal expansion may be quite different along the different crystallographic directions, as for example along the c and a axes for Zn and Cd.

are ratios of such derivatives:

$$\beta = -\frac{(\partial^2 F/\partial V \partial T)_{TV}}{V(\partial^2 F/\partial V^2)_T}, \tag{1.36}$$

$$\gamma = \frac{V(\partial^2 F/\partial V \partial T)_{TV}}{T(\partial^2 F/\partial T^2)_V}. \tag{1.37}$$

Relations Among Thermodynamic Functions

The relations discussed here are of primary importance in transforming experimental data for more convenient comparison with theoretical calculations.

In order to calculate $C_P - C_V$, write dS considering S as a function of T and V, and then as a function of T and P, to get

$$dS = (\partial S/\partial T)_V\, dT + (\partial S/\partial V)_T\, dV = (\partial S/\partial T)_P\, dT + (\partial S/\partial P)_T\, dP. \quad (1.38)$$

Multiplying through by T, dividing by dT, and taking constant pressure ($dP = 0$) gives

$$C_V + T(\partial S/\partial V)_T(\partial V/\partial T)_P = C_P, \quad (1.39)$$

or

$$C_P - C_V = T(\partial S/\partial V)_T(\partial V/\partial T)_P = T(\partial P/\partial T)_V(\partial V/\partial T)_P$$
$$= -T(\partial P/\partial V)_T[(\partial V/\partial T)_P]^2,$$

and finally

$$C_P - C_V = TV\beta^2/k_T. \quad (1.40)$$

In order to calculate $k_T - k_S$, write dV considering V as a function of P and T, and then as a function of P and S; proceeding in a manner similar to the above derivation leads to the result

$$k_T - k_S = TV\beta^2/C_P. \quad (1.41)$$

The difference between the adiabatic and isothermal bulk moduli is

$$B_S - B_T = k_S^{-1} - k_T^{-1} = B_T B_S(k_T - k_S). \quad (1.42)$$

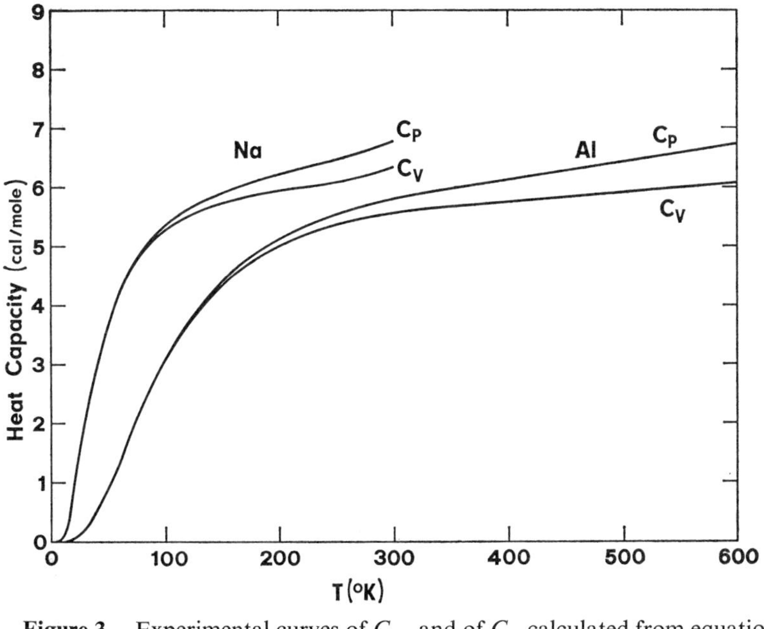

Figure 3. Experimental curves of C_P, and of C_V calculated from equation (1.40), as functions of T at $P = 0$; C_P and $C_V \to 0$ as $T \to 0$, and $C_P > C_V > 0$ for $T \neq 0$.

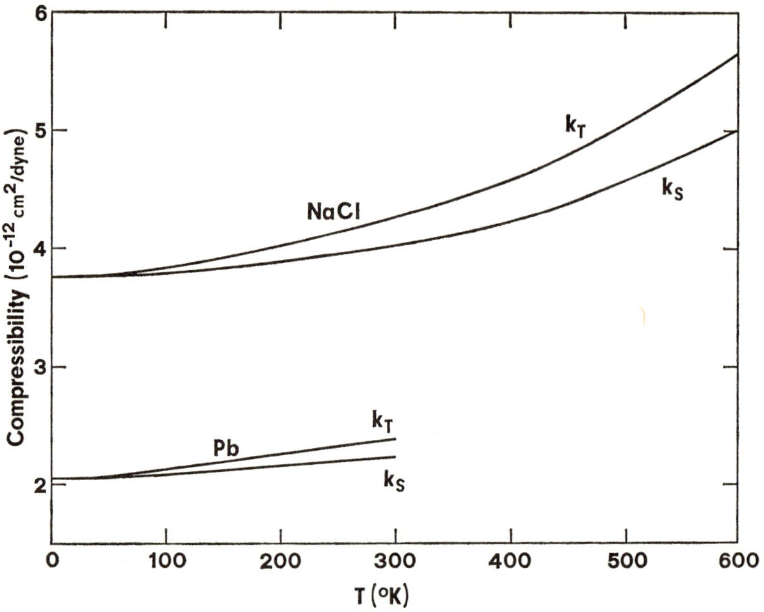

Figure 4. Experimental curves of k_T and k_S as functions of T at $P = 0$. The temperature derivatives of k_T and k_S approach zero as T approaches zero, and $k_T > k_S > 0$ for $T \neq 0$, $k_T = k_S > 0$ at $T = 0$.

The characteristic temperature-dependence of C_P and C_V at zero pressure is shown in Figure 3, and that of k_T and k_S at zero pressure is shown in Figure 4.

The ratio of heat capacities is obtained by equating the two expressions for $TV\beta^2$ which are given by (1.40) and (1.41).

$$TV\beta^2 = C_P(k_T - k_S) = k_T(C_P - C_V).$$

Dividing this by $C_P k_T$ gives

$$\frac{C_P}{C_V} = \frac{k_T}{k_S} = \frac{B_S}{B_T}. \tag{1.43}$$

There are several useful Maxwell type of relations among the thermodynamic functions. For example, from the definition (1.28) of β,

$$(\partial\beta/\partial P)_T = -V^{-2}(\partial V/\partial P)_T(\partial V/\partial T)_P + V^{-1}(\partial^2 V/\partial P\partial T)_{TP}, \tag{1.44}$$

and from the definition (1.26) of k_T,

$$(\partial k_T/\partial T)_P = V^{-2}(\partial V/\partial T)_P(\partial V/\partial P)_T - V^{-1}(\partial^2 V/\partial T\partial P)_{PT}. \tag{1.45}$$

Comparing (1.44) and (1.45) leads to the result

$$(\partial\beta/\partial P)_T = -(\partial k_T/\partial T)_P. \tag{1.46}$$

From (1.32) for C_V as a second derivative of F, and from (1.10) for P as a first derivative of F, it follows that

$$(\partial C_V/\partial V)_T = T(\partial^2 P/\partial T^2)_V. \tag{1.47}$$

The counterpart of this last equation, with V and P interchanged, is derived as follows:

$$\begin{aligned}(\partial C_P/\partial P)_T &= T(\partial^2 S/\partial P\partial T)_{TP} = T[\partial/\partial T(\partial S/\partial P)_T]_P \\ &= -T[\partial/\partial T(\partial V/\partial T)_P]_P,\end{aligned}$$

or

$$(\partial C_P/\partial P)_T = -T(\partial^2 V/\partial T^2)_P. \tag{1.48}$$

An alternate form for the pressure derivative of C_P may be obtained by noting, from the definition of β,

$$\begin{aligned}(\partial\beta/\partial T)_P &= -V^{-2}[(\partial V/\partial T)_P]^2 + V^{-1}(\partial^2 V/\partial T^2)_P \\ &= -\beta^2 - (1/TV)(\partial C_P/\partial P)_T,\end{aligned} \tag{1.49}$$

where in the last equality we used (1.48). Rewriting (1.49) gives the desired result:

$$(\partial C_P/\partial P)_T = -TV[\beta^2 + (\partial\beta/\partial T)_P]. \tag{1.50}$$

If W is any thermodynamic function, the volume and pressure derivatives at constant temperature may be related by writing

$$(\partial W/\partial V)_T = (\partial W/\partial P)_T(\partial P/\partial V)_T,$$

or

$$(\partial W/\partial V)_T = -(B_T/V)(\partial W/\partial P)_T. \tag{1.51}$$

To relate the temperature derivatives at constant volume and pressure, consider $W = W(V,T)$ and write

$$dW = (\partial W/\partial V)_T\, dV + (\partial W/\partial T)_V\, dT.$$

Divide by dT and take constant pressure, to get

$$(\partial W/\partial T)_P = (\partial W/\partial V)_T(\partial V/\partial T)_P + (\partial W/\partial T)_V;$$

then with (1.51) the desired result is

$$(\partial W/\partial T)_V = (\partial W/\partial T)_P + \beta B_T(\partial W/\partial P)_T. \tag{1.52}$$

Finally, we calculate the pressure derivatives of $k_T - k_S$, and of $B_S - B_T$, in a form which can be evaluated from experimental data. For abbreviation, write

$$\Delta k = k_T - k_S;$$
$$\Delta B = B_S - B_T. \tag{1.53}$$

Differentiating the expression (1.41) for Δk gives

$$\left(\frac{\partial \Delta k}{\partial P}\right)_T = \Delta k\left[\frac{1}{V}\left(\frac{\partial V}{\partial P}\right)_T + \frac{2}{\beta}\left(\frac{\partial \beta}{\partial P}\right)_T - \frac{1}{C_P}\left(\frac{\partial C_P}{\partial P}\right)_T\right]. \tag{1.54}$$

The first term in square brackets is $-k_T$; furthermore, with the aid of (1.46) for $(\partial\beta/\partial P)_T$ and (1.50) for $(\partial C_P/\partial P)_T$, we can write (1.54) as

$$\left(\frac{\partial \Delta k}{\partial P}\right)_T = \Delta k\left\{-k_T - \frac{2}{\beta}\left(\frac{\partial k_T}{\partial T}\right)_P + \Delta k\left[1 + \frac{1}{\beta^2}\left(\frac{\partial \beta}{\partial T}\right)_P\right]\right\}. \tag{1.55}$$

For the pressure derivative of ΔB, it is convenient to define the dimensionless quantities X_T and X_S:

$$X_T = B_T\,\Delta k,$$
$$X_S = B_S\,\Delta k. \tag{1.56}$$

Then according to (1.42),

$$B_S = B_T + B_T X_S = B_T(1 + X_S),$$

or

$$B_T = B_S/(1 + X_S), \tag{1.57}$$

and using (1.42) again gives

$$\Delta B = B_S X_S/(1 + X_S). \tag{1.58}$$

Differentiation of this equation leads directly to

$$\left(\frac{\partial \Delta B}{\partial P}\right)_T = \frac{1}{(1 + X_S)^2}\left[X_S(2 + X_S)\left(\frac{\partial B_S}{\partial P}\right)_T + B_S^2\left(\frac{\partial \Delta k}{\partial P}\right)_T\right]. \tag{1.59}$$

Equation (1.59) is convenient when the measured value of $(\partial B_S/\partial P)_T$ is available; $(\partial \Delta k/\partial P)_T$ is to be calculated from (1.55). If the measured value of $(\partial B_T/\partial P)_T$ is available, one can proceed alternately and write (1.42) as

$$B_T = B_S(1 - X_T),$$

or

$$B_S = B_T/(1 - X_T),$$

and upon using (1.42) again,

$$\Delta B = B_T X_T/(1 - X_T). \tag{1.60}$$

Differentiation of this equation then gives

$$\left(\frac{\partial \Delta B}{\partial P}\right)_T = \frac{1}{(1 - X_T)^2}\left[X_T(2 - X_T)\left(\frac{\partial B_T}{\partial P}\right)_T + B_T^2\left(\frac{\partial \Delta k}{\partial P}\right)_T\right]. \tag{1.61}$$

Starting with the definition (1.30) of the Grüneisen parameter γ, and proceeding along lines similar to the above derivations, we can express the volume derivative of γ in terms of readily measured quantities as

$$\left(\frac{\partial \ln \gamma}{\partial \ln V}\right)_T = \frac{B_T}{\beta}\left(\frac{\partial k_T}{\partial T}\right)_P - \frac{X_T}{\beta^2}\left(\frac{\partial \beta}{\partial T}\right)_P + (1 - X_T)\left[1 - \left(\frac{\partial B_S}{\partial P}\right)_T\right]. \tag{1.62}$$

Stability

In the preceding discussion we have been concerned with thermodynamic processes along equilibrium lines. The question of stability concerns virtual variations, i.e., variations of the system away from equilibrium. Consider a thermodynamic system at equilibrium. Now if the thermodynamic variables are changed so as to bring the system out of equilibrium, a virtual process, then there should be a thermodynamic restoring force which tends to move the system back to equilibrium. The presence of such a restoring force means the original equilibrium configuration is stable. The original equilibrium configuration is unstable, of course, if a thermodynamic force develops which tends to move the system further away from equilibrium.

The general thermodynamic stability requirement is that the entropy is maximum, with respect to all virtual variations, at equilibrium:

$$d^2S < 0, \quad \text{thermodynamic stability.} \tag{1.63}$$

An equivalent statement of thermodynamic stability* is that the internal energy is minimum, with respect to all virtual variations, at equilibrium:

$$d^2U > 0. \tag{1.64}$$

Now according to (1.7), U can be considered a function of V and S, and for any infinitesimal variation we can write

$$d^2U = \left(\frac{\partial^2 U}{\partial S^2}\right)_V (dS)^2 + 2\left(\frac{\partial^2 U}{\partial S \partial V}\right)_{VS} dS dV + \left(\frac{\partial^2 U}{\partial V^2}\right)_S (dV)^2. \tag{1.65}$$

The stability condition is that the right-hand side must be positive for any dS and dV, except of course the trivial case $dS = dV = 0$.

The right-hand side of (1.65) is a homogeneous quadratic form in the

* H. B. Callen, *Thermodynamics*, John Wiley & Sons Inc., New York, 1960.

two variables dS and dV. Consider a general homogeneous quadratic form in the real variables y_α,

$$\sum_{\alpha\alpha'} M_{\alpha\alpha'} y_\alpha y_{\alpha'}, \tag{1.66}$$

where the matrix of coefficients $M_{\alpha\alpha'}$ is real symmetric. The quadratic form is said to be positive definite if, for all values of the y_α except all $y_\alpha = 0$, the form is positive. The matrix $\mathbf{M}$ is said to be positive definite if all its eigenvalues are positive. According to a theorem of linear algebra,

$$\sum_{\alpha\alpha'} M_{\alpha\alpha'} y_\alpha y_{\alpha'} > 0 \quad \text{if and only if} \quad \mathbf{M} > 0, \tag{1.67}$$

where >0 is to be read positive definite. To prove (1.67), let μ_s and $\mathbf{w}_s$ be the eigenvalues and eigenvectors, respectively, of $\mathbf{M}$:

$$\sum_{\alpha\alpha'} w_{\alpha s} M_{\alpha\alpha'} w_{\alpha' s'} = \mu_s \delta_{ss'}. \tag{1.68}$$

Equation (1.68) expresses the diagonalization of $\mathbf{M}$, and since $\mathbf{M}$ is real symmetric the μ_s are real and the $\mathbf{w}_s$ are complete in the space of the variables. The completeness means that the y_α can be expanded as linear functions of the $\mathbf{w}_s$:

$$y_\alpha = \sum_s c_s w_{\alpha s}. \tag{1.69}$$

Substituting (1.69) into the quadratic form (1.66), and using the diagonalization equation (1.68), gives

$$\sum_{\alpha\alpha'} M_{\alpha\alpha'} y_\alpha y_{\alpha'} = \sum_s c_s^2 \mu_s. \tag{1.70}$$

Now if $\mathbf{M} > 0$, all $\mu_s > 0$ by definition; hence, the right-hand side of (1.70) is positive, so the homogeneous quadratic form is positive definite. To prove the converse, choose the arbitrary y_α so that all $c_s = 0$ except one, say $c_{s'}$. Then from (1.70) the quadratic form is positive only if $\mu_{s'} > 0$, and this argument can be extended to each eigenvalue μ_s in turn. This proves (1.67). The condition $\mathbf{M} > 0$ can also be stated as the condition that each of the principal minors of $\mathbf{M}$ is positive.

Returning to the thermodynamic stability condition, the matrix of the coefficients of the homogeneous quadratic form (1.65) must be positive definite:

$$\begin{pmatrix} a & c \\ c & b \end{pmatrix} > 0, \tag{1.71}$$

where

$$a = (\partial^2 U/\partial S^2)_V, \tag{1.72}$$

$$b = (\partial^2 U/\partial V^2)_S, \tag{1.73}$$

$$c = (\partial^2 U/\partial S\,\partial V)_{VS}. \tag{1.74}$$

The eigenvalues of the matrix (1.71) are μ_1 and μ_2; the sum and product of these are the trace and determinant, respectively, of the matrix. Stability requires

$$\mu_1 + \mu_2 = a + b > 0, \tag{1.75}$$

$$\mu_1\mu_2 = ab - c^2 > 0. \tag{1.76}$$

It is easily seen that these conditions are satisfied if and only if

$$a > 0, \qquad b > 0, \qquad ab > c^2. \tag{1.77}$$

We will discuss the interpretation of these three conditions in turn.

The first condition in (1.77) is $(\partial^2U/\partial S^2)_V > 0$; in view of the relation (1.9) for T and the definition (1.23) of C_V, this is

$$(\partial T/\partial S)_V = T/C_V > 0. \tag{1.78}$$

Since $T > 0$, this implies $C_V > 0$, and since $C_V = (\partial Q/\partial T)_V$ this means that the temperature of the crystal must increase if heat is added to the crystal while the volume is held constant. The second condition in (1.77) is $(\partial^2U/\partial V^2)_S > 0$; in view of the relation (1.8) for P and the definition (1.25) of k_S, this is

$$-(\partial P/\partial V)_S = 1/Vk_S > 0. \tag{1.79}$$

Since $V > 0$, this implies $k_S > 0$, and this means that the crystal volume must decrease as the pressure is increased while the entropy is held constant.

To study the third stability condition in (1.77), it is convenient to write ab as

$$\left(\frac{\partial^2U}{\partial S^2}\right)_V\left(\frac{\partial^2U}{\partial V^2}\right)_S = -\left(\frac{\partial T}{\partial S}\right)_V\left(\frac{\partial P}{\partial V}\right)_S = \frac{T}{Vk_SC_V}. \tag{1.80}$$

The coefficient c may be evaluated as

$$\left(\frac{\partial^2U}{\partial S\partial V}\right)_{VS} = \left(\frac{\partial T}{\partial V}\right)_S = -\left(\frac{\partial T}{\partial S}\right)_V\left(\frac{\partial S}{\partial V}\right)_T = -\frac{T\beta B_T}{C_V}, \tag{1.81}$$

where the expression (1.29) for βB_T was used. Now the condition $ab > c^2$ is

$$\frac{T}{Vk_SC_V} > \left(\frac{T\beta}{k_TC_V}\right)^2. \tag{1.82}$$

Multiplying (1.82) by the positive quantity $k_T^2k_S(VC_V/T)$ gives

$$k_T^2 > (TV\beta^2k_S/C_V),$$

and from (1.43)

$$k_T^2 > (TV\beta^2k_T/C_P),$$

and from (1.41)

$$k_T^2 > k_T(k_T - k_S) = k_T^2 - k_Tk_S.$$

Since $k_S > 0$, this last result implies $k_T > 0$. Finally, since $C_V > 0$, equation (1.40) for $C_P - C_V$ then implies $C_P > 0$. Again referring to the equations for $C_P - C_V$, and for $k_T - k_S$, and noting that C_P, C_V, k_T, and k_S have all been shown to be positive, we can summarize the thermodynamic stability conditions quite simply by the results

$$C_P > C_V > 0, \tag{1.83}$$

$$k_T > k_S > 0. \tag{1.84}$$

2. STRESS–STRAIN VARIABLES

Homogeneous Strain

The basic equilibrium thermodynamics of the preceding section is here generalized to the case of anisotropic stresses and strains. It should be emphasized that the results of Section 1 are still correct in terms of pressure and volume variables, even for anisotropic crystals.

For thermodynamic calculations, the crystal is considered to be a homogeneous, anisotropic elastic medium. The applied stresses are uniform, i.e., constant on a given crystal surface, and the resulting strains are homogeneous i.e., uniform throughout the crystal. Consider in particular the finite strain from an arbitrary initial configuration, which corresponds to arbitrary applied initial stress, to a new final configuration. Such strain is presumably brought about by the application of additional stresses. The position of a small element of mass is denoted by **X** in the initial configuration, and by **x** in the final configuration.* For homogeneous strains, the vectors **X** and **x** are related by a linear transformation according to

$$x_i = \sum_j \alpha_{ij} X_j, \tag{2.1}$$

where the indices i,j represent Cartesian coordinates and take on the values x,y,z or 1,2,3. The inverse transformation is

$$X_i = \sum_j \xi_{ij} x_j; \tag{2.2}$$

obviously the matrices $\boldsymbol{\alpha}$ and $\boldsymbol{\xi}$ are inverse to each other:

$$\sum_j \alpha_{ij} \xi_{jk} = \delta_{ik}. \tag{2.3}$$

* On occasion we will use the symbols **X** and **x** to denote functional evaluation in the initial and final configuration, respectively, as for example $V(\mathbf{X})$ and $V(\mathbf{x})$. In addition we will refer to the "configuration **X**," or the "configuration **x**," with obvious meaning.

It also follows, from the definitions of the transformations, that

$$\begin{aligned} \alpha_{ij} &= (\partial x_i/\partial X_j); \\ \xi_{ij} &= (\partial X_i/\partial x_j). \end{aligned} \tag{2.4}$$

For homogeneous strains, α_{ij} and ξ_{ij} are constants, independent of the location in the crystal.

The displacement in the strain from **X** to **x** is the vector **u**, defined by

$$u_i = x_i - X_i. \tag{2.5}$$

The displacement gradients are

$$u_{ij} = (\partial u_i/\partial X_j); \tag{2.6}$$

differentiation of (2.5) with respect to X_j and comparison with (2.4) leads to the relation between the transformation coefficients and the displacement gradients:

$$\alpha_{ij} = \delta_{ij} + u_{ij}. \tag{2.7}$$

The u_{ij} are, of course, also constants for a homogeneous strain.

Another measure of the strain from **X** to **x** is the change of the distance between any two elements of mass in the crystal. From (2.1) and (2.7),

$$\sum_i x_i^2 = \sum_{ijk} \alpha_{ij}\alpha_{ik} X_j X_k = \sum_i X_i^2 + 2\sum_{ij} \eta_{ij} X_i X_j, \tag{2.8}$$

where the Lagrangian strain parameters η_{ij} are

$$\eta_{ij} = \tfrac{1}{2}\Big(u_{ij} + u_{ji} + \sum_k u_{ki}u_{kj}\Big). \tag{2.9}$$

Since in calculating the length of **x** in (2.8) the origin of coordinates is arbitrary, the result can be interpreted in terms of the vector $\Delta\mathbf{X}$ in the initial configuration and $\Delta\mathbf{x}$ in the final configuration, between any two elements of mass in the crystal:

$$|\Delta\mathbf{x}|^2 - |\Delta\mathbf{X}|^2 = 2\sum_{ij} \eta_{ij}\Delta X_i \Delta X_j. \tag{2.10}$$

It is seen from (2.9) that the η_{ij} are symmetric, i.e., $\eta_{ij} = \eta_{ji}$, and with the aid of (2.7),

$$\eta_{ij} = \tfrac{1}{2}\Big(\sum_k \alpha_{ki}\alpha_{kj} - \delta_{ij}\Big). \tag{2.11}$$

The final configuration of the crystal may be completely specified in terms of the initial configuration and either the transformation coefficients α_{ij} or the displacement gradients u_{ij}; the Lagrangian strains η_{ij} do not give a complete description, however, since they contain no information about rotations of the crystal.

There are two useful identities which reflect the geometry of strain; these are

$$V(\mathbf{x})/V(\mathbf{X}) = J = \det[\alpha_{ij}]; \tag{2.12}$$

$$(\partial J/\partial \alpha_{ij})_{\alpha'} = \xi_{ji} J. \tag{2.13}$$

In (2.12), J is defined as the determinant of $\boldsymbol{\alpha}$, and in (2.13) the subscript α' means all other α_{kl} are to be held constant in differentiating with respect to α_{ij}. To prove the first identity, let L_{ij} and l_{ij} be the j Cartesian component of the i edge of the crystal in the configurations $\mathbf{X}$ and $\mathbf{x}$, respectively; then

$$V(\mathbf{x}) = \det[l_{ij}] = \det[(\alpha L)_{ij}] = \det[\alpha_{ij}]V(\mathbf{X}).$$

To prove the second identity, expand the determinant J in cofactors and differentiate to get

$$\partial J/\partial \alpha_{ij} = \alpha^{ij}, \tag{2.14}$$

where α^{ij} is the cofactor of α_{ij}. A well-known property of cofactors is

$$\sum_j \alpha_{kj}\alpha^{ij} = J\delta_{ik};$$

this is so since if $k = i$ it is just the expansion of J in cofactors, and if $k \neq i$ it is the expansion of a determinant whose k,j rows are equal and hence it is zero. But the solution of this last set of equations, for any i,j, is

$$\alpha^{ij} = \xi_{ji} J, \tag{2.15}$$

as can be seen with the aid of (2.3). Hence (2.14) and (2.15) together prove Jacobi's identity (2.13).

Stresses and Elastic Constants

The applied stress is represented by the tensor τ_{ij}, and is taken to be symmetric, so that there is no net torque on the crystal. The combined first and second laws of thermodynamics states that $dU = TdS - đW$; in order to evaluate this it is necessary to calculate the work $đW$ done by the crystal against the applied stresses, in changing the configuration from $\mathbf{X}$ to $\mathbf{X} + \Delta\mathbf{X}$. This is in strict analogy to $đW = PdV$ for the pressure–volume case.

Let the surface of the crystal in configuration $\mathbf{X}$ be $\mathbf{S}$, with a surface element denoted by $d\mathbf{S}$. The i component of force on $d\mathbf{S}$, due to the applied stress, is

$$f_i = \sum_j \tau_{ij} dS_j. \tag{2.16}$$

Let the displacement gradients corresponding to the strain $\Delta\mathbf{X}$ be Δu_{ij}. The displacement of $d\mathbf{S}$ in the i direction is then

$$\Delta X_i = \sum_k \Delta u_{ik} X_k. \tag{2.17}$$

The work done by the crystal against the stress applied to $d\mathbf{S}$ is then

$$-\sum_i f_i \Delta X_i = -\sum_{ijk} \tau_{ij} dS_j \Delta u_{ik} X_k; \tag{2.18}$$

note that we are using the customary convention that positive stresses are directed outward from the crystal surface. The total work done by the crystal is just the integral of this expression over the surface **S**. This surface integral is transformed to a volume integral by Gauss' theorem and evaluated as

$$\begin{aligned} \Delta W &= -\int_{\mathbf{S}} \sum_{ijk} \tau_{ij} \Delta u_{ik} X_k \, dS_j = -\int_V \sum_{ij} \tau_{ij} \Delta u_{ij} \, dV \\ &= -\sum_{ij} \tau_{ij} \Delta u_{ij} V(\mathbf{X}), \end{aligned} \tag{2.19}$$

since τ_{ij} and Δu_{ij} are constants. Finally, since τ_{ij} is symmetric, the antisymmetric part of Δu_{ij} gives no contribution to the Σ_{ij} in (2.19), so Δu_{ij} may be replaced by $\Delta \eta_{ij}$ to leading order, according to (2.9). Writing the work (2.19) in differential form then gives

$$dW = -V \sum_{ij} \tau_{ij} d\eta_{ij}. \tag{2.20}$$

The combined first and second laws of thermodynamics for stress–strain variables is therefore

$$dU = TdS + V \sum_{ij} \tau_{ij} d\eta_{ij}. \tag{2.21}$$

Since $F = U - TS$, the differential of F is

$$dF = -SdT + V \sum_{ij} \tau_{ij} d\eta_{ij}. \tag{2.22}$$

It is convenient to consider U as a function of the configuration and the entropy; then the dependent variables are the stresses and the temperature. It is also convenient to take F as a function of the configuration and the temperature, with the dependent variables being the stresses and the entropy. These dependences are denoted as

$$\begin{aligned} U &= U(\mathbf{x}, S), \\ F &= F(\mathbf{x}, T). \end{aligned} \tag{2.23}$$

The state functions U and F must remain unchanged if the crystal and the system of applied stresses are translated, or rotated, together. In terms of the configuration dependence of U and F, this means that the state functions must be invariant with respect to translation or rotation, without deformation, of the crystal. Translational invariance is obviously satisfied, since the origin of coordinates is irrelevant in our specification of the crystal configuration. The rotational invariance implies that the state functions depend on the final configuration **x** only through the initial configuration **X** and the rotation-independent strains η_{ij}; the functional dependences may therefore be written

$$U(\mathbf{x},S) = U(\mathbf{X},\eta_{ij},S),$$
$$F(\mathbf{x},T) = F(\mathbf{X},\eta_{ij},T), \tag{2.24}$$

where the η_{ij} are the Lagrangian strain parameters from **X** to **x**.

Expressions for the dependent variables are obtained from the combined first and second laws. From (2.21) for dU, it follows that

$$\tau_{ij} = V^{-1}(\partial U/\partial \eta_{ij})_{S\eta'}; \tag{2.25}$$

$$T = (\partial U/\partial S)_{\eta}. \tag{2.26}$$

From (2.22) for dF, it follows that

$$\tau_{ij} = V^{-1}(\partial F/\partial \eta_{ij})_{T\eta'}; \tag{2.27}$$

$$S = -(\partial F/\partial T)_{\eta}. \tag{2.28}$$

The subscript η' means that all other η_{kl} are to be held constant while differentiating with respect to η_{ij}, and the subscript η means that all η_{ij} are held constant, i.e., the differentiation is carried out at constant configuration. The equations for the stresses are to be evaluated at the initial configuration **X**, i.e., at $\eta_{ij} = 0$, since in the derivation leading to (2.20) for the work dW, the stresses were evaluated at **X**. This means, for example, that in order to calculate the variation of stress with strain, one must calculate $\tau_{ij}(\mathbf{X} + \Delta\mathbf{X}) - \tau_{ij}(\mathbf{X})$ from (2.25) or (2.27). A more general derivation* leads to the stress evaluated at an arbitrary final configuration **x**:

$$\tau_{ij}(\mathbf{x}) = V(\mathbf{x})^{-1} \sum_{kl} \alpha_{ik}\alpha_{jl}(\partial U/\partial \eta_{kl})_{S\eta'}, \tag{2.29}$$

$$= V(\mathbf{x})^{-1} \sum_{kl} \alpha_{ik}\alpha_{jl}(\partial F/\partial \eta_{kl})_{T\eta'}; \tag{2.30}$$

here the strains from **X** to **x** are measured by α_{ij} and η_{ij} and the functions on the right-hand sides are to be evaluated at **x**. We will, however, continue to base our work on the equations which were derived here, namely (2.25) and (2.27) for the stresses.

The above equations for the dependent variables are obvious generalizations of the corresponding equations of Section 1. In particular, in (1.9) for T and (1.11) for S, the constant volume condition is replaced now by a

* D. C. Wallace, in *Solid State Physics*, edited by H. Ehrenreich, F. Seitz, and D. Turnbull, Academic Press Inc., New York, 1970, Vol. 25, p. 301.

constant configuration condition. In (1.8) and (1.10) for P, the volume derivative is replaced by a strain derivative. The difference in sign in the equations for P, and those for τ_{ij}, is due to the sign convention: A positive pressure is exerted inward on the crystal, while a positive stress is exerted outward. If the stress is an isotropic pressure, for example, one has $\tau_{ij} = -P\delta_{ij}$. Again the equations (2.25) and (2.27) are considered as equilibrium conditions, since they determine the anisotropic forces on the crystal at equilibrium.

The elastic constants of second, third, and higher order are defined as strain derivatives of the state functions of second, third, and higher order, respectively. The adiabatic elastic constants of second, third, and fourth order are

$$C^S_{ijkl} = V^{-1}(\partial^2 U/\partial\eta_{ij}\partial\eta_{kl})_{S\eta'}, \tag{2.31}$$

$$C^S_{ijklmn} = V^{-1}(\partial^3 U/\partial\eta_{ij}\partial\eta_{kl}\partial\eta_{mn})_{S\eta'}, \tag{2.32}$$

$$C^S_{ijklmnpq} = V^{-1}(\partial^4 U/\partial\eta_{ij}\partial\eta_{kl}\partial\eta_{mn}\partial\eta_{pq})_{S\eta'}, \tag{2.33}$$

and the isothermal elastic constants are

$$C^T_{ijkl} = V^{-1}(\partial^2 F/\partial\eta_{ij}\partial\eta_{kl})_{T\eta'}, \tag{2.34}$$

$$C^T_{ijklmn} = V^{-1}(\partial^3 F/\partial\eta_{ij}\partial\eta_{kl}\partial\eta_{mn})_{T\eta'}, \tag{2.35}$$

$$C^T_{ijklmnpq} = V^{-1}(\partial^4 F/\partial\eta_{ij}\partial\eta_{kl}\partial\eta_{mn}\partial\eta_{pq})_{T\eta'}. \tag{2.36}$$

These elastic constants are defined at an arbitrary configuration $\mathbf{X}$, where the strains η_{ij} are measured from $\mathbf{X}$ and the derivatives are to be evaluated at $\mathbf{X}$, i.e., at $\eta_{ij} = 0$. Since rotational invariance requires that the state functions depend on a final configuration $\mathbf{x}$ only through $\mathbf{X}$ and η_{ij}, according to (2.24), then U and F may be expanded in powers of the strains η_{ij}:

$$U(\mathbf{X},\eta_{ij},S) = U(\mathbf{X},S) + V\sum_{ij}\tau_{ij}\eta_{ij} + \tfrac{1}{2}V\sum_{ijkl}C^S_{ijkl}\eta_{ij}\eta_{kl} + \cdots; \tag{2.37}$$

$$F(\mathbf{X},\eta_{ij},T) = F(\mathbf{X},T) + V\sum_{ij}\tau_{ij}\eta_{ij} + \tfrac{1}{2}V\sum_{ijkl}C^T_{ijkl}\eta_{ij}\eta_{kl} + \cdots. \tag{2.38}$$

In view of the definitions of the elastic constants as multiple strain derivatives of the state functions, and also since the strains η_{ij} are symmetric, the elastic constants have complete Voigt symmetry. The Voigt symmetry means that the elastic constants are invariant under interchange of the indices i and j which correspond to a strain η_{ij}, and they are also invariant under interchange of the pairs ij and kl which correspond to two strains η_{ij} and η_{kl}. For example, for either adiabatic or isothermal constants,

$$C_{ijkl} = C_{jikl} = C_{klij} = \cdots. \tag{2.39}$$

It is therefore convenient to use Voigt notation, in which a pair of Cartesian indices ij is replaced by a single index α, according to the scheme

$$\begin{array}{lllllll} ij = 11 & 22 & 33 & 32 \text{ or } 23 & 31 \text{ or } 13 & 21 \text{ or } 12 \\ \alpha = 1 & 2 & 3 & 4 & 5 & 6 \end{array} \tag{2.40}$$

In Voigt notation, the adiabatic or isothermal elastic constants are completely symmetric:

$$\begin{aligned} C_{\alpha\beta} &= C_{\beta\alpha}; \\ C_{\alpha\beta\gamma} &= C_{\alpha\gamma\beta} = C_{\beta\alpha\gamma} = \cdots; \\ C_{\alpha\beta\gamma\delta} &= C_{\alpha\beta\delta\gamma} = C_{\beta\alpha\delta\gamma} = \cdots. \end{aligned} \tag{2.41}$$

Stress–Strain Relations

In order to calculate the variation of stress with strain, we need to consider two configurations, say $\mathbf{X}$ and $\bar{\mathbf{X}}$, and calculate $\tau_{ij} - \bar{\tau}_{ij}$. We will first establish some relations involving the two configurations, considering only symmetric strains, i.e., no rotations, from $\bar{\mathbf{X}}$ to $\mathbf{X}$. Let the transformation coefficients from $\bar{\mathbf{X}}$ to $\mathbf{X}$ be a_{ij},

$$a_{ij} = a_{ji} = (\partial X_i/\partial \bar{X}_j), \tag{2.42}$$

and let the corresponding Lagrangian strain parameters be n_{ij},

$$n_{ij} = \tfrac{1}{2}\left(\sum_k a_{ki}a_{kj} - \delta_{ij}\right). \tag{2.43}$$

Inversion of (2.43) as a power series in n_{ij} gives

$$a_{ij} = \delta_{ij} + n_{ij} - \tfrac{1}{2}\sum_k n_{ki}n_{kj} + \cdots; \tag{2.44}$$

this may be verified by using (2.44) in (2.43). The ratio of the crystal volumes in the two configurations is, from (2.12),

$$V/\bar{V} = \det[a_{ij}] = 1 + \sum_i n_{ii} + \cdots. \tag{2.45}$$

Now let the Lagrangian strains from $\bar{\mathbf{X}}$ to an arbitrary final configuration $\mathbf{x}$ be $\bar{\eta}_{ij}$, and those from $\mathbf{X}$ to $\mathbf{x}$ be η_{ij}. Then from (2.11) for Lagrangian strains, it follows that

$$\begin{aligned} \bar{\eta}_{ij} + \tfrac{1}{2}\delta_{ij} &= \tfrac{1}{2}\sum_k \frac{\partial x_k}{\partial \bar{X}_i}\frac{\partial x_k}{\partial \bar{X}_j} \\ &= \tfrac{1}{2}\sum_k \sum_{rs} \frac{\partial x_k}{\partial X_r}\frac{\partial X_r}{\partial \bar{X}_i}\frac{\partial x_k}{\partial X_s}\frac{\partial X_s}{\partial \bar{X}_j} \\ &= \sum_{rs} (\eta_{rs} + \tfrac{1}{2}\delta_{rs})a_{ri}a_{sj}. \end{aligned} \tag{2.46}$$

Differentiating (2.46) gives the result

$$(\partial\bar{\eta}_{ij}/\partial\eta_{rs})_{\eta'} = a_{ri}a_{sj}. \tag{2.47}$$

If the strain from $\bar{\mathbf{X}}$ to $\mathbf{X}$ is carried out adiabatically, $\tau_{ij} - \bar{\tau}_{ij}$ may be calculated by evaluating (2.25) for τ_{ij} at $\mathbf{X}$, as follows.

$$\begin{aligned} \tau_{ij} &= V^{-1}(\partial U/\partial\eta_{ij})_{S\eta'}, \quad \text{at } \mathbf{X}; \\ &= \bar{V}^{-1}(\bar{V}/V)\sum_{kl}(\partial U/\partial\bar{\eta}_{kl})_{S\bar{\eta}'}(\partial\bar{\eta}_{kl}/\partial\eta_{ij})_{\eta'}; \end{aligned} \tag{2.48}$$

and from the strain expansion of U, equation (2.37), and (2.47) for $(\partial\bar{\eta}_{kl}/\partial\eta_{ij})_{\eta'}$,

$$\tau_{ij} = (\bar{V}/V)\sum_{kl} a_{ik}a_{jl}\left\{\bar{\tau}_{kl} + \sum_{mn}\bar{C}^S_{klmn}n_{mn} + \cdots\right\}, \tag{2.49}$$

where the strain n_{mn} appears as a result of evaluation at $\mathbf{X}$. With the expansions (2.45) for $V/\bar{V}$ and (2.44) for a_{ij}, (2.49) may be evaluated to first order in the strains n_{ij} from $\bar{\mathbf{X}}$ to $\mathbf{X}$ to give

$$\tau_{ij} = \bar{\tau}_{ij} + \sum_{kl} n_{kl}[-\bar{\tau}_{ij}\delta_{kl} + \bar{\tau}_{il}\delta_{jk} + \bar{\tau}_{jk}\delta_{il} + \bar{C}^S_{ijkl}]. \tag{2.50}$$

Equation (2.50) is the desired relation between the variation of stress and the variation of strain, to first order in the strains and for adiabatic processes. We can now return to our original notation and summarize the results in terms of the stress–strain coefficients B_{ijkl} as follows. For adiabatic strains

$$\tau_{ij}(\mathbf{x},S) = \tau_{ij}(\mathbf{X},S) + \sum_{kl} B^S_{ijkl}\eta_{kl} + \cdots, \tag{2.51}$$

and for isothermal strains

$$\tau_{ij}(\mathbf{x},T) = \tau_{ij}(\mathbf{X},T) + \sum_{kl} B^T_{ijkl}\eta_{kl} + \cdots, \tag{2.52}$$

where

$$B^S_{ijkl} = \tfrac{1}{2}(\tau_{il}\delta_{jk} + \tau_{jl}\delta_{ik} + \tau_{ik}\delta_{jl} + \tau_{jk}\delta_{il} - 2\tau_{ij}\delta_{kl}) + C^S_{ijkl}, \tag{2.53}$$

$$B^T_{ijkl} = \tfrac{1}{2}(\tau_{il}\delta_{jk} + \tau_{jl}\delta_{ik} + \tau_{ik}\delta_{jl} + \tau_{jk}\delta_{il} - 2\tau_{ij}\delta_{kl}) + C^T_{ijkl}. \tag{2.54}$$

Since k,l are summed in (2.50), k and l were interchanged in the quantity in brackets in (2.50) to obtain the symmetric forms shown in (2.53) and (2.54). In (2.51) and (2.52) the B_{ijkl} are to be evaluated at $\mathbf{X}$, the η_{ij} measure the symmetric strain from $\mathbf{X}$ to $\mathbf{x}$, and $+\cdots$ indicates terms of second and higher order in the η_{ij}. Finally, it should be recalled that rotations of the crystal are not included in these stress–strain relations.

The stress–strain derivatives may be written down directly from (2.51) and (2.52); evaluated at $\mathbf{X}$, where $\eta_{ij} = 0$, these are

$$(\partial\tau_{ij}/\partial\eta_{kl})_{S\eta'} = B^S_{ijkl}, \tag{2.55}$$

$$(\partial\tau_{ij}/\partial\eta_{kl})_{T\eta'} = B^T_{ijkl}. \tag{2.56}$$

The elastic compliances S_{ijkl} are components of the tensor inverse to the tensor of the B_{ijkl}; it is convenient to define the inverse in the following symmetric way, for either adiabatic or isothermal coefficients.

$$\sum_{kl} B_{ijkl}S_{klmn} = \sum_{kl} S_{ijkl}B_{klmn} = \tfrac{1}{2}(\delta_{im}\delta_{jn} + \delta_{in}\delta_{jm}). \tag{2.57}$$

The stress–strain derivatives (2.55) and (2.56) may then be inverted to give

$$(\partial\eta_{ij}/\partial\tau_{kl})_{S\tau'} = S^S_{ijkl}, \tag{2.58}$$

$$(\partial\eta_{ij}/\partial\tau_{kl})_{T\tau'} = S^T_{ijkl}. \tag{2.59}$$

The stress–strain coefficients (or elastic stiffnesses) B_{ijkl}, and the compliances S_{ijkl}, are symmetric in i,j and in k,l, but not necessarily in ij, kl.

The compressibility, which measures the variation of volume with pressure, can be calculated from the compliances. The adiabatic compressibility, for example, is

$$k_S = -V^{-1}(\partial V/\partial P)_S.$$

But from the geometry of homogeneous strain, in particular from (2.12), this volume derivative evaluated at **X** is just

$$k_S = -(\partial J/\partial P)_S. \tag{2.60}$$

Now $J = V(\mathbf{x})/V(\mathbf{X})$, and is a function only of the symmetric strains η_{ij}; by chain-rule differentiation (2.60) may be written

$$k_S = -\sum_{ijkl}(\partial J/\partial\eta_{ij})_{\eta'}(\partial\eta_{ij}/\partial\tau_{kl})_{S\tau'}(\partial\tau_{kl}/\partial P). \tag{2.61}$$

With the help of Jacobi's identity (2.13), and the relation (2.11) between η_{ij} and α_{ij}, it follows that

$$(\partial J/\partial\eta_{ij})_{\eta'} = \delta_{ij}, \quad \text{at } \mathbf{X}. \tag{2.62}$$

If the isotropic pressure P is varied while the nonisotropic part of the stress is held constant, then $d\tau_{kl} = -dP\delta_{kl}$, or

$$(\partial\tau_{kl}/\partial P) = -\delta_{kl}. \tag{2.63}$$

With these results, and with (2.58) for the adiabatic compliances, (2.61) becomes

$$k_S = \sum_{ij} S^S_{iijj}. \tag{2.64}$$

The above derivation carried through for isothermal strains gives

$$k_T = \sum_{ij} S^T_{iijj}. \tag{2.65}$$

Since the elastic stiffnesses and compliances are symmetric in their first two Cartesian indices, and also in their last two, they may be written in

Voigt notation. In other words, all the independent B_{ijkl}, or S_{ijkl}, are contained in the set $B_{\alpha\beta}$, or $S_{\alpha\beta}$. The 6 × 6 matrices $[B_{\alpha\beta}]$ and $[S_{\alpha\beta}]$ are not generally symmetric, however:

$$B_{\alpha\beta} \neq B_{\beta\alpha}, \qquad S_{\alpha\beta} \neq S_{\beta\alpha}, \quad \text{in general.} \tag{2.66}$$

The S_{ijkl} are defined as inverses to the B_{ijkl}, according to (2.57). It is also convenient to define the $B_{\alpha\beta}$ as equal to the corresponding B_{ijkl}, and to define the $S_{\alpha\beta}$ as inverses to the $B_{\alpha\beta}$:

$$\sum_{\beta} B_{\alpha\beta} S_{\beta\gamma} = \sum_{\beta} S_{\alpha\beta} B_{\beta\gamma} = \delta_{\alpha\gamma}. \tag{2.67}$$

When this is done, the $S_{\alpha\beta}$ are *not* equal to the corresponding S_{ijkl}, simply because the number of times a given $S_{\alpha\beta}$ appears in the sum in (2.67) is in some cases less than the number of times the corresponding S_{ijkl} appears in the sum in (2.57). It is easy to show the following relations.*

$$S_{ijkl} = S_{\alpha\beta}, \quad \text{if } \alpha,\beta = 1,2,3;$$
$$S_{ijkl} = \tfrac{1}{2} S_{\alpha\beta}, \quad \text{if } \alpha = 1,2,3, \ \beta = 4,5,6, \ \text{and if } \alpha = 4,5,6, \ \beta = 1,2,3;$$
$$S_{ijkl} = \tfrac{1}{4} S_{\alpha\beta}, \quad \text{if } \alpha,\beta = 4,5,6. \tag{2.68}$$

In terms of Voigt notation compliances, the compressibilities are

$$k_S = \sum_{\alpha,\beta=1}^{3} S_{\alpha\beta}^{S}, \tag{2.69}$$

$$k_T = \sum_{\alpha,\beta=1}^{3} S_{\alpha\beta}^{T}. \tag{2.70}$$

Thermodynamic Functions

We continue the generalization of the equilibrium thermodynamics of Section 1 by defining the symmetric thermal strain tensor,

$$\beta_{ij} = (\partial \eta_{ij}/\partial T)_{\tau}; \tag{2.71}$$

the symmetric thermal stress tensor,

$$b_{ij} = (\partial \tau_{ij}/\partial T)_{\eta}; \tag{2.72}$$

the heat capacity at constant configuration,

$$C_{\eta} = T(\partial S/\partial T)_{\eta} = (\partial U/\partial T)_{\eta}; \tag{2.73}$$

and the heat capacity at constant stress,

$$C_{\tau} = T(\partial S/\partial T)_{\tau}. \tag{2.74}$$

* D. C. Wallace, in *Solid State Physics*, edited by H. Ehrenreich, F. Seitz, and D. Turnbull, Academic Press Inc., New York, 1970, Vol. 25, p. 301.

The thermal strains β_{ij} are the thermal expansion coefficients for crystals of arbitrary symmetry, and the thermal stresses b_{ij} are coefficients which constitute a generalization of the function $-\beta B_T$. If the stress is isotropic pressure, $C_\tau = C_P$. However, $C_\eta \neq C_V$ in general, since in defining C_V it is assumed that as T is varied, P is also varied in such a way as to maintain constant V, and this does not correspond to constant η_{ij} in general for anisotropic crystals. It should be noted, however, that C_V is a unique function of the thermodynamic variables. The thermal expansion coefficient at constant stress is evaluated in a manner similar to the compressibility calculation.

$$\begin{aligned}\beta &= V^{-1}(\partial V/\partial T)_\tau \\ &= (\partial J/\partial T)_\tau, \quad \text{at } \mathbf{X} \\ &= \sum_{ij} (\partial J/\partial \eta_{ij})_{\eta'} (\partial \eta_{ij}/\partial T)_\tau,\end{aligned}$$

or with (2.62) and the definition (2.71) of the thermal strains,

$$\beta = \sum_i \beta_{ii}. \tag{2.75}$$

This holds, of course, for arbitrary stress, including isotropic pressure.

To relate the thermal stresses and strains, consider τ_{ij} as a function of the η_{ij} and T, and write

$$d\tau_{ij} = \sum_{kl} (\partial \tau_{ij}/\partial \eta_{kl})_{T\eta'} d\eta_{kl} + (\partial \tau_{ij}/\partial T)_\eta dT. \tag{2.76}$$

Dividing (2.76) by dT and taking constant stresses gives

$$\begin{aligned}0 &= \sum_{kl} (\partial \tau_{ij}/\partial \eta_{kl})_{T\eta'} (\partial \eta_{kl}/\partial T)_\tau + (\partial \tau_{ij}/\partial T)_\eta \\ &= \sum_{kl} B_{ijkl}^T \beta_{kl} + b_{ij},\end{aligned}$$

or

$$b_{ij} = -\sum_{kl} B_{ijkl}^T \beta_{kl}. \tag{2.77}$$

Inversion of (2.77) with the elastic compliances gives

$$\beta_{ij} = -\sum_{kl} S_{ijkl}^T b_{kl}. \tag{2.78}$$

The quantities most accessible to experimental determination are the elastic stiffnesses B_{ijkl}^T and the thermal strains β_{ij}; the thermal stresses b_{ij} can then be calculated from (2.77).

The Maxwell equations are obtained by differentiating the dependent variables, and comparing the various derivatives in the form of second derivatives of the state functions. For example, from (2.25)–(2.28) it is seen that

$$(\partial \tau_{ij}/\partial S)_\eta = V^{-1}(\partial T/\partial \eta_{ij})_{S\eta'}; \tag{2.79}$$

$$(\partial \tau_{ij}/\partial T)_\eta = -V^{-1}(\partial S/\partial \eta_{ij})_{T\eta'}. \tag{2.80}$$

If W is any thermodynamic function, the stress and strain derivatives may be related by writing

$$(\partial W/\partial \eta_{ij})_{T\eta'} = \sum_{kl} (\partial W/\partial \tau_{kl})_{T\tau'} (\partial \tau_{kl}/\partial \eta_{ij})_{T\eta'},$$

or

$$(\partial W/\partial \eta_{ij})_{T\eta'} = \sum_{kl} B^T_{klij} (\partial W/\partial \tau_{kl})_{T\tau'}. \tag{2.81}$$

This is a generalization of (1.51). The temperature derivatives at constant strain and stress may be related by writing

$$dW = \sum_{ij} (\partial W/\partial \tau_{ij})_{T\tau'} d\tau_{ij} + (\partial W/\partial T)_\tau dT;$$

dividing by dT, taking constant strain, and using the definition (2.72) of the b_{ij} gives

$$(\partial W/\partial T)_\eta = (\partial W/\partial T)_\tau + \sum_{ij} b_{ij} (\partial W/\partial \tau_{ij})_{T\tau'}. \tag{2.82}$$

This is a generalization of (1.52), and it illustrates the interpretation of the thermal stresses b_{ij} as a generalization of the quantity $-\beta B_T$.

From the stress–strain derivative (2.56) it is obvious that

$$(\partial b_{ij}/\partial \eta_{kl})_{T\eta'} = (\partial B^T_{ijkl}/\partial T)_\eta, \tag{2.83}$$

and from the strain–stress derivative (2.59),

$$(\partial \beta_{ij}/\partial \tau_{kl})_{T\tau'} = (\partial S^T_{ijkl}/\partial T)_\tau. \tag{2.84}$$

In order to evaluate the temperature derivative in (2.83) from the measured temperature derivative of B^T_{ijkl} at constant stress, one may use (2.82) to write

$$(\partial B^T_{ijkl}/\partial T)_\eta = (\partial B^T_{ijkl}/\partial T)_\tau + \sum_{mn} b_{mn} (\partial B^T_{ijkl}/\partial \tau_{mn})_{T\tau'}. \tag{2.85}$$

In order to evaluate the temperature derivative in (2.84) from the measured temperature derivative of B^T_{ijkl} at constant stress, we differentiate the inverse relation (2.57) with respect to T at constant stress to obtain

$$\sum_{kl} [(\partial S^T_{ijkl}/\partial T)_\tau B^T_{klmn} + S^T_{ijkl} (\partial B^T_{klmn}/\partial T)_\tau] = 0. \tag{2.86}$$

Again using the inverse relation, this may be solved to give

$$(\partial S^T_{ijkl}/\partial T)_\tau = -\sum_{mnpq} S^T_{ijmn} (\partial B^T_{mnpq}/\partial T)_\tau S^T_{pqkl}. \tag{2.87}$$

The difference between adiabatic and isothermal stress–strain coefficients may be derived from (2.76) for $d\tau_{ij}$, by dividing by $d\eta_{kl}$ and taking constant S,η' to obtain

$$(\partial\tau_{ij}/\partial\eta_{kl})_{S\eta'} = (\partial\tau_{ij}/\partial\eta_{kl})_{T\eta'} + (\partial\tau_{ij}/\partial T)_{\eta}(\partial T/\partial\eta_{kl})_{S\eta'},$$

or

$$B^S_{ijkl} = B^T_{ijkl} + b_{ij}(\partial T/\partial\eta_{kl})_{S\eta'}. \tag{2.88}$$

The last term is evaluated with the Maxwell equation (2.79) and the definition (2.73) of C_η as

$$\begin{aligned}(\partial T/\partial\eta_{kl})_{S\eta'} &= V(\partial\tau_{kl}/\partial S)_{\eta}\\ &= V(\partial\tau_{kl}/\partial T)_{\eta}(\partial T/\partial S)_{\eta}\\ &= (VT/C_\eta)b_{kl}.\end{aligned} \tag{2.89}$$

Then (2.88) is simply

$$B^S_{ijkl} - B^T_{ijkl} = (TV/C_\eta)b_{ij}b_{kl}. \tag{2.90}$$

This is the generalization of (1.42) for $B_S - B_T$. In addition, since the elastic constants C_{ijkl} and the stiffnesses B_{ijkl} differ only by components of the stress, according to (2.53) for adiabatic coefficients and (2.54) for isothermal coefficients, then we have

$$C^S_{ijkl} - C^T_{ijkl} = B^S_{ijkl} - B^T_{ijkl}. \tag{2.91}$$

The difference in heat capacities $C_\tau - C_\eta$ is derived as follows. Consider S a function of η_{ij} and T, and then a function of τ_{ij} and T, and write

$$\begin{aligned}dS &= (\partial S/\partial T)_\eta dT + \sum_{ij}(\partial S/\partial\eta_{ij})_{T\eta'}d\eta_{ij}\\ &= (\partial S/\partial T)_\tau dT + \sum_{ij}(\partial S/\partial\tau_{ij})_{T\tau'}d\tau_{ij}.\end{aligned} \tag{2.92}$$

Multiply (2.92) by T/dT and take constant τ_{ij} to obtain

$$C_\eta + T\sum_{ij}(\partial S/\partial\eta_{ij})_{T\eta'}(\partial\eta_{ij}/\partial T)_\tau = C_\tau;$$

then with the Maxwell equation (2.80) this is

$$C_\tau - C_\eta = -TV\sum_{ij} b_{ij}\beta_{ij}, \tag{2.93}$$

or

$$C_\tau - C_\eta = TV\sum_{ijkl} B^T_{ijkl}\beta_{ij}\beta_{kl}. \tag{2.94}$$

This is the generalization of (1.40) for $C_P - C_V$. The ratio of heat capacities is found by expressing the quantity $TVb_{ij}\sum_{kl} b_{kl}\beta_{kl}$ by means of (2.90) and also by (2.93), and equating the results:

$$C_\eta \sum_{kl} (B^S_{ijkl} - B^T_{ijkl})\beta_{kl} = -\, b_{ij}(C_\tau - C_\eta);$$

now dividing by $C_\eta b_{ij}$ and using the relation (2.77) between the b_{ij} and β_{kl},

$$\frac{C_\tau}{C_\eta} = \frac{\sum_{kl} B^S_{ijkl}\beta_{kl}}{\sum_{kl} B^T_{ijkl}\beta_{kl}}. \tag{2.95}$$

This is the generalization of (1.43) for C_P/C_V.

We can also calculate the difference $C_V - C_\eta$ from the preceding results. In fact, considering the stress to be isotropic pressure, $C_\tau = C_P$, and (2.93) minus (1.40) gives

$$C_V - C_\eta = -(TV/k_T)\left[\beta^2 + k_T \sum_{ij} b_{ij}\beta_{ij}\right]. \tag{2.96}$$

This can be written in a more symmetric form by defining a compressibility tensor

$$k^T_{ij} = \sum_k S^T_{kkij}; \tag{2.97}$$

then the principal components of the compressibility are k^T_{ii} and $k_T = \Sigma_i\, k^T_{ii}$. Furthermore, the thermal expansion coefficient at constant stress is, from (2.75) and (2.78),

$$\beta = -\sum_{ij} k^T_{ij} b_{ij}, \tag{2.98}$$

and finally

$$C_V - C_\eta = (TV/k_T) \sum_{ij} b_{ij}(\beta k^T_{ij} - k_T\beta_{ij}). \tag{2.99}$$

The Grüneisen parameter γ may be written, from its definition (1.30) and the relation (1.29) for βB_T, as

$$\gamma = (V/C_V)(\partial S/\partial V)_T; \tag{2.100}$$

this may be generalized to a tensor γ_{ij} for stress–strain variables,

$$\gamma_{ij} = (1/C_\eta)(\partial S/\partial \eta_{ij})_{T\eta'}. \tag{2.101}$$

From the Maxwell equation (2.80) it is seen that γ_{ij} is simply related to b_{ij}:

$$\gamma_{ij} = -(V/C_\eta) b_{ij}. \tag{2.102}$$

In terms of the thermal strains,

$$\gamma_{ij} = (V/C_\eta) \sum_{kl} B^T_{ijkl}\beta_{kl} = (V/C_\tau) \sum_{kl} B^S_{ijkl}\beta_{kl}, \tag{2.103}$$

where we used (2.95) for the second equality. Also since $C_\eta = (\partial U/\partial T)_\eta$, we can rewrite (2.102) in the form

$$\gamma_{ij} = -V(\partial\tau_{ij}/\partial U)_\eta. \tag{2.104}$$

This equation provides a possible way of measuring γ_{ij} directly, by observing the anisotroic stress $d\tau_{ij}$ produced by a constant configuration change in the internal energy dU [see discussion following (1.31)].

For theoretical calculations, it is convenient to write the thermodynamic functions as derivatives of the state functions. The temperature, entropy, stresses, and elastic constants are already expressed as such derivatives in (2.25)–(2.28) and (2.31–(2.36). From these results and the definitions of C_η and b_{ij}, it is obvious that

$$C_\eta = -T(\partial^2 F/\partial T^2)_\eta, \tag{2.105}$$

$$b_{ij} = V^{-1}(\partial^2 F/\partial T\partial\eta_{ij})_{\eta,T\eta'}. \tag{2.106}$$

Again the β_{ij} and γ_{ij} are more complicated, being ratios of state function derivatives.

Crystal Symmetries

The second-order elastic constants may be arranged in a 6 × 6 matrix in Voigt notation. Since this matrix is symmetric, the maximum number of independent $C_{\alpha\beta}$ is 21 for adiabatic or isothermal constants. For a given crystal symmetry, certain of these 21 elastic constants may be equal, or otherwise simply related, since they represent derivatives of the state functions with respect to strains which are crystallographically equivalent. Thus the crystal symmetry may further reduce the number of independent second-order elastic constants; the same observations apply also to the set of third-order elastic constants $C_{\alpha\beta\gamma}$, the set of fourth-order elastic constants $C_{\alpha\beta\gamma\delta}$, and so on. In addition, the equations of thermoelasticity are invariant with respect to inversion of the coordinate system. Therefore, in considering the effects of crystal symmetry, we are concerned only with the 11 Laue groups, since all classes in a given group have a common array of elastic constants of each order. The Laue groups and the number of independent second-, third-, and fourth-order elastic constants for each, as well as for isotropic materials, are listed in Table 1. The scheme of independent second-order elastic constants for each group is listed in Table 2. These tables are valid for a crystal in the presence of arbitrary applied stress, but the crystal symmetry must be chosen to correspond to the stressed configuration.

Consider for example the crystal groups cubic, hexagonal, orthorhombic, and tetragonal I, with no shear stresses:

$$\tau_{ij} = \delta_{ij}\tau_{ii}. \tag{2.107}$$

Table 1. Laue groups and the number of independent elastic constants. The second column gives the Hermann–Mauguin symbol for each class in the group, and the last three columns give the number of independent second-, third-, and fourth-order elastic constants, respectively.[a]

Laue group	Point groups	Crystal system	$C_{\alpha\beta}$	$C_{\alpha\beta\gamma}$	$C_{\alpha\beta\gamma\delta}$
N	$1, \bar{1}$	Triclinic	21	56	126
M	$m, 2, \frac{2}{m}$	Monoclinic	13	32	70
0	$2mm, 222, \frac{2}{m}\frac{2}{m}\frac{2}{m}$	Orthorhombic	9	20	42
T II	$4, \bar{4}, \frac{4}{m}$	Tetragonal	7	16	36
T I	$4mm, \bar{4}2m, 422, \frac{4}{m}\frac{2}{m}\frac{2}{m}$	Tetragonal	6	12	25
R II	$3, \bar{3}$	Rhombohedral	7	20	42
R I	$3m, 32, \bar{3}\frac{2}{m}$	Rhombohedral	6	14	28
H II	$\bar{6}, 6, \frac{6}{m}$	Hexagonal	5	12	24
H I	$\bar{6}2m, 6mm, 622, \frac{6}{m}\frac{2}{m}\frac{2}{m}$	Hexagonal	5	10	19
C II	$23, \frac{2}{m}\bar{3}$	Cubic	3	8	14
C I	$\bar{4}3m, 432, \frac{4}{m}\bar{3}\frac{2}{m}$	Cubic	3	6	11
I		Isotropic	2	3	4

[a] From D. C. Wallace, in *Solid State Physics*, edited by H. Ehrenreich, F. Seitz, and D. Turnbull, Academic Press Inc., New York, 1970, Vol. 25, p. 301.

For adiabatic or isothermal coefficients, we can write

$$B_{ijkl} = C_{ijkl} + \Delta_{ijkl}, \tag{2.108}$$

where Δ_{ijkl} is given by (2.53) or (2.54). In Voigt notation the matrix $[\Delta_{\alpha\beta}]$ has nonzero elements in the upper left 3×3 matrix, and also nonzero Δ_{44}, Δ_{55}, and Δ_{66}. For the crystal symmetries considered, the matrix $[C_{\alpha\beta}]$ has nonzero elements in the same locations, and hence so do the matrices

Table 2. Scheme of independent second-order elastic constants for each group. Only the subscripts $\alpha\beta$ for each $C_{\alpha\beta}$ are listed. The scheme is the same for the two hexagonal groups (denoted as H) and the two cubic groups (denoted as C).[a]

N	M	O	T II	T I	R II	R I	H	C	I
11	11	11	11	11	11	11	11	11	11
12	12	12	12	12	12	12	12	12	12
13	13	13	13	13	13	13	13	12	12
14	0	0	0	0	14	14	0	0	0
15	15	0	0	0	15	0	0	0	0
16	0	0	16	0	0	0	0	0	0
22	22	22	11	11	11	11	11	11	11
23	23	23	13	13	13	13	13	12	12
24	0	0	0	0	−14	−14	0	0	0
25	25	0	0	0	−15	0	0	0	0
26	0	0	−16	0	0	0	0	0	0
33	33	33	33	33	33	33	33	11	11
34	0	0	0	0	0	0	0	0	0
35	35	0	0	0	0	0	0	0	0
36	0	0	0	0	0	0	0	0	0
44	44	44	44	44	44	44	44	44	44[b]
45	0	0	0	0	0	0	0	0	0
46	46	0	0	0	−15	0	0	0	0
55	55	55	44	44	44	44	44	44	44[b]
56	0	0	0	0	14	14	0	0	0
66	66	66	66	66	66[c]	66[c]	66[c]	44	44[b]

[a] From D. C. Wallace, in *Solid State Physics*, edited by H. Ehrenreich, F. Seitz, and D. Turnbull, Academic Press Inc., New York, 1970, Vol. 25, p. 301.
[b] $C_{44} = \frac{1}{2}(C_{11} - C_{12})$
[c] $C_{66} = \frac{1}{2}(C_{11} - C_{12})$

$[B_{\alpha\beta}]$ and $[S_{\alpha\beta}]$. This tells us that

$$\beta_{ij} = \delta_{ij}\beta_{ii},$$
$$b_{ij} = \delta_{ij}b_{ii}, \tag{2.109}$$
$$k_{ij}^T = \delta_{ij}k_{ii}^T.$$

Under these conditions many of the thermoelastic equations are greatly simplified; for example,

$$B_{ijkl}^S - B_{ijkl}^T = \delta_{ij}\delta_{kl}(TV/C_\eta)b_{ii}b_{kk}; \tag{2.110}$$
$$C_V - C_\eta = (TV/k_T)\sum_i b_{ii}(\beta k_{ii}^T - k_T\beta_{ii}). \tag{2.111}$$

It is instructive to transfer these equations to Voigt notation:

$$B^S_{\alpha\beta} - B^T_{\alpha\beta} = \delta_\alpha \delta_\beta (TV/C_\eta) b_\alpha b_\beta; \tag{2.112}$$

$$C_V - C_\eta = (TV/k_T) \sum_\alpha \delta_\alpha b_\alpha (\beta k^T_\alpha - k_T \beta_\alpha); \tag{2.113}$$

where

$$\delta_\alpha = 1, \quad \alpha = 1,2,3; \qquad \delta_\alpha = 0, \quad \alpha = 4,5,6. \tag{2.114}$$

For a cubic crystal, or isotropic material, under isotropic pressure P,

$$\tau_{ij} = -P\delta_{ij}, \tag{2.115}$$

and the various thermoelastic tensors simplify even further:

$$\begin{aligned} \beta_{ij} &= \tfrac{1}{3}\beta\delta_{ij}, \\ b_{ij} &= -\beta B_T \delta_{ij}, \\ k^T_{ij} &= \tfrac{1}{3}k_T \delta_{ij}. \end{aligned} \tag{2.116}$$

For this example we find

$$B^S_{ijkl} - B^T_{ijkl} = \delta_{ij}\delta_{kl}(B_S - B_T); \tag{2.117}$$

$$C_V - C_\eta = 0; \tag{2.118}$$

$$C_\tau - C_\eta = C_P - C_V, \tag{2.119}$$

where $C_P - C_V$ is given by (1.40). In Voigt notation (2.117) is

$$\begin{aligned} &B^S_{11} - B^T_{11} = B^S_{12} - B^T_{12} = B_S - B_T; \\ &B^S_{44} - B^T_{44} = 0. \end{aligned} \tag{2.120}$$

When one wishes to treat problems relating to a crystal of prescribed symmetry in the presence of applied stress, or zero stress, the procedure is to construct the matrix $[C_{\alpha\beta}]$ from Table 2, then $[B_{\alpha\beta}]$ with the help of (2.53) or (2.54), and invert $[B_{\alpha\beta}]$ to get $[S_{\alpha\beta}]$; this gives all the second-order elastic coefficients required for treating homogeneous strain problems. The inversion of $[B_{\alpha\beta}]$ is discussed in detail elsewhere;* certain results of use here are as follows. For a cubic crystal under isotropic pressure, and for adiabatic or isothermal cases,

$$\begin{aligned} k &= 3(S_{11} + 2S_{12}), \\ B &= \tfrac{1}{3}(B_{11} + 2B_{12}) = \tfrac{1}{3}(C_{11} + 2C_{12} + P); \end{aligned} \tag{2.121}$$

* D. C. Wallace, in *Solid State Physics*, edited by H. Ehrenreich, F. Seitz, and D. Turnbull, Academic Press Inc., New York, 1970, Vol. 25, p. 301.

for tetragonal I or hexagonal at zero stress,

$$\begin{aligned} k_{11} &= k_{22} = k_{\perp}, \qquad k_{33} = k_{\parallel}; \\ k &= k_{\parallel} + 2k_{\perp}; \\ k_{\parallel} &= (C_{11} + C_{12} - 2C_{13})/C, \\ k_{\perp} &= (C_{33} - C_{13})/C, \\ C &= C_{33}(C_{11} + C_{12}) - 2C_{13}^2. \end{aligned} \tag{2.122}$$

3. WAVE PROPAGATION

Equation of Motion

In order to treat elastic wave propagation it is necessary to make a further generalization of the homogeneous strain concept of the preceding section. The initial configuration **X** still represents a crystal in the presence of uniform stress as a homogeneous elastic medium. The final configuration **x**, however, now corresponds to an elastic wave propagating in this medium, so the strain parameters α_{ij}, u_{ij}, and η_{ij} are no longer constants but vary with the time t and the location **X**. If we consider only elastic waves of very long wavelength, such that the wavelength is much greater than the range of effective forces in the crystal, then in any region whose size is of the order of the range of forces, the deformation is approximately homogeneous and the equations of Section 2 can be used. The range of effective forces in crystals should always be finite, even for Coulomb forces as in ionic crystals and metals, because charge neutrality is maintained over small regions. We will therefore work in the limit of long-wavelength elastic waves. For shorter wavelengths it is necessary to take into account the detailed structure of the forces in the crystal; this leads us into lattice dynamics, which is the subject of the following chapters.

Let us define the state function densities by script letters $\mathscr{U}$ and $\mathscr{F}$; these are densities per unit *initial* volume. Rotational invariance still holds locally, so in correspondence to (2.24)

$$\begin{aligned} \mathscr{U}(\mathbf{x},S) &= \mathscr{U}(\mathbf{X},\eta_{ij},S), \\ \mathscr{F}(\mathbf{x},T) &= \mathscr{F}(\mathbf{X},\eta_{ij},T), \end{aligned} \tag{3.1}$$

where $\eta_{ij} = \eta_{ij}(\mathbf{X},t)$ are the Lagrangian strains from **X** to **x**. For adiabatic motion, for example, the density of the Lagrangian per unit unitial volume is

$$\mathscr{L} = \tfrac{1}{2}\rho(\mathbf{X}) \sum_i \dot{x}_i^2 - \mathscr{U}(\mathbf{x},S), \tag{3.2}$$

where $\rho(\mathbf{X})$ is the (constant) density of the elastic medium in the initial configuration. Lagrange's equations of motion in the absence of body forces are

$$\frac{d}{dt}\frac{\partial \mathscr{L}}{\partial \dot{x}_i} + \sum_k \frac{\partial}{\partial X_k}\frac{\partial \mathscr{L}}{\partial \alpha_{ik}} = 0, \tag{3.3}$$

or

$$\rho(\mathbf{X})\ddot{x}_i - \sum_k (\partial/\partial X_k)(\partial \mathscr{U}/\partial \alpha_{ik})_{S\alpha'} = 0. \tag{3.4}$$

Since $\mathscr{U}$ depends on the strain only through the η_{ij}, according to (3.1), we have

$$(\partial \mathscr{U}/\partial \alpha_{ik})_{S\alpha'} = \sum_{lm} (\partial \mathscr{U}/\partial \eta_{lm})_{S\eta'}(\partial \eta_{lm}/\partial \alpha_{ik})_{\alpha'}. \tag{3.5}$$

Also from (2.11) relating η_{ij} and α_{ij},

$$(\partial \eta_{lm}/\partial \alpha_{ik})_{\alpha'} = \tfrac{1}{2}(\alpha_{il}\delta_{km} + \alpha_{im}\delta_{kl}). \tag{3.6}$$

With (3.6), and noting the symmetry of η_{kl}, the equation of motion (3.4) is

$$\rho(\mathbf{X})\ddot{x}_i = \sum_{kl} \frac{\partial}{\partial X_k}\alpha_{il}\left(\frac{\partial \mathscr{U}}{\partial \eta_{kl}}\right)_{S\eta'}. \tag{3.7}$$

For small-amplitude wave propagation the right-hand side of (3.7) is linearized by evaluating the derivatives at $\mathbf{X}$, where $\alpha_{ij} = \delta_{ij}$ and $\eta_{ij} = 0$. From the internal energy expansion (2.37), considered as a local equation with $\mathscr{U} = U/V$,

$$(\partial \mathscr{U}/\partial \eta_{kl})_{S\eta'} = \tau_{kl} + \sum_{mn} C^S_{klmn}\eta_{mn} + \cdots, \tag{3.8}$$

and from the definition (2.4) of α_{ij},

$$(\partial \alpha_{il}/\partial X_k)_{\alpha'} = (\partial \alpha_{ik}/\partial X_l)_{\alpha'} = (\partial^2 x_i/\partial X_l \partial X_k). \tag{3.9}$$

With (3.9) and (3.8), and differentiating η_{mn} in (3.8) by means of (3.6) and (3.9), the right side of the equation of motion (3.7) evaluated at $\mathbf{X}$ is

$$\sum_{jkl} (\partial^2 x_k/\partial X_j \partial X_l)[\tau_{jl}\delta_{ik} + C^S_{ijkl}]. \tag{3.10}$$

For isothermal motion the potential in the Lagrangian density $\mathscr{L}$ is $\mathscr{F}$ in place of $\mathscr{U}$, and the elastic constants in (3.10) are then C^T_{ijkl} in place of C^S_{ijkl}. Introducing the wave propagation coefficients A_{ijkl}, the equation of motion for adiabatic or isothermal propagation, evaluated at $\mathbf{X}$, is

$$\rho\ddot{x}_i = \sum_{jkl} A_{ijkl}(\partial^2 x_k/\partial X_j \partial X_l), \tag{3.11}$$

where for adiabatic propagation one uses

$$A^S_{ijkl} = \tau_{jl}\delta_{ik} + C^S_{ijkl}, \tag{3.12}$$

and for isothermal propagation one uses

$$A^T_{ijkl} = \tau_{jl}\delta_{ik} + C^T_{ijkl}. \tag{3.13}$$

From these definitions of the propagation coefficients, it is obvious that

$$A^S_{ijkl} - A^T_{ijkl} = C^S_{ijkl} - C^T_{ijkl}, \tag{3.14}$$

and this may be evaluated from (2.90) and (2.91).

The wave propagation coefficients also appear as the second-order coefficients in the expansion of the state functions in powers of the displacement gradients u_{ij}. This may be shown by starting with the expansions (2.37) and (2.38) for U and F, respectively, putting in (2.9) for η_{ij} in terms of the u_{ij}, and collecting terms of the same order. For F, for example, the expansion is

$$F(\mathbf{x},T) = F(\mathbf{X},T) + V\sum_{ij}\tau_{ij}u_{ij} + \tfrac{1}{2}V\sum_{ijkl}A^T_{ijkl}u_{ij}u_{kl} + \cdots. \tag{3.15}$$

Now it should be noted that the original expansion (2.38) of F in powers of η_{ij} contains the rotational invariance condition, and the Voigt symmetry of the elastic constants C^T_{ijkl}, which appear in that expansion, simply reflect the rotational invariance. The rotational invariance of (3.15) is not obvious, since the u_{ij} are not symmetric in general, but it is still contained in the symmetry properties of the coefficients. These symmetry properties are

$$\tau_{ij} = \tau_{ji}, \tag{3.16}$$

and from (3.12) and (3.13) and the Voigt symmetry of the C_{ijkl},

$$A_{ijkl} = A_{klij}, \tag{3.17}$$

$$A_{ijkl} - \tau_{jl}\delta_{ik} = A_{jikl} - \tau_{il}\delta_{jk}, \tag{3.18}$$

for adiabatic or isothermal coefficients. Finally it should be noted that the wave-propagation coefficients do not have sufficient symmetry to be written in Voigt notation, unless the stresses vanish.

Elastic Waves

The linearized equation of motion (3.11) is a harmonic equation, and the solutions are plane elastic waves. The displacement in a plane wave of frequency ω and wave vector $\mathbf{k}$ is*

$$\mathbf{x} - \mathbf{X} = \mathbf{w}\sin(\mathbf{k}\cdot\mathbf{X} - \omega t), \tag{3.19}$$

where $\mathbf{w}$ gives the direction and magnitude of the displacement. With this

* We always use (wave vector) = (2π/wavelength) and ω = angular frequency.

displacement the equation of motion is

$$\rho\omega^2 w_i = \sum_{jkl} A_{ijkl} k_j k_l w_k, \tag{3.20}$$

or with $\hat{\mathbf{k}}$ a unit vector in the direction of $\mathbf{k}$ and v the wave velocity,

$$\hat{\mathbf{k}} = \mathbf{k}/|\mathbf{k}|, \qquad v = \omega/|\mathbf{k}|, \tag{3.21}$$

the equation of motion is

$$\rho v^2 w_i = \sum_{jkl} A_{ijkl} \hat{k}_j \hat{k}_l w_k. \tag{3.22}$$

Here the distinction between adiabatic and isothermal propagation has been dropped, since the equations cover either case. It is noted from (3.20) that $\omega^2 \propto k^2$; hence, the elastic waves are dispersionless, i.e., phase and group velocities are the same. This result is correct, of course, only for long-wavelength waves, which is the region of validity of our theory. In addition, it should be clearly recognized that the externally applied stresses τ_{ij} enter explicitly the wave-propagation coefficients, according to their definitions (3.12) and (3.13); hence, the externally applied stresses make an explicit contribution to the elastic wave velocities. The physical interpretation of this is as follows. If one generates a wave by applying a time-varying stress to the surface of a crystal, say with a transducer, then the initial applied stress works alternately with and against the transducer in producing displacements at the crystal surface. Alternatively, the initial applied stress can be considered as a body force throughout the crystal, providing restoring forces for the displacements which occur when a wave is propagating in the crystal.

The equations of motion (3.20) and (3.22) are eigenvalue–eigenvector equations; for a given direction of $\mathbf{k}$ each has three solutions, which will be denoted by $s = 1,2,3$. We define the real symmetric 3×3 propagation matrices $\mathbf{L}(\hat{\mathbf{k}})$ by

$$L_{ik} = \sum_{jl} A_{ijkl} \hat{k}_j \hat{k}_l. \tag{3.23}$$

The eigenvalues and eigenvectors of $\mathbf{L}$ are ρv_s^2 and $\mathbf{w}_s$, respectively, and the eigenvalue–eigenvector equation of motion is

$$\rho v_s^2 w_{is} = \sum_k L_{ik} w_{ks}. \tag{3.24}$$

The eigenvectors are chosen to be normalized, and since $\mathbf{L}$ is real symmetric they are orthonormal and complete:

$$\sum_i w_{is} w_{is'} = \delta_{ss'}, \quad \text{orthonormality}; \tag{3.25}$$

$$\sum_s w_{is} w_{i's} = \delta_{ii'}, \quad \text{completeness}. \tag{3.26}$$

The diagonalization of **L** is expressed by multiplying (3.24) by $w_{is'}$ and summing on i to obtain

$$\sum_{ij} w_{is} L_{ij} w_{js'} = \rho v_s^2 \delta_{ss'}, \quad \text{diagonalization.} \tag{3.27}$$

There are certain simplifications if the applied stress is isotropic pressure.

$$\tau_{ij} = -P\delta_{ij}, \tag{3.28}$$

and from (2.53) or (2.54),

$$B_{ijkl} = -P(\delta_{jl}\delta_{ik} + \delta_{il}\delta_{jk} - \delta_{ij}\delta_{kl}) + C_{ijkl}, \tag{3.29}$$

and from (3.12) or (3.13),

$$A_{ijkl} = -P\delta_{jl}\delta_{ik} + C_{ijkl}. \tag{3.30}$$

In the propagation matrix **L**, on account of the Σ_{jl} in (3.23), only the symmetric combination $A_{ijkl} + A_{ilkj}$ appears, and from (3.29) and (3.30) it is seen that

$$A_{ijkl} + A_{ilkj} = B_{ijkl} + B_{ilkj}. \tag{3.31}$$

Therefore, for isotropic pressure,

$$L_{ik} = \sum_{jl} B_{ijkl} \hat{k}_j \hat{k}_l, \tag{3.32}$$

and the measured elastic wave velocities may be interpreted directly in terms of stress–strain coefficients B_{ijkl}. Furthermore, these coefficients can be written in Voigt notation, and for isotropic pressure they have Voigt symmetry:

$$B_{\alpha\beta} = B_{\beta\alpha}, \quad \text{isotropic pressure.} \tag{3.33}$$

In fact, the propagation matrices may be simply expressed in terms of the elastic constants, for from (3.29) and (3.32),

$$L_{ik} = \sum_{jl} C_{ijkl} \hat{k}_j \hat{k}_l - P\delta_{ik}, \tag{3.34}$$

since $\Sigma_j \hat{k}_j \hat{k}_j = 1$. The eigenvalues are formally written

$$\rho v_s^2 = \sum_{ijkl} w_{is} C_{ijkl} \hat{k}_j \hat{k}_l w_{ks} - P; \tag{3.35}$$

this result shows clearly how the applied pressure appears explicitly in the elastic wave velocities of a crystal under pressure.

It is instructive to work out the propagation matrices for a few particular crystals. For a tetragonal I crystal under isotropic pressure, there are six independent $B_{\alpha\beta}$:

$$
\begin{aligned}
&B_{11} = B_{22} = C_{11} - P,\\
&B_{33} = C_{33} - P,\\
&B_{12} = B_{21} = C_{12} + P,\\
&B_{13} = B_{23} = B_{31} = B_{32} = C_{13} + P,\\
&B_{44} = B_{55} = C_{44} - P,\\
&B_{66} = C_{66} - P.
\end{aligned}
\tag{3.36}
$$

Elements of the propagation matrix are easily worked out for any direction of **k** from (3.32), to give

$$
\begin{aligned}
&L_{11} = B_{11}\hat{k}_1^2 + B_{66}\hat{k}_2^2 + B_{44}\hat{k}_3^2,\\
&L_{22} = B_{66}\hat{k}_1^2 + B_{11}\hat{k}_2^2 + B_{44}\hat{k}_3^2,\\
&L_{33} = B_{44}(\hat{k}_1^2 + \hat{k}_2^2) + B_{33}\hat{k}_3^2,\\
&L_{12} = L_{21} = (B_{12} + B_{66})\hat{k}_1\hat{k}_2,\\
&L_{13} = L_{31} = (B_{13} + B_{44})\hat{k}_1\hat{k}_3,\\
&L_{23} = L_{32} = (B_{13} + B_{44})\hat{k}_2\hat{k}_3.
\end{aligned}
\tag{3.37}
$$

For a hexagonal crystal the results are the same, with the additional condition $C_{66} = \frac{1}{2}(C_{11} - C_{12})$.

For a cubic crystal under isotropic pressure there are only three independent $B_{\alpha\beta}$:

$$
\begin{aligned}
&B_{11} = B_{22} = B_{33} = C_{11} - P,\\
&B_{12} = B_{23} = B_{31} = \cdots = C_{12} + P,\\
&B_{44} = B_{55} = B_{66} = C_{44} - P.
\end{aligned}
\tag{3.38}
$$

The propagation matrix elements are represented by

$$
\begin{aligned}
&L_{ii} = B_{11}\hat{k}_i^2 + B_{44}(1 - \hat{k}_i^2),\\
&L_{ij} = (B_{12} + B_{44})\hat{k}_i\hat{k}_j, \quad i \neq j.
\end{aligned}
\tag{3.39}
$$

In our discussion of elastic waves we have neglected attenuation and interactions among the waves. Thus the present elastic waves are normal modes of thermoelastic motion, and have infinite lifetimes; this is in fact a pretty good approximation to the situation in most real crystals, since the lifetime of an elastic wave is generally very long compared to its period of vibration. It is well known, however, that under certain resonance conditions, a traveling elastic wave will generate a harmonic at twice the frequency, and two interacting elastic waves will generate sum- and difference-frequency waves. These anharmonic effects are related to third- and higher-order elastic constants, and may be studied by keeping such terms in the expansion of the equation of motion through equation (3.8).*

Stability

For the crystal to be stable against all elastic homogeneous deformations, the energies of all the long-wavelength elastic waves must be positive:

$$[\omega_s(\hat{\mathbf{k}})]^2 > 0, \quad \text{for all } \hat{\mathbf{k}}, s. \tag{3.40}$$

Equivalently, the propagation matrix must be positive definite for every direction of $\mathbf{k}$:

$$\mathbf{L}(\hat{\mathbf{k}}) > 0, \quad \text{for all } \hat{\mathbf{k}}. \tag{3.41}$$

This must be true for adiabatic and isothermal waves. For isothermal waves, for example, the stability requirement is the same as $d^2F > 0$ for all virtual strains at constant T. To prove this, refer to the displacement–gradient expansion (3.15) for F to obtain for infinitesimal u_{ij}

$$d^2F_T = V \sum_{ijkl} A^T_{ijkl} u_{ij} u_{kl}; \tag{3.42}$$

since u_{ij} are arbitrary, they may be written $u_{ij} = y_i \hat{k}_j$, where $\mathbf{y}$ and $\hat{\mathbf{k}}$ are arbitrary vectors (with $\hat{\mathbf{k}}$ of unit length). Then identifying the isothermal propagation matrix $\mathbf{L}^T$ by means of the definition (3.23),

$$d^2F_T = V \sum_{ik} L^T_{ik} y_i y_k. \tag{3.43}$$

From the theorem (1.67), d^2F_T is positive definite if and only if $\mathbf{L}^T$ is positive definite for every $\hat{\mathbf{k}}$. Thus we have shown that the free energy is minimum with respect to arbitrary virtual homogeneous strains at constant temperature if and only if all long-wavelength isothermal elastic waves have positive energies. The same arguments hold for adiabatic waves, so the stability requirements (3.40) or (3.41) are equivalent to

$$\begin{aligned} d^2F_T &> 0, \\ d^2U_S &> 0. \end{aligned} \tag{3.44}$$

The general stability results summarized in (1.83) and (1.84) must still hold, of course.

The adiabatic or isothermal propagation coefficients A_{ijkl} can be written as a symmetric 9×9 matrix, and the stability condition (3.44) is that each of these matrices (the adiabatic and the isothermal) must be positive definite. When the stress applied to the crystal is zero, $A_{ijkl} = C_{ijkl}$, and the stability requirement is that each symmetric 9×9 matrix of C_{ijkl} (the adiabatic and

* D. C. Wallace, in *Solid State Physics*, edited by H. Ehrenreich, F. Seitz, and D. Turnbull, Academic Press Inc., New York, 1970, Vol. 25, p. 301.

the isothermal) must be positive definite. Further, according to Table 2, for the crystal groups cubic, hexagonal, orthorhombic, and tetragonal I, the 9×9 matrix of C_{ijkl} has nonzero elements only in the three 3×3 matrices along the diagonal, so that the 9×9 matrix of C_{ijkl} is positive definite if and only if the 6×6 matrix of $C_{\alpha\beta}$ (Voigt notation) is positive definite. Hence for these crystal groups and at zero applied stress, the stability condition is

$$[C_{\alpha\beta}] > 0, \tag{3.45}$$

for adiabatic and for isothermal $C_{\alpha\beta}$.

Consider for example a cubic crystal at zero stress, where

$$[C_{\alpha\beta}] = \left(\begin{array}{ccc|ccc} C_{11} & C_{12} & C_{12} & & & \\ C_{12} & C_{11} & C_{12} & & 0 & \\ C_{12} & C_{12} & C_{11} & & & \\ \hline & & & C_{44} & 0 & 0 \\ & 0 & & 0 & C_{44} & 0 \\ & & & 0 & 0 & C_{44} \end{array}\right). \tag{3.46}$$

The eigenvalues μ_α of $[C_{\alpha\beta}]$ are easily found to be

$$\begin{aligned} \mu_1 &= C_{11} + 2C_{12}, \\ \mu_2 = \mu_3 &= C_{11} - C_{12} \\ \mu_4 = \mu_5 = \mu_6 &= C_{44}. \end{aligned} \tag{3.47}$$

Stability then requires, for isothermal and adiabatic constants,

$$C_{11} + 2C_{12} > 0, \qquad C_{11} - C_{12} > 0, \qquad C_{44} > 0; \tag{3.48}$$

note that the first condition here requires k_T and k_S both positive, according to (2.121).

The example of tetragonal I or hexagonal at zero stress is more interesting; here

$$[C_{\alpha\beta}] = \left(\begin{array}{ccc|ccc} C_{11} & C_{12} & C_{13} & & & \\ C_{12} & C_{11} & C_{13} & & 0 & \\ C_{13} & C_{13} & C_{33} & & & \\ \hline & & & C_{44} & 0 & 0 \\ & 0 & & 0 & C_{44} & 0 \\ & & & 0 & 0 & C_{66} \end{array}\right). \tag{3.49}$$

The eigenvalues μ_α are

$$\mu_1 = C_{11} - C_{12},$$

$$\mu_2 \text{ and } \mu_3 = \tfrac{1}{2}\{(C_{11} + C_{12} + C_{33}) \pm [(C_{11} + C_{12} - C_{33})^2 + 8C_{13}^2]^{1/2}\},$$

$$\mu_4 = \mu_5 = C_{44}, \tag{3.50}$$

$$\mu_6 = C_{66}.$$

The two eigenvalues μ_2 and μ_3 correspond to the + and − signs here; they are both positive if and only if $\mu_2 + \mu_3$ and $\mu_2\mu_3$ are both positive. Stability then requires

$$C_{11} - C_{12} > 0, \tag{3.51a}$$

$$C_{11} + C_{12} + C_{33} > 0, \quad \text{from } \mu_2 + \mu_3 > 0, \tag{3.51b}$$

$$C = (C_{11} + C_{12})C_{33} - 2C_{13}^2 > 0, \quad \text{from } \mu_2\mu_3 > 0, \tag{3.51c}$$

$$C_{44} > 0, \tag{3.51d}$$

$$C_{66} > 0. \tag{3.51e}$$

From (3.51c),

$$(C_{11} + C_{12})C_{33} > 0, \tag{3.52}$$

and this, combined with (3.51b), gives

$$(C_{11} + C_{12}) > 0, \quad C_{33} > 0, \tag{3.53}$$

and the first of these, together with (3.51a), gives

$$C_{11} > 0. \tag{3.54}$$

These conditions further prove that the compressibility is positive, since $2(C_{13} - C_{33})^2$ must always be positive, and writing this out

$$2C_{13}^2 - 4C_{13}C_{33} + 2C_{33}^2 > 0, \tag{3.55}$$

or in view of (3.51c)

$$(C_{11} + C_{12})C_{33} - 4C_{13}C_{33} + 2C_{33}^2 > 0, \tag{3.56}$$

and finally since $C_{33} > 0$,

$$(C_{11} + C_{12} + 2C_{33} - 4C_{13}) > 0. \tag{3.57}$$

Since $k = k_\parallel + 2k_\perp$, given by (2.122), this last result along wtih (3.51c) implies that $k > 0$, and this holds of course for k_T and k_S. It is interesting to note that $k_\parallel$ and $k_\perp$ are not both required to be positive for stability.

When the stress applied to the crystal is not zero, the stability conditions are most conveniently worked out in terms of the propagation coefficients, either from (3.41) or from (3.44).

4. VOLUME CORRECTIONS

Expansions for Small Pressure

The derivations of the preceding sections were carried out for materials in the presence of arbitrary applied stress. In most laboratory experiments, however, the stress is small compared to the elastic constants, or equivalently, the Lagrangian strains η_{ij} from the zero-stress configuration are small compared to 1. It is therefore useful to study the region of small stress in more detail.

We begin with the case of pressure–volume variables. Slater* has used two expansions, one for the volume as a function of P and T:

$$(V_0 - V)/V_0 = -a_0(T) + a_1(T)P - a_2(T)P^2 + \cdots, \tag{4.1}$$

and one for the pressure as a function of V and T:

$$P(V,T) = P_0(T) + P_1(T)[(V_0 - V)/V_0] + P_2(T)[(V_0 - V)/V_0]^2 + \cdots. \tag{4.2}$$

Here $V = V(P,T)$, and $V_0 = V(P = 0, T = 0)$. The coefficients a_α and P_α in these expansions are functions only of T, and may be evaluated from $P = 0$ experimental results. From (4.1), if $P = 0$, $a_0(T)$ is just $(V - V_0)/V_0$, the relative change in volume from $T = 0$. Likewise from (4.2), $P_0(T)$ is just the pressure required to reduce the volume to V_0. Bridgman's† expansion of the volume as a function of P and T is

$$V(P,T) = V(0,T)[1 - A_1(T)P + A_2(T)P^2 - \cdots]. \tag{4.3}$$

These three expansions are not independent, of course, but are simply related. One can rewrite (4.1) as

$$V_0 - V(P,T) = V_0[-a_0 + a_1P - a_2P^2 + \cdots], \tag{4.4}$$

and evaluate this at $P = 0$ as

$$V_0 - V(0,T) = -a_0V_0. \tag{4.5}$$

* J. C. Slater, *Introduction to Chemical Physics*, McGraw-Hill Book Co. Inc., New York, 1939.

† P. W. Bridgman, *The Physics of High Pressure*, G. Bell and Sons Ltd., London, 1949.

Then (4.5) minus (4.4) is

$$V(P,T) - V(0,T) = (1 + a_0)^{-1}V(0,T)[-a_1P + a_2P^2 - \cdots], \quad (4.6)$$

and comparing this with Bridgman's equation (4.3) gives

$$A_1 = a_1/(1 + a_0), \qquad A_2 = a_2/(1 + a_0), \quad \text{etc.} \quad (4.7)$$

In addition, the pressure expansion (4.2) may be inverted to a volume expansion and then related to (4.1); we will do this only to the *second order* in $(V_0 - V)/V_0$. Write (4.2) as

$$0 = -P + P_0 + P_1[(V_0 - V)/V_0] + P_2[(V_0 - V)/V_0]^2, \quad (4.8)$$

put in (4.1) for $(V_0 - V)/V_0$ to order P^2, and set the coefficient of each power of P equal to zero to obtain

$$\text{zeroth order:} \qquad P_0 - a_0P_1 + a_0^2P_2 = 0; \quad (4.9)$$

$$\text{first order:} \qquad -1 + a_1P_1 - 2a_0a_1P_2 = 0; \quad (4.10)$$

$$\text{second order:} \qquad -a_2P_1 + a_1^2P_2 + 2a_0a_2P_2 = 0. \quad (4.11)$$

Solving for P_0, P_1, and P_2 gives

$$P_0 = (a_0/a_1) + (a_0^2a_2/a_1^3), \quad (4.12)$$

$$P_1 = (1/a_1) + 2(a_0a_2/a_1^3), \quad (4.13)$$

$$P_2 = (a_2/a_1^3). \quad (4.14)$$

It is also useful to relate the expansion coefficients to directly measured properties; we will do this for the coefficients a_α of the Slater expansion (4.1), and use the notation

$$a'_\alpha = da_\alpha/dT, \qquad a''_\alpha = d^2a_\alpha/dT^2, \quad \text{etc.} \quad (4.15)$$

First of all at zero pressure

$$a_0 = e^I - 1, \quad (4.16)$$

where

$$I = I(T) = \int_0^T \beta(T')\, dT'. \quad (4.17)$$

The proof is as follows.

$$dV = (\partial V/\partial T)_P dT = V\beta dT, \quad \text{at} \quad P = 0;$$

therefore $\beta dT = dV/V$, and

$$I = \int_0^T \beta\, dT' = \int_{V_0}^V dV'/V' = \ln(V/V_0),$$

so that

$$(V/V_0) - 1 = a_0 = e^I - 1. \tag{4.18}$$

Differentiating (4.18) gives

$$(\partial V/\partial T)_P = V_0 a_0',$$

and since $V_0 = V/(1 + a_0)$, the thermal expansion coefficient is

$$\beta = a_0'/(1 + a_0), \quad \text{at any } T \text{ and } \quad P = 0. \tag{4.19}$$

In the same way we find from (4.1) that

$$k_T = a_1/(1 + a_0), \quad \text{at any } T \text{ and } \quad P = 0. \tag{4.20}$$

Differentiating (4.19) and (4.20) with respect to T at constant P leads to

$$(\partial\beta/\partial T)_P + \beta^2 = a_0''/(1 + a_0), \quad P = 0; \tag{4.21}$$

$$(\partial k_T/\partial T)_P + \beta k_T = a_1'/(1 + a_0), \quad P = 0. \tag{4.22}$$

Finally it is easy to show from the above equations that

$$B_T^{-2}[(\partial B_T/\partial P)_T + 1] = 2a_2/(1 + a_0) = 2A_2, \quad P = 0. \tag{4.23}$$

Corrections to Fixed Volume

It is often a great simplification in theoretical work to restrict calculations to conditions of fixed configuration of the crystal. For this reason we will develop equations by which thermodynamic functions measured at $P = 0$ can be corrected to correspond to the fixed configuration of the crystal at $T = 0$, $P = 0$. (We could, of course, correct to some *other* configuration by means of the same equations.) This procedure removes the configuration-dependence, leaving only the explicit temperature-dependence, of the thermodynamic functions. The natural way to work out such corrections is in the form of power series in the strains relating the two configurations.

For the case of pressure–volume variables the expansion parameter is a_0, and the volume corrections are easily worked out in terms of the Slater coefficients a_α. For the isothermal compressibility, for example,

$$k_T(V) = k_T(V_0) + \int_{V_0}^{V} \left(\frac{\partial k_T}{\partial V}\right)_T dV, \tag{4.24}$$

and transforming to a pressure integral,

$$k_T(V) = k_T(V_0) + \int_{P_0}^{0} \left(\frac{\partial k_T}{\partial P}\right)_T dP, \tag{4.25}$$

since $P(V_0) = P_0$ and $P(V) = 0$. Now from the Slater expansion (4.1),

$$V = V_0[1 + a_0 - a_1P + a_2P^2 - \cdots];$$
$$-(\partial V/\partial P)_T = V_0[a_1 - 2a_2P + 3a_3P^2 - \cdots];$$

so that

$$k_T = \frac{[a_1 - 2a_2P + 3a_3P^2 - \cdots]}{[1 + a_0 - a_1P + a_2P^2 - \cdots]}. \tag{4.26}$$

Now (4.26) is expanded for small a_0 and P, keeping terms to, say, the second order in both (i.e., to order a_0^2, P^2, a_0P), and differentiated with respect to P to obtain $(\partial k_T/\partial P)_T$. Then the integral in (4.25) is carried out and evaluated with the aid of (4.12) for P_0, and the volume correction is obtained correct to the second order in a_0:

$$\begin{aligned} k_T(V_0) = k_T(V) &+ a_0[a_1 - 2(a_2/a_1)] \\ &+ a_0^2[3(a_3/a_1^2) - 2(a_2^2/a_1^3) - a_1] + \cdots, \end{aligned} \tag{4.27}$$

where $+ \cdots$ indicates terms of third and higher order in a_0.

In a similar way the volume corrections for the other thermodynamic functions may be worked out as power series in a_0. Results correct to order a_0^2 are

$$\beta(V_0) = \beta(V) + a_0[a_0' - (a_1'/a_1)] + a_0^2[(a_2'/a_1^2) - (a_1'a_2/a_1^3) - a_0'] + \cdots; \tag{4.28}$$

$$S(V_0) = S(V) - a_0V_0(a_0'/a_1) - \tfrac{1}{2}a_0^2V_0[2(a_0'a_2/a_1^3) - (a_1'/a_1^2)] + \cdots; \tag{4.29}$$

and results correct to order a_0 are

$$C_V(V_0) = C_V(V) - a_0V_0T[(a_0''/a_1) - 2(a_0'a_1'/a_1^2) + 2(a_0')^2(a_2/a_1^3)] + \cdots; \tag{4.30}$$

$$\beta B_T(V_0) = \beta B_T(V) + a_0[2(a_0'a_2/a_1^3) - (a_1'/a_1^2)] + \cdots. \tag{4.31}$$

It should be noted that all the quantities on the right-hand sides of these equations are evaluated at $P = 0$.

It is useful to express these volume corrections in terms of directly measurable quantities. If W is any thermodynamic function, we can write

$$W(V_0,T) = W(V,T) - a_0V_0(\partial W/\partial V)_T + \tfrac{1}{2}a_0^2V_0^2(\partial^2 W/\partial V^2)_T + \cdots, \tag{4.32}$$

where the derivatives are evaluated at V,T. For the adiabatic and isothermal bulk moduli, the first volume derivative is conveniently transformed to a pressure derivative with the aid of the general relation (1.51), and then

$$B_S(V_0) = B_S(V) + a_0 B_T(\partial B_S/\partial P)_T + \cdots; \tag{4.33}$$

$$B_T(V_0) = B_T(V) + a_0 B_T(\partial B_T/\partial P)_T + \cdots. \tag{4.34}$$

For the function βB_T, we note from (1.33) for B_T and (1.35) for βB_T that

$$(\partial \beta B_T/\partial V)_T = -V^{-1}(\partial B_T/\partial T)_V, \tag{4.35}$$

and the constant-volume temperature derivative of B_T is transformed by the general relation (1.52) to give finally

$$\beta B_T(V_0) = \beta B_T(V) + a_0[(\partial B_T/\partial T)_P + \beta B_T(\partial B_T/\partial P)_T] + \cdots. \tag{4.36}$$

The entropy volume correction is easily evaluated from the general expansion (4.32), since by the Maxwell equation (1.29)

$$(\partial S/\partial V)_T = \beta B_T,$$

and therefore

$$(\partial^2 S/\partial V^2)_T = (\partial \beta B_T/\partial V)_T.$$

Then with the aid of (4.35) or (4.36),

$$S(V_0) = S(V) - a_0 V_0 \beta B_T - \tfrac{1}{2} a_0^2 V_0[(\partial B_T/\partial T)_P + \beta B_T(\partial B_T/\partial P)_T] + \cdots. \tag{4.37}$$

For the heat capacity we can combine (1.47) and (1.29) to write

$$(\partial C_V/\partial V)_T = T(\partial^2 P/\partial T^2)_V = T(\partial \beta B_T/\partial T)_V, \tag{4.38}$$

and this is transformed by the general relation (1.52) to

$$(\partial C_V/\partial V)_T = T[(\partial \beta B_T/\partial T)_P + \beta B_T(\partial \beta B_T/\partial P)_T]. \tag{4.39}$$

Carrying out the differentiation and noting from (1.46) that

$$(\partial \beta/\partial P)_T = B_T^{-2}(\partial B_T/\partial T)_P, \tag{4.40}$$

the heat capacity volume correction is

$$C_V(V_0) = C_V(V) - a_0 V_0 T[\beta^2 B_T(\partial B_T/\partial P)_T + B_T(\partial \beta/\partial T)_P + 2\beta(\partial B_T/\partial T)_P] + \cdots. \tag{4.41}$$

Finally, let us evaluate the volume correction for the thermal expansion coefficient, to second order in a_0. The first volume derivative is

$$(\partial \beta/\partial V)_T = -(B_T/V)(\partial \beta/\partial P)_T = -(VB_T)^{-1}(\partial B_T/\partial T)_P, \tag{4.42}$$

where (4.40) gives the second equality. The second volume derivative is conveniently evaluated by writing

$$(\partial^2\beta/\partial V^2)_T = -(B_T/V)[\partial(\partial\beta/\partial V)_T/\partial P]_T, \tag{4.43}$$

and by evaluating the expression in square brackets by differentiating the last term in (4.42). When this is done the volume correction for β is

$$\beta(V_0) = \beta(V) + a_0V_0(VB_T)^{-1}(\partial B_T/\partial T)_P - \tfrac{1}{2}(a_0V_0/V)^2\{B_T^{-1}(\partial B_T/\partial T)_P \times [(\partial B_T/\partial P)_T - 1] - (\partial^2 B_T/\partial P\partial T)_{TP}\}.$$

Since $V_0/V = 1 - a_0 + \cdots$, this is, correct to order a_0^2,

$$\beta(V_0) = \beta(V) + a_0B_T^{-1}(\partial B_T/\partial T)_P + \tfrac{1}{2}a_0^2\{(\partial^2 B_T/\partial P\partial T)_{TP} - B_T^{-1}(\partial B_T/\partial T)_P[(\partial B_T/\partial P)_T + 1]\}. \tag{4.44}$$

Some formal simplification may be achieved by using (4.23) to introduce the Bridgman coefficient A_2, with the result

$$\beta(V_0) = \beta(V) + a_0B_T^{-1}(\partial B_T/\partial T)_P + a_0^2B_T(\partial A_2B_T/\partial T)_P + \cdots. \tag{4.45}$$

Expansions for Small Stress

The preceding results can be used to evaluate thermodynamic functions at the fixed volume V_0, for any temperature. For a cubic crystal the fixed volume corresponds to a fixed configuration; for crystals of lower symmetry, however, the fixed volume corresponds to a configuration which is a unique function of temperature, but which is still not generally a constant configuration. The correction of thermodynamic functions to conditions of constant configuration is a straightforward generalization of the volume correction.

Let $\mathbf{X}$ be the configuration at any temperature and zero stress, and let $\mathbf{X}_0$ be that at zero temperature and zero stress. The strain from $\mathbf{X}$ to $\mathbf{X}_0$ is $\eta_{ij}(T)$, and the corresponding stress is $\tau_{ij}(T)$. In other words at any temperature, in order to bring the zero-stress configuration $\mathbf{X}$ to the fixed configuration $\mathbf{X}_0$, the crystal must undergo the strain η_{ij} measured from $\mathbf{X}$, and this may be accomplished by applying the stresses τ_{ij}. From the stress–strain relation (2.52), and its inverse,

$$\tau_{ij} = \sum_{kl} C_{ijkl}^T\eta_{kl} + \cdots, \tag{4.46}$$

$$\eta_{ij} = \sum_{kl} S_{ijkl}^T\tau_{kl} + \cdots, \tag{4.47}$$

where C_{ijkl}^T and S_{ijkl}^T are evaluated at zero stress, and we are considering isothermal strains. The second-order terms in these expansions are easily worked out; the present discussion will be limited to first-order strain corrections. If W is any thermodynamic function, the formal strain correction is

$$W(\mathbf{X}_0) = W(\mathbf{X}) + \sum_{ij} (\partial W/\partial \eta_{ij})_{T\eta'}\eta_{ij} + \cdots \tag{4.48}$$

$$= W(\mathbf{X}) + \sum_{ijkl} (\partial W/\partial \eta_{ij})_{T\eta'} S^T_{ijkl}\tau_{kl} + \cdots . \tag{4.49}$$

In order to calculate the corrections from experimental data, it is sometimes more convenient to introduce the stress derivatives of W:

$$W(\mathbf{X}_0) = W(\mathbf{X}) + \sum_{ij} (\partial W/\partial \tau_{ij})_{T\tau'}\tau_{ij} + \cdots \tag{4.50}$$

$$= W(\mathbf{X}) + \sum_{ijkl} (\partial W/\partial \tau_{ij})_{T\tau'} C^T_{ijkl}\eta_{kl} + \cdots . \tag{4.51}$$

These expansions are applicable at any temperature, and the coefficients of η_{ij} or τ_{ij} on the right-hand sides are to be evaluated at zero stress.

The first step in evaluating the strain corrections is to derive the variation of second-order elastic constants with strain; this may be done in the same manner as the derivation in Section 2 of the variation of the stress with strain. For isothermal elastic constants for example, consider two configurations $\bar{\mathbf{X}}$ and $\mathbf{X}$ connected by the Lagrangian strains n_{ij}, and calculate in analogy to equation (2.48) for the stresses:

$$\begin{aligned} C^T_{ijkl} &= V^{-1}(\partial^2 F/\partial \eta_{ij}\partial \eta_{kl})_{T\eta'}, \quad \text{at } \mathbf{X}; \\ &= (\bar{V}/V) \sum_{mnpq} a_{im}a_{jn}a_{kp}a_{lq}\{\bar{C}^T_{mnpq} + \sum_{rs} \bar{C}^T_{mnpqrs}n_{rs} + \cdots\}. \end{aligned} \tag{4.52}$$

This may be evaluated with the aid of (2.45) for $V/\bar{V}$ and (2.44) for a_{ij}, to give an expression for $C^T_{ijkl} - \bar{C}^T_{ijkl}$ to first order in the n_{ij}. This expression is then applied to the present case of configurations $\mathbf{X}$ and $\mathbf{X}_0$ connected by the strains η_{ij}, with the result

$$\begin{aligned} C^T_{ijkl}(\mathbf{X}_0) &= C^T_{ijkl}(\mathbf{X}) + \sum_{pq} C^T_{ijklpq}\eta_{pq} \\ &\quad + \sum_{p} [-C^T_{ijkl}\eta_{pp} + C^T_{ijkp}\eta_{lp} + C^T_{ijpl}\eta_{kp} \\ &\quad + C^T_{ipkl}\eta_{jp} + C^T_{pjkl}\eta_{ip}] + \cdots , \end{aligned} \tag{4.53}$$

where all the elastic constants on the right are evaluated at zero stress. When this derivation is carried through for adiabatic elastic constants, the result is slightly different:

$$\begin{aligned} C^S_{ijkl}(\mathbf{X}_0) &= C^S_{ijkl}(\mathbf{X}) + \sum_{pq} C^{ST}_{ijklpq}\eta_{pq} \\ &\quad + \sum_{p} [-C^S_{ijkl}\eta_{pp} + C^S_{ijkp}\eta_{lp} + C^S_{ijpl}\eta_{kp} \\ &\quad + C^S_{ipkl}\eta_{jp} + C^S_{pjkl}\eta_{ip}] + \cdots , \end{aligned} \tag{4.54}$$

where the mixed third-order elastic constant appears because the strain is taken at constant temperature, and is defined by

$$C_{ijklpq}^{ST} = V^{-1}[\partial(\partial^2 U/\partial\eta_{ij}\partial\eta_{kl})_{S\eta'}/\partial\eta_{pq}]_{T\eta'}. \tag{4.55}$$

The differences between adiabatic, isothermal, and mixed third-order elastic constants are generally small enough to be neglected; the differences can be evaluated from experimental data if they are needed.

The strain derivatives of the elastic constants, evaluated at zero stress, are written down from (4.53) and (4.54). For example,

$$\begin{aligned}(\partial C_{ijkl}^{T}/\partial\eta_{pq})_{T\eta'} = C_{ijklpq}^{T} + \tfrac{1}{2}[&-2C_{ijkl}^{T}\delta_{pq} + C_{ijkp}^{T}\delta_{lq} + C_{ijkq}^{T}\delta_{lp} \\ &+ C_{ijpl}^{T}\delta_{kq} + C_{ijql}^{T}\delta_{kp} + C_{ipkl}^{T}\delta_{jq} + C_{iqkl}^{T}\delta_{jp} \\ &+ C_{pjkl}^{T}\delta_{iq} + C_{qjkl}^{T}\delta_{ip}].\end{aligned} \tag{4.56}$$

The stress derivatives are then given by chain-rule differentiation:

$$(\partial C_{ijkl}^{T}/\partial\tau_{rs})_{T\tau'} = \sum_{pq} (\partial C_{ijkl}^{T}/\partial\eta_{pq})_{T\eta'} S_{pqrs}^{T}. \tag{4.57}$$

Another useful relation is, from (2.54) for the stress–strain coefficients,

$$\begin{aligned}(\partial B_{ijkl}^{T}/\partial\tau_{rs})_{T\tau'} = (\partial C_{ijkl}^{T}/\partial\tau_{rs})_{T\tau'} + \tfrac{1}{4}[&(\delta_{ir}\delta_{ls} + \delta_{is}\delta_{lr})\delta_{jk} + (\delta_{jr}\delta_{ls} + \delta_{js}\delta_{lr})\delta_{ik} \\ &+ (\delta_{ir}\delta_{ks} + \delta_{is}\delta_{kr})\delta_{jl} + (\delta_{jr}\delta_{ks} + \delta_{js}\delta_{kr})\delta_{il} \\ &- 2(\delta_{ir}\delta_{js} + \delta_{is}\delta_{jr})\delta_{kl}].\end{aligned} \tag{4.58}$$

We can now proceed with the thermodynamic corrections to fixed configuration. The strain derivatives of the thermal stress tensor b_{ij} are given by (2.83) and (2.85), and when these are evaluated at zero stress with the aid of (4.58) the result leads to

$$\begin{aligned}b_{ij}(\mathbf{X}_0) = b_{ij}(\mathbf{X}) &+ \sum_{p} [b_{ip}\eta_{jp} + b_{jp}\eta_{ip} - b_{ij}\eta_{pp}] \\ &+ \sum_{kl} \Big\{(\partial C_{ijkl}^{T}/\partial T)_{\tau} + \sum_{pq} b_{pq}(\partial C_{ijkl}^{T}/\partial\tau_{pq})_{T\tau'}\Big\}\eta_{kl} + \cdots.\end{aligned} \tag{4.59}$$

The stress derivatives of the thermal strain tensor β_{ij} are given by (2.84), and the general strain expansion (4.51) can be used to write

$$\beta_{ij}(\mathbf{X}_0) = \beta_{ij}(\mathbf{X}) + \sum_{klmn} (\partial S_{ijkl}^{T}/\partial T)_{\tau} C_{klmn}^{T}\eta_{mn} + \cdots. \tag{4.60}$$

The strain derivatives of the entropy are given by the Maxwell equation (2.80), which may be written

$$(\partial S/\partial\eta_{ij})_{T\eta'} = -Vb_{ij}; \tag{4.61}$$

then to first order in the strains

$$S(\mathbf{X}_0) = S(\mathbf{X}) - V \sum_{ij} b_{ij}\eta_{ij} + \cdots. \tag{4.62}$$

Finally, the strain derivatives of the heat capacity $C_\eta = T(\partial S/\partial T)_\eta$ may be calculated with the aid of (4.61) as

$$(\partial C_\eta/\partial \eta_{ij})_{T\eta'} = -TV(\partial b_{ij}/\partial T)_\eta \tag{4.63}$$

so that

$$C_\eta(\mathbf{X}_0) = C_\eta(\mathbf{X}) - TV \sum_{ij} (\partial b_{ij}/\partial T)_\eta \eta_{ij} + \cdots. \tag{4.64}$$

The constant-strain temperature derivative of b_{ij} is more conveniently expressed with the general relation (2.82) as

$$(\partial b_{ij}/\partial T)_\eta = (\partial b_{ij}/\partial T)_\tau + \sum_{kl} b_{kl}(\partial b_{ij}/\partial \tau_{kl})_{T\tau'}, \tag{4.65}$$

and the stress derivative is transformed to a strain derivative

$$(\partial b_{ij}/\partial \tau_{kl})_{T\tau'} = \sum_{mn} (\partial b_{ij}/\partial \eta_{mn})_{T\eta'} S^T_{mnkl}. \tag{4.66}$$

The strain derivative in (4.66) can at last be evaluated from (4.59) for $b_{ij}(\mathbf{X}_0) - b_{ij}(\mathbf{X})$.

5. APPROXIMATE THEORIES

Equation of State

As a result of the equilibrium conditions, the stresses and the configuration are related at any given temperature; this relation is the equation of state and may be written formally as

$$f(\tau_{ij};\mathbf{X},\eta_{ij};T) = 0. \tag{5.1}$$

Our discussion here will be restricted to the case of pressure–volume variables, and will consider in particular the isothermal P–V relation, which may be written

$$P = P(V,T), \quad \text{or} \quad V = (P,T). \tag{5.2}$$

The Slater equations (4.1) and (4.2), and the Bridgman equation (4.3), are examples of P–V relations valid for a given crystal phase (we note that the P–V relation is nonanalytic at phase boundaries); these equations are infinite series, however, and it is desirable to find approximate P–V relations which can be expressed in simple form. One such approximate equation of state is the Murnaghan equation, which will be discussed here; there are many others.

The Murnaghan equation is based on the assumption that the isothermal bulk modulus is a linear function of pressure, at any temperature; that is,

$$B_T(P,T) = \bar{B} + \bar{B}'P, \tag{5.3}$$

where the abbreviations are

$$\bar{B} = B_T(P = 0,T); \qquad \bar{B}' = (\partial B_T/\partial P)_{T;P=0}. \tag{5.4}$$

Since $B_T = -V(\partial P/\partial V)_T$, the approximation (5.3) is

$$-dP/(\bar{B} + \bar{B}'P) = dV/V, \quad \text{at constant } T. \tag{5.5}$$

Integrating (5.5) at constant T gives one form of the Murnaghan equation,

$$\ln(\bar{V}/V) = (1/\bar{B}')\ln[(\bar{B} + \bar{B}'P)/\bar{B}], \tag{5.6}$$

where the abbreviations are

$$V = V(P,T); \qquad \bar{V} = V(P = 0,T). \tag{5.7}$$

It is convenient to introduce the dimensionless pressure p, defined by

$$p = P/\bar{B}; \tag{5.8}$$

then the Murnaghan equation is simply

$$\ln(\bar{V}/V) = (1/\bar{B}')\ln(1 + \bar{B}'p), \tag{5.9}$$

or

$$p = (1/\bar{B}')[(\bar{V}/V)^{\bar{B}'} - 1]. \tag{5.10}$$

If (5.9) is expanded as a power series in p, it is "exact" only to the second order. In order to see more clearly the nature of the Murnaghan approximation, let us write out the correct expansion of $V/\bar{V}$, to the fourth order in p. The general expansion is

$$V = \bar{V} + (\partial V/\partial P)_T P + \tfrac{1}{2}(\partial^2 V/\partial P^2)_T P^2 + \cdots, \tag{5.11}$$

where the derivatives are evaluated at $P = 0$. From the definition of B_T, $(\partial V/\partial P)_T$ is $-VB_T$ at any P, and continued differentiation of this gives the coefficients of (5.11), and

$$\begin{aligned} V/\bar{V} = 1 - p &+ \tfrac{1}{2}(1 + \bar{B}')p^2 - \tfrac{1}{6}[1 + 3\bar{B}' + 2(\bar{B}')^2 - \bar{B}\bar{B}'']p^3 \\ &+ \tfrac{1}{24}[1 + 6\bar{B}' + 11(\bar{B}')^2 + 6(\bar{B}')^3 - 4\bar{B}\bar{B}'' \\ &- 6\bar{B}\bar{B}'\bar{B}'' + \bar{B}^2\bar{B}''']p^4 + \cdots. \end{aligned} \tag{5.12}$$

But according to (5.3) the Murnaghan approximation consists of setting $\bar{B}'', \bar{B}''', \cdots$, equal to zero. Thus the Murnaghan equation (5.9) for $V/\bar{V}$ must be the same as the expansion (5.12) carried to infinite order, with all the coefficients $\bar{B}'', \bar{B}''', \cdots$, being omitted. From dimensional arguments we might expect $\bar{B}', \bar{B}\bar{B}'', \bar{B}^2\bar{B}''', \cdots$, to be all of the same order of magnitude for a given crystal. If we take a representative value of 5, and set* $\bar{B}' = 5$, $\bar{B}\bar{B}'' = -5$, $\bar{B}^2\bar{B}''' = 5$, then the Murnaghan approximation gives the coefficient of p^2 correctly in the expansion (5.12) of $V/\bar{V}$, the coefficient of p^3 in error by -7%, and the coefficient of p^4 in error by -14%. Therefore, if p is sufficiently small so that the volume expansion converges quite rapidly, say $p \leqslant 0.1$, then the Murnaghan equation is expected to be a good approximation.

In a recent paper Anderson† has shown that the Murnaghan equation, based on $\bar{B}$ and $\bar{B}'$ measured at zero pressure, fits the experimental data quite well up to the highest pressures measured, for many materials at room temperature. This includes compressions as high as $V/\bar{V} \sim 0.7$ and $p \sim 1$.

Einstein Approximation

The Einstein approximation‡ begins to consider the atomic nature of crystals. For a crystal containing N atoms there are $3N$ degrees of motional freedom, and the normal modes of vibration are a set of $3N$ waves whose frequencies are ω_κ, $\kappa = 1,2,\cdots,3N$. These waves obey boson statistics, and the crystal free energy is

$$F = \Phi_0 + \tfrac{1}{2}\sum_\kappa \hbar\omega_\kappa + KT\sum_\kappa \ln(1 - e^{-\hbar\omega_\kappa/KT}), \tag{5.13}$$

where Φ_0 is the total potential of the static crystal and K is Boltzmann's constant. Here we are neglecting the conduction-electron free energy in the case of metals, and the effects of the interactions among the vibrational waves (anharmonicity). The Einstein approximation consists of letting all the frequencies be equal:

$$\omega_\kappa = \omega_{\mathrm{E}}, \quad \text{for all } \kappa. \tag{5.14}$$

The Einstein temperature Θ_{E} is defined by

$$\hbar\omega_{\mathrm{E}} = K\Theta_{\mathrm{E}}, \tag{5.15}$$

and depends only on the configuration of the crystal.

The thermodynamic functions are conveniently expressed in terms of Θ_{E}. The free energy (5.13) is

$$F = \Phi_0 + \tfrac{3}{2}NK\Theta_{\mathrm{E}} + 3NKT\ln(1 - e^{-\Theta_{\mathrm{E}}/T}); \tag{5.16}$$

the entropy and heat capacity are then

$$S = 3NK[(\Theta_{\mathrm{E}}/T)(e^{\Theta_{\mathrm{E}}/T} - 1)^{-1} - \ln(1 - e^{-\Theta_{\mathrm{E}}/T})], \tag{5.17}$$

$$C_V = 3NK(\Theta_{\mathrm{E}}/T)^2 e^{\Theta_{\mathrm{E}}/T}(e^{\Theta_{\mathrm{E}}/T} - 1)^{-2}. \tag{5.18}$$

* In all our model calculations we have found $\bar{B}\bar{B}'' < 0$; see, e.g., Chapter 8.

† O. L. Anderson, *J. Phys. Chem. Solids* **27,** 547 (1966).

‡ A. Einstein, *Ann. Phys.* **22,** 180 (1907).

In the high-temperature limit $T \gg \Theta_{\mathrm{E}}$, the heat capacity approaches $3NK$; this gives a qualitative accounting of the Dulong–Petit empirical rule, $C_V \approx 3NK$ at room temperature. In the low-temperature limit the Einstein heat capacity approaches zero, not with the experimentally observed T^3-dependence, however, but with the dependence $(\Theta_{\mathrm{E}}/T)^2 e^{-\Theta_{\mathrm{E}}/T}$.

For the Einstein approximation the entropy is a function of Θ_{E}/T; the temperature-dependence of S is explicit and the configuration-dependence is through that of Θ_{E}. For this functional dependence, the Grüneisen parameter is simply expressed by the following calculation.

$$S = S(\Theta_{\mathrm{E}}/T); \tag{5.19}$$

therefore

$$C_V = T(\partial S/\partial T)_V = -(\Theta_{\mathrm{E}}/T)S', \tag{5.20}$$

and from (1.29),

$$\beta B_T = (\partial S/\partial V)_T = T^{-1}(d\Theta_{\mathrm{E}}/dV)S', \tag{5.21}$$

where the abbreviation is $S' = dS(x)/dx$. Then from the definition (1.30) of γ,

$$\gamma = -(d \ln \Theta_{\mathrm{E}}/d \ln V). \tag{5.22}$$

Therefore for the Einstein approximation, γ is independent of T and depends only on the crystal configuration. Integration of (5.22) gives

$$\Theta_{\mathrm{E}} \propto V^{-\gamma}, \tag{5.23}$$

giving the volume-dependence of Θ_{E}.

An interesting example of an Einstein model is

$$\omega_{\mathrm{E}} = \langle \omega^2 \rangle^{1/2}, \tag{5.24}$$

where $\langle \omega^2 \rangle$ is the average over all the lattice waves κ of ω_κ^2. This is a convenient model, since for a prescribed crystal and set of interactions among the atoms, $\langle \omega^2 \rangle$ and its volume and strain derivatives are easily calculated (see, e.g., Chapter 3). It should be emphasized, however, that the Einstein approximation is of value only for estimates of thermodynamic functions; the correct lattice-wave sums are easily evaluated for a prescribed crystal model.

Debye Approximation

The Debye approximation* treats the crystal as an isotropic elastic medium. The elastic waves are presumed to be quantized in the volume V of the crystal, and are denoted by the polarization $s = 1,2,3$ and the wave vector $\mathbf{k}$. These waves correspond to the dispersionless waves of Section 3,

* P. Debye, *Ann. Phys.* **39**, 789 (1912).

where for an isotropic medium there are always two transverse and one longitudinal polarization and for each polarization the wave velocity is independent of the direction of **k**. The frequencies of the normal vibrational modes are then

$$\omega_s(\mathbf{k}) = v_s\,|\mathbf{k}|, \tag{5.25}$$

where v_s are the wave velocities. According to the volume quantization, the number of allowed states up to $|\mathbf{k}|$, for each s, is $\frac{4}{3}\pi\,|\mathbf{k}|^3\,V$. Then from (5.25), the number of allowed states up to ω, for each s, is $\frac{4}{3}\pi v_s^{-3}\omega^3 V$. The total number of states with frequencies equal to or less than ω is therefore

$$n(\omega) = \tfrac{4}{3}\pi\omega^3 V \sum_s v_s^{-3}. \tag{5.26}$$

Now we follow the customary procedure and cut off the allowed frequencies at a common maximum frequency ω_D, defined such that the total number of allowed states is $3N$ for a crystal containing N atoms. In keeping with the volume quantization, it would be more appropriate to cut off the allowed states at a common $|\mathbf{k}|$, thus giving different maximum frequencies for longitudinal and transverse waves. The choice of the cut off is unimportant, however, since it affects the short-wavelength waves, where the Debye approximation is not appropriate for a crystal in any case.

The Debye frequency ω_D is therefore defined by setting $n(\omega_\mathrm{D}) = 3N$:

$$3N = \tfrac{4}{3}\pi\omega_\mathrm{D}^3 V \sum_s v_s^{-3}. \tag{5.27}$$

The number of allowed states with frequencies between ω and $\omega + d\omega$ is $g(\omega)d\omega$, where the distribution function $g(\omega)$ is given by

$$g(\omega) = dn(\omega)/d\omega = 4\pi\omega^2 V \sum_s v_s^{-3} = 9N\omega^2/\omega_\mathrm{D}^3\,. \tag{5.28}$$

To evaluate thermodynamic functions for the Debye approximation, in particular the harmonic free energy (5.13), we transform sums over the normal modes κ to integrals over ω, according to

$$\sum_\kappa f(\omega_\kappa) = \int_0^{\omega_\mathrm{D}} f(\omega)g(\omega)\,d\omega. \tag{5.29}$$

It is also convenient to introduce the Debye temperature Θ_D, which is a function only of the crystal volume, and is defined by

$$K\Theta_\mathrm{D} = \hbar\omega_\mathrm{D}. \tag{5.30}$$

The harmonic free energy (5.13) is then

$$F = \Phi_0 + 9N\left\{\tfrac{1}{8}K\Theta_\mathrm{D} + KT(T/\Theta_\mathrm{D})^3\int_0^{\Theta_\mathrm{D}/T} \zeta^2 \ln(1 - e^{-\zeta})\,d\zeta\right\}. \tag{5.31}$$

To calculate the heat capacity in the Debye approximation, it is simplest to calculate $C_V = -T(\partial^2 F/\partial T^2)_V$ from (5.13),

$$C_V = K \sum_{\kappa} (\hbar\omega_\kappa/KT)^2 e^{\hbar\omega_\kappa/KT}(e^{\hbar\omega_\kappa/KT} - 1)^{-2}, \tag{5.32}$$

and then transform to an integral according to (5.29) to obtain

$$C_V = 9NK(T/\Theta_D)^3 \int_0^{\Theta_D/T} \zeta^4 e^\zeta (e^\zeta - 1)^{-2}\, d\zeta. \tag{5.33}$$

In the low-temperature limit $\Theta_D/T \gg 1$, and since the integrand goes as $\zeta^4 e^{-\zeta}$ for large ζ, the integral can be extended to ∞ and evaluated as

$$\int_0^\infty \zeta^4 e^\zeta (e^\zeta - 1)^{-2}\, d\zeta = 4\pi^4/15. \tag{5.34}$$

Then at low temperatures,

$$C_V = 3NK(4\pi^4/5)(T/\Theta_D)^3; \tag{5.35}$$

this gives the correct (experimentally observed) temperature-dependence of the crystal heat capacity at low temperatures, with the omission of conduction-electron contributions in the case of metals. In the high-temperature limit, expansion of (5.32) for $KT \gg \hbar\omega_\kappa$, all κ, gives

$$C_V = 3NK - \tfrac{1}{12}K \sum_{\kappa} (\hbar\omega_\kappa/KT)^2 + \cdots; \tag{5.36}$$

transformation of the sum to an integral according to (5.29) then leads to

$$C_V = 3NK[1 - \tfrac{1}{20}(\Theta_D/T)^2 + \cdots]. \tag{5.37}$$

Here the $+\cdots$ represents terms of order higher than 2 in (Θ_D/T); this equation shows how the Debye heat capacity approaches $3NK$ at high temperatures.

Neglecting electronic and anharmonic effects, the Debye C_V has the same temperature-dependence at low T, and approaches the same limit at high T, as does the accurate lattice dynamical theory. The Debye approximation therefore provides a sensitive way to represent experimental C_V data as a function of T, in the form of the effective experimental Θ_D as a function of T. In the Debye approximation, of course, Θ_D is independent of T at fixed volume; the effective Θ_D for the experimental heat capacity is defined by (5.33), where the left-hand side is the experimental C_V at any temperature. Some curves of Θ_D *vs* T, for experimental heat capacities corrected to fixed volume and with the electronic contributions removed, are shown in Figure 5.

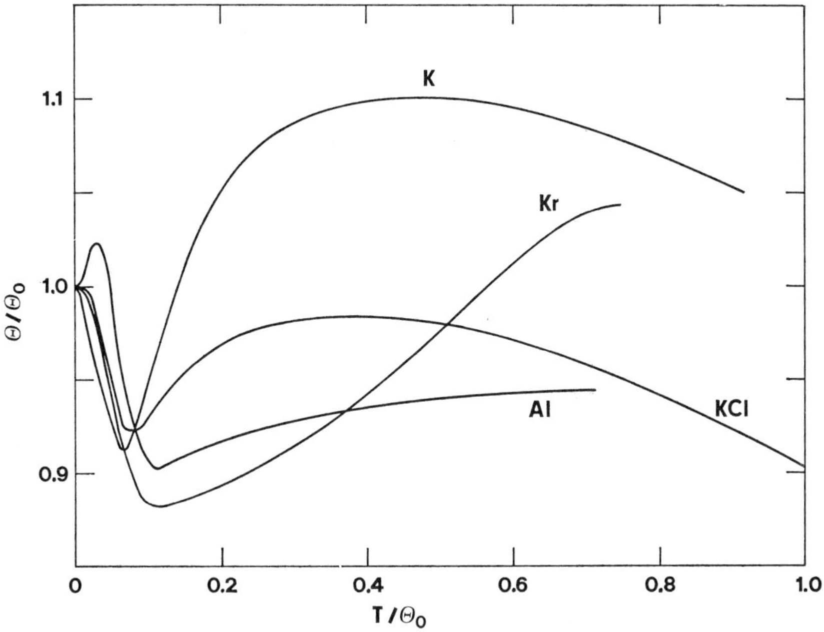

Figure 5. Effective experimental Debye temperatures Θ, for C_V corrected to the fixed volume V_0, and with the electronic heat capacity subtracted out for K and Al. The value Θ_0 of Θ at $T = 0$ is 71.9°K for Kr, 90.6°K for K, 235°K for KCl, and 428°K for Al. The temperature variations of these effective Debye temperatures are caused by lattice-wave dispersion and interactions.

The temperature variation of these experimental Θ_D values results from the dispersion of the lattice waves and from anharmonicity; the theory of these effects is discussed in the following chapters.

Just as in the Einstein approximation, the Debye approximation gives the entropy as a function of (Θ_D/T); therefore, according to (5.22) and (5.23),

$$\gamma = -(d \ln \Theta_D/d \ln V), \tag{5.38}$$

$$\Theta_D \propto V^{-\gamma}. \tag{5.39}$$

One can also define an effective experimental Debye temperature for the entropy, or for any thermodynamic function, in the same way as that defined above for C_V. Again because of dispersion and anharmonicity of the lattice waves, such effective Debye temperatures are not constant in temperature, and are generally different from one another for a given material.

Grüneisen Approximations

We will discuss two different approximations made by Grünesien* to explain approximately the zero-pressure equation of state, i.e., the temperature variation of the lattice configuration at $P = 0$. The first approximation was employed to calculate the thermal strain parameters β_{ij} for the hexagonal metals Zn and Cd. In Voigt notation, β_α, for hexagonal or tetragonal I crystals the thermal strains have the symmetry

$$\beta_1 = \beta_2 = \beta_\perp, \qquad \beta_3 = \beta_\parallel, \qquad \beta_4 = \beta_5 = \beta_6 = 0; \tag{5.40}$$

the same symmetry holds for the thermal stresses b_α and the Grüneisen parameters γ_α, defined in Section 2. Now the equation (2.103) for the Grüneisen parameters γ_{ij} may be inverted to write

$$\beta_{ij} = (C_\eta/V) \sum_{kl} S^T_{ijkl}\gamma_{kl}; \tag{5.41}$$

writing this in Voigt notation for the two independent β_α for hexagonal or tetragonal I crystals at zero stress gives

$$\beta_1 = (C_\eta/V)[(S^T_{11} + S^T_{12})\gamma_1 + S^T_{13}\gamma_3]; \tag{5.42}$$

$$\beta_3 = (C_\eta/V)[2S^T_{13}\gamma_1 + S^T_{33}\gamma_3]. \tag{5.43}$$

Grüneisen approximated the quantities $\gamma_1 C_\eta$ and $\gamma_3 C_\eta$, each by a Debye formula, with a different effective Grüneisen parameter and Debye temperature for each of the two principal directions. In other words,

$$\gamma_1 C_\eta = \bar{\gamma}_1 C(\Theta_1/T), \tag{5.44}$$

$$\gamma_3 C_\eta = \bar{\gamma}_3 C(\Theta_3/T), \tag{5.45}$$

where $\bar{\gamma}_1$, $\bar{\gamma}_3$, Θ_1, and Θ_3 are constant parameters and $C(\Theta/T)$ is the Debye heat capacity function (5.33). With these four parameters suitably adjusted, and the measured $S^T_{\alpha\beta}$ (assumed temperature-independent), Grüneisen obtained reasonably good fits of (5.42) and (5.43) to the measured β_1 and β_3 for Zn and Cd. The two principal β_α have quite different temperature-dependence and magnitude for each of these metals; in fact $\beta_\perp$ is negative at low temperatures for both.

For the case of pressure–volume variables, the counterpart of Grüneisen's first approximation is $\gamma C_V = \bar{\gamma} C(\Theta/T)$, where $\bar{\gamma}$ is a constant and $C(\Theta/T)$ is again the Debye heat capacity. It is popular nowadays to reinterpret this as the approximation $\gamma =$ constant; it makes little difference since for most materials either approximation is roughly of the same accuracy. Several

* E. Grüneisen, in *Handbuch der Physik*, edited by H. Geiger and K. Scheel, Springer-Verlag, Berlin, 1926, Vol. 10, p. 1.

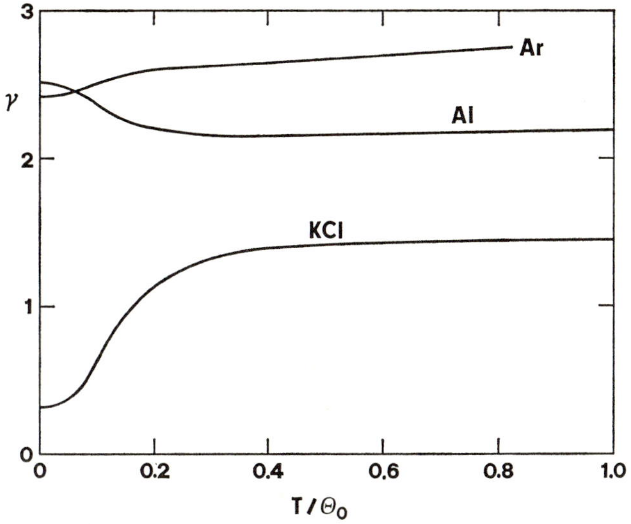

Figure 6. Experimental curves of $\gamma = V\beta B_T/C_V$ at $P = 0$, where the electronic contribution was subtracted out for Al. The temperature variations are mainly due to lattice-wave dispersion and to the volume-dependence of γ. The large variation of γ with T for KCl is typical of the alkali halides. The value of Θ_0 is 92.0°K for Ar, 428°K for Al, and 235°K for KCl.

curves of the experimental γ as a function of temperature and at zero pressure, calculated from

$$\gamma = V\beta B_T/C_V$$

where the conduction-electron contribution has been subtracted out at low temperatures, are shown in Figure 6. The temperature variation of these Grüneisen parameters is due primarily to the dispersion of the lattice waves and to the volume-dependence of γ.

Grüneisen's second approximation was applied to the calculation of $a_0(T)$ at $P = 0$, where, as in Section 4,

$$a_0(T) = (V - V_0)/V_0, \quad \text{at} \quad P = 0. \tag{5.46}$$

This approximation is discussed by writing the free energy in the form

$$F(V,T) = F_0(V) + F_T(V,T), \tag{5.47}$$

where for the thermal contribution F_T,

$$F_T = 0 \quad \text{at} \quad T = 0. \tag{5.48}$$

For pressure–volume variables the equilibrium condition is

$$P = -(\partial F/\partial V)_T = -(dF_0/dV) - (\partial F_T/\partial V)_T; \tag{5.49}$$

multiplying by V gives

$$PV + V(dF_0/dV) = \hat{\gamma}U, \tag{5.50}$$

where U is the internal energy and

$$\hat{\gamma} = -(V/U)(\partial F_T/\partial V)_T. \tag{5.51}$$

The reason for introducing $\hat{\gamma}$ is that it is approximately constant in temperature; in fact if $\hat{\gamma}$ *is* independent of T, then $\hat{\gamma} = \gamma$. To show this, differentiate (5.50) with respect to T at constant V:

$$V(\partial P/\partial T)_V = \hat{\gamma}C_V, \quad \text{for } \hat{\gamma} \text{ independent of } T; \tag{5.52}$$

then from (1.31) it is seen that $\hat{\gamma} = \gamma$.

We proceed, without making assumptions about $\hat{\gamma}$, by applying (5.50) at $P = 0$ and any T,

$$\hat{\gamma}U = V(dF_0/dV), \quad \text{at} \quad P = 0. \tag{5.53}$$

The right-hand side is expanded into

$$\begin{aligned}\hat{\gamma}U = {} & [V(dF_0/dV)]_0 + (V - V_0)[(dF_0/dV) + V(d^2F_0/dV^2)]_0 \\ & + \tfrac{1}{2}(V - V_0)^2\,[2(d^2F_0/dV^2) + V(d^3F_0/dV^3)]_0 + \cdots,\end{aligned} \tag{5.54}$$

where the notation $[\ ]_0$ means to evaluate at V_0. To evaluate the expansion coefficients, use the $T = 0$ equilibrium condition from (5.49),

$$(dF_0/dV) = -P, \quad \text{at} \quad T = 0; \tag{5.55}$$

then repeated volume differentiation gives

$$(d^2F_0/dV^2) = B_0/V, \tag{5.56}$$

$$(d^3F_0/dV^3) = -(B_0/V^2)(1 + B_0'), \tag{5.57}$$

where B_0 and B_0' are, respectively, the bulk modulus and its pressure derivative at $T = 0$. Then the expansion (5.54) is

$$\hat{\gamma}U = a_0V_0B_0 + \tfrac{1}{2}a_0^2V_0B_0(1 - B_0') + \cdots, \quad \text{at} \quad P = 0. \tag{5.58}$$

Now $a_0(T)$ may be expressed by inverting the series (5.58). To first order this is

$$a_0 = \hat{\gamma}U/V_0B_0 + \cdots. \tag{5.59}$$

To get a correct second-order expression for a_0, put (5.59) in for the a_0^2 term in (5.58), giving

$$\hat{\gamma}U = a_0V_0B_0[1 + \tfrac{1}{2}(\hat{\gamma}U/V_0B_0)(1 - B_0') + \cdots], \tag{5.60}$$

or solving for a_0,

$$a_0 = U/(Q - bU + \cdots), \tag{5.61}$$

where

$$Q = V_0 B_0/\hat{\gamma}, \qquad b = \tfrac{1}{2}(B_0' - 1). \tag{5.62}$$

Equation (5.61) is the form given by Grüneisen. If all higher-order terms are kept, this equation is exact, but it is not of much use since $\hat{\gamma} = \hat{\gamma}(V,T)$ is a complicated function, and so are the higher-order terms. However, two circumstances make the equation a useful approximation, namely (a) $\hat{\gamma}$ is nearly independent of T, so that Q is also, and (b) the series in the denominator converges rapidly. Therefore the approximation

$$a_0 = U/(Q - bU) \tag{5.63}$$

has been found to give a fairly good fit to measured values of a_0 *vs* T at $P = 0$, when Q and b are taken as adjustable constant parameters. Extension to include another parameter c,

$$a_0 = U/(Q - bU + cU^2), \tag{5.64}$$

should give improved fits to experimental data up to larger values of a_0.

2

THE CRYSTAL POTENTIAL

Here the crystal is considered to be a finite array of interacting ions or atoms in the presence of externally applied mechanical forces, and the total potential energy is studied as a function of the configuration of the array. Since the vibrational energy of a crystal is generally small compared to its potential energy, the crystal potential is a first approximation to the free energy or to the internal energy, so the treatment here is the mechanical analog of the thermoelastic calculations of Chapter 1. The potential is independent of T and S, however, so the mechanical elastic constants depend only on the configuration of the crystal, and do not distinguish between adiabatic and isothermal processes. The kinetic energy of the moving atoms is added to the potential in Chapter 3, to give the total Hamiltonian and hence the normal vibrational modes of the crystal.

6. PROPERTIES OF POTENTIAL ENERGY COEFFICIENTS

Potential Energy Expansion

The ions of the crystal are presumed to interact with one another, and to be acted upon by externally applied forces and fields. In the presence of these interactions the static (nonvibrating) crystal assumes an equilibrium configuration, i.e., one in which the net force on each ion vanishes. We label the ions by the letters M,N, $\cdots$, and continue to use subscripts $i,j,\cdots$, for Cartesian indices. The equilibrium positions of the ions are given by the vectors **R**(M), and displacements from equilibrium are denoted by **U**(M). At the moment, it is not necessary to specify the different kinds of atoms which might appear in a given unit cell of the crystal.*

The potential energy of the crystal, due to the interactions among the ions in a given configuration, is Φ; this is presumed to be an analytic function of the positions of the ions, and hence it may be expanded in the displacements from an arbitrary initial configuration:

$$\begin{aligned}\Phi = \Phi_0 &+ \sum_{\mathrm{M}}\sum_{i}\Phi_i(\mathrm{M})U_i(\mathrm{M}) + \tfrac{1}{2}\sum_{\mathrm{MN}}\sum_{ij}\Phi_{ij}(\mathrm{M,N})U_i(\mathrm{M})U_j(\mathrm{N})\\ &+ (1/3!)\sum_{\mathrm{MNP}}\sum_{ijk}\Phi_{ijk}(\mathrm{M,N,P})U_i(\mathrm{M})U_j(\mathrm{N})U_k(\mathrm{P})\\ &+ (1/4!)\sum_{\mathrm{MNPQ}}\sum_{ijkl}\Phi_{ijkl}(\mathrm{M,N,P,Q})U_i(\mathrm{M})U_j(\mathrm{N})U_k(\mathrm{P})U_l(\mathrm{Q}) + \cdots.\end{aligned} \tag{6.1}$$

Here Φ_0 is the potential of interaction among the ions when they are all located at their initial equilibrium positions **R**(M). This expansion defines the potential energy coefficients $\Phi_i(\mathrm{M}), \Phi_{ij}(\mathrm{M,N}), \cdots$; these are obviously derivatives of the potential, evaluated at the initial configuration:

$$\Phi_i(\mathrm{M}) = \partial\Phi/\partial U_i(\mathrm{M}), \tag{6.2}$$

$$\Phi_{ij}(\mathrm{M,N}) = \partial^2\Phi/\partial U_i(\mathrm{M})\partial U_j(\mathrm{N}), \tag{6.3}$$

and so on. From their definition the potential energy coefficients are obviously symmetric in their index pairs Mi,Nj, $\cdots$.

$$\Phi_{ij}(\mathrm{M,N}) = \Phi_{ji}(\mathrm{N,M}), \tag{6.4}$$

$$\Phi_{ijk}(\mathrm{M,N,P}) = \Phi_{jik}(\mathrm{N,M,P}) = \Phi_{ikj}(\mathrm{M,P,N}) = \cdots, \tag{6.5}$$

$$\Phi_{ijkl}(\mathrm{M,N,P,Q}) = \Phi_{jikl}(\mathrm{N,M,P,Q}) = \cdots. \tag{6.6}$$

Finally, since the initial configuration is arbitrary, the expansion of Φ may be taken about different initial configurations on different occasions; then it should be recognized that Φ_0 and the other potential energy coefficients are functions of the initial configuration.

The potential energy expansion is fundamental to the atomic theory of crystals, and several remarks about its validity are in order. First, the potential Φ is supposed to represent the entire energy of the crystal except for kinetic energy of the ions; this means that Φ includes the total energy of the crystal electrons. Imagine that Φ is calculated for a sequence of different configurations of the ions, with the electrons in their ground state in each case. Then certainly Φ should be an analytic function of the positions

* Starting with equation (6.51), each ion will be labeled by two indices, with M,N, $\cdots$, denoting the unit cell and $\mu,\nu, \cdots$, denoting the ion in a given cell; until then a single index for each ion is sufficient.

of the ions, except along lines in configuration space where the nature of the electronic ground state changes, such as where a band begins to fill or empty; at such points Φ is nonanalytic but continuous. The potential energy expansion should therefore be appropriate for describing homogeneous deformations from the initial configuration, as in the present chapter, with the recognition that Φ may be nonanalytic at boundaries between regions where it is analytic. In addition, Φ has been assumed to depend on the positions of the ions, and no other variables. This should also be true since the electronic ground state should depend only on the positions of the ions. If one considers the core electrons as belonging to each ion, as for example in sodium chloride composed of Na^+ and Cl^-, then the crystal potential depends on the polarizations of the cores; these polarizations, however, are in turn unique functions of the ion positions. There are exceptions, as for example in magnetic materials where the crystal potential depends on the directions of the localized (core) spins; such a dependence leads to spin waves and spin wave–phonon interactions, which are not included in our work.

Second, the potential energy expansion is presumed in Chapter 3 to be the appropriate potential for describing the *motion* of the ions. For this application one has to rely on the adiabatic approximation,* which says essentially that the kinetic energy involved in the motion of the ions is sufficiently small so that excitations of the crystal electrons from their ground state (induced by the ion kinetic energy) are a negligible effect. For metals this is a rather poorer approximation than for nonmetals, since metals have empty electronic states with energies essentially continuous with the ground state energy. In any case, however, one can begin with the adiabatic approximation, and then treat nonadiabatic effects by perturbation theory (see, e.g., Chapter 5).

Finally from physical reasons, the total crystal potential should not really depend on the positions of all the ions, but only on their relative positions, given by the distances $|\mathbf{R}(M) + \mathbf{U}(M) - \mathbf{R}(N) - \mathbf{U}(N)|$. This is true, of course, but the mathematical formulation in terms of the positions $\mathbf{R}(M) + \mathbf{U}(M)$ is more convenient; the correct physics is then recovered with subsidiary conditions of translational and rotational invariance, which are discussed in the following section. Further, in complete analogy with the thermoelastic theory of Section 3, the propagation of long-wavelength waves in the crystal can be discussed in terms of the equations derived for homogeneous deformations, provided the wavelengths are long compared to the (finite) effective range of interactions among the ions in the crystal. This brings up the interesting point of volume-dependent energies. If Φ contains a

* M. Born and K. Huang, *Dynamical Theory of Crystal Lattices*, Clarendon Press, Oxford, 1954.

term that is dependent only on the volume, or the macroscopic configuration, of the crystal, then this term contributes to the variation of the crystal potential with homogeneous deformation. In calculating the normal vibrational modes of the crystal, however, with the boundary conditions of fixed macroscopic configuration, it may appear that such a contribution to Φ will drop out completely. This is not the case if the theory is done correctly, since at any location in a very long wavelength wave the deformation is homogeneous over the effective range of interactions of the ions, and hence a term depending only on the macroscopic configuration of the crystal will contribute to the energy of the wave. These points are considered in more detail in the following chapters.

Up to now we have considered the external forces only implicitly, in that they are important in determining the equilibrium configuration of the crystal. The force applied to ion M is $\mathbf{f}(\mathrm{M})$; these forces are presumably derived from a potential, such as an electric field. Let $W(\mathbf{R}(\mathrm{M}))$ be the work done by the externally applied forces when they are brought from the value zero to the arbitrary initial values $\mathbf{f}(\mathrm{M})$. Then the total potential of the system of crystal plus externally applied forces is Ψ, given by

$$\Psi(\mathbf{R}(\mathrm{M})) = \Phi(\mathbf{R}(\mathrm{M})) - W(\mathbf{R}(\mathrm{M})). \tag{6.7}$$

The potential Ψ represents a closed system, and in any real process in which the equilibrium configuration of the crystal is changed, by changing the externally applied forces, Ψ is conserved. For a virtual process, in which the crystal is deformed while the externally applied forces are held constant, Ψ is not conserved. Since the work done by the external forces in a virtual displacement is

$$\sum_{\mathrm{M}} \sum_{i} f_i(\mathrm{M}) U_i(\mathrm{M}), \tag{6.8}$$

the total system potential may be expanded for virtual displacements with the aid of the expansion (6.1) of Φ:

$$\Psi = \Psi_0 + \sum_{\mathrm{M}} \sum_{i} [\Phi_i(\mathrm{M}) - f_i(\mathrm{M})] U_i(\mathrm{M}) + \tfrac{1}{2} \sum_{\mathrm{MN}} \sum_{ij} \Phi_{ij}(\mathrm{M,N}) U_i(\mathrm{M}) U_j(\mathrm{N}) + \cdots. \tag{6.9}$$

Before proceeding with the mathematical analysis, there is another point which should be discussed. For our finite crystal model, we will always assume that the equilibrium positions of the ions correspond to a lattice which is perfectly periodic throughout the bulk of the crystal, with departure from periodicity near the surface, roughly within the range of effective interactions among the ions. We will also assume that surface effects are negligible compared to bulk effects, and will eliminate surface effects from our calculations either by neglecting the surface distortions, or equivalently by evaluating expressions in the perfectly periodic interior of the crystal.

It may be thought that surface effects are automatically eliminated by considering a finite portion of a crystal within an infinite lattice, but this is not the case. For, in order to calculate the properties of the subcrystal, one needs the potential Φ due only to the interactions of the ions *within* the subcrystal. The interactions of these ions with the rest of the infinite lattice are in fact just external forces, applied to the subcrystal ions near the surface, within the range of effective interactions. These external forces, which ensure that the equilibrium configuration of the subcrystal is perfectly periodic throughout, are still a surface effect which must be eliminated, or neglected, in the subsequent theory.

In fact the infinite lattice model has led to a "paradox" regarding the equilibrium conditions. Consider a subcrystal inside an infinite lattice, and write the potential χ of the ions in this subcrystal due to *all* the interactions involving these ions:

$$\chi = \chi_0 + \sum_{\mathrm{M}} \sum_{i} \chi_i(\mathrm{M}) U_i(\mathrm{M}) + \cdots.$$

Obviously the total force on each ion vanishes at equilibrium, hence the coefficients $\chi_i(\mathrm{M})$ must vanish. If now χ is interpreted as the potential of the subcrystal alone, then the mechanical stresses on the subcrystal are given by derivatives of χ with respect to homogeneous strains, and since homogeneous strains are constructed of linear combinations of the ion displacements $\mathbf{U}(\mathrm{M})$, the stresses evaluated at equilibrium are in turn linear functions of the coefficients $\chi_i(\mathrm{M})$. (Details of the homogeneous deformation calculations are presented in the following section.) This argument has led to the conclusion that the infinite lattice model requires a second equilibrium condition, namely that the stresses vanish at equilibrium.* This conclusion is incorrect, since χ is not the potential of the subcrystal alone, but is the potential of the total system of subcrystal plus externally applied forces, the external forces arising from interactions of the ions within the subcrystal with those outside. Thus χ corresponds to Ψ, given by (6.7), and the strain derivatives of χ are not stresses. The mechanical stresses need not vanish, and they can be calculated from the strain derivatives of Φ, where surface effects may be neglected, or equivalently they can be calculated by summing the externally applied surface forces.

EQUILIBRIUM AND INVARIANCE

The equilibrium condition is that the total force on each ion must vanish when all ions are located at their equilibrium positions $\mathbf{R}(\mathrm{M})$. The i component of the total force on ion M is just $-\partial\Psi/\partial U_i(\mathrm{M})$, so the equilibrium

* M. Born and K. Huang, *Dynamical Theory of Crystal Lattices*, Clarendon Press, Oxford, 1954.

condition is

$$\partial\Psi/\partial U_i(\mathrm{M}) = 0, \quad \text{at all} \quad \mathbf{U}(\mathrm{M}) = 0, \tag{6.10}$$

or

$$\Phi_i(\mathrm{M}) - f_i(\mathrm{M}) = 0, \quad \text{for all M},i. \tag{6.11}$$

As a result of this condition, Ψ is a stationary function with respect to all virtual displacements of the ions from equilibrium. Neither Φ nor W is similarly stationary in general. The macroscopic equilibrium conditions for the crystal are that the total externally applied force, and the total externally applied torque, should vanish:

$$\sum_{\mathrm{M}} f_i(\mathrm{M}) = 0, \quad \text{for all } i; \tag{6.12}$$

$$\sum_{\mathrm{M}} f_i(\mathrm{M})R_j(\mathrm{M}) \text{ is symmetric in } i,j, \quad \text{for all } i,j. \tag{6.13}$$

For physical reasons these conditions must be contained in the general equilibrium condition (6.11), since the collection of ions representing the crystal should experience no net total force, and no net total torque, as a result of the interactions among the ions; that is,

$$\sum_{\mathrm{M}} \Phi_i(\mathrm{M}) = 0, \quad \text{for all } i; \tag{6.14}$$

$$\sum_{\mathrm{M}} \Phi_i(\mathrm{M})R_j(\mathrm{M}) \text{ is symmetric in } i,j, \quad \text{for all } i,j. \tag{6.15}$$

In view of (6.11), these conditions are equivalent to (6.12) and (6.13).

The invariance conditions require that the total system potential Ψ is invariant under translation or rotation of the entire system of ions and external forces. This is equivalent to requiring that the crystal potential Φ is invariant under translation or rotation, without deformation, of the crystal alone. To discuss these conditions it is necessary to derive the transformation properties of the potential energy coefficients, i.e., their variation with variations of the initial configuration. Consider two initial equilibrium configurations in which the ions are located at $\mathbf{R}(\mathrm{M})$ and $\bar{\mathbf{R}}(\mathrm{M})$, respectively, and where the corresponding potential energy coefficients are $\Phi_i(\mathrm{M})$ and $\bar{\Phi}_i(\mathrm{M})$, with similar notation for higher-order coefficients. Let the displacements from $\mathbf{R}(\mathrm{M})$ to $\bar{\mathbf{R}}(\mathrm{M})$ be $\mathbf{V}(\mathrm{M})$, so that

$$\bar{\mathbf{R}}(\mathrm{M}) = \mathbf{R}(\mathrm{M}) + \mathbf{V}(\mathrm{M}). \tag{6.16}$$

Now with the arbitrary displacements $\mathbf{U}(\mathrm{M})$ referred to the same coordinate system throughout, the first potential energy coefficient is, from (6.2),

$$\bar{\Phi}_i(\mathrm{M}) = \partial\Phi/\partial U_i(\mathrm{M}), \quad \text{at all} \quad U_i(\mathrm{M}) = V_i(\mathrm{M}), \tag{6.17}$$

or

$$\bar{\Phi}_i(\mathrm{M}) = \Phi_i(\mathrm{M}) + \sum_{\mathrm{N}}\sum_{j} \Phi_{ij}(M,\mathrm{N})V_j(\mathrm{N}) + \cdots. \tag{6.18}$$

Similarly for the higher-order potential coefficients, it follows that

$$\overline{\Phi}_{ij}(\mathrm{M,N}) = \Phi_{ij}(\mathrm{M,N}) + \sum_{\mathrm{P}} \sum_{k} \Phi_{ijk}(\mathrm{M,N,P}) V_k(\mathrm{P}) + \cdots, \tag{6.19}$$

$$\overline{\Phi}_{ijk}(\mathrm{M,N,P}) = \Phi_{ijk}(\mathrm{M,N,P}) + \sum_{\mathrm{Q}} \sum_{l} \Phi_{ijkl}(\mathrm{M,N,P,Q}) V_l(\mathrm{Q}) + \cdots, \tag{6.20}$$

$$\overline{\Phi}_{ijkl}(\mathrm{M,N,P,Q}) = \Phi_{ijkl}(\mathrm{M,N,P,Q}) + \cdots. \tag{6.21}$$

Now let the system undergo the translation $\boldsymbol{\epsilon}$, i.e.,

$$V_i(\mathrm{M}) = \epsilon_i, \quad \text{for all M}, \tag{6.22}$$

where ϵ_i are arbitrary. In this case the externally applied forces remain unchanged, and the expansion of Ψ for virtual displacements about the new equilibrium positions $\overline{\mathbf{R}}(\mathrm{M})$ is, according to (6.9),

$$\overline{\Psi} = \overline{\Psi}_0 + \sum_{\mathrm{M}} \sum_{i} [\overline{\Phi}_i(\mathrm{M}) - f_i(\mathrm{M})] U_i(\mathrm{M}) + \tfrac{1}{2} \sum_{\mathrm{MN}} \sum_{ij} \overline{\Phi}_{ij}(\mathrm{M,N}) U_i(\mathrm{M}) U_j(\mathrm{N}) + \cdots. \tag{6.23}$$

The translational invariance condition requires that this be the same function of the ion displacements as the original expansion (6.9), and since the displacements are arbitrary, each potential coefficient in the two expansions (6.9) and (6.23) must be equal. Writing this out for the first-order potential coefficients,

$$\Phi_i(\mathrm{M}) - f_i(\mathrm{M}) = \overline{\Phi}_i(\mathrm{M}) - f_i(\mathrm{M}),$$

or with $\overline{\Phi}_i(\mathrm{M})$ given by (6.18) and with the translation (6.22),

$$\Phi_i(\mathrm{M}) = \Phi_i(\mathrm{M}) + \sum_{\mathrm{N}} \sum_{j} \Phi_{ij}(\mathrm{M,N}) \epsilon_j + \cdots. \tag{6.24}$$

Finally the ϵ_j are arbitrary, so the coefficient of each power of ϵ_j in (6.24) must vanish, leading to the translational invariance condition

$$\sum_{\mathrm{N}} \Phi_{ij}(\mathrm{M,N}) = 0, \quad \text{for all } \mathrm{M}, i, j, \tag{6.25}$$

and from the symmetry (6.4) of $\Phi_{ij}(\mathrm{M,N})$ in its index pairs,

$$\sum_{\mathrm{M}} \Phi_{ij}(\mathrm{M,N}) = 0, \quad \text{for all } \mathrm{N}, i, j. \tag{6.26}$$

We note that the corresponding conditions on the coefficients of higher powers of ϵ_i are contained in the first-order conditions for successive potential coefficients listed below.

Applying the translational invariance condition to the higher-order potential coefficients in the two expansions (6.9) and (6.23) gives the following set of conditions.

$$\sum_{\mathrm{M}} \Phi_{ijk}(\mathrm{M,N,P}) = \sum_{\mathrm{N}} \Phi_{ijk}(\mathrm{M,N,P}) = \sum_{\mathrm{P}} \Phi_{ijk}(\mathrm{M,N,P}) = 0, \tag{6.27}$$

$$\sum_{\mathrm{M}} \Phi_{ijkl}(\mathrm{M,N,P,Q}) = \sum_{\mathrm{N}} \Phi_{ijkl}(\mathrm{M,N,P,Q}) = \cdots = 0. \tag{6.28}$$

These equations are valid for arbitrary initial configuration of the crystal.

The rotational invariance conditions may be derived by rotating the system of ions plus external forces, and the coordinate system, and requiring the system potential to be the same function of the ion displacements as it was originally. For this purpose it is sufficient to consider infinitesimal rotations $\boldsymbol{\omega}$, and work to first order in $\boldsymbol{\omega}$. Under the rotation, the ion positions $\mathbf{R}(\mathrm{M})$ go to $\bar{\mathbf{R}}(\mathrm{M})$, given by

$$\bar{R}_i(\mathrm{M}) = \sum_{i'} (\delta_{ii'} + \omega_{ii'}) R_{i'}(\mathrm{M}), \tag{6.29}$$

where for infinitesimal rotation

$$\omega_{ii'} = -\omega_{i'i}. \tag{6.30}$$

The ion displacements (6.16) are therefore

$$V_i(\mathrm{M}) = \sum_{i'} \omega_{ii'} R_{i'}(\mathrm{M}). \tag{6.31}$$

The transformed potential coefficients are given by (6.18)–(6.21), with the transformation displacements given by (6.31); for example,

$$\bar{\Phi}_i(\mathrm{M}) = \Phi_i(\mathrm{M}) + \sum_{\mathrm{N}} \sum_{jj'} \Phi_{ij}(\mathrm{M,N}) \omega_{jj'} R_{j'}(\mathrm{N}), \tag{6.32}$$

$$\bar{\Phi}_{ij}(\mathrm{M,N}) = \Phi_{ij}(\mathrm{M,N}) + \sum_{\mathrm{P}} \sum_{kk'} \Phi_{ijk}(\mathrm{M,N,P}) \omega_{kk'} R_{k'}(\mathrm{P}), \tag{6.33}$$

to first order in $\omega_{ii'}$. The external force $\mathbf{f}(\mathrm{M})$ is still applied to ion M, but should be rotated to the new direction $\bar{\mathbf{f}}(\mathrm{M})$, given by the vector transformation (6.29):

$$\bar{f}_i(\mathrm{M}) = \sum_{i'} (\delta_{ii'} + \omega_{ii'}) f_{i'}(\mathrm{M}). \tag{6.34}$$

Now the total system potential, expressed as an expansion in the displacements $\mathbf{U}(\mathrm{M})$ of the ions from the rotated equilibrium positions $\bar{\mathbf{R}}(\mathrm{M})$ is

$$\begin{aligned} \bar{\Psi} = \bar{\Psi}_0 &+ \sum_{\mathrm{M}} \sum_i [\bar{\Phi}_i(\mathrm{M}) - \bar{f}_i(\mathrm{M})] U_i(\mathrm{M}) \\ &+ \tfrac{1}{2} \sum_{\mathrm{MN}} \sum_{ij} \bar{\Phi}_{ij}(\mathrm{M,N}) U_i(\mathrm{M}) U_j(\mathrm{N}) + \cdots. \end{aligned} \tag{6.35}$$

These displacements $\mathbf{U}(\mathrm{M})$ are still measured in the original coordinate system; in order to maintain the coordinate system the same relative to the crystal, the coordinate system is also rotated through the angle $\boldsymbol{\omega}$. The displacements $\mathbf{U}(\mathrm{M})$ are denoted $\hat{\mathbf{U}}(\mathrm{M})$ when measured in the rotated coordinate system, and the transformation is just the inverse of the vector transformation (6.29):

$$\hat{U}_i(\mathrm{M}) = \sum_{i'} (\delta_{ii'} - \omega_{ii'}) U_{i'}(\mathrm{M}); \tag{6.36}$$

inverting this to first order in $\omega_{ii'}$ gives

$$U_i(\mathrm{M}) = \sum_{i'} (\delta_{ii'} + \omega_{ii'}) \hat{U}_{i'}(\mathrm{M}). \tag{6.37}$$

This expression may be used in the potential expansion (6.35) to give the expansion of the total system potential in terms of the displacements $\hat{\mathbf{U}}(\mathrm{M})$ of the ions from the rotated equilibrium positions and measured in the rotated coordinate system. This expansion is written

$$\begin{aligned}\hat{\Psi} = \hat{\Psi}_0 &+ \sum_{\mathrm{M}} \sum_{i} [\hat{\Phi}_i(\mathrm{M}) - \hat{f}_i(\mathrm{M})] \hat{U}_i(\mathrm{M}) \\ &+ \tfrac{1}{2} \sum_{\mathrm{MN}} \sum_{ij} \hat{\Phi}_{ij}(\mathrm{M,N}) \hat{U}_i(\mathrm{M}) \hat{U}_j(N) + \cdots,\end{aligned} \tag{6.38}$$

and the coefficients are, to first order in $\boldsymbol{\omega}$,

$$\begin{aligned}\hat{\Phi}_i(\mathrm{M}) - \hat{f}_i(\mathrm{M}) = {}& \Phi_i(\mathrm{M}) - f_i(\mathrm{M}) \\ &+ \sum_{i'} \Phi_{i'}(\mathrm{M}) \omega_{i'i} + \sum_{\mathrm{N}} \sum_{jj'} \Phi_{ij}(\mathrm{M,N}) R_{j'}(\mathrm{N}) \omega_{jj'},\end{aligned} \tag{6.39}$$

$$\begin{aligned}\hat{\Phi}_{ij}(\mathrm{M,N}) = {}& \Phi_{ij}(\mathrm{M,N}) + \sum_{i'} \Phi_{i'j}(\mathrm{M,N}) \omega_{i'i} + \sum_{j'} \Phi_{ij'}(\mathrm{M,N}) \omega_{j'j} \\ &+ \sum_{\mathrm{P}} \sum_{kk'} \Phi_{ijk}(\mathrm{M,N,P}) R_{k'}(\mathrm{P}) \omega_{kk'},\end{aligned} \tag{6.40}$$

$$\begin{aligned}\hat{\Phi}_{ijk}(\mathrm{M,N,P}) = {}& \Phi_{ijk}(\mathrm{M,N,P}) + \sum_{i'} \Phi_{i'jk}(\mathrm{N,N,P}) \omega_{i'i} \\ &+ \sum_{j'} \Phi_{ij'k}(\mathrm{M,N,P}) \omega_{j'j} + \sum_{k'} \Phi_{ijk'}(\mathrm{M,N,P}) \omega_{k'k} \\ &+ \sum_{\mathrm{Q}} \sum_{ll'} \Phi_{ijkl}(\mathrm{M,N,P,Q}) R_{l'}(\mathrm{Q}) \omega_{ll'},\end{aligned} \tag{6.41}$$

and so on.

Finally, rotational invariance requires each of the coefficients in the expansion (6.38) to be the same as the corresponding coefficient in the original expansion (6.9), since the description is unchanged when the system and the coordinate system are rotated together. Thus,

$$\hat{\Phi}_i(\mathrm{M}) - \hat{f}_i(\mathrm{M}) = \Phi_i(\mathrm{M}) - f_i(\mathrm{M}), \tag{6.42}$$

$$\hat{\Phi}_{ij}(\mathrm{M,N}) = \Phi_{ij}(\mathrm{M,N}), \tag{6.43}$$

$$\hat{\Phi}_{ijk}(\mathrm{M,N,P}) = \Phi_{ijk}(\mathrm{M,N,P}), \tag{6.44}$$

and so on. This obviously requires the vanishing of the terms that are linear in $\omega_{ii'}$ in (6.39)–(6.41), and since $\omega_{ii'}$ is antisymmetric according to (6.30), such requirements can be expressed in terms of the symmetry of the coefficients of $\omega_{ii'}$. That is,

$$\sum_{ii'} f_{ii'}\omega_{ii'} = 0$$

implies that

$$f_{ii'} \text{ is symmetric in } i,i'.$$

Then the rotational invariance conditions, to first order in $\boldsymbol{\omega}$, are

$$\sum_{\mathrm{N}} \Phi_{ij}(\mathrm{M,N})R_k(\mathrm{N}) + \Phi_j(\mathrm{M})\delta_{ik} \quad \text{is symmetric in } j,k; \tag{6.45}$$

$$\sum_{\mathrm{P}} \Phi_{ijk}(\mathrm{M,N,P})R_l(\mathrm{P}) + \Phi_{kj}(\mathrm{M,N})\delta_{il} + \Phi_{ik}(\mathrm{M,N})\delta_{jl} \quad \text{is symmetric in } k,l; \tag{6.46}$$

$$\sum_{\mathrm{Q}} \Phi_{ijkl}(\mathrm{M,N,P,Q})R_m(\mathrm{Q}) + \Phi_{ljk}(\mathrm{M,N,P})\delta_{im} + \Phi_{ilk}(\mathrm{M,N,P})\delta_{jm} + \Phi_{ijl}(\mathrm{M,N,P})\delta_{km} \quad \text{is symmetric in } l,m. \tag{6.47}$$

The equilibrium and invariance conditions place restrictions on the potential energy coefficients, and these conditions must be satisfied by any physically acceptable model for interactions among the ions in a crystal. In addition the equilibrium and invariance conditions are of basic importance in formal theoretical derivations.

Stability

The equilibrium configuration of ions plus external forces is a stable equilibrium if the total system potential Ψ is minimum with respect to arbitrary small virtual displacements of the ions from equilibrium. The system potential is expanded for virtual displacements in (6.9), and in view of the equilibrium condition (6.11) this is

$$\Psi = \Psi_0 + \tfrac{1}{2}\sum_{\mathrm{MN}}\sum_{ij}\Phi_{ij}(\mathrm{M,N})U_i(\mathrm{M})U_j(\mathrm{N}) + \cdots. \tag{6.48}$$

The stability condition is that the homogeneous quadratic form in (6.48) is positive definite, i.e., positive for any values of the $U_i(\mathrm{M})$ except all $U_i(\mathrm{M}) = 0$. Let us introduce the indices $\alpha,\beta,\cdots$, where α stands for the pair Mi; then the stability condition is written

$$\sum_{\alpha\beta} \Phi_{\alpha\beta} U_\alpha U_\beta > 0, \tag{6.49}$$

and from the theorem (1.67) this quadratic form is positive definite if and only if the matrix of coefficients is positive definite:

$$[\Phi_{\alpha\beta}] > 0. \tag{6.50}$$

The stability condition encompasses all possible displacements of the ions, and hence all the motional degrees of freedom of the crystal. As a result of the stability condition, all normal lattice waves are required to have positive energies. If the ion displacements are restricted to correspond to a homogeneous strain, then (6.49) becomes a homogeneous quadratic form in the strain parameters, like (3.42), and the stability condition is equivalent to requiring all long-wavelength acoustic lattice waves to have positive energies, in analogy with (3.40). These points are studied in more detail in subsequent pages.

Lattice Symmetry Properties

The system is now specialized to a collection of ions representing a finite crystal of initial volume V, plus initial externally applied forces. There are N_0 unit cells in the crystal and n ions in each unit cell. It now requires two indices to specify a given ion: the unit cell will be denoted by M,N,P, $\cdots$, and the particular ion in a given unit cell will be denoted by $\mu,\nu,\pi,\cdots$. The initial equilibrium positions of the ions are $\mathbf{R}(\mathrm{M}\mu)$, which can be written

$$\mathbf{R}(\mathrm{M}\mu) = \mathbf{R}(\mathrm{M}) + \mathbf{R}(\mu), \tag{6.51}$$

where $\mathbf{R}(\mathrm{M})$ is the location of a reference point in unit cell M and $\mathbf{R}(\mu)$ is the location relative to the reference point of ion μ within a unit cell. Similarly the displacements of the ions from equilibrium are $\mathbf{U}(\mathrm{M}\mu)$. Extension of the preceding notation is obvious; for example, the expansion of the crystal potential Φ is

$$\Phi = \Phi_0 + \sum_{\mathrm{M}\mu} \sum_i \Phi_i(\mathrm{M}\mu) U_i(\mathrm{M}\mu) + \tfrac{1}{2} \sum_{\mathrm{MN}\mu\nu} \sum_{ij} \Phi_{ij}(\mathrm{M}\mu,\mathrm{N}\nu) U_i(\mathrm{M}\mu) U_j(\mathrm{N}\nu) + \cdots. \tag{6.52}$$

Of course for a primitive lattice there is only one ion per unit cell, all ions are the same, and the indices $\mu,\nu,\cdots$, can be dropped and the preceding notation is recovered.

In the interior of the crystal, the equilibrium configuration of the lattice is perfectly periodic, so the potential energy coefficients remain unchanged if the lattice, or the coordinate system, is translated by an arbitrary unit cell vector $\mathbf{R}(\mathrm{Q})$. This is expressed by writing

$$\Phi_i(\mathrm{M}\mu) = \Phi_i(\mathrm{M} + \mathrm{Q},\mu), \tag{6.53}$$

$$\Phi_{ij}(\mathrm{M}\mu,\mathrm{N}\nu) = \Phi_{ij}(\mathrm{M} + \mathrm{Q},\mu;\mathrm{N} + \mathrm{Q},\nu), \tag{6.54}$$

$$\Phi_{ijk}(\mathrm{M}\mu,\mathrm{N}\nu,\mathrm{P}\pi) = \Phi_{ijk}(\mathrm{M} + \mathrm{Q},\mu;\mathrm{N} + \mathrm{Q},\nu;\mathrm{P} + \mathrm{Q},\pi), \tag{6.55}$$

and so on. The origin of coordinates will be taken at a unit cell in the interior, such that $\mathbf{R}(\mathrm{M}) = 0$ for $\mathrm{M} = 0$. Then the potential energy coefficients coupling any ions in the interior can be related to those coupling the ions in the unit cell at the origin, since from the above equations with $\mathbf{R}(\mathrm{Q}) = -\mathbf{R}(\mathrm{M})$, or $\mathbf{R}(\mathrm{Q}) = -\mathbf{R}(\mathrm{N})$,

$$\Phi_i(\mathrm{M}\mu) = \Phi_i(0\mu), \tag{6.56}$$

$$\Phi_{ij}(\mathrm{M}\mu,\mathrm{N}\nu) = \Phi_{ij}(0\mu;\mathrm{N} - \mathrm{M},\nu) = \Phi_{ij}(\mathrm{M} - \mathrm{N},\mu;0\nu), \tag{6.57}$$

$$\Phi_{ijk}(\mathrm{M}\mu,\mathrm{N}\nu,\mathrm{P}\pi) = \Phi_{ijk}(0\mu;\mathrm{N} - \mathrm{M},\nu;\mathrm{P} - \mathrm{M},\pi) = \cdots. \tag{6.58}$$

There is a special set of inversion symmetry properties of the potential energy coefficients, whenever each ion in the crystal is a center of inversion symmetry. Consider a primitive lattice, which always has this inversion symmetry, and write the total system potential expansion

$$\begin{aligned}\Psi = \Psi_0 &+ \tfrac{1}{2}\sum_{\mathrm{MN}}\sum_{ij}\Phi_{ij}(\mathrm{M,N})U_i(\mathrm{M})U_j(\mathrm{N}) \\ &+ \tfrac{1}{6}\sum_{\mathrm{MNP}}\sum_{ijk}\Phi_{ijk}(\mathrm{M,N,P})U_i(\mathrm{M})U_j(\mathrm{N})U_k(\mathrm{P}) + \cdots.\end{aligned} \tag{6.59}$$

Now if the coordinate system is inverted, each $\mathbf{R}(\mathrm{M}) \rightarrow \mathbf{R}(-\mathrm{M}) = -\mathbf{R}(\mathrm{M})$, and $\Phi_{ij}(\mathrm{M,N}) \rightarrow \Phi_{ij}(-\mathrm{M},-\mathrm{N})$, and if the $\mathbf{U}(\mathrm{M})$ are expressed in the inverted coordinates $\mathbf{U}(\mathrm{M}) \rightarrow -\mathbf{U}(\mathrm{M})$, and the total system potential is

$$\begin{aligned}\Psi = \Psi_0 &+ \tfrac{1}{2}\sum_{\mathrm{MN}}\sum_{ij}\Phi_{ij}(-\mathrm{M},-\mathrm{N})U_i(\mathrm{M})U_j(\mathrm{N}) \\ &- \tfrac{1}{6}\sum_{\mathrm{MNP}}\sum_{ijk}\Phi_{ijk}(-\mathrm{M},-\mathrm{N},-\mathrm{P})U_i(\mathrm{M})U_j(\mathrm{N})U_k(\mathrm{P}) + \cdots.\end{aligned} \tag{6.60}$$

But this must be the same function of the arbitrary ion displacements $\mathbf{U}(\mathrm{M})$ as the initial expansion (6.59); hence each of the coefficients in the two expansions must be equal, so that

$$\Phi_{ij}(\mathrm{M,N}) = \Phi_{ij}(-\mathrm{M},-\mathrm{N}), \tag{6.61}$$

$$\Phi_{ijk}(\mathrm{M,N,P}) = -\Phi_{ijk}(-\mathrm{M},-\mathrm{N},-\mathrm{P}), \tag{6.62}$$

$$\Phi_{ijkl}(\mathrm{M,N,P,Q}) = \Phi_{ijkl}(-\mathrm{M},-\mathrm{N},-\mathrm{P},-\mathrm{Q}), \tag{6.63}$$

and so on.

In addition (6.61) can be written, for $\mathbf{R}(\mathrm{M}) = 0$,

$$\Phi_{ij}(0,\mathrm{P}) = \Phi_{ij}(0,-\mathrm{P}), \tag{6.64}$$

and (6.57) for a primitive lattice gives

$$\begin{aligned} \Phi_{ij}(\mathrm{M},\mathrm{N}) &= \Phi_{ij}(0,\mathrm{N}-\mathrm{M}), \\ \Phi_{ij}(\mathrm{N},\mathrm{M}) &= \Phi_{ij}(0,\mathrm{M}-\mathrm{N}). \end{aligned} \tag{6.65}$$

Comparing (6.64) and (6.65), and including the general symmetry (6.4) of $\Phi_{ij}(\mathrm{M},\mathrm{N})$ in its index pairs, gives

$$\Phi_{ij}(\mathrm{M},\mathrm{N}) = \Phi_{ij}(\mathrm{N},\mathrm{M}) = \Phi_{ji}(\mathrm{M},\mathrm{N}). \tag{6.66}$$

Thus for a primitive lattice $\Phi_{ij}(\mathrm{M},\mathrm{N})$ is symmetric in i,j and in M,N. It should be noted that all these symmetry relations, (6.53)–(6.58) and (6.61)–(6.66), were derived for the perfectly periodic interior of the crystal.

Generalization of the inversion symmetry properties to nonprimitive lattices in which each ion is at a center of inversion symmetry is obvious from the preceding discussion. Consider the unit cell $\mathrm{M} = 0$ in the interior of the crystal, and for the moment put the origin of coordinates at the position of ion μ, so that $\mathbf{R}(\mu) = 0$. Then there are two equivalent ions located at $\mathbf{R}(\mathrm{N}\nu)$ and $-\mathbf{R}(\mathrm{N}\nu) = \mathbf{R}(-\mathrm{N}-\nu)$; another two equivalent ions located at $\mathbf{R}(\mathrm{P}\pi)$ and $-\mathbf{R}(\mathrm{P}\pi)$; and so on. The potential energy coefficients coupling these ions to the one at 0μ are related by

$$\Phi_{ij}(0\mu,\mathrm{N}\nu) = \Phi_{ij}(0\mu,-\mathrm{N}-\nu), \quad \text{with} \quad \mathbf{R}(\mu) = 0; \tag{6.67}$$

$$\Phi_{ijk}(0\mu,\mathrm{N}\nu,\mathrm{P}\pi) = -\Phi_{ijk}(0\mu,-\mathrm{N}-\nu,-\mathrm{P}-\pi), \quad \text{with} \quad \mathbf{R}(\mu) = 0; \tag{6.68}$$

and these must hold for each μ in turn.

Finally we should comment further on the elimination of surface effects; the general idea is to transform all results so that they may be evaluated in the interior. For example, the translational invariance condition (6.25) is independent of the cell index M when evaluated in the interior, and may be written

$$\sum_{\mathrm{N}\nu} \Phi_{ij}(0\mu,\mathrm{N}\nu) = 0, \quad \text{for all } \mu,i,j\text{; nonprimitive lattice;} \tag{6.69}$$

$$\sum_{\mathrm{N}} \Phi_{ij}(0,\mathrm{N}) = 0, \quad \text{for all } i,j\text{; primitive lattice.} \tag{6.70}$$

The rotational invariance condition (6.45) is also independent of the cell index M when evaluated in the interior, and may be written for a nonprimitive lattice,

$$\sum_{\mathrm{N}\nu} \Phi_{ij}(0\mu,\mathrm{N}\nu) R_k(\mathrm{N}\nu) + \Phi_j(0\mu)\delta_{ik} \quad \text{is symmetric in } j,k; \tag{6.71}$$

or for a primitive lattice,

$$\sum_{\mathrm{N}} \Phi_{ij}(0,\mathrm{N})R_k(\mathrm{N}) + \Phi_j(0)\delta_{ik} \quad \text{is symmetric in } j,k. \tag{6.72}$$

It is presumed that these sums over lattice points converge in the region of perfect periodicity of the lattice.

7. QUADRATIC HOMOGENEOUS DEFORMATION

Strain Expansions of the Potential

For simplicity the external forces will now be restricted to surface forces, representing arbitrary mechanical stresses applied to the crystal. Thus, for example, electric and magnetic fields are not included, so the discussion does not cover such effects as electrostriction and magnetostriction. Since the $\mathbf{f}(\mathrm{M}\mu)$ are applied only near the crystal surface, and since the equilibrium condition (6.11) holds for all ions,

$$f_i(\mathrm{M}\mu) = \Phi_i(\mathrm{M}\mu) = 0, \quad \text{for all } \mathrm{M},\mu,i \text{ in the interior.} \tag{7.1}$$

The theory presented throughout this book is easily extended to include the effects of externally applied fields which penetrate the crystal. The interesting example of a primitive lattice in the presence of gravity is a trivial extension of our results. The case of a piezoelectric crystal in the presence of an electric field is more complicated, but straightforward.

Since the crystal is in the presence of arbitrary mechanical stress, the calculations here are the mechanical analog of the stress–strain thermoelasticity of Section 2. The crystal potential may be considered to be an approximation to the thermodynamic state functions:

$$\Phi \approx F; \qquad \Phi \approx U. \tag{7.2}$$

In this approximation, referred to as the potential approximation, the stresses and elastic constants are given by strain derivatives of Φ. One should always keep in mind that the quantities so calculated are only the mechanical approximations to the corresponding thermodynamic quantities.

From the initial equilibrium configuration, let the ions undergo a homogeneous deformation given by the displacements

$$U_i(\mathrm{M}\mu) = S_i(\mu) + \sum_j u_{ij}R_j(\mathrm{M}\mu). \tag{7.3}$$

Here the u_{ij} are the displacement gradients introduced in Section 2, and the vectors $\mathbf{S}(\mu)$ are the sublattice displacements which occur during the homogeneous strain. This deformation is explicitly assumed to be brought about by additional forces $\mathbf{g}(M\mu)$ applied to the ions near the crystal surface, while the initial forces $\mathbf{f}(M\mu)$ are held constant. Because of the lattice distortion near the crystal surface, the displacement gradients may not be constant in the surface region, but this will be neglected in the process of neglecting surface effects, so we may consider the u_{ij} as constants. The sublattice displacements must be considered as dependent variables, and are to be eliminated in favor of the independent displacement gradients. When this is done, the crystal potential Φ may be expanded as

$$\Phi = \Phi_0 + V \sum_{ij} \tilde{A}_{ij} u_{ij} + \tfrac{1}{2} V \sum_{ijkl} \tilde{A}_{ijkl} u_{ij} u_{kl} + \cdots. \tag{7.4}$$

This equation defines formally the $\tilde{A}$ coefficients in terms of the potential energy coefficients. Henceforth in this section we will not consider terms of higher order than quadratic in the strain parameters.

It is convenient to use the displacement gradients, since the ion displacements are given by the linear relation (7.3). Nevertheless we know from rotational invariance, just as for the state functions in Section 2, that the crystal potential must depend only on the symmetric finite strain parameters η_{ij}, given by

$$\eta_{ij} = \tfrac{1}{2}\left(u_{ij} + u_{ji} + \sum_k u_{ki} u_{kj}\right). \tag{7.5}$$

Therefore the crystal potential can be expanded in the η_{ij}:

$$\Phi = \Phi_0 + V \sum_{ij} \tilde{C}_{ij} \eta_{ij} + \tfrac{1}{2} V \sum_{ijkl} \tilde{C}_{ijkl} \eta_{ij} \eta_{kl}. \tag{7.6}$$

Again, this equation defines formally the $\tilde{C}$ coefficients, and because of the symmetry of the η_{ij}, these coefficients must have complete Voigt symmetry:

$$\tilde{C}_{ij} = \tilde{C}_{ji}; \tag{7.7}$$

$$\tilde{C}_{ijkl} = \tilde{C}_{jikl} = \tilde{C}_{klij} = \cdots. \tag{7.8}$$

By using (7.5) to transform (7.6) to an expansion in the u_{ij}, and comparing the coefficients with those of (7.4), it is easily found that

$$\tilde{A}_{ij} = \tilde{C}_{ij}, \tag{7.9}$$

$$\tilde{A}_{ijkl} = \tilde{C}_{jl} \delta_{ik} + \tilde{C}_{ijkl}. \tag{7.10}$$

From these relations, and the symmetries (7.7) and (7.8) of the $\tilde{C}$ coefficients, it is obvious that

$$\tilde{A}_{ij} = \tilde{A}_{ji}; \tag{7.11}$$

$$\tilde{A}_{ijkl} = \tilde{A}_{klij}; \tag{7.12}$$

$$\tilde{A}_{ijkl} + \tilde{A}_{il} \delta_{jk} = \tilde{A}_{jikl} + \tilde{A}_{jl} \delta_{ik}. \tag{7.13}$$

The $\tilde{C}_{ij}$ and $\tilde{C}_{ijkl}$ are the mechanical analogs of the stresses and second-order elastic constants, respectively, of Section 2, and the $\tilde{A}_{ijkl}$ are the mechanical analogs of the wave-propagation coefficients of Section 3.

Apart from special symmetries which result for particular crystal structures, the maximum general symmetry of the $\tilde{C}$ coefficients of each order is the Voigt symmetry. Since the $\tilde{A}$ and $\tilde{C}$ coefficients are not independent, this Voigt symmetry must be contained also in the $\tilde{A}$ coefficients. For the first-order coefficients $\tilde{A}_{ij}$ the Voigt symmetry is contained in (7.11); for the $\tilde{A}_{ijkl}$ it is contained in (7.11)–(7.13).

Primitive Lattice

Since there is only one ion per unit cell, the homogeneous deformation (7.3) is

$$U_i(\mathrm{M}) = \sum_j u_{ij} R_j(\mathrm{M}), \tag{7.14}$$

and the expansion (6.1) of the crystal potential is

$$\Phi = \Phi_0 + \sum_{\mathrm{M}} \sum_{ij} \Phi_i(\mathrm{M}) R_j(\mathrm{M}) u_{ij} + \tfrac{1}{2} \sum_{\mathrm{MN}} \sum_{ijkl} \Phi_{ij}(\mathrm{M,N}) R_k(\mathrm{M}) R_l(\mathrm{N}) u_{ik} u_{jl} + \cdots. \tag{7.15}$$

Comparing this with (7.4), the $\tilde{A}$ coefficients are seen to be

$$\tilde{A}_{ij} = V^{-1} \sum_{\mathrm{M}} \Phi_i(\mathrm{M}) R_j(\mathrm{M}), \tag{7.16}$$

$$\tilde{A}_{ijkl} = V^{-1} \sum_{\mathrm{MN}} \Phi_{ik}(\mathrm{M,N}) R_j(\mathrm{M}) R_l(\mathrm{N}). \tag{7.17}$$

It may appear that these coefficients depend on the location of the origin of coordinates, but this is not the case, since from the equilibrium condition (6.14), which incidentally may also be looked upon as a translational invariance condition,

$$\sum_{\mathrm{M}} \Phi_i(\mathrm{M}) R_j(\mathrm{M}) = \sum_{\mathrm{M}} \Phi_i(\mathrm{M}) [R_j(\mathrm{M}) - R_j], \tag{7.18}$$

where R_j are components of an arbitrary vector. Similarly, with the aid of (6.25) and (6.26), it follows that $\tilde{A}_{ijkl}$ is independent of the origin of coordinates.

The symmetries (7.11)–(7.13) of the $\tilde{A}$ coefficients must be satisfied on account of the definition of these coefficients by the expansion (7.4). It is nevertheless instructive to relate these symmetries to the equilibrium and invariance conditions of Section 6, through the expressions (7.16) and (7.17) for the $\tilde{A}$ coefficients. The equilibrium conditions (6.13) or (6.15), which may also be looked upon as rotational invariance conditions, are

$$\sum_{\mathrm{M}} \Phi_i(\mathrm{M}) R_j(\mathrm{M}) = \sum_{\mathrm{M}} \Phi_j(\mathrm{M}) R_i(\mathrm{M});$$

this is just the symmetry (7.11), $\tilde{A}_{ij} = \tilde{A}_{ji}$. Also, the symmetry (6.4) of $\Phi_{ij}(\text{M,N})$ in its index pairs allows (7.17) to be rewritten by interchanging these pairs and then relabeling M and N to obtain

$$\sum_{\text{MN}} \Phi_{ik}(\text{M,N})R_j(\text{M})R_l(\text{N}) = \sum_{\text{MN}} \Phi_{ki}(\text{M,N})R_l(\text{M})R_j(\text{N});$$

this is just the symmetry (7.12), $\tilde{A}_{ijkl} = \tilde{A}_{klij}$. Finally, one can multiply the rotational invariance condition (6.45) by a component of $\mathbf{R}(\text{M})$ and sum over M to get

$$\sum_{\text{MN}} \Phi_{ik}(\text{M,N})R_j(\text{M})R_l(\text{N}) + \sum_{\text{M}} \Phi_k(\text{M})R_j(\text{M})\delta_{il} \quad \text{is symmetric in } k,l.$$

From (7.16) and (7.17) this is

$$\tilde{A}_{ijkl} + \tilde{A}_{kj}\delta_{il} \quad \text{is symmetric in } k,l; \tag{7.19}$$

interchanging ij with kl, and using (7.12), gives the symmetry (7.13).

We now wish to eliminate surface effects. The mechanical stresses $\tilde{A}_{ij} = \tilde{C}_{ij}$ contain essential surface contributions when written in the form (7.16), since by (7.1) the $\Phi_i(\text{M})$ vanish in the interior of the crystal. In other words, (7.16) is just the mechanical stress tensor represented as a sum of surface forces. Surface effects can be eliminated from $\tilde{A}_{ij}$ by first calculating Φ per unit volume evaluated in the interior, and then differentiating with respect to strain; this is done for central potentials in Section 9. Surface effects can be eliminated from $\tilde{A}_{ijkl}$ by taking the combination symmetric in j,l and evaluating in the interior. Defining this combination as $\hat{A}_{ikjl}$, we have

$$\hat{A}_{ikjl} = \tfrac{1}{2}(\tilde{A}_{ijkl} + \tilde{A}_{ilkj}), \tag{7.20}$$

and from (7.17),

$$\hat{A}_{ikjl} = \tfrac{1}{2}V^{-1} \sum_{\text{MN}} \Phi_{ik}(\text{M,N})[R_j(\text{M})R_l(\text{N}) + R_l(\text{M})R_j(\text{N})]$$
$$= -\tfrac{1}{2}V^{-1} \sum_{\text{MN}} \Phi_{ik}(\text{M,N})[R_j(\text{N}) - R_j(\text{M})][R_l(\text{N}) - R_l(\text{M})].$$

The last equality follows with the translational invariance conditions (6.25) and (6.26). Now the last Σ_{N} is the same for all ions M in the interior, so surface effects are eliminated by evaluating at $\mathbf{R}(\text{M}) = 0$ and multiplying by the number of cells N_0 to account for the Σ_{M}. If the volume per unit cell is V_C, then

$$V_C = V/N_0, \tag{7.21}$$

and the expression for $\hat{A}_{ikjl}$ evaluated in the interior is

$$\hat{A}_{ikjl} = -\tfrac{1}{2}V_C^{-1} \sum_{\text{N}} \Phi_{ik}(0,\text{N})R_j(\text{N})R_l(\text{N}). \tag{7.22}$$

There are three important symmetry properties of the $\hat{A}$ coefficients; these three together are again equivalent to the Voigt symmetry of the second-order elastic constants:

$$\hat{A}_{ikjl} = \hat{A}_{iklj}, \tag{7.23}$$

$$\hat{A}_{ikjl} = \hat{A}_{kijl}, \tag{7.24}$$

$$\hat{A}_{ikjl} + \tilde{C}_{ik}\delta_{jl} = \hat{A}_{jlik} + \tilde{C}_{jl}\delta_{ik}. \tag{7.25}$$

The first of these is obvious from the definition (7.20) of $\hat{A}_{ikjl}$. The second follows from (7.22) for $\hat{A}_{ikjl}$, along with the primitive lattice symmetry (6.66) of the $\Phi_{ij}(\mathrm{M,N})$. To derive the last symmetry (7.25), write out (7.19) as

$$\tilde{A}_{ijkl} + \tilde{A}_{kj}\delta_{il} = \tilde{A}_{ijlk} + \tilde{A}_{jl}\delta_{ik};$$

subtract (7.13) from this and interchange k,l in the result to get

$$\tilde{A}_{ijkl} + \tilde{A}_{ik}\delta_{jl} = \tilde{A}_{jilk} + \tilde{A}_{jl}\delta_{ik};$$

then take the combination of this equation which is symmetric in j,l

$$\tfrac{1}{2}(\tilde{A}_{ijkl} + \tilde{A}_{ilkj}) + \tilde{A}_{ik}\delta_{jl} = \tfrac{1}{2}(\tilde{A}_{jilk} + \tilde{A}_{lijk}) + \tilde{A}_{jl}\delta_{ik}.$$

On the right-hand side, $\tilde{A}_{lijk}$ can be replaced by $\tilde{A}_{jkli}$, by (7.12), and this equation is then (7.25) when the definition (7.20) of the $\hat{A}$ coefficients is used.

The next step is to calculate the elastic constants $\tilde{C}_{ijkl}$ in terms of the $\hat{A}$ coefficients. From (7.10),

$$\tilde{C}_{ijkl} = \tilde{A}_{ijkl} - \tilde{C}_{jl}\delta_{ik}, \tag{7.26}$$

and taking the combination symmetric in j,l gives, with (7.20) for $\hat{A}_{ikjl}$,

$$\tfrac{1}{2}(\tilde{C}_{ijkl} + \tilde{C}_{ilkj}) = \hat{A}_{ikjl} - \tilde{C}_{jl}\delta_{ik}. \tag{7.27}$$

To solve this for the elastic constants, interchange i and j to get

$$\tfrac{1}{2}(\tilde{C}_{jikl} + \tilde{C}_{jlki}) = \hat{A}_{jkil} - \tilde{C}_{il}\delta_{jk}; \tag{7.28}$$

then, in this, interchange i and k to get

$$\tfrac{1}{2}(\tilde{C}_{jkil} + \tilde{C}_{jlik}) = \hat{A}_{jikl} - \tilde{C}_{kl}\delta_{ij}. \tag{7.29}$$

Now take (7.27) plus (7.28) minus (7.29) and use the Voigt symmetry of $\tilde{C}_{ijkl}$, to obtain the result

$$\tilde{C}_{ijkl} = \hat{A}_{ikjl} + \hat{A}_{jkil} - \hat{A}_{ijkl} - \tilde{C}_{jl}\delta_{ik} - \tilde{C}_{il}\delta_{jk} + \tilde{C}_{kl}\delta_{ij}. \tag{7.30}$$

In the third $\hat{A}$ coefficient on the right we also interchanged i and j, in view of (7.24). The Voigt symmetry (7.8) of the $\tilde{C}_{ijkl}$ must be satisfied on account of the definition of these coefficients by the potential expansion (7.6); the right-hand side of (7.30) must also have complete Voigt symmetry, and the relations (7.23)–(7.25) along with $\tilde{C}_{ij} = \tilde{C}_{ji}$ are just the necessary and sufficient conditions to ensure this symmetry. The proof is left as an exercise.

Equation (7.30) is a convenient formal expression for the elastic constants, in terms of the initial applied stresses $\tilde{C}_{ij}$, and the $\hat{A}_{ijkl}$ from which surface contributions have been eliminated. Mechanical approximation to the stress–strain coefficients is, from (2.53) or (2.54),

$$\tilde{B}_{ijkl} = \hat{A}_{ikjl} + \hat{A}_{jkil} - \hat{A}_{ijkl} + \tfrac{1}{2}(\tilde{C}_{ik}\delta_{jl} - \tilde{C}_{il}\delta_{jk} + \tilde{C}_{jk}\delta_{il} - \tilde{C}_{jl}\delta_{ik} + 2\tilde{C}_{kl}\delta_{ij} - 2\tilde{C}_{ij}\delta_{kl}), \tag{7.31}$$

and mechanical approximation to the propagation coefficients is, from (3.12) or (3.13) or from (7.10),

$$\tilde{A}_{ijkl} = \hat{A}_{ikjl} + \hat{A}_{jkil} - \hat{A}_{ijkl} - \tilde{C}_{il}\delta_{jk} + \tilde{C}_{kl}\delta_{ij}. \tag{7.32}$$

These equations simplify further when the mechanical stress is zero, or an isotropic pressure.

Nonprimitive Lattice

The homogeneous deformation is now given by (7.3)

$$U_i(\mathrm{M}\mu) = \mathrm{S}_i(\mu) + \sum_j u_{ij} R_j(\mathrm{M}\mu), \tag{7.33}$$

including the sublattice displacements $\mathbf{S}(\mu)$. The expansion (6.52) of the crystal potential is then

$$\begin{aligned}\Phi = \Phi_0 &+ \sum_{\mathrm{M}\mu}\sum_i \Phi_i(\mathrm{M}\mu)\left[S_i(\mu) + \sum_j u_{ij} R_j(\mathrm{M}\mu)\right] \\ &+ \tfrac{1}{2}\sum_{\mathrm{MN}\mu\nu}\sum_{ij}\Phi_{ij}(\mathrm{M}\mu,\mathrm{N}\nu)\Big\{S_i(\mu)S_j(\nu) \\ &\qquad + 2S_i(\mu)\sum_k u_{jk}R_k(\mathrm{N}\nu) + \sum_{kl} u_{ik}u_{jl}R_k(\mathrm{M}\mu)R_l(\mathrm{N}\nu)\Big\}.\end{aligned} \tag{7.34}$$

In writing (7.34), two cross terms linear in $\mathbf{S}(\mu)$ have been combined by interchanging indices to yield the middle term in braces.

The first step is to eliminate the sublattice displacements in favor of the strain parameters u_{ij}. This is done by requiring that the homogeneously deformed configuration is still an equilibrium configuration, i.e., the net force on each ion vanishes. The initial equilibrium condition is (6.11), which reads

$$f_i(\mathrm{M}\mu) = \Phi_i(\mathrm{M}\mu), \quad \text{initial configuration.} \tag{7.35}$$

Now the homogeneous deformation is brought about by additional surface forces $\mathbf{g}(\mathrm{M}\mu)$, and under the deformation the potential coefficients $\Phi_i(\mathrm{M}\mu)$ transform to $\overline{\Phi}_i(\mathrm{M}\mu)$, given in general by (6.18), so the final equilibrium condition is

$$f_i(\mathrm{M}\mu) + g_i(\mathrm{M}\mu) = \overline{\Phi}_i(\mathrm{M}\mu), \quad \text{deformed configuration.} \tag{7.36}$$

With $\overline{\Phi}_i(\mathrm{M}\mu)$ given by (6.18), where the ion displacements are (7.33), the final equilibrium condition is

$$f_i(\mathrm{M}\mu) + g_i(\mathrm{M}\mu)$$
$$= \Phi_i(\mathrm{M}\mu) + \sum_{\mathrm{N}\nu}\sum_j \Phi_{ij}(\mathrm{M}\mu,\mathrm{N}\nu)\left[S_j(\nu) + \sum_k u_{jk}R_k(\mathrm{N}\nu)\right] + \cdots. \tag{7.37}$$

It will be necessary to solve this for $S_j(\nu)$ only to first order in the u_{ij}. In view of the initial equilibrium condition (7.35), $f_i(\mathrm{M}\mu)$ and $\Phi_i(\mathrm{M}\mu)$ cancel from (7.37). In addition we will eliminate surface effects by evaluating (7.37) in the interior, where the surface forces $\mathbf{g}(\mathrm{M}\mu)$ vanish; the result can be written as a set of inhomogeneous equations for the $S_j(\nu)$:

$$\sum_{\mathrm{N}\nu}\sum_j \Phi_{ij}(0\mu,\mathrm{N}\nu)S_j(\nu) = -\sum_{\mathrm{N}\nu}\sum_{jk} \Phi_{ij}(0\mu,\mathrm{N}\nu)R_k(\mathrm{N}\nu)u_{jk}. \tag{7.38}$$

The homogeneous equations, obtained by setting the right-hand side equal to zero, have as solution an arbitrary vector $\mathbf{T}$ independent of ν, since

$$\sum_j T_j \sum_{\mathrm{N}\nu} \Phi_{ij}(0\mu,\mathrm{N}\nu) = 0, \tag{7.39}$$

by the translational invariance condition (6.69) evaluated in the interior. Thus the sublattice displacements determined by (7.38) are arbitrary to within a vector $\mathbf{T}$; this is a direct result of eliminating surface effects from (7.37). However, the *relative* sublattice displacements are determined by (7.38), and any uniform sublattice displacement $\mathbf{T}$ is a translation of the entire crystal and is guaranteed by translational invariance to contribute nothing to our calculations.

The condition of solubility for the inhomogeneous equations is that the homogeneous solution $\mathbf{T}$ is orthogonal to the inhomogeneous term:

$$\sum_{\mathrm{N}\mu\nu}\sum_{ijk} T_i\Phi_{ij}(0\mu,\mathrm{N}\nu)R_k(\mathrm{N}\nu)u_{jk} = 0, \tag{7.40}$$

or

$$\sum_{ijk} T_i u_{jk} \sum_{\mathrm{N}\mu\nu} \Phi_{ij}(0\mu,\mathrm{N}\nu)R_k(\mathrm{N}\nu) = 0. \tag{7.41}$$

This condition is satisfied with the neglect of surface effects, since by the translational invariance condition (6.25) and (6.26) we can write

$$\sum_{\mathrm{MN}\mu\nu} \Phi_{ij}(\mathrm{M}\mu,\mathrm{N}\nu)[R_k(\mathrm{N}\nu) - R_k(\mathrm{M}\mu)] = 0, \tag{7.42}$$

and here the $\Sigma_{\mathrm{N}\mu\nu}$ is the same for all cells M in the interior, so it can be evaluated at M = 0 to give

$$\sum_{N\mu\nu} \Phi_{ij}(0\mu,N\nu)[R_k(N\nu) - R_k(\mu)] = 0. \qquad (7.43)$$

But the term involving $R_k(\mu)$ vanishes by translational invariance, so

$$\sum_{N\mu\nu} \Phi_{ij}(0\mu,N\nu)R_k(N\nu) = 0, \qquad (7.44)$$

and (7.41) is satisfied.

On the left-hand side of (7.38), the matrix of coefficients $\Sigma_N \Phi_{ij}(0\mu,N\nu)$ is a $3n \times 3n$ real symmetric matrix, symmetric in the index pairs μi and νj, where n is the number of ions per unit cell. Since the solutions for $S_j(\nu)$ are arbitrary within added vector components T_j, the inhomogeneous equations (7.38) are not all independent, and the matrix of coefficients is singular. The arbitrariness of the solutions may be removed by setting the sublattice displacement zero for any desired sublattice, say sublattice σ:

$$S_i(\sigma) = 0, \quad i = 1,2,3; \qquad (7.45)$$

this of course corresponds to a particular choice of the homogeneous solution T_i. Then the inhomogeneous equations (7.38) are considered only for $\mu,\nu \neq \sigma$, and the $(3n - 3) \times (3n - 3)$ matrix of coefficients $\Sigma_N \Phi_{ij}(0\mu,N\nu)$ is in general no longer singular and may be inverted by a real symmetric $(3n - 3) \times (3n - 3)$ matrix $\mathbf{\Gamma}$,

$$\sum_{\mu i} \Gamma_{ki}(\pi\mu) \sum_{N} \Phi_{ij}(0\mu,N\nu) = \delta_{kj}\delta_{\pi\nu}, \quad \pi,\mu,\nu \neq \sigma. \qquad (7.46)$$

With this relation the inhomogeneous equations are solved for $S_i(\mu)$, $\mu \neq \sigma$, to give

$$S_i(\mu) = -\sum_{N\nu\pi}\sum_{jkl} \Gamma_{ij}(\mu\nu)\Phi_{jk}(0\nu,N\pi)R_l(N\pi)u_{kl}. \qquad (7.47)$$

Finally it is convenient to include the index σ in our calculations by setting

$$\Gamma_{ij}(\mu\nu) = 0, \quad \mu \text{ or } \nu = \sigma; \qquad (7.48)$$

then the solution (7.47) is always applicable, and gives $S_i(\sigma) = 0$. The important property for general derivations is the symmetry of $\mathbf{\Gamma}$,

$$\Gamma_{ij}(\mu\nu) = \Gamma_{ji}(\nu\mu), \qquad (7.49)$$

which includes $\mu,\nu = \sigma$.

We will abbreviate the solution (7.47) for the sublattice displacements by introducing the coefficients

$$X_{i,jk}(\mu) = -\sum_{N\nu\pi}\sum_{l} \Gamma_{il}(\mu\nu)\Phi_{lj}(0\nu,N\pi)R_k(N\pi); \qquad (7.50)$$

then the $S_i(\mu)$ are related to the strains u_{ij} by

$$S_i(\mu) = \sum_{jk} X_{i,jk}(\mu) u_{jk}. \tag{7.51}$$

The X coefficients have the important symmetry

$$X_{i,jk}(\mu) = X_{i,kj}(\mu). \tag{7.52}$$

To prove this we simply note that (7.50) contains the sum

$$\sum_{\mathrm{N}\pi} \Phi_{ij}(0\nu,\mathrm{N}\pi) R_k(\mathrm{N}\pi),$$

and since we are taking $\Phi_j(0\nu) = 0$ according to (7.1), the rotational invariance condition (6.71) requires this sum to be symmetric in j,k.

The next step is to eliminate the sublattice displacements from the expansion (7.34) of the crystal potential. The first term involving $S_i(\mu)$ in (7.34) may be evaluated in the interior by setting the cell index $\mathrm{M} = 0$ and multiplying by the number of cells N_0 to account for the Σ_{M}:

$$\sum_{\mathrm{M}\mu} \sum_{i} \Phi_i(\mathrm{M}\mu) S_i(\mu) = N_0 \sum_{\mu} \sum_{i} \Phi_i(0\mu) S_i(\mu) = 0, \tag{7.53}$$

since $\Phi_i(0\mu) = 0$. It is because this term vanishes that the $S_i(\mu)$ need to be determined only to first order in the u_{ij}. Eliminating surface effects from the term of second order in $S_i(\mu)$ in (7.34), and replacing one of the $S_i(\mu)$ factors by means of the equation (7.38), gives

$$\begin{aligned} \tfrac{1}{2} \sum_{\mathrm{M\,N}\mu\nu} \sum_{ij} \Phi_{ij}(\mathrm{M}\mu,\mathrm{N}\nu) S_i(\mu) S_j(\nu) &= \tfrac{1}{2} N_0 \sum_{\mathrm{N}\mu\nu} \sum_{ij} \Phi_{ij}(0\mu,\mathrm{N}\nu) S_i(\mu) S_j(\nu) \\ &= -\tfrac{1}{2} N_0 \sum_{\mathrm{N}\mu\nu} \sum_{ijk} \Phi_{ij}(0\mu,\mathrm{N}\nu) R_k(\mathrm{N}\nu) S_i(\mu) u_{jk}. \end{aligned} \tag{7.54}$$

Surface effects may also be eliminated from the cross term in $S_i(\mu) u_{jk}$ in (7.34), to obtain

$$\sum_{\mathrm{MN}\mu\nu} \sum_{ijk} \Phi_{ij}(\mathrm{M}\mu,\mathrm{N}\nu) R_k(\mathrm{N}\nu) S_i(\mu) u_{jk} = N_0 \sum_{\mathrm{N}\mu\nu} \sum_{ijk} \Phi_{ij}(0\mu,\mathrm{N}\nu) R_k(\mathrm{N}\nu) S_i(\mu) u_{jk}. \tag{7.55}$$

Thus (7.54) cancels half of this contribution. Now with the help of (7.53)–(7.55), and putting the solution (7.51) for $S_i(\mu)$ into the term (7.55), the potential expansion to second order in the displacement gradients is

$$\begin{aligned} \Phi = \Phi_0 &+ \sum_{ij} \sum_{\mathrm{M}\mu} \Phi_i(\mathrm{M}\mu) R_j(\mathrm{M}\mu) u_{ij} \\ &+ \tfrac{1}{2} \sum_{ijkl} \Big\{ N_0 \sum_{\mathrm{N}\mu\nu} \sum_{m} \Phi_{mk}(0\mu,\mathrm{N}\nu) R_l(\mathrm{N}\nu) X_{m,ij}(\mu) \\ &\qquad + \sum_{\mathrm{MN}\mu\nu} \Phi_{ik}(\mathrm{M}\mu,\mathrm{N}\nu) R_j(\mathrm{M}\mu) R_l(\mathrm{N}\nu) \Big\} u_{ij} u_{kl}. \end{aligned} \tag{7.56}$$

The $\tilde{A}$ coefficients are defined by (7.4) as the coefficients in the expansion of Φ in powers of u_{ij}; comparison of that equation with (7.56) gives

$$\tilde{A}_{ij} = V^{-1} \sum_{M\mu} \Phi_i(M\mu) R_j(M\mu); \tag{7.57}$$

$$\begin{aligned}\tilde{A}_{ijkl} = V^{-1}\Big\{ & N_0 \sum_{N\mu\nu} \sum_m \Phi_{mk}(0\mu,N\nu) R_l(N\nu) X_{m,ij}(\mu) \\ & + \sum_{MN\mu\nu} \Phi_{ik}(M\mu,N\nu) R_j(M\mu) R_l(N\nu)\Big\}.\end{aligned} \tag{7.58}$$

Again the mechanical stresses $\tilde{A}_{ij} = \tilde{C}_{ij}$ contain essential surface contributions when written in the form (7.57). Surface contributions have already been eliminated from the first term in $\tilde{A}_{ijkl}$, and just as for primitive lattices surface effects may be eliminated from the second term by taking the combination symmetric in j,l and evaluating in the interior. The treatment of the second term in $\tilde{A}_{ijkl}$ is as follows, for the combination symmetric in j,l.

$$\begin{aligned}&\tfrac{1}{2} \sum_{MN\mu\nu} \Phi_{ik}(M\mu,N\nu)[R_j(M\mu)R_l(N\nu) + R_l(M\mu)R_j(N\nu)] \\ &\qquad = -\tfrac{1}{2} \sum_{MN\mu\nu} \Phi_{ik}(M\mu,N\nu)[R_j(N\nu) - R_j(M\mu)][R_l(N\nu) - R_l(M\mu)],\end{aligned}$$

from the translational invariance conditions (6.25) and (6.26). The $\Sigma_{N\mu\nu}$ is the same for all cells M in the interior, so surface effects are eliminated by evaluating this at $M = 0$ and multiplying by N_0 to account for the Σ_M, giving the term

$$-\tfrac{1}{2} N_0 \sum_{N\mu\nu} \Phi_{ik}(0\mu,N\nu)[R_j(N\nu) - R_j(\mu)][R_l(N\nu) - R_l(\mu)].$$

With this result, and the volume per unit cell $V_C = V/N_0$, the $\hat{A}_{ikjl}$ are written with surface effects eliminated:

$$\hat{A}_{ikjl} = \tfrac{1}{2}(\tilde{A}_{ijkl} + \tilde{A}_{ilkj}), \tag{7.59}$$

where

$$\begin{aligned}\hat{A}_{ikjl} = {} & \tfrac{1}{2} V_C^{-1} \sum_{N\mu\nu} \sum_m \Phi_{mk}(0\mu,N\nu)[R_l(N\nu) X_{m,ij}(\mu) + R_j(N\nu) X_{m,il}(\mu)] \\ & - \tfrac{1}{2} V_C^{-1} \sum_{N\mu\nu} \Phi_{ik}(0\mu,N\nu)[R_j(N\nu) - R_j(\mu)][R_l(N\nu) - R_l(\mu)].\end{aligned} \tag{7.60}$$

In addition the lattice point vector components $R_l(N\nu)$ in the first line of (7.60) may be replaced by $R_l(N\nu) - R_l(\mu)$, if it is convenient to do so for computational purposes, since the added term in $R_l(\mu)$ is zero by the translational invariance condition (6.69).

The three symmetry properties (7.23)–(7.25) for the $\hat{A}_{ikjl}$ were proved for a primitive lattice; it is now possible to prove these properties for a nonprimitive lattice as well. The first is

$$\hat{A}_{ikjl} = \hat{A}_{iklj}; \tag{7.61}$$

this follows from the definition (7.59) of the $\hat{A}_{ikjl}$. The second symmetry is

$$\hat{A}_{ikjl} = \hat{A}_{kijl}; \tag{7.62}$$

to prove this we need to examine in detail the terms in (7.60). Define

$$f_{ijkl} = \tfrac{1}{2}V_C^{-1} \sum_{N\mu\nu} \sum_{m} \Phi_{mk}(0\mu,N\nu)R_l(N\nu)X_{m,ij}(\mu). \tag{7.63}$$

From the symmetry (7.52) of the X coefficients, it is obvious that

$$f_{ijkl} = f_{jikl}. \tag{7.64}$$

Now writing out (7.63) with the definition (7.50) of the X coefficients,

$$f_{ijkl} = -\tfrac{1}{2}V_C^{-1} \sum_{N\mu\nu} \sum_{P\pi\rho} \sum_{mn} \Phi_{mk}(0\mu,N\nu)R_l(N\nu)\Gamma_{mn}(\mu\pi)\Phi_{ni}(0\pi,P\rho)R_j(P\rho). \tag{7.65}$$

If this expression is symmetric in i,j, it is obviously also symmetric in k,l, so

$$f_{ijkl} = f_{ijlk}. \tag{7.66}$$

Finally, in (7.65) interchange m with n, μ with π, and $N\nu$ with $P\rho$; in view of the symmetry (7.49) of $\Gamma_{mn}(\mu\pi)$ the result is unchanged. But the result corresponds precisely to interchanging i with k and j with l, simultaneously, so

$$f_{ijkl} = f_{klij}. \tag{7.67}$$

We can now prove the symmetry (7.62). Denote the first line of (7.60) for $\hat{A}_{ikjl}$ by $\hat{A}^{(1)}_{ikjl}$; from (7.63) this is

$$\begin{aligned} \hat{A}^{(1)}_{ikjl} &= f_{ijkl} + f_{ilkj} \\ &= f_{klij} + f_{kjil} \\ &= \hat{A}^{(1)}_{kijl}, \end{aligned} \tag{7.68}$$

where for the second equality we used the symmetry (7.67). To prove (7.62) for the last line of (7.60), calculate as follows.

$$\begin{aligned} &\sum_{N\mu\nu} \Phi_{ik}(0\mu,N\nu)[R_j(N\nu) - R_j(\mu)][R_l(N\nu) - R_l(\mu)] \\ &= \sum_{N\mu\nu} \Phi_{ik}(0\mu,N\nu)[R_j(N) + R_j(\nu) - R_j(\mu)][R_l(N) + R_l(\nu) - R_l(\mu)], \end{aligned} \tag{7.69}$$

and since for every $\mathbf{R}(N)$ there is a $\mathbf{R}(-N) = -\mathbf{R}(N)$,

$$= \sum_{N\mu\nu} \Phi_{ik}(0\mu,-N\nu)[-R_j(N) + R_j(\nu) - R_j(\mu)][-R_l(N) + R_l(\nu) - R_l(\mu)],$$

and changing the sign of each square bracket and using the lattice displacement symmetry (6.57),

$$= \sum_{N\mu\nu} \Phi_{ik}(N\mu,0\nu)[R_j(N) - R_j(\nu) + R_j(\mu)][R_l(N) - R_l(\nu) + R_l(\mu)],$$

and interchanging μ and ν, then finally using $\Phi_{ik}(N\nu,0\mu) = \Phi_{ki}(0\mu,N\nu)$,

$$= \sum_{N\mu\nu} \Phi_{ki}(0\mu,N\nu)[R_j(N) - R_j(\mu) + R_j(\nu)][R_l(N) - R_l(\mu) + R_l(\nu)]. \quad (7.70)$$

Thus we have shown that this term is the same when k and i are interchanged, and the symmetry (7.62) is proved for $\hat{A}_{ikjl}$.

The third symmetry of the $\hat{A}_{ikjl}$ is

$$\hat{A}_{ikjl} + \tilde{C}_{ik}\delta_{jl} = \hat{A}_{jlik} + \tilde{C}_{jl}\delta_{ik}. \quad (7.71)$$

Again write $\hat{A}^{(1)}_{ikjl}$ for the first line in (7.60), and from (7.63)

$$\hat{A}^{(1)}_{ikjl} = f_{ijkl} + f_{ilkj};$$

with the symmetry (7.67) this is

$$\hat{A}^{(1)}_{ikjl} = f_{ijkl} + f_{kjil};$$

and with the symmetries (7.64) and (7.66)

$$\hat{A}^{(1)}_{ikjl} = f_{jilk} + f_{jkli} = \hat{A}^{(1)}_{jlik}. \quad (7.72)$$

Now denote the second line of (7.60) by $\hat{A}^{(2)}_{ikjl}$, and write this as the combination symmetric in j,l of the second line of (7.58):

$$\hat{A}^{(2)}_{ikjl} = \tfrac{1}{2}(\tilde{A}^{(2)}_{ijkl} + \tilde{A}^{(2)}_{ilkj}); \quad (7.73)$$

$$\tilde{A}^{(2)}_{ijkl} = V^{-1} \sum_{MN\mu\nu} \Phi_{ik}(M\mu,N\nu)R_j(M\mu)R_l(N\nu). \quad (7.74)$$

In (7.74) use $\Phi_{ik}(M\mu,N\nu) = \Phi_{ki}(N\nu,M\mu)$; then interchange $M\mu$ with $N\nu$ to show

$$\tilde{A}^{(2)}_{ijkl} = \tilde{A}^{(2)}_{klij}. \quad (7.75)$$

Also, the rotational invariance condition (6.45) for a nonprimitive lattice is

$$\sum_{N\nu} \Phi_{ik}(M\mu,N\nu)R_l(N\nu) + \Phi_k(M\mu)\delta_{il} \quad \text{is symmetric in } k,l.$$

Multiply this by $V^{-1}R_j(M\mu)$ and sum over $M\mu$, and compare with (7.74) for $\tilde{A}^{(2)}_{ijkl}$ and (7.57) for $\tilde{A}_{ij} = \tilde{C}_{ij}$, to show that

$$\tilde{A}^{(2)}_{ijkl} + \tilde{C}_{kj}\delta_{il} \quad \text{is symmetric in } k,l. \quad (7.76)$$

Just as in the calculation following (7.25), the symmetries (7.75) and (7.76) along with $\tilde{C}_{ij} = \tilde{C}_{ji}$, and the definition (7.73) of $\hat{A}^{(2)}_{ikjl}$, are sufficient to show that

$$\hat{A}^{(2)}_{ikjl} + \tilde{C}_{ik}\delta_{jl} = \hat{A}^{(2)}_{jlik} + \tilde{C}_{jl}\delta_{ik}. \tag{7.77}$$

Finally, (7.72) and (7.77) together prove the symmetry (7.71) for $\hat{A}_{ikjl}$.

From this point the solution for the elastic constants is the same as for a primitive lattice; in particular, the elastic constants $\tilde{C}_{ijkl}$ are given by (7.30), the stress–strain coefficients $\tilde{B}_{ijkl}$ by (7.31), and the wave-propagation coefficients by (7.32). The three symmetry properties (7.61), (7.62), and (7.71) for the $\hat{A}_{ikjl}$, along with $\tilde{C}_{ij} = \tilde{C}_{ji}$, are again just the necessary and sufficient conditions that the theoretical expression for $\tilde{C}_{ijkl}$ have complete Voigt symmetry. For any lattice in an initial configuration corresponding to arbitrary externally applied homogeneous surface forces, the $\hat{A}_{ikjl}$ with surface effects eliminated are given by (7.60).

It is well known that the sublattice displacements during a homogeneous deformation are zero for any crystal in which each ion is at a center of inversion symmetry in the initial equilibrium configuration. This can be proved by means of the inversion symmetry properties of the potential energy coefficients. The right-hand side of the equations (7.38) for the $S_j(\nu)$ may be written

$$-\sum_{N\nu}\sum_{jk}\Phi_{ij}(0\mu,N\nu)[R_k(N\nu) - R_k(\mu)]u_{jk},$$

where the added term in $R_k(\mu)$ vanishes by translational invariance. But the $\Sigma_{N\nu}$ counts ions at $\pm[R_k(N\nu) - R_k(\mu)]$, for which $\Phi_{ij}(0\mu,N\nu)$ is the same by the inversion symmetry (6.67), and hence the sum vanishes for any i,j,k. Therefore, each and every $S_i(\mu)$ must vanish. Furthermore this argument is not limited to first order in the strains u_{ij}. By constructing an arbitrary finite deformation from a sequence of infinitesimal strains and applying the argument at each step, which may be done since the initial equilibrium configuration is arbitrary in the present theory, it follows that the sublattice displacements $\mathbf{S}(\mu)$ must vanish for finite homogeneous deformations, provided of course that Φ remains analytic for the configurations considered. Besides crystals in which each ion is at a center of inversion symmetry, there may be other nonprimitive lattices where in reality, or perhaps for a given theoretical model, the sublattice displacements are zero; in any such case the coefficients $X_{i,jk}(\mu)$ are all zero, and the preceding equations are greatly simplified.

Second-Order Sublattice Displacements

Let us reconsider, in more general terms, the equilibrium condition for the lattice in the homogeneously deformed configuration. Since all externally applied forces are surface forces, the equilibrium condition in the interior of the crystal for the arbitrary initial configuration is

$$\Phi_i(0\mu) = 0, \quad \text{for all } \mu, i; \tag{7.78}$$

likewise in the interior for the homogeneously deformed configuration,

$$\overline{\Phi}_i(0\mu) = 0, \quad \text{for all } \mu, i. \tag{7.79}$$

If the ion displacements corresponding to the homogeneous deformation are $\mathbf{U}(M\mu)$, then $\overline{\Phi}_i(0\mu)$ may be expanded as [see, e.g., (6.18)]

$$\begin{aligned}\overline{\Phi}_i(0\mu) = \Phi_i(0\mu) &+ \sum_{N\nu}\sum_{j} \Phi_{ij}(0\mu, N\nu)U_j(N\nu) \\ &+ \tfrac{1}{2}\sum_{NP\nu\pi}\sum_{jk} \Phi_{ijk}(0\mu, N\nu, P\pi)U_j(N\nu)U_k(P\pi) + \cdots.\end{aligned} \tag{7.80}$$

Now the independent variables describing a homogeneous deformation are the displacement gradients u_{ij}, and the equilibrium condition allows us to determine the sublattice displacements $S_i(\mu)$ in terms of the u_{ij}. The solution for $S_i(\mu)$, to first order in the u_{ij}, has already been found in (7.51); we will now proceed to the second-order solution. Let us expand the sublattice displacements as a series,

$$S_i(\mu) = S_i^{(1)}(\mu) + S_i^{(2)}(\mu) + \cdots, \tag{7.81}$$

where $S_i^{(1)}(\mu)$ and $S_i^{(2)}(\mu)$ are, respectively, of first and second order in the u_{ij}. The ion displacements $\mathbf{U}(M\mu)$ corresponding to a homogeneous deformation are given by (7.3), and may be similarly expanded in powers of the u_{ij}:

$$U_i^{(1)}(M\mu) = S_i^{(1)}(\mu) + \sum_j u_{ij}R_j(M\mu), \tag{7.82}$$

$$U_i^{(2)}(M\mu) = S_i^{(2)}(\mu), \tag{7.83}$$

and so on. But the solution for the first-order sublattice displacements is given by (7.51), and with this solution (7.82) may be written

$$U_i^{(1)}(M\mu) = \sum_{jk} Y_{i,jk}(M\mu)u_{jk}, \tag{7.84}$$

where

$$Y_{i,jk}(M\mu) = R_k(M\mu)\delta_{ij} + X_{i,jk}(\mu). \tag{7.85}$$

The equilibrium condition (7.79) is that $\overline{\Phi}_i(0\mu)$ must vanish, and since the u_{ij} are arbitrary strain parameters, the coefficient of each term in u_{ij} in the expansion (7.80) must separately vanish. It is sufficient here to write down the complete equilibrium condition in each order in the strains, as follows.

$$\Phi_i(0\mu) = 0, \quad \text{for all } \mu,i; \tag{7.86}$$

$$\sum_{N\nu}\sum_{j} \Phi_{ij}(0\mu,N\nu)U_j^{(1)}(N\nu) = 0, \quad \text{for all } \mu,i; \tag{7.87}$$

$$\sum_{N\nu}\sum_{j} \Phi_{ij}(0\mu,N\nu)S_j^{(2)}(\nu) + \tfrac{1}{2}\sum_{NP\nu\pi}\sum_{jk}\Phi_{ijk}(0\mu,N\nu,P\pi)U_j^{(1)}(N\nu)U_k^{(1)}(P\pi) = 0, \quad \text{for all } \mu,i; \tag{7.88}$$

and so on. The first of these is, of course, satisfied by the initial equilibrium condition (7.78). The second is just the inhomogeneous equation (7.38) for the first-order sublattice displacements, and is satisfied because $U_j^{(1)}(N\nu)$ is given by (7.84) in terms of the solution for the first-order sublattice displacements.

The second-order equilibrium condition (7.88) is a set of inhomogeneous equations for the $S_j^{(2)}(\nu)$, and the solution is obtained from the inverse relation (7.46) as

$$S_i^{(2)}(\mu) = -\tfrac{1}{2}\sum_{NP\nu\pi\rho}\sum_{jkl}\Gamma_{ij}(\mu\nu)\Phi_{jkl}(0\nu,N\pi,P\rho)U_k^{(1)}(N\pi)U_l^{(1)}(P\rho). \tag{7.89}$$

This expresses $\mathbf{S}^{(2)}(\mu)$ as a function quadratic in the strains, since from (7.84) the $\mathbf{U}^{(1)}(M\mu)$ are linear in the u_{ij}. Incidentally, for a crystal in which each ion is at a center of inversion symmetry, we showed, following (7.77), that all $S_i^{(1)}(\mu)$ must vanish. Then from (7.82), the first-order ion displacements are simply $U_i^{(1)}(M\mu) = \Sigma_j\, u_{ij}R_j(M\mu)$, and with the inversion symmetry (6.68) of $\Phi_{ijk}(0\mu,N\nu,P\pi)$ it follows that the second term in (7.88) vanishes for arbitrary u_{ij}. Hence it is verified for such crystals that all $S_i^{(2)}(\mu)$ also vanish by symmetry.

We also wish to discuss the use of an alternate "phase factor" in describing homogeneous deformations of nonprimitive lattices. The ion displacements corresponding to a homogeneous deformation have been written in (7.3),

$$U_i(M\mu) = S_i(\mu) + \sum_j u_{ij}R_j(M\mu);$$

these same displacements may be written instead in a manner which provides greater mathematical convenience on occasion, as

$$U_i(M\mu) = \bar{S}_i(\mu) + \sum_j u_{ij}R_j(M). \tag{7.90}$$

In the description (7.90) the sublattice displacements are obviously different from before, being given by

$$\bar{S}_i(\mu) = S_i(\mu) + \sum_j u_{ij}R_j(\mu). \tag{7.91}$$

This description is perhaps less direct in the physical sense, because for the example of a NaCl type of lattice for which $S_i(\mu)$ are zero by symmetry, the $\bar{S}_i(\mu)$ are not zero. In the treatment of long-wavelength acoustic waves, the two different sublattice displacements $\mathbf{S}(\mu)$ and $\bar{\mathbf{S}}(\mu)$ simply correspond to the use of different phase factors in describing the waves.

All the above homogeneous deformation calculations can be carried out with the displacements (7.90), and the results are the same after the sublattice displacements are eliminated. In particular, the matrix of coefficients $\Gamma_{ij}(\mu\nu)$ is defined exactly as before, by (7.46) and (7.48). The coefficients $X_{i,jk}(\mu)$ are changed from (7.50) to

$$\bar{X}_{i,jk}(\mu) = -\sum_{\mathrm{N}\nu\pi}\sum_{l} \Gamma_{il}(\mu\nu)\Phi_{lj}(0\nu,\mathrm{N}\pi)R_k(\mathrm{N}), \tag{7.92}$$

and the first-order sublattice displacements $\bar{S}_i^{(1)}(\mu)$ are given by an equation analogous to (7.51):

$$\bar{S}_i^{(1)}(\mu) = \sum_{jk} \bar{X}_{i,jk}(\mu)u_{jk}. \tag{7.93}$$

The first-order ion displacements corresponding to a homogeneous deformation were written as a linear combination of the strains in (7.84); these same displacements are still given by the same combination of the strains, namely

$$U_i^{(1)}(\mathrm{M}\mu) = \sum_{jk} Y_{i,jk}(\mathrm{M}\mu)u_{jk},$$

but when we are using the alternate phase factor the Y coefficients may be written

$$Y_{i,jk}(\mathrm{M}\mu) = R_k(\mathrm{M})\delta_{ij} + \bar{X}_{i,jk}(\mu). \tag{7.94}$$

Finally with the alternate phase factor, the coefficients $\hat{A}_{ikjl}$ with surface effects eliminated are found to be

$$\begin{aligned}\hat{A}_{ikjl} = \tfrac{1}{2}V_C^{-1}\sum_{\mathrm{N}\mu\nu}\sum_{m}\Phi_{mk}(0\mu,\mathrm{N}\nu)[R_l(\mathrm{N})\bar{X}_{m,ij}(\mu) + R_j(\mathrm{N})\bar{X}_{m,il}(\mu)]\\ - \tfrac{1}{2}V_C^{-1}\sum_{\mathrm{N}\mu\nu}\Phi_{ik}(0\mu,\mathrm{N}\nu)R_j(\mathrm{N})R_l(\mathrm{N}).\end{aligned} \tag{7.95}$$

The algebra may be checked by showing that this expression is equal to the previous result (7.60) for $\hat{A}_{ikjl}$; this procedure is an interesting exercise.

Wave Propagation

The treatment of long-wavelength acoustic waves is just the mechanical analog of the treatment in Section 3 of long-wavelength thermoelastic waves. The equation of motion may be derived from Lagrange's equation (3.3), where the potential in the Lagrangian density (3.2) is now the crystal potential density $V^{-1}\Phi$. For plane wave solutions, the eigenvalue–eigenvector equation corresponding to (3.20)–(3.24) may be written

$$\rho\omega_s^2 w_{is} = \sum_{jkl} \tilde{A}_{ijkl} k_j k_l w_{ks}, \quad s = 1,2,3; \tag{7.96}$$

where $\mathbf{k}$ is the wave vector, ω_s and $\mathbf{w}_s$ are, respectively, the frequencies and eigenvectors of the three independent waves for each $\mathbf{k}$, and ρ is the density of the crystal in the initial configuration. If the mass of one unit cell is M_C, then

$$\rho = M_C/V_C. \tag{7.97}$$

Because of the Σ_{jl} in (7.96), only the combination of $\tilde{A}_{ijkl}$ symmetric in j,l contributes to the sum, so we have the important result

$$\rho\omega_s^2 w_{is} = \sum_{jkl} \hat{A}_{ikjl} k_j k_l w_{ks}, \tag{7.98}$$

or with (7.97) for ρ,

$$M_C\omega_s^2 w_{is} = \sum_{jkl} V_C \hat{A}_{ikjl} k_j k_l w_{ks}. \tag{7.99}$$

Therefore by (7.98) or (7.99), surface effects, including the externally applied stresses, are completely eliminated from the theoretical expressions for the long-wavelength acoustic wave frequencies.

Mechanical stability of the crystal against excitation of long-wavelength acoustic waves requires the energy of all such waves to be positive, or

$$[\omega_s(\mathbf{k})]^2 > 0, \quad \text{for all } \mathbf{k}, s. \tag{7.100}$$

This is the same as saying that the propagation matrices are positive definite:

$$\left[\sum_{jl} \hat{A}_{ikjl} k_j k_l\right] > 0, \quad \text{for all } \mathbf{k}, \tag{7.101}$$

and by the same kind of calculation which leads to (3.44), this is equivalent to

$$d^2\Phi = V \sum_{ijkl} \tilde{A}_{ijkl} u_{ij} u_{kl} > 0. \tag{7.102}$$

Equation (7.102) is just the homogeneous deformation part of the complete stability requirement (6.49).

8. THIRD-ORDER ELASTIC CONSTANTS

Third-Order Strain Coefficients

The formal theoretical calculation of third-order elastic constants is a straightforward extension of the procedures in Section 7 for calculating second-order elastic constants. The crystal potential may be expanded in powers of the displacement gradients u_{ij}, defining the theoretically simple $\tilde{A}$

coefficients,

$$\Phi = \Phi_0 + V \sum_{ij} \tilde{A}_{ij} u_{ij} + \tfrac{1}{2} V \sum_{ijkl} \tilde{A}_{ijkl} u_{ij} u_{kl} + \tfrac{1}{6} V \sum_{ijklmn} \tilde{A}_{ijklmn} u_{ij} u_{kl} u_{mn} + \cdots \tag{8.1}$$

Because of rotational invariance of the crystal potential, this potential may also be expanded in powers of the symmetric finite strains η_{ij}, defining the $\tilde{C}$ coefficients,

$$\Phi = \Phi_0 + V \sum_{ij} \tilde{C}_{ij} \eta_{ij} + \tfrac{1}{2} V \sum_{ijkl} \tilde{C}_{ijkl} \eta_{ij} \eta_{kl} + \tfrac{1}{6} V \sum_{ijklmn} \tilde{C}_{ijklmn} \eta_{ij} \eta_{kl} \eta_{mn} + \cdots . \tag{8.2}$$

Without considering additional symmetries due to particular crystal symmetries, the maximum general symmetry of the $\tilde{C}$ coefficients is the complete Voigt symmetry, which follows from their definition through the expansion (8.2). For the third-order elastic constants, the Voigt symmetry is

$$\tilde{C}_{ijklmn} = \tilde{C}_{jiklmn} = \tilde{C}_{klijmn} = \tilde{C}_{mnklij} = \cdots . \tag{8.3}$$

Again the expansions (8.1) and (8.2) are not independent; the two sets of coefficients up to second order are related by (7.9) and (7.10), and for the third-order coefficients one finds

$$\tilde{A}_{ijklmn} = \tilde{C}_{ijln} \delta_{km} + \tilde{C}_{kljn} \delta_{im} + \tilde{C}_{jlmn} \delta_{ik} + \tilde{C}_{ijklmn}. \tag{8.4}$$

From the definition of the $\tilde{A}$ coefficients through the expansion (8.1), these coefficients must be symmetric in their index pairs:

$$\tilde{A}_{ijklmn} = \tilde{A}_{klijmn} = \tilde{A}_{mnklij} = \cdots . \tag{8.5}$$

In addition, because of the relation (8.4) between $\tilde{A}_{ijklmn}$ and the $\tilde{C}$ coefficients, and because of the complete Voigt symmetry of the latter, the $\tilde{A}_{ijklmn}$ satisfy a rotational invariance symmetry, which is obviously

$$\tilde{A}_{ijklmn} + \tilde{C}_{klin} \delta_{jm} + \tilde{C}_{ilmn} \delta_{jk} = \tilde{A}_{jiklmn} + \tilde{C}_{kljn} \delta_{im} + \tilde{C}_{jlmn} \delta_{ik}. \tag{8.6}$$

The two symmetries (8.5) and (8.6) of the $\tilde{A}_{ijklmn}$ are completely equivalent to the Voigt symmetry (8.3) of the $\tilde{C}_{ijklmn}$, when the relation (8.4) between these coefficients and the Voigt symmetry of $\tilde{C}_{ijkl}$ are taken into account.

Just as for the second-order coefficients, a combination which allows the elimination of surface effects is

$$\hat{A}_{ikjlmn} = \tfrac{1}{2}(\tilde{A}_{ijklmn} + \tilde{A}_{ilkjmn}); \tag{8.7}$$

note the asymmetry of $\hat{A}_{ikjlmn}$ in the first four indices as compared to the last two. We could define a more symmetric set of coefficients, but the combination (8.7) is sufficient for the elimination of surface effects, as will be shown in the detailed calculations below. Now again, because the $\hat{A}$ coefficients are linearly dependent on the $\tilde{A}$ coefficients, and hence in turn on the $\tilde{C}$ coefficients, the $\hat{A}_{ikjlmn}$ must satisfy a set of symmetry relations which are equivalent to the maximum general symmetry of the third-order coefficients, namely the Voigt symmetry of the $\tilde{C}_{ijklmn}$.

From the definition (8.7) it is obvious that

$$\hat{A}_{ikjlmn} = \hat{A}_{ikljmn}; \tag{8.8}$$

also from (8.7) and the symmetry (8.5) it follows that

$$\hat{A}_{ikjlmn} = \hat{A}_{kijlmn}. \tag{8.9}$$

The symmetry involving interchange of the last two indices of $\hat{A}_{ikjlmn}$ may be written

$$\begin{aligned}\hat{A}_{ikjlmn} &+ \tfrac{1}{2}(\tilde{C}_{ijml} + \tilde{C}_{ilmj})\delta_{kn} + \tfrac{1}{2}(\tilde{C}_{klmj} + \tilde{C}_{kjml})\delta_{in} \\ &= \hat{A}_{ikjlnm} + \tfrac{1}{2}(\tilde{C}_{ijnl} + \tilde{C}_{ilnj})\delta_{km} + \tfrac{1}{2}(\tilde{C}_{klnj} + \tilde{C}_{kjnl})\delta_{im};\end{aligned} \tag{8.10}$$

to show this, interchange ij with mn in (8.6) to obtain

$$\tilde{A}_{ijklmn} + \tilde{C}_{klmj}\delta_{in} + \tilde{C}_{mlij}\delta_{kn} = \tilde{A}_{ijklnm} + \tilde{C}_{klnj}\delta_{im} + \tilde{C}_{nlij}\delta_{km}, \tag{8.11}$$

and then use the definition (8.7) of $\hat{A}_{ikjlmn}$. Finally by straightforward manipulation, one can verify the symmetry of $\hat{A}_{ikjlmn}$ with respect to interchange of the first two index pairs:

$$\begin{aligned}\hat{A}_{ikjlmn} + \tilde{C}_{ikmn}\delta_{jl} + \tfrac{1}{2}(\tilde{C}_{ilkn} + \tilde{C}_{klin})\delta_{jm} + \tfrac{1}{2}(\tilde{C}_{ijkn} + \tilde{C}_{kjin})\delta_{lm} \\ \text{is symmetric in } ik \leftrightarrow jl.\end{aligned} \tag{8.12}$$

There is one more independent symmetry of the $\hat{A}_{ikjlmn}$, namely that involving the interchange of ik with mn, or jl with mn; this may be written down from (8.13) below.

The third-order elastic constants may be expressed in terms of $\hat{A}_{ikjlmn}$ with the aid of (8.4) and (8.7), as

$$\begin{aligned}\tfrac{1}{2}(\tilde{C}_{ijklmn} + \tilde{C}_{ilkjmn}) = \hat{A}_{ikjlmn} &- \tfrac{1}{2}(\tilde{C}_{ijln}\delta_{km} + \tilde{C}_{kljn}\delta_{im} + \tilde{C}_{jlmn}\delta_{ik} \\ &+ \tilde{C}_{iljn}\delta_{km} + \tilde{C}_{kjln}\delta_{im} + \tilde{C}_{ljmn}\delta_{ik}).\end{aligned} \tag{8.13}$$

This may be solved for the third-order elastic constants by a procedure similar to that leading to (7.30) for the second-order elastic constants; the result is

$$\begin{aligned}\tilde{C}_{ijklmn} = \hat{A}_{ikjlmn} &+ \hat{A}_{jkilmn} - \hat{A}_{ijklmn} + \tilde{C}_{klmn}\delta_{ij} - \tilde{C}_{jlmn}\delta_{ik} - \tilde{C}_{ilmn}\delta_{jk} \\ &- \tfrac{1}{2}(2\tilde{C}_{ijln} + \tilde{C}_{iljn} + \tilde{C}_{jlin})\delta_{km} + \tfrac{1}{2}(\tilde{C}_{jlkn} - \tilde{C}_{kljn})\delta_{im} \\ &+ \tfrac{1}{2}(\tilde{C}_{ilkn} - \tilde{C}_{klin})\delta_{jm}.\end{aligned} \tag{8.14}$$

The second-order elastic constants here may be calculated from the $\hat{A}_{ikjl}$ and the components $\tilde{C}_{ij}$ of the applied stress, by (7.30). We will now derive the theoretical expressions for the $\hat{A}_{ikjlmn}$.

Primitive Lattice

From a glance at (7.16) and (7.17), it is obvious that the third-order coefficients for a primitive lattice are

$$\tilde{A}_{ijklmn} = V^{-1} \sum_{\mathrm{MNP}} \Phi_{ikm}(\mathrm{M,N,P}) R_j(\mathrm{M}) R_l(\mathrm{N}) R_n(\mathrm{P}). \tag{8.15}$$

From the translational invariance condition (6.27), it follows that the $\tilde{A}_{ijklmn}$ are independent of the location of the origin of coordinates. The combination of (8.15) symmetric in j,l gives $\hat{A}_{ikjlmn}$ according to (8.7),

$$\hat{A}_{ikjlmn} = \tfrac{1}{2} V^{-1} \sum_{\mathrm{MNP}} \Phi_{ikm}(\mathrm{M,N,P})[R_j(\mathrm{M}) R_l(\mathrm{N}) + R_l(\mathrm{M}) R_j(\mathrm{N})] R_n(\mathrm{P}),$$

and in view of the translational invariance (6.27) this may be written

$$\hat{A}_{ikjlmn} = -\tfrac{1}{2} V^{-1} \sum_{\mathrm{MNP}} \Phi_{ikm}(\mathrm{M,N,P})[R_j(\mathrm{N}) - R_j(\mathrm{M})] \times [R_l(\mathrm{N}) - R_l(\mathrm{M})][R_n(\mathrm{P}) - R_n(\mathrm{M})]. \tag{8.16}$$

Here the Σ_{NP} is the same for all ions M in the interior, so surface effects are eliminated by evaluating at M = 0 and multiplying by N_0 to give

$$\hat{A}_{ikjlmn} = -\tfrac{1}{2} V_C^{-1} \sum_{\mathrm{NP}} \Phi_{ikm}(0,\mathrm{N,P}) R_j(\mathrm{N}) R_l(\mathrm{N}) R_n(\mathrm{P}). \tag{8.17}$$

This result clearly shows the asymmetry of $\hat{A}_{ikjlmn}$ in the last two indices as compared to the first four.

It is always useful to check the theoretical calculations by verifying the expected symmetry of the coefficients. Let us do this here simply by proving the symmetries (8.5) and (8.6) of the $\tilde{A}_{ijklmn}$. Because of the general symmetry (6.5) of the $\Phi_{ikm}(\mathrm{M,N,P})$, we can interchange the pairs Mi and Nk, and rewrite (8.15) as

$$\tilde{A}_{ijklmn} = V^{-1} \sum_{\mathrm{MNP}} \Phi_{kim}(\mathrm{N,M,P}) R_j(\mathrm{M}) R_l(\mathrm{N}) R_n(\mathrm{P});$$

then by interchanging M and N in the sum it is shown that

$$\tilde{A}_{ijklmn} = \tilde{A}_{klijmn}.$$

Similar interchanges of other index pairs establish the complete symmetry (8.5).

The rotational invariance condition (6.46) can be written

$$\sum_{\mathrm{P}} \Phi_{ikm}(\mathrm{M,N,P})R_n(\mathrm{P}) + \Phi_{mk}(\mathrm{M,N})\delta_{in} + \Phi_{im}(\mathrm{M,N})\delta_{kn} \quad \text{is symmetric in } m,n. \tag{8.18}$$

Multiply this by $V^{-1}R_j(\mathrm{M})R_l(\mathrm{N})$ and sum over M and N to get

$$V^{-1}\sum_{\mathrm{MNP}} \Phi_{ikm}(\mathrm{M,N,P})R_j(\mathrm{M})R_l(\mathrm{N})R_n(\mathrm{P}) + V^{-1}\sum_{\mathrm{MN}}\Phi_{mk}(\mathrm{M,N})R_j(\mathrm{M})R_l(\mathrm{N})\delta_{in}$$
$$+ V^{-1}\sum_{\mathrm{MN}}\Phi_{im}(\mathrm{M,N})R_j(\mathrm{M})R_l(\mathrm{N})\delta_{kn} \quad \text{is symmetric in } m,n. \tag{8.19}$$

With the expressions (7.17) and (8.15) for $\tilde{A}$ coefficients, this is

$$\tilde{A}_{ijklmn} + \tilde{A}_{mjkl}\delta_{in} + \tilde{A}_{ijml}\delta_{kn} \quad \text{is symmetric in } m,n. \tag{8.20}$$

Introducing the second-order elastic constants by means of (7.10) leads to

$$\tilde{A}_{ijklmn} + \tilde{C}_{mjkl}\delta_{in} + \tilde{C}_{ijml}\delta_{kn} \quad \text{is symmetric in } m,n, \tag{8.21}$$

since the additional term in the stresses is

$$\tilde{C}_{jl}(\delta_{km}\delta_{in} + \delta_{kn}\delta_{im}),$$

and this term is separately symmetric in m,n. Finally, interchanging ij with mn, and using the symmetry (8.5) of $\tilde{A}_{ijklmn}$ in its index pairs, establishes the rotational invariance symmetry (8.6).

Nonprimitive Lattice

In order to calculate the third- and higher-order elastic constants for a nonprimitive lattice from the method of homogeneous deformation, it is necessary in principle to calculate the sublattice displacements $\mathbf{S}(\mu)$ to second and higher order in the strains u_{ij}. It turns out, however, that the terms in $\mathbf{S}(\mu)$ of second order in the u_{ij} drop out in the calculation of third-order elastic constants. Let us write the sublattice displacements as an expansion in powers of u_{ij}, as (7.81)

$$S_i(\mu) = S_i^{(1)}(\mu) + S_i^{(2)}(\mu) + \cdots, \tag{8.22}$$

and the crystal potential as an expansion in powers of the ion displacements $U_i(\mathrm{M}\mu)$,

$$\Phi = \Phi_0 + \Phi_1 + \Phi_2 + \Phi_3 + \cdots. \tag{8.23}$$

Each of these contributions to Φ will now be studied in turn, keeping all contributions to third order in the strains u_{ij}.

The functions Φ_1 and Φ_2 were already written out in the expansion (7.34); from that equation

$$\Phi_1 = \sum_{\mathrm{M}\mu}\sum_i \Phi_i(M\mu)[S_i(\mu) + \sum_j u_{ij}R_j(M\mu)]. \tag{8.24}$$

Here the $S_i(\mu)$ are given by (8.22) as a power series in u_{ij}, but to any order in that expansion the quantity $\Sigma_\mu \Sigma_i \Phi_i(M\mu)S_i(\mu)$ is independent of M for unit cells M in the interior, so surface effects may be eliminated by evaluating at M = 0. Then since $\Phi_i(0\mu) = 0$,

$$\Phi_1 = \sum_{M\mu} \sum_{ij} \Phi_i(M\mu)R_j(M\mu)u_{ij} = V \sum_{ij} \tilde{A}_{ij}u_{ij}, \tag{8.25}$$

just as in Section 7, and the $S_i(\mu)$ drop out of Φ_1.

To discuss Φ_2 it is convenient to use the notation of (7.82)–(7.84), and write for the ion displacements

$$U_i(M\mu) = \sum_{jk} Y_{i,jk}(M\mu)u_{jk} + S_i^{(2)}(\mu) + \cdots, \tag{8.26}$$

where

$$Y_{i,jk}(M\mu) = R_k(M\mu)\delta_{ij} + X_{i,jk}(\mu). \tag{8.27}$$

The first term in (8.26) is linear in the strains, the second term $S_i^{(2)}(\mu)$ is quadratic, and so on. The terms in Φ_2 up to third order in the u_{ij} are then

$$\begin{aligned}\Phi_2 &= \tfrac{1}{2} \sum_{MN\mu\nu} \sum_{ij} \Phi_{ij}(M\mu,N\nu)U_i(M\mu)U_j(N\nu) \\ &= \tfrac{1}{2} \sum_{MN\mu\nu} \sum_{ijklmn} \Phi_{ij}(M\mu,N\nu)Y_{i,kl}(M\mu)Y_{j,mn}(N\nu)u_{kl}u_{mn} \\ &\quad + \sum_{MN\mu\nu} \sum_{ijkl} \Phi_{ij}(M\mu,N\nu)S_i^{(2)}(\mu)Y_{j,kl}(N\nu)u_{kl},\end{aligned} \tag{8.28}$$

where the two terms linear in $S_i^{(2)}(\mu)$ and $S_j^{(2)}(\nu)$ were combined by interchanging indices. Now with the neglect or surface effects the last line of (8.28) vanishes, since the $\Sigma_{N\nu} \Sigma_j$ is independent of M for all cells M in the interior, and again by the solution of the equations (7.38) for the first-order sublattice displacements [compare (7.87)],

$$\sum_{N\nu} \sum_{j} \Phi_{ij}(M\mu,N\nu)Y_{j,kl}(N\nu) = 0, \quad \text{for} \quad M = 0. \tag{8.29}$$

Therefore the second-order sublattice displacements $S_i^{(2)}(\mu)$ drop out of Φ_2, and only the first line of (8.28) remains. Then Φ_2 is

$$\Phi_2 = \tfrac{1}{2}V \sum_{ijkl} \tilde{A}_{ijkl}u_{ij}u_{kl},$$

where from the first line of (8.28)

$$\tilde{A}_{klmn} = V^{-1} \sum_{MN\mu\nu} \sum_{ij} \Phi_{ij}(M\mu,N\nu)Y_{i,kl}(M\mu)Y_{j,mn}(N\nu). \tag{8.30}$$

Further, the $Y_{i,kl}(M\mu)$ is composed of two terms according to (8.27); for the term involving $X_{i,kl}(\mu)$, the $\Sigma_{N\nu} \Sigma_j$ in (8.30) may be evaluated in the interior, and vanishes by (8.29), so with the remaining term involving $R_l(\mathrm{M}\mu)$ we have

$$\tilde{A}_{klmn} = V^{-1} \sum_{\mathrm{MN}\mu\nu} \sum_{j} \Phi_{kj}(\mathrm{M}\mu,\mathrm{N}\nu) R_l(\mathrm{M}\mu) Y_{j,mn}(\mathrm{N}\nu). \tag{8.31}$$

By interchanging indices this can also be written

$$\tilde{A}_{klmn} = V^{-1} \sum_{\mathrm{MN}\mu\nu} \sum_{i} \Phi_{im}(\mathrm{M}\mu,\mathrm{N}\nu) Y_{i,kl}(\mathrm{M}\mu) R_n(\mathrm{N}\nu). \tag{8.32}$$

Either of the expressions (8.31) or (8.32) is the same as (7.58).

The terms in Φ of third order in the ion displacements are

$$\Phi_3 = \tfrac{1}{6} \sum_{\mathrm{MNP}\mu\nu\pi} \sum_{ijk} \Phi_{ijk}(\mathrm{M}\mu,\mathrm{N}\nu,\mathrm{P}\pi) U_i(\mathrm{M}\mu) U_j(\mathrm{N}\nu) U_k(\mathrm{P}\pi). \tag{8.33}$$

Here it is only necessary to express the $U_i(\mathrm{M}\mu)$ to first order in the strains u_{ij}, which is the first term in (8.26). Then Φ_3 is

$$\Phi_3 = \tfrac{1}{6} V \sum_{ijklmn} \tilde{A}_{ijklmn} u_{ij} u_{kl} u_{mn},$$

where

$$\tilde{A}_{ijklmn} = V^{-1} \sum_{\mathrm{MNP}\mu\nu\pi} \sum_{i'j'k'} \Phi_{i'j'k'}(\mathrm{M}\mu,\mathrm{N}\nu,\mathrm{P}\pi) \times Y_{i',ij}(\mathrm{M}\mu) Y_{j',kl}(\mathrm{N}\nu) Y_{k',mn}(\mathrm{P}\pi). \tag{8.34}$$

Before eliminating surface effects we will outline the procedure for verifying the two symmetry properties (8.5) and (8.6) of $\tilde{A}_{ijklmn}$. First interchange $\mathrm{M}\mu i'$ with $\mathrm{N}\nu j'$ in (8.34); then, since $\Phi_{i'j'k'}(\mathrm{M}\mu,\mathrm{N}\nu,\mathrm{P}\pi)$ remains the same according to the general symmetry (6.5), this shows that

$$\tilde{A}_{ijklmn} = \tilde{A}_{klijmn}.$$

By interchanging other index sets, the complete symmetry (8.5) is verified. With (8.34) and (8.27) the $\tilde{A}_{ijklmn}$ may be separated into two contributions as follows.

$$\tilde{A}_{ijklmn} = \tilde{A}^{(1)}_{ijklmn} + \tilde{A}^{(2)}_{ijklmn}; \tag{8.35}$$

$$\tilde{A}^{(1)}_{ijklmn} = V^{-1} \sum_{\mathrm{MNP}\mu\nu\pi} \sum_{i'j'k'} \Phi_{i'j'k'}(\mathrm{M}\mu,\mathrm{N}\nu,\mathrm{P}\pi) Y_{i',ij}(\mathrm{M}\mu) Y_{j',kl}(\mathrm{N}\nu) X_{k',mn}(\pi); \tag{8.36}$$

$$\tilde{A}^{(2)}_{ijklmn} = V^{-1} \sum_{\mathrm{MNP}\mu\nu\pi} \sum_{i'j'} \Phi_{i'j'm}(\mathrm{M}\mu,\mathrm{N}\nu,\mathrm{P}\pi) Y_{i',ij}(\mathrm{M}\mu) Y_{j',kl}(\mathrm{N}\nu) R_n(\mathrm{P}\pi). \tag{8.37}$$

The rotational invariance condition (6.46) may be written

$$\sum_{\mathrm{P}\pi} \Phi_{i'j'm}(\mathrm{M}\mu,\mathrm{N}\nu,\mathrm{P}\pi) R_n(\mathrm{P}\pi) + \Phi_{mj'}(\mathrm{M}\mu,\mathrm{N}\nu)\delta_{i'n} + \Phi_{i'm}(\mathrm{M}\mu,\mathrm{N}\nu)\delta_{j'n} \quad \text{is symmetric in } m,n. \tag{8.38}$$

Multiplying by $V^{-1}Y_{i',ij}(\mathrm{M}\mu)Y_{j',kl}(\mathrm{N}\nu)$, summing over $\mathrm{M}\mu i'$ and $\mathrm{N}\nu j'$, and using (8.29) and identifying $\tilde{A}_{ijkl}$ from the expressions (8.31) and (8.32), converts (8.38) to

$$\tilde{A}^{(2)}_{ijklmn} + \tilde{A}_{klmj}\delta_{in} + \tilde{A}_{mlij}\delta_{kn} \quad \text{is symmetric in } m,n. \tag{8.39}$$

Introducing the elastic constants $\tilde{C}_{ijkl}$ by the relation (7.10), and noting that the extra term in the stresses $\tilde{C}_{lj}(\delta_{km}\delta_{in} + \delta_{kn}\delta_{im})$ is separately symmetric in m,n,

$$\tilde{A}^{(2)}_{ijklmn} + \tilde{C}_{klmj}\delta_{in} + \tilde{C}_{mlij}\delta_{kn} \quad \text{is symmetric in } m,n. \tag{8.40}$$

Finally from the symmetry (7.52) of $X_{k',mn}(\pi)$, the expression (8.36) for $\tilde{A}^{(1)}_{ijklmn}$ is symmetric in m,n, and this together with (8.40) allows verification of the rotational symmetry (8.6) of $\tilde{A}_{ijklmn}$.

When the definition (8.27) of $Y_{i,jk}(\mathrm{M}\mu)$ is used and (8.34) for $\tilde{A}_{ijklmn}$ is written out, surface effects can be eliminated from all but one term. The result is

$$\begin{aligned}
\tilde{A}_{ijklmn} &= V_C^{-1} \sum_{\mathrm{NP}\mu\nu\pi} \sum_{i'j'k'} \Phi_{i'j'k'}(0\mu,\mathrm{N}\nu,\mathrm{P}\pi)X_{i',ij}(\mu)X_{j',kl}(\nu)X_{k',mn}(\pi) \\
&+ V_C^{-1} \sum_{\mathrm{NP}\mu\nu\pi} \sum_{i'j'} \{\Phi_{i'j'm}(0\mu,\mathrm{N}\nu,\mathrm{P}\pi)X_{i',ij}(\mu)X_{j',kl}(\nu)R_n(\mathrm{P}\pi) \\
&\quad + \Phi_{i'j'k}(0\mu,\mathrm{N}\nu,\mathrm{P}\pi)X_{i',ij}(\mu)X_{j',mn}(\nu)R_l(\mathrm{P}\pi) \\
&\quad + \Phi_{i'j'i}(0\mu,\mathrm{N}\nu,\mathrm{P}\pi)X_{i',kl}(\mu)X_{j',mn}(\nu)R_j(\mathrm{P}\pi)\} \\
&+ V_C^{-1} \sum_{\mathrm{NP}\mu\nu\pi} \sum_{i'} \{\Phi_{i'km}(0\mu,\mathrm{N}\nu,\mathrm{P}\pi)X_{i',ij}(\mu)R_l(\mathrm{N}\nu)R_n(\mathrm{P}\pi) \\
&\quad + \Phi_{i'im}(0\mu,\mathrm{N}\nu,\mathrm{P}\pi)X_{i',kl}(\mu)R_j(\mathrm{N}\nu)R_n(\mathrm{P}\pi) \\
&\quad + \Phi_{i'ik}(0\mu,\mathrm{N}\nu,\mathrm{P}\pi)X_{i',mn}(\mu)R_j(\mathrm{N}\nu)R_l(\mathrm{P}\pi)\} \\
&+ V^{-1} \sum_{\mathrm{MNP}\mu\nu\pi} \Phi_{ikm}(\mathrm{M}\mu,\mathrm{N}\nu,\mathrm{P}\pi)R_j(\mathrm{M}\mu)R_l(\mathrm{N}\nu)R_n(\mathrm{P}\pi).
\end{aligned} \tag{8.41}$$

For the last term in (8.41), surface effects are eliminated in $\hat{A}_{ikjlmn}$ by taking the combination symmetric in j,l according to the definition (8.7). This combination is

$$\begin{aligned}
&\tfrac{1}{2}V^{-1} \sum_{\mathrm{MNP}\mu\nu\pi} \Phi_{ikm}(\mathrm{M}\mu,\mathrm{N}\nu,\mathrm{P}\pi)[R_j(\mathrm{M}\mu)R_l(\mathrm{N}\nu) + R_l(\mathrm{M}\mu)R_j(\mathrm{N}\nu)]R_n(\mathrm{P}\pi) \\
&\qquad = -\tfrac{1}{2}V_C^{-1} \sum_{\mathrm{NP}\mu\nu\pi} \Phi_{ikm}(0\mu,\mathrm{N}\nu,\mathrm{P}\pi)[R_j(\mathrm{N}\nu) - R_j(\mu)] \\
&\qquad\quad \times [R_l(\mathrm{N}\nu) - R_l(\mu)][R_n(\mathrm{P}\pi) - R_n(\mu)].
\end{aligned} \tag{8.42}$$

The third-order elastic constants $\tilde{C}_{ijklmn}$ are then given by (8.14). In any crystal where the first-order sublattice displacements are zero, the $X_{i,jk}(\mu)$ are zero, and $\hat{A}_{ikjlmn}$ is given simply by (8.42).

9. CENTRAL POTENTIALS

Potential Energy Coefficients

Consider a model in which each pair of ions $M\mu$ and $N\nu$ interact through the central potential $\tilde{\varphi}(M\mu,N\nu)$. The tilde means that the ions are in arbitrary positions $\mathbf{R}(M\mu) + \mathbf{U}(M\mu)$, not necessarily the equilibrium positions $\mathbf{R}(M\mu)$. The form of the potential may depend on the type of ions μ,ν which are coupled, but for a given pair μ,ν the potential depends only on the distance between the ions. This may be expressed by writing

$$\tilde{\varphi}(M\mu,N\nu) = \tilde{\varphi}_{\mu\nu}(|\mathbf{R}(M\mu) + \mathbf{U}(M\mu) - \mathbf{R}(N\nu) - \mathbf{U}(N\nu)|). \tag{9.1}$$

To denote evaluation of $\tilde{\varphi}$, and later on its derivatives, at the equilibrium positions, we simply remove the tilde:

$$\varphi(M\mu,N\nu) = \varphi_{\mu\nu}(|\mathbf{R}(M\mu) - \mathbf{R}(N\nu)|). \tag{9.2}$$

The total crystal potential is

$$\Phi = \tfrac{1}{2} \sum_{MN\mu\nu}{}' \tilde{\varphi}(M\mu,N\nu), \tag{9.3}$$

where Σ' means the terms $M\mu = N\nu$ are omitted.

Such a model is useful in some cases as an approximation for real crystals. For alkali halides, central potentials give a good approximation to the static crystal potential Φ and its strain derivatives, but they do not give a good account of vibrational properties overall. For simple metals a local pseudopotential perturbation theory leads to an expression for Φ as a volume-dependent term plus a sum of central potentials among ion pairs. The interactions among rare gas atoms are mainly central potentials, with a small contribution from noncentral forces. The central-potential model is also useful for illustrating details of the formal theory.

The potential energy coefficients may be calculated by expanding (9.3) in powers of the ion displacements $\mathbf{U}(M\mu)$. It is convenient to consider φ a function of distance squared, and use the notation (suppressing all indices)

$$\begin{aligned} \varphi &= \varphi(R^2), \\ \varphi' &= d\varphi/dR^2, \\ \varphi'' &= d^2\varphi/d(R^2)^2, \end{aligned} \tag{9.4}$$

and so on. With this schematic notation, we can write

$$\tilde{\varphi}(|\mathbf{R} + \mathbf{U}|^2) = \varphi(R^2) + \Delta\varphi'(R^2) + \tfrac{1}{2}\Delta^2\varphi''(R^2) + \cdots, \tag{9.5}$$

where

$$\Delta = 2\mathbf{R} \cdot \mathbf{U} + \mathbf{U} \cdot \mathbf{U}. \tag{9.6}$$

If each central potential in the sum (9.3) over all pairs is expanded by (9.5), and terms of each order in the displacements $\mathbf{U}(M\mu)$ are collected together, then the potential energy coefficients are obtained by comparison of this expansion with the general expansion (6.1). [They can also be obtained, of course, by direct differentiation of the total Φ given by (9.3), with the definitions (6.2) and (6.3).] Potential energy coefficients which couple three or more *different* ions are always zero for central potentials, since they are two-body potentials. With all $\varphi, \varphi', \varphi'', \cdots$, evaluated at the (arbitrary) initial equilibrium configuration, corresponding to locating the ions at $\mathbf{R}(M\nu)$, and with the abbreviation

$$R_i(M\mu,N\nu) = -R_i(N\nu,M\mu) = R_i(M\mu) - R_i(N\nu), \tag{9.7}$$

the results are as follows.

$$\Phi_0 = \tfrac{1}{2} \sum_{MN\mu\nu}{}' \varphi(M\mu,N\nu), \quad \text{at } \mathbf{R}(M\mu); \tag{9.8}$$

$$\Phi_i(M\mu) = 2 \sum_{N\nu}{}' \varphi'(M\mu,N\nu) R_i(M\mu,N\nu); \tag{9.9}$$

and for $M\mu \neq N\nu$,

$$\Phi_{ij}(M\mu,N\nu)$$
$$= -2\varphi'(M\mu,N\nu)\delta_{ij} - 4\varphi''(M\mu,N\nu)R_i(M\mu,N\nu)\, R_j(M\mu,N\nu); \tag{9.10}$$

$$\Phi_{ijk}(M\mu,M\mu,N\nu)$$
$$= -4\varphi''(M\mu,N\nu)[R_i(M\mu,N\nu)\delta_{jk} + R_j(M\mu,N\nu)\delta_{ik} + R_k(M\mu,N\nu)\delta_{ij}]$$
$$- 8\varphi'''(M\mu,N\nu)R_i(M\mu,N\nu)\, R_j(M\mu,N\nu)R_k(M\mu,N\nu); \tag{9.11}$$

$$\Phi_{ijkl}(M\mu,M\mu,M\mu,N\nu)$$
$$= -4\varphi''(M\mu,N\nu)[\delta_{ij}\delta_{kl} + \delta_{ik}\delta_{jl} + \delta_{il}\delta_{jk}]$$
$$\begin{aligned} - 8\varphi'''(M\mu,N\nu)[&R_i(M\mu,N\nu)R_j(M\mu,N\nu)\delta_{kl} \\ &+ R_i(M\mu,N\nu)R_k(M\mu,N\nu)\delta_{jl} \\ &+ R_i(M\mu,N\nu)R_l(M\mu,N\nu)\delta_{jk} \\ &+ R_j(M\mu,N\nu)R_k(M\mu,N\nu)\delta_{il} \\ &+ R_j(M\mu,N\nu)R_l(M\mu,N\nu)\delta_{ik} \\ &+ R_k(M\mu,N\nu)R_l(M\mu,N\nu)\delta_{ij}] \end{aligned} \tag{9.12}$$
$$- 16\varphi''''(M\mu,N\nu)R_i(M\mu,N\nu)R_j(M\mu,N\nu)R_k(M\mu,N\nu)R_l(M\mu,N\nu).$$

All the independent potential energy coefficients up to fourth order are given above; other coefficients are simply related to those given. The "self-coupling" coefficients like $\Phi_{ij}(\mathrm{M}\mu,\mathrm{M}\mu)$ are determined so as to satisfy the translational invariance conditions (6.25)–(6.28). These are

$$\Phi_{ij}(\mathrm{M}\mu,\mathrm{M}\mu) = -\sum_{\mathrm{N}\nu}{}' \Phi_{ij}(\mathrm{M}\mu,\mathrm{N}\nu); \tag{9.13}$$

$$\Phi_{ijk}(\mathrm{M}\mu,\mathrm{M}\mu,\mathrm{M}\mu) = -\sum_{\mathrm{N}\nu}{}' \Phi_{ijk}(\mathrm{M}\mu,\mathrm{M}\mu,\mathrm{N}\nu); \tag{9.14}$$

$$\Phi_{ijkl}(\mathrm{M}\mu,\mathrm{M}\mu,\mathrm{M}\mu,\mathrm{M}\mu) = -\sum_{\mathrm{N}\nu}{}' \Phi_{ijkl}(\mathrm{M}\mu,\mathrm{M}\mu,\mathrm{M}\mu,\mathrm{N}\nu). \tag{9.15}$$

Also, according to (9.10)–(9.12), the potential energy coefficients are completely symmetric in their Cartesian indices:

$$\Phi_{ij}(\mathrm{M}\mu,\mathrm{N}\nu) = \Phi_{ji}(\mathrm{M}\mu,\mathrm{N}\nu); \tag{9.16}$$

$$\Phi_{ijk}(\mathrm{M}\mu,\mathrm{M}\mu,\mathrm{N}\nu) = \Phi_{jik}(\mathrm{M}\mu,\mathrm{M}\mu,\mathrm{N}\nu) = \Phi_{ikj}(\mathrm{M}\mu,\mathrm{M}\mu,\mathrm{N}\nu) = \cdots; \tag{9.17}$$

$$\Phi_{ijkl}(\mathrm{M}\mu,\mathrm{M}\mu,\mathrm{M}\mu,\mathrm{N}\nu) = \Phi_{jikl}(\mathrm{M}\mu,\mathrm{M}\mu,\mathrm{M}\mu,\mathrm{N}\nu) = \cdots. \tag{9.18}$$

Then from the general symmetries (6.4)–(6.6) of the potential coefficients in their index pairs, it follows that they are completely symmetric in their ion indices:

$$\Phi_{ij}(\mathrm{M}\mu,\mathrm{N}\nu) = \Phi_{ij}(\mathrm{N}\nu,\mathrm{M}\mu); \tag{9.19}$$

$$\Phi_{ijk}(\mathrm{M}\mu,\mathrm{M}\mu,\mathrm{N}\nu) = \Phi_{ijk}(\mathrm{M}\mu,\mathrm{N}\nu,\mathrm{M}\mu) = \Phi_{ijk}(\mathrm{N}\nu,\mathrm{M}\mu,\mathrm{M}\mu); \tag{9.20}$$

$$\Phi_{ijkl}(\mathrm{M}\mu,\mathrm{M}\mu,\mathrm{M}\mu,\mathrm{N}\nu) = \Phi_{ijkl}(\mathrm{M}\mu,\mathrm{M}\mu,\mathrm{N}\nu,\mathrm{M}\mu) = \cdots. \tag{9.21}$$

Finally, coefficients such as $\Phi_{ijk}(\mathrm{M}\mu,\mathrm{N}\nu,\mathrm{N}\nu)$ may be related to $\Phi_{ijk}(\mathrm{M}\mu,\mathrm{M}\mu,\mathrm{N}\nu)$ either by direct derivation of the coefficients, or by use of translational invariance. For example, from (6.27),

$$\sum_{\mathrm{P}\pi} \Phi_{ijk}(\mathrm{M}\mu,\mathrm{N}\nu,\mathrm{P}\pi) = 0,$$

and for a two-body potential this gives

$$\Phi_{ijk}(\mathrm{M}\mu,\mathrm{N}\nu,\mathrm{M}\mu) + \Phi_{ijk}(\mathrm{M}\mu,\mathrm{N}\nu,\mathrm{N}\nu) = 0,$$

or

$$\Phi_{ijk}(\mathrm{M}\mu,\mathrm{N}\nu,\mathrm{N}\nu) = -\Phi_{ijk}(\mathrm{M}\mu,\mathrm{M}\mu,\mathrm{N}\nu). \tag{9.22}$$

Extension to the fourth-order coefficients gives

$$\Phi_{ijkl}(\mathrm{M}\mu,\mathrm{N}\nu,\mathrm{N}\nu,\mathrm{N}\nu) = -\Phi_{ijkl}(\mathrm{M}\mu,\mathrm{M}\mu,\mathrm{N}\nu,\mathrm{N}\nu) = \Phi_{ijkl}(\mathrm{M}\mu,\mathrm{M}\mu,\mathrm{M}\mu,\mathrm{N}\nu). \tag{9.23}$$

Now let us examine the equilibrium condition (6.11):

$$\Phi_i(M\mu) - f_i(M\mu) = 0, \quad \text{for all } M,\mu,i. \tag{9.24}$$

The coefficients $\Phi_i(M\mu)$ are given by (9.9); they are the same for any cell M in the interior, and so may be evaluated at M = 0. Then since $f_i(0\mu) = 0$, the equilibrium condition in the interior is

$$\Phi_i(0\mu) = -2\sum_{N\nu}{}' \varphi'(0\mu,N\nu)[R_i(N\nu) - R_i(\mu)] = 0, \quad \text{for all } \mu,i. \tag{9.25}$$

For any crystal which has inversion symmetry, whether or not the center of inversion is a lattice site, the $\Sigma_{N\nu}$ in (9.25) vanishes by symmetry, and hence the equilibrium condition in the interior is satisfied for any macroscopic configuration of the crystal. This includes all primitive lattices, all alkali halide lattices, diamond, hexagonal close-packed (hcp), etc. In other cases the condition (9.25) presumably places a restriction on the $\varphi'(0\mu,N\nu)$. Near the surface of the crystal, where the lattice is not perfectly periodic, the $\Phi_i(M\mu)$ no longer vanish by symmetry, and the equilibrium condition (9.24) determines the distortion in the surface region.

The invariance conditions must be trivially satisfied for central potentials, since these potentials are indifferent to the location and orientation of the crystal. The translational invariance conditions are satisfied identically, since they are used in the definitions (9.13)–(9.15) of the self-coupling coefficients. The rotational invariance condition (6.45), for example, is

$$\sum_{N\nu} \Phi_{ij}(M\mu,N\nu)R_k(N\nu) + \Phi_j(M\mu)\delta_{ik} \quad \text{is symmetric in } j,k. \tag{9.26}$$

In the first term, $R_k(N\nu)$ can be replaced by $R_k(N\nu) - R_k(M\mu)$, since the term in $R_k(M\mu)$ vanishes by translational invariance, and with the notation (9.7) we can rewrite the quantity in (9.26) as

$$-\sum_{N\nu} \Phi_{ij}(M\mu,N\nu)R_k(M\mu,N\nu) + \Phi_j(M\mu)\delta_{ik}. \tag{9.27}$$

Now with the results (9.9) and (9.10) for the potential coefficients this is

$$\begin{aligned} 2\sum_{N\nu}{}' \varphi'(M\mu\, N\nu)[R_k(M\mu,N\nu)\delta_{ij} + R_j(M\mu,N\nu)\delta_{ik}] \\ + 4\sum_{N\nu}{}' \varphi''(M\mu,N\nu)R_i(M\mu,N\nu)R_j(M\mu,N\nu)R_k(M\mu,N\nu). \end{aligned} \tag{9.28}$$

This expression is obviously symmetric in j,k, and (9.26) is satisfied.

The condition of lattice stability against arbitrary small displacements of the ions, represented by (6.49) or (6.50), is not necessarily satisfied by central potentials in general, and must be checked for a given crystal structure and prescribed central potential.

ELASTIC CONSTANTS

It is of interest to investigate sublattice displacements for the central-potential model. The first-order sublattice displacements are proportional to the inhomogeneous part (right-hand side) of (7.38), which may be written

$$-\sum_{N\nu}\sum_{jk}\Phi_{ij}(0\mu,N\nu)R_k(N\nu)u_{jk} = -\sum_{jk}u_{jk}\sum_{N\nu}\Phi_{ij}(0\mu,N\nu)[R_k(N\nu) - R_k(\mu)]. \tag{9.29}$$

The term in $R_k(\mu)$ was added for convenience, and it vanishes by translational invariance. With the abbreviation (9.7) for $R_k(0\mu,N\nu)$ and with (9.10) for $\Phi_{ij}(0\mu,N\nu)$, the right side of (9.29) is

$$\begin{aligned}-\sum_{jk}u_{jk}\sum_{N\nu}&[2\varphi'(0\mu,N\nu)R_k(0\mu,N\nu)\delta_{ij}\\&+4\varphi''(0\mu,N\nu)R_i(0\mu,N\nu)R_j(0\mu,N\nu)R_k(0\mu,N\nu)].\end{aligned} \tag{9.30}$$

The $\Sigma_{N\nu}$ in (9.30) vanishes for every μ for any crystal in which each ion is at a center of inversion symmetry, as for example, NaCl and CsCl lattices; this is as it should be from the general results of Section 7. However, the $\Sigma_{N\nu}$ in (9.30) also vanishes for other symmetries, as for example, diamond or hcp (where all ions are the same); thus, central potentials predict zero sublattice displacements in these lattices. This represents a failure of the central-potential model, since such crystals generally have a sublattice displacement accompanying a homogeneous shear deformation.

Let us now consider only those crystals for which the sublattice displacements during a homogeneous deformation are zero. With this information, and with any prescribed model for the interactions among the ions, one can bypass the formal theory of Section 7 and calculate the elastic constants directly; we shall illustrate this for the central-potential model. Corresponding to the homogeneous strains u_{ij}, the ion displacements in the perfectly periodic interior of the crystal are

$$U_i(M\mu) = \sum_j u_{ij}R_j(M\mu), \quad \text{for} \quad S_i(\mu) = 0. \tag{9.31}$$

At the same time the squared distance between two ions becomes

$$\begin{aligned}&|\mathbf{R}(M\mu) + \mathbf{U}(M\mu) - \mathbf{R}(N\nu) - \mathbf{U}(N\nu)|^2\\&\quad = |\mathbf{R}(M\mu) - \mathbf{R}(N\nu)|^2 + 2\sum_{ij}\eta_{ij}R_i(M\mu,N\nu)R_j(M\mu,N\nu),\end{aligned} \tag{9.32}$$

according to (2.10), where η_{ij} are the Lagrangian strains. Then the central potential between two ions in the interior varies with the strains η_{ij} as

$$\tilde{\varphi}(\mathrm{M}\mu,\mathrm{N}\nu) = \varphi(\mathrm{M}\mu,\mathrm{N}\nu) + 2\varphi'(\mathrm{M}\mu,\mathrm{N}\nu)\sum_{ij}\eta_{ij}R_i(\mathrm{M}\mu,\mathrm{N}\nu)R_j(\mathrm{M}\mu,\mathrm{N}\nu)$$
$$+ 2\varphi''(\mathrm{M}\mu,\mathrm{N}\nu)\sum_{ijkl}\eta_{ij}\eta_{kl}R_i(\mathrm{M}\mu,\mathrm{N}\nu)R_j(\mathrm{M}\mu,\mathrm{N}\nu)R_k(\mathrm{M}\mu,\mathrm{N}\nu)R_l(\mathrm{M}\mu,\mathrm{N}\nu) + \cdots. \tag{9.33}$$

The crystal potential Φ is given by (9.3) as the sum of central potentials; when surface effects are neglected the sum can be evaluated in the interior as

$$\Phi = \tfrac{1}{2}N_0 \sum_{\mathrm{N}\mu\nu}{}' \tilde{\varphi}(0\mu,\mathrm{N}\nu). \tag{9.34}$$

Then with (9.33), the crystal potential may be expanded directly in powers of the Lagrangian strains.

$$\Phi = \Phi_0 + N_0 \sum_{N\mu\nu}{}' \sum_{ij}\varphi'(0\mu,\mathrm{N}\nu)R_i(0\mu,\mathrm{N}\nu)R_j(0\mu,\mathrm{N}\nu)\eta_{ij}$$
$$+ N_0 \sum_{\mathrm{N}\mu\nu}{}' \sum_{ijkl}\varphi''(0\mu,\mathrm{N}\nu)R_i(0\mu,\mathrm{N}\nu)R_j(0\mu,\mathrm{N}\nu) \tag{9.35}$$
$$\times R_k(0\mu,\mathrm{N}\nu)R_l(0\mu,\mathrm{N}\nu)\eta_{ij}\eta_{kl} + \cdots.$$

The mechanical stresses $\tilde{C}_{ij}$ and elastic constants $\tilde{C}_{ijkl}, \tilde{C}_{ijklmn}, \cdots$, are obtained by comparing (9.35) with the general expansion

$$\Phi = \Phi_0 + V\sum_{ij}\tilde{C}_{ij}\eta_{ij} + \tfrac{1}{2}V\sum_{ijkl}\tilde{C}_{ijkl}\eta_{ij}\eta_{kl} + \cdots. \tag{9.36}$$

Let us use the further abbreviation

$$R_i \equiv R_i(0\mu,\mathrm{N}\nu); \tag{9.37}$$

then the results up to fourth-order elastic constants are

$$\tilde{C}_{ij} = V_C^{-1} \sum_{\mathrm{N}\mu\nu}{}' \varphi'(0\mu,\mathrm{N}\nu)R_iR_j, \tag{9.38}$$

$$\tilde{C}_{ijkl} = 2V_C^{-1} \sum_{\mathrm{N}\mu\nu}{}' \varphi''(0\mu,\mathrm{N}\nu)R_iR_jR_kR_l \tag{9.39}$$

$$\tilde{C}_{ijklmn} = 4V_C^{-1} \sum_{\mathrm{N}\mu\nu}{}' \varphi'''(0\mu,\mathrm{N}\nu)R_iR_jR_kR_lR_mR_n, \tag{9.40}$$

$$\tilde{C}_{ijklmnpq} = 8V_C^{-1} \sum_{\mathrm{N}\mu\nu}{}' \varphi''''(0\mu,\mathrm{N}\nu)R_iR_jR_kR_lR_mR_nR_pR_q. \tag{9.41}$$

From these results it is seen that the elastic constants for central potentials not only have Voigt symmetry, but are *completely* symmetric in their Cartesian indices.

It is instructive to evaluate the general results of Sections 7 and 8 for the case of central potentials and zero sublattice displacements, and show how the above results are recovered. Starting with (7.57) for $\tilde{A}_{ij}$, and using (9.9) for $\Phi_i(\text{M}\mu)$, the mechanical stresses are calculated as follows.

$$\begin{aligned}\tilde{A}_{ij} = \tilde{C}_{ij} &= V^{-1}\sum_{\text{M}\mu}\Phi_i(\text{M}\mu)R_j(\text{M}\mu)\\ &= 2V^{-1}\sum_{\text{MN}\mu\nu}{}' \varphi'(\text{M}\mu,\text{N}\nu)R_i(\text{M}\mu,\text{N}\nu)R_j(\text{M}\mu), \qquad (9.42)\\ &= V^{-1}\sum_{\text{MN}\mu\nu}{}' \varphi'(\text{M}\mu,\text{N}\nu)R_i(\text{M}\mu,\text{N}\nu)R_j(\text{M}\mu,\text{N}\nu),\end{aligned}$$

where in the last step we interchanged $\text{M}\mu$ with $\text{N}\nu$ in half the quantity and used $R_i(\text{M}\mu,\text{N}\nu) = -R_i(\text{N}\nu,\text{M}\mu)$. But in the last line of (9.42) the $\Sigma_{\text{N}\mu\nu}$ is the same for all M in the interior, so surface effects can be eliminated by evaluating at $\text{M} = 0$ and multiplying by N_0. In this way (9.38) for $\tilde{C}_{ij}$ is reproduced. It is interesting to note that contributions to the first line in (9.42) arise entirely from ions in the surface region, since $\Phi_i(\text{M}\mu) = 0$ for ions in the interior, while in the last line of (9.42) surface contributions are negligible, and may be eliminated!

With zero sublattice displacements, the $\hat{A}_{ikjl}$ can be written down from (7.60), and transformed with (9.10) for $\Phi_{ij}(\text{M}\mu,\text{N}\nu)$, as follows.

$$\begin{aligned}\hat{A}_{ikjl} &= -\tfrac{1}{2}V_C^{-1}\sum_{\text{N}\mu\nu}{}' \Phi_{ik}(0\mu,\text{N}\nu)R_j(0\mu,\text{N}\nu)R_l(0\mu,\text{N}\nu),\\ &= V_C^{-1}\sum_{\text{N}\mu\nu}{}' [\varphi'(0\mu,\text{N}\nu)R_jR_l\delta_{ik} + 2\varphi''(0\mu,\text{N}\nu)R_iR_jR_kR_l], \qquad (9.43)\\ &= \tilde{C}_{jl}\delta_{ik} + 2V_C^{-1}\sum_{\text{N}\mu\nu}{}' \varphi''(0\mu,\text{N}\nu)R_iR_jR_kR_l.\end{aligned}$$

Here we are again using the abbreviation (9.37), and the prime on $\Sigma'_{N\mu\nu}$ is introduced in the first line because $R_i(\text{M}\mu,\text{M}\mu) = 0$ and $\Phi_{ik}(\text{M}\mu,\text{M}\mu)$ is finite. The general expression for $\tilde{C}_{ijkl}$, as a combination of $\hat{A}_{ikjl}$ and $\tilde{C}_{ij}$ coefficients, is given by (7.30). By rewriting (9.43) as

$$\hat{A}_{ikjl} - \tilde{C}_{jl}\delta_{ik} = 2V_C^{-1}\sum_{\text{N}\mu\nu}{}' \varphi''(0\mu,\text{N}\nu)R_iR_jR_kR_l, \qquad (9.44)$$

it is immediately seen that (7.30) again gives the result (9.39) for $\tilde{C}_{ijkl}$. In a similar way it is verified that the general equation (8.14) reproduces the result (9.40) for $\tilde{C}_{ijklmn}$.

Finally, consider the cubic lattices NaCl or CsCl, or any of the primitive cubic lattices [simple cubic, body-centered cubic (bcc), face-centered cubic (fcc)]. For these cases the elastic constant indices x, y, and z are equivalent, the lattice sums in (9.38)–(9.41) vanish unless the Cartesian indices are equal in pairs, so there is only one independent stress component, two independent second-order elastic constants, three independent third-order elastic constants, and so on. The initial stress must be isotropic pressure:

$$\tilde{C}_{ij} = -P\delta_{ij}, \tag{9.45}$$

$$P = -V_C^{-1} \sum_{N\mu\nu}{}' \varphi'(0\mu,N\nu)R_x^2; \tag{9.46}$$

this is a result of requiring cubic symmetry in the initial configuration. In Voigt notation, the two independent second-order elastic constants are

$$\tilde{C}_{11} = 2V_C^{-1} \sum_{N\mu\nu}{}' \varphi''(0\mu,N\nu)R_x^4, \tag{9.47}$$

$$\tilde{C}_{12} = \tilde{C}_{44} = 2V_C^{-1} \sum_{N\mu\nu}{}' \varphi''(0\mu,N\nu)R_x^2R_y^2; \tag{9.48}$$

the three independent third-order elastic constants are

$$\tilde{C}_{111} = 4V_C^{-1} \sum_{N\mu\nu}{}' \varphi'''(0\mu,N\nu)R_x^6, \tag{9.49}$$

$$\tilde{C}_{112} = \tilde{C}_{166} = 4V_C^{-1} \sum_{N\mu\nu}{}' \varphi'''(0\mu,N\nu)R_x^4R_y^2, \tag{9.50}$$

$$\tilde{C}_{123} = \tilde{C}_{144} = \tilde{C}_{456} = 4V_C^{-1} \sum_{N\mu\nu}{}' \varphi'''(0\mu,N\nu)R_x^2R_y^2R_z^2; \tag{9.51}$$

and the four independent fourth-order elastic constants are

$$\tilde{C}_{1111} = 8V_C^{-1} \sum_{N\mu\nu}{}' \varphi''''(0\mu,N\nu)R_x^8, \tag{9.52}$$

$$\tilde{C}_{1112} = \tilde{C}_{1166} = 8V_C^{-1} \sum_{N\mu\nu}{}' \varphi''''(0\mu,N\nu)R_x^6R_y^2 \tag{9.53}$$

$$\tilde{C}_{1122} = \tilde{C}_{1266} = \tilde{C}_{4444} = 8V_C^{-1} \sum_{N\mu\nu}{}' \varphi''''(0\mu,N\nu)R_x^4R_y^4, \tag{9.54}$$

$$\tilde{C}_{1123} = \tilde{C}_{1144} = \tilde{C}_{1244} = \tilde{C}_{1456} = \tilde{C}_{4466} = 8V_C^{-1} \sum_{N\mu\nu}{}' \varphi''''(0\mu,N\nu)R_x^4R_y^2R_z^2. \tag{9.55}$$

The above relations among the elastic constants of each order are the cubic-symmetry Cauchy relations; in particular these are (9.48) for second-order elastic constants, (9.50) and (9.51) for third-order constants, and (9.53)–(9.55) for fourth-order constants. In general these Cauchy relations will be destroyed if noncentral forces are included in the crystal potential. In addition the Cauchy relations will generally be destroyed for the complete adiabatic and isothermal elastic constants, even at zero temperature, on account of the temperature-dependent vibrational contributions to the thermodynamic state functions (see Chapter 4). Both these effects contribute to the departure from Cauchy relations of the measured elastic constants in real crystals. It should also be clearly recognized that the Cauchy relations apply only to the elastic constants. The three independent stress–strain coefficients for a cubic crystal under isotropic pressure, for example, are

$$B_{11} = C_{11} - P, \qquad B_{12} = C_{12} + P, \qquad B_{44} = C_{44} - P. \tag{9.56}$$

Hence the potential approximation result $\tilde{C}_{12} = \tilde{C}_{44}$ does *not* imply $\tilde{B}_{12} = \tilde{B}_{44}$, unless $P = 0$.

All the results of this section may be applied to any primitive lattice simply by dropping the indices $\mu, \nu, \cdots$, which denote different ions in each unit cell.

3

LATTICE DYNAMICS

A crystal of vibrating ions is described as a set of weakly interacting phonons, and different formulations of the phonon eigenvalue–eigenvector problem are related. It is shown for any lattice in any initial equilibrium configuration that the long-wavelength acoustic phonons are the same as long-wavelength sound waves in the potential approximation; i.e., the methods of homogeneous deformation and long waves are in agreement. The phonons (their operators and energies) are renormalized to leading order in phonon–phonon interactions, and the energy levels of the crystal are calculated to that order. Self-consistent phonons are described in first-order random-phase approximation.

10. HARMONIC PHONONS

The Lattice Vibration Problem

In a real crystal the ions are in a state of continual motion, with each ion constantly exchanging kinetic and potential energy. This motion is the statistical–mechanical description of the heat energy of a crystal. Before thermodynamic functions can be calculated for the crystal, it is first necessary to find the quantum energy levels associated with this mechanical motion of the ions. The kinetic energy of all the ions in the crystal is

$$KE = \frac{1}{2} \sum_{N\nu} \sum_{i} M_\nu [\dot{U}_i(N\nu)]^2, \tag{10.1}$$

where M_ν is the mass of an ion of type ν and $\dot{U}(N\nu)$ is the instantaneous velocity of ion $N\nu$. The total Hamiltonian for the system of crystal plus externally applied forces is $KE + \Psi$, where Ψ is the total system potential of Section 6. To get the energy levels of the crystal alone for arbitrary initial configuration we need to subtract Ψ_0, the work done by the external forces in bringing the crystal to its initial equilibrium configuration. Therefore the crystal Hamiltonian is

$$\mathscr{H} = KE + \Psi - \Psi_0. \tag{10.2}$$

The adiabatic approximation is implicit in this Hamiltonian, since there is no reference to the electronic coordinates or velocities. The adiabatic approximation is quite good for all crystals as far as the total energy and other total equilibrium thermodynamic functions are concerned; more detailed discussion of the adiabatic approximation, and of nonadiabatic effects in metals, is provided in Chapter 5.

In (6.9) the potential Ψ is expanded in powers of the virtual displacements of the ions from their equilibrium positions; with this expansion the Hamiltonian may be written

$$\begin{aligned}\mathscr{H} = \Phi_0 &+ \tfrac{1}{2}\sum_{N\nu}\sum_{i} M_\nu[\dot{U}_i(N\nu)]^2 + \tfrac{1}{2}\sum_{MN\mu\nu}\sum_{ij}\Phi_{ij}(M\mu,N\nu)U_i(M\mu)U_j(N\nu)\\ &+ (1/3!)\sum_{MNP\mu\nu\pi}\sum_{ijk}\Phi_{ijk}(M\mu,N\nu,P\pi)U_i(M\mu)U_j(N\nu)U_k(P\pi)\\ &+ (1/4!)\sum_{MNPQ\mu\nu\pi\rho}\sum_{ijkl}\Phi_{ijkl}(M\mu,N\nu,P\pi,Q\rho)\\ &\times U_i(M\mu)U_j(N\nu)U_k(P\pi)U_l(Q\rho) + \cdots .\end{aligned} \tag{10.3}$$

The variables of this mechanical problem are the positions and velocities of all the ions, and these variables satisfy the well-known commutation relations*

$$[M_\nu\dot{U}_i(N\nu),U_j(P\pi)] = -i\hbar\delta_{NP}\delta_{\nu\pi}\delta_{ij}; \tag{10.4}$$

$$[\dot{U}_i(N\nu),\dot{U}_j(P\pi)] = [U_i(N\nu),U_j(P\pi)] = 0. \tag{10.5}$$

Now the problem is completely defined, by the Hamiltonian (10.3) and the commutators (10.4) and (10.5). In order to assist in solving the problem we will make two approximations, both of which are very much in keeping with the nature of real crystals, namely the harmonic-perturbation approximation, and the neglect of surface effects (the latter includes incidentally the periodic boundary condition).

* P. A. M. Dirac, *The Principles of Quantum Mechanics*, Clarendon Press, Oxford, 1947, Third Edition.

The first approximation is based on the fact that during the vibrations of the ions in a real crystal, the displacements from equilibrium are generally small compared to equilibrium interionic distances. This approximation is not appropriate for solid He and H_2, where the vibrational displacements are quite large because of the small masses and weak forces; for these crystals it is more appropriate to begin with the self-consistent phonon description of Section 14. In addition the small-displacement approximation may begin to break down for many materials at very high temperatures; for the vast majority of crystals, however, the approximation is quite good for temperatures up to a few times the Debye temperature. We will therefore consider the small parameter ϵ as an expansion parameter for the problem, defined by

$$\epsilon = \frac{\langle U^2 \rangle^{1/2}}{R}, \tag{10.6}$$

where $\langle U^2 \rangle$ is an average displacement squared and R is an average nearest-neighbor distance. The definition of ϵ is obviously an order-of-magnitude definition, which is all we require. The terms in $\mathscr{H}$ can now be ordered qualitatively in powers of ϵ as

$$\mathscr{H} = \Phi_0 + \mathscr{H}_2 + \mathscr{H}_3 + \mathscr{H}_4 + \cdots, \tag{10.7}$$

where from (10.3)

$$\mathscr{H}_2 = \tfrac{1}{2} \sum_{\mathrm{N}\nu} \sum_{i} M_\nu [\dot{U}_i(\mathrm{N}\nu)]^2 + \tfrac{1}{2} \sum_{\mathrm{MN}\mu\nu} \sum_{ij} \Phi_{ij}(\mathrm{M}\mu,\mathrm{N}\nu) U_i(\mathrm{M}\mu) U_j(\mathrm{N}\nu), \tag{10.8}$$

$$\mathscr{H}_3 = (1/3!) \sum_{\mathrm{MNP}\mu\nu\pi} \sum_{ijk} \Phi_{ijk}(\mathrm{M}\mu,\mathrm{N}\nu,\mathrm{P}\pi) U_i(\mathrm{M}\mu) U_j(\mathrm{N}\nu) U_k(\mathrm{P}\pi), \tag{10.9}$$

$$\mathscr{H}_4 = (1/4!) \sum_{\mathrm{MNPQ}\mu\nu\pi\rho} \sum_{ijkl} \Phi_{ijkl}(\mathrm{M}\mu,N\nu,\mathrm{P}\pi,\mathrm{Q}\rho) U_i(\mathrm{M}\mu) U_j(\mathrm{N}\nu) U_k(\mathrm{P}\pi) U_l(\mathrm{Q}\rho). \tag{10.10}$$

The harmonic-perturbation approximation consists of solving the homogeneous quadratic Hamiltonian $\mathscr{H}_2$ for its normal vibrational modes, and then treating all higher-order terms $\mathscr{H}_3, \mathscr{H}_4, \cdots$, by perturbation theory.

The distortion of the crystal near its surface may be neglected either by assuming that the lattice is perfectly periodic everywhere, including the surface region, or else by considering a finite subcrystal in the interior of an infinite lattice, which corresponds to applying external forces in the surface region designed to bring the finite subcrystal to perfect periodicity everywhere. Either of these pictures may be taken as a basis for the following calculations. Let the edges of each identical unit cell be the three vectors $\mathbf{a}_\alpha$, $\alpha = 1,2,3$, so the unit cell volume is

$$V_C = \mathbf{a}_1 \cdot (\mathbf{a}_2 \times \mathbf{a}_3). \tag{10.11}$$

Further let the crystal be constructed of a three-dimensional array of $L \times L \times L$ unit cells, where $L^3 = N_0$, so the set of unit cell vectors (also called the primitive lattice) is

$$\mathbf{R}(\mathrm{M}) = M_1\mathbf{a}_1 + M_2\mathbf{a}_2 + M_3\mathbf{a}_3, \tag{10.12}$$

where M_1,M_2,M_3 are integers in a range L, say from $-\frac{1}{2}L$ to $+\frac{1}{2}L$. Now to the same degree of accuracy to which surface effects may be neglected, the motion of the ions may be restricted by applying boundary conditions on the displacements of the surface ions. The solution of the problem of motion is simplified by using the periodic boundary condition, requiring displacements of ions on opposite faces of the crystal to be the same:

$$U_i(\mathrm{M}\mu) = U_i(\mathrm{M} + \mathrm{L},\mu), \tag{10.13}$$

where the ion $\mathrm{M}\mu$ is on one crystal surface and $\mathrm{M} + \mathrm{L},\mu$ is an equivalent ion on the opposite surface.

We now seek a complete set of wave vectors $\mathbf{k}$ such that an arbitrary displacement of all unit cells in the crystal, subject to the periodic boundary condition and without regard to internal displacements of the ions in each unit cell, may be analyzed in terms of the complete set of plane waves $e^{i\mathbf{k}\cdot\mathbf{R}(\mathrm{M})}$. In other words, an arbitrary function $f(\mathbf{R}(\mathrm{M}))$ of the discrete set of N_0 vectors $\mathbf{R}(\mathrm{M})$ may be represented as a discrete Fourier transform by a sum over N_0 plane waves:

$$f(\mathbf{R}(\mathrm{M})) = N_0^{-1/2} \sum_{\mathbf{k}} g(\mathbf{k}) e^{i\mathbf{k}\cdot\mathbf{R}(\mathrm{M})}, \quad \text{for each } \mathbf{R}(\mathrm{M}). \tag{10.14}$$

The periodic boundary condition then requires the phase of each allowed plane wave to be periodic in the edge of the crystal, or

$$e^{iL\mathbf{k}\cdot\mathbf{a}_1} = e^{iL\mathbf{k}\cdot\mathbf{a}_2} = e^{iL\mathbf{k}\cdot\mathbf{a}_3} = 1, \quad \text{for all } \mathbf{k}. \tag{10.15}$$

The complete set of $\mathbf{k}$ vectors is conveniently described in terms of the inverse lattice. The inverse unit cell is the cell whose edges are the three vectors $\mathbf{b}_\beta$, $\beta = 1,2,3$, defined so that

$$\mathbf{a}_\alpha \cdot \mathbf{b}_\beta = 2\pi\delta_{\alpha\beta}. \tag{10.16}$$

With the aid of (10.11) for the direct-lattice unit-cell volume, the inverse-lattice unit cell vectors may be written

$$\begin{aligned} \mathbf{b}_1 &= (2\pi/V_C)(\mathbf{a}_2 \times \mathbf{a}_3), \\ \mathbf{b}_2 &= (2\pi/V_C)(\mathbf{a}_3 \times \mathbf{a}_1), \\ \mathbf{b}_3 &= (2\pi/V_C)(\mathbf{a}_1 \times \mathbf{a}_2). \end{aligned} \tag{10.17}$$

The inverse (reciprocal) lattice is defined by the set of vectors

$$\mathbf{Q}(\mathrm{M}) = M_1\mathbf{b}_1 + M_2\mathbf{b}_2 + M_3\mathbf{b}_3, \tag{10.18}$$

where M_1,M_2,M_3 are integers (positive, negative, or zero). The inverse lattice may be taken to be finite or infinite, but in any case is perfectly periodic according to (10.18), and there should be no confusion from using the same index M to denote direct-lattice vectors **R**(M) and reciprocal-lattice vectors **Q**(M). We also note that the inverse lattice depends only on the primitive lattice of the crystal, and is independent of the structure of each unit cell, i.e., it is independent of the ion position vectors **R**(μ). Hence for example, since the primitive lattice is fcc for both fcc and diamond lattices, the inverse lattice is the same for both, namely bcc. Finally, for any **Q**(M) and any **R**(N),

$$\mathbf{Q}(\mathrm{M}) \cdot \mathbf{R}(\mathrm{N}) = 2\pi(M_1N_1 + M_2N_2 + M_3N_3) = 2\pi \text{ (integer)}, \tag{10.19}$$

which follows from (10.16).

The periodic boundary condition (10.15) will be satisfied by any wave vector **k** of the form

$$\mathbf{k} = L^{-1}(K_1\mathbf{b}_1 + K_2\mathbf{b}_2 + K_3\mathbf{b}_3), \tag{10.20}$$

where K_1,K_2,K_3 are any integers. Furthermore, from (10.19)

$$e^{i(\mathbf{k}+\mathbf{Q})\cdot\mathbf{R}(\mathrm{M})} = e^{i\mathbf{k}\cdot\mathbf{R}(\mathrm{M})} \tag{10.21}$$

for any inverse-lattice vector **Q**, so all the independent plane waves are obtained by taking **k** in one unit cell of the inverse lattice. This corresponds to taking K_1,K_2,K_3 as integers in a range L, and since $L^3 = N_0$, there are just N_0 vectors **k** in the complete set. In addition, because of the point symmetry of the inverse lattice, the Fourier-transformed equations for the ion motion are essentially the same for **k** vectors related by point-group operations, so that in detailed calculations it is necessary to consider only one **k** in each group, and hence only a relatively small representative portion of the inverse unit cell. To make use of this symmetry it is convenient (but not necessary) to take the allowed **k** as those in the first Brillouin zone, rather than in one inverse unit cell. The first zone is constructed as follows: from an inverse-lattice point **Q**(M) draw lines to all other **Q**(N), construct planes which are perpendicular bisectors of these lines, and take the smallest polyhedron of these planes which encloses **Q**(M). It is easily verified that the volume of the first zone is the same as the inverse unit-cell volume, and that the set of **k** vectors in each of these two volumes is the same modulo inverse-lattice vectors. It should also be noted that in summing over all **k**, if a point on a zone surface (or inverse unit-cell surface) is counted, then the equivalent point on the opposite surface is not counted, since the two vectors differ exactly by a **Q**(M). The Brillouin zones for several simple lattices are pictured in Appendix 1.

We can now complete the discussion of the formal properties of discrete Fourier transforms. The orthonormality of the set of plane waves is expressed by the relation

$$N_0^{-1} \sum_{\mathrm{M}} e^{i\mathbf{k}\cdot\mathbf{R}(\mathrm{M})} = \delta(\mathbf{k}), \tag{10.22}$$

where $\delta(\mathbf{k}) = 0$ unless $\mathbf{k}$ is an inverse-lattice vector $\mathbf{Q}$, and $\delta(\mathbf{Q}) = 1$. This must hold since the Σ_{M} remains the same if the lattice is translated by any primitive-lattice vector $\mathbf{R}(\mathrm{N})$; but this translation replaces each $\mathbf{R}(\mathrm{M})$ by $\mathbf{R}(\mathrm{M}) + \mathbf{R}(\mathrm{N})$ and hence multiplies the expression by $e^{i\mathbf{k}\cdot\mathbf{R}(\mathrm{N})}$, and the requirement $e^{i\mathbf{k}\cdot\mathbf{R}(\mathrm{N})} = 1$ is equivalent to the requirement $\mathbf{k} =$ any $\mathbf{Q}$, including of course $\mathbf{Q} = 0$. Equation (10.22) is seen as the orthogonality of the plane waves $\mathbf{k}$ and $\mathbf{k}'$ when it is written

$$N_0^{-1} \sum_{\mathrm{M}} e^{i(\mathbf{k}-\mathbf{k}')\cdot\mathbf{R}(\mathrm{M})} = \delta(\mathbf{k} - \mathbf{k}'). \tag{10.23}$$

This also has the familiar interpretation as the requirement for conservation of "crystal momentum" in a process in which $\mathbf{k} \to \mathbf{k}'$, where if $\mathbf{k}' - \mathbf{k} = 0$ the process is normal, and if $\mathbf{k}' - \mathbf{k} = \mathbf{Q}(\mathrm{M}) \neq 0$ it is an Umklapp process. Note that there are no Umklapp processes involving only two phonons (lattice waves) since for all allowed $\mathbf{k},\mathbf{k}'$ in the Brillouin zone, $\mathbf{k} - \mathbf{k}'$ can never equal a nonzero $\mathbf{Q}(\mathrm{M})$; there are of course three-, four-, and many-phonon Umklapp processes, and electron–phonon Umklapp processes as well.

The general Fourier transform (10.14) may be inverted by multiplying by $e^{-i\mathbf{k}'\cdot\mathbf{R}(\mathrm{M})}$, summing on M, and using the orthogonality (10.23) to give

$$g(\mathbf{k}) = N_0^{-1/2} \sum_{\mathrm{M}} f(\mathbf{R}(\mathrm{M}))e^{-i\mathbf{k}\cdot\mathbf{R}(\mathrm{M})}. \tag{10.24}$$

But $f(\mathbf{R}(\mathrm{N}))$ can be reconstructed from this expression for the Fourier coefficients $g(\mathbf{k})$, according to (10.14), for any $\mathbf{R}(\mathrm{N})$:

$$\begin{aligned} f(\mathbf{R}(\mathrm{N})) &= N_0^{-1/2} \sum_{\mathbf{k}} g(\mathbf{k})e^{i\mathbf{k}\cdot\mathbf{R}(\mathrm{N})} \\ &= \sum_{\mathrm{M}} f(\mathbf{R}(\mathrm{M}))N_0^{-1} \sum_{\mathbf{k}} e^{i\mathbf{k}\cdot[\mathbf{R}(\mathrm{N})-\mathbf{R}(\mathrm{M})]}. \end{aligned} \tag{10.25}$$

Since this must hold for every $\mathbf{R}(\mathrm{N})$, it implies the completeness relation

$$N_0^{-1} \sum_{\mathbf{k}} e^{i\mathbf{k}\cdot[\mathbf{R}(\mathrm{N})-\mathbf{R}(\mathrm{M})]} = \delta(\mathbf{R}(\mathrm{N}) - \mathbf{R}(\mathrm{M})), \tag{10.26}$$

or for a general primitive-lattice vector $\mathbf{R}(\mathrm{M})$,

$$N_0^{-1} \sum_{\mathbf{k}} e^{i\mathbf{k}\cdot\mathbf{R}(\mathrm{M})} = \delta(\mathbf{R}(\mathrm{M})), \tag{10.27}$$

where $\delta(\mathbf{R}(\mathrm{M}))$ is zero if $\mathbf{R}(\mathrm{M}) \neq 0$ and $\delta(0) = 1$.

There is one more important property of the allowed $\mathbf{k}$ vectors; from (10.12) for the primitive-lattice vectors $\mathbf{R}(\mathrm{M})$ and (10.20) for the wave vectors $\mathbf{k}$, and with (10.16) for $\mathbf{a}_\alpha \cdot \mathbf{b}_\beta$ it follows that

$$\mathbf{k} \cdot \mathbf{R}(\mathrm{M}) = 2\pi L^{-1}(K_1 M_1 + K_2 M_2 + K_3 M_3), \tag{10.28}$$

and it also follows from the geometry that (10.28) is *independent of the macroscopic configuration of the crystal.* As a special result, $\mathbf{k} \cdot \mathbf{R}(\mathrm{M})$ is invariant under homogeneous deformation of the crystal; the same is not true, of course, for $\mathbf{k} \cdot \mathbf{R}(\mathrm{M}\mu)$ because of the existence of sublattice displacements in general.

Dynamical Matrices

We are now prepared to find the normal-mode solutions of the harmonic Hamiltonian $\mathscr{H}_2$; these normal modes are traveling (or standing) vibrational waves in the crystal, called phonons. There are as many independent phonons as there are degrees of motional freedom in the crystal, namely $3nN_0$ for N_0 unit cells with n ions in each. Diagonalization of $\mathscr{H}_2$ requires simultaneous diagonalization of the kinetic energy KE and the quadratic crystal potential Φ_2, according to (10.8) for $\mathscr{H}_2$. This procedure represents a well-known problem in classical mechanics (or linear algebra), and may be cast in the form of an ordinary eigenvalue problem, or a generalized eigenvalue problem. We will use the ordinary eigenvalue problem formulation here because it corresponds to the most common usage by physicists; several different (but equivalent) formulations are discussed in the following section.

At the heart of the diagonalization of $\mathscr{H}_2$ is the diagonalization of N_0 dynamical matrices $\mathbf{D}(\mathbf{k})$, whose elements are given by

$$D_{ij}(\nu\pi,\mathbf{k}) = (M_\nu M_\pi)^{-1/2} N_0^{-1} \sum_{\mathrm{NP}} \Phi_{ij}(\mathrm{N}\nu,\mathrm{P}\pi) e^{i\mathbf{k}\cdot[\mathbf{R}(\mathrm{P}\pi)-\mathbf{R}(\mathrm{N}\nu)]}, \tag{10.29}$$

where it will be recalled that M_ν is the mass of an ion of type ν; $\mathbf{D}(\mathbf{k})$ is a $3n \times 3n$ matrix whose rows and columns are labeled by the index pairs νi and πj, respectively, and it is obvious from the definition (10.29) that

$$D_{ij}(\nu\pi,-\mathbf{k}) = D_{ij}(\nu\pi,\mathbf{k})^*, \tag{10.30}$$

or

$$\mathbf{D}(-\mathbf{k}) = \mathbf{D}(\mathbf{k})^*. \tag{10.31}$$

Furthermore, $\mathbf{D}(\mathbf{k})$ is Hermitian; to show this write the elements of $\mathbf{D}(\mathbf{k})^\dagger$,

$$D_{ji}(\pi\nu,\mathbf{k})^* = (M_\nu M_\pi)^{-1/2} N_0^{-1} \sum_{\mathrm{NP}} \Phi_{ji}(\mathrm{N}\pi,\mathrm{P}\nu) e^{-i\mathbf{k}\cdot[\mathbf{R}(\mathrm{P}\nu)-\mathbf{R}(\mathrm{N}\pi)]}.$$

Now interchanging N and P, and then using $\Phi_{ji}(\mathrm{P}\pi,\mathrm{N}\nu) = \Phi_{ij}(\mathrm{N}\nu,\mathrm{P}\pi)$ according to (6.4), this is

$$\begin{aligned} D_{ji}(\pi\nu,\mathbf{k})^* &= (M_\nu M_\pi)^{-1/2} N_0^{-1} \sum_{\mathrm{NP}} \Phi_{ij}(\mathrm{N}\nu,\mathrm{P}\pi) e^{-i\mathbf{k}\cdot[\mathbf{R}(\mathrm{N}\nu)-\mathbf{R}(\mathrm{P}\pi)]} \\ &= D_{ij}(\nu\pi,\mathbf{k}), \end{aligned} \tag{10.32}$$

where the last equality represents (10.29). Hence,

$$\mathbf{D}(\mathbf{k})^{\dagger} = \mathbf{D}(\mathbf{k}). \tag{10.33}$$

Since $\mathbf{D}(\mathbf{k})$ is Hermitian, it is diagonalized by a unitary matrix of eigenvectors $\mathbf{w}(\mathbf{k}s)$, where $s = 1,2,\cdots,3n$, and the eigenvalues $[\omega(\mathbf{k}s)]^2$ are real. As we shall see later, each phonon is characterized by a wave vector $\mathbf{k}$ and polarization s, its frequency is $\omega(\mathbf{k}s)$, and its eigenvector $\mathbf{w}(\mathbf{k}s)$ describes the direction and relative magnitude of the corresponding displacements of the ions in a unit cell. The components of $\mathbf{w}(\mathbf{k}s)$ are written $w_i(\nu,ks)$, and the diagonalization of $\mathbf{D}(\mathbf{k})$ and the orthonormality and completeness of the eigenvectors are written out in the following equations.

Diagonalization:

$$\sum_{\nu i \pi j} w_i(\nu,\mathbf{k}s)^* D_{ij}(\nu\pi,\mathbf{k}) w_j(\pi,\mathbf{k}s') = [\omega(\mathbf{k}s)]^2 \delta_{ss'}; \tag{10.34}$$

Orthonormality:

$$\sum_{\nu i} w_i(\nu,\mathbf{k}s)^* w_i(\nu,\mathbf{k}s') = \delta_{ss'}; \tag{10.35}$$

Completeness:

$$\sum_{s} w_i(\nu,\mathbf{k}s)^* w_j(\pi,\mathbf{k}s) = \delta_{\nu\pi}\delta_{ij}. \tag{10.36}$$

Also since $\mathbf{D}(-\mathbf{k}) = \mathbf{D}(\mathbf{k})^*$ according to (10.31), the eigenvectors may be chosen so that $\mathbf{w}(-\mathbf{k}) = \mathbf{w}(\mathbf{k})^*$, or

$$w_i(\nu,-\mathbf{k}s) = w_i(\nu,\mathbf{k}s)^*, \tag{10.37}$$

and since $[\omega(\mathbf{k}s)]^2$ are real, it also follows that

$$[\omega(-\mathbf{k}s)]^2 = [\omega(\mathbf{k}s)]^2. \tag{10.38}$$

Another property of the dynamical matrices is that surface effects may be eliminated directly from the defining equation (10.29), since there the Σ_{P} is the same for all cells N in the interior. Evaluating at $\mathrm{N} = 0$ and multiplying by N_0 to account for the Σ_{N} gives

$$D_{ij}(\nu\pi,\mathbf{k}) = (M_\nu M_\pi)^{-1/2} \sum_{\mathrm{P}} \Phi_{ij}(0\nu,\mathrm{P}\pi) e^{i\mathbf{k}\cdot[\mathbf{R}(\mathrm{P}\pi)-\mathbf{R}(\nu)]}. \tag{10.39}$$

Now let us introduce the phonon coordinates $q(\mathbf{k}s)$ and define the linear transformation of the ion displacements

$$U_i(\mathrm{N}\nu) = (N_0 M_\nu)^{-1/2} \sum_{\mathbf{k}s} q(\mathbf{k}s) e^{i\mathbf{k}\cdot\mathbf{R}(\mathrm{N}\nu)} w_i(\nu,\mathbf{k}s); \tag{10.40}$$

the time derivative of this is

$$\dot{U}_i(\mathrm{N}\nu) = (N_0 M_\nu)^{-1/2} \sum_{\mathbf{k}s} \dot{q}(\mathbf{k}s) e^{i\mathbf{k}\cdot\mathbf{R}(\mathrm{N}\nu)} w_i(\nu,\mathbf{k}s). \tag{10.41}$$

With (10.41) the total crystal kinetic energy (10.1) is

$$KE = \tfrac{1}{2} \sum_{N\nu} \sum_{i} M_\nu [\dot{U}_i(N\nu)]^2$$
$$= \tfrac{1}{2} \sum_{\mathbf{k}\mathbf{k}'ss'} \dot{q}(\mathbf{k}s)\dot{q}(\mathbf{k}'s')\delta(\mathbf{k} + \mathbf{k}') \sum_{\nu i} w_i(\nu,\mathbf{k}s) w_i(\nu,\mathbf{k}'s'), \tag{10.42}$$

where the $\delta(\mathbf{k} + \mathbf{k}')$ replaces a primitive-lattice sum, according to its definition (10.22). Since $\delta(\mathbf{k} + \mathbf{k}')$ requires $\mathbf{k}' = -\mathbf{k}$, the $\Sigma_{\nu i}$ in (10.42) becomes $\delta_{ss'}$ from (10.35) and (10.37), and the kinetic energy is simply

$$KE = \tfrac{1}{2} \sum_{\mathbf{k}s} \dot{q}(\mathbf{k}s)\dot{q}(-\mathbf{k}s). \tag{10.43}$$

With the transformation (10.40) the quadratic crystal potential is

$$\Phi_2 = \tfrac{1}{2} \sum_{NP\nu\pi} \sum_{ij} \Phi_{ij}(N\nu,P\pi) U_i(N\nu) U_j(P\pi)$$
$$= \tfrac{1}{2} \sum_{\mathbf{k}\mathbf{k}'ss'} q(\mathbf{k}s) q(\mathbf{k}'s') \Big\{ N_0^{-1} \sum_{NP\nu\pi} \sum_{ij} (M_\nu M_\pi)^{-1/2} \Phi_{ij}(N\nu,P\pi) \tag{10.44}$$
$$\times\, e^{i[\mathbf{k}\cdot\mathbf{R}(N\nu) + \mathbf{k}'\cdot\mathbf{R}(P\pi)]} w_i(\nu,\mathbf{k}s) w_j(\pi,\mathbf{k}'s') \Big\}.$$

The Fourier transform of second-order potential energy coefficients appearing in (10.44) can be rewritten as follows.

$$N_0^{-1}(M_\nu M_\pi)^{-1/2} \sum_{NP} \Phi_{ij}(N\nu,P\pi) e^{i[\mathbf{k}\cdot\mathbf{R}(N\nu) + \mathbf{k}'\cdot\mathbf{R}(P\pi)]}$$
$$= N_0^{-1}(M_\nu M_\pi)^{-1/2} \sum_{N} e^{i(\mathbf{k}+\mathbf{k}')\cdot\mathbf{R}(N)} \sum_{P} \Phi_{ij}(N\nu,P\pi) e^{i\mathbf{k}'\cdot[\mathbf{R}(P\pi) - \mathbf{R}(N\nu)]} e^{i(\mathbf{k}+\mathbf{k}')\cdot\mathbf{R}(\nu)},$$

and for all cells N in the interior, the Σ_P is independent of N, and so may be evaluated at N = 0 by neglecting surface effects. Then since the Σ_N is $N_0\delta(\mathbf{k} + \mathbf{k}')$, the above quantity is

$$\delta(\mathbf{k} + \mathbf{k}')(M_\nu M_\pi)^{-1/2} \sum_{P} \Phi_{ij}(0\nu,P\pi) e^{i\mathbf{k}'\cdot[\mathbf{R}(P\pi) - \mathbf{R}(\nu)]} = \delta(\mathbf{k} + \mathbf{k}') D_{ij}(\nu\pi,\mathbf{k}'),$$

according to (10.39) for $\mathbf{D}(\mathbf{k}')$ with surface effects eliminated. With this last result the expression (10.44) for Φ_2 is

$$\Phi_2 = \tfrac{1}{2} \sum_{\mathbf{k}'ss'} q(-\mathbf{k}'s) q(\mathbf{k}'s') \sum_{\nu i\pi j} D_{ij}(\nu\pi,\mathbf{k}') w_i(\nu,-\mathbf{k}'s) w_j(\pi,\mathbf{k}'s'), \tag{10.45}$$

and finally with the diagonalization (10.34)

$$\Phi_2 = \tfrac{1}{2} \sum_{\mathbf{k}s} [\omega(\mathbf{k}s)]^2 q(\mathbf{k}s) q(-\mathbf{k}s). \tag{10.46}$$

The momentum $p(\mathbf{k}s)$ conjugate to $q(\mathbf{k}s)$ is

$$p(\mathbf{k}s) = \dot{q}(-\mathbf{k}s); \tag{10.47}$$

this is shown by calculating commutators as follows. The phonon coordinate transformation (10.40) may be inverted with the orthonormality relations (10.23) and (10.35) to give

$$q(\mathbf{k}s) = N_0^{-1/2} \sum_{N\nu} \sum_{i} M_\nu^{1/2} U_i(N\nu) e^{-i\mathbf{k}\cdot\mathbf{R}(N\nu)} w_i(\nu,-\mathbf{k}s), \tag{10.48}$$

and taking the time derivative to calculate $p(\mathbf{k}s)$ by (10.47),

$$p(\mathbf{k}s) = N_0^{-1/2} \sum_{N\nu} \sum_{i} M_\nu^{1/2} \dot{U}_i(N\nu) e^{i\mathbf{k}\cdot\mathbf{R}(N\nu)} w_i(\nu,\mathbf{k}s). \tag{10.49}$$

The commutators are then calculated from the position and momentum commutators (10.4) and (10.5), and the orthonormality relations, to give

$$[p(\mathbf{k}s),q(\mathbf{k}'s')] = -i\hbar\delta(\mathbf{k} - \mathbf{k}')\delta_{ss'}; \tag{10.50}$$

$$[p(\mathbf{k}s),p(\mathbf{k}'s')] = [q(\mathbf{k}s),q(\mathbf{k}'s')] = 0. \tag{10.51}$$

These results show that $q(\mathbf{k}s)$ and $p(\mathbf{k}s)$ are conjugate coordinates and momenta, and further that all the phonons are statistically independent. It is also seen by taking the complex conjugate of (10.48) and using (10.37), that

$$q(-\mathbf{k}s) = q(\mathbf{k}s)^*; \qquad \dot{q}(-\mathbf{k}s) = \dot{q}(\mathbf{k}s)^*. \tag{10.52}$$

Finally we can construct the harmonic Hamiltonian $\mathscr{H}_2$ from its parts, namely (10.43) for KE and (10.46) for Φ_2; putting in the momenta by (10.47) this is

$$\mathscr{H}_2 = KE + \Phi_2 = \tfrac{1}{2} \sum_{\mathbf{k}s} \{|p(\mathbf{k}s)|^2 + [\omega(\mathbf{k}s)]^2 \, |q(\mathbf{k}s)|^2\}. \tag{10.53}$$

Thus $\mathscr{H}_2$ is a set of $3nN_0$ independent harmonic oscillator Hamiltonians, with each oscillator (phonon) labeled by the indices $\mathbf{k}s$ and having the frequency $\omega(\mathbf{k}s)$. It is clear from the phonon transformation (10.40) of the ion displacements that the phonon $\mathbf{k}s$ has wave vector $\mathbf{k}$ and gives rise to displacements of the ions in a unit cell whose directions and relative magnitudes are given by the eigenvector components $w_i(\nu,\mathbf{k}s)$.

The condition that the crystal be stable against all possible (small) displacements of the ions was written in (6.49) as the requirement that the second-order crystal potential Φ_2 be positive definite. In the diagonal form (10.46) of Φ_2, the stability condition is

$$\sum_{\mathbf{k}s} [\omega(\mathbf{k}s)]^2 \, |q(\mathbf{k}s)|^2 > 0. \tag{10.54}$$

Since the $q(\mathbf{k}s)$ are independent variables, and since $|q(\mathbf{k}s)|^2 \geq 0$, the positive definite condition (10.54) requires

$$[\omega(\mathbf{k}s)]^2 > 0, \quad \text{for all } \mathbf{k},s. \tag{10.55}$$

Alternatively one can argue that when a phonon $\mathbf{k}s$ is excited, the time-dependence of the state is given by the factor $e^{-i\omega(\mathbf{k}s)t}$, and if $[\omega(\mathbf{k}s)]^2$ is negative, then $\omega(\mathbf{k}s)$ will be imaginary, and the state wave function will increase exponentially in the positive or negative time domain. Therefore, stability of the crystal against excitation of all phonons again requires (10.55). We will always take the positive square root for $\omega(\mathbf{k}s)$, so that as convention

$$\omega(\mathbf{k}s) > 0. \tag{10.56}$$

There are several more useful results of the eigenvalue–eigenvector problem defined by (10.34)–(10.36). The diagonalization equation (10.34) can be inverted with the aid of the completeness (10.36) to give

$$D_{ij}(\nu\pi,\mathbf{k}) = \sum_s w_i(\nu,\mathbf{k}s)[\omega(\mathbf{k}s)]^2 w_j(\pi,\mathbf{k}s)^*. \tag{10.57}$$

Let us define the inverse dynamical matrix as $\mathbf{E}(\mathbf{k}) = \mathbf{D}(\mathbf{k})^{-1}$, where the elements of $\mathbf{E}(\mathbf{k})$ satisfy the inverse relation

$$\sum_{\nu j} D_{ij}(\mu\nu,\mathbf{k}) E_{jk}(\nu\pi,\mathbf{k}) = \delta_{\mu\pi}\delta_{ik}. \tag{10.58}$$

Then we can write

$$E_{ij}(\nu\pi,\mathbf{k}) = \sum_s w_i(\nu,\mathbf{k}s)[\omega(\mathbf{k}s)]^{-2} w_j(\pi,\mathbf{k}s)^*; \tag{10.59}$$

this may be verified by using the orthonormality (10.35) and completeness (10.36) to show that the right-hand sides of (10.57) and (10.59) satisfy the inverse relation (10.58).

The sum of the eigenvalues of $\mathbf{D}(\mathbf{k})$ is just

$$\sum_s [\omega(\mathbf{k}s)]^2 = \operatorname{Tr} \mathbf{D}(\mathbf{k}) = \sum_{\nu i} D_{ii}(\nu\nu,\mathbf{k}), \tag{10.60}$$

which of course follows from (10.34) and (10.36). From (10.39) for $\mathbf{D}(\mathbf{k})$,

$$D_{ii}(\nu\nu,\mathbf{k}) = M_\nu^{-1} \sum_{\mathrm{P}} \Phi_{ii}(0\nu,\mathrm{P}\nu) e^{i\mathbf{k}\cdot\mathbf{R}(\mathrm{P})}, \tag{10.61}$$

and this leads to a particularly simple result for the average of all phonon frequencies squared:

$$\langle[\omega(\mathbf{k}s)]^2\rangle = (3nN_0)^{-1} \sum_{\mathbf{k}s} [\omega(\mathbf{k}s)]^2 = (3n)^{-1} \sum_{\nu i} M_\nu^{-1}\Phi_{ii}(0\nu,0\nu), \tag{10.62}$$

where the completeness (10.27) of the $\mathbf{k}$ vectors was used. The self-coupling coefficient in (10.62) is given by the translational invariance condition (6.25) in terms of other second-order potential energy coefficients as

$$\Phi_{ii}(0\nu,0\nu) = -\sum_{\mathrm{P}\pi}{}' \Phi_{ii}(0\nu,\mathrm{P}\pi). \tag{10.63}$$

Interacting Phonon Description

For further theoretical work, especially involving effects of interactions among the phonons, it is convenient to transform the lattice vibration problem into a description in terms of phonon creation and annihilation operators. These are denoted $A(\mathbf{k}s)^\dagger$ and $A(\mathbf{k}s)$ for creation and annihilation operators, respectively, and they are defined by

$$A(\mathbf{k}s) = [2\hbar\omega(\mathbf{k}s)]^{-1/2}[\omega(\mathbf{k}s)q(\mathbf{k}s) + ip(-\mathbf{k}s)], \qquad (10.64)$$

$$A(\mathbf{k}s)^\dagger = [2\hbar\omega(\mathbf{k}s)]^{-1/2}[\omega(\mathbf{k}s)q(-\mathbf{k}s) - ip(\mathbf{k}s)]. \qquad (10.65)$$

Note that these operators are dimensionless, and also $A(\mathbf{k}s)^\dagger = A(\mathbf{k}s)^*$. Solving (10.64) and (10.65) for the phonon coordinates and momenta gives

$$q(\mathbf{k}s) = [\hbar/2\omega(\mathbf{k}s)]^{1/2}[A(\mathbf{k}s) + A(-\mathbf{k}s)^\dagger], \qquad (10.66)$$

$$p(\mathbf{k}s) = i[\hbar\omega(\mathbf{k}s)/2]^{1/2}[A(\mathbf{k}s)^\dagger - A(-\mathbf{k}s)]. \qquad (10.67)$$

The creation and annihilation operator commutators may be calculated directly from the definitions (10.64) and (10.65), and the coordinate and momentum commutators (10.50) and (10.51); this gives

$$[A(\mathbf{k}s),A(\mathbf{k}'s')^\dagger] = \delta(\mathbf{k} - \mathbf{k}')\delta_{ss'}; \qquad (10.68)$$

$$[A(\mathbf{k}s),A(\mathbf{k}'s')] = [A(\mathbf{k}s)^\dagger,A(\mathbf{k}'s')^\dagger] = 0. \qquad (10.69)$$

Thus the $A(\mathbf{k}s)$ and $A(\mathbf{k}s)^\dagger$ represent a set of $3nN_0$ independent bosons.

The harmonic Hamiltonian $\mathcal{H}_2$ is given as a sum of phonon momenta and coordinates in (10.53), and may be transformed to creation and annihilation operators with the aid of (10.66)–(10.69):

$$\mathcal{H}_2 = \sum_{\mathbf{k}s} \hbar\omega(\mathbf{k}s)[A(\mathbf{k}s)^\dagger A(\mathbf{k}s) + \tfrac{1}{2}]. \qquad (10.70)$$

In order to transform the higher-order potential energy terms $\mathcal{H}_3$ and $\mathcal{H}_4$ of (10.9) and (10.10), we write the ion-displacement transformation (10.40) in terms of $A(\mathbf{k}s)^\dagger$ and $A(\mathbf{k}s)$ with the aid of (10.66), as follows.

$$U_i(\mathrm{N}\nu) = (\hbar/2N_0M_\nu)^{1/2}\sum_{\mathbf{k}s} [\omega(\mathbf{k}s)]^{-1/2}[A(\mathbf{k}s) + A(-\mathbf{k}s)^\dagger]e^{i\mathbf{k}\cdot\mathbf{R}(\mathrm{N}\nu)}w_i(\nu,\mathbf{k}s). \qquad (10.71)$$

Then $\mathcal{H}_3$ and $\mathcal{H}_4$ are written

$$\mathcal{H}_3 = \sum_{\mathbf{k}\mathbf{k}'\mathbf{k}''}\sum_{ss's''} \Phi(\mathbf{k}s,\mathbf{k}'s',\mathbf{k}''s'')[A(\mathbf{k}s) + A(-\mathbf{k}s)^\dagger] \times [A(\mathbf{k}'s') + A(-\mathbf{k}'s')^\dagger][A(\mathbf{k}''s'') + A(-\mathbf{k}''s'')^\dagger]; \qquad (10.72)$$

$$\mathcal{H}_4 = \sum_{\mathbf{k}\mathbf{k}'\mathbf{k}''\mathbf{k}'''}\sum_{ss's''s'''} \Phi(\mathbf{k}s,\mathbf{k}'s',\mathbf{k}''s'',\mathbf{k}'''s''')[A(\mathbf{k}s) + A(-\mathbf{k}s)^\dagger] \times [A(\mathbf{k}'s') + A(-\mathbf{k}'s')^\dagger][A(\mathbf{k}''s'') + A(-\mathbf{k}''s'')^\dagger] \times [A(\mathbf{k}'''s''') + A(-\mathbf{k}'''s''')^\dagger]. \qquad (10.73)$$

Here we have put all the numerical factors into the transformed potential energy coefficients, which are given by

$$\begin{aligned}\Phi(\mathbf{k}s,\mathbf{k}'s',\mathbf{k}''s'') = {} & (1/3!)(\hbar/2N_0)^{3/2}[\omega(\mathbf{k}s)\omega(\mathbf{k}'s')\omega(\mathbf{k}''s'')]^{-1/2} \\ & \times \sum_{\mathrm{MNP}\mu\nu\pi} \sum_{ijk} \Phi_{ijk}(\mathrm{M}\mu,\mathrm{N}\nu,\mathrm{P}\pi)(M_\mu M_\nu M_\pi)^{-1/2} \\ & \times e^{i\mathbf{k}\cdot\mathbf{R}(\mathrm{M}\mu)} e^{i\mathbf{k}'\cdot\mathbf{R}(\mathrm{N}\nu)} e^{i\mathbf{k}''\cdot\mathbf{R}(\mathrm{P}\pi)} \\ & \times w_i(\mu,\mathbf{k}s)w_j(\nu,\mathbf{k}'s')w_k(\pi,\mathbf{k}''s'');\end{aligned} \tag{10.74}$$

$$\begin{aligned}\Phi(\mathbf{k}s,\mathbf{k}'s',\mathbf{k}''s'',\mathbf{k}'''s''') = {} & (1/4!)(\hbar/2N_0)^{2}[\omega(\mathbf{k}s)\omega(\mathbf{k}'s')\omega(\mathbf{k}''s'')\omega(\mathbf{k}'''s''')]^{-1/2} \\ & \times \sum_{\mathrm{MNPQ}\mu\nu\pi\rho} \sum_{ijkl} \Phi_{ijkl}(\mathrm{M}\mu,\mathrm{N}\nu,\mathrm{P}\pi,\mathrm{Q}\rho)(M_\mu M_\nu M_\pi M_\rho)^{-1/2} \\ & \times e^{i\mathbf{k}\cdot\mathbf{R}(\mathrm{M}\mu)} e^{i\mathbf{k}'\cdot\mathbf{R}(\mathrm{N}\nu)} e^{i\mathbf{k}''\cdot\mathbf{R}(\mathrm{P}\pi)} e^{i\mathbf{k}'''\cdot\mathbf{R}(\mathrm{Q}\rho)} \\ & \times w_i(\mu,\mathbf{k}s)w_j(\nu,\mathbf{k}'s')w_k(\pi,\mathbf{k}''s'')w_l(\rho,\mathbf{k}'''s''').\end{aligned} \tag{10.75}$$

The higher-order many-phonon terms $\mathscr{H}_5,\mathscr{H}_6,\cdots$, will not be considered explicitly in our work. The reason for keeping both $\mathscr{H}_3$ and $\mathscr{H}_4$ is that $\mathscr{H}_3$ contributes to the total energy in second-order perturbation and $\mathscr{H}_4$ in first-order perturbation, so to this leading order $\mathscr{H}_3$ and $\mathscr{H}_4$ are expected to be about equally important in the total crystal energy. $\mathscr{H}_3$ and $\mathscr{H}_4$ describe three- and four-phonon interactions, respectively, and the transformed potential energy coefficients contain information about the magnitudes of all such interactions. Because of the general symmetries (6.5) and (6.6) of the potential energy coefficients in their index sets $\mathrm{M}\mu i,\mathrm{N}\nu j,\cdots$, it is immediately seen that the transformed coefficients (10.74) and (10.75) are symmetric in $\mathbf{k}s,\mathbf{k}'s',\cdots$, or

$$\Phi(\mathbf{k}s,\mathbf{k}'s',\mathbf{k}''s'') = \Phi(\mathbf{k}'s',\mathbf{k}s,\mathbf{k}''s'') = \Phi(\mathbf{k}s,\mathbf{k}''s'',\mathbf{k}'s') = \cdots, \tag{10.76}$$

$$\Phi(\mathbf{k}s,\mathbf{k}'s',\mathbf{k}''s'',\mathbf{k}'''s''') = \Phi(\mathbf{k}'s',\mathbf{k}s,\mathbf{k}''s'',\mathbf{k}'''s''') = \cdots. \tag{10.77}$$

In addition, it follows from the definitions (10.74) and (10.75), and from (10.37) and (10.38), that

$$\Phi(-\mathbf{k}s,-\mathbf{k}'s',-\mathbf{k}''s'') = \Phi(\mathbf{k}s,\mathbf{k}'s',\mathbf{k}''s'')^*; \tag{10.78}$$

$$\Phi(-\mathbf{k}s,-\mathbf{k}'s',-\mathbf{k}''s'',-\mathbf{k}'''s''') = \Phi(\mathbf{k}s,\mathbf{k}'s',\mathbf{k}''s'',\mathbf{k}'''s''')^*. \tag{10.79}$$

Surface effects can be eliminated directly from the transformed potential coefficients; consider (10.74) for $\Phi(\mathbf{k}s,\mathbf{k}'s',\mathbf{k}''s'')$, which contains the factor

$$\sum_{\mathrm{MNP}} \Phi_{ijk}(\mathrm{M}\mu,\mathrm{N}\nu,\mathrm{P}\pi) e^{i\mathbf{k}\cdot\mathbf{R}(\mathrm{M})} e^{i\mathbf{k}'\cdot\mathbf{R}(\mathrm{N})} e^{i\mathbf{k}''\cdot\mathbf{R}(\mathrm{P})}.$$

By the lattice-vector translation symmetry (6.58) this may be written

$$\sum_{\mathrm{M}} e^{i(\mathbf{k}+\mathbf{k}'+\mathbf{k}'')\cdot\mathbf{R}(\mathrm{M})} \sum_{\mathrm{NP}} \Phi_{ijk}(0\mu;\mathrm{N}-M,\nu;\mathrm{P}-\mathrm{M},\pi) \times e^{i\mathbf{k}'\cdot[\mathbf{R}(\mathrm{N})-\mathbf{R}(\mathrm{M})]} e^{i\mathbf{k}''\cdot[\mathbf{R}(\mathrm{P})-\mathbf{R}(\mathrm{M})]},$$

and by letting N − M be N and P − M be P in the sum, it is

$$\sum_{\mathrm{M}} e^{i(\mathbf{k}+\mathbf{k}'+\mathbf{k}'')\cdot\mathbf{R}(\mathrm{M})} \sum_{\mathrm{NP}} \Phi_{ijk}(0\mu,\mathrm{N}\nu,\mathrm{P}\pi) e^{i\mathbf{k}'\cdot\mathbf{R}(\mathrm{N})} e^{i\mathbf{k}''\cdot\mathbf{R}(\mathrm{P})}. \tag{10.80}$$

This procedure eliminates surface effects from the transformed potential coefficients, and it also shows, because the Σ_{M} in (10.80) is $N_0\delta(\mathbf{k}+\mathbf{k}'+\mathbf{k}'')$, the important conclusion

$$\Phi(\mathbf{k}s,\mathbf{k}'s',\mathbf{k}''s'') \quad \text{contains} \quad \delta(\mathbf{k}+\mathbf{k}'+\mathbf{k}''), \tag{10.81}$$

$$\Phi(\mathbf{k}s,\mathbf{k}'s',\mathbf{k}''s'',\mathbf{k}'''s''') \quad \text{contains} \quad \delta(\mathbf{k}+\mathbf{k}'+\mathbf{k}''+\mathbf{k}'''). \tag{10.82}$$

The properties (10.81) and (10.82) are just the requirements for conservation of total "crystal momentum" in three- and four-phonon processes, respectively. Unlike two-phonon terms, Umklapp processes are allowed in three- and four-phonon interactions [see discussion following (10.23)].

The notation is shortened considerably by introducing the phonon index κ to stand for the pair $\mathbf{k}s$, according to

$$\kappa = \mathbf{k}s; \qquad -\kappa = -\mathbf{k}s. \tag{10.83}$$

Then the total Hamiltonian for the system of interacting phonons is

$$\mathscr{H} = \Phi_0 + \mathscr{H}_2 + \mathscr{H}_3 + \mathscr{H}_4, \tag{10.84}$$

$$\mathscr{H}_2 = \sum_{\kappa} \hbar\omega_\kappa(A_\kappa^\dagger A_\kappa + \tfrac{1}{2}), \tag{10.85}$$

$$\mathscr{H}_3 = \sum_{\kappa\kappa'\kappa''} \Phi_{\kappa\kappa'\kappa''}(A_\kappa + A_{-\kappa}^\dagger)(A_{\kappa'} + A_{-\kappa'}^\dagger)(A_{\kappa''} + A_{-\kappa''}^\dagger), \tag{10.86}$$

$$\mathscr{H}_4 = \sum_{\kappa\kappa'\kappa''\kappa'''} \Phi_{\kappa\kappa'\kappa''\kappa'''}(A_\kappa + A_{-\kappa}^\dagger)(A_{\kappa'} + A_{-\kappa'}^\dagger) \times (A_{\kappa''} + A_{-\kappa''}^\dagger)(A_{\kappa'''} + A_{-\kappa'''}^\dagger). \tag{10.87}$$

The phonon commutators are

$$[A_\kappa, A_{\kappa'}^\dagger] = \delta_{\kappa\kappa'}, \tag{10.88}$$

$$[A_\kappa, A_{\kappa'}] = [A_\kappa^\dagger, A_{\kappa'}^\dagger] = 0; \tag{10.89}$$

these, together with the expressions (10.84)–(10.87) for $\mathscr{H}$, completely define the interacting phonon problem. The coefficients $\Phi_{\kappa\kappa'\kappa''}$ and $\Phi_{\kappa\kappa'\kappa''\kappa'''}$ are of course still given by (10.74) and (10.75), and they are completely symmetric in their indices $\kappa,\kappa',\cdots$.

In terms of the smallness parameter ϵ defined by (10.6), the orders of magnitude of the quantities entering $\mathscr{H}$ are as follows.

$$
\begin{aligned}
A_\kappa, A_\kappa^\dagger &\sim 1, \\
\hbar\omega_\kappa &\sim \epsilon^2, \\
\sum_{\kappa'\kappa''} \Phi_{\kappa\kappa'\kappa''} &\sim \epsilon\hbar\omega_\kappa, \\
\sum_{\kappa'\kappa''\kappa'''} \Phi_{\kappa\kappa'\kappa''\kappa'''} &\sim \epsilon^2\hbar\omega_\kappa.
\end{aligned}
\tag{10.90}
$$

These orders of ϵ serve as the basis for the perturbation treatment of the phonon interaction terms. We could also consider $\sqrt{\hbar}$ as the expansion parameter; in fact each quantity in (10.90) is of the same order in $\sqrt{\hbar}$ as in ϵ, since according to (10.74) and (10.75), $\Phi_{\kappa\kappa'\kappa''} \sim \hbar^{3/2}$ and $\Phi_{\kappa\kappa'\kappa''\kappa'''} \sim \hbar^2$. The orders of thermodynamic functions are simply related to the ordering procedure here, as is discussed in Chapter 5.

It should also be recognized that the use of the periodic boundary condition implies that the macroscopic configuration of the crystal is fixed. However, since the theory is valid for arbitrary macroscopic configuration, corresponding to arbitrary externally applied uniform surface forces, the calculations may be repeated at any desired macroscopic configuration, or equivalently any result may be differentiated with respect to homogeneous strains. In this case the phonon frequencies ω_κ and all the potential coefficients Φ_0, $\Phi_{\kappa\kappa'\kappa''}$, and $\Phi_{\kappa\kappa'\kappa''\kappa'''}$ are functions of the macroscopic configuration of the crystal, while the phonon commutators are of course constant.

Finally, it should be noted that the allowed **k** vectors are densely distributed in the inverse-lattice space, there being N_0 values of **k** in the volume $(2\pi)^3/V_C$ of one inverse-lattice cell. This corresponds to a density of points equal to $N_0V_C/(2\pi)^3 = V/(2\pi)^3$. As a result of the dense distribution all $\Sigma_{\mathbf{k}}$, such as those appearing in the Hamiltonian contributions (10.85)–(10.87), may be transformed to $\int d\mathbf{k}$. The transformation to an integral is useful for evaluating thermodynamic functions in the low-temperature region, as will be shown in Section 18; in other cases the sums may be evaluated directly by picking representative points in the Brillouin zone, as discussed in Appendix 1.

11. EIGENVALUE PROBLEMS

Generalized Eigenvalue Problem

Historically, generalized eigenvalue problems arose in the study of small oscillations of mechanical systems. Lattice dynamics is, of course, a mechanical small-oscillation problem, the quantum aspects being unimportant in solving for the normal modes. We are going to discuss several different procedures for solving the same lattice dynamics problem; it is important to understand the essential equivalence of these procedures, and it is quite useful to be able to work with any of the different transformations since they each have some advantage for formal derivations or for computations.

The elegant generalized eigenvalue-problem solution of small-oscillation problems has been discussed by Goldstein.* Let us use the index β for a complete set of ion indices and a Cartesian index, while κ still stands for the phonon indices:

$$\beta = \mathrm{M}\mu i; \qquad \kappa = \mathbf{k}s; \qquad \beta \text{ and } \kappa = 1,2,\cdots,3nN_0. \tag{11.1}$$

The harmonic Hamiltonian of Section 10 is written simply as

$$\mathscr{H}_2 = KE + \Phi_2; \tag{11.2}$$

$$KE = \tfrac{1}{2}\sum_{\beta\beta'} T_{\beta\beta'}\dot{U}_\beta \dot{U}_{\beta'}; \tag{11.3}$$

$$\Phi_2 = \tfrac{1}{2}\sum_{\beta\beta'} \Phi_{\beta\beta'} U_\beta U_{\beta'}. \tag{11.4}$$

Here $T_{\beta\beta'}$ is a real diagonal matrix whose elements are the ion masses M_β,

$$T_{\beta\beta'} = M_\beta \delta_{\beta\beta'}, \tag{11.5}$$

and $\Phi_{\beta\beta'}$ is a real symmetric matrix whose elements are the second-order potential energy coefficients,

$$\Phi_{\beta\beta'} = \Phi_{\beta'\beta} = \Phi^*_{\beta\beta'}. \tag{11.6}$$

The normal modes of $\mathscr{H}_2$ may be found either by the transformation method or by the equation of motion method, both of which will be presented.

In the transformation method one defines a linear transformation of the ion displacements U_β to phonon coordinates q_κ, as

$$U_\beta = \sum_\kappa v_{\beta\kappa} q_\kappa, \tag{11.7}$$

and by taking the time derivative

$$\dot{U}_\beta = \sum_\kappa v_{\beta\kappa} \dot{q}_\kappa. \tag{11.8}$$

Here $v_{\beta\kappa}$ are dimensionless complex transformation coefficients, and for the lattice-vibration problem they may be chosen to satisfy

$$v_{\beta,-\kappa} = v^*_{\beta\kappa}. \tag{11.9}$$

* H. Goldstein, *Classical Mechanics*, Addison-Wesley Publishing Co. Inc., Cambridge, Mass., 1950.

Again for the lattice-dynamics problem, either by inverting the transformations (11.7) and (11.8), or by noting that for every κ there is a $-\kappa$ in Σ_κ, (11.9) implies

$$q_{-\kappa} = q_\kappa^*, \qquad \dot{q}_{-\kappa} = \dot{q}_\kappa^*. \tag{11.10}$$

Now the transformation coefficients $v_{\beta\kappa}$ are to be chosen so as to simultaneously diagonalize KE and Φ_2; we note that such coefficients can indeed be found and that they form a nonunitary matrix. The requirement that the kinetic energy be diagonal is the condition of orthonormalization:

$$\sum_{\beta\beta'} v_{\beta\kappa} T_{\beta\beta'} v_{\beta'\kappa'} = \sum_{\beta} M_\beta v_{\beta\kappa} v_{\beta\kappa'} = M_C \delta_{\kappa,-\kappa'}, \tag{11.11}$$

where we have arbitrarily normalized to the mass per unit cell M_C (remember $v_{\beta\kappa}$ are supposed to be dimensionless). The requirement that the potential energy be diagonal is the diagonalization condition:

$$\sum_{\beta\beta'} v_{\beta\kappa} \Phi_{\beta\beta'} v_{\beta'\kappa'} = M_C \omega_\kappa^2 \delta_{\kappa,-\kappa'}. \tag{11.12}$$

With (11.11) and (11.12), the harmonic Hamiltonian (11.2) becomes

$$\mathscr{H}_2 = \tfrac{1}{2} \sum_\kappa (M_C \dot{q}_\kappa \dot{q}_{-\kappa} + M_C \omega_\kappa^2 q_\kappa q_{-\kappa}). \tag{11.13}$$

Now q_κ has dimensions of a position coordinate, and the momentum conjugate to q_κ is $p_\kappa = M_C \dot{q}_{-\kappa}$. Then by (11.10), the Hamiltonian is again a sum of $3nN_0$ independent harmonic oscillator Hamiltonians, with frequencies ω_κ:

$$\mathscr{H}_2 = \tfrac{1}{2} \sum_\kappa \{M_C^{-1} |p_\kappa|^2 + M_C \omega_\kappa^2 |q_\kappa|^2\}. \tag{11.14}$$

The transformation coefficients are completely determined by the orthonormality (11.11) and the diagonalization (11.12). They are assumed to be linearly independent, and the completeness relation is obtained by multiplying (11.11) by $v_{\beta',-\kappa'}$ and summing over κ'.

$$\sum_\beta M_\beta v_{\beta\kappa} \sum_{\kappa'} v_{\beta\kappa'} v_{\beta',-\kappa'} = \sum_{\kappa'} M_C \delta_{\kappa,-\kappa'} v_{\beta',-\kappa'} = M_C v_{\beta'\kappa},$$

and since this must hold for every $v_{\beta'\kappa}$ the completeness relation is

$$M_\beta \sum_{\kappa'} v_{\beta\kappa'} v_{\beta',-\kappa'} = M_C \delta_{\beta\beta'}. \tag{11.15}$$

The equation of motion method is just as easy. The kinetic and potential energies were written in (11.3) and (11.4), and from these the Lagrangian equations of motion are

$$\sum_{\beta'} (T_{\beta\beta'} \ddot{U}_{\beta'} + \Phi_{\beta\beta'} U_{\beta'}) = 0, \tag{11.16}$$

or

$$M_\beta \ddot{U}_\beta + \sum_{\beta'} \Phi_{\beta\beta'} U_{\beta'} = 0. \tag{11.17}$$

This last is Newton's equation of motion, and the same result is obtained from the Hamiltonian equations of motion

$$\ddot{U}_\beta = (i/\hbar)[\mathscr{H}_2,\dot{U}_\beta], \tag{11.18}$$

where this may be evaluated from the momentum-position commutators

$$[M_\beta \dot{U}_\beta, U_{\beta'}] = -i\hbar\delta_{\beta\beta'}. \tag{11.19}$$

Assume $3nN_0$ oscillatory solutions labeled by the index κ:

$$U_\beta = c v_{\beta\kappa} e^{-i\omega_\kappa t}, \tag{11.20}$$

where c is an arbitrary constant of dimensions the same as U_β, and $v_{\beta\kappa}$ are to be determined. With these solutions the equations of motion (11.17) become

$$\sum_{\beta'} \Phi_{\beta\beta'} v_{\beta'\kappa} = \omega_\kappa^2 M_\beta v_{\beta\kappa}. \tag{11.21}$$

This is the set of eigenvalue–eigenvector equations for the generalized eigenvalue problem, and they completely determine the eigenvalues $M_C\omega_\kappa^2$ and (apart from normalization) the eigenvectors $v_{\beta\kappa}$.

The orthonormalization is derived in the customary way as follows. Multiply (11.21) by $v_{\beta,-\kappa'}$ and sum over β to get

$$\sum_{\beta\beta'} v_{\beta,-\kappa'} \Phi_{\beta\beta'} v_{\beta'\kappa} = \omega_\kappa^2 \sum_\beta M_\beta v_{\beta,-\kappa'} v_{\beta\kappa}; \tag{11.22}$$

in this replace κ by $-\kappa'$, κ' by $-\kappa$, interchange β and β', and use $\Phi_{\beta\beta'} = \Phi_{\beta'\beta}$ and $\omega_{-\kappa}^2 = \omega_\kappa^2$ to get

$$\sum_{\beta\beta'} v_{\beta'\kappa} \Phi_{\beta\beta'} v_{\beta,-\kappa'} = \omega_{\kappa'}^2 \sum_\beta M_\beta v_{\beta\kappa} v_{\beta,-\kappa'}. \tag{11.23}$$

The left-hand sides of (11.22) and (11.23) are the same, so the difference of these two equations is

$$(\omega_\kappa^2 - \omega_{\kappa'}^2) \sum_\beta M_\beta v_{\beta,-\kappa'} v_{\beta\kappa} = 0. \tag{11.24}$$

If $\kappa \neq \kappa'$, $\omega_\kappa^2 - \omega_{\kappa'}^2 \neq 0$ (except for degeneracy), so (11.24) gives the orthonormalization condition

$$\sum_\beta M_\beta v_{\beta,-\kappa'} v_{\beta\kappa} = M_C \delta_{\kappa\kappa'}, \tag{11.25}$$

arbitrarily normalized to M_C. This can also be chosen to hold regardless of degeneracies among the frequencies ω_κ. The orthonormalization (11.25) is the same as (11.11) obtained in the transformation method by requiring the kinetic energy to be diagonal; furthermore, with (11.25) and the eigenvalue–eigenvector equation (11.21), the diagonalization condition (11.12) of the transformation method may be reconstructed.

Our results are easily written down in the complete notation of lattice dynamics; for example,

$$\Phi_{\beta\beta'} = \Phi_{ij}(\mathrm{M}\mu,\mathrm{N}\nu). \tag{11.26}$$

The transformation coefficients $v_{\beta\kappa}$ may be based on either of two different phase factors, namely $e^{i\mathbf{k}\cdot\mathbf{R}(\mathrm{N}\nu)}$ or $e^{i\mathbf{k}\cdot\mathbf{R}(\mathrm{N})}$; the first is more physical since it describes the phase variation within each unit cell, while the second is more convenient mathematically. Nevertheless, the final solutions for the motion of the ions are identical in each case, the difference in phase factors being taken up by the eigenvectors.

For the first phase factor the transformation coefficients are

$$v_{\beta\kappa} = N_0^{-1/2} e^{i\mathbf{k}\cdot\mathbf{R}(\mathrm{N}\nu)} v_j(\nu,\mathbf{k}s); \qquad \beta = \mathrm{N}\nu j, \quad \kappa = \mathbf{k}s. \tag{11.27}$$

The orthonormalization (11.11) is then written

$$N_0^{-1} \sum_{\mathrm{N}\nu} \sum_j M_\nu e^{i(\mathbf{k}+\mathbf{k}')\cdot\mathbf{R}(\mathrm{N}\nu)} v_j(\nu,\mathbf{k}s) v_j(\nu,\mathbf{k}'s') = M_C \delta(\mathbf{k}+\mathbf{k}')\delta_{ss'}. \tag{11.28}$$

The left-hand side contains $N_0^{-1} \Sigma_{\mathrm{N}} e^{i(\mathbf{k}+\mathbf{k}')\cdot\mathbf{R}(\mathrm{N})} = \delta(\mathbf{k}+\mathbf{k}')$, according to (10.22), and with this the factor $e^{i(\mathbf{k}+\mathbf{k}')\cdot\mathbf{R}(\nu)} = 1$ and the orthonormalization becomes

$$\sum_{\nu j} M_\nu v_j(\nu,\mathbf{k}s) v_j(\nu,-\mathbf{k}s') = M_C \delta_{ss'}. \tag{11.29}$$

The diagonalization (11.12) is written

$$\begin{aligned} N_0^{-1} \sum_{\mathrm{NP}\nu\pi} \sum_{ij} \Phi_{ij}(\mathrm{N}\nu,\mathrm{P}\pi) e^{i[\mathbf{k}\cdot\mathbf{R}(\mathrm{N}\nu)+\mathbf{k}'\cdot\mathbf{R}(\mathrm{P}\pi)]} v_i(\nu,\mathbf{k}s) v_j(\pi,\mathbf{k}'s') \\ = M_C[\omega(\mathbf{k}s)]^2 \delta(\mathbf{k}+\mathbf{k}')\delta_{ss'}. \end{aligned} \tag{11.30}$$

Just as in Section 10, following (10.44), the Σ_{NP} may be rewritten and surface effects eliminated as follows.

$$\begin{aligned} &N_0^{-1} \sum_{\mathrm{NP}} \Phi_{ij}(\mathrm{N}\nu,\mathrm{P}\pi) e^{i[\mathbf{k}\cdot\mathbf{R}(\mathrm{N}\nu)+\mathbf{k}'\cdot\mathbf{R}(\mathrm{P}\pi)]} \\ &= N_0^{-1} \sum_{\mathrm{N}} e^{i(\mathbf{k}+\mathbf{k}')\cdot\mathbf{R}(\mathrm{N})} \sum_{\mathrm{P}} \Phi_{ij}(\mathrm{N}\nu,\mathrm{P}\pi) e^{i\mathbf{k}'\cdot[\mathbf{R}(\mathrm{P}\pi)-\mathbf{R}(\mathrm{N}\nu)]} e^{i(\mathbf{k}+\mathbf{k}')\cdot\mathbf{R}(\nu)} \\ &= \delta(\mathbf{k}+\mathbf{k}') \sum_{\mathrm{P}} \Phi_{ij}(0\nu,\mathrm{P}\pi) e^{i\mathbf{k}'\cdot[\mathbf{R}(\mathrm{P}\pi)-\mathbf{R}(\nu)]}. \end{aligned} \tag{11.31}$$

This brings us to the definition of a new dynamical matrix for the generalized eigenvalue problem, which we will call $\mathbf{G}(\mathbf{k})$; we write the elements as

$$G_{ij}(\nu\pi,\mathbf{k}) = N_0^{-1} \sum_{\mathrm{NP}} \Phi_{ij}(\mathrm{N}\nu,\mathrm{P}\pi) e^{i\mathbf{k}\cdot[\mathbf{R}(\mathrm{P}\pi)-\mathbf{R}(\mathrm{N}\nu)]}, \tag{11.32}$$

or with surface effects eliminated,

$$G_{ij}(\nu\pi,\mathbf{k}) = \sum_{\mathrm{P}} \Phi_{ij}(0\nu,\mathrm{P}\pi) e^{i\mathbf{k}\cdot[\mathbf{R}(\mathrm{P}\pi)-\mathbf{R}(\nu)]}. \tag{11.33}$$

Then with (11.31) and (11.33), the diagonalization of Φ_2 given by (11.30) reduces to the diagonalization of the dynamical matrices:

$$\sum_{vi\pi j} v_i(\nu,-\mathbf{k}s)G_{ij}(\nu\pi,\mathbf{k})v_j(\pi,\mathbf{k}s') = M_C[\omega(\mathbf{k}s)]^2\delta_{ss'}. \tag{11.34}$$

Now the transformation coefficients $v_i(\nu,\mathbf{k}s)$ are completely determined by the orthonormalization (11.29) and the diagonalization (11.34), and just as in Section 10 the following relations are obvious [see (10.31), (10.33), (10.37), and (10.38)].

$$\mathbf{G}(-\mathbf{k}) = \mathbf{G}(\mathbf{k})^*, \tag{11.35}$$

$$\mathbf{G}(\mathbf{k})^\dagger = \mathbf{G}(\mathbf{k}), \tag{11.36}$$

$$\mathbf{v}(-\mathbf{k}) = \mathbf{v}(\mathbf{k})^*, \tag{11.37}$$

$$[\omega(-\mathbf{k}s)]^2 = [\omega(\mathbf{k}s)]^2. \tag{11.38}$$

The equivalent statement of the problem as an eigenvalue–eigenvector equation, or an equation of motion, is written out from (11.21) and reduced to the form

$$\sum_{\pi j} G_{ij}(\nu\pi,\mathbf{k})v_j(\pi,\mathbf{k}s) = M_\nu[\omega(\mathbf{k}s)]^2 v_i(\nu,\mathbf{k}s). \tag{11.39}$$

The completeness relation (11.15) becomes

$$M_\nu \sum_{s} v_i(\nu,-\mathbf{k}s)v_j(\pi,\mathbf{k}s) = M_C\delta_{\nu\pi}\delta_{ij}. \tag{11.40}$$

With the second phase factor $e^{i\mathbf{k}\cdot\mathbf{R}(\mathrm{N})}$ instead of $e^{i\mathbf{k}\cdot\mathbf{R}(\mathrm{N}\nu)}$, the transformation coefficients are

$$v_{\beta\kappa} = N_0^{-1/2}e^{i\mathbf{k}\cdot\mathbf{R}(\mathrm{N})}\bar{v}_j(\nu,\mathbf{k}s); \qquad \beta = \mathrm{N}\nu j, \quad \kappa = \mathbf{k}s. \tag{11.41}$$

Comparison with (11.27) gives

$$\bar{v}_j(\nu,\mathbf{k}s) = e^{i\mathbf{k}\cdot\mathbf{R}(\nu)}v_j(\nu,\mathbf{k}s). \tag{11.42}$$

The dynamical matrix $\mathbf{G}(\mathbf{k})$ is correspondingly replaced by $\overline{\mathbf{G}}(\mathbf{k})$, where

$$\overline{\mathbf{G}}_{ij}(\nu\pi,\mathbf{k}) = N_0^{-1}\sum_{\mathrm{NP}}\Phi_{ij}(\mathrm{N}\nu,\mathrm{P}\pi)e^{i\mathbf{k}\cdot[\mathbf{R}(\mathrm{P})-\mathbf{R}(\mathrm{N})]}. \tag{11.43}$$

The preceding results still hold as follows.
Diagonalization:

$$\sum_{vi\pi j} \bar{v}_i(\nu,-\mathbf{k}s)\overline{\mathbf{G}}_{ij}(\nu\pi,\mathbf{k})\bar{v}_j(\pi,\mathbf{k}s') = M_C[\omega(\mathbf{k}s)]^2\delta_{ss'}; \tag{11.44}$$

Orthonormality:

$$\sum_{vj} M_\nu\bar{v}_j(\nu,-\mathbf{k}s)\bar{v}_j(\nu,\mathbf{k}s') = M_C\delta_{ss'}, \tag{11.45}$$

Completeness:
$$M_\nu \sum_s \bar{v}_i(\nu,-\mathbf{k}s)\bar{v}_j(\pi,\mathbf{k}s) = M_C \delta_{\nu\pi}\delta_{ij}. \tag{11.46}$$

This new phase factor corresponds to writing the ion displacements during a homogeneous deformation by (7.90), whereas the first phase factor $e^{i\mathbf{k}\cdot\mathbf{R}(\mathbf{N}\nu)}$ corresponds to the equation before (7.90).

Ordinary Eigenvalue Problem

A simple redefinition of the transformation diagonalizing $\mathscr{H}_2$ leads to an ordinary eigenvalue problem (diagonalization of one Hermitian matrix) instead of a generalized eigenvalue problem (simultaneous diagonalization of two Hermitian matrices). Let us first look at this situation in a matrix formulation, defining the $3nN_0 \times 3nN_0$ matrices $\mathbf{T}$ and $\mathbf{\Phi}$ by

$$\mathbf{T} = [T_{\beta\beta'}], \qquad \mathbf{\Phi} = [\Phi_{\beta\beta'}]. \tag{11.47}$$

The generalized eigenvalue problem (11.21) is

$$\mathbf{\Phi}\mathbf{v}_\kappa = \omega_\kappa^2 \mathbf{T}\mathbf{v}_\kappa. \tag{11.48}$$

There are several ways to transform this to an ordinary eigenvalue problem, and we take the procedure which gives the results of Section 10; $\mathbf{T}$ is real symmetric (in fact diagonal) and is nonsingular, so $\mathbf{T}^{-1}$, $\mathbf{T}^{1/2}$, and $\mathbf{T}^{-1/2}$ exist and are real symmetric. Multiplying (11.48) by $\mathbf{T}^{-1/2}$ and inserting $\mathbf{T}^{-1/2}\mathbf{T}^{1/2} = 1$, gives

$$(\mathbf{T}^{-1/2}\mathbf{\Phi}\mathbf{T}^{-1/2})\mathbf{T}^{1/2}\mathbf{v}_\kappa = \omega_\kappa^2 \mathbf{T}^{1/2}\mathbf{v}_\kappa. \tag{11.49}$$

Defining the real symmetric matrix $\boldsymbol{\chi}$ and the eigenvectors $\mathbf{w}_\kappa$ by

$$\boldsymbol{\chi} = \mathbf{T}^{-1/2}\mathbf{\Phi}\mathbf{T}^{-1/2}, \tag{11.50}$$

$$\mathbf{w}_\kappa = \mathbf{T}^{1/2}\mathbf{v}_\kappa, \tag{11.51}$$

then (11.49) becomes the ordinary eigenvalue problem

$$\boldsymbol{\chi}\mathbf{w}_\kappa = \omega_\kappa^2 \mathbf{w}_\kappa. \tag{11.52}$$

The important result is that the frequencies ω_κ are the same in the generalized problem (11.48) and the ordinary problem (11.52); the normalization of the $\mathbf{w}_\kappa$ will be chosen as unity in keeping with Section 10.

In other words, the kinetic energy is diagonal but not a multiple of the unit matrix, and by simply including the ion masses in the linear transformations the kinetic energy becomes in essence a unit matrix. Let us do this explicitly by going back to the transformations (11.7) and (11.8) and writing them (with new q_κ and $\dot{q}_\kappa$) as

$$U_\beta = M_\beta^{-1/2} \sum_\kappa w_{\beta\kappa} q_\kappa, \tag{11.53}$$

$$\dot{U}_\beta = M_\beta^{-1/2} \sum_\kappa w_{\beta\kappa} \dot{q}_\kappa. \tag{11.54}$$

It is convenient (but not necessary) to take the $w_{\beta\kappa}$ as dimensionless, so the q_κ are generalized coordinates of dimension $M_\beta^{1/2}U_\beta$. The ordinary eigenvalue problem (11.52) is

$$\sum_{\beta'} (M_\beta M_{\beta'})^{-1/2}\Phi_{\beta\beta'}w_{\beta'\kappa} = \omega_\kappa^2 w_{\beta\kappa}. \tag{11.55}$$

Since $(M_\beta M_{\beta'})^{-1/2}\Phi_{\beta\beta'}$ is real symmetric, the $w_{\beta\kappa}$ are unitary (or satisfy the orthonormalization),

$$\sum_{\beta} w_{\beta\kappa}^* w_{\beta\kappa'} = \delta_{\kappa\kappa'}; \tag{11.56}$$

with this the eigenvalue–eigenvector equation (11.55) is transformed to diagonal form

$$\sum_{\beta\beta'} w_{\beta\kappa}^*(M_\beta M_{\beta'})^{-1/2}\Phi_{\beta\beta'}w_{\beta'\kappa'} = \omega_\kappa^2\delta_{\kappa\kappa'}, \tag{11.57}$$

and the completeness relation is

$$\sum_{\kappa} w_{\beta\kappa}^* w_{\beta'\kappa} = \delta_{\beta\beta'}. \tag{11.58}$$

For the lattice-vibration problem it is always possible to choose

$$w_{\beta,-\kappa} = w_{\beta\kappa}^*, \tag{11.59}$$

and this implies $q_{-\kappa} = q_\kappa^*$, $\dot{q}_{-\kappa} = \dot{q}_\kappa^*$. Finally with (11.53)–(11.59), the Hamiltonian $\mathscr{H}_2$ of (11.2)–(11.4) is again a sum of independent harmonic oscillator Hamiltonians

$$\mathscr{H}_2 = \tfrac{1}{2}\sum_{\kappa}\{|p_\kappa|^2 + \omega_\kappa^2\,|q_\kappa|^2\},$$

where $p_\kappa = \dot{q}_{-\kappa}$ is the generalized momentum conjugate to q_κ.

If now the eigenvectors $w_{\beta\kappa}$ are chosen as

$$w_{\beta\kappa} = N_0^{-1/2}e^{i\mathbf{k}\cdot\mathbf{R}(\mathrm{N}\nu)}w_j(\nu,\mathbf{k}s); \qquad \beta = \mathrm{N}\nu j, \quad \kappa = \mathbf{k}s, \tag{11.60}$$

then all the results of Section 10 are reproduced. In particular the diagonalization (11.57) reduces to (10.34) for diagonalization of the dynamical matrices, the orthonormality (11.56) reduces to (10.35), and the completeness (11.58) reduces to (10.36). It is also useful to be able to work with the equation of motion (or eigenvalue–eigenvector equation); from (11.55) this reduces to

$$\sum_{\pi j} D_{ij}(\nu\pi,\mathbf{k})w_j(\pi,\mathbf{k}s) = [\omega(\mathbf{k}s)]^2 w_i(\nu,\mathbf{k}s), \tag{11.61}$$

where the dynamical matrix is the same as (10.29).

Once again it is possible to choose an alternate phase factor $e^{i\mathbf{k}\cdot\mathbf{R}(\mathrm{N})}$ in place of $e^{i\mathbf{k}\cdot\mathbf{R}(\mathrm{N}\nu)}$. With this new phase factor the eigenvector components $w_{\beta\kappa}$ are

$$w_{\beta\kappa} = N_0^{-1/2}e^{i\mathbf{k}\cdot\mathbf{R}(\mathrm{N})}\bar{w}_j(\nu,\mathbf{k}s); \qquad \beta = \mathrm{N}\nu j, \quad \kappa = \mathbf{k}s, \tag{11.62}$$

and comparison with (11.60) gives

$$\bar{w}_j(\nu,\mathbf{k}s) = e^{i\mathbf{k}\cdot\mathbf{R}(\nu)} w_j(\nu,\mathbf{k}s). \tag{11.63}$$

The dynamical matrices appearing in the diagonalization problem are now $\bar{\mathbf{D}}(\mathbf{k})$, and their elements are

$$\bar{D}_{ij}(\nu\pi,\mathbf{k}) = (M_\nu M_\pi)^{-1}\, {}^2N_0^{-1} \sum_{\mathrm{NP}} \Phi_{ij}(\mathrm{N}\nu,\mathrm{P}\pi) e^{i\mathbf{k}\cdot[\mathbf{R}(\mathrm{P})-\mathbf{R}(\mathrm{N})]}; \tag{11.64}$$

this form is particularly useful for calculating strain derivatives of the phonon frequencies $\omega(\mathbf{k}s)$, since $\mathbf{k}\cdot\mathbf{R}(\mathrm{M})$ is invariant under homogeneous deformation of the crystal. The diagonalization (10.34), orthonormality (10.35), completeness (10.36), and the eigenvalue–eigenvector equation (11.61) all hold with $\mathbf{D}(\mathbf{k})$ and $\mathbf{w}(\mathbf{k}s)$ replaced, respectively, by $\bar{\mathbf{D}}(\mathbf{k})$ and $\bar{\mathbf{w}}(\mathbf{k}s)$; the eigenvalues $[\omega(\mathbf{k}s)]^2$ are of course unchanged. In addition, if the entire Hamiltonian is transformed with the new phase factors, i.e., by using (11.62) instead of (11.60), the final results are identical to those of Section 10. In particular, $\mathscr{H}$ is still given by (10.84)–(10.87) in terms of phonon creation and annihilation operators, $\Phi_{\kappa\kappa'\kappa''}$ and $\Phi_{\kappa\kappa'\kappa''\kappa'''}$ are still given by (10.74) and (10.75) with each factor $e^{i\mathbf{k}\cdot\mathbf{R}(\mu)} w_i(\mu,\mathbf{k}s)$ being replaced by $\bar{w}_i(\mu,\mathbf{k}s)$, which is the same according to (11.63).

Comparison of (10.29) for $\mathbf{D}(\mathbf{k})$ and (11.64) for $\bar{\mathbf{D}}(\mathbf{k})$ shows that these matrices are unitarily equivalent, since

$$D_{ij}(\nu\pi,\mathbf{k}) = e^{-i\mathbf{k}\cdot[\mathbf{R}(\nu)-\mathbf{R}(\pi)]}\, \bar{D}_{ij}(\nu\pi,\mathbf{k}), \tag{11.65}$$

or

$$\mathbf{D}(\mathbf{k}) = \mathbf{S}(\mathbf{k})^\dagger \bar{\mathbf{D}}(\mathbf{k}) \mathbf{S}(\mathbf{k}), \tag{11.66}$$

where

$$S_{ij}(\nu\pi,\mathbf{k}) = e^{i\mathbf{k}\cdot\mathbf{R}(\nu)} \delta_{\nu\pi}\delta_{ij}. \tag{11.67}$$

It follows that $\mathbf{S}(\mathbf{k})$ is unitary, i.e., $\mathbf{S}(\mathbf{k})^\dagger = \mathbf{S}(\mathbf{k})^{-1}$, since

$$[\mathbf{S}(\mathbf{k})^\dagger \mathbf{S}(\mathbf{k})]_{\mu i,\pi l} = \sum_{\nu j} S_{ji}(\nu\mu,\mathbf{k})^* S_{jl}(\nu\pi,\mathbf{k}) = \delta_{\mu\pi}\delta_{il}. \tag{11.68}$$

Finally we observe that one can easily transform between the generalized and the ordinary eigenvalue-problem formulations in any of the subsequent theory. The eigenvectors are related by (11.51) except for normalization; since $w_{\beta\kappa}$ are normalized to unity and $v_{\beta\kappa}$ to M_C, the correct relation is $w_{\beta\kappa} = (M_\beta/M_C)^{1/2} v_{\beta\kappa}$, or writing this out

$$w_i(\nu,\mathbf{k}s) = (M_\nu/M_C)^{1/2} v_i(\nu,\mathbf{k}s). \tag{11.69}$$

With this relation it is seen that the ordinary and generalized orthonormality equations, (10.35) and (11.29), respectively, are equivalent, and so are the ordinary and generalized completeness relations, (10.36) and (11.40), respectively. Furthermore from (10.29) for $\mathbf{D}(\mathbf{k})$ and (11.32) for $\mathbf{G}(\mathbf{k})$ it follows that

$$D_{ij}(\nu\pi,\mathbf{k}) = (M_\nu M_\pi)^{-1/2} G_{ij}(\nu\pi,\mathbf{k}), \tag{11.70}$$

and then the ordinary diagonalization (10.34) and the generalized diagonalization (11.34) are also equivalent. If the generalized eigenvalue problem formulation is used, the complete Hamiltonian is again identical with (10.84)–(10.87), with each factor $M_\mu^{-1/2} w_i(\mu,\mathbf{k}s)$ in $\Phi_{\kappa\kappa'\kappa''}$ and $\Phi_{\kappa\kappa'\kappa''\kappa'''}$ being replaced by $M_C^{-1/2} v_i(\mu,\mathbf{k}s)$, which is the same according to (11.69).

Eigenvalue Derivatives

Here we derive the expression for the first derivative of an eigenvalue with respect to an arbitrary variable, and apply the result to the calculation of the phonon Grüneisen parameters. Consider an ordinary eigenvalue problem corresponding to a Hermitian matrix $\mathbf{D} = [D_{\alpha\alpha'}]$ whose eigenvectors and eigenvalues are $\mathbf{w}_s$ and ω_s, respectively, where s has the same number of values as α. Let the matrix elements $D_{\alpha\alpha'}$ be dependent on a parameter x and make an expansion about x_0 as follows.

$$D_{\alpha\alpha'}(x) = D_{\alpha\alpha'}(x_0) + (x - x_0)(\partial D_{\alpha\alpha'}/\partial x)_0 + \cdots, \tag{11.71}$$

where the partial derivative means other parameters are to be held constant, and $(\partial D_{\alpha\alpha'}/\partial x)_0$ is evaluated at x_0. If the eigenvectors are normalized to unity, the perturbation expansion of the eigenvalues is

$$\omega_s(x) = \omega_s(x_0) + \sum_{\alpha\alpha'} w_{\alpha s}^*(x - x_0)(\partial D_{\alpha\alpha'}/\partial x)_0 w_{\alpha' s} + \cdots. \tag{11.72}$$

The derivative of $\omega_s(x)$ with respect to x evaluated at x_0, denoted $(\partial\omega_s/\partial x)_0$, is given exactly by differentiating the first-order term in (11.72), since higher-order terms vanish at x_0:

$$(\partial\omega_s/\partial x)_0 = \sum_{\alpha\alpha'} w_{\alpha s}^*(\partial D_{\alpha\alpha'}/\partial x)_0 w_{\alpha' s}. \tag{11.73}$$

Note that we have used nothing more than first-order perturbation theory. In other contexts the result (11.73) is known as the Hellmann–Feynman theorem.* The second derivative $(\partial^2\omega_s/\partial x^2)_0$ can of course be obtained from second-order perturbation theory.

The result (11.73) provides us with a very useful and practical way of computing the phonon Grüneisen parameters; in fact when a detailed model calculation is being carried out, these parameters may be obtained as easily as the phonon frequencies themselves. Let us define generalized phonon Grüneisen parameters by

$$\gamma_{ij}(\mathbf{k}s) = -[\omega(\mathbf{k}s)]^{-1}[\partial\omega(\mathbf{k}s)/\partial\eta_{ij}]_{\eta'}, \tag{11.74}$$

* H. Hellmann, *Einfuhrung in die Quantenchemie*, Franz Deuticke, Leipzig, 1937; R. P. Feynman, *Phys. Rev.* **56**, 340 (1939).

where as usual the η_{ij} are Lagrangian strain parameters. Since the eigenvalues of the dynamical matrix are $[\omega(\mathbf{k}s)]^2$, it is convenient to rewrite (11.74) as

$$\gamma_{ij}(\mathbf{k}s) = -\frac{1}{2[\omega(\mathbf{k}s)]^2}\frac{\partial[\omega(\mathbf{k}s)]^2}{\partial\eta_{ij}}. \tag{11.75}$$

Further, it is convenient to calculate these derivatives from the dynamical matrices $\bar{\mathbf{D}}(\mathbf{k})$ rather than $\mathbf{D}(\mathbf{k})$, since as we mentioned previously the phase factor in $\bar{\mathbf{D}}(\mathbf{k})$ does not depend on the η_{ij}. Since the eigenvalues and normalized eigenvectors of $\bar{\mathbf{D}}(\mathbf{k})$ are $[\omega(\mathbf{k}s)]^2$ and $\bar{\mathbf{w}}(\mathbf{k}s)$, respectively, it follows from (11.73) and (11.75) that

$$\gamma_{kl}(\mathbf{k}s) = -\tfrac{1}{2}[\omega(\mathbf{k}s)]^{-2}\sum_{\nu i\pi j}\bar{w}_i(\nu,\mathbf{k}s)^*[\partial\bar{D}_{ij}(\nu\pi,\mathbf{k})/\partial\eta_{kl}]_{\eta'}\bar{w}_j(\pi,\mathbf{k}s). \tag{11.76}$$

Also from (11.64) for $\bar{\mathbf{D}}(\mathbf{k})$,

$$[\partial\bar{D}_{ij}(\nu\pi,\mathbf{k})/\partial\eta_{kl}]_{\eta'} = (M_\nu M_\pi)^{-1/2}N_0^{-1}\sum_{\mathrm{NP}}[\partial\Phi_{ij}(\mathrm{N}\nu,\mathrm{P}\pi)/\partial\eta_{kl}]_{\eta'}e^{i\mathbf{k}\cdot[\mathbf{R}(\mathrm{P})-\mathbf{R}(\mathrm{N})]}. \tag{11.77}$$

For computational purposes it is convenient to eliminate surface effects from (11.77) by setting $\mathrm{N}=0$, $\mathbf{R}(\mathrm{N})=0$, and replacing $N_0^{-1}\Sigma_{\mathrm{N}}$ by 1. Ordinarily the strain derivative of $\Phi_{ij}(0\nu,\mathrm{P}\pi)$ can be evaluated directly for a given model; a formal expression may be obtained in terms of third-order potential energy coefficients. The general equation for variation of second-order potential energy coefficients with ion displacements is (6.19), and the first-order ion displacements corresponding to a homogeneous deformation of the crystal are abbreviated in (7.84). It follows from these equations that

$$[\partial\Phi_{ij}(0\nu,\mathrm{P}\pi)/\partial\eta_{kl}]_{\eta'} = \tfrac{1}{2}\sum_{Q\rho}\sum_{m}\Phi_{ijm}(0\nu,\mathrm{P}\pi,\mathrm{Q}\rho)[Y_{m,kl}(\mathrm{Q}\rho)+Y_{m,lk}(\mathrm{Q}\rho)], \tag{11.78}$$

where $Y_{m,kl}(\mathrm{Q}\rho)$ are given by (7.85). In any case the evaluation of the dynamical matrix strain derivatives (11.77) is essentially no more difficult than evaluation of the same kind of lattice sums for the dynamical matrix.

For cubic crystals which are constrained to remain cubic, e.g., by limiting the applied stress to arbitrary isotropic pressure and allowing for arbitrary temperature, the only macroscopic strain variable is the volume, and the appropriate phonon Grüneisen parameters are the volume derivatives,

$$\gamma(\mathbf{k}s) = -[V/\omega(\mathbf{k}s)][d\omega(\mathbf{k}s)/dV]. \tag{11.79}$$

These parameters are also important for a crystal of any symmetry, in, e.g., the theory of volume thermal expansion, where the volume variation is not unique and has to be specified in terms of strain variations η_{ij}. In any case (11.79) is given by (11.76) and (11.77) with $\partial/\partial\eta_{kl}$ replaced by $V d/dV$.

Detailed lattice dynamical expressions for the first- and second-strain derivatives of the phonon frequencies are derived in Section 17.

12. LONG-WAVELENGTH ACOUSTIC PHONONS

Primitive Lattice

Acoustic phonons are defined as those for which the frequency $\omega(\mathbf{k}s)$ goes to zero as $\mathbf{k} \to 0$. For each direction of $\mathbf{k}$ there are three acoustic phonon branches, denoted by $s = 1,2,3$ or $s =$ acoustic. The object of the present section is to find the harmonic acoustic phonon frequencies in the long-wavelength (small $\mathbf{k}$) region, and compare these to the frequencies of the mechanical sound waves discussed in Section 7. The procedure is called the method of long waves, and consists of expanding the phonon eigenvalue-eigenvector equation for small $|\mathbf{k}|$ along a fixed direction of $\mathbf{k}$ and solving for $[\omega(\mathbf{k}s)]^2$ to second order in $|\mathbf{k}|$.* Because of the inversion symmetry of the primitive lattice, the term of order $|\mathbf{k}|$ vanishes and the leading contribution to $[\omega(\mathbf{k}s)]^2$ is of order $|\mathbf{k}|^2$. Then in the long-wavelength limit $\omega(\mathbf{k}s) \propto |\mathbf{k}|$ for acoustic phonons, and it will be shown that the velocities of these long-wavelength phonons are the same as the mechanical sound-wave velocities. In the process an important relation is found between long-wavelength acoustic phonon eigenvectors and the sublattice displacements which occur during a homogeneous deformation of the crystal. The calculations are valid for a crystal in an arbitrary initial equilibrium configuration, corresponding to arbitrary uniform stresses applied to the crystal surface. In addition the elimination of surface effects is essentially simpler in the method of long waves than in the method of homogeneous deformation, since the macroscopic configuration of the crystal is fixed (by the periodic boundary conditions) and surface effects may be eliminated from the dynamical matrices at the start of the calculation.

We begin with the trivial case of a primitive lattice. The dynamical matrices with surface effects eliminated are written from (10.39),

$$D_{ij}(\mathbf{k}) = M_C^{-1} \sum_{\mathrm{N}} \Phi_{ij}(0,\mathrm{N}) e^{i\mathbf{k}\cdot\mathbf{R}(\mathrm{N})}, \tag{12.1}$$

where M_C is the mass of one ion and also the mass of one unit cell (neglecting electron compared to nuclear masses). For a primitive lattice, $D(\mathbf{k})$ is real, since for every N in Σ_{N} there is a $-\mathrm{N}$ for which $\mathbf{R}(-\mathrm{N}) = -\mathbf{R}(\mathrm{N})$ and

* M. Born and K. Huang, *Dynamical Theory of Crystal Lattices*, Clarendon Press, Oxford, 1954.

$\Phi_{ij}(0,-\mathrm{N}) = \Phi_{ij}(0,\mathrm{N})$, according to the lattice vector translation symmetry (6.64). Therefore (12.1) may be written

$$D_{ij}(\mathbf{k}) = M_C^{-1} \sum_{\mathrm{N}} \Phi_{ij}(0,\mathrm{N})\cos[\mathbf{k}\cdot\mathbf{R}(\mathrm{N})], \tag{12.2}$$

and because of the translational invariance condition (6.25) which requires $\Sigma_{\mathrm{N}}\, \Phi_{ij}(0,\mathrm{N}) = 0$, this may also be written

$$D_{ij}(\mathbf{k}) = M_C^{-1} \sum_{\mathrm{N}}{}' \Phi_{ij}(0,\mathrm{N})\{\cos[\mathbf{k}\cdot\mathbf{R}(\mathrm{N})] - 1\}. \tag{12.3}$$

This last is a particularly useful form for computational purposes. Since according to (10.33) $\mathbf{D}(\mathbf{k})$ is Hermitian, and since it is real for a primitive lattice it must also be symmetric; the symmetry follows also from (6.66) for the symmetry of $\Phi_{ij}(\mathrm{M},\mathrm{N})$ in its Cartesian indices. Hence

$$\mathbf{D}(\mathbf{k}) = \text{real symmetric } 3 \times 3, \tag{12.4}$$

for a primitive lattice. For each $\mathbf{k}$ there are three phonons $s = 1,2,3$, and the eigenvalue–eigenvector equation (11.61) is

$$\sum_j D_{ij}(\mathbf{k})w_j(\mathbf{k}s) = [\omega(\mathbf{k}s)]^2 w_i(\mathbf{k}s). \tag{12.5}$$

The eigenvectors $\mathbf{w}(\mathbf{k}s)$ can be chosen real,

$$w_i(\mathbf{k}s) = w_i(\mathbf{k}s)^* = w_i(-\mathbf{k}s), \tag{12.6}$$

and the orthonormality and completeness relations (10.35) and (10.36) simplify for a primitive lattice to

$$\sum_i w_i(\mathbf{k}s)w_i(\mathbf{k}s') = \delta_{ss'}, \quad \text{orthonormality;} \tag{12.7}$$

$$\sum_s w_i(\mathbf{k}s)w_j(\mathbf{k}s) = \delta_{ij}, \quad \text{completeness.} \tag{12.8}$$

Expansion of the dynamical matrix (12.3) in powers of $\mathbf{k}$ gives

$$D_{ij}(\mathbf{k}) = -\tfrac{1}{2}M_C^{-1} \sum_{\mathrm{N}}{}' \Phi_{ij}(0,\mathrm{N})[\mathbf{k}\cdot\mathbf{R}(\mathrm{N})]^2 + \cdots, \tag{12.9}$$

where $+\cdots$ indicates terms of order $\mathbf{k}^4$ and so on. This shows immediately that $[\omega(\mathbf{k}s)]^2$ is of leading order $\mathbf{k}^2$, and hence all branches are acoustic for a primitive lattice. Let us rewrite the eigenvalue–eigenvector equation (12.5), to leading order $\mathbf{k}^2$ with (12.9) and with $\mathbf{k}\cdot\mathbf{R}(\mathrm{N})$ written out as a Cartesian index sum, to obtain

$$M_C[\omega(\mathbf{k}s)]^2 w_i(\mathbf{k}s) = -\tfrac{1}{2} \sum_{\mathrm{N}}{}' \sum_{jmn} \Phi_{ij}(0,\mathrm{N})R_m(\mathrm{N})R_n(\mathrm{N})k_m k_n w_j(\mathbf{k}s). \tag{12.10}$$

Now the Σ_{N}' can be replaced by Σ_{N} because $\Phi_{ij}(0,0)$ is presumably finite while $\mathbf{R}(0) = 0$, and then with (7.22) for $\hat{A}_{ikjl}$,

$$-\tfrac{1}{2}\sum_{\mathrm{N}}{}' \Phi_{ij}(0,\mathrm{N})R_m(\mathrm{N})R_n(\mathrm{N}) = V_C \hat{A}_{ijmn}.$$

Then (12.10) is, for each direction of $\mathbf{k}$ and to leading order in $|\mathbf{k}|$,

$$M_C[\omega(\mathbf{k}s)]^2 w_i(\mathbf{k}s) = \sum_{jkl} V_C \hat{A}_{ikjl} k_j k_l w_k(\mathbf{k}s), \tag{12.11}$$

which is identical to the eigenvalue–eigenvector equation (7.99) for long-wavelength mechanical sound waves.

Nonprimitive Lattice

The long-waves calculation is a good deal more complicated for nonprimitive lattices than for primitive lattices, the complications being associated with the sublattice displacements. Consider a fixed direction of $\mathbf{k}$ and for small $\mathbf{k}$ let

$$\hat{\mathbf{k}} = \mathbf{k}/|\mathbf{k}|; \qquad \lambda = |\mathbf{k}|. \tag{12.12}$$

Here $\hat{\mathbf{k}}$ is a dimensionless unit vector and λ is a variable small parameter with dimension the same as $\mathbf{k}$. The acoustic phonon frequencies and eigenvectors may be formally expanded for small λ as

$$\omega(\mathbf{k}s) = 0 + \lambda\omega^{(1)}(s) + \cdots, \tag{12.13}$$

$$w_i(\nu,\mathbf{k}s) = w_i^{(0)}(\nu,s) + i\lambda w_i^{(1)}(\nu,s) + \tfrac{1}{2}\lambda^2 w_i^{(2)}(\nu,s) + \cdots. \tag{12.14}$$

In (12.13) the frequencies are *assumed* to go to zero as $\lambda \to 0$, and we will see if such solutions can be found. The dynamical matrix with surface effects eliminated is given by (10.39), and this may also be expanded in powers of λ. When this is done the eigenvalue–eigenvector equation (11.61) may be written as a power series in λ; keeping all terms which will contribute to order λ^2 the result is

$$\begin{aligned}\sum_{\mathrm{P}}\sum_{\pi j}(M_\nu M_\pi)^{-1/2}\Phi_{ij}(0\nu,\mathrm{P}\pi)\{1 &+ i\lambda\hat{\mathbf{k}}\cdot[\mathbf{R}(P\pi) - \mathbf{R}(\nu)] \\ &- \tfrac{1}{2}\lambda^2[\hat{\mathbf{k}}\cdot[\mathbf{R}(\mathrm{P}\pi) - \mathbf{R}(\nu)]]^2\} \\ \times\ \{w_j^{(0)}(\pi,s) &+ i\lambda w_j^{(1)}(\pi,s) + \tfrac{1}{2}\lambda^2 w_j^{(2)}(\pi,s)\} \\ &= \lambda^2[\omega^{(1)}(s)]^2 w_i^{(0)}(\nu,s).\end{aligned} \tag{12.15}$$

We recall from Section 11 that the eigenvalues and eigenvectors (except for normalization) are completely determined by this equation; the object is to solve the problem to order λ^2 so as to determine the first-order frequencies $\omega^{(1)}(s)$ appearing on the right-hand side.

The zeroth-order equation, i.e., terms in (12.15) of zeroth order in λ, is

$$\sum_{\mathrm{P}}\sum_{\pi j}(M_\nu M_\pi)^{-1/2}\Phi_{ij}(0\nu,\mathrm{P}\pi)w_j^{(0)}(\pi,s) = 0. \tag{12.16}$$

This equation has three independent nonvanishing solutions

$$w_j^{(0)}(\pi,s) = (M_\pi/M_C)^{1/2}y_j(s), \qquad s = 1,2,3, \tag{12.17}$$

where the $\mathbf{y}(s)$ are three mutually orthogonal dimensionless unit vectors. To show these are solutions, use them in (12.16) to obtain

$$(M_C M_\nu)^{-1/2} \sum_j y_j(s) \sum_{P\pi} \Phi_{ij}(0\nu,P\pi) = 0, \tag{12.18}$$

and this is satisfied since the $\Sigma_{P\pi}$ vanishes by the translational invariance condition (6.69) evaluated in the interior [compare (7.39)]. Since there are only three mutually orthogonal vectors in Cartesian space, there are only three acoustic branches. The solutions (12.17) satisfy the zeroth order orthonormality [from (10.35)] since

$$\sum_{\pi j} w_j^{(0)}(\pi,s)w_j^{(0)}(\pi,s') = \sum_\pi (M_\pi/M_C) \sum_j y_j(s)y_j(s') = \delta_{ss'}, \tag{12.19}$$

but of course they are not complete [according to (10.36)] since there are only 3 out of the set of $3n$ eigenvectors. The orientation of the vector set $\mathbf{y}(s)$ is arbitrary until the second-order equation is solved.

The terms in (12.15) of first order in λ are

$$\sum_{P\pi} \sum_j (M_\nu M_\pi)^{-1/2}\Phi_{ij}(0\nu,P\pi)\{i\lambda w_j^{(1)}(\pi,s) + i\lambda\hat{\mathbf{k}} \cdot [\mathbf{R}(P\pi) - \mathbf{R}(\nu)]w_j^{(0)}(\pi,s)\} = 0. \tag{12.20}$$

Now we divide out $i\lambda M_\nu^{-1/2}$, put in the zeroth-order solution (12.17), multiply through by $M_C^{1/2}$, and write this as a set of inhomogeneous equations for $w_j^{(1)}(\pi,s)$:

$$\begin{aligned}\sum_{P\pi} \sum_j (M_C/M_\pi)^{1/2}\Phi_{ij}(0\nu,P\pi)w_j^{(1)}(\pi,s) \\ = -\sum_{P\pi} \sum_j \Phi_{ij}(0\nu,P\pi)\hat{\mathbf{k}} \cdot [\mathbf{R}(P\pi) - \mathbf{R}(\nu)]y_j(s).\end{aligned} \tag{12.21}$$

On the right-hand side the term in $\mathbf{R}(\nu)$ vanishes by translational invariance, according to (6.69) again; with the abbreviation

$$z_j(\pi,s) = (M_C/M_\pi)^{1/2}w_j^{(1)}(\pi,s) \tag{12.22}$$

in the left-hand side and with $\hat{\mathbf{k}} \cdot \mathbf{R}(P\pi)$ written out as a Cartesian index sum in the right-hand side, the first-order equation (12.21) is

$$\sum_{P\pi} \sum_j \Phi_{ij}(0\nu,P\pi)z_j(\pi,s) = -\sum_{P\pi} \sum_{jl} \Phi_{ij}(0\nu,P\pi)R_l(P\pi)y_j(s)\hat{k}_l. \tag{12.23}$$

This is now exactly the same set of equations for $z_j(\pi,s)$ as the equations (7.38) for the sublattice displacements $S_j(\pi)$, where here the homogeneous strain parameters u_{jl} are replaced by $y_j(s)\hat{k}_l$.

The solution of (12.23) for $z_j(\pi,s)$ proceeds just like the solution of the inhomogeneous equations (7.38). In particular the homogeneous equations, obtained by setting the right-hand side of (12.23) equal to zero, have three independent solutions in the form of three mutually orthogonal vectors $\mathbf{T}(s)$, since with $z_j(\pi,s) = T_j(s)$ the left side of (12.23) vanishes by translational invariance. These particular solutions represent uniform translations of the crystal, and will contribute nothing to subsequent calculations because of translational invariance. The solubility conditions are that the homogeneous solutions $\mathbf{T}(s)$ are orthogonal to the right side of (12.23), and these conditions are satisfied with the neglect of surface effects by (7.44). The solutions for $z_j(\pi,s)$ are then given by (7.51) for each s:

$$z_j(\pi,s) = \sum_{mn} X_{j,mn}(\pi) y_m(s) \hat{k}_n, \tag{12.24}$$

where the particular solution $T_j(s)$ has been dropped. This is the important relation between long-wavelength acoustic phonon eigenvectors and the homogeneous deformation sublattice displacements. Before discussing the physical meaning of this result, let us proceed to the second-order eigenvalue–eigenvector equation.

The terms in (12.15) of second order in λ are as follows.

$$\begin{aligned}\sum_{\mathrm{P}} \sum_{\pi j} (M_\nu M_\pi)^{-1/2} \Phi_{ij}(0\nu,\mathrm{P}\pi) \\ \times \{\tfrac{1}{2}\lambda^2 w_j^{(2)}(\pi,s) - \lambda^2 \hat{\mathbf{k}} \cdot [\mathbf{R}(\mathrm{P}\pi) - \mathbf{R}(\nu)] w_j^{(1)}(\pi,s) \\ - \tfrac{1}{2}\lambda^2 [\hat{\mathbf{k}} \cdot [\mathbf{R}(\mathrm{P}\pi) - \mathbf{R}(\nu)]]^2 w_j^{(0)}(\pi,s)\} \\ = \lambda^2 [\omega^{(1)}(s)]^2 w_i^{(0)}(\nu,s).\end{aligned} \tag{12.25}$$

Equation (12.25) may be considered as a set of inhomogeneous equations for $w_j^{(2)}(\pi,s)$; after dividing out λ^2 it may be written

$$\begin{aligned}\tfrac{1}{2}\sum_{\mathrm{P}} \sum_{\pi j} (M_\nu M_\pi)^{-1/2} \Phi_{ij}(0\nu,\mathrm{P}\pi) w_j^{(2)}(\pi,s) \\ = \sum_{\mathrm{P}} \sum_{\pi j} (M_\nu M_\pi)^{-1/2} \Phi_{ij}(0\nu,\mathrm{P}\pi) \{\hat{\mathbf{k}} \cdot [\mathbf{R}(\mathrm{P}\pi) - \mathbf{R}(\nu)] w_j^{(1)}(\pi,s) \\ + \tfrac{1}{2}[\hat{\mathbf{k}} \cdot [\mathbf{R}(\mathrm{P}\pi) - \mathbf{R}(\nu)]]^2 w_j^{(0)}(\pi,s)\} + [\omega^{(1)}(s)]^2 w_i^{(0)}(\nu,s).\end{aligned} \tag{12.26}$$

Here the homogeneous equations again have three independent nonvanishing solutions of the form $M_\pi^{1/2} T_j(s)$ where $\mathbf{T}(s)$ are arbitrary vectors [since $\Sigma_{\mathrm{P}\pi} \Phi_{ij}(0\nu,\mathrm{P}\pi)$ vanishes by translational invariance], and the condition of solubility for the inhomogeneous equations is that the homogeneous solutions are orthogonal to the inhomogeneous term:

$$\sum_{\nu i} M_\nu^{1/2} T_i(s)[\text{right-hand side of (12.26)}] = 0, \qquad s = 1,2,3. \tag{12.27}$$

Since $T_i(s)$ are arbitrary, the solubility condition (12.27) requires

$$\sum_{\nu} M_\nu^{1/2}[\text{right-hand side of (12.26)}] = 0, \qquad s = 1,2,3. \tag{12.28}$$

Written out in detail, (12.28) is

$$\begin{aligned}\sum_{\nu} M_\nu^{1/2}[\omega^{(1)}(s)]^2 w_i^{(0)}(\nu,s) &= -\sum_{P\nu\pi}\sum_{j} M_\pi^{-1/2}\Phi_{ij}(0\nu,P\pi)\{\hat{\mathbf{k}}\cdot[\mathbf{R}(P\pi)-\mathbf{R}(\nu)]w_j^{(1)}(\pi,s)\\ &\qquad + \tfrac{1}{2}[\hat{\mathbf{k}}\cdot[\mathbf{R}(P\pi)-\mathbf{R}(\nu)]]^2 w_j^{(0)}(\pi,s)\}.\end{aligned} \tag{12.29}$$

It now remains to put the zeroth-order eigenvectors (12.17), and first-order eigenvectors given by (12.22) and (12.24), into equation (12.29). The term involving $w_j^{(1)}(\pi,s)$ is, with (12.22) for $z_j(\pi,s)$,

$$-M_C^{-1/2}\sum_{P\nu\pi}\sum_{j}\Phi_{ij}(0\nu,P\pi)\hat{\mathbf{k}}\cdot[\mathbf{R}(P\pi)-\mathbf{R}(\nu)]z_j(\pi,s). \tag{12.30}$$

It is convenient to rewrite this term by replacing $\Phi_{ij}(0\nu,P\pi)$ by $\Phi_{ji}(P\pi,0\nu)$, with the symmetry (6.4), and then replacing this by $\Phi_{ji}(0\pi,-P\nu)$, with the symmetry (6.57). Then the quantity (12.30) is

$$-M_C^{-1/2}\sum_{P\nu\pi}\sum_{j}\Phi_{ji}(0\pi,-P\nu)\hat{\mathbf{k}}\cdot[\mathbf{R}(P)+\mathbf{R}(\pi)-\mathbf{R}(\nu)]z_j(\pi,s),$$

and since for every primitive lattice $\mathbf{R}(P)$ in Σ_P there is a $\mathbf{R}(-P) = -\mathbf{R}(P)$, this is

$$-M_C^{-1/2}\sum_{P\nu\pi}\sum_{j}\Phi_{ji}(0\pi,P\nu)\hat{\mathbf{k}}\cdot[-\mathbf{R}(P)+\mathbf{R}(\pi)-\mathbf{R}(\nu)]z_j(\pi,s),$$

and interchanging π and ν inside the sum gives

$$M_C^{-1/2}\sum_{P\nu\pi}\sum_{j}\Phi_{ji}(0\nu,P\pi)\hat{\mathbf{k}}\cdot[\mathbf{R}(P\pi)-\mathbf{R}(\nu)]z_j(\nu,s). \tag{12.31}$$

Now the term in $\mathbf{R}(\nu)$ can be dropped because it vanishes by translational invariance, and with (12.24) for $z_j(\pi,s)$ this is

$$M_C^{-1/2}\sum_{P\nu\pi}\sum_{jklm}\Phi_{mi}(0\nu,P\pi)R_j(P\pi)X_{m,kl}(\nu)\hat{k}_j\hat{k}_l y_k(s).$$

Finally this term can be written with the coefficients f_{ijkl} defined by (7.63) as

$$M_C^{-1/2}\sum_{jkl} 2V_C f_{klij}\hat{k}_j\hat{k}_l y_k(s). \tag{12.32}$$

With (12.17) for $w_j^{(0)}(\pi,s)$, the last term in (12.29) is written out as

$$-\tfrac{1}{2}M_C^{-1/2}\sum_{P\nu\pi}\sum_{jkl}\Phi_{ik}(0\nu,P\pi)[R_j(P\pi)-R_j(\nu)][R_l(P\pi)-R_l(\nu)]\hat{k}_j\hat{k}_l y_k(s), \tag{12.33}$$

and the left-hand side of (12.29) is

$$M_C^{-1/2}\sum_{\nu} M_\nu[\omega^{(1)}(s)]^2 y_i(s) = M_C^{1/2}[\omega^{(1)}(s)]^2 y_i(s). \tag{12.34}$$

Now we can restore the acoustic phonon frequencies $\omega(\mathbf{k}s)$ by means of (12.13) and the wave vectors $\mathbf{k}$ by (12.12), gather up the terms (12.32)–(12.34), and rewrite the solubility condition (12.29) as an eigenvalue–eigenvector equation for $\omega(\mathbf{k}s)$ and $\mathbf{y}(s)$:

$$\begin{aligned} M_C[\omega(\mathbf{k}s)]^2 y_i(s) = \sum_{jkl} \Big\{ & V_C(f_{klij} + f_{kjil}) \\ & - \tfrac{1}{2} \sum_{\mathrm{P}\nu\pi} \Phi_{ik}(0\nu,\mathrm{P}\pi)[R_j(\mathrm{P}\pi) - R_j(\nu)][R_l(\mathrm{P}\pi) - R_l(\nu)] \Big\} k_j k_l y_k(s). \end{aligned} \tag{12.35}$$

Here the first term on the right has been written as the symmetric combination $(f_{klij} + f_{kjil})$ because the Σ_{jl} ensures that only this combination contributes to the sum; the last term on the right is already symmetric in j,l. But from (7.60) for $\hat{A}_{ikjl}$, and with the symmetry (7.68) for the first line of $\hat{A}_{ikjl}$ which contains the f_{ijkl}, we have for (12.35)

$$M_C[\omega(\mathbf{k}s)]^2 y_i(s) = \sum_{jkl} V_C \hat{A}_{ikjl} k_j k_l y_k(s). \tag{12.36}$$

It should be remembered that this equation of motion is correct only to leading order in $|\mathbf{k}|$, namely to order $|\mathbf{k}|^2$. It is interesting to note that in the process of solving the long-waves equation of motion we have transformed to the more natural description corresponding to a generalized eigenvalue problem [compare (12.17) and (12.22) with (11.69)].

Long-Waves Homogeneous-Deformation Equivalence

For any lattice in an arbitrary initial equilibrium configuration, corresponding to arbitrary applied uniform surface stresses which may be positive or negative (outward or inward on the crystal surface), there are three acoustic phonon branches for which $\omega(\mathbf{k}s) \propto |\mathbf{k}|$ for very small $|\mathbf{k}|$, and where it should be kept in mind that the proportionality constant (the velocity) depends on the direction of $\mathbf{k}$. To this leading order in $|\mathbf{k}|$ the equation of motion for the acoustic phonons is (12.11) for primitive lattices and (12.36) for nonprimitive lattices, and in either case is identical to (7.99) for long-wavelength mechanical sound waves as derived by the method of homogeneous deformation. Since the equation of motion is an eigenvalue–eigenvector equation, it completely determines the frequencies and (apart from normalization) the eigenvectors, hence the long-wavelength acoustic phonons are identical to the long-wavelength mechanical sound waves. The homogeneous-deformation theory is valid only for long wavelengths, long compared to the effective range of interactions among the ions, while of course lattice dynamics is valid for all allowed wavelengths, and when the phonon wavelength becomes of order or less than the range of interactions, then $\omega(\mathbf{k}s)$ will depart from being linear with $|\mathbf{k}|$.

An arbitrary ion displacement is written as a sum over all phonon coordinates in (10.40); accordingly the displacement of ion Nν in the j direction, due to a phonon $\mathbf{k}s$, is proportional to

$$(M_C/M_\nu)^{1/2}e^{i\mathbf{k}\cdot\mathbf{R}(\mathrm{N}\nu)}w_j(\nu,\mathbf{k}s), \qquad (12.37)$$

where we multiplied by $M_C^{1/2}$ for convenience. For an acoustic phonon ($s = 1,2,3$) in the limit of long wavelength, i.e., to zeroth order in $|\mathbf{k}|$, the phase factor in (12.37) is 1 and $(M_C/M_\nu)^{1/2}w_j(\nu,\mathbf{k}s)$ is $y_j(s)$, so the displacement is a uniform translation of the crystal. The first-order terms in the small $\mathbf{k}$ expansion of (12.37) are

$$(M_C/M_\nu)^{1/2}[i\mathbf{k}\cdot\mathbf{R}(\mathrm{N}\nu)w_j^{(0)}(\nu,s) + i\lambda w_j^{(1)}(\nu,s)]. \qquad (12.38)$$

Now in (12.38) we divide out $i\lambda$, replace $w_j^{(0)}(\nu,s)$ by (12.17) and $w_j^{(1)}(\nu,s)$ by (12.22) and (12.24), and express the ion displacement due to the long-wavelength acoustic phonon $\mathbf{k}s$ as proportional to

$$\sum_i \hat{k}_i R_i(\mathrm{N}\nu)y_j(s) + \sum_{mn} X_{j,mn}(\nu)y_m(s)\hat{k}_n = \sum_{mn} Y_{j,mn}(\mathrm{N}\nu)y_m(s)\hat{k}_n,$$

where $Y_{j,mn}(\mathrm{N}\nu)$ is given by (7.85). But according to (7.84) this is exactly the first-order displacement $U_j^{(1)}(\mathrm{N}\nu)$, due to a homogeneous deformation with displacement gradients given by

$$u_{mn} = y_m(s)\hat{k}_n. \qquad (12.39)$$

Thus it is clear that a long-wavelength acoustic phonon produces a homogeneous deformation at any location in the crystal, and the long-wavelength acoustic phonon eigenvectors $w_j^{(1)}(\nu,s)$ are related to the first-order sublattice displacements by

$$(M_C/M_\nu)^{1/2}w_j^{(1)}(\nu,s) = \sum_{mn} X_{j,mn}(\nu)y_m(s)\hat{k}_n = S_j^{(1)}(\nu). \qquad (12.40)$$

The time-dependence of any harmonic phonon state is of course $e^{-i\omega t}$.

The use of the alternate phase factors $e^{i\mathbf{k}\cdot\mathbf{R}(\mathrm{N})}$ in place of $e^{i\mathbf{k}\cdot\mathbf{R}(\mathrm{N}\nu)}$ in describing the phonons was discussed in Section 11. It is now obvious that for the long-wavelength acoustic phonons this alternate phase factor is equivalent to the alternate "phase factor" used in (7.90) to define the ion displacements corresponding to a homogeneous deformation. Components of the eigenvectors corresponding to the alternate phase factors are $\bar{w}_j(\nu,\mathbf{k}s)$, given by (11.63), and the above calculations can all be carried through to given, in place of (12.40),

$$(M_C/M_\nu)^{1/2}\bar{w}_j^{(1)}(\nu,s) = \sum_{mn} \bar{X}_{j,mn}(\nu)y_m(s)\hat{k}_n = \bar{S}_j^{(1)}(\nu), \qquad (12.41)$$

where $\bar{X}_{j,mn}(\nu)$ are given by (7.92).

For a crystal with n ions per unit cell there are also $3n - 3$ phonon branches labeled by $s = 4,5,\cdots,3n$, for which the $\omega(\mathbf{k}s)$ approach nonzero values as $|\mathbf{k}| \to 0$; these are the optic branches. The method of long waves can also be used to find the frequencies of the long-wavelength optic phonons.* For these phonons the homogeneous strains are zero, but the sublattice displacements are not, and at $|\mathbf{k}| = 0$ the motion corresponds to the sublattices remaining rigid and vibrating with respect to one another. An explicit separation of the long-wavelength dynamical matrices into acoustic and optic submatrices is shown in Section 15 for the example of two like atoms per unit cell with central potential interactions.

Acoustic Phonon Velocities

For thermodynamic calculations it will be convenient to define the long-wavelength acoustic phonon velocities $c(\hat{\mathbf{k}}s)$, which depend on the direction but not on the magnitude of $\mathbf{k}$, according to

$$\omega(\mathbf{k}s) = c(\hat{\mathbf{k}}s)\,|\mathbf{k}|, \quad \text{for small } |\mathbf{k}| \text{ and } s = \text{acoustic}. \tag{12.42}$$

In view of the equation of motion (12.11) for a primitive lattice, or (12.36) for a nonprimitive lattice, the quantities $M_C[\omega(\mathbf{k}s)]^2$ for long-wavelength acoustic phonons are eigenvalues of the matrices $\Sigma_{jl}\, V_C \hat{A}_{ikjl} k_j k_l$; therefore the squared velocities $[c(\hat{\mathbf{k}}s)]^2$ are eigenvalues of the real symmetric 3×3 matrices $\mathbf{d}(\hat{\mathbf{k}})$, where for each $\hat{\mathbf{k}}$,

$$d_{ik}(\hat{\mathbf{k}}) = (V_C/M_C) \sum_{jl} \hat{A}_{ikjl} \hat{k}_j \hat{k}_l. \tag{12.43}$$

Also for a given direction $\hat{\mathbf{k}}$, the eigenvectors of $\mathbf{d}(\hat{\mathbf{k}})$ are obviously $\mathbf{y}(s)$, according to (12.36). Since the eigenvectors are different for each direction $\hat{\mathbf{k}}$, we will denote them generally by $\mathbf{y}(\hat{\mathbf{k}}s)$, and write the diagonalization of $\mathbf{d}(\mathbf{k})$ as

$$\sum_{ij} y_i(\hat{\mathbf{k}}s) d_{ij}(\hat{\mathbf{k}}) y_j(\hat{\mathbf{k}}s') = [c(\hat{\mathbf{k}}s)]^2 \delta_{ss'}. \tag{12.44}$$

The eigenvectors satisfy the usual orthonormality and completeness relations in three-dimensional space; in fact since $\mathbf{d}(\hat{\mathbf{k}})$ is real symmetric, the $\mathbf{y}(\hat{\mathbf{k}}s)$ form a real orthogonal matrix. For a primitive lattice, since all branches are acoustic, it follows that

$$\mathbf{d}(\hat{\mathbf{k}}) = \lim_{|\mathbf{k}| \to 0} \frac{\mathbf{D}(\mathbf{k})}{|\mathbf{k}|^2}; \tag{12.45}$$

$$\mathbf{y}(\hat{\mathbf{k}}s) = \lim_{|\mathbf{k}| \to 0} \mathbf{w}(\mathbf{k}s), \quad \text{primitive lattice}. \tag{12.46}$$

* M. Born and K. Huang, *Dynamical Theory of Crystal Lattices*, Clarendon Press, Oxford, 1954.

It also follows for any lattice that in the potential approximation, defined by (7.2), the acoustic phonon velocities $c(\hat{\mathbf{k}}s)$ are equal to the thermodynamic sound velocities v_s discussed in Section 3:

$$c(\hat{\mathbf{k}}s) = v_s(\hat{\mathbf{k}}), \quad \text{potential approximation.} \tag{12.47}$$

The phonon Grüneisen parameters are defined by (11.74) and (11.79); the strain and volume derivatives of the phonon velocities $c(\hat{\mathbf{k}}s)$ have an essentially different meaning from the Grüneisen parameters, so we will define them with a different symbol, σ, as follows.

$$\sigma_{ij}(\hat{\mathbf{k}}s) = -[c(\hat{\mathbf{k}}s)]^{-1}[\partial c(\hat{\mathbf{k}}s)/\partial \eta_{ij}]_{\eta'}; \tag{12.48}$$

$$\sigma(\hat{\mathbf{k}}s) = -[V/c(\hat{\mathbf{k}}s)][dc(\hat{\mathbf{k}}s)/dV]. \tag{12.49}$$

Just as in the definition (11.79) of the phonon Grüneisen parameters $\gamma(\mathbf{k}s)$, the volume variation in (12.49) is not unique for anisotropic crystals, and it is understood to be prescribed in terms of a set of strains η_{ij}. With the aid of the diagonalization equation (12.44) which gives $[c(\hat{\mathbf{k}}s)]^2$ as eigenvalues of $\mathbf{d}(\hat{\mathbf{k}})$, and the eigenvalue derivative theorem (11.73), the σ parameters may be calculated from

$$\sigma_{kl}(\hat{\mathbf{k}}s) = -\tfrac{1}{2}[c(\hat{\mathbf{k}}s)]^{-2} \sum_{ij} y_i(\hat{\mathbf{k}}s)[\partial d_{ij}(\hat{\mathbf{k}})/\partial \eta_{kl}]_{\eta'} y_j(\hat{\mathbf{k}}s); \tag{12.50}$$

$$\sigma(\hat{\mathbf{k}}s) = -\tfrac{1}{2}[c(\hat{\mathbf{k}}s)]^{-2} \sum_{ij} y_i(\hat{\mathbf{k}}s)\{V d[d_{ij}(\hat{\mathbf{k}})]/dV\} y_j(\hat{\mathbf{k}}s). \tag{12.51}$$

At this point an interpretive discussion of the various eigenvalue strain derivatives is in order. Consider first the phonon Grüneisen parameters $\gamma_{ij}(\mathbf{k}s)$ and $\gamma(\mathbf{k}s)$, which are, respectively, logarithmic strain and volume derivatives of the phonon frequency $\omega(\mathbf{k}s)$; these are calculated from appropriate derivatives of the dynamical matrix $\mathbf{D}(\mathbf{k})$. Now each $\mathbf{D}(\mathbf{k})$ is characterized by an allowed $\mathbf{k}$ vector, a particular $\mathbf{k}$ of the complete set of N_0 such vectors, and under a homogeneous deformation of the lattice each allowed $\mathbf{k}$ varies with the strain. As it was pointed out in Section 10, following (10.28), if we consider a given $\mathbf{k}$ and watch the product $\mathbf{k} \cdot \mathbf{R}(\mathrm{N})$ for any $\mathbf{R}(\mathrm{N})$, we find that although both $\mathbf{k}$ and $\mathbf{R}(\mathrm{N})$ vary under a homogeneous deformation, this product is invariant. This of course simplifies the calculation of strain and volume derivatives of $\mathbf{D}(\mathbf{k})$ [particularly $\bar{\mathbf{D}}(\mathbf{k})$, equation (11.64)], but it should always be remembered that when we talk about the variation of $\bar{\mathbf{D}}(\mathbf{k})$ with strain we are following a given allowed $\mathbf{k}$ vector, and this $\mathbf{k}$ is varying with the strain. From another point of view, the calculated $\gamma_{ij}(\mathbf{k}s)$ and $\gamma(\mathbf{k}s)$ evaluated at $\eta_{ij} = 0$ correspond to the same $\mathbf{k}$ as does $\omega(\mathbf{k}s)$ evaluated at $\eta_{ij} = 0$, but the strain and volume derivatives correctly account for the variation of this given $\mathbf{k}$ with the η_{ij}. It should also be noted that in all our calculations the initial equilibrium configuration, from which the η_{ij} are measured, is arbitrary.

The situation is essentially different in the logarithmic strain and volume derivatives of the velocity $c(\hat{\mathbf{k}}s)$, namely $\sigma_{ij}(\hat{\mathbf{k}}s)$ and $\sigma(\hat{\mathbf{k}}s)$, respectively. Here the derivatives are carried out for a fixed direction $\hat{\mathbf{k}}$, i.e., $\hat{\mathbf{k}}$ does not vary with the strain. Therefore the strain variation of each $c(\hat{\mathbf{k}}s)$ does *not* correspond to a given allowed **k** (a given phonon), since both the direction and magnitude of an allowed **k** will generally vary with the strain. Hence the relation between the phonon Grüneisen parameters for long-wavelength acoustic phonons and the σ parameters is *not* the same as the simple relation (12.42) between the phonon frequencies and velocities. Again the initial equilibrium configuration at which the σ parameters are evaluated may be chosen as desired.

13. RENORMALIZED PHONONS

Operator-Renormalization Method

In preparation for calculating the partition function and other thermodynamic functions of crystals, we now turn to evaluation of the quantum energy levels of the interacting phonon system. The phonon–phonon interactions will be treated only to the leading order of perturbation in which they contribute to the system energy levels, namely to order $\epsilon^2\hbar\omega_\kappa$. A particularly simple approach to this calculation is provided by the operator-renormalization method; the basic idea is to obtain improved phonon creation and annihilation operators representing "dressed" phonons which do not interact with one another in first-order perturbation, and then calculate the energies of these phonons in second-order perturbation (order $\epsilon^2\hbar\omega_\kappa$). The operator renormalization method is a general procedure for solving many-body perturbation problems,* and in order to discuss electron–phonon interactions in Chapter 5 we will present the basic theory for a system of interacting bosons and fermions.

The starting point is the relation between the properties of a creation operator and its commutator with the Hamiltonian $\mathscr{H}$. Suppose for example there exist a number of operators $\theta_i^\dagger$ for which

$$[\mathscr{H},\theta_i^\dagger] = \omega_i\theta_i^\dagger, \tag{13.1}$$

where ω_i are real positive numbers. Then if ψ is an eigenfunction of $\mathscr{H}$ with eigenvalue (energy) E,

$$\mathscr{H}\psi = E\psi, \tag{13.2}$$

* D. C. Wallace, *Phys. Rev.* **152,** 247 (1966).

and operating on ψ with the commutator $[\mathscr{H},\theta_i^\dagger]$ and using (13.1) and (13.2) gives

$$[\mathscr{H},\theta_i^\dagger]\psi = \mathscr{H}\theta_i^\dagger\psi - \theta_i^\dagger\mathscr{H}\psi = \omega_i\theta_i^\dagger\psi,$$

or

$$\mathscr{H}\theta_i^\dagger\psi = (E + \omega_i)\theta_i^\dagger\psi. \tag{13.3}$$

This last equation tells us that each of the functions $\theta_i^\dagger\psi$ which is not zero is an eigenfunction of $\mathscr{H}$, with eigenvalue $E + \omega_i$. Obviously, repeated operations by the creation operators $\theta_i^\dagger$ produce functions which are either zero or are eigenfunctions of $\mathscr{H}$. The problem of finding the eigenfunctions of $\mathscr{H}$ has therefore been replaced by the problem of finding the creation operators $\theta_i^\dagger$ which satisfy (13.1). Solution of (13.1) alone does not necessarily provide the complete solution to the problem, however, because of the following circumstances. First, one needs a starting eigenfunction ψ from which to construct the new eigenfunctions $\theta_i^\dagger\psi$; the starting eigenfunction may be chosen as the ground state of $\mathscr{H}$ for convenience. In terms of the energies, we are saying that (13.1) determines only the excitation energies ω_i, but not the ground-state energy. Second, for calculating thermodynamic functions one needs the complete set of energy levels of $\mathscr{H}$, or at least all those in some energy range for special calculations, and it is correspondingly necessary to find the complete set of creation operators, or at least some subset thereof. Finally, some of the functions $\theta_i^\dagger\psi$ might be zero, and hence not eigenfunctions of $\mathscr{H}$, and this must be recognized in order to properly count the allowed states.

The operator-renormalization procedure may be developed in a straightforward way for perturbation problems. A perturbation expansion of $\mathscr{H}$ is written $\mathscr{H}_0 + \mathscr{H}_1 + \cdots$, and there are no difficulties if one has available the complete set of solutions of $\mathscr{H}_0$. In particular let us assume that $\mathscr{H}_0$ describes a system of bosons whose creation and annihilation operators are $A_{0\kappa}^\dagger$ and $A_{0\kappa}$, respectively, and a system of fermions whose creation and annihilation operators are $C_{0\lambda}^\dagger$ and $C_{0\lambda}$, respectively. The indices κ and λ are the quantum numbers of the bosons and fermions, respectively, and $\mathscr{H}_0$ is

$$\mathscr{H}_0 = G_0 + \sum_\kappa \omega_{0\kappa}A_{0\kappa}^\dagger A_{0\kappa} + \sum_\lambda \omega_{0\lambda}C_{0\lambda}^\dagger C_{0\lambda}, \tag{13.4}$$

where G_0 is the zeroth-order ground-state energy and $\omega_{0\kappa},\omega_{0\lambda}$ are the single-particle energies. These particles are assumed to be all independent (non-interacting in zeroth order) so that their operators satisfy the following relations.

$$[A_{0\kappa},A_{0\kappa'}^\dagger] = \delta_{\kappa\kappa'}, \qquad [A_{0\kappa},A_{0\kappa'}] = 0; \tag{13.5}$$

$$\{C_{0\lambda},C_{0\lambda'}^\dagger\} = \delta_{\lambda\lambda'}, \qquad \{C_{0\lambda},C_{0\lambda'}\} = 0; \tag{13.6}$$

$$[A_{0\kappa},C_{0\lambda}] = [A_{0\kappa},C_{0\lambda}^\dagger] = [A_{0\kappa}^\dagger,C_{0\lambda}] = [A_{0\kappa}^\dagger,C_{0\lambda}^\dagger] = 0. \tag{13.7}$$

The notation is $[A,B]$ for a commutator, and $\{A,B\}$ for an anticommutator. A single state vector of $\mathscr{H}_0$ is denoted by $|\cdots n_\kappa \cdots n_\lambda \cdots (0)\rangle$, where $n_\kappa = 0,1,2,\cdots$, is the number of bosons in the single-particle state κ, and $n_\lambda = 0,1$ is the number of fermions in the single-particle state λ, while the (0) is used to denote a zeroth-order state, i.e., an eigenvector of $\mathscr{H}_0$.

We can always arrange to take the zeroth-order ground state as the particle vacuum $|\cdots 0_\kappa \cdots 0_\lambda \cdots (0)\rangle = |0(0)\rangle$; this means that the system bosons and fermions describe *excitations* above the ground state, and for real conserved particles such as electrons in metals the energy of the filled Fermi sea is contained in G_0. Any annihilation operator operates on the vacuum to give zero:

$$A_{0\kappa}\,|0(0)\rangle = C_{0\lambda}\,|0(0)\rangle = 0. \tag{13.8}$$

Indeed these equations for the complete set of annihilation operators uniquely determine (apart from normalization) the ground state $|0(0)\rangle$. The complete set of orthonormal state vectors is

$$|\cdots n_\kappa \cdots n_\lambda \cdots (0)\rangle = \prod_\kappa (n_\kappa!)^{-1/2}(A_{0\kappa}^\dagger)^{n\kappa}\prod_\lambda (C_{0\lambda}^\dagger)^{n\lambda}\,|0(0)\rangle, \tag{13.9}$$

where the ordering of the set of $A_{0\kappa}^\dagger$ operators is arbitrary since they all commute, but the ordering of the $C_{0\lambda}^\dagger$ operators is fixed according to some (any) prescribed arrangement of the quantum numbers $\lambda,\lambda',\cdots$. The result of operating on a state vector by any of the single-particle operators is then given by the following equations.

$$A_{0\kappa}^\dagger\,|\cdots n_\kappa \cdots n_\lambda \cdots (0)\rangle = (n_\kappa + 1)^{1/2}\,|\cdots n_\kappa + 1 \cdots n_\lambda \cdots (0)\rangle; \tag{13.10}$$

$$A_{0\kappa}\,|\cdots n_\kappa \cdots n_\lambda \cdots (0)\rangle = n_\kappa^{1/2}\,|\cdots n_\kappa - 1 \cdots n_\lambda \cdots (0)\rangle; \tag{13.11}$$

$$C_{0\lambda}^\dagger\,|\cdots n_\kappa \cdots n_\lambda \cdots (0)\rangle = (1 - n_\lambda)(-1)^{p\lambda}\,|\cdots n_\kappa \cdots 1_\lambda \cdots (0)\rangle; \tag{13.12}$$

$$C_{0\lambda}\,|\cdots n_\kappa \cdots n_\lambda \cdots (0)\rangle = n_\lambda(-1)^{p\lambda}\,|\cdots n_\kappa \cdots 0_\lambda \cdots (0)\rangle; \tag{13.13}$$

here p_λ is the sum of $n_{\lambda'}$ which stand to the left of the n_λ position. It then follows that the state vectors are eigenvectors of the number operators $A_{0\kappa}^\dagger A_{0\kappa}$ and $C_{0\lambda}^\dagger C_{0\lambda}$, with eigenvalues n_κ and n_λ, respectively:

$$A_{0\kappa}^\dagger A_{0\kappa}\,|\cdots n_\kappa \cdots n_\lambda \cdots (0)\rangle = n_\kappa\,|\cdots n_\kappa \cdots n_\lambda \cdots (0)\rangle; \tag{13.14}$$

$$C_{0\lambda}^\dagger C_{0\lambda}\,|\cdots n_\kappa \cdots n_\lambda \cdots (0)\rangle = n_\lambda\,|\cdots n_\kappa \cdots n_\lambda \cdots (0)\rangle. \tag{13.15}$$

In addition the number operators, together with $\mathscr{H}_0$, form a complete set of commuting operators.

It is clear from (13.4) for $\mathscr{H}_0$, and (13.5)–(13.7) for the particle commutators and anticommutators, that the creation operators all satisfy the Hamiltonian commutator equation (13.1) in zeroth order:

$$[\mathscr{H}_0,A_{0\kappa}^\dagger] = \omega_{0\kappa}A_{0\kappa}^\dagger; \tag{13.16}$$

$$[\mathscr{H}_0,C_{0\lambda}^\dagger] = \omega_{0\lambda}C_{0\lambda}^\dagger. \tag{13.17}$$

It is further clear that the $A_{0\kappa}^\dagger$ and $C_{0\lambda}^\dagger$ are *all* the creation operators, and that operating on the vacuum they give *all* the eigenvectors. Hence if there is a perturbation which causes interactions among the zeroth-order particles, each of the zeroth-order states is altered in perturbation theory, or equivalently each of the zeroth-order particle operators is altered (renormalized), and the set of renormalized creation operators contains *all* the creation operators of $\mathscr{H}$.

Let us consider the effects of a perturbation formally by writing

$$\mathscr{H} = \mathscr{H}_0 + \mathscr{H}_1 + \mathscr{H}_2 + \cdots, \tag{13.18}$$

where in terms of the expansion parameter ϵ the orders are

$$\mathscr{H}_1 \sim \epsilon\mathscr{H}_0, \qquad \mathscr{H}_2 \sim \epsilon^2\mathscr{H}_0, \cdots. \tag{13.19}$$

Suppose each creation operator is corrected to order m by constructing a series of operators of each order; the corrected creation operators are denoted $\tilde{A}_\kappa^\dagger$ and $\tilde{C}_\lambda^\dagger$ and may be written

$$\tilde{A}_\kappa^\dagger = A_{0\kappa}^\dagger + A_{1\kappa}^\dagger + \cdots + A_{m\kappa}^\dagger, \tag{13.20}$$

$$\tilde{C}_\lambda^\dagger = C_{0\lambda}^\dagger + C_{1\lambda}^\dagger + \cdots + C_{m\lambda}^\dagger, \tag{13.21}$$

where

$$A_{0\kappa}^\dagger, C_{0\lambda}^\dagger \sim 1; \qquad A_{1\lambda}^\dagger, C_{1\lambda}^\dagger \sim \epsilon; \quad \cdots. \tag{13.22}$$

These corrected operators are designed to satisfy the Hamiltonian commutator equations (13.1) to order m; that is,

$$[\mathscr{H},\tilde{A}_\kappa^\dagger] = \Omega_\kappa\tilde{A}_\kappa^\dagger + O(\epsilon^{m+1}), \tag{13.23}$$

$$[\mathscr{H},\tilde{C}_\lambda^\dagger] = \Omega_\lambda\tilde{C}_\lambda^\dagger + O(\epsilon^{m+1}), \tag{13.24}$$

where the corrected particle energies Ω_κ and Ω_λ are real positive numbers. The renormalized operators are also required to satisfy the appropriate commutators and anticommutators to order m:

$$[\tilde{A}_\kappa,\tilde{A}_{\kappa'}^\dagger] = \delta_{\kappa\kappa'} + O(\epsilon^{m+1}), \qquad [\tilde{A}_\kappa,\tilde{A}_{\kappa'}] = O(\epsilon^{m+1}); \tag{13.25}$$

$$\{\tilde{C}_\lambda,\tilde{C}_{\lambda'}^\dagger\} = \delta_{\lambda\lambda'} + O(\epsilon^{m+1}), \qquad \{\tilde{C}_\lambda,\tilde{C}_{\lambda'}\} = O(\epsilon^{m+1}); \tag{13.26}$$

$$[\tilde{A}_\kappa,\tilde{C}_\lambda] = O(\epsilon^{m+1}), \qquad [\tilde{A}_\kappa,\tilde{C}_\lambda^\dagger] = O(\epsilon^{m+1}). \tag{13.27}$$

Note that these commutator and anticommutator requirements are equivalent to requiring that the renormalized single-particle state vectors are orthonormal to order m.

Equations (13.23)–(13.27) describe a complete solution correct to order m, where of course m can be $0,1,\cdots$. This description is just the same as the zeroth-order description, except the renormalized energy levels are different, and now the commutators and anticommutators are not exact but have "errors" of order $m + 1$. Hence all the equations describing the zeroth-order system are now given correctly to order m in terms of renormalized particles. In particular if $|0(m)\rangle$ is the ground-state vector correct to order m, then

$$\tilde{A}_\kappa\,|0(m)\rangle = O(\epsilon^{m+1}), \qquad \tilde{C}_\lambda\,|0(m)\rangle = O(\epsilon^{m+1}), \tag{13.28}$$

and in fact these conditions determine $|0(m)\rangle$ uniquely, apart from normalization. Further, the Hamiltonian is diagonal to order m in terms of renormalized particles:

$$\mathscr{H} = G(m) + \sum_\kappa \Omega_\kappa \tilde{A}_\kappa^\dagger \tilde{A}_\kappa + \sum_\lambda \Omega_\lambda \tilde{C}_\lambda^\dagger \tilde{C}_\lambda + O(\epsilon^{m+1}), \tag{13.29}$$

where $G(m)$ is the ground-state energy correct to order m, i.e.,

$$G(m) = G_0 + G_1 + \cdots + G_m. \tag{13.30}$$

The state vectors $|\cdots n_\kappa \cdots n_\lambda \cdots (m)\rangle$ are orthonormal to order m, and represent states composed of renormalized particles just like the zeroth-order representation (13.9); in addition, the transformations (13.10)–(13.15) for renormalized particle operators and renormalized state vectors all hold to order m.

The renormalized particle energies are also obtained as a power series in the expansion parameter; that is,

$$\Omega_\kappa = \omega_{0\kappa} + \omega_{1\kappa} + \cdots + \omega_{m\kappa}, \tag{13.31}$$

$$\Omega_\lambda = \omega_{0\lambda} + \omega_{1\lambda} + \cdots + \omega_{m\lambda}, \tag{13.32}$$

where

$$\omega_{1\kappa} \sim \epsilon\omega_{0\kappa}, \qquad \omega_{1\lambda} \sim \epsilon\omega_{0\lambda}, \qquad \cdots. \tag{13.33}$$

It is possible to calculate the next higher order contributions $\omega_{m+1,\kappa}$ and $\omega_{m+1,\lambda}$ to the particle energies, without finding the $(m + 1)$-order contributions to the creation operators. For the bosons, for example, assume $A_{m+1,\kappa}^\dagger$ exists such that the Hamiltonian commutator equations are satisfied to order $m + 1$:

$$[\mathscr{H},\tilde{A}_\kappa^\dagger + A_{m+1,\kappa}^\dagger] = (\Omega_\kappa + \omega_{m+1,\kappa})(\tilde{A}_\kappa^\dagger + A_{m+1,\kappa}^\dagger) + O(\epsilon^{m+2}), \tag{13.34}$$

where $\tilde{A}_\kappa^\dagger$ satisfies (13.23). Because of (13.23), the terms in (13.34) of order $0,1,\cdots,m$ are already satisfied, and the terms of order $m+1$ are

$$\begin{aligned}[\mathscr{H}_0,A^\dagger_{m+1,\kappa}] + [\mathscr{H}_1,A^\dagger_{m\kappa}] + \cdots + [\mathscr{H}_{m+1},A^\dagger_{0\kappa}] \\ = \omega_{0\kappa}A^\dagger_{m+1,\kappa} + \omega_{1\kappa}A^\dagger_{m\kappa} + \cdots + \omega_{m+1,\kappa}A^\dagger_{0\kappa}.\end{aligned} \tag{13.35}$$

This operator equation may be evaluated in any representation; let us take the particular zeroth-order matrix element

$$\langle\cdots n_\kappa + 1 \cdots n_\lambda \cdots (0)|\,[\text{equation (13.35)}]\,|\cdots n_\kappa \cdots n_\lambda \cdots (0)\rangle.$$

Then the terms containing $A^\dagger_{m+1,\kappa}$ cancel, and if $\omega_{m+1,\kappa}$ is treated as a number, the result is

$$\begin{aligned}\langle\cdots n_{\kappa+1} \cdots n_\lambda \cdots (0)|\,[\mathscr{H}_1,A^\dagger_{m\kappa}] + [\mathscr{H}_2,A^\dagger_{m-1,\kappa}] + \cdots + [\mathscr{H}_{m+1},A^\dagger_{0\kappa}] \\ - \omega_{1\kappa}A^\dagger_{m\kappa} - \omega_{2\kappa}A^\dagger_{m-1,\kappa} - \cdots - \omega_{m\kappa}A^\dagger_{1\kappa}\,|\cdots n_\kappa \cdots n_\lambda \cdots (0)\rangle \\ = (n_\kappa + 1)^{1/2}\omega_{m+1,\kappa}.\end{aligned} \tag{13.36}$$

For the fermions the procedure is quite similar. Assume $C^\dagger_{m+1,\lambda}$ exists such that

$$[\mathscr{H},\tilde{C}^\dagger_\lambda + C^\dagger_{m+1,\lambda}] = (\Omega_\lambda + \omega_{m+1,\lambda})(\tilde{C}^\dagger_\lambda + C^\dagger_{m+1,\lambda}) + O(\epsilon^{m+2}), \tag{13.37}$$

where $\tilde{C}^\dagger_\lambda$ satisfies (13.24). The terms of order $0,1,\cdots,m$ are already satisfied by (13.24), and the terms of order $m+1$ in the matrix element

$$\langle\cdots n_\kappa \cdots 1_\lambda \cdots (0)|\,[\text{equation (13.37)}]\,|\cdots n_\kappa \cdots 0_\lambda \cdots (0)\rangle$$

give

$$\begin{aligned}\langle\cdots n_\kappa \cdots 1_\lambda \cdots (0)|\,[\mathscr{H}_1,C^\dagger_{m\lambda}] + [\mathscr{H}_2,C^\dagger_{m-1,\lambda}] + \cdots + [\mathscr{H}_{m+1},C^\dagger_{0\lambda}] \\ - \omega_{1\lambda}C^\dagger_{m\lambda} - \omega_{2\lambda}C^\dagger_{m-1,\lambda} - \cdots - \omega_{m\lambda}C^\dagger_{1\lambda}\,|\cdots n_\kappa \cdots 0_\lambda \cdots (0)\rangle \\ = (-1)^{p\lambda}\omega_{m+1,\lambda}.\end{aligned} \tag{13.38}$$

If the renormalized particles are found to order m, where m may be $0,1,\cdots$, then (13.36) and (13.38) give the correct $(m+1)$-order contributions to the particle energies without requiring the $(m+1)$-order solutions. If the energies $\omega_{m+1,\kappa}$ and $\omega_{m+1,\lambda}$ so calculated are *not* numbers, but depend on the state vectors in the matrix elements (i.e., depend on the n_κ and n_λ), then the renormalized particles are not independent in order $m+1$. In that case the truly independent normal modes of $\mathscr{H}$ in order $m+1$ must be specified by quantum numbers other than the original κ and λ of the zeroth-order system. In addition, if the Ω_κ and Ω_λ are real numbers to order m, then the $\omega_{m+1,\kappa}$ given by (13.36) and $\omega_{m+1,\lambda}$ given by (13.38) are also real. In carrying out the renormalization procedure in real problems, however, it is necessary to introduce complex energies in order to avoid divergencies on the real axis, and when the renormalized particles are not independent in order $m + 1$ then the energies $\omega_{m+1,\kappa}$ and $\omega_{m+1,\lambda}$ are complex. The general procedure of the operator-renormalization method, and its basic simplicity, are illustrated by the treatment of the interacting phonon system, which follows. The original reference* may be consulted for further remarks of a general nature.

First-Order Phonon Operators

We return now to the system of interacting phonons, and continue with our original notation, which is slightly different from the preceding discussion of the operator-renormalization method. The system is described by (10.84)–(10.89) as follows.

$$\mathscr{H} = \Phi_0 + \mathscr{H}_2 + \mathscr{H}_3 + \mathscr{H}_4, \tag{13.39}$$

where $\mathscr{H}_2$ represents a set of noninteracting phonons

$$\mathscr{H}_2 = G_2 + \sum_{\kappa} \hbar\omega_\kappa A_\kappa^\dagger A_\kappa, \tag{13.40}$$

$$G_2 = \tfrac{1}{2} \sum_{\kappa} \hbar\omega_\kappa, \tag{13.41}$$

and $\mathscr{H}_3$ and $\mathscr{H}_4$ represent phonon–phonon interactions

$$\mathscr{H}_3 = \sum_{\kappa\kappa'\kappa''} \Phi_{\kappa\kappa'\kappa''}(A_\kappa + A_{-\kappa}^\dagger)(A_{\kappa'} + A_{-\kappa'}^\dagger)(A_{\kappa''} + A_{-\kappa''}^\dagger), \tag{13.42}$$

$$\mathscr{H}_4 = \sum_{\kappa\kappa'\kappa''\kappa'''} \Phi_{\kappa\kappa'\kappa''\kappa'''}(A_\kappa + A_{-\kappa}^\dagger)(A_{\kappa'} + A_{-\kappa'}^\dagger)(A_{\kappa''} + A_{-\kappa''}^\dagger)(A_{\kappa'''} + A_{-\kappa'''}^\dagger). \tag{13.43}$$

Thus the zeroth-order phonon energies and creation operators are $\hbar\omega_\kappa$ and $A_\kappa^\dagger$, respectively, $\mathscr{H}_3$ is of order $\epsilon\mathscr{H}_2$, and $\mathscr{H}_4$ is of order $\epsilon^2\mathscr{H}_2$. The zeroth-order phonon–operator commutators are

$$[A_\kappa, A_{\kappa'}^\dagger] = \delta_{\kappa\kappa'}, \qquad [A_\kappa, A_{\kappa'}] = 0; \tag{13.44}$$

and from these it follows that

$$[\mathscr{H}_2, A_k^\dagger] = \hbar\omega_\kappa A_\kappa^\dagger. \tag{13.45}$$

The operator-renormalization procedure requires the calculation of a series of commutators, and these can all be evaluated from the commutators (13.44) and their Hermitian conjugates. A useful relation, for any operators A, B, and C, is

$$[A,BC] = B[A,C] + [A,B]C; \tag{13.46}$$

* D. C. Wallace, *Phys. Rev.* **152,** 247 (1966).

this may be verified by writing out all the terms in each side, and it is easily extended to products of several operators, as for example,

$$[A,BCD] = BC[A,D] + B[A,C]D + [A,B]CD. \tag{13.47}$$

The first step is to calculate the first-order phonon energies $\hbar\omega_{1\kappa}$; with the zeroth-order state vectors denoted simply by $|\cdots n_\kappa \cdots\rangle$, the general equation (13.36) gives

$$(n_\kappa + 1)^{1/2}\hbar\omega_{1\kappa} = \langle\cdots n_\kappa + 1 \cdots|\,[\mathscr{H}_3,A_\kappa^\dagger]\,|\cdots n_\kappa \cdots\rangle. \tag{13.48}$$

The commutator in (13.48) is calculated from (13.42) for $\mathscr{H}_3$ and is

$$[\mathscr{H}_3,A_\kappa^\dagger] = 3\sum_{\kappa'\kappa''}\Phi_{\kappa\kappa'\kappa''}(A_{\kappa'} + A_{-\kappa'}^\dagger)(A_{\kappa''} + A_{-\kappa''}^\dagger), \tag{13.49}$$

where we have combined three terms by using the symmetry (10.76) of $\Phi_{\kappa\kappa'\kappa''}$ in its indices. Now (13.49) couples only states in which the total of all occupation numbers changes by 0 or 2, so the matrix element (13.48) vanishes. Hence,

$$\hbar\omega_{1\kappa} = 0. \tag{13.50}$$

The next step is to find first-order corrections $B_\kappa^\dagger$ to the phonon-creation operators, where

$$B_\kappa^\dagger \sim \epsilon, \tag{13.51}$$

such that the renormalized operators $A_\kappa^\dagger + B_\kappa^\dagger$ are good creation operators of $\mathscr{H}$ to one order higher than the bare phonon operators $A_\kappa^\dagger$. In other words, we want to satisfy the Hamiltonian commutator equations to one order higher than the corresponding leading-order equations (13.45), so we write

$$[\mathscr{H}_2 + \mathscr{H}_3,A_\kappa^\dagger + B_\kappa^\dagger] = \hbar\omega_\kappa(A_\kappa^\dagger + B_\kappa^\dagger) + O(\epsilon^2\hbar\omega_\kappa), \tag{13.52}$$

where $\hbar\omega_{1\kappa}$ has been dropped since it vanishes by (13.50). The terms of order $\hbar\omega_\kappa$ are satisfied by (13.45); the terms of order $\epsilon\hbar\omega_\kappa$ give a set of operator equations which determine the $B_\kappa^\dagger$. In order to solve these equations it will be necessary to replace $\hbar\omega_\kappa$ on the right-hand side of (13.52) by $\hbar\omega_\kappa + i\gamma_\kappa$, where γ_κ are positive real infinitesimal numbers; then the terms of order $\epsilon\hbar\omega_\kappa$ in (13.52) are

$$[\mathscr{H}_2,B_\kappa^\dagger] + [\mathscr{H}_3A_\kappa^\dagger] = (\hbar\omega_\kappa + i\gamma_\kappa)B_\kappa^\dagger. \tag{13.53}$$

The introduction of the infinitesimals γ_κ into (13.53) is simply a device for avoiding divergencies. In our operator calculations γ_κ can be taken arbitrarily small, in particular $\gamma_\kappa \ll \hbar\omega_\kappa$ for every κ; note that we are not including the three modes of uniform translation of the crystal, for which $\hbar\omega_\kappa = 0$. After the operator calculations are completed and thermal averages are taken, as in Chapter 4, the limit $\gamma_\kappa \to 0$ can be taken to obtain not only the temperature-dependent renormalized phonon energies, but their lifetimes as well. The sign of $i\gamma_\kappa$ in (13.53) was chosen to produce positive lifetimes in the positive time domain.

According to (13.49) the commutator $[\mathscr{H}_3, A_\kappa^\dagger]$ is quadratic in phonon operators, so we expect each term in (13.53) to be quadratic in phonon operators. Therefore $B_\kappa^\dagger$ should have the general form

$$\begin{aligned} B_\kappa^\dagger = \sum_{\kappa'\kappa''} \Phi_{\kappa\kappa'\kappa''}(\alpha_{\kappa\kappa'\kappa''}A_{\kappa'}A_{\kappa''} + \beta_{\kappa\kappa'\kappa''}A_{\kappa'}A_{-\kappa''}^\dagger \\ + \eta_{\kappa\kappa'\kappa''}A_{-\kappa'}^\dagger A_{\kappa''} + \zeta_{\kappa\kappa'\kappa''}A_{-\kappa'}^\dagger A_{-\kappa''}^\dagger), \end{aligned} \tag{13.54}$$

where the coefficients $\alpha_{\kappa\kappa'\kappa''}$, $\beta_{\kappa\kappa'\kappa''}$, $\eta_{\kappa\kappa'\kappa''}$, $\zeta_{\kappa\kappa'\kappa''}$ are to be determined so as to satisfy (13.53). The expression (13.54) includes all nontrivial contributions to $B_\kappa^\dagger$, since any other contribution, say $\tilde{B}_\kappa^\dagger$, will be required by (13.53) to satisfy

$$[\mathscr{H}_2, \tilde{B}_\kappa^\dagger] = (\hbar\omega_\kappa + i\gamma_\kappa)\tilde{B}_\kappa^\dagger,$$

and can in no way help to remove the leading-order phonon–phonon interactions. With (13.54), a straightforward calculation gives

$$\begin{aligned} [\mathscr{H}_2, B_\kappa^\dagger] = \sum_{\kappa'\kappa''} \Phi_{\kappa\kappa'\kappa''} \\ \times \{-\alpha_{\kappa\kappa'\kappa''}(\hbar\omega_{\kappa'} + \hbar\omega_{\kappa''})A_{\kappa'}A_{\kappa''} - \beta_{\kappa\kappa'\kappa''}(\hbar\omega_{\kappa'} - \hbar\omega_{\kappa''})A_{\kappa'}A_{-\kappa''}^\dagger \\ + \eta_{\kappa\kappa'\kappa''}(\hbar\omega_{\kappa'} - \hbar\omega_{\kappa''})A_{-\kappa'}^\dagger A_{\kappa''} + \zeta_{\kappa\kappa'\kappa''}(\hbar\omega_{\kappa'} + \hbar\omega_{\kappa''})A_{-\kappa'}^\dagger A_{-\kappa''}^\dagger\}, \end{aligned} \tag{13.55}$$

where we have used $\hbar\omega_\kappa = \hbar\omega_{-\kappa}$. Now with the contributions (13.49), (13.54), and (13.55), equation (13.53) can be written as

$$\begin{aligned} \sum_{\kappa'\kappa''} \Phi_{\kappa\kappa'\kappa''}\{[3 - \alpha_{\kappa\kappa'\kappa''}(\hbar\omega_\kappa + \hbar\omega_{\kappa'} + \hbar\omega_{\kappa''} + i\gamma_\kappa)]A_{\kappa'}A_{\kappa''} \\ + [3 - \beta_{\kappa\kappa'\kappa''}(\hbar\omega_\kappa + \hbar\omega_{\kappa'} - \hbar\omega_{\kappa''} + i\gamma_\kappa)]A_{\kappa'}A_{-\kappa''}^\dagger \\ + [3 - \eta_{\kappa\kappa'\kappa''}(\hbar\omega_\kappa - \hbar\omega_{\kappa'} + \hbar\omega_{\kappa''} + i\gamma_\kappa)]A_{-\kappa'}^\dagger A_{\kappa''} \\ + [3 - \zeta_{\kappa\kappa'\kappa''}(\hbar\omega_\kappa - \hbar\omega_{\kappa'} - \hbar\omega_{\kappa''} + i\gamma_\kappa)]A_{-\kappa'}^\dagger A_{-\kappa''}^\dagger\} = 0. \end{aligned} \tag{13.56}$$

An operator is zero if and only if it produces zero when operating on any state function, so (13.56) is satisfied if and only if the coefficient of every operator term vanishes. This requirement then determines the coefficients in the $B_\kappa^\dagger$ operators as follows.

$$\begin{aligned} \alpha_{\kappa\kappa'\kappa''} &= 3(\hbar\omega_\kappa + \hbar\omega_{\kappa'} + \hbar\omega_{\kappa''} + i\gamma_\kappa)^{-1}, \\ \beta_{\kappa\kappa'\kappa''} &= 3(\hbar\omega_\kappa + \hbar\omega_{\kappa'} - \hbar\omega_{\kappa''} + i\gamma_\kappa)^{-1}, \\ \eta_{\kappa\kappa'\kappa''} &= 3(\hbar\omega_\kappa - \hbar\omega_{\kappa'} + \hbar\omega_{\kappa''} + i\gamma_\kappa)^{-1}, \\ \zeta_{\kappa\kappa'\kappa''} &= 3(\hbar\omega_\kappa - \hbar\omega_{\kappa'} - \hbar\omega_{\kappa''} + i\gamma_\kappa)^{-1}. \end{aligned} \tag{13.57}$$

The first-order corrections $B_\kappa^\dagger$ to the phonon creation operators are now determined by (13.54) and (13.57) so that the renormalized operators $A_\kappa^\dagger + B_\kappa^\dagger$ satisfy their role as creation operators of $\mathscr{H}$ to order $\epsilon\hbar\omega_\kappa$. The renormalized phonons should also be independent bosons to the same order, and this requires the following commutator relations to be satisfied.

$$[A_\kappa + B_\kappa, A_{\kappa'}^\dagger + B_{\kappa'}^\dagger] = \delta_{\kappa\kappa'} + O(\epsilon^2); \tag{13.58}$$

$$[A_\kappa + B_\kappa, A_{\kappa'} + B_{\kappa'}] = O(\epsilon^2). \tag{13.59}$$

The zeroth-order terms in these equations are already satisfied by the zeroth-order phonon–operator commutators (13.44), and the first-order terms are

$$[A_\kappa, B_{\kappa'}^\dagger] + [B_\kappa, A_{\kappa'}^\dagger] = 0, \tag{13.60}$$

$$[A_\kappa, B_{\kappa'}] + [B_\kappa, A_{\kappa'}] = 0. \tag{13.61}$$

Note that B_κ is the Hermitian conjugate of (13.54), and since $\Phi^*_{\kappa\kappa'\kappa''} = \Phi_{-\kappa,-\kappa',-\kappa''}$ according to (10.78), B_κ is

$$B_\kappa = \sum_{\kappa'\kappa''} \Phi_{-\kappa,-\kappa',-\kappa''}(\alpha^*_{\kappa\kappa'\kappa''} A^\dagger_{\kappa''} A^\dagger_{\kappa'} + \beta^*_{\kappa\kappa'\kappa''} A_{-\kappa''} A^\dagger_{\kappa'} + \eta^*_{\kappa\kappa'\kappa''} A^\dagger_{\kappa''} A_{-\kappa'} + \zeta^*_{\kappa\kappa'\kappa''} A_{-\kappa''} A_{-\kappa'}). \tag{13.62}$$

Without loss of generality, $\gamma_\kappa = \gamma_{-\kappa}$ can be taken, so that from (13.57) the coefficients $\alpha_{\kappa\kappa'\kappa''}$, $\beta_{\kappa\kappa'\kappa''}$, $\eta_{\kappa\kappa'\kappa''}$, and $\zeta_{\kappa\kappa'\kappa''}$ remain the same when the sign of any κ, κ', or κ'' is changed. Then letting $\kappa' \to -\kappa''$ and $\kappa'' \to -\kappa'$ inside the sum in (13.62), B_κ may be written

$$B_\kappa = \sum_{\kappa'\kappa''} \Phi_{-\kappa\kappa'\kappa''}(\alpha^*_{\kappa\kappa''\kappa'} A^\dagger_{-\kappa'} A^\dagger_{-\kappa''} + \beta^*_{\kappa\kappa''\kappa'} A_{\kappa'} A^\dagger_{-\kappa''} + \eta^*_{\kappa\kappa''\kappa'} A^\dagger_{-\kappa'} A_{\kappa''} + \zeta^*_{\kappa\kappa''\kappa'} A_{\kappa'} A_{\kappa''}). \tag{13.63}$$

Now with $B_\kappa^\dagger$ given by (13.54) and B_κ given by (13.63), the first-order commutators are as follows. Equation (13.60) is

$$\sum_{\kappa''} \Phi_{-\kappa\kappa'\kappa''}\{(\beta^*_{\kappa\kappa''\kappa'} + \eta^*_{\kappa\kappa'\kappa''} + \zeta_{\kappa'\kappa''\kappa} + \zeta_{\kappa'\kappa\kappa''})A^\dagger_{-\kappa''} + (\beta_{\kappa'\kappa''\kappa} + \eta_{\kappa'\kappa\kappa''} + \zeta^*_{\kappa\kappa'\kappa''} + \zeta^*_{\kappa\kappa''\kappa'})A_{\kappa''}\} = 0; \tag{13.64}$$

and (13.61) is

$$\sum_{\kappa''} \Phi_{-\kappa,-\kappa',\kappa''}\{(\alpha^*_{\kappa'\kappa\kappa''} + \alpha^*_{\kappa'\kappa''\kappa} - \alpha^*_{\kappa\kappa'\kappa''} - \alpha^*_{\kappa\kappa''\kappa'})A^\dagger_{-\kappa''} + (\beta^*_{\kappa'\kappa\kappa''} + \eta^*_{\kappa'\kappa''\kappa} - \beta^*_{\kappa\kappa'\kappa''} - \eta^*_{\kappa\kappa''\kappa'})A_{\kappa''}\} = 0. \tag{13.65}$$

The coefficient of every operator in (13.64) and (13.65) must vanish, so that

$$\beta^*_{\kappa\kappa''\kappa'} + \eta^*_{\kappa\kappa'\kappa''} + \zeta_{\kappa'\kappa''\kappa} + \zeta_{\kappa'\kappa\kappa''} = 0, \tag{13.66}$$

$$\beta_{\kappa'\kappa''\kappa} + \eta_{\kappa'\kappa\kappa''} + \zeta^*_{\kappa\kappa'\kappa''} + \zeta^*_{\kappa\kappa''\kappa'} = 0, \tag{13.67}$$

$$\alpha^*_{\kappa'\kappa\kappa''} + \alpha^*_{\kappa'\kappa''\kappa} - \alpha^*_{\kappa\kappa'\kappa''} - \alpha^*_{\kappa\kappa''\kappa'} = 0, \tag{13.68}$$

$$\beta^*_{\kappa'\kappa\kappa''} + \eta^*_{\kappa'\kappa''\kappa} - \beta^*_{\kappa\kappa'\kappa''} - \eta^*_{\kappa\kappa''\kappa'} = 0. \tag{13.69}$$

It is easily verified that the conditions (13.66)–(13.69) hold if and only if γ_κ are the same for all κ; henceforth this will be taken:

$$\gamma_\kappa = \gamma, \quad \text{for all } \kappa. \tag{13.70}$$

With this the renormalized-phonon-creation operators $A^\dagger_\kappa + B^\dagger_\kappa$ satisfy the independent boson commutators to first order in ϵ, as given by (13.58) and (13.59).

SECOND-ORDER PHONON ENERGIES

The second-order contributions to the phonon energies are $\hbar\omega_{2\kappa} \sim \epsilon^2\hbar\omega_\kappa$, and these may be calculated from the general equation (13.36). Since $\hbar\omega_{1\kappa} = 0$ by (13.50), the proper expression in the present notation is

$$(n_\kappa + 1)^{1/2}\hbar\omega_{2\kappa} = \langle\cdots n_\kappa + 1\cdots|\,[\mathscr{H}_3,B^\dagger_\kappa] + [\mathscr{H}_4,A^\dagger_\kappa]\,|\cdots n_\kappa\cdots\rangle. \tag{13.71}$$

The commutator $[\mathscr{H}_4,A^\dagger_\kappa]$ is calculated from (13.43) for $\mathscr{H}_4$ and is

$$[\mathscr{H}_4,A^\dagger_\kappa] = 4\sum_{\kappa'\kappa''\kappa'''}\Phi_{\kappa\kappa'\kappa''\kappa'''}(A_{\kappa'} + A^\dagger_{-\kappa'}) \times (A_{\kappa''} + A^\dagger_{-\kappa''})(A_{\kappa'''} + A^\dagger_{-\kappa'''}), \tag{13.72}$$

where we have combined four terms by using the symmetry (10.77) of $\Phi_{\kappa\kappa'\kappa''\kappa'''}$ in its indices. Now the only terms in (13.72) which will contribute to the matrix element (13.71) are those containing an $A^\dagger_\kappa$ and a product $A^\dagger_{\kappa'}A_{\kappa'}$ for any κ'; picking out all such terms gives the effective operator $[\mathscr{H}_4,A^\dagger_\kappa]$ for the matrix element:

$$\begin{aligned}[\mathscr{H}_4,A^\dagger_\kappa] \to 4\sum_{\kappa'}\{&\Phi_{\kappa,\kappa',-\kappa,-\kappa'}(A_{\kappa'}A^\dagger_\kappa A^\dagger_{\kappa'} + A^\dagger_{-\kappa'}A^\dagger_\kappa A_{-\kappa'})\\ &+ \Phi_{\kappa,\kappa',-\kappa',-\kappa}(A_{\kappa'}A^\dagger_{\kappa'}A^\dagger_\kappa + A^\dagger_{-\kappa'}A_{-\kappa'}A^\dagger_\kappa)\\ &+ \Phi_{\kappa,-\kappa,\kappa',-\kappa'}(A^\dagger_\kappa A_{\kappa'}A^\dagger_{\kappa'} + A^\dagger_\kappa A^\dagger_{-\kappa'}A_{-\kappa'})\}.\end{aligned} \tag{13.73}$$

In order to evaluate the matrix element it is convenient to commute the operators to normal order; for example,

$$A_{\kappa'}A^\dagger_\kappa A^\dagger_{\kappa'} = A^\dagger_\kappa A^\dagger_{\kappa'}A_{\kappa'} + A^\dagger_\kappa + A^\dagger_{\kappa'}\delta_{\kappa\kappa'}. \tag{13.74}$$

The term here involving $\delta_{\kappa\kappa'}$ can be dropped, since it removes the $\Sigma_{\kappa'}$ in (13.73), and hence it is of relative order N_0^{-1}, since there are of order N_0 terms in $\Sigma_{\kappa'}$. When all operators are in normal order, the matrix elements are evaluated according to [see (13.10) and (13.14)]

$$\langle\cdots n_\kappa + 1 \cdots| A_\kappa^\dagger |\cdots n_\kappa \cdots\rangle = (n_\kappa + 1)^{1/2}, \tag{13.75}$$

$$\langle\cdots n_\kappa + 1 \cdots| A_\kappa^\dagger A_{\kappa'}^\dagger A_{\kappa'} |\cdots n_\kappa \cdots\rangle = (n_\kappa + 1)^{1/2} n_{\kappa'}. \tag{13.76}$$

Then making use of the symmetry of $\Phi_{\kappa\kappa'\kappa''\kappa'''}$ in its indices, the contribution to $\hbar\omega_{2\kappa}$ from $[\mathscr{H}_4, A_\kappa^\dagger]$ in (13.71) is

$$\hbar\omega_{2\kappa}(\text{quartic}) = 12 \sum_{\kappa'} \Phi_{\kappa,-\kappa,\kappa',-\kappa'}(n_{\kappa'} + n_{-\kappa'} + 1). \tag{13.77}$$

Evaluation of the matrix element involving $[\mathscr{H}_3, B_\kappa^\dagger]$ in (13.71) proceeds in the same manner as the above calculation of the quartic contribution to $\hbar\omega_{2\kappa}$. Calculation of $[\mathscr{H}_3, B_\kappa^\dagger]$ is straightforward but lengthy; again we pick out all operator terms of the form $A_\kappa^\dagger A_{\kappa'}^\dagger A_{\kappa'}$ for any κ', commute to normal order and drop terms containing $\delta_{\kappa\kappa'}$ as of relative order N_0^{-1}, and use the symmetry of $\Phi_{\kappa\kappa'\kappa''}$ in its indices to group together the resulting terms. In addition we use the symmetries

$$\beta_{\kappa\kappa'\kappa''} = \eta_{\kappa\kappa''\kappa'}, \tag{13.78}$$

$$\zeta_{\kappa\kappa'\kappa''} = \zeta_{\kappa\kappa''\kappa'}, \tag{13.79}$$

which follow from (13.57). The result for the contribution to $\hbar\omega_{2\kappa}$ from $[\mathscr{H}_3, B_\kappa^\dagger]$ in (13.71) is

$$\begin{aligned} \hbar\omega_{2\kappa}&(\text{cubic}) \\ &= -6 \sum_{\kappa'\kappa''} \{\Phi_{\kappa\kappa'\kappa''}\Phi_{-\kappa,-\kappa',-\kappa''}[\alpha_{\kappa\kappa'\kappa''}(n_{\kappa'} + n_{\kappa''} + 1) + \beta_{\kappa\kappa'\kappa''}(n_{-\kappa''} - n_{\kappa'}) \\ &\quad + \eta_{\kappa\kappa'\kappa''}(n_{-\kappa'} - n_{\kappa''}) - \zeta_{\kappa\kappa'\kappa''}(n_{-\kappa'} + n_{-\kappa''} + 1)] \\ &\quad + \Phi_{\kappa,-\kappa,\kappa''}\Phi_{\kappa',-\kappa',-\kappa''}(\eta_{\kappa\kappa\kappa''} - \zeta_{\kappa\kappa\kappa''})(n_{\kappa'} + n_{-\kappa'} + 1)\}. \end{aligned} \tag{13.80}$$

The renormalized excitation energies have the following significance. If the number of renormalized phonons in each state κ' is $n_{\kappa'}$, then the energy required to add one renormalized phonon κ to the system, i.e., to increase n_κ to $n_\kappa + 1$, is

$$\hbar\omega_\kappa + \hbar\omega_{2\kappa}(\text{quartic}) + \hbar\omega_{2\kappa}(\text{cubic}), \tag{13.81}$$

correct to order $\epsilon^2\hbar\omega_\kappa$.

The renormalization procedure is now complete to the order required here, and it is instructive to discuss the renormalized phonon picture of the lattice-dynamics problem. To begin with the bare (zeroth-order) phonons are created by the boson operators $A_\kappa^\dagger$, and these phonons interact via three-phonon interactions in first order in ϵ (the Hamiltonian term $\mathscr{H}_3$), and via four-phonon interactions in second order in ϵ (the Hamiltonian term $\mathscr{H}_4$). The renormalized phonons are created by the operators $A_\kappa^\dagger + B_\kappa^\dagger$, and these have interactions in second order in ϵ, but not in first order. Also, the renormalized phonons are not exact bosons, but are bosons only to order ϵ, according to the renormalized-operator commutators (13.58) and (13.59). The physical picture of a renormalized phonon is apparent from the first-order creation operator $B_\kappa^\dagger$ of (13.54): in addition to the bare phonon κ, the first-order renormalized phonon carries with it a cloud of bare phonon pairs, phonon–hole pairs, and hole pairs, and each pair has net crystal "momentum" κ, i.e., has wave vector $\mathbf{k} \pm \mathbf{Q}$. [The associated bare phonons correspond to creation operators, and the holes correspond to annihilation operators, in (13.54).] The associated cloud is of order ϵ compared to the initial bare phonon, and is designed just right to screen out first-order interactions between the renormalized phonons.

Both the quartic and cubic contributions to $\hbar\omega_{2\kappa}$, given by (13.77) and (13.80), respectively, are of second order *relative* to $\hbar\omega_\kappa$, i.e., are of order $\epsilon^2\hbar\omega_\kappa$. These expressions for $\hbar\omega_{2\kappa}$ are not strictly numbers, since each depends on the occupation numbers of all the phonons, represented by the set of n_κ. This means that the interactions among the phonons cannot be renormalized away in second order, and that true normal modes of the lattice-dynamics problem which are independent to the order $\epsilon^2\hbar\omega_\kappa$ cannot be described by the quantum numbers κ. The operator-renormalization procedure can still be carried to higher orders, so as to calculate the higher-order contributions to the interaction energies, but this will not be necessary here.

It will be recalled from (10.81) and (10.82) that $\Phi_{\kappa\kappa'\kappa''}$ and $\Phi_{\kappa\kappa'\kappa''\kappa'''}$ contain "momentum"-conserving δ functions. The δ function is automatically satisfied in each contribution to (13.77) for $\hbar\omega_{2\kappa}$(quartic), since $\Phi_{\kappa,-\kappa,\kappa',-\kappa'}$ contains $\delta(\mathbf{k} - \mathbf{k} + \mathbf{k}' - \mathbf{k}') = \delta(0) = 1$, and the $\Sigma_{\kappa'}$ in (13.77) does not include Umklapp processes. The δ function *does* restrict the sums in (13.80) for $\hbar\omega_{2\kappa}$(cubic); the terms $\Phi_{\kappa\kappa'\kappa''}\Phi_{-\kappa,-\kappa',-\kappa''} = |\Phi_{\kappa\kappa'\kappa''}|^2$ contain $\delta(\mathbf{k} + \mathbf{k}' + \mathbf{k}'')$ and allow Umklapp processes, and the terms $\Phi_{\kappa,-\kappa,\kappa''}\Phi_{\kappa',-\kappa',-\kappa''}$ contain a $\delta(\mathbf{k}'')$ and do not include Umklapp processes. In these latter terms [last line of (13.80)], $\mathbf{k}'' = 0$ is required, and for the acoustic modes the value $\mathbf{k}'' = 0$ should always be omitted since these are modes of uniform translation of the crystal. Therefore the $\Sigma_{\kappa''} = \Sigma_{\mathbf{k}''s''}$ should include only the optic modes $s'' = 4,5,\cdots,3n$, so the last line of (13.80) appears only for nonprimitive lattices.

Total System Energy Levels

The total system energy levels $\mathscr{E} = \mathscr{E}(\cdots n_\kappa \cdots)$ may be written as a series in the expansion parameter ϵ,

$$\mathscr{E} = \Phi_0 + \mathscr{E}_2 + \mathscr{E}_4. \tag{13.82}$$

The ground-state energy corresponds to all $n_\kappa = 0$, i.e., no phonons, and is written

$$\mathscr{E}_G = \Phi_0 + G_2 + G_4. \tag{13.83}$$

It is convenient to calculate first the ground-state energy $\mathscr{E}_G$, and then to calculate the energy due to excitation of the renormalized phonons, to obtain the complete energy levels $\mathscr{E}$.

The second-order (harmonic) ground-state energy is given by (13.41),

$$G_2 = \tfrac{1}{2} \sum_\kappa \hbar\omega_\kappa; \tag{13.84}$$

this is the well-known zero-point vibrational energy of the zeroth-order (harmonic) phonons. The fourth-order ground state energy G_4 is easily calculated by perturbation theory according to

$$G_4 = \langle 0| \mathscr{H}_4 |0\rangle + \sum_{(\cdots n_\kappa \cdots)}' \frac{|\langle \cdots n_\kappa \cdots | \mathscr{H}_3 |0\rangle|^2}{[G_2 - \mathscr{E}_2(\cdots n_\kappa \cdots)]}, \tag{13.85}$$

where $|0\rangle$ is the harmonic phonon vacuum $|\cdots 0_\kappa \cdots\rangle$ and the first-order term $\langle 0| \mathscr{H}_3 |0\rangle$ has been dropped, since $\mathscr{H}_3$ has no diagonal elements. To calculate $\langle 0| \mathscr{H}_4 |0\rangle$, the expression (13.43) for $\mathscr{H}_4$ is multiplied out, all terms are dropped except those containing two creation and two annihilation operators (diagonal terms), and any remaining term with an annihilation operator on the right is dropped, since

$$A_\kappa |0\rangle = 0. \tag{13.86}$$

At this point the matrix element looks like

$$\begin{aligned}\langle 0| \mathscr{H}_4 |0\rangle = \sum_{\kappa\kappa'\kappa''\kappa'''} \Phi_{\kappa\kappa'\kappa''\kappa'''} \langle 0| &A_\kappa A_{\kappa'} A^\dagger_{-\kappa''} A^\dagger_{-\kappa'''} \\ &+ A_\kappa A^\dagger_{-\kappa'} A_{\kappa''} A^\dagger_{-\kappa'''} + A^\dagger_{-\kappa} A_{\kappa'} A_{\kappa''} A^\dagger_{-\kappa'''} |0\rangle.\end{aligned} \tag{13.87}$$

Now the operators in (13.87) are commuted to normal order with the commutators (13.44), and terms with an annihilation operator on the right are again dropped by (13.86). Since $\langle 0 \mid 0\rangle = 1$, the result is

$$\langle 0| \mathscr{H}_4 |0\rangle = \sum_{\kappa\kappa'\kappa''\kappa'''} \Phi_{\kappa\kappa'\kappa''\kappa'''}(\delta_{\kappa,-\kappa''}\delta_{\kappa',-\kappa'''} + \delta_{\kappa,-\kappa'''}\delta_{\kappa',-\kappa''} + \delta_{\kappa,-\kappa'}\delta_{\kappa'',-\kappa'''}),$$

or in view of the symmetry of $\Phi_{\kappa\kappa'\kappa''\kappa'''}$ in its indices,

$$\langle 0|\mathscr{H}_4|0\rangle = 3\sum_{\kappa\kappa'} \Phi_{\kappa,-\kappa,\kappa',-\kappa'}. \tag{13.88}$$

To evaluate the second-order perturbation term involving $\mathscr{H}_3$ in (13.85), the expression (13.42) for $\mathscr{H}_3$ is multiplied out, the operators are commuted to normal order, and terms with an annihilation operator on the right are dropped to give the result

$$\mathscr{H}_3|0\rangle = \sum_{\kappa\kappa'\kappa''} \Phi_{\kappa\kappa'\kappa''}(3A^\dagger_{-\kappa}\delta_{\kappa',-\kappa''} + A^\dagger_{-\kappa}A^\dagger_{-\kappa'}A^\dagger_{-\kappa''})|0\rangle. \tag{13.89}$$

Three terms linear in creation operators have been grouped together here by interchanging indices inside the sum. The $A^\dagger_{-\kappa}$ operator in (13.89) will couple $|0\rangle$ to a state $\langle\cdots n_\kappa\cdots|$ which has only one phonon, say κ^*, while all other $n_{\kappa'}$ are zero. Then

$$\langle\cdots 1_{\kappa*}\cdots|\sum_{\kappa\kappa'\kappa''}\Phi_{\kappa\kappa'\kappa''}3A^\dagger_{-\kappa}\,\delta_{\kappa',-\kappa''}|0\rangle = 3\sum_{\kappa'}\Phi_{-\kappa*,\kappa',-\kappa'}.$$

The energy denominator is

$$[G_2 - \mathscr{E}_2(\cdots 1_{\kappa*}\cdots)]^{-1} = -(\hbar\omega_{\kappa*})^{-1},$$

and the $\Sigma'_{(\cdots n_\kappa\cdots)}$ over all possible combinations of the set of n_κ in $\langle\cdots n_\kappa\cdots|$ reduces to

$$\sum_{\kappa*}\left|3\sum_{\kappa'}\Phi_{-\kappa*,\kappa',-\kappa'}\right|^2(-\hbar\omega_{\kappa*})^{-1} = -9\sum_{\kappa\kappa'\kappa''}\Phi_{\kappa,-\kappa,\kappa''}\Phi_{\kappa',-\kappa',-\kappa''}(\hbar\omega_{\kappa''})^{-1}. \tag{13.90}$$

The same procedure is used to evaluate the $A^\dagger_{-\kappa}A^\dagger_{-\kappa'}A^\dagger_{-\kappa''}$ terms in (13.89), and the contribution of these terms to (13.85) for G_4 is found to be

$$-6\sum_{\kappa\kappa'\kappa''}|\Phi_{\kappa\kappa'\kappa''}|^2(\hbar\omega_\kappa + \hbar\omega_{\kappa'} + \hbar\omega_{\kappa''})^{-1}. \tag{13.91}$$

The complete result for G_4 is then the sum of (13.88), (13.90), and (13.91):

$$G_4 = 3\sum_{\kappa\kappa'}\Phi_{\kappa,-\kappa,\kappa',-\kappa'} - 18\hbar^{-1}\sum_{\kappa\kappa'\kappa''}\{\tfrac{1}{3}|\Phi_{\kappa\kappa'\kappa''}|^2(\omega_\kappa+\omega_{\kappa'}+\omega_{\kappa''})^{-1} + \tfrac{1}{2}\Phi_{\kappa,-\kappa,\kappa''}\Phi_{\kappa',-\kappa',-\kappa''}(\omega_{\kappa''})^{-1}\}. \tag{13.92}$$

Again the last term in (13.92) contains a $\delta(\mathbf{k}'')$ and should be included only for s'' = optic modes, and hence only for nonprimitive lattices.

The excitation energies are obtained simply by adding up the energies of the renormalized phonons. The zeroth-order phonons are all independent, so the energy of the set of n_κ phonons in each phonon state κ is just

$$\mathscr{E}_2 - G_2 = \sum_\kappa n_\kappa\hbar\omega_\kappa, \tag{13.93}$$

or

$$\mathscr{E}_2 = \sum_{\kappa} (n_\kappa + \tfrac{1}{2})\hbar\omega_\kappa. \tag{13.94}$$

This result corresponds to replacing each $A_\kappa^\dagger A_\kappa$ in $\mathscr{H}_2$ by its eigenvalue n_κ.

Because of the second-order interactions among the renormalized phonons, the total excitation energy $\mathscr{E}_4 - G_4$ is *not* simply the sum of the single-phonon excitation energies; that is,

$$\mathscr{E}_4 - G_4 \neq \sum_{\kappa} n_\kappa \hbar\omega_{2\kappa}. \tag{13.95}$$

This is because each $\hbar\omega_{2\kappa}$ depends on all the occupation numbers $n_{\kappa'}$, and if we start from the vacuum and create renormalized phonons one at a time, the excitation energies $\hbar\omega_{2\kappa}$ change with each added phonon. Consider the terms in $\hbar\omega_{2\kappa}$ which contain a sum like $\Sigma_{\kappa'} S_{\kappa\kappa'} n_{\kappa'}$, where $S_{\kappa\kappa'}$ are numerical coefficients. By going through the mental exercise of creating one renormalized phonon at a time until the set corresponding to the numbers n_κ is obtained, it is easily verified that the total excitation energy from such a contribution is $\frac{1}{2}\Sigma_{\kappa\kappa'} S_{\kappa\kappa'} n_\kappa n_{\kappa'}$. In other words, since this energy contribution is a sum of two-phonon interactions, a factor of $\frac{1}{2}$ is needed to avoid counting each interaction twice. In the same way, if the excitation energies are calculated out to the order of three-phonon interactions, i.e., if they contain terms like $\Sigma_{\kappa'\kappa''} S_{\kappa\kappa'\kappa''} n_{\kappa'} n_{\kappa''}$, then the total excitation energy is

$$(1/3!)\, \Sigma_{\kappa\kappa'\kappa''} S_{\kappa\kappa'\kappa''} n_\kappa n_{\kappa'} n_{\kappa''},$$

and so on for higher-order interactions. This is a general rule for calculating the total energy of a system of interacting particles, and it is emphasized here since the point has not been universally recognized. For the constant terms in $\hbar\omega_{2\kappa}$, i.e., those independent of $n_{\kappa'}$, the counting factor is of course 1. Note also that the counting factor might appear to be incorrect for the diagonal terms ($n_\kappa n_\kappa$ terms), but these give a relative contribution of order N_0^{-1} to the total energy and hence are unimportant. In fact additional contributions to such diagonal terms were consistently dropped in the derivation of $\hbar\omega_{2\kappa}$, since they were of relative order N_0^{-1}.

Now the total fourth-order excitation energy is written down directly from (13.77) for $\hbar\omega_{2\kappa}$(quartic) and (13.80) for $\hbar\omega_{2\kappa}$(cubic), with the counting factor of $\frac{1}{2}$ properly included:

$$\begin{aligned} \mathscr{E}_4 - G_4 &= 6 \sum_{\kappa\kappa'} \Phi_{\kappa,-\kappa,\kappa',-\kappa'} (n_{\kappa'} + n_{-\kappa'} + 2) n_\kappa \\ &- 3 \sum_{\kappa\kappa'\kappa''} \{ |\Phi_{\kappa\kappa'\kappa''}|^2 [\alpha_{\kappa\kappa'\kappa''}(n_{\kappa'} + n_{\kappa''} + 2) n_\kappa + \beta_{\kappa\kappa'\kappa''}(n_{-\kappa''} - n_{\kappa'}) n_\kappa \\ &\qquad + \eta_{\kappa\kappa'\kappa''}(n_{-\kappa'} - n_{\kappa''}) n_\kappa - \zeta_{\kappa\kappa'\kappa''}(n_{-\kappa'} + n_{-\kappa''} + 2) n_\kappa] \\ &\qquad + \Phi_{\kappa,-\kappa,\kappa''} \Phi_{\kappa',-\kappa',-\kappa''} (\eta_{\kappa\kappa\kappa''} - \zeta_{\kappa\kappa\kappa''})(n_{\kappa'} + n_{-\kappa'} + 2) n_\kappa \}. \end{aligned} \tag{13.96}$$

With this the system energy levels are calculated to fourth order in ϵ, i.e., to order $\epsilon^2\hbar\omega_\kappa$, and we are prepared to calculate the partition function directly from these energy levels.

14. SELF-CONSISTENT PHONONS

Statistical Perturbation Method

In several cases of practical interest it is not appropriate to treat the phonon interaction effects by perturbation theory. For solid helium at moderate pressures, for example, the bare phonon ω_κ^2 are generally negative, the anharmonic effects being sufficiently large to make the renormalized squared frequencies positive. In addition there are many crystals which undergo ferroelectriclike phase changes in which one or more "soft" optic modes (Cochran modes), or in some cases one or more "soft" acoustic modes, have frequencies which become very small at the transition. The theoretical picture is that at temperatures above the transition temperature, certain of the harmonic ω_κ^2 are negative but the renormalized phonons are stabilized by anharmonicity, and as the temperature is decreased the phonon energy renormalization decreases and these soft mode frequencies become smaller, so the crystal becomes unstable toward a displacive or otherwise deformation-related transition. We note in passing that the transition will generally take place before any of the renormalized phonon frequencies go to zero. Finally even for normal crystals, where the harmonic ω_κ^2 are all positive, the anharmonic frequency renormalization increases with temperature at high temperatures, and if the crystal does not melt first, the frequency shifts and widths will become of the same order of magnitude as the bare ω_κ. Although the main emphasis of this book is the study of leading-order anharmonic effects in thermodynamic properties of crystals, we will briefly discuss the self-consistent phonon theory which is presently being developed for application to the large-anharmonicity aspects of lattice dynamics.*, †

Again the operator-renormalization technique provides a particularly simple approach. We will use the statistical perturbation method, which is a general method for approximating statistical averages in a many-body problem.‡ To begin, suppose that one or more approximate creation operators $\theta_i^\dagger$ are found which satisfy the Hamiltonian commutator equations

$$[\mathscr{H},\theta_i^\dagger] = \omega_i\theta_i^\dagger + R_i^\dagger, \qquad (14.1)$$

* P. Choquard, *The Anharmonic Crystal*, W. A. Benjamin Inc., New York, 1967.

† N. R. Werthamer, *Am. J. Phys.* **37**, 763 (1969).

‡ D. C. Wallace, *Phys. Rev.* **152**, 261 (1966).

where ω_i are real positive numbers and the remainders $R_i^\dagger$ are "small" operators. The $\theta_i^\dagger$ and θ_i need not represent good bosons or fermions; in other words, we are not concerned about the commutators and anticommutators among these operators. Further, the energies ω_i and the operators $\theta_i^\dagger$ may be temperature-dependent from the beginning. The $R_i^\dagger$ are to be small in the sense that they give small contributions to the statistical averages which it is desired to calculate. The Hermitian conjugate of (14.1) is

$$[\mathscr{H},\theta_i] = -\omega_i\theta_i - R_i, \tag{14.2}$$

and to the extent that the remainders may be neglected the $\theta_i^\dagger$ create excitations of energy ω_i, while the θ_i annihilate excitations of energy ω_i.

If the remainders are neglected completely, it is possible to obtain a simple expression for the statistical averages $\langle\theta_i^\dagger\chi\rangle$, where χ is any operator. To do this, write (14.1) with the remainder omitted,

$$[\mathscr{H},\theta_i^\dagger] = \omega_i\theta_i^\dagger, \tag{14.3}$$

or

$$\mathscr{H}\theta_i^\dagger = \theta_i^\dagger(\mathscr{H} + \omega_i). \tag{14.4}$$

Now operate on (14.4) with $\mathscr{H}$ to obtain

$$\mathscr{H}^2\theta_i^\dagger = \mathscr{H}\theta_i^\dagger(\mathscr{H} + \omega_i) = \theta_i^\dagger(\mathscr{H} + \omega_i)^2, \tag{14.5}$$

where the last equality follows with (14.4) again. Hence by induction it follows that

$$\mathscr{H}^n\theta_i^\dagger = \theta_i^\dagger(\mathscr{H} + \omega_i)^n, \quad n = 0,1, \cdots. \tag{14.6}$$

Then if α is any constant,

$$\begin{aligned} e^{\alpha\mathscr{H}}\theta_i^\dagger &= \sum_{n=0}^{\infty}(n!)^{-1}\alpha^n\mathscr{H}^n\theta_i^\dagger = \sum_{n=0}^{\infty}(n!)^{-1}\alpha^n\theta_i^\dagger(\mathscr{H} + \omega_i)^n \\ &= \theta_i^\dagger e^{\alpha(\mathscr{H}+\omega_i)}. \end{aligned} \tag{14.7}$$

The partition function Z is given by

$$Z = \mathrm{Tr}\, e^{-\mathscr{H}/KT}, \tag{14.8}$$

and the statistical average $\langle\theta_i^\dagger\chi\rangle$ is defined by

$$\langle\theta_i^\dagger\chi\rangle = Z^{-1}\,\mathrm{Tr}\,\theta_i^\dagger\chi\, e^{-\mathscr{H}/KT}. \tag{14.9}$$

By the cyclic permutation theorem for traces, (14.9) may be written

$$\langle\theta_i^\dagger\chi\rangle = Z^{-1}\,\mathrm{Tr}\,\chi\, e^{-\mathscr{H}/KT}\theta_i^\dagger, \tag{14.10}$$

and with (14.7), where $\alpha = -(KT)^{-1}$,

$$\langle\theta_i^\dagger\chi\rangle = Z^{-1}\,\mathrm{Tr}\,\chi\theta_i^\dagger\, e^{-(\mathscr{H}+\omega_i)/KT}. \tag{14.11}$$

Finally, this may be written in the simple form

$$\langle\theta_i^\dagger\chi\rangle = e^{-\omega_i/KT}\langle\chi\theta_i^\dagger\rangle. \tag{14.12}$$

It is useful to write (14.12) in a form containing operator commutators or anticommutators. This is done by transforming (14.12) to

$$[e^{\omega_i/KT} \pm 1]\langle\theta_i^\dagger\chi\rangle = \langle\chi\theta_i^\dagger \pm \theta_i^\dagger\chi\rangle, \tag{14.13}$$

and from this it follows that

$$\langle\theta_i^\dagger\chi\rangle = [e^{\omega_i/KT} - 1]^{-1}\langle[\chi,\theta_i^\dagger]\rangle; \tag{14.14}$$

$$\langle\theta_i^\dagger\chi\rangle = [e^{\omega_i/KT} + 1]^{-1}\langle\{\chi,\theta_i^\dagger\}\rangle. \tag{14.15}$$

Equations (14.14) and (14.15) are the zeroth-order basic equations of the statistical perturbation method; zeroth-order because they do not take account of the remainder operators $R_i^\dagger$. These equations are useful because the commutators or anticommutators on the right-hand sides are generally simpler operators than are the products appearing on the left-hand sides. Take, for example, the case of bosons, for which the creation and annihilation operators satisfy the commutators

$$[\theta_i,\theta_{i'}^\dagger] = \delta_{ii'}; \qquad [\theta_i^\dagger,\theta_{i'}^\dagger] = 0. \tag{14.16}$$

Then (14.14) with $\chi = \theta_{i'}$ gives

$$\langle\theta_i^\dagger\theta_{i'}\rangle = \delta_{ii'}[e^{\omega_i/KT} - 1]^{-1}, \tag{14.17}$$

and with $\chi = \theta_{i'}^\dagger$,

$$\langle\theta_i^\dagger\theta_{i'}^\dagger\rangle = 0. \tag{14.18}$$

The result (14.17) gives the well-known statistical average boson occupation number $\langle\theta_i^\dagger\theta_i\rangle = [e^{\omega_i/KT} - 1]^{-1}$, and it is interesting to note that we have not required the complete set of boson operators and energies to find this result, but only the commutators (14.3) and (14.16) involving the single-boson operators $\theta_i^\dagger$ and θ_i.

Consider now the case of fermions, for which the creation and annihilation operators satisfy the anticommutators

$$\{\theta_i,\theta_{i'}^\dagger\} = \delta_{ii'}; \qquad \{\theta_i^\dagger,\theta_{i'}^\dagger\} = 0. \tag{14.19}$$

Then (14.15) with $\chi = \theta_{i'}$ gives

$$\langle\theta_i^\dagger\theta_{i'}\rangle = \delta_{ii'}[e^{\omega_i/KT} + 1]^{-1}, \tag{14.20}$$

and with $\chi = \theta_{i'}^\dagger$,

$$\langle\theta_i^\dagger\theta_{i'}^\dagger\rangle = 0. \tag{14.21}$$

Again (14.20) gives the statistical average fermion occupation number $\langle\theta_i^\dagger\theta_i\rangle = [e^{\omega_i/KT} + 1]^{-1}$, and no information about any other particles in the system was required to obtain this result. It should also be noted that the basic equations (14.14) and (14.15) hold for a grand canonical statistical average, with ω_i replaced by $\omega_i - \mu$ where μ is the chemical potential.

The contribution of $R_i^\dagger$ to the statistical average $\langle\theta_i^\dagger\chi\rangle$ can be calculated by a procedure similar to the derivation of the zeroth-order expression (14.12) for $\langle\theta_i^\dagger\chi\rangle$. The Hamiltonian commutator equation (14.1) can be written

$$\mathscr{H}\theta_i^\dagger = \theta_i^\dagger(\mathscr{H} + \omega_i) + R_i^\dagger, \tag{14.22}$$

and operating on this with $\mathscr{H}$ gives

$$\begin{aligned}\mathscr{H}^2\theta_i^\dagger &= \mathscr{H}\theta_i^\dagger(\mathscr{H} + \omega_i) + \mathscr{H}R_i^\dagger\\ &= \theta_i^\dagger(\mathscr{H} + \omega_i)^2 + R_i^\dagger(\mathscr{H} + \omega_i) + \mathscr{H}R_i^\dagger,\end{aligned} \tag{14.23}$$

where the last equality follows with (14.22) again. It is then easy to show by induction that

$$\mathscr{H}^n\theta_i^\dagger = \theta_i^\dagger(\mathscr{H} + \omega_i)^n + \sum_{p=0}^{n-1}\mathscr{H}^p R_i^\dagger(\mathscr{H} + \omega_i)^{n-p-1}, \quad n = 1,2,\cdots. \tag{14.24}$$

Now since

$$e^{-\mathscr{H}/KT} = \sum_{n=0}^{\infty}(n!)^{-1}(-KT)^{-n}\mathscr{H}^n,$$

equation (14.24) may be used to show

$$e^{-\mathscr{H}/KT}\theta_i^\dagger = \theta_i^\dagger\, e^{-(\mathscr{H}+\omega_i)/KT} + \sum_{n=1}^{\infty}(n!)^{-1}(-KT)^{-n}\sum_{p=0}^{n-1}\mathscr{H}^p R_i^\dagger(\mathscr{H} + \omega_i)^{n-p-1}. \tag{14.25}$$

The statistical average $\langle\theta_i^\dagger\chi\rangle$ is defined by (14.9), and with (14.25) this becomes

$$\begin{aligned}\langle\theta_i^\dagger\chi\rangle &= e^{-\omega_i/KT}\langle\chi\theta_i^\dagger\rangle\\ &\quad + Z^{-1}\,\mathrm{Tr}\,\chi\sum_{n=1}^{\infty}(n!)^{-1}(-KT)^{-n}\sum_{p=0}^{n-1}\mathscr{H}^p R_i^\dagger(\mathscr{H} + \omega_i)^{n-p-1}.\end{aligned} \tag{14.26}$$

This last is the basic equation for $\langle\theta_i^\dagger\chi\rangle$ with $R_i^\dagger$ included. The philosophy of the statistical perturbation method is that the last term in (14.26), which we call the perturbation term, may be approximated since $R_i^\dagger$ is small. A particularly useful average is $\langle\theta_i^\dagger\theta_i\rangle$, obtained by setting $\chi = \theta_i$, and for this the perturbation term is easily evaluated as a power series in $R_i^\dagger$ as follows. Equation (14.2) for $[\mathscr{H},\theta_i]$ is written

$$\theta_i\mathscr{H} = (\mathscr{H} + \omega_i)\theta_i + R_i, \tag{14.27}$$

and again by induction it follows that

$$\theta_i \mathscr{H}^n = (\mathscr{H} + \omega_i)^n \theta_i + O(R_i) \quad n = 0,1,\cdots, \tag{14.28}$$

where it will not be necessary to include the details of the term of order R_i. Note to zeroth order, i.e., neglecting the $O(R_i)$, (14.28) is just the Hermitian conjugate of (14.6). The last term in (14.26), with $\chi = \theta_i$, is

$$Z^{-1}\,\mathrm{Tr}\,\theta_i \sum_{n=1}^{\infty} (n!)^{-1}(-KT)^{-n} \sum_{p=0}^{n-1} \mathscr{H}^p R_i^\dagger (\mathscr{H} + \omega_i)^{n-p-1},$$

and permuting θ_i inside the trace it is

$$Z^{-1}\,\mathrm{Tr} \sum_{n=1}^{\infty} (n!)^{-1}(-KT)^{-n} \sum_{p=0}^{n-1} \mathscr{H}^p R_i^\dagger (\mathscr{H} + \omega_i)^{n-p-1}\theta_i;$$

then with (14.28) it is

$$Z^{-1}\,\mathrm{Tr} \sum_{n=1}^{\infty} (n!)^{-1}(-KT)^{-n} \sum_{p=0}^{n-1} \mathscr{H}^p R_i^\dagger \theta_i \mathscr{H}^{n-p-1} + O(R_i^\dagger R_i),$$

and finally permuting $\mathscr{H}^p$ inside the trace this quantity becomes

$$Z^{-1}\,\mathrm{Tr} \sum_{n=1}^{\infty} (n!)^{-1}(-KT)^{-n} \sum_{p=0}^{n-1} R_i^\dagger \theta_i \mathscr{H}^{n-1} + O(R_i^\dagger R_i). \tag{14.29}$$

Now the sum on p contains n terms which are all the same, so Σ_p can be replaced by n and (14.29) is

$$\begin{aligned} -(KT)^{-1}Z^{-1}\,\mathrm{Tr}\,R_i^\dagger\theta_i \sum_{n=1}^{\infty} [(n-1)!]^{-1}(-KT)^{-(n-1)}\mathscr{H}^{n-1} + O(R_i^\dagger R_i) \\ = -(KT)^{-1}Z^{-1}\,\mathrm{Tr}\,R_i^\dagger\theta_i e^{-\mathscr{H}/KT} + O(R_i^\dagger R_i) \quad (14.30) \\ = -(KT)^{-1}\langle R_i^\dagger\theta_i\rangle + O(R_i^\dagger R_i). \end{aligned}$$

The basic equation (14.26) may now be written

$$\langle\theta_i^\dagger\theta_i\rangle = e^{-\omega_i/KT}\langle\theta_i\theta_i^\dagger\rangle - (KT)^{-1}\langle R_i^\dagger\theta_i\rangle + O(R_i^\dagger R_i). \tag{14.31}$$

The perturbation term has been evaluated to order $R_i^\dagger$, with a second-order term in $R_i^\dagger R_i$ left over. The second-order term is easily evaluated, but this will not be required here and henceforth that term will be neglected. It is convenient to transform (14.31) to a form containing commutators or anticommutators, by writing

$$[e^{\omega_i/KT} \pm 1]\langle\theta_i^\dagger\theta_i\rangle = \langle\theta_i\theta_i^\dagger \pm \theta_i^\dagger\theta_i\rangle - (KT)^{-1}\,e^{\omega_i/KT}\langle R_i^\dagger\theta_i\rangle. \tag{14.32}$$

This result may now be put in the form of the first-order basic equations of the statistical perturbation method, which are

$$\langle\theta_i^\dagger\theta_i\rangle = [e^{\omega_i/KT} - 1]^{-1}\langle[\theta_i,\theta_i^\dagger]\rangle - (KT)^{-1}\, e^{\omega_i/KT}[e^{\omega_i/KT} - 1]^{-1}\langle R_i^\dagger\theta_i\rangle; \tag{14.33}$$

$$\langle\theta_i^\dagger\theta_i\rangle = [e^{\omega_i/KT} + 1]^{-1}\langle\{\theta_i,\theta_i^\dagger\}\rangle - (KT)^{-1}\, e^{\omega_i/KT}[e^{\omega_i/KT} + 1]^{-1}\langle R_i^\dagger\theta_i\rangle. \tag{14.34}$$

These equations are useful for calculating the averages $\langle\theta_i^\dagger\theta_i\rangle$ correct to first order in the $R_i^\dagger$, and also to that order the averages $\langle R_i^\dagger\theta_i\rangle$ may be evaluated to zeroth order by means of the Hermitian conjugate zeroth-order equations (14.14) and (14.15):

$$\langle\chi^\dagger\theta_i\rangle = [e^{\omega_i/KT} - 1]^{-1}\langle[\theta_i,\chi^\dagger]\rangle; \tag{14.35}$$

$$\langle\chi^\dagger\theta_i\rangle = [e^{\omega_i/KT} + 1]^{-1}\langle\{\theta_i,\chi^\dagger\}\rangle. \tag{14.36}$$

A further procedure of the statistical perturbation method consists of introducing the renormalized energies Ω_i, defined by

$$\Omega_i = \omega_i + \delta\omega_i, \tag{14.37}$$

where $\delta\omega_i$ is presumably of the same order as statistical averages of operators linear in $R_i^\dagger$ or R_i. The renormalized energies may be determined so as to make the first-order perturbation terms in (14.33) or (14.34) vanish identically. To do this, rewrite the Hamiltonian commutator equation (14.1) by introducing Ω_i to find

$$[\mathscr{H},\theta_i^\dagger] = \Omega_i\theta_i^\dagger + (R_i^\dagger - \delta\omega_i\theta_i^\dagger). \tag{14.38}$$

Now derivation of the basic equations above depended only on the commutator equation (14.1), so in view of (14.38) all those derivations are now valid with ω_i replaced by Ω_i, and $R_i^\dagger$ replaced by $R_i^\dagger - \delta\omega_i\theta_i^\dagger$. In particular, (14.33) becomes

$$\begin{aligned}\langle\theta_i^\dagger\theta_i\rangle = {} & [e^{\Omega_i/KT} - 1]^{-1}\langle[\theta_i,\theta_i^\dagger]\rangle \\ & - (KT)^{-1}e^{\Omega_i/KT}[e^{\Omega_i/KT} - 1]^{-1}\langle(R_i^\dagger - \delta\omega_i\theta_i^\dagger)\theta_i\rangle.\end{aligned} \tag{14.39}$$

The first-order perturbation term in (14.39) vanishes if

$$\langle(R_i^\dagger - \delta\omega_i\theta_i^\dagger)\theta_i\rangle = 0, \tag{14.40}$$

and solving this for $\delta\omega_i$ gives

$$\delta\omega_i = \frac{\langle R_i^\dagger\theta_i\rangle}{\langle\theta_i^\dagger\theta_i\rangle}. \tag{14.41}$$

Thus with $\delta\omega_i$ given by (14.41), the first-order basic equation (14.39) reduces to a zeroth-order basic equation with renormalized energies Ω_i:

$$\langle\theta_i^\dagger\theta_i\rangle = [e^{\Omega_i/KT} - 1]^{-1}\langle[\theta_i,\theta_i^\dagger]\rangle. \tag{14.42}$$

The same procedure goes through for the anticommutator form (14.34), giving

$$\langle\theta_i^\dagger\theta_i\rangle = [e^{\Omega_i/KT} + 1]^{-1}\langle\{\theta_i,\theta_i^\dagger\}\rangle, \tag{14.43}$$

where $\delta\omega_i$ is again given by (14.41).

In principle the coupled equations (14.41) and (14.42), or (14.41) and (14.43), must be solved self-consistently. However, all the above calculations were carried only to first order in the $R_i^\dagger$ and R_i operators, so to this order it is consistent to evaluate the averages $\langle\theta_i^\dagger\theta_i\rangle$ from (14.42) or (14.43) alone, with $\delta\omega_i$ evaluated from (14.41) in zeroth order. With (14.35) and (14.36) for $\langle R_i^\dagger\theta_i\rangle$, and with (14.14) and (14.15) for $\langle\theta_i^\dagger\theta_i\rangle$, zeroth-order evaluation of (14.41) is

$$\delta\omega_i = \frac{\langle[\theta_i,R_i^\dagger]\rangle}{\langle[\theta_i,\theta_i^\dagger]\rangle}, \tag{14.44}$$

or

$$\delta\omega_i = \frac{\langle\{\theta_i,R_i^\dagger\}\rangle}{\langle\{\theta_i,\theta_i^\dagger\}\rangle}. \tag{14.45}$$

It should also be noted that $\delta\omega_i$ has been constructed specifically to remove first-order terms in the particular average $\langle\theta_i^\dagger\theta_i\rangle$; in order to remove first-order terms in the average $\langle\theta_i^\dagger\chi\rangle$, $\delta\omega_i$ will in general depend on χ.

The statistical perturbation method works entirely with operator equations, and does not require knowledge of the system wave functions, or how these wave functions are transformed by the operators. Further discussion of the general method, and application to the Heisenberg ferromagnet problem, is provided in the original reference.* We will now apply the method to derive the self-consistent phonon theory.

First-Order Self-Consistent Phonons

For simplicity, the discussion will be restricted to a primitive lattice. The Hamiltonian of the crystal in the presence of externally applied stresses is (10.2), and neglecting surface effects the potential $\Psi - \Psi_0 = \Phi_0 + \Phi_2 + \cdots$ is just the crystal potential $\Phi = \Phi_0 + \Phi_1 + \Phi_2 + \cdots$, since Φ_1 has contributions only from the ions in the surface region. Therefore for arbitrary macroscopic configuration of the crystal and with surface effects neglected,

$$\mathscr{H} = KE + \Phi. \tag{14.46}$$

* D. C. Wallace, *Phys. Rev.* **152,** 261 (1966).

Let us introduce an effective harmonic potential Θ and write

$$\mathscr{H} = (KE + \Theta) + (\Phi - \Theta) = \mathscr{H}_{\rm H} + (\Phi - \Theta). \tag{14.47}$$

Here $\mathscr{H}_{\rm H}$ is an effective harmonic Hamiltonian, and Θ is to be determined so that the "perturbation" $\Phi - \Theta$ is small in the statistical sense. For a primitive lattice the kinetic and potential contributions to $\mathscr{H}_{\rm H}$ are

$$KE = \tfrac{1}{2}\sum_{\rm N}\sum_{i} M_C[\dot{U}_i({\rm N})]^2, \tag{14.48}$$

$$\Theta = \tfrac{1}{2}\sum_{\rm MN}\sum_{ij} \Theta_{ij}({\rm M,N})U_i({\rm M})U_j({\rm N}), \tag{14.49}$$

where as usual M_C is the mass of one ion (or one unit cell) and $\Theta_{ij}({\rm M,N})$ are the effective potential energy coefficients.

It will be useful to derive a simple relation involving the commutators of the crystal potential with the ion velocities. The position and momentum commutators (10.4) and (10.5) may be written

$$[U_i({\rm N}),\dot{U}_j({\rm P})] = (i\hbar/M_C)\delta_{\rm NP}\delta_{ij}, \tag{14.50}$$

$$[U_i({\rm N}),U_j({\rm P})] = 0. \tag{14.51}$$

Since Φ is a general Maclaurin series in the ion displacements, it follows that

$$[\Phi,\dot{U}_i({\rm N})] = (i\hbar/M_C)[\partial\Phi/\partial U_i({\rm N})], \tag{14.52}$$

$$[\Phi,U_i({\rm N})] = 0. \tag{14.53}$$

Repeated application of (14.52) gives

$$[[\Phi,\dot{U}_i({\rm N})],\dot{U}_j({\rm P})] = (i\hbar/M_C)^2[\partial^2\Phi/\partial U_i({\rm N})\partial U_j({\rm P})], \tag{14.54}$$

and so on. These results also hold for Θ, which is quadratic in the ion displacements according to (14.49); in particular,

$$[[\Theta,\dot{U}_i({\rm N})],\dot{U}_j({\rm P})] = (i\hbar/M_C)^2\Theta_{ij}({\rm M,N}). \tag{14.55}$$

The harmonic Hamiltonian is diagonalized by the transformations of Section 10. The transformation from the ion displacements $\mathbf{U}({\rm N})$ to the phonon coordinates $q(\mathbf{k}s)$ is (10.40):

$$U_i({\rm N}) = (N_0M_C)^{-1/2}\sum_{\mathbf{k}s} q(\mathbf{k}s)\, e^{i\mathbf{k}\cdot\mathbf{R}({\rm N})}W_i(\mathbf{k}s), \tag{14.56}$$

and the transformation to creation and annihilation operators $A(\mathbf{k}s)$ and $A(\mathbf{k}s)^\dagger$ is (10.66):

$$q(\mathbf{k}s) = [\hbar/2\Omega(\mathbf{k}s)]^{1/2}[A(\mathbf{k}s) + A(-\mathbf{k}s)^\dagger], \tag{14.57}$$

where we are using effective eigenvectors $\mathbf{W}(\mathbf{k}s)$ in (14.56), and effective frequencies $\Omega(\mathbf{k}s)$ in (14.57), appropriate for the effective potential Θ. The effective dynamical matrices are denoted $\mathscr{D}(\mathbf{k})$, and corresponding to (10.29) their elements are

$$\mathscr{D}_{ij}(\mathbf{k}) = (N_0 M_C)^{-1} \sum_{\mathrm{NP}} \Theta_{ij}(\mathrm{N,P}) e^{i\mathbf{k}\cdot[\mathbf{R}(\mathrm{P})-\mathbf{R}(\mathrm{N})]}. \tag{14.58}$$

The diagonalization, orthonormalization, and completeness relations are, respectively,

$$\sum_{ij} W_i(-\mathbf{k}s)\mathscr{D}_{ij}(\mathbf{k})W_j(\mathbf{k}s') = [\Omega(\mathbf{k}s)]^2\delta_{ss'}; \tag{14.59}$$

$$\sum_{i} W_i(-\mathbf{k}s)W_i(\mathbf{k}s') = \delta_{ss'}; \tag{14.60}$$

$$\sum_{s} W_i(-\mathbf{k}s)W_j(\mathbf{k}s) = \delta_{ij}. \tag{14.61}$$

With these transformations the harmonic Hamiltonian $\mathscr{H}_{\mathrm{H}}$ is diagonalized, and the total Hamiltonian (14.47) becomes

$$\mathscr{H} = \sum_{\mathbf{k}s} \hbar\Omega(\mathbf{k}s)[A(\mathbf{k}s)^{\dagger}A(\mathbf{k}s) + \tfrac{1}{2}] + (\Phi - \Theta), \tag{14.62}$$

where the phonon operators have the usual commutators

$$[A(\mathbf{k}s), A(\mathbf{k}'s')^{\dagger}] = \delta_{\mathbf{k}\mathbf{k}'}\delta_{ss'}, \tag{14.63}$$

$$[A(\mathbf{k}s), A(\mathbf{k}'s')] = 0. \tag{14.64}$$

We are now ready to derive the self-consistent phonon frequencies. From (14.62)–(14.64), the Hamiltonian commutator equation is

$$[\mathscr{H}, A(\mathbf{k}s)^{\dagger}] = \hbar\Omega(\mathbf{k}s)A(\mathbf{k}s)^{\dagger} + R(\mathbf{k}s)^{\dagger}, \tag{14.65}$$

where

$$R(\mathbf{k}s)^{\dagger} = [\Phi - \Theta, A(\mathbf{k}s)^{\dagger}]. \tag{14.66}$$

Corresponding to the remainders (14.66), the phonon energy shifts $\delta\hbar\Omega(\mathbf{k}s)$ of (14.44) are

$$\delta\hbar\Omega(\mathbf{k}s) = \frac{\langle[A(\mathbf{k}s), R(\mathbf{k}s)^{\dagger}]\rangle}{\langle[A(\mathbf{k}s), A(\mathbf{k}s)^{\dagger}]\rangle} = \langle[A(\mathbf{k}s), R(\mathbf{k}s)^{\dagger}]\rangle, \tag{14.67}$$

where the denominator is 1 by (14.63). With (14.66) this may be written

$$\delta\hbar\Omega(\mathbf{k}s) = -\langle[[\Phi - \Theta, A(\mathbf{k}s)^{\dagger}], A(\mathbf{k}s)]\rangle, \tag{14.68}$$

and the object is to choose Θ such that each energy shift vanishes:

$$\delta\hbar\Omega(\mathbf{k}s) = 0, \quad \text{for all } \mathbf{k}, s. \tag{14.69}$$

Evaluation of (14.68) is straightforward. Inversion of (14.57) and its time derivative gives [see, e.g., (10.64) and (10.65)]

$$\begin{aligned} A(\mathbf{k}s)^{\dagger} &= [2\hbar\Omega(\mathbf{k}s)]^{-1/2}[\Omega(\mathbf{k}s)q(-\mathbf{k}s) - i\dot{q}(-\mathbf{k}s)], \\ A(\mathbf{k}s) &= [2\hbar\Omega(\mathbf{k}s)]^{-1/2}[\Omega(\mathbf{k}s)q(\mathbf{k}s) + i\dot{q}(\mathbf{k}s)], \end{aligned} \tag{14.70}$$

and inversion of (14.56) and its time derivative gives [see, e.g., (10.48) and (10.49)]

$$q(\mathbf{k}s) = (M_C/N_0)^{1/2} \sum_{\mathrm{N}i} U_i(\mathrm{N})\, e^{-i\mathbf{k}\cdot\mathbf{R}(\mathrm{N})} W_i(-\mathbf{k}s),$$
$$\dot{q}(\mathbf{k}s) = (M_C/\mathrm{N}_0)^{1/2} \sum_{\mathrm{N}i} \dot{U}_i(\mathrm{N})\, e^{-i\mathbf{k}\cdot\mathbf{R}(\mathrm{N})} W_i(-\mathbf{k}s). \tag{14.71}$$

Now we put (14.70) into (14.68), and drop all terms in $q(\mathbf{k}s)$ since by (14.71) these operators are linear in $U_i(\mathrm{N})$ and hence they commute with $\Phi - \Theta$. We then replace the $\dot{q}(\mathbf{k}s)$ operators according to (14.71), and obtain

$$\delta\hbar\Omega(\mathbf{k}s) = -[M_C/2\hbar\Omega(\mathbf{k}s)]N_0^{-1} \sum_{\mathrm{NP}} \sum_{ij} \langle[[\Phi - \Theta, \dot{U}_i(\mathrm{N})], \dot{U}_j(\mathrm{P})]\rangle$$
$$\times\, e^{i\mathbf{k}\cdot[\mathbf{R}(\mathrm{N})-\mathbf{R}(\mathrm{P})]} W_i(\mathbf{k}s) W_j(-\mathbf{k}s). \tag{14.72}$$

If this is to vanish for each and every $\mathbf{k}s$, then each term in $\Sigma_{\mathrm{NP}}\, \Sigma_{ij}$ must separately vanish, so

$$\langle[[\Phi - \Theta, \dot{U}_i(\mathrm{N})], \dot{U}_j(\mathrm{P})]\rangle = 0, \quad \text{for all N,P,}i,j, \tag{14.73}$$

Finally with the aid of (14.54) and (14.55), the double commutator is converted to a second derivative with respect to displacements, and (14.73) determines the effective harmonic potential coefficients as statistical averages of second derivatives of the true crystal potential:

$$\Theta_{ij}(\mathrm{N,P}) = \langle \partial^2\Phi/\partial U_i(\mathrm{N})\partial U_j(\mathrm{P})\rangle. \tag{14.74}$$

Therefore the effective harmonic potential coefficients, and hence also the self-consistent frequencies, the eigenvectors, and the phonon operators, are all temperature-dependent.

It is fair to take statistical averages such as (14.74) in the effective harmonic representation, that is, the trace may be evaluated by summing over diagonal matrix elements between the effective harmonic phonon states. When this is done, only terms in Φ of even order in the ion displacements contribute to (14.74), so that

$$\Theta_{ij}(\mathrm{M,N}) = \Phi_{ij}(\mathrm{M,N}) + \tfrac{1}{2} \sum_{\mathrm{PQ}} \sum_{kl} \Phi_{ijkl}(\mathrm{M,N,P,Q})\langle U_k(\mathrm{P})U_l(\mathrm{Q})\rangle + \cdots. \tag{14.75}$$

It is not meant to imply here that the series converges rapidly, as would be the case for an ordinary perturbation problem, but simply that the terms odd in the $U_i(\mathrm{M})$ average out and hence do not contribute to $\Theta_{ij}(\mathrm{M,N})$. Now the effective phonon frequencies are obtained by the diagonalization (14.59), and the potential coefficients appearing in the effective dynamical matrices $\mathscr{D}(\mathbf{k})$ are given by (14.74), so in principle these two equations (really sets of equations) must be solved self-consistently. It is clear from the basic equations of the statistical perturbation method, however, and especially from the interpretation of the energy shifts (14.44) and (14.45), that the self-consistent phonon calculation is *not* a nonperturbative result, but is correct to first order in the remainder operators given by (14.66).

FIRST-ORDER STATISTICAL AVERAGES

We will use, without derivation, the general results of statistical mechanics for a system of harmonic phonons (bosons); the appropriate derivations are provided in Section 16. The harmonic partition function is

$$Z_{\mathrm{H}} = \mathrm{Tr}\, e^{-\mathscr{H}_{\mathrm{H}}/KT}; \tag{14.76}$$

the harmonic free energy is

$$F_{\mathrm{H}} = -KT \ln Z_{\mathrm{H}}; \tag{14.77}$$

and the harmonic density matrix is

$$\rho_{\mathrm{H}} = Z_{\mathrm{H}}^{-1}\, e^{-\mathscr{H}_{\mathrm{H}}/KT}. \tag{14.78}$$

The harmonic statistical average of any operator χ is defined by

$$\langle \chi \rangle_{\mathrm{H}} = \mathrm{Tr}\, \chi e^{-\mathscr{H}_{\mathrm{H}}/KT}. \tag{14.79}$$

The lattice-dynamics Hamiltonian $\mathscr{H}$ is given by (14.47), and if $\Phi - \Theta$ is treated as a perturbation, then the crystal free energy F correct to first order in $\Phi - \Theta$ is given by

$$F = F_{\mathrm{H}} + \langle \Phi - \Theta \rangle_{\mathrm{H}}. \tag{14.80}$$

An alternate form of the free energy is often used, namely

$$F = \mathrm{Tr}\, [\rho_{\mathrm{H}}(\mathscr{H} + KT \ln \rho_{\mathrm{H}})], \tag{14.81}$$

and it is instructive to show that this is identical to (14.80). To do this one simply uses (14.76)–(14.79) and transforms (14.81) as follows.

$$\begin{aligned} &\mathrm{Tr}\, [\rho_{\mathrm{H}}(\mathscr{H} + KT \ln \rho_{\mathrm{H}})] \\ &= Z_{\mathrm{H}}^{-1}\, \mathrm{Tr}\, [e^{-\mathscr{H}_{\mathrm{H}}/KT}(\mathscr{H} + KT \ln e^{-\mathscr{H}_{\mathrm{H}}/KT} - KT \ln Z_{\mathrm{H}})] \\ &= Z_{\mathrm{H}}^{-1}\, \mathrm{Tr}\, [e^{-\mathscr{H}_{\mathrm{H}}/KT}(\mathscr{H} - \mathscr{H}_{\mathrm{H}})] - Z_{\mathrm{H}}^{-1} KT \ln Z_{\mathrm{H}}\, \mathrm{Tr}\, e^{-\mathscr{H}_{\mathrm{H}}/KT} \\ &= \langle \mathscr{H} - \mathscr{H}_{\mathrm{H}} \rangle_{\mathrm{H}} - KT \ln Z_{\mathrm{H}}, \end{aligned} \tag{14.82}$$

and since $\mathscr{H} - \mathscr{H}_{\mathrm{H}} = \Phi - \Theta$ according to (14.47), and the last term is F_{H} by (14.77), the result is

$$\mathrm{Tr}\, [\rho_{\mathrm{H}}(\mathscr{H} + KT \ln \rho_{\mathrm{H}})] = F_{\mathrm{H}} + \langle \Phi - \Theta \rangle_{\mathrm{H}}. \tag{14.83}$$

To proceed with evaluation of the thermodynamic functions, it is useful to transform the kinetic and potential energies *separately* to forms involving the effective phonon creation and annihilation operators. The KE is given by (14.48), and Θ is given by (14.49); these are easily transformed to

$$KE = \tfrac{1}{4}\sum_{\mathbf{k}s} \hbar\Omega(\mathbf{k}s)[2A(\mathbf{k}s)^{\dagger}A(\mathbf{k}s) + 1 - A(\mathbf{k}s)^{\dagger}A(-\mathbf{k}s)^{\dagger} - A(-\mathbf{k}s)A(\mathbf{k}s)]; \tag{14.84}$$

$$\Theta = \tfrac{1}{4}\sum_{\mathbf{k}s} \hbar\Omega(\mathbf{k}s)[2A(\mathbf{k}s)^{\dagger}A(\mathbf{k}s) + 1 + A(\mathbf{k}s)^{\dagger}A(-\mathbf{k}s)^{\dagger} + A(-\mathbf{k}s)A(\mathbf{k}s)]. \tag{14.85}$$

The sum of these is of course $\mathscr{H}_{\mathrm{H}}$:

$$\mathscr{H}_{\mathrm{H}} = \sum_{\mathbf{k}s} \hbar\Omega(\mathbf{k}s)[A(\mathbf{k}s)^{\dagger}A(\mathbf{k}s) + \tfrac{1}{2}]. \tag{14.86}$$

Now since the $A(\mathbf{k}s)^{\dagger}$ are exact creation operators of $\mathscr{H}_{\mathrm{H}}$, and since they satisfy the boson commutators (14.63) and (14.64), then their harmonic statistical averages are given exactly by the boson zeroth-order equations (14.17) and (14.18).

$$\langle A(\mathbf{k}s)^{\dagger}A(\mathbf{k}'s')\rangle_{\mathrm{H}} = [e^{\hbar\Omega(\mathbf{k}s)/KT} - 1]^{-1}\delta_{\mathbf{k}\mathbf{k}'}\delta_{ss'}; \tag{14.87}$$

$$\langle A(ks)^{\dagger}A(\mathbf{k}'s')^{\dagger}\rangle_{\mathrm{H}} = \langle A(\mathbf{k}s)A(\mathbf{k}'s')\rangle_{\mathrm{H}} = 0. \tag{14.88}$$

With these results it is easily verified that (14.84)–(14.86) give

$$\langle KE\rangle_{\mathrm{H}} = \langle\Theta\rangle_{\mathrm{H}} = \tfrac{1}{2}\langle\mathscr{H}_{\mathrm{H}}\rangle_{\mathrm{H}} = \tfrac{1}{4}\sum_{\mathbf{k}s} \hbar\Omega(\mathbf{k}s)\coth[\hbar\Omega(\mathbf{k}s)/2KT], \tag{14.89}$$

where we have used

$$[e^{\hbar\Omega(\mathbf{k}s)/KT} - 1]^{-1} + \tfrac{1}{2} = \tfrac{1}{2}\coth[\hbar\Omega(\mathbf{k}s)/2KT]. \tag{14.90}$$

The internal energy U is just the statistical average of $\mathscr{H}$, and with a harmonic average this can be written

$$U = \langle KE\rangle_{\mathrm{H}} + \langle\Phi\rangle_{\mathrm{H}} = \tfrac{1}{4}\sum_{\mathbf{k}s} \hbar\Omega(\mathbf{k}s)\coth[\hbar\Omega(\mathbf{k}s)/2KT] + \langle\Phi\rangle_{\mathrm{H}}, \tag{14.91}$$

or alternatively,

$$U = \langle\mathscr{H}_{\mathrm{H}}\rangle_{\mathrm{H}} + \langle\Phi - \Theta\rangle_{\mathrm{H}} = \tfrac{1}{2}\sum_{\mathbf{k}s} \hbar\Omega(\mathbf{k}s)\coth[\hbar\Omega(\mathbf{k}s)/2KT] + \langle\Phi - \Theta\rangle_{\mathrm{H}}. \tag{14.92}$$

The harmonic free energy (14.77) is (see Section 16)

$$\begin{aligned} F_{\mathrm{H}} &= \sum_{\mathbf{k}s} \{\tfrac{1}{2}\hbar\Omega(\mathbf{k}s) + KT\ln[1 - e^{-\hbar\Omega(\mathbf{k}s)/KT}]\} \\ &= KT\sum_{\mathbf{k}s} \ln\{2\sinh[\hbar\Omega(\mathbf{k}s)/2KT]\}, \end{aligned} \tag{14.93}$$

and the total free energy correct to first order in $\Phi - \Theta$ is, from (14.80),

$$F = F_{\mathrm{H}} + \langle\Phi\rangle_{\mathrm{H}} - \tfrac{1}{4}\sum_{\mathbf{k}s} \hbar\Omega(\mathbf{k}s)\coth[\hbar\Omega(\mathbf{k}s)/2KT]. \tag{14.94}$$

It should always be remembered, especially when deriving other thermodynamic functions from the above equations, that the self-consistent frequencies $\Omega(\mathbf{k}s)$ depend on the temperature.

Finally it is of interest to evaluate the effective harmonic frequencies $\Omega(\mathbf{k}s)$ for the case of an ordinary perturbation problem. Here the effective harmonic potential coefficients may be written from (14.75),

$$\Theta_{ij}(\mathrm{M,N}) = \Phi_{ij}(\mathrm{M,N}) + \delta\Phi_{ij}(\mathrm{M,N}), \tag{14.95}$$

where

$$\delta\Phi_{ij}(\mathrm{M,N}) = \tfrac{1}{2}\sum_{\mathrm{PQ}}\sum_{kl}\Phi_{ijkl}(\mathrm{M,N,P,Q})\langle U_k(\mathrm{P})U_l(\mathrm{Q})\rangle. \tag{14.96}$$

The effective dynamical matrix (14.58) is then given by

$$\mathscr{D}_{ij}(\mathbf{k}) = D_{ij}(\mathbf{k}) + \delta D_{ij}(\mathbf{k}), \tag{14.97}$$

where $D_{ij}(\mathbf{k})$ are the ordinary dynamical matrix elements of (10.29) and

$$\delta D_{ij}(\mathbf{k}) = (N_0 M_C)^{-1}\sum_{\mathrm{NP}}\delta\Phi_{ij}(\mathrm{N,P})e^{i\mathbf{k}\cdot[\mathbf{R}(\mathrm{P})-\mathbf{R}(\mathrm{N})]}. \tag{14.98}$$

Now the eigenvalues and eigenvectors of $\mathbf{D}(\mathbf{k})$ are $[\omega(\mathbf{k}s)]^2$ and $\mathbf{w}(\mathbf{k}s)$, respectively, and with $\delta\mathbf{D}(\mathbf{k})$ treated as a perturbation to first order, the diagonalization of (14.97) reads

$$[\Omega(\mathbf{k}s)]^2 = [\omega(\mathbf{k}s)]^2 + \sum_{ij} w_i(-\mathbf{k}s)\delta D_{ij}(\mathbf{k})w_j(\mathbf{k}s). \tag{14.99}$$

To evaluate the last term in (14.99), we begin with (14.96); the quantities $\langle U_k(\mathrm{P})U_l(\mathrm{Q})\rangle$ are calculated by using (14.56) and (14.57) to transform to $A(\mathbf{k}s)^\dagger$ and $A(\mathbf{k}s)$ operators, and then using (14.87) and (14.88) for the averages. The result for $\delta\Phi_{ij}(\mathrm{M,N})$ is then put into (14.98), and this into (14.99), to obtain

$$[\Omega(\mathbf{k}s)]^2 = [\omega(\mathbf{k}s)]^2 + 24\hbar^{-1}\omega(\mathbf{k}s) \times \sum_{\mathbf{k}'s'}\Phi(\mathbf{k}s,-\mathbf{k}s,\mathbf{k}'s',-\mathbf{k}'s')\langle 2A(\mathbf{k}'s')^\dagger A(\mathbf{k}'s') + 1\rangle, \tag{14.100}$$

where $\Phi(\mathbf{k}s,-\mathbf{k}s,\mathbf{k}'s',-\mathbf{k}'s')$ is given by (10.75). Taking the square root of (14.100) to first order in the perturbation term, and multiplying by $\hbar$, gives

$$\hbar\Omega(\mathbf{k}s) = \hbar\omega(\mathbf{k}s) + \hbar\delta\omega(\mathbf{k}s), \tag{14.101}$$

where

$$\hbar\delta\omega(\mathbf{k}s) = 12\sum_{\mathbf{k}'s'}\Phi(\mathbf{k}s,-\mathbf{k}s,\mathbf{k}'s',-\mathbf{k}'s')\langle 2A(\mathbf{k}'s')^\dagger A(\mathbf{k}'s') + 1\rangle. \tag{14.102}$$

This temperature-dependent shift in the self-consistent phonon energy, to first order in the ordinary anharmonic perturbation theory, is just the statistical average of $\hbar\omega_{2\kappa}$(quartic) given by (13.77). The leading-order energy shift from the terms in Φ which are odd in the ion displacements, namely $\hbar\omega_{2\kappa}$(cubic) given by (13.80), may be expected to be roughly the same magnitude as the quartic contribution.

Interactions Among Self-Consistent Phonons

In first- (and higher-) order perturbation, the odd terms in the crystal potential Φ cause interactions among the first-order self-consistent phonons discussed above. In addition, the even terms in Φ cause interactions in second- and higher-order perturbation. There are several ways to try to include the odd terms in the calculation. One way is to keep the second-order term in the basic equation (14.31), i.e., the term of order $R_i^\dagger R_i$, and determine the self-consistent energies to second order. An alternate procedure, which we will discuss briefly, is to introduce an effective anharmonic potential Θ_{A} and write the total Hamiltonian as

$$\mathscr{H} = \mathscr{H}_{\mathrm{A}} + (\Phi - \Theta_{\mathrm{A}}). \tag{14.103}$$

Here $\Theta_{\mathrm{A}} = \Theta_2 + \Theta_3$, with Θ_2 a harmonic potential like (14.49) and

$$\Theta_3 = (1/3!) \sum_{\mathrm{MNP}} \sum_{ijk} \Theta_{ijk}(\mathrm{M,N,P}) U_i(\mathrm{M}) U_j(\mathrm{N}) U_k(\mathrm{P}), \tag{14.104}$$

and the effective anharmonic Hamiltonian $\mathscr{H}_{\mathrm{A}}$ is

$$\mathscr{H}_{\mathrm{A}} = KE + \Theta_{\mathrm{A}}. \tag{14.105}$$

Now the harmonic part of $\mathscr{H}_{\mathrm{A}}$ is diagonalized in terms of the phonons $A_\kappa^\dagger$, and the renormalization operators $B_\kappa^\dagger$ are constructed according to (13.54), but based on the transformed potential coefficients $\Theta_{\kappa\kappa'\kappa''}$ instead of the $\Phi_{\kappa\kappa'\kappa''}$, and the Hamiltonian commutator equations read

$$[\mathscr{H}, A_\kappa^\dagger + B_\kappa^\dagger] = \hbar\Omega_\kappa(A_\kappa^\dagger + B_\kappa^\dagger) + R_\kappa^\dagger, \tag{14.106}$$

where

$$R_\kappa^\dagger = [\Theta_3, B_\kappa^\dagger] + [\Phi - \Theta_{\mathrm{A}}, A_\kappa^\dagger + B_\kappa^\dagger] + i\gamma B_\kappa^\dagger. \tag{14.107}$$

Equation (14.107) is still exact, and there are several points to be noted. First, the $i\gamma B_\kappa^\dagger$ term can ordinarily be dropped since γ is an infinitesimal, and statistical averages involving this term will ordinarily be of order γ. Second, the renormalized phonons $A_\kappa^\dagger + B_\kappa^\dagger$ are not good bosons, since their commutators are

$$[A_\kappa + B_\kappa, A_{\kappa'}^\dagger + B_{\kappa'}^\dagger] = \delta_{\kappa\kappa'} + [B_\kappa, B_{\kappa'}^\dagger], \tag{14.108}$$

$$[A_\kappa + B_\kappa, A_{\kappa'} + B_{\kappa'}] = [B_\kappa, B_{\kappa'}]. \tag{14.109}$$

The departure of the commutators (14.108) and (14.109) from boson commutators is the kinematical interaction among the renormalized phonons;

the presence of this interaction causes no difficulty in the statistical perturbation method. Finally the first-order energy shifts (14.44) are now written

$$\delta\hbar\Omega_\kappa = \frac{\langle[A_\kappa + B_\kappa, R_\kappa^\dagger]\rangle}{\langle[A_\kappa + B_\kappa, A_\kappa^\dagger + B_\kappa^\dagger]\rangle}, \tag{14.110}$$

and the optimum determination of the potential coefficients in Θ_A is such as to make these energy shifts vanish. A simpler procedure is to determine the potential coefficients in some less optimum way; for example, one might choose $\Theta_{ij}(M,N)$ according to (14.74) and by analogy

$$\Theta_{ijk}(M,N,P) = \langle \partial^3\Phi/\partial U_i(M)\partial U_j(N)\partial U_k(P)\rangle, \tag{14.111}$$

and then evaluate the energy shifts from (14.110).

In any case the statistical perturbation method offers a great deal of flexibility in seeking a suitable approximate solution. It should also be noted that if the equation (14.110) is evaluated for an ordinary perturbation problem, the correct statistical average of $\hbar\omega_{2\kappa}$(cubic) is recovered in the energy shifts from the contribution $[\Theta_3, B_\kappa^\dagger]$ to $R_\kappa^\dagger$ in (14.107).

15. CENTRAL POTENTIALS

Dynamical Matrices

The general formula for dynamical matrices in the ordinary-eigenvalue-problem formulation is (10.29), and for central potentials the second-order potential energy coefficients are given by (9.10). For a primitive lattice the dynamical matrices are further simplified in (12.2) and (12.3), and with central potentials the result is

$$D_{ij}(\mathbf{k}) = M_C^{-1} \sum_N{}' \,[2\varphi'(0,N)\delta_{ij} + 4\varphi''(0,N)R_i(N)R_j(N)]\{1 - \cos[\mathbf{k}\cdot\mathbf{R}(N)]\}. \tag{15.1}$$

In addition, for a primitive lattice, the matrices $\mathbf{d}(\hat{\mathbf{k}})$ whose eigenvalues are $[c(\hat{\mathbf{k}}s)]^2$ are written in (12.45), and the small $\mathbf{k}$ limit of (15.1) gives directly

$$d_{ik}(\hat{\mathbf{k}}) = M_C^{-1} \sum_N{}' \sum_{jl} [\varphi'(0,N)\delta_{ik} + 2\varphi''(0,N)\, R_i(N)R_k(N)]\, R_j(N)R_l(N)\hat{k}_j\hat{k}_l. \tag{15.2}$$

Recall that $c(\hat{\mathbf{k}}s)$ are the long-wavelength acoustic phonon velocities, and for a primitive lattice all branches are acoustic. Comparison of (15.2) with the stresses and elastic constants calculated for central potentials in the potential approximation [equations (9.38) and (9.39), respectively] allows us to write

$$d_{ik}(\hat{\mathbf{k}}) = (V_C/M_C) \sum_{jl} [\tilde{C}_{jl}\delta_{ik} + \tilde{C}_{ijkl}]\hat{k}_j\hat{k}_l, \tag{15.3}$$

which of course is the same as the general expression (12.43).

Consider now a lattice which has two identical atoms per unit cell, and a center of inversion symmetry (presumably halfway between the two equilibrium sites in a unit cell), as for example, diamond or hcp. In this case the central potential which couples all pairs of ions is the same function, i.e., a single central potential function $\varphi(R^2)$ describes all the interactions. If the two ions in each unit cell are labeled 0,1 in place of the general indices $\mu,\nu,\cdots$, then the single central potential provides the following simplification.

$$\begin{aligned} \varphi(00,\mathrm{N}0) &= \varphi(01,\mathrm{N}1); \\ \varphi'(00,\mathrm{N}0) &= \varphi'(01,\mathrm{N}1); \quad \text{etc.} \end{aligned} \tag{15.4}$$

Also with the definition $\mathbf{R}(\mathrm{M}\mu,\mathrm{N}\nu) = \mathbf{R}(\mathrm{M}\mu) - \mathbf{R}(\mathrm{N}\nu)$, as introduced in (9.7),

$$\mathbf{R}(00,\mathrm{N}0) = \mathbf{R}(01,\mathrm{N}1), \tag{15.5}$$

and with (15.4) and (15.5) it is seen that the second-order potential energy coefficients (9.10) for central potentials satisfy

$$\Phi_{ij}(00,\mathrm{N}0) = \Phi_{ij}(01,\mathrm{N}1). \tag{15.6}$$

We will take the origin of coordinates at the equilibrium position of ion 0 in unit cell 0, so that

$$\begin{aligned} \mathbf{R}(00) &= 0, \\ \mathbf{R}(\mathrm{N}0) &= \mathbf{R}(\mathrm{N}), \\ \mathbf{R}(\mathrm{N}1) &= \mathbf{R}(\mathrm{N}) + \mathbf{R}(1), \end{aligned} \tag{15.7}$$

where $\mathbf{R}(1)$ is the vector form equilibrium site 0 to site 1, i.e., $\mathbf{R}(1)$ is the basis vector. All the independent $\Phi_{ij}(\mathrm{M}\mu,\mathrm{N}\nu)$ are included in the set $\Phi_{ij}(00,\mathrm{N}\nu)$, and from (9.10) these are

$$\Phi_{ij}(00,\mathrm{N}0) = -2\varphi'(00,\mathrm{N}0)\delta_{ij} - 4\varphi''(00,\mathrm{N}0)R_i(\mathrm{N})R_j(\mathrm{N}), \tag{15.8}$$

$$\Phi_{ij}(00,\mathrm{N}1) = -2\varphi'(00,\mathrm{N}1)\delta_{ij} - 4\varphi''(00,\mathrm{N}1)R_i(\mathrm{N}1)R_j(\mathrm{N}1). \tag{15.9}$$

The mass of each ion is now $\frac{1}{2}M_C$, where M_C is still the mass per unit cell, and the dynamical matrices with surface effects eliminated are written from (10.39),

$$D_{ij}(\mu\nu,\mathbf{k}) = (2/M_C) \sum_{N} \Phi_{ij}(0\mu,\mathrm{N}\nu)e^{i\mathbf{k}\cdot[\mathbf{R}(\mathrm{N}\nu)-\mathbf{R}(\mu)]}. \tag{15.10}$$

In view of (15.6) it follows that

$$D_{ij}(00,\mathbf{k}) = D_{ij}(11,\mathbf{k}), \tag{15.11}$$

and because the $\Phi_{ij}(00,\mathrm{N}0)$ are symmetric in i,j according to (15.8),

$$D_{ij}(00,\mathbf{k}) = D_{ji}(00,\mathbf{k}). \tag{15.12}$$

Also the Hermitian property (10.32) of the dynamical matrices requires

$$D_{ij}(01,\mathbf{k}) = D_{ji}(10,\mathbf{k})^*, \tag{15.13}$$

and again because the $\Phi_{ij}(00,\mathrm{N}1)$ are symmetric in i,j by (15.9),

$$D_{ij}(01,\mathbf{k}) = D_{ji}(01,\mathbf{k}). \tag{15.14}$$

The relations (15.11)–(15.14) allow the dynamical matrix for each $\mathbf{k}$ to be written in terms of two 3×3 matrices $\mathbf{f}(\mathbf{k})$ and $\mathbf{g}(\mathbf{k})$ as follows.

$$\mathbf{D}(\mathbf{k}) = \begin{pmatrix} \mathbf{f}(\mathbf{k}) & \mathbf{g}(\mathbf{k}) \\ \mathbf{g}(\mathbf{k})^* & \mathbf{f}(\mathbf{k}) \end{pmatrix}, \tag{15.15}$$

where $\mathbf{f}(\mathbf{k})$ is the real symmetric matrix whose elements are

$$f_{ij}(\mathbf{k}) = D_{ij}(00,\mathbf{k}) = (2/M_C) \sum_{\mathrm{N}} \Phi_{ij}(00,\mathrm{N}0) e^{i\mathbf{k}\cdot\mathbf{R}(\mathrm{N})}, \tag{15.16}$$

and $\mathbf{g}(\mathbf{k})$ is the complex symmetric matrix whose elements are

$$g_{ij}(\mathbf{k}) = D_{ij}(01,\mathbf{k}) = (2/M_C) \sum_{\mathrm{N}} \Phi_{ij}(00,\mathrm{N}1) e^{i\mathbf{k}\cdot\mathbf{R}(\mathrm{N}1)}. \tag{15.17}$$

Note that the Σ_{N} in (15.16) contains the self-coupling coefficient $\Phi_{ij}(00,00)$; according to (9.13) this coefficient is given as a sum over all other $\Phi_{ij}(00,\mathrm{N}\nu)$ so as to satisfy the translational invariance condition:

$$\Phi_{ij}(00,00) = -\sum_{\mathrm{N}}{}' \Phi_{ij}(00,\mathrm{N}0) - \sum_{\mathrm{N}} \Phi_{ij}(00,\mathrm{N}1). \tag{15.18}$$

Then the elements of $\mathbf{f}(\mathbf{k})$ in (15.16) become

$$f_{ij}(\mathbf{k}) = (2/M_C)\left\{\sum_{\mathrm{N}}{}' \Phi_{ij}(00,\mathrm{N}0)\{\cos[\mathbf{k}\cdot\mathbf{R}(\mathrm{N})] - 1\} - \sum_{\mathrm{N}} \Phi_{ij}(00,\mathrm{N}1)\right\}. \tag{15.19}$$

It is also convenient to separate $\mathbf{g}(\mathbf{k})$ into its real and imaginary parts, as

$$\mathbf{g}(\mathbf{k}) = \mathscr{R}\mathbf{g}(\mathbf{k}) + i\mathscr{I}\mathbf{g}(\mathbf{k}), \tag{15.20}$$

where $\mathscr{R}\mathbf{g}$ and $\mathscr{I}\mathbf{g}$ are each real symmetric, and have the matrix elements

$$\mathscr{R}g_{ij}(\mathbf{k}) = (2/M_C) \sum_{\mathrm{N}} \Phi_{ij}(00,\mathrm{N}1)\cos[\mathbf{k}\cdot\mathbf{R}(\mathrm{N}1)], \tag{15.21}$$

$$\mathscr{I}g_{ij}(\mathbf{k}) = (2/M_C) \sum_{\mathrm{N}} \Phi_{ij}(00,\mathrm{N}1)\sin[\mathbf{k}\cdot\mathbf{R}(\mathrm{N}1)]. \tag{15.22}$$

For a given central potential φ, the dynamical matrices are easily computed from equations (15.19)–(15.22), with the second-order potential coefficients given by (15.8) and (15.9).

The Hermitian dynamical matrix (15.15) may be transformed to a real symmetric matrix by a simple unitary transformation. Suppressing the index $\mathbf{k}$, the diagonalization of $\mathbf{D}$ in matrix notation is

$$\mathbf{w}^\dagger \mathbf{D} \mathbf{w} = \boldsymbol{\omega}^2, \tag{15.23}$$

where $\mathbf{w}$ is the 6×6 matrix of eigenvectors of $\mathbf{D}$, and $\boldsymbol{\omega}^2$ represents a diagonal matrix whose elements are the six eigenvalues of $\mathbf{D}$. If $\mathrm{S_a}$ is a unitary matrix, i.e., $\mathrm{S_a^{-1}} = \mathrm{S_a^\dagger}$, then (15.23) may be rewritten

$$\mathbf{w}^\dagger \mathrm{S_a^{-1}} \mathrm{S_a} \mathbf{D} \mathrm{S_a^{-1}} \mathrm{S_a} \mathbf{w} = (\mathrm{S_a}\mathbf{w})^\dagger (\mathrm{S_a} \mathbf{D} \mathrm{S_a^{-1}})(\mathrm{S_a}\mathbf{w}) = \boldsymbol{\omega}^2. \tag{15.24}$$

In words, the unitarily transformed dynamical matrix is $\mathrm{S_a}\mathbf{D}\mathrm{S_a^{-1}}$, its eigenvector matrix is $\mathrm{S_a}\mathbf{w}$, and its eigenvalues are of course the same as those of $\mathbf{D}$. The unitary matrix needed here is

$$\mathrm{S_a} = \frac{1}{\sqrt{2}} \begin{pmatrix} \mathbf{I} & i\mathbf{I} \\ i\mathbf{I} & \mathbf{I} \end{pmatrix}, \tag{15.25}$$

where $\mathbf{I}$ is the 3×3 unit matrix. Obviously, $\mathrm{S_a}$ is unitary, since

$$\mathrm{S_a^\dagger} = \frac{1}{\sqrt{2}} \begin{pmatrix} \mathbf{I} & -i\mathbf{I} \\ -i\mathbf{I} & \mathbf{I} \end{pmatrix}, \tag{15.26}$$

and

$$\mathrm{S_a^\dagger} \mathrm{S_a} = \mathrm{S_a} \mathrm{S_a^\dagger} = \begin{pmatrix} \mathbf{I} & 0 \\ 0 & \mathbf{I} \end{pmatrix}. \tag{15.27}$$

With the real and imaginary parts of $\mathbf{g}$ written out, the dynamical matrix $\mathbf{D}$ of (15.15) is

$$\mathbf{D} = \begin{pmatrix} \mathbf{f} & \mathscr{R}\mathbf{g} + i\mathscr{I}\mathbf{g} \\ \mathscr{R}\mathbf{g} - i\mathscr{I}\mathbf{g} & \mathbf{f} \end{pmatrix}, \tag{15.28}$$

and the transformed matrix is found to be

$$\mathrm{S_a}\mathbf{D}\mathrm{S_a^{-1}} = \begin{pmatrix} \mathbf{f} + \mathscr{I}\mathbf{g} & \mathscr{R}\mathbf{g} \\ \mathscr{R}\mathbf{g} & \mathbf{f} - \mathscr{I}\mathbf{g} \end{pmatrix}. \tag{15.29}$$

Needless to say, the real symmetric matrix (15.29) is more convenient for computational purposes than is the Hermitian matrix (15.15). The form (15.29) holds for each $\mathbf{k}$ vector, and the eigenvalues of (15.29) are of course $[\omega(\mathbf{k}s)]^2$, $s = 1,2,\cdots,6$. The eigenvectors of (15.29) are $\mathrm{S_a}\mathbf{w}(\mathbf{k}s)$, and the eigenvectors $\mathbf{w}(\mathbf{k}s)$ of $\mathbf{D}(\mathbf{k})$ can always be recovered if needed by

$$\mathbf{w}(\mathbf{k}s) = \mathrm{S_a^{-1}} \mathrm{S_a} \mathbf{w}(\mathbf{k}s). \tag{15.30}$$

In addition, if it is desired to compute the inverse $\mathbf{E}$ of $\mathbf{D}$, this can always be obtained from the inverse of the transformed dynamical matrix by

$$\mathbf{E} = \mathbf{D}^{-1} = \mathbf{S}_a^{-1}(\mathbf{S}_a\mathbf{D}\mathbf{S}_a^{-1})^{-1}\mathbf{S}_a. \tag{15.31}$$

Long Waves

The trivial case of a primitive lattice is discussed above, and the matrices $\mathbf{d}(\hat{\mathbf{k}})$ whose eigenvalues are $[c(\hat{\mathbf{k}}s)]^2$ are given in (15.2). For the example of two like atoms per unit cell, in a lattice which has inversion symmetry, it is possible to block-diagonalize the dynamical matrices in the long-wavelength limit, and obtain separately the 3×3 matrices for the acoustic and optic branches. This calculation exemplifies a very useful general procedure, which is not restricted to the example discussed here.

We consider a fixed direction of $\mathbf{k}$, and let λ be the magnitude of $\mathbf{k}$, so that

$$\mathbf{k} = \lambda\hat{\mathbf{k}}. \tag{15.32}$$

The matrices $\mathbf{f}$ and $\mathbf{g}$ are expanded for small λ to order λ^2, as

$$\mathbf{f} = \mathbf{f}_0 + \tfrac{1}{2}\lambda^2\mathbf{f}_2; \tag{15.33}$$

$$\mathscr{R}\mathbf{g} = -\mathbf{f}_0 + \tfrac{1}{2}\lambda^2\mathbf{g}_2; \tag{15.34}$$

$$\mathscr{I}\mathbf{g} = \lambda\mathbf{g}_1. \tag{15.35}$$

For a given $\hat{\mathbf{k}}$, these real symmetric 3×3 expansion matrices are obtained by expanding (15.19), (15.21), and (15.22):

$$f_{0,ij} = -(2/M_C)\sum_{\mathrm{N}} \Phi_{ij}(00,\mathrm{N}1), \tag{15.36}$$

$$f_{2,ij} = -(2/M_C)\sum_{\mathrm{N}}{}' \Phi_{ij}(00,\mathrm{N}0)[\hat{\mathbf{k}}\cdot\mathbf{R}(\mathrm{N})]^2, \tag{15.37}$$

$$g_{1,ij} = (2/M_C)\sum_{\mathrm{N}} \Phi_{ij}(00\,\mathrm{N}1)[\hat{\mathbf{k}}\cdot\mathbf{R}(\mathrm{N}1)], \tag{15.38}$$

$$g_{2,ij} = -(2/M_C)\sum_{\mathrm{N}} \Phi_{ij}(00,\mathrm{N}1)[\hat{\mathbf{k}}\cdot\mathbf{R}(\mathrm{N}1)]^2. \tag{15.39}$$

To order λ^2, the real symmetric form (15.29) of the dynamical matrix is

$$\mathbf{S}_a\mathbf{D}\mathbf{S}_a^{-1} = \begin{pmatrix} \mathbf{f}_0 + \lambda\mathbf{g}_1 + \tfrac{1}{2}\lambda^2\mathbf{f}_2 & -\mathbf{f}_0 + \tfrac{1}{2}\lambda^2\mathbf{g}_2 \\ -\mathbf{f}_0 + \tfrac{1}{2}\lambda^2\mathbf{g}_2 & \mathbf{f}_0 - \lambda\mathbf{g}_1 + \tfrac{1}{2}\lambda^2\mathbf{f}_2 \end{pmatrix}. \tag{15.40}$$

The block-diagonalization procedure is similar to the method of long waves described in Section 12. The real symmetric 6×6 matrix (15.40) is diagonalized by a real orthogonal matrix, say $\mathbf{S}$; note that orthogonal is the real counterpart of unitary, and corresponds to $\tilde{\mathbf{S}} = \mathbf{S}^{-1}$, where $\tilde{\mathbf{S}}$ is the transpose of $\mathbf{S}$. We therefore write down a 6×6 matrix $\mathbf{S}$ in terms of 3×3 submatrices,

$$\mathbf{S} = \begin{pmatrix} \mathbf{a} & \mathbf{b} \\ \mathbf{c} & \mathbf{d} \end{pmatrix},$$

then write each of the submatrices as a power series in λ to order λ^2, and finally require $\mathbf{S}$ to be orthogonal and $\mathbf{S}(\mathbf{S}_a\mathbf{D}\mathbf{S}_a^{-1})\mathbf{S}^{-1}$ to be block-diagonal to zeroth order in λ, then to first order in λ, and finally to second order in λ. The detailed calculation is recommended as an instructive exercise.

It is convenient to describe the results in the form of two successive orthogonal transformations. The first transformation is

$$\mathbf{S}_b = \frac{1}{\sqrt{2}}\begin{pmatrix} \mathbf{I} & \mathbf{I} \\ -\mathbf{I} & \mathbf{I} \end{pmatrix}; \tag{15.41}$$

it is obvious that

$$\tilde{\mathbf{S}}_b = \frac{1}{\sqrt{2}}\begin{pmatrix} \mathbf{I} & -\mathbf{I} \\ \mathbf{I} & \mathbf{I} \end{pmatrix} = \mathbf{S}_b^{-1}. \tag{15.42}$$

With the real symmetric dynamical matrix (15.40), the further transformed matrix, correct to order λ^2, is found to be

$$\mathbf{S}_b\mathbf{S}_a\mathbf{D}\mathbf{S}_a^{-1}\mathbf{S}_b^{-1} = \begin{pmatrix} \frac{1}{2}\lambda^2(\mathbf{f}_2 + \mathbf{g}_2) & -\lambda\mathbf{g}_1 \\ -\lambda\mathbf{g}_1 & 2\mathbf{f}_0 + \frac{1}{2}\lambda^2(\mathbf{f}_2 - \mathbf{g}_2) \end{pmatrix}. \tag{15.43}$$

To define the second transformation, let us introduce the 3×3 matrix $\mathbf{h}_1$,

$$\mathbf{h}_1 = \mathbf{f}_0^{-1}\mathbf{g}_1. \tag{15.44}$$

Since $\mathbf{f}_0$ is real symmetric, so is $\mathbf{f}_0^{-1}$, and since $\mathbf{g}_1$ is also real symmetric, it follows that

$$\tilde{\mathbf{h}}_1 = \mathbf{g}_1\mathbf{f}_0^{-1}. \tag{15.45}$$

The second orthogonal transformation is

$$\mathbf{S}_c = \begin{pmatrix} \mathbf{I} - \frac{1}{8}\lambda^2\tilde{\mathbf{h}}_1\mathbf{h}_1 & \frac{1}{2}\lambda\tilde{\mathbf{h}}_1 \\ -\frac{1}{2}\lambda\mathbf{h}_1 & \mathbf{I} - \frac{1}{8}\lambda^2\mathbf{h}_1\tilde{\mathbf{h}}_1 \end{pmatrix}. \tag{15.46}$$

It is easily verified that $\mathbf{S}_c$ is orthogonal to order λ^2, i.e.,

$$\tilde{\mathbf{S}}_c\mathbf{S}_c = \begin{pmatrix} \mathbf{I} & 0 \\ 0 & \mathbf{I} \end{pmatrix} + O(\lambda^4). \tag{15.47}$$

Then from (15.43), the block-diagonal form of the dynamical matrix, correct to order λ^2, is

$$\mathbf{S}_c\mathbf{S}_b\mathbf{S}_a\mathbf{D}\mathbf{S}_a^{-1}\mathbf{S}_b^{-1}\mathbf{S}_c^{-1}$$

$$= \frac{1}{2}\begin{pmatrix} \lambda^2(\mathbf{f}_2 + \mathbf{g}_2 - \mathbf{g}_1\mathbf{f}_0^{-1}\mathbf{g}_1) & 0 \\ 0 & 4\mathbf{f}_0 + \lambda^2[\mathbf{f}_2 - \mathbf{g}_2 + \frac{1}{2}(\mathbf{f}_0^{-1}\mathbf{g}_1\mathbf{g}_1 + \mathbf{g}_1\mathbf{g}_1\mathbf{f}_0^{-1})] \end{pmatrix}. \tag{15.48}$$

Equation (15.48) constitutes a complete separation of the dynamical matrix into acoustic and optic phonon submatrices in the long-wavelength limit. For each direction $\hat{\mathbf{k}}$ the real symmetric 3×3 matrix $\mathbf{d}(\hat{\mathbf{k}})$, whose eigenvalues are the acoustic phonon squared velocities $[c(\hat{\mathbf{k}}s)]^2$, $s = 1,2,3$, is

$$\mathbf{d}(\hat{\mathbf{k}}) = \tfrac{1}{2}(\mathbf{f}_2 + \mathbf{g}_2 - \mathbf{g}_1\mathbf{f}_0^{-1}\mathbf{g}_1). \tag{15.49}$$

Also for each direction $\hat{\mathbf{k}}$, the real symmetric 3×3 matrix whose eigenvalues are the optic phonon squared frequencies $[\omega(\mathbf{k}s)]^2$, $s = 4,5,6$, is

$$2\mathbf{f}_0 + \tfrac{1}{2}\lambda^2[\mathbf{f}_2 - \mathbf{g}_2 + \tfrac{1}{2}(\mathbf{f}_0^{-1}\mathbf{g}_1\mathbf{g}_1 + \mathbf{g}_1\mathbf{g}_1\mathbf{f}_0^{-1})]. \tag{15.50}$$

The sum of $[c(\hat{\mathbf{k}}s)]^2$ for a given $\hat{\mathbf{k}}$ is just the trace of $\mathbf{d}(\hat{\mathbf{k}})$,

$$\sum_{s=1}^{3}[c(s)]^2 = \tfrac{1}{2}\,\mathrm{Tr}(\mathbf{f}_2 + \mathbf{g}_2 - \mathbf{g}_1\mathbf{f}_0^{-1}\mathbf{g}_1), \tag{15.51}$$

and the sum of the optic eigenvalues $[\omega(\mathbf{k}s)]^2$ for a given $\hat{\mathbf{k}}$ is the trace of (15.50),

$$\sum_{s=4}^{6}[\omega(s)]^2 = 2\,\mathrm{Tr}\,\mathbf{f}_0 + \tfrac{1}{2}\lambda^2\,\mathrm{Tr}(\mathbf{f}_2 - \mathbf{g}_2 + \mathbf{g}_1\mathbf{f}_0^{-1}\mathbf{g}_1). \tag{15.52}$$

To obtain the last term in (15.52), we permuted $\mathbf{g}_1$ inside the trace. Finally the total sum of $[\omega(\mathbf{k}s)]^2$ for a given $\hat{\mathbf{k}}$ is, from (15.51) and (15.52),

$$\sum_{s=1}^{6}[\omega(s)]^2 = 2\,\mathrm{Tr}(\mathbf{f}_0 + \tfrac{1}{2}\lambda^2\mathbf{f}_2), \tag{15.53}$$

since of course $[\omega(s)]^2 = \lambda^2[c(s)]^2$ for the acoustic modes. Note that (15.53) is the same as the trace of the original real symmetric dynamical matrix (15.40).

Eigenvalue Derivatives

The calculation of phonon Grüneisen parameters will be illustrated for a primitive lattice with central potential interactions among the ions. Let us drop the ion indices M,N,· · ·, for the moment, and use the schematic notation $\varphi(R^2)$ for the central potential, to calculate some required derivatives. Under a homogeneous deformation with displacement gradients u_{ij}, and corresponding Lagrangian strains η_{ij}, the ion displacements are $U_i = \Sigma_j u_{ij}R_j$, so the symmetric strain derivatives of the ion position vector components are

$$(\partial R_i/\partial \eta_{kl})_{\eta'} = \tfrac{1}{2}(R_k\delta_{il} + R_l\delta_{ik}), \tag{15.54}$$

where here and in the following the derivatives are evaluated at the initial configuration, i.e., at $\eta_{ij} = 0$. Also under this deformation, the R^2 vary by the amount $2\Sigma_{ij}\ \eta_{ij}R_iR_j$ [see, e.g., (2.10)], so that

$$(\partial R^2/\partial \eta_{ij})_{\eta'} = 2R_iR_j. \tag{15.55}$$

The strain derivatives of $\varphi(R^2)$, $\varphi'(R^2)$, etc., are then calculated by chain-rule differentiation, as follows.

$$[\partial\varphi(R^2)/\partial\eta_{ij}]_{\eta'} = [\partial\varphi(R^2)/\partial R^2](\partial R^2/\partial\eta_{ij})_{\eta'} = 2\varphi'(R^2)R_iR_j; \tag{15.56}$$

$$[\partial\varphi'(R^2)/\partial\eta_{ij}]_{\eta'} = 2\varphi''(R^2)R_iR_j; \tag{15.57}$$

$$[\partial\varphi''(R^2)/\partial\eta_{ij}]_{\eta'} = 2\varphi'''(R^2)R_iR_j. \tag{15.58}$$

Consider also an infinitesimal isotropic volume change δV, in which each position vector component R_i changes by the amount $\frac{1}{3}(\delta V/V)R_i$. Note that this is a volume change without shear, and for a cubic crystal it corresponds to the volume change resulting from a change in the temperature or in the isotropic pressure. The following derivatives, evaluated at $\delta V = 0$, may then be calculated.

$$V(dR_i/dV) = \tfrac{1}{3}R_i; \tag{15.59}$$

$$V(dR^2/dV) = \tfrac{2}{3}R^2; \tag{15.60}$$

$$V[d\varphi'(R^2)/dV] = \tfrac{2}{3}\varphi''(R^2)R^2, \tag{15.61}$$

$$V[d\varphi''(R^2)/dV] = \tfrac{2}{3}\varphi'''(R^2)R^2. \tag{15.62}$$

The derivatives (15.54)–(15.62) hold for any primitive lattice, and also for a nonprimitive lattice in the case of zero sublattice displacements.

For a primitive lattice with central potential interactions among the ions, $\mathbf{D}(\mathbf{k})$ is given by (15.1), and the strain derivatives are easily evaluated with the aid of (15.54), (15.57), and (15.58). With the abbreviations

$$\mathbf{R} \equiv \mathbf{R}(\mathrm{N}), \qquad R_i \equiv R_i(\mathrm{N}), \tag{15.63}$$

the results are

$$\begin{aligned}&[\partial D_{ij}(\mathbf{k})/\partial\eta_{kl}]_{\eta'}\\ &= M_C^{-1}\sum_{\mathrm{N}}{}'\{2\varphi''(0,\mathrm{N})[2R_kR_l\delta_{ij} + R_iR_k\delta_{jl} + R_iR_l\delta_{jk} + R_jR_k\delta_{il} + R_jR_l\delta_{ik}]\\ &\qquad + 8\varphi'''(0,\mathrm{N})R_iR_jR_kR_l\}[1 - \cos(\mathbf{k}\cdot\mathbf{R})].\end{aligned} \tag{15.64}$$

The generalized phonon Grüneisen parameters are defined by (11.74), and it is shown in (11.76) how they may be calculated from the strain derivatives of the dynamical matrices. For a primitive lattice, (11.76) is simply

$$\gamma_{kl}(\mathbf{k}s) = -\tfrac{1}{2}[\omega(\mathbf{k}s)]^{-2} \sum_{ij} w_i(\mathbf{k}s)[\partial D_{ij}(\mathbf{k})/\partial\eta_{kl}]_{\eta'} w_j(\mathbf{k}s). \tag{15.65}$$

The volume derivative of (15.1) for $\mathbf{D}(\mathbf{k})$ is easily evaluated with the aid of (15.59)–(15.62); again with the abbreviations (15.63) the result is

$$V[dD_{ij}(\mathbf{k})/dV] = \tfrac{2}{3}M_C^{-1} \sum_{\mathrm{N}}{}' \{2\varphi''(0,\mathrm{N})[R^2\delta_{ij} + 2R_iR_j] + 4\varphi'''(0,\mathrm{N})R^2R_iR_j\}[1 - \cos(\mathbf{k}\cdot\mathbf{R})]. \tag{15.66}$$

The ordinary (volume) phonon Grüneisen parameters are defined by (11.79), and they may be calculated from a formula similar to (15.65), namely

$$\gamma(\mathbf{k}s) = -\tfrac{1}{2}[\omega(\mathbf{k}s)]^{-2} \sum_{ij} w_i(\mathbf{k}s)[V dD_{ij}(\mathbf{k})/dV] w_j(\mathbf{k}s). \tag{15.67}$$

For a fixed direction $\hat{\mathbf{k}}$, the strain derivatives of the matrix $\mathbf{d}(\hat{\mathbf{k}})$ of (15.2) are evaluated with the aid of (15.54), (15.57), and (15.58), to give

$$\begin{aligned}
[\partial d_{ik}(\hat{\mathbf{k}})/\partial\eta_{mn}]_{\eta'} = M_C^{-1} \sum_{\mathrm{N}}{}' \sum_{jl} \{&\varphi'(0,\mathrm{N})\delta_{ik}[R_jR_m\delta_{ln} + R_jR_n\delta_{lm}] \\
&+ \varphi''(0,\mathrm{N})[2R_jR_lR_mR_n\delta_{ik} + 2R_iR_jR_k(R_m\delta_{ln} + R_n\delta_{lm}) \\
&+ R_iR_jR_l(R_m\delta_{kn} + R_n\delta_{km}) + R_jR_kR_l(R_m\delta_{in} + R_n\delta_{im})] \\
&+ 4\varphi'''(0,\mathrm{N})R_iR_jR_kR_lR_mR_n\}\hat{k}_j\hat{k}_l,
\end{aligned} \tag{15.68}$$

where some terms symmetric in j,l have been combined inside the Σ_{jl}, and the abbreviations (15.63) have been used. The phonon velocity strain derivatives $\sigma_{ij}(\hat{\mathbf{k}}s)$ are defined in (12.48), and it is shown in (12.50) how they may be calculated from the strain derivatives of $\mathbf{d}(\hat{\mathbf{k}})$, given for the present model by (15.68). The volume derivative of $\mathbf{d}(\hat{\mathbf{k}})$ of (15.2) is evaluated with the aid of (15.59)–(15.62), to give

$$\begin{aligned}
V d[d_{ij}(\hat{\mathbf{k}})]/dV
&= \tfrac{2}{3}M_C^{-1} \sum_{\mathrm{N}}{}' \sum_{jl} \{\varphi'(0,\mathrm{N})R_jR_l\delta_{ik} + \varphi''(0,\mathrm{N})[R^2R_jR_l\delta_{ik} + 4R_iR_jR_kR_l] \\
&\quad + 2\varphi'''(0,\mathrm{N})R^2R_iR_jR_kR_l\}\hat{k}_j\hat{k}_l.
\end{aligned} \tag{15.69}$$

Again the phonon velocity volume derivative $\sigma(\hat{\mathbf{k}}s)$, defined by (12.49), may be calculated from (15.69) for the present model, according to (12.51).

For the example of two like atoms per unit cell with central potential interactions among the ions, the phonon Grüneisen parameters may be calculated from the real symmetric dynamical matrices (15.29). Furthermore, in the long-wavelength limit the various eigenvalue derivatives may be calculated from the acoustic and optic mode matrices (15.49) and (15.50), respectively.

4

PHONON THERMODYNAMICS

In this chapter the Helmholtz free energy is calculated to leading order in the anharmonicity, which is fourth order in ϵ. The other thermodynamic functions, including the directly measurable renormalized phonon frequencies and lifetimes, are also derived. It is shown that at $T = 0$ the sound waves and long-wavelength renormalized acoustic phonons are identical, and that to leading order in the anharmonicity the $T = 0$ Debye temperature calculated from the measured sound velocities is the same as that obtained from the low-temperature entropy or heat capacity. Explicit expressions are obtained for the first- and second-strain derivatives of the harmonic phonon frequencies. The temperature- and configuration-dependences of thermodynamic functions are discussed for the limits of low and high temperatures. Some of the pitfalls involved in trying to calculate thermodynamic functions from measured phonon frequencies are pointed out, and remedies are described.

16. ANHARMONIC PERTURBATION EXPANSION

Helmholtz Free Energy

In the preceding sections the *mechanical* motion of the ions in a crystal has been studied, and the solutions to the mechanical problem of motion provide the basis for *statistical mechanical* calculations in the present section. For a given equilibrium configuration of the crystal, the partition function Z can be calculated from the crystal energy levels, and the Helmholtz free energy is then given by

$$F = -KT \ln Z. \tag{16.1}$$

Then certain thermodynamic quantities which are obtained by differentiating F with respect to T can be obtained for this configuration (e.g., the internal energy U, the entropy S, and the heat capacity at constant configuration C_η). To obtain thermodynamic forces, which depend on the variation of F with respect to configuration (e.g., the stresses and elastic constants), it is necessary in principle to find the crystal energy levels, then Z, and finally F for a new configuration which is infinitesimally removed from the original configuration. Thus it is necessary in principle to have the mechanical solutions available for various equilibrium configurations of the crystal; this is why the externally applied mechanical forces were introduced in Chapter 2, and the subsequent lattice statics and lattice-dynamics calculations were carried out for arbitrary initial configuration. In practice the free energy may be considered a function of the temperature and of the macroscopic configuration of the crystal, through the configuration-dependence of the energy levels, and hence configuration derivatives of F may be evaluated directly. It must be emphasized that if our mechanical calculations had been restricted to a single configuration, such as, for example, that corresponding to vanishing stresses, then in principle we would not be allowed to differentiate F with respect to strains. It should also be noted that the externally applied mechanical forces are envisioned to be positive or negative, and they allow the crystal configuration to be chosen quite arbitrarily. Finally our calculations are restricted to regions in which the crystal potential Φ is analytic, as discussed in Section 6.

The canonical partition function is the sum over all the system energy levels $\mathscr{E}$,

$$Z = \sum e^{-\mathscr{E}/KT}. \tag{16.2}$$

The lattice-dynamics energy levels are written as a perturbation expansion in (13.82), i.e., $\mathscr{E} = \Phi_0 + \mathscr{E}_2 + \mathscr{E}_4$, and we will now treat $\mathscr{E}_4$ as small and calculate Z to first order in $\mathscr{E}_4$. The harmonic partition function is defined by

$$Z_{\mathrm{H}} = \sum e^{-\mathscr{E}_2/KT}; \tag{16.3}$$

evaluation of (16.2) for Z then proceeds as follows.

$$\begin{aligned} Z &= e^{-\Phi_0/KT} \sum e^{-(\mathscr{E}_2+\mathscr{E}_4)/KT} \\ &= e^{-\Phi_0/KT} \sum e^{-\mathscr{E}_2/KT}[1 - (\mathscr{E}_4/KT)] \\ &= e^{-\Phi_0/KT} Z_{\mathrm{H}}\{1 - Z_{\mathrm{H}}^{-1} \sum (\mathscr{E}_4/KT) e^{-\mathscr{E}_2/KT}\} \\ &= e^{-\Phi_0/KT} Z_{\mathrm{H}}[1 - (KT)^{-1}\langle \mathscr{E}_4 \rangle_{\mathrm{H}}]. \end{aligned} \tag{16.4}$$

The quantity $\langle \mathscr{E}_4 \rangle_H$ is defined by the last step in (16.4), according to

$$\langle \mathscr{E}_4 \rangle_H = \sum \mathscr{E}_4 e^{-\mathscr{E}_2/KT}, \tag{16.5}$$

and this is, of course, a harmonic statistical average of $\mathscr{E}_4$. Although the expansion (16.4) of Z is an expansion in powers of $\mathscr{E}_4$, and hence in powers of the number of unit cells N_0 in the crystal, the corresponding free energy expansion below has every term proportional to N_0, as it should.

With Z given by (16.4), the free energy expression (16.1) becomes

$$\begin{aligned} F &= \Phi_0 - KT \ln Z_H - KT \ln [1 - (KT)^{-1} \langle \mathscr{E}_4 \rangle_H] \\ &= \Phi_0 - KT \ln Z_H + \langle \mathscr{E}_4 \rangle_H, \end{aligned} \tag{16.6}$$

again to first order in $\mathscr{E}_4$. Let us define the harmonic free energy F_H and the leading-order anharmonic free energy F_A, and write our perturbation expansion of the free energy in the form

$$F = \Phi_0 + F_H + F_A, \tag{16.7}$$

where

$$F_H = -KT \ln Z_H, \tag{16.8}$$

$$F_A = \langle \mathscr{E}_4 \rangle_H. \tag{16.9}$$

The detailed expressions for F_H and F_A will now be derived.

The sum (16.3) over all harmonic energy levels $\mathscr{E}_2$ is written

$$Z_H = \sum_{(\cdots n_\kappa \cdots)} e^{-\mathscr{E}_2(\cdots n_\kappa \cdots)/KT}, \tag{16.10}$$

where all possible sets of the numbers n_κ are included, and $\mathscr{E}_2$ is given by (13.94):

$$\mathscr{E}_2(\cdots n_\kappa \cdots) = \sum_\kappa \hbar\omega_\kappa(n_\kappa + \tfrac{1}{2}). \tag{16.11}$$

Evaluation of (16.10) is as follows.

$$\begin{aligned} Z_H &= \sum_{(\cdots n_\kappa \cdots)} e^{-\Sigma_\kappa \hbar\omega_\kappa(n_\kappa + \frac{1}{2})/KT} \\ &= \sum_{(\cdots n_\kappa \cdots)} \prod_\kappa e^{-\hbar\omega_\kappa(n_\kappa + \frac{1}{2})/KT} \\ &= \prod_\kappa e^{-\frac{1}{2}\hbar\omega_\kappa/KT} \sum_{n=0}^{\infty} e^{-n\hbar\omega_\kappa/KT} \\ &= \prod_\kappa e^{-\frac{1}{2}\hbar\omega_\kappa/KT}[1 - e^{-\hbar\omega_\kappa/KT}]^{-1}. \end{aligned} \tag{16.12}$$

Then the harmonic free energy (16.8) is

$$F_H = -KT \sum_\kappa \{\ln e^{-\frac{1}{2}\hbar\omega_\kappa/KT} + \ln[1 - e^{-\hbar\omega_\kappa/KT}]^{-1}\},$$

or

$$F_{\mathrm{H}} = \sum_{\kappa} \{\tfrac{1}{2}\hbar\omega_{\kappa} + KT \ln[1 - e^{-\hbar\omega_{\kappa}/KT}]\}. \qquad (16.13)$$

Harmonic statistical averages of functions of the phonon occupation numbers, e.g., $\langle n_{\kappa}\rangle_{\mathrm{H}}$, $\langle n_{\kappa}n_{\kappa'}\rangle_{\mathrm{H}}$, etc., may be derived by calculations similar to the above calculation of F_{H}. It is simpler, however, to use the statistical perturbation method of Section 14. Since the harmonic phonon-creation operators $A_{\kappa}^{\dagger}$ are exact creation operators of $\mathscr{H}_2$, according to (13.45), the zeroth-order basic equations of the statistical perturbation method apply exactly, and for the harmonic phonons (14.14) reads

$$\langle A_{\kappa}^{\dagger}\chi\rangle_{\mathrm{H}} = [e^{\hbar\omega_{\kappa}/KT} - 1]^{-1}\langle[\chi,A_{\kappa}^{\dagger}]\rangle_{\mathrm{H}}, \qquad (16.14)$$

where χ is any operator. Now the harmonic phonons are good bosons, so $[A_{\kappa},A_{\kappa}^{\dagger}] = 1$ and n_{κ} is the eigenvalue of $A_{\kappa}^{\dagger}A_{\kappa}$, and (16.14) gives

$$\langle n_{\kappa}\rangle_{\mathrm{H}} = \langle A_{\kappa}^{\dagger}A_{\kappa}\rangle_{\mathrm{H}} = [e^{\hbar\omega_{\kappa}/KT} - 1]^{-1}. \qquad (16.15)$$

It will be convenient to use the abbreviation $\bar{n}_{\kappa}$ for the statistical average phonon occupation number (16.15), and since $\omega_{\kappa} = \omega_{-\kappa}$ we have

$$\bar{n}_{\kappa} = \bar{n}_{-\kappa} = [e^{\hbar\omega_{\kappa}/KT} - 1]^{-1}. \qquad (16.16)$$

The average $\langle n_{\kappa}n_{\kappa'}\rangle_{\mathrm{H}}$ is calculated in the same way:

$$\begin{aligned}\langle n_{\kappa}n_{\kappa'}\rangle_{\mathrm{H}} &= \langle A_{\kappa}^{\dagger}A_{\kappa}A_{\kappa'}^{\dagger}A_{\kappa'}\rangle_{\mathrm{H}} = \bar{n}_{\kappa}\langle[A_{\kappa}A_{\kappa'}^{\dagger}A_{\kappa'},A_{\kappa}^{\dagger}]\rangle_{\mathrm{H}} \\ &= \bar{n}_{\kappa}\langle A_{\kappa'}^{\dagger}A_{\kappa'} + (A_{\kappa'}^{\dagger}A_{\kappa'} + 1)\delta_{\kappa\kappa'}\rangle \qquad (16.17) \\ &= \bar{n}_{\kappa}\bar{n}_{\kappa'} + \bar{n}_{\kappa}(\bar{n}_{\kappa'} + 1)\delta_{\kappa\kappa'}.\end{aligned}$$

According to (16.9) the anharmonic free energy is just the average $\langle\mathscr{E}_4\rangle_{\mathrm{H}}$, and $\mathscr{E}_4$ may be written down from (13.96) for $\mathscr{E}_4 - G_4$ and (13.92) for G_4. The statistical average then involves terms like $\langle n_{\kappa}\rangle_{\mathrm{H}}$ and $\langle n_{\kappa}n_{\kappa'}\rangle_{\mathrm{H}}$, which are given by (16.15) and (16.17), respectively, and further since these factors appear inside sums over κ,κ' the $\delta_{\kappa\kappa'}$ term in (16.17) may be dropped, since it leads to contributions to $\langle\mathscr{E}_4\rangle_{\mathrm{H}}$ which are of relative order N_0^{-1}. Therefore $\langle\mathscr{E}_4\rangle_{\mathrm{H}}$ is just $\mathscr{E}_4$ with each n_{κ} replaced by $\bar{n}_{\kappa}$; then since $\bar{n}_{-\kappa} = \bar{n}_{\kappa}$ by (16.16), the result is

$$\begin{aligned}\langle\mathscr{E}_4\rangle_{\mathrm{H}} = {} & 12\sum_{\kappa\kappa'}\Phi_{\kappa,-\kappa,\kappa',-\kappa'}[(\bar{n}_{\kappa'} + 1)\bar{n}_{\kappa} + \tfrac{1}{4}] \\ & - 3\sum_{\kappa\kappa'\kappa''}\{|\Phi_{\kappa\kappa'\kappa''}|^2\,[\alpha_{\kappa\kappa'\kappa''}(\bar{n}_{\kappa'} + \bar{n}_{\kappa''} + 2)\bar{n}_{\kappa} + \beta_{\kappa\kappa'\kappa''}(\bar{n}_{\kappa''} - \bar{n}_{\kappa'})\bar{n}_{\kappa} \\ & \quad + \eta_{\kappa\kappa'\kappa''}(\bar{n}_{\kappa'} - \bar{n}_{\kappa''})\bar{n}_{\kappa} - \zeta_{\kappa\kappa'\kappa''}(\bar{n}_{\kappa'} + \bar{n}_{\kappa''} + 2)\bar{n}_{\kappa} \qquad (16.18) \\ & \quad + 2(\hbar\omega_{\kappa} + \hbar\omega_{\kappa'} + \hbar\omega_{\kappa''})^{-1}] \\ & + \Phi_{\kappa,-\kappa,\kappa''}\Phi_{\kappa',-\kappa',-\kappa''}[2(\eta_{\kappa\kappa\kappa''} - \zeta_{\kappa\kappa\kappa''})(\bar{n}_{\kappa'} + 1)\bar{n}_{\kappa} + 3(\hbar\omega_{\kappa''})^{-1}]\}.\end{aligned}$$

The complex coefficients $\alpha_{\kappa\kappa'\kappa''}$, $\beta_{\kappa\kappa'\kappa''}$, $\eta_{\kappa\kappa'\kappa''}$, and $\zeta_{\kappa\kappa'\kappa''}$ are given in (13.57), and now that all the singular occupation numbers n_κ have been replaced by the smoothly varying functions $\bar{n}_\kappa$, we can take the limit $\gamma \to 0$ in these coefficients. This limit produces real and imaginary contributions to each sum; for example, if f_κ is any function, the general relation is

$$\lim_{\gamma\to 0^+} \sum_\kappa \frac{f_\kappa}{(\omega_\kappa \pm i\gamma)} = \sum_\kappa \frac{f_\kappa}{(\omega_\kappa)_p} \mp i\pi \sum_\kappa f_\kappa \delta(\omega_\kappa), \tag{16.19}$$

where the notation $\Sigma_\kappa f_\kappa/(\omega_\kappa)_p$ represents the principal part of the sum (or integral). Note that we are dealing with three-dimensional principal part sums, in the sums over **k** vectors.

The limit $\gamma \to 0$ is taken in each term in (16.18) for $\langle \mathscr{E}_4 \rangle_{\mathrm{H}}$, and the imaginary terms are discarded since they relate to the lifetimes of the system states, and are of no interest in the free energy calculation. Then making use of the symmetry of each $\Phi_{\kappa\kappa'\kappa''}$ in its indices, and interchanging indices freely inside the sums, several of the terms can be grouped together as shown in the following relations for the limit $\gamma \to 0$.

$$\begin{aligned} \alpha_{\kappa\kappa'\kappa''}(\bar{n}_{\kappa'} + \bar{n}_{\kappa''} + 2)\bar{n}_\kappa &\to 3(\hbar\omega_\kappa + \hbar\omega_{\kappa'} + \hbar\omega_{\kappa''})_p^{-1}(\bar{n}_{\kappa'} + \bar{n}_{\kappa''} + 2)\bar{n}_\kappa \\ &\to 6(\hbar\omega_\kappa + \hbar\omega_{\kappa'} + \hbar\omega_{\kappa''})_p^{-1}(\bar{n}_{\kappa'}\bar{n}_\kappa + \bar{n}_\kappa); \end{aligned} \tag{16.20}$$

$$\begin{aligned} \zeta_{\kappa\kappa'\kappa''}(\bar{n}_{\kappa'} + \bar{n}_{\kappa''} + 2)\bar{n}_\kappa &\to 3(\hbar\omega_\kappa - \hbar\omega_{\kappa'} - \hbar\omega_{\kappa''})_p^{-1}(\bar{n}_{\kappa'} + \bar{n}_{\kappa''} + 2)\bar{n}_\kappa \\ &\to -6(\hbar\omega_\kappa + \hbar\omega_{\kappa'} - \hbar\omega_{\kappa''})_p^{-1}(\bar{n}_\kappa\bar{n}_{\kappa''} + \bar{n}_{\kappa''}); \end{aligned} \tag{16.21}$$

$$\begin{aligned} (\beta_{\kappa\kappa'\kappa''} - \eta_{\kappa\kappa'\kappa''})(\bar{n}_{\kappa''} - \bar{n}_{\kappa'})\bar{n}_\kappa &\to [3(\hbar\omega_\kappa + \hbar\omega_{\kappa'} - \hbar\omega_{\kappa''})_p^{-1} \\ &\quad - 3(\hbar\omega_\kappa - \hbar\omega_{\kappa'} + \hbar\omega_{\kappa''})_p^{-1}](\bar{n}_{\kappa''} - \bar{n}_{\kappa'})\bar{n}_\kappa \\ &\to 6(\hbar\omega_\kappa + \hbar\omega_{\kappa'} - \hbar\omega_{\kappa''})_p^{-1}\,(\bar{n}_{\kappa''}\bar{n}_\kappa - \bar{n}_{\kappa'}\bar{n}_\kappa); \end{aligned} \tag{16.22}$$

$$(\eta_{\kappa\kappa\kappa''} - \zeta_{\kappa\kappa\kappa''})(\bar{n}_{\kappa'} + 1)\bar{n}_\kappa \to 6(\hbar\omega_{\kappa''})_p^{-1}(\bar{n}_{\kappa'}\bar{n}_\kappa + \bar{n}_\kappa). \tag{16.23}$$

In addition, by interchanging indices in the quartic contribution, one obtains the symmetric combination

$$(\bar{n}_{\kappa'} + 1)\bar{n}_\kappa + \tfrac{1}{4} \to (\bar{n}_{\kappa'} + \tfrac{1}{2})(\bar{n}_\kappa + \tfrac{1}{2}). \tag{16.24}$$

Finally, the anharmonic free energy is the real part of (16.18) for $\langle \mathscr{E}_4 \rangle_{\mathrm{H}}$, and with the contributions (16.20)–(16.24) the result is

$$F_{\mathrm{A}} = 12 \sum_{\kappa\kappa'} \Phi_{\kappa,-\kappa,\kappa',-\kappa'}(\bar{n}_\kappa + \tfrac{1}{2})(\bar{n}_{\kappa'} + \tfrac{1}{2})$$

$$- \frac{18}{\hbar} \sum_{\kappa\kappa'\kappa''} \left\{ |\Phi_{\kappa\kappa'\kappa''}|^2 \left[\frac{(\bar{n}_\kappa \bar{n}_{\kappa'} + \bar{n}_\kappa + \frac{1}{3})}{(\omega_\kappa + \omega_{\kappa'} + \omega_{\kappa''})_p} + \frac{(2\bar{n}_\kappa \bar{n}_{\kappa''} - \bar{n}_\kappa \bar{n}_{\kappa'} + \bar{n}_{\kappa''})}{(\omega_\kappa + \omega_{\kappa'} - \omega_{\kappa''})_p} \right] \right.$$

$$\left. + 2\Phi_{\kappa,-\kappa,\kappa''} \Phi_{\kappa',-\kappa',-\kappa''} \frac{(\bar{n}_\kappa \bar{n}_{\kappa'} + \bar{n}_\kappa + \frac{1}{4})}{(\omega_{\kappa''})_p} \right\}. \quad (16.25)$$

As long as all the harmonic phonon frequencies ω_κ are positive, and of course with the uniform translational modes corresponding to $\omega_\kappa = 0$ omitted from the sums, the denominators $\omega_\kappa + \omega_{\kappa'} + \omega_{\kappa''}$ and $\omega_{\kappa''}$ in (16.25) never vanish, and the sum is equal to its principal part for these terms (i.e., the principal part designation can be dropped for these terms). In addition, the last term in (16.25) is to be included only for nonprimitive lattices, since in the $\Sigma_{\kappa''} = \Sigma_{\mathbf{k}''s''}$ only the s'' = optic branches are to be included. The total contribution to F_{A} from the fourth-order ground-state energy G_4 is contained in the numbers $\frac{1}{4}$ in the quartic contribution, and $\frac{1}{3}$ and $\frac{1}{4}$ in the cubic contribution.

The crystal free energy correct to leading order in phonon–phonon interactions is $F = \Phi_0 + F_{\mathrm{H}} + F_{\mathrm{A}}$, and is a function of the crystal configuration and of the temperature. The function Φ_0 depends only on the configuration; F_{H} is given by (16.13), the temperature-dependence is explicit, while the configuration-dependence is through that of the ω_κ; and F_{A} is given by (16.25), with the temperature-dependence contained in the $\bar{n}_\kappa$, and the configuration-dependence contained in the $\Phi_{\kappa\kappa'\kappa''}$, $\Phi_{\kappa\kappa'\kappa''\kappa'''}$, and ω_κ.

Phonon Frequency Shifts and Widths

As it was mentioned in Section 13, the energy required to create a renormalized phonon depends on the occupation numbers of all the phonons in the system; this energy is $\hbar\omega_\kappa + \hbar\omega_{2\kappa}$, according to (13.81). Now inelastic neutron-scattering experiments measure the statistical average of the energy required to create (or annihilate) a normal mode of the crystal associated with a particular **k** vector; hence these experiments measure $\langle \hbar\omega_\kappa + \hbar\omega_{2\kappa} \rangle$. Of course $\langle \hbar\omega_\kappa \rangle = \hbar\omega_\kappa$, and to leading order the average energy renormalization may be calculated as the harmonic average $\langle \hbar\omega_{2\kappa} \rangle_{\mathrm{H}}$. The detailed contributions to $\hbar\omega_{2\kappa}$ were written in (13.77) and (13.80), and the harmonic statistical average is obtained by replacing each n_κ with $\bar{n}_\kappa$. Then the limit $\gamma \to 0$ is taken to obtain the real and imaginary parts, according to (16.19), and the result is written

$$\langle \hbar\omega_{2\kappa}\rangle_{\mathrm{H}} = \hbar\Delta_\kappa - i\hbar\Gamma_\kappa, \tag{16.26}$$

where it is easily verified from (13.77) and (13.80), that

$$\begin{aligned}
\hbar\Delta_\kappa = {} & 12 \sum_{\kappa'} \Phi_{\kappa,-\kappa,\kappa',-\kappa'}(2\bar{n}_{\kappa'} + 1) \\
& - \frac{18}{\hbar} \sum_{\kappa'\kappa''} \Bigg\{ |\Phi_{\kappa\kappa'\kappa''}|^2 \Bigg[\frac{(\bar{n}_{\kappa'} + \bar{n}_{\kappa''} + 1)}{(\omega_\kappa + \omega_{\kappa'} + \omega_{\kappa''})_p} + \frac{(\bar{n}_{\kappa''} - \bar{n}_{\kappa'})}{(\omega_\kappa + \omega_{\kappa'} - \omega_{\kappa''})_p} \\
& + \frac{(\bar{n}_{\kappa'} - \bar{n}_{\kappa''})}{(\omega_\kappa - \omega_{\kappa'} + \omega_{\kappa''})_p} - \frac{(\bar{n}_{\kappa'} + \bar{n}_{\kappa''} + 1)}{(\omega_\kappa - \omega_{\kappa'} - \omega_{\kappa''})_p} \Bigg] \\
& + 2\Phi_{\kappa,-\kappa,\kappa''}\Phi_{\kappa',-\kappa',-\kappa''} \frac{(2\bar{n}_{\kappa'} + 1)}{(\omega_{\kappa''})_p} \Bigg\};
\end{aligned} \tag{16.27}$$

$$\begin{aligned}
\hbar\Gamma_\kappa = {} & 18\pi\hbar^{-1} \sum_{\kappa'\kappa''} |\Phi_{\kappa\kappa'\kappa''}|^2 \{(\bar{n}_{\kappa'} + \bar{n}_{\kappa''} + 1)\delta(\omega_\kappa - \omega_{\kappa'} - \omega_{\kappa''}) \\
& + (\bar{n}_{\kappa'} - \bar{n}_{\kappa''})[\delta(\omega_\kappa + \omega_{\kappa'} - \omega_{\kappa''}) - \delta(\omega_\kappa - \omega_{\kappa'} + \omega_{\kappa''})]\}.
\end{aligned} \tag{16.28}$$

In (16.28) the terms involving $\delta(\omega_\kappa + \omega_{\kappa'} + \omega_{\kappa''})$ and $\delta(\omega_{\kappa''})$ were dropped, since these δ functions are never satisfied.

Since the time-dependence of a renormalized phonon state is $e^{-i(\omega_\kappa + 2\omega_\kappa)t}$, the statistical average renormalized phonon lifetime is Γ_κ^{-1}. Further, if the anharmonicity is small, i.e., if the leading-order perturbation treatment is reasonably accurate, then the inelastic neutron-scattering lineshape is Lorentzian, with a peak at $\hbar\Omega_\kappa$ given by

$$\hbar\Omega_\kappa = \hbar(\omega_\kappa + \Delta_\kappa), \tag{16.29}$$

and a full width at half maximum equal to $2\hbar\Gamma_\kappa$. A significant departure from Lorentzian lineshapes in a neutron-scattering experiment would indicate higher-order anharmonic effects; the corresponding analysis may be given in terms of the phonon spectral function, which may be derived by a self-consistent operator renormalization. We note that the frequency shifts Δ_κ and the half-widths Γ_κ are thermodynamic quantities in the true thermodynamic sense. Additional contributions to the phonon energy shifts and widths, due to the electron–phonon interactions in metals, are derived in Section 25.

It is convenient to separate the temperature-dependent and temperature-independent contributions to the frequency shifts Δ_κ; for this purpose we define the coefficients θ_κ and $\theta_{\kappa\kappa'}$, which depend only on the crystal configuration:

$$\theta_\kappa = 12\hbar^{-1}\sum_{\kappa'}\Phi_{\kappa,-\kappa,\kappa',-\kappa'}$$
$$- 18\hbar^{-2}\sum_{\kappa'\kappa''}\{|\Phi_{\kappa\kappa'\kappa''}|^2\,[(\omega_\kappa + \omega_{\kappa'} + \omega_{\kappa''})_p^{-1} - (\omega_\kappa - \omega_{\kappa'} - \omega_{\kappa''})_p^{-1}]$$
$$+ 2\Phi_{\kappa,-\kappa,\kappa''}\Phi_{\kappa',-\kappa',-\kappa''}(\omega_{\kappa''})_p^{-1}\}; \tag{16.30}$$

$$\theta_{\kappa\kappa'} = 24\hbar^{-1}\Phi_{\kappa,-\kappa,\kappa',-\kappa'}$$
$$- 36\hbar^{-2}\sum_{\kappa''}\{|\Phi_{\kappa\kappa'\kappa''}|^2\,[(\omega_\kappa + \omega_{\kappa'} + \omega_{\kappa''})_p^{-1} - (\omega_\kappa + \omega_{\kappa'} - \omega_{\kappa''})_p^{-1}$$
$$+ (\omega_\kappa - \omega_{\kappa'} + \omega_{\kappa''})_p^{-1} - (\omega_\kappa - \omega_{\kappa'} - \omega_{\kappa''})_p^{-1}] \tag{16.31}$$
$$+ 2\Phi_{\kappa,-\kappa,\kappa''}\Phi_{\kappa',-\kappa',-\kappa''}(\omega_{\kappa''})_p^{-1}\}.$$

From (16.31) it is easily verified that

$$\theta_{\kappa\kappa'} = \theta_{\kappa'\kappa}; \tag{16.32}$$

to show this requires the symmetry of $\Phi_{\kappa\kappa'\kappa''}$ in its indices, and requires letting $\kappa'' \to -\kappa''$ with $\omega_{-\kappa''} = \omega_{\kappa''}$ inside the sum for the last term in (16.31). Another interesting relation is

$$\theta_\kappa = \tfrac{1}{2}\sum_{\kappa'}\theta_{\kappa\kappa'}; \tag{16.33}$$

this follows since the two middle terms in the square bracket in (16.31) cancel when summed over κ'.

Now Δ_κ is written out in (16.27), with terms in $\bar{n}_{\kappa'}$ and in $\bar{n}_{\kappa''}$; if in each $\bar{n}_{\kappa''}$ term we interchange κ' and κ'' inside the $\Sigma_{\kappa'\kappa''}$, then each average phonon occupation number appears as $\bar{n}_{\kappa'}$, and with $\theta_{\kappa\kappa'}$ given by (16.31) it is seen that

$$\Delta_\kappa = \sum_{\kappa'}\theta_{\kappa\kappa'}(\bar{n}_{\kappa'} + \tfrac{1}{2}). \tag{16.34}$$

Further, with (16.33) this is

$$\Delta_\kappa = \theta_\kappa + \sum_{\kappa'}\theta_{\kappa\kappa'}\bar{n}_{\kappa'}. \tag{16.35}$$

The temperature-dependence of (16.34) and (16.35) is contained in the $\bar{n}_{\kappa'}$.

The total fourth-order excitation energy of the interacting phonon system, namely $\mathscr{E}_4 - G_4$, was calculated in (13.96) by summing the single renormalized phonon energy shifts, with the two-phonon interaction terms multiplied by $\frac{1}{2}$ to avoid double counting. By the same procedure the real part of $\langle\mathscr{E}_4 - G_4\rangle_{\mathrm{H}}$, which is $F_{\mathrm{A}} - G_4$ according to (16.9), may be calculated by summing the statistical average shifts $\hbar\Delta_\kappa$ of (16.35), again with a $\frac{1}{2}$ counting factor, to give

$$F_{\mathrm{A}} = G_4 + \sum_{\kappa} \hbar\theta_{\kappa}\bar{n}_{\kappa} + \tfrac{1}{2}\sum_{\kappa\kappa'} \hbar\theta_{\kappa\kappa'}\bar{n}_{\kappa}\bar{n}_{\kappa'}. \tag{16.36}$$

This equation agrees with (16.25), where G_4 is given by (13.92). Alternately (16.36) may be written as the sum of energy shifts $\hbar\Delta_{\kappa}$, with a double counting correction subtracted:

$$F_{\mathrm{A}} = G_4 + \sum_{\kappa} \hbar\Delta_{\kappa}\bar{n}_{\kappa} - \tfrac{1}{2}\sum_{\kappa\kappa'} \hbar\theta_{\kappa\kappa'}\bar{n}_{\kappa}\bar{n}_{\kappa'}. \tag{16.37}$$

These last two equations for F_{A} are the same by virtue of (16.35). We note that G_4 cannot be expressed as a simple function of the θ_{κ} and $\theta_{\kappa\kappa'}$.

Thermodynamic Functions at Constant Configuration

Let us first consider thermodynamic functions which are composed of the free energy and its constant-configuration temperature derivatives; these functions are defined in Sections 1 and 2. Some useful preliminary relations are obtained from the definition (16.16) of $\bar{n}_{\kappa}$, as follows. Inversion of (16.16) gives

$$e^{\hbar\omega_{\kappa}/KT} - 1 = 1/\bar{n}_{\kappa}; \tag{16.38}$$

then

$$e^{\hbar\omega_{\kappa}/KT} = (\bar{n}_{\kappa} + 1)/\bar{n}_{\kappa}. \tag{16.39}$$

Differentiating (16.16) with respect to T, and using (16.39), leads to

$$(\partial\bar{n}_{\kappa}/\partial T)_{\eta} = (\hbar\omega_{\kappa}/KT^2)(\bar{n}_{\kappa}^2 + \bar{n}_{\kappa}); \tag{16.40}$$

$$(\partial^2\bar{n}_{\kappa}/\partial T^2)_{\eta} = [(\hbar\omega_{\kappa}/KT^2)(2\bar{n}_{\kappa} + 1) - (2/T)](\partial\bar{n}_{\kappa}/\partial T)_{\eta}. \tag{16.41}$$

As usual the subscript η means constant configuration, and this in turn means constant ω_{κ}.

Now $F = \Phi_0 + F_{\mathrm{H}} + F_{\mathrm{A}}$, according to (16.7), and obviously

$$(\partial\Phi_0/\partial T)_{\eta} = 0. \tag{16.42}$$

We will use subscripts H and A to denote contributions arising from F_{H} or its temperature derivatives, and from F_{A} or its temperature derivatives, respectively. The entropy is

$$S = -(\partial F/\partial T)_{\eta} = S_{\mathrm{H}} + S_{\mathrm{A}}, \tag{16.43}$$

where from (16.13) for F_{H},

$$S_{\mathrm{H}} = K\sum_{\kappa}\{(\hbar\omega_{\kappa}/KT)\bar{n}_{\kappa} - \ln[1 - e^{-\hbar\omega_{\kappa}/KT}]\}, \tag{16.44}$$

and from (16.36) for F_{A} along with the symmetry $\theta_{\kappa\kappa'} = \theta_{\kappa'\kappa}$,

$$S_{\mathrm{A}} = -\sum_{\kappa}\left[\hbar\theta_{\kappa} + \sum_{\kappa'} \hbar\theta_{\kappa\kappa'}\bar{n}_{\kappa'}\right](\partial\bar{n}_{\kappa}/\partial T)_{\eta}. \tag{16.45}$$

In terms of Δ_κ given by (16.35), this last equation is simplified to

$$S_{\mathrm{A}} = -\sum_{\kappa} \hbar\Delta_\kappa(\partial \bar{n}_\kappa/\partial T)_\eta. \tag{16.46}$$

An alternate form for S_{H} is

$$S_{\mathrm{H}} = K \sum_{\kappa} [(\bar{n}_\kappa + 1)\ln(\bar{n}_\kappa + 1) - \bar{n}_\kappa \ln \bar{n}_\kappa]; \tag{16.47}$$

this is shown to be equal to (16.44) by the following calculation.

$$\begin{aligned}(\bar{n}_\kappa + 1)\ln(\bar{n}_\kappa + 1) - \bar{n}_\kappa \ln \bar{n}_\kappa &= \bar{n}_\kappa \ln[(\bar{n}_\kappa + 1)/\bar{n}_\kappa] + \ln(\bar{n}_\kappa + 1)\\ &= \bar{n}_\kappa \ln[e^{\hbar\omega_\kappa/KT}] + \ln[\bar{n}_\kappa e^{\hbar\omega_\kappa/KT}]\\ &= \bar{n}_\kappa(\hbar\omega_\kappa/KT) - \ln[(1/\bar{n}_\kappa)e^{-\hbar\omega_\kappa/KT}]\\ &= \bar{n}_\kappa(\hbar\omega_\kappa/KT) - \ln[1 - e^{-\hbar\omega_\kappa/KT}],\end{aligned} \tag{16.48}$$

where we used (16.39) and then (16.38). The harmonic entropy (16.47) is a function only of the $3nN_0$ statistical average occupation numbers $\bar{n}_\kappa$, one for each harmonic phonon. The function (16.47) is the maximum entropy, and hence the equilibrium value of the entropy, for a set of independent bosons with energies $\hbar\omega_\kappa$. In addition, $S_{\mathrm{H}} + S_{\mathrm{A}}$ is given, correct to first order in the frequency shifts Δ_κ, which is the order of (16.46) for S_{A}, by the same formula as S_{H}, only with each ω_κ replaced by $\Omega_\kappa = \omega_\kappa + \Delta_\kappa$. Thus to order Δ_κ, the renormalized phonons have the same thermodynamic distribution as a set of independent bosons with energies $\hbar\Omega_\kappa$; this result is in keeping with the properties of the renormalized phonons as derived in Section 13. The anharmonic contributions to other thermodynamic functions are *not* obtained simply by replacing ω_κ with Ω_κ in the corresponding harmonic formulas; a full discussion of this situation is provided in Section 20.

Another noteworthy point is that the natural thermodynamic variables of our calculations are the crystal configuration and the temperature, and the functional dependence on these variables is easily identified. The condition of constant entropy, however, is a very complicated condition in general, and since S must be considered a dependent variable it is a nontrivial matter to carry out a constant S derivative. In particular, constant S does *not* imply that all $\bar{n}_\kappa$ are constant, as has often been assumed in the past. We therefore stress the fact that all lattice-dynamical thermodynamic functions can be calculated, and should be calculated, from the Helmholtz free energy as a function of the crystal configuration and temperature.

To proceed with the calculation of lattice-dynamical expressions for the thermodynamic functions, the internal energy is

$$U = F + TS = \Phi_0 + U_{\mathrm{H}} + U_{\mathrm{A}}. \tag{16.49}$$

The notation is $U_{\rm H} = F_{\rm H} + TS_{\rm H}$, and from (16.13) for $F_{\rm H}$ and (16.44) for $S_{\rm H}$,

$$U_{\rm H} = \sum_{\kappa} \hbar\omega_{\kappa}(\bar{n}_{\kappa} + \tfrac{1}{2}). \tag{16.50}$$

Similarly $U_{\rm A} = F_{\rm A} + TS_{\rm A}$, and from (16.37) for $F_{\rm A}$ and (16.46) for $S_{\rm A}$ a convenient form for $U_{\rm A}$ is

$$U_{\rm A} = G_4 + \sum_{\kappa} \hbar\Delta_{\kappa}[\bar{n}_{\kappa} - T(\partial\bar{n}_{\kappa}/\partial T)_{\eta}] - \tfrac{1}{2}\sum_{\kappa\kappa'} \hbar\theta_{\kappa\kappa'}\bar{n}_{\kappa}\bar{n}_{\kappa'}. \tag{16.51}$$

The internal energy is, of course, the statistical average of the Hamiltonian, and (16.50) for $U_{\rm H}$ is obviously the harmonic average of $\mathscr{H}_{\rm H} = \mathscr{H}_2$ [see, e.g., (10.85) or (13.40) for $\mathscr{H}_2$]. It should be noted, however, that the statistical average of $\mathscr{H}$ correct to fourth order is *not* simply the harmonic average $\langle \mathscr{E}_2 + \mathscr{E}_4\rangle_{\rm H}$, so that $U_{\rm A}$ is *not* the average $\langle\mathscr{E}_4\rangle_{\rm H}$ of the perturbation energy. Indeed, in leading-order perturbation the statistical average of the perturbation energy goes into F, according to (16.9), and the internal energy U has an added TS term. This is a general result of perturbation theory, and again the point is mentioned here because it has not been universally recognized.

The heat capacity at constant configuration is

$$C_{\eta} = T(\partial S/\partial T)_{\eta} = -T(\partial^2 F/\partial T^2)_{\eta} = (\partial U/\partial T)_{\eta}, \tag{16.52}$$

and is written as a harmonic plus an anharmonic contribution,

$$C_{\eta} = C_{\eta \rm H} + C_{\eta \rm A}. \tag{16.53}$$

It then follows that

$$C_{\eta\rm H} = \sum_{\kappa} \hbar\omega_{\kappa}(\partial\bar{n}_{\kappa}/\partial T)_{\eta}; \tag{16.54}$$

$$C_{\eta\rm A} = -T\sum_{\kappa} \hbar\Delta_{\kappa}(\partial^2\bar{n}_{\kappa}/\partial T^2)_{\eta} - T\sum_{\kappa\kappa'} \hbar\theta_{\kappa\kappa'}(\partial\bar{n}_{\kappa}/\partial T)_{\eta}(\partial\bar{n}_{\kappa'}/\partial T)_{\eta}. \tag{16.55}$$

In Section 10 we introduced the expansion parameter ϵ, defined by (10.6), to denote the orders of the various contributions to the lattice-dynamics Hamiltonian. Since the (dimensionless) $\bar{n}_{\kappa}$ are of order 1, we use the convention $\hbar\omega_{\kappa}/KT \sim 1$, or

$$\hbar\omega_{\kappa} \sim KT \sim \epsilon^2, \tag{16.56}$$

so that all the harmonic functions $F_{\rm H}$, $S_{\rm H}$, $U_{\rm H}$, $C_{\eta\rm H}$ are second-order quantities, and the corresponding anharmonic functions are fourth-order quantities. By this convention, differentiation with respect to T does not change the order of a quantity.

Stresses and Elastic Constants

Let us now consider the thermodynamic functions which contain strain derivatives of the free energy at constant temperature. An abbreviated notation for the first- and second-strain derivatives of the harmonic phonon frequencies is

$$\gamma_{\kappa,ij} = -\omega_\kappa^{-1}(\partial\omega_\kappa/\partial\eta_{ij})_{\eta'}; \tag{16.57}$$

$$\xi_{\kappa,ijkl} = \omega_\kappa^{-1}(\partial^2\omega_\kappa/\partial\eta_{ij}\partial\eta_{kl})_{\eta'}; \tag{16.58}$$

note that $\gamma_{\kappa,ij}$ are the same as the phonon Grüneisen parameters $\gamma_{ij}(\mathbf{k}s)$ defined by (11.74). Detailed lattice-dynamical expressions for the first- and second-strain derivatives of ω_κ are calculated in Section 17 [see (17.62) and (17.63)]. Now $(\partial\omega_\kappa/\partial\eta_{ij})_{\eta'}$ is calculated from third-order potential energy coefficients, and so is a first-order anharmonic quantity; likewise $(\partial^2\omega_\kappa/\partial\eta_{ij}\partial\eta_{kl})_{\eta'}$ is second-order anharmonic. The appropriate ordering of the γ_κ and ξ_κ parameters is specified by

$$\hbar\omega_\kappa\gamma_{\kappa,ij} \sim \epsilon^3, \qquad \hbar\omega_\kappa\xi_{\kappa,ijkl} \sim \epsilon^4. \tag{16.59}$$

Hence differentiation with respect to strain (or volume) increases the order of a thermodynamic quantity by one. This ordering procedure is in agreement with the customary designation of the elastic constants C_{ijkl} as second order, C_{ijklmn} as third order, and so on.

Some useful preliminary relations are obtained from (16.16) for $\bar{n}_\kappa$, together with the alternate form (16.39). Direct calculation gives

$$(\partial\bar{n}_\kappa/\partial\omega_\kappa)_T = -(\hbar/KT)(\bar{n}_\kappa^2 + \bar{n}_\kappa). \tag{16.60}$$

With this it follows that

$$\begin{aligned}(\partial\bar{n}_\kappa/\partial\eta_{ij})_{T\eta'} &= (\partial\bar{n}_\kappa/\partial\omega_\kappa)_T(\partial\omega_\kappa/\partial\eta_{ij})_{\eta'} \\ &= (\hbar\omega_\kappa/KT)(\bar{n}_\kappa^2 + \bar{n}_\kappa)\gamma_{\kappa,ij} \\ &= T(\partial\bar{n}_\kappa/\partial T)_\eta\gamma_{\kappa,ij},\end{aligned} \tag{16.61}$$

where we will continue to use $(\partial\bar{n}_\kappa/\partial T)_\eta$, given by (16.40), as an abbreviation.

The stresses τ_{ij} and isothermal elastic constants C_{ijkl}^T are given by (2.27) and (2.34), respectively, as

$$\tau_{ij} = V^{-1}(\partial F/\partial\eta_{ij})_{T\eta'}; \tag{16.62}$$

$$C_{ijkl}^T = V^{-1}(\partial^2 F/\partial\eta_{ij}\partial\eta_{kl})_{T\eta'}. \tag{16.63}$$

Since $F = \Phi_0 + F_{\mathrm{H}} + F_{\mathrm{A}}$ by (16.7), the leading contributions to τ_{ij} and C_{ijkl}^T arise from strain derivatives of Φ_0; these are calculated in Section 7 and are

$$\tilde{C}_{ij} = V^{-1}(\partial\Phi_0/\partial\eta_{ij})_{\eta'}; \tag{16.64}$$

$$\tilde{C}_{ijkl} = V^{-1}(\partial^2\Phi_0/\partial\eta_{ij}\partial\eta_{kl})_{\eta'}. \tag{16.65}$$

For a given crystal model, the $\tilde{C}$ coefficients are most easily calculated by direct differentiation of Φ_0 with surface effects eliminated, as illustrated in Section 9 for central potentials. Note that $\tilde{C}_{ij}$ is first order, and $\tilde{C}_{ijkl}$ is second order. The next contributions to τ_{ij} and C^T_{ijkl} arise from strain derivatives of F_H; from (16.13) for F_H the results are

$$V^{-1}(\partial F_H/\partial \eta_{ij})_{T\eta'} = -V^{-1} \sum_{\kappa} \hbar\omega_\kappa \gamma_{\kappa,ij}(\bar{n}_\kappa + \tfrac{1}{2}); \tag{16.66}$$

$$V^{-1}(\partial^2 F_H/\partial \eta_{ij}\partial \eta_{kl})_{T\eta'} = V^{-1} \sum_{\kappa} \hbar\omega_\kappa [\xi_{\kappa,ijkl}(\bar{n}_\kappa + \tfrac{1}{2}) - \gamma_{\kappa,ij}\gamma_{\kappa,kl} T(\partial \bar{n}_\kappa/\partial T)_\eta]. \tag{16.67}$$

Note that (16.66) is a third-order quantity, and (16.67) is fourth order. The anharmonic free energy F_A will give fifth- and sixth-order contributions to the stresses and elastic constants, respectively, and these contributions will not be calculated explicitly here. Thus, to third order the stresses are

$$\tau_{ij} = \tilde{C}_{ij} - V^{-1} \sum_{\kappa} \hbar\omega_\kappa \gamma_{\kappa,ij}(\bar{n}_\kappa + \tfrac{1}{2}), \tag{16.68}$$

and to fourth order the isothermal elastic constants are

$$C^T_{ijkl} = \tilde{C}_{ijkl} + V^{-1} \sum_{\kappa} \hbar\omega_\kappa [\xi_{\kappa,ijkl}(\bar{n}_\kappa + \tfrac{1}{2}) - \gamma_{\kappa,ij}\gamma_{\kappa,kl} T(\partial \bar{n}_\kappa/\partial T)_\eta]. \tag{16.69}$$

The thermal stress tensor b_{ij} is defined by (2.72), and is calculated from (16.68):

$$b_{ij} = (\partial \tau_{ij}/\partial T)_\eta = -V^{-1} \sum_{\kappa} \hbar\omega_\kappa \gamma_{\kappa,ij}(\partial \bar{n}_\kappa/\partial T)_\eta. \tag{16.70}$$

Note that the first-order term $\tilde{C}_{ij}$ in τ_{ij} drops out because it is independent of T, and as a result b_{ij} is a third-order quantity. The difference between adiabatic and isothermal elastic constants was expressed in terms of the b_{ij} in (2.90) and (2.91), and from (16.70) the explicit lattice-dynamics expression is

$$C^S_{ijkl} - C^T_{ijkl} = (T/VC_\eta) \sum_{\kappa} \hbar\omega_\kappa \gamma_{\kappa,ij}(\partial \bar{n}_\kappa/\partial T)_\eta \sum_{\kappa'} \hbar\omega_{\kappa'} \gamma_{\kappa',kl}(\partial \bar{n}_{\kappa'}/\partial T)_\eta. \tag{16.71}$$

Now because C_η is a quantity of lowest order second order, (16.71) is a fourth-order quantity, and hence the adiabatic–isothermal difference is of the same order as the temperature-dependent contribution in (16.69) to the C^T_{ijkl}. Also, to the order to which our equations are correct, namely to fourth order, C_η in (16.71) may be replaced by $C_{\eta H}$ of (16.54).

We noted in Section 7 that the adiabatic and isothermal elastic constants of the potential approximation are the same, and depend only on the crystal configuration. Now we find that the strain derivatives of F_H give the leading contributions to the explicit temperature-dependence of the elastic constants, as well as the adiabatic–isothermal difference. Furthermore these (fourth-order) contributions will generally destroy the Cauchy relations which are found in the potential approximation, as for example, the cubic-crystal Cauchy relations for central potentials found in Section 9. It is not possible to determine the sign of $C_{ijkl}^S - C_{ijkl}^T$ from the general expression (16.71), without evaluating it in detail, but it is obvious that this difference goes to zero as $T \to 0$. From the above equations the stresses and the adiabatic and isothermal elastic constants may be calculated for any crystal configuration and temperature, and similar equations are easily written for the higher-order elastic constants.

If it is desired to carry out theoretical calculations of quantities related to the thermal-expansion tensor, and to compare these calculations with experimental results, the simplest procedure from the theoretical point of view is to calculate b_{ij} from (16.70). The experimental b_{ij} are easily obtained from the measured thermal strains β_{ij} and the isothermal stress–strain coefficients B_{ijkl}^T, through the relation (2.77). Recall that at zero stress the B_{ijkl}^T are equal to the isothermal elastic constants C_{ijkl}^T. Since the measured elastic constants are generally available to good accuracy, the experimental b_{ij} may generally be determined as accurately as the measured β_{ij}. If, on the other hand, it is desirable to calculate β_{ij} directly, then one needs to calculate b_{ij} and also the S_{ijkl}^T, and compute β_{ij} from the relation (2.78); note that in this case the leading contribution to S_{ijkl}^T is $\tilde{S}_{ijkl}$ defined in the potential approximation. An alternate but equally valid procedure for calculating β_{ij} directly is as follows for the example of zero stress. At various configurations and temperatures the free energy is calculated at least to harmonic order, namely $\Phi_0 + F_{\text{H}}$, and then at each temperature the free energy is minimized as a function of the configurational parameters, as for example, c and a for a hcp lattice. This gives the zero-stress configuration as a function of temperature, and the thermal strains at zero stress may be obtained by numerical differentiation with respect to temperature.

Pressure–Volume Variables

When the stress is isotropic pressure P, the independent variables may be taken to be the crystal volume V and the temperature T. When considering noncubic crystals, it must always be kept in mind that a variation dT at constant V does not generally correspond to constant configuration, but corresponds merely to V being held constant by a variation dP along with dT; in other words, $dV = 0$ does not necessarily imply $d\eta_{ij} = 0$. Also, a volume variation dV at constant T is not generally a uniform volume variation, but corresponds to the particular configuration variation resulting from a dP which brings about the dV. Furthermore, for a given crystal, the way in which the configuration changes in a constant T or a constant V process depends on the initial values of T and V. These considerations are all implicit in the following calculations for the case of pressure and volume variables.

The volume derivatives of the harmonic phonon frequencies are denoted simply as

$$\gamma_\kappa = -(V/\omega_\kappa)(d\omega_\kappa/dV); \tag{16.72}$$

$$\xi_\kappa = (V^2/\omega_\kappa)(d^2\omega_\kappa/dV^2). \tag{16.73}$$

Again γ_κ is the ordinary phonon Grüneisen parameter $\gamma(\mathbf{k}s)$ defined by (11.79), and the orders of these volume derivatives are the same as the corresponding strain derivatives, equation (16.59):

$$\hbar\omega_\kappa\gamma_\kappa \sim \epsilon^3; \qquad \hbar\omega_\kappa\xi_\kappa \sim \epsilon^4. \tag{16.74}$$

The volume derivative of $\bar{n}_\kappa$ is, with the aid of (16.60),

$$V(\partial\bar{n}_\kappa/\partial V)_T = (\hbar\omega_\kappa/KT)(\bar{n}_\kappa + \bar{n}_\kappa)\gamma_\kappa. \tag{16.75}$$

The pressure and the isothermal bulk modulus are given by (1.10) and (1.33), as

$$P = -(\partial F/\partial V)_T; \tag{16.76}$$

$$B_T = V(\partial^2 F/\partial V^2)_T. \tag{16.77}$$

Again the leading-order terms arising from volume derivatives of Φ_0 are denoted by tildes, according to

$$\tilde{P} = -(d\Phi_0/dV); \tag{16.78}$$

$$\tilde{B} = V(d^2\Phi_0/dV^2). \tag{16.79}$$

Including the contributions from volume derivatives of F_{H}, given by (16.13), we find the pressure correct to third order is

$$P = \tilde{P} + V^{-1}\sum_\kappa \hbar\omega_\kappa\gamma_\kappa(\bar{n}_\kappa + \tfrac{1}{2}), \tag{16.80}$$

and the isothermal bulk modulus correct to fourth order is

$$B_T = \tilde{B} + V^{-1}\sum_\kappa \hbar\omega_\kappa[\xi_\kappa(\bar{n}_\kappa + \tfrac{1}{2}) - \gamma_\kappa^2(\hbar\omega_\kappa/KT)(\bar{n}_\kappa^2 + \bar{n}_\kappa)]. \tag{16.81}$$

For a particular crystal structure and a model for the interactions among the ions, it is possible to eliminate surface effects from the expression for Φ_0, and calculate $\tilde{P}$ and $\tilde{B}$ directly from (16.78) and (16.79).

The thermal expansion coefficient β is most conveniently calculated in the form of the quantity βB_T; from (1.35) this is

$$\beta B_T = -(\partial^2 F/\partial V\partial T)_{TV} = (\partial P/\partial T)_V. \tag{16.82}$$

For a noncubic crystal, the simple quantity $(\partial P/\partial T)_V$ is generously endowed with complications; P is given by (16.80) and the quantities $\tilde{P}$, ω_κ, and γ_κ all vary with T at constant V, since constant V is not constant configuration. For a cubic crystal, however, as long as the stress is isotropic pressure, the configuration is determined solely by V (i.e., $dV = 0$ implies $d\eta_{ij} = 0$), and by applying (16.82) to (16.80), βB_T is found to be

$$\beta B_T = V^{-1} \sum_\kappa \hbar\omega_\kappa \gamma_\kappa (\partial \bar{n}_\kappa / \partial T)_V, \quad \text{cubic crystal.} \tag{16.83}$$

Again this is a third-order quantity, analogous to b_{ij} of (16.70). For many noncubic crystals, (16.83) should be a good approximation to βB_T since the constant-volume temperature derivatives of $\tilde{P}$, ω_κ, and γ_κ should give contributions small compared to those retained in (16.83).

The difference between adiabatic and isothermal bulk moduli is given by (1.42), and with $k_T - k_S$ taken from (1.41) and with $B_S/C_P = B_T/C_V$ from (1.43), that difference may be written

$$B_S - B_T = (TV/C_V)(\beta B_T)^2. \tag{16.84}$$

This is a fourth-order quantity, analogous to (16.71) for $C_{ijkl}^S - C_{ijkl}^T$. For a cubic crystal under isotropic pressure, $C_V = C_\eta$; for a noncubic crystal, C_V has to be calculated from C_η by (2.96) or (2.99), but the difference $C_V - C_\eta$ should be small compared to C_V for most crystals.

17. SOUND WAVES AND ACOUSTIC PHONONS

A Theorem

From the calculations of Sections 7 and 12, we were able to conclude that in leading (harmonic) order the adiabatic sound waves, the isothermal sound waves, and the long-wavelength acoustic phonons are all the same, and propagate as mechanical waves in a crystal. When anharmonicity is included, however, the three waves are generally different, and propagate as adiabatic waves, isothermal waves, and phonons, respectively. Now the difference between adiabatic and isothermal propagation coefficients vanishes at $T = 0$. This is not surprising since the very requirement $T = 0$ is an isothermal condition. In fact we conclude that under the condition $T = 0$ the long-wavelength acoustic phonons must be just isothermal sound waves, and hence all three types of waves are identical at $T = 0$. In other words, at $T = 0$ the potential which governs the motion of renormalized phonons is the ground-state energy $\mathscr{E}_G$, and this is the same as the potential which governs the motion of sound waves, namely F at $T = 0$.

In the present section we carry out a detailed calculation of the long-wavelength renormalized acoustic phonon frequencies to second order in the anharmonicity (the energies to order ϵ^4), and also the sound wave frequencies to the same order. In the process it is shown that the long-wavelength acoustic phonon frequencies are still proportional to the wave vector magnitude $|\mathbf{k}|$, for any lattice and at any temperature. General expressions are derived for the first- and second-strain derivatives of the harmonic phonon frequencies, and our $T = 0$ theorem is confirmed to leading order in the anharmonicity. It is shown that for a given sound wave at finite temperatures, the adiabatic velocity is greater than the isothermal velocity, and the isothermal velocity is likely to be less than the acoustic phonon velocity.

Long-Wavelength Potential Energy Coefficients

The first step in the analysis is the examination of the transformed potential energy coefficients $\Phi(\mathbf{k}s,\mathbf{k}'s',\mathbf{k}''s'')$ and $\Phi(\mathbf{k}s,-\mathbf{k}s,\mathbf{k}'s',-\mathbf{k}'s')$ in the limit of long wavelength for one of the wave vectors $\mathbf{k}$, and with the corresponding s = acoustic. The calculations facing us are rather tedious, and it is convenient (but not necessary) to use the alternate phase factors $e^{i\mathbf{k}\cdot\mathbf{R}(\mathrm{N})}$ in place of $e^{i\mathbf{k}\cdot\mathbf{R}(\mathrm{N}\nu)}$. This variation is explained in detail in Section 11; in particular the transformed potential coefficients are given by (10.74) and (10.75), and the use of alternate phase factors means we simply introduce the $\bar{\mathbf{w}}(\mathbf{k}s)$ according to (11.63):

$$\bar{w}_j(\nu,\mathbf{k}s) = e^{i\mathbf{k}\cdot\mathbf{R}(\nu)} w_j(\nu,\mathbf{k}s). \tag{17.1}$$

Now $\Phi(\mathbf{k}s,\mathbf{k}'s',\mathbf{k}''s'')$ contains a $\delta(\mathbf{k} + \mathbf{k}' + \mathbf{k}'')$ according to (10.81), and inside a sum over $\mathbf{k}''$ we can write

$$\Phi(\mathbf{k}s,\mathbf{k}'s',\mathbf{k}''s'') = \delta(\mathbf{k} + \mathbf{k}' + \mathbf{k}'')\Phi(\mathbf{k}s,\mathbf{k}'s',-\mathbf{k} - \mathbf{k}'s''). \tag{17.2}$$

Also from the symmetry (10.76) of $\Phi(\mathbf{k}s,\mathbf{k}'s',\mathbf{k}''s'')$ in its index pairs,

$$\Phi(\mathbf{k}s,\mathbf{k}'s',-\mathbf{k} - \mathbf{k}'s'') = \Phi(-\mathbf{k} - \mathbf{k}'s'',\mathbf{k}'s',\mathbf{k}s). \tag{17.3}$$

This coefficient is now written, from the definition (10.74) and with the replacement (17.1), with the indices in the order given on the right of (17.3),

$$\begin{aligned}\Phi(\mathbf{k}s,\mathbf{k}'s',-\mathbf{k} - \mathbf{k}'s'') &= (1/3!)(\hbar/2N_0)^{3/2}[\omega(\mathbf{k}s)\omega(\mathbf{k}'s')\omega(\mathbf{k} + \mathbf{k}'s'')]^{-1/2} \\ &\times \sum_{\mathrm{MNP}\mu\nu\pi}\sum_{ijk}\Phi_{ijk}(\mathrm{M}\mu,\mathrm{N}\nu,\mathrm{P}\pi)(M_\mu M_\nu M_\pi)^{-1/2}e^{-i(\mathbf{k}+\mathbf{k}')\cdot\mathbf{R}(\mathrm{M})} \\ &\times e^{i\mathbf{k}'\cdot\mathbf{R}(\mathrm{N})}e^{i\mathbf{k}\cdot\mathbf{R}(\mathrm{P})}\bar{w}_i(\mu,-\mathbf{k} - \mathbf{k}'s'')\bar{w}_j(\nu,\mathbf{k}'s')\bar{w}_k(\pi,\mathbf{k}s).\end{aligned} \tag{17.4}$$

Surface effects are eliminated from (17.4) by setting M = 0 with $\mathbf{R}(\mathrm{M}) = 0$, and replacing Σ_M by N_0. In addition the complex conjugate of (17.4) is obtained by replacing each $\mathbf{k},\mathbf{k}'$ by $-\mathbf{k},-\mathbf{k}'$, according to the general result (10.78) for $\Phi(\mathbf{k}s,\mathbf{k}'s',\mathbf{k}''s'')^*$. Then the detailed expression for the absolute square of (17.4) is

$$
\begin{aligned}
&|\Phi(\mathbf{k}s,\mathbf{k}'s',-\mathbf{k}-\mathbf{k}'s'')|^2 \\
&\quad = (\hbar^3/288)[\omega(\mathbf{k}s)\omega(\mathbf{k}'s')\omega(\mathbf{k}+\mathbf{k}'s'')]^{-1}N_0^{-1} \\
&\quad \times \sum_{\mathrm{NN'PP'}} \sum_{\mu\mu'\nu\nu'\pi\pi'} \sum_{ii'jj'kk'} \Phi_{ijk}(0\mu,\mathrm{N}\nu,\mathrm{P}\pi)\Phi_{i'j'k'}(0\mu',\mathrm{N}'\nu',\mathrm{P}'\pi') \\
&\quad \times (M_\mu M_{\mu'} M_\nu M_{\nu'} M_\pi M_{\pi'})^{-1/2} e^{i\mathbf{k}'\cdot[\mathbf{R}(\mathrm{N})-\mathbf{R}(\mathrm{N}')]} e^{i\mathbf{k}\cdot[\mathbf{R}(\mathrm{P})-\mathbf{R}(\mathrm{P}')]} \\
&\quad \times \bar{w}_i(\mu,-\mathbf{k}-\mathbf{k}'s'')\bar{w}_{i'}(\mu',\mathbf{k}+\mathbf{k}'s'') \\
&\quad \times \bar{w}_j(\nu,\mathbf{k}'s')\bar{w}_{j'}(\nu',-\mathbf{k}'s')\bar{w}_k(\pi,\mathbf{k}s)\bar{w}_{k'}(\pi',-\mathbf{k}s).
\end{aligned}
\tag{17.5}
$$

For the fourth-order coefficient $\Phi(\mathbf{k}s,-\mathbf{k}s,\mathbf{k}'s',-\mathbf{k}'s')$ the δ function is automatically satisfied, and by the symmetry (10.77) of this coefficient in its index pairs,

$$
\Phi(\mathbf{k}s,-\mathbf{k}s,\mathbf{k}'s',-\mathbf{k}'s') = \Phi(-\mathbf{k}'s',\mathbf{k}'s',\mathbf{k}s,-\mathbf{k}s). \tag{17.6}
$$

The coefficient on the right of (17.6) is written down from its definition (10.75), the replacement (17.1) is made, and surface effects are eliminated to give the result

$$
\begin{aligned}
&\Phi(\mathbf{k}s,-\mathbf{k}s,\mathbf{k}'s',-\mathbf{k}'s') \\
&\quad = (\hbar^2/96)[\omega(\mathbf{k}s)\omega(\mathbf{k}'s')]^{-1}N_0^{-1} \sum_{\mathrm{NPQ}} \sum_{\mu\nu\pi\rho} \sum_{ijkl} \Phi_{ijkl}(0\mu,\mathrm{N}\nu,\mathrm{P}\pi,\mathrm{Q}\rho) \\
&\quad \times (M_\mu M_\nu M_\pi M_\rho)^{-1/2} e^{i\mathbf{k}'\cdot\mathbf{R}(\mathrm{N})} e^{i\mathbf{k}\cdot[\mathbf{R}(\mathrm{P})-\mathbf{R}(\mathrm{Q})]} \\
&\quad \times \bar{w}_i(\mu,-\mathbf{k}'s')\bar{w}_j(\nu,\mathbf{k}'s')\bar{w}_k(\pi,\mathbf{k}s)\bar{w}_l(\rho,-\mathbf{k}s).
\end{aligned}
\tag{17.7}
$$

There is one more third-order coefficient which is of interest, namely $\Phi(\mathbf{k}s,-\mathbf{k}s,\mathbf{k}''s'')$ for s'' = optic. This coefficient contains $\delta(\mathbf{k}'')$, so we set $\mathbf{k}''=0$ and follow the same procedure as above to obtain

$$
\begin{aligned}
\Phi(\mathbf{k}s,-\mathbf{k}s,0s'') &= \tfrac{1}{6}(\hbar/2)^{3/2}[\omega(\mathbf{k}s)]^{-1}[\omega(0s'')]^{-1/2}N_0^{-1/2} \\
&\quad \times \sum_{\mathrm{NP}\mu\nu\pi} \sum_{ijk} \Phi_{ijk}(0\mu,\mathrm{N}\nu,\mathrm{P}\pi)(M_\mu M_\nu M_\pi)^{-1/2} \\
&\quad \times e^{i\mathbf{k}\cdot[\mathbf{R}(\mathrm{N})-\mathbf{R}(\mathrm{P})]}\bar{w}_i(\mu,0s'')\bar{w}_j(\nu,\mathbf{k}s)\bar{w}_k(\pi,-\mathbf{k}s).
\end{aligned}
\tag{17.8}
$$

Each of the coefficients (17.5), (17.7), and (17.8) may now be expanded for small $\mathbf{k}$ with s = acoustic. For the fourth-order coefficient (17.7), the quantity to be expanded is

$$
e^{i\mathbf{k}\cdot[\mathbf{R}(\mathrm{P})-\mathbf{R}(\mathrm{Q})]}\bar{w}_k(\pi,\mathbf{k}s)\bar{w}_l(\rho,-\mathbf{k}s). \tag{17.9}
$$

The phase factor expands into

$$
1 + i\mathbf{k}\cdot[\mathbf{R}(\mathrm{P})-\mathbf{R}(\mathrm{Q})] - \tfrac{1}{2}[\mathbf{k}\cdot[\mathbf{R}(\mathrm{P})-\mathbf{R}(\mathrm{Q})]]^2 - \cdots, \tag{17.10}
$$

and the eigenvector components are expanded as in the method of long waves [see, e.g., (12.14); recall s = acoustic],

$$\bar{w}_j(\nu,\pm\mathbf{k}s) = \bar{w}_j^{(0)}(\nu,s) \pm i\,|\mathbf{k}|\,\bar{w}_j^{(1)}(\nu,s) + \tfrac{1}{2}\,|\mathbf{k}|^2\,\bar{w}_j^{(2)}(\nu,s) + \cdots. \tag{17.11}$$

It may be seen from (17.1) that $\bar{\mathbf{w}} = \mathbf{w}$ to zeroth order in $|\mathbf{k}|$; hence from the zeroth-order long-waves solution (12.17) for $w_j^{(0)}(\nu,s)$ it follows that

$$\bar{w}_j^{(0)}(\nu,s) = (M_\nu/M_C)^{1/2}y_j(s). \tag{17.12}$$

In addition the first-order long-waves eigenvectors are related to the homogeneous-deformation sublattice displacements, and when we are using the alternate phase factors the relation is (12.41):

$$\bar{w}_j^{(1)}(\nu,s) = (M_\nu/M_C)^{1/2}\sum_{mn}\bar{X}_{j,mn}(\nu)y_m(s)\hat{k}_n. \tag{17.13}$$

The complete expansion of the quantity (17.9) contains one zeroth-order term, three first-order terms, and six second-order terms. Because $\bar{w}_j^{(0)}(\nu,s)$ depends on ν only through the factor $M_\nu^{1/2}$, as shown in (17.12), many of these terms vanish since they contain the sums

$$\sum_{\mathrm{P}\pi}\Phi_{ijkl}(0\mu,\mathrm{N}\nu,\mathrm{P}\pi,\mathrm{Q}\rho) = \sum_{\mathrm{Q}\rho}\Phi_{ijkl}(0\mu,\mathrm{N}\nu,\mathrm{P}\pi,\mathrm{Q}\rho) = 0; \tag{17.14}$$

this is, of course, just the translational invariance condition (6.28). To order $|\mathbf{k}|^2$ the only terms in the expansion of (17.9) which do not sum to zero are

$$|\mathbf{k}|^2\,[\hat{\mathbf{k}}\cdot\mathbf{R}(\mathrm{P})\bar{w}_k^{(0)}(\pi,s) + \bar{w}_k^{(1)}(\pi,s)][\hat{\mathbf{k}}\cdot\mathbf{R}(\mathrm{Q})\bar{w}_l^{(0)}(\rho,s) + \bar{w}_l^{(1)}(\rho,s)]. \tag{17.15}$$

It is convenient at this point to define the zeroth-order (i.e., independent of the magnitude of $\mathbf{k}$) coefficients $H_j(\mathrm{N}\nu,\hat{\mathbf{k}}s)$, according to

$$H_j(\mathrm{N}\nu,\hat{\mathbf{k}}s) = \hat{\mathbf{k}}\cdot\mathbf{R}(\mathrm{N})y_j(s) + \sum_{mn}\bar{X}_{j,mn}(\nu)y_m(s)\hat{k}_n. \tag{17.16}$$

With the aid of (17.12) and (17.13) it then follows that each expression in brackets in (17.15) is like

$$\hat{\mathbf{k}}\cdot\mathbf{R}(\mathrm{Q})\bar{w}_l^{(0)}(\rho,s) + \bar{w}_l^{(1)}(\rho,s) = (M_\rho/M_C)^{1/2}H_l(\mathrm{Q}\rho,\hat{\mathbf{k}}s). \tag{17.17}$$

Furthermore $H_j(\mathrm{N}\nu,\hat{\mathbf{k}}s)$ is proportional to the total first-order displacement of ion Nν in direction j due to the local homogeneous deformation caused by the long-wavelength acoustic phonon $\mathbf{k}s$. To show this we note from (12.39) that the displacement gradients induced by the phonon are proportional to

$$u_{mn} = y_m(s)\hat{k}_n, \tag{17.18}$$

and then from (7.84) the total first-order ion displacements are

$$U_j^{(1)}(\mathrm{N}\nu) = \sum_{mn}Y_{j,mn}(\mathrm{N}\nu)y_m(s)\hat{k}_n. \tag{17.19}$$

From (7.94) the Y coefficients are

$$Y_{j,mn}(\mathrm{N}\nu) = R_n(\mathrm{N})\delta_{jm} + \bar{X}_{j,mn}(\nu), \tag{17.20}$$

and with this it is obvious that

$$\sum_{mn} Y_{j,mn}(\mathrm{N}\nu)y_m(s)\hat{k}_n = H_j(\mathrm{N}\nu,\hat{\mathbf{k}}s). \tag{17.21}$$

This last equation is the key to relating renormalized acoustic phonons and sound waves.

The small $\mathbf{k}$ expansion of the fourth-order coefficient (17.7) may now be concluded as follows. The contributing terms (17.15) in the phase factor and eigenvector expansion are replaced by H coefficients with (17.17), and then inserted into (17.7) to obtain

$$\begin{aligned}
&\Phi(\mathbf{k}s,-\mathbf{k}s,\mathbf{k}'s',-\mathbf{k}'s') \\
&\quad = (\hbar^2/96M_C)[\omega(\mathbf{k}s)\omega(\mathbf{k}'s')]^{-1}N_0^{-1} \\
&\quad\quad \times \sum_{\mathrm{NPQ}} \sum_{\mu\nu\pi\rho} \sum_{ijkl} \Phi_{ijkl}(0\mu,\mathrm{N}\nu,\mathrm{P}\pi,\mathrm{Q}\rho)(M_\mu M_\nu)^{-1/2} \\
&\quad\quad \times e^{i\mathbf{k}'\cdot\mathbf{R}(\mathrm{N})}\bar{w}_i(\mu,-\mathbf{k}'s')\bar{w}_j(\nu,\mathbf{k}'s')\,|\mathbf{k}|^2\, H_k(\mathrm{P}\pi,\hat{\mathbf{k}}s)H_l(\mathrm{Q}\rho,\hat{\mathbf{k}}s).
\end{aligned} \tag{17.22}$$

This expression is understood to be correct only to leading order in $|\mathbf{k}|$. Since (17.22) contains a factor $[\omega(\mathbf{k}s)]^{-1}$, and since for small $\mathbf{k}$ and $s =$ acoustic $\omega(\mathbf{k}s) = c(\hat{\mathbf{k}}s)\,|\mathbf{k}|$, by (12.42), then (17.22) is of leading order $|\mathbf{k}|$:

$$\Phi(\mathbf{k}s,-\mathbf{k}s,\mathbf{k}'s',-\mathbf{k}'s') \propto |\mathbf{k}|, \quad \text{for small } \mathbf{k} \text{ and } s = \text{acoustic}. \tag{17.23}$$

In the very same way the coefficients (17.5) and (17.8) are expanded for small $\mathbf{k}$; to leading order $\omega(\mathbf{k} + \mathbf{k}'s'')$ and $\bar{w}_i(\mu,-\mathbf{k} - \mathbf{k}'s'')$ in (17.5) may be replaced by $\omega(\mathbf{k}'s'')$ and $\bar{w}_i(\mu,-\mathbf{k}'s'')$, respectively, and the results are

$$\begin{aligned}
&|\Phi(\mathbf{k}s,\mathbf{k}'s',-\mathbf{k} - \mathbf{k}'s'')|^2 \\
&\quad = (\hbar^3/288M_C)[\omega(\mathbf{k}s)\omega(\mathbf{k}'s')\omega(\mathbf{k}'s'')]^{-1}N_0^{-1} \\
&\quad\quad \times \sum_{\mathrm{NN'PP'}} \sum_{\mu\mu'\nu\nu'\pi\pi'} \sum_{ii'jj'kk'} \Phi_{ijk}(0\mu,\mathrm{N}\nu,\mathrm{P}\pi)\Phi_{i'j'k'}(0\mu',\mathrm{N}'\nu',\mathrm{P}'\pi') \\
&\quad\quad \times (M_\mu M_{\mu'} M_\nu M_{\nu'})^{-1/2} e^{i\mathbf{k}'\cdot[\mathbf{R}(\mathrm{N})-\mathbf{R}(\mathrm{N}')]}\bar{w}_i(\mu,-\mathbf{k}'s'')\bar{w}_{i'}(\mu',\mathbf{k}'s'') \\
&\quad\quad \times \bar{w}_j(\nu,\mathbf{k}'s')\bar{w}_{j'}(\nu',-\mathbf{k}'s')\,|\mathbf{k}|^2\, H_k(\mathrm{P}\pi,\hat{\mathbf{k}}s)H_{k'}(\mathrm{P}'\pi',\hat{\mathbf{k}}s);
\end{aligned} \tag{17.24}$$

$$\begin{aligned}
&\Phi(\mathbf{k}s,-\mathbf{k}s,0s'') \\
&\quad = (1/6M_C)(\hbar/2)^{3/2}[\omega(\mathbf{k}s)]^{-1}[\omega(0s'')]^{-1/2}N_0^{-1/2} \\
&\quad\quad \times \sum_{\mathrm{NP}\mu\nu\pi} \sum_{ijk} \Phi_{ijk}(0\mu,\mathrm{N}\nu,\mathrm{P}\pi)M_\mu^{-1/2}\bar{w}_i(\mu,0s'') \\
&\quad\quad \times |\mathbf{k}|^2\, H_j(\mathrm{N}\nu,\hat{\mathbf{k}}s)H_k(\mathrm{P}\pi,\hat{\mathbf{k}}s).
\end{aligned} \tag{17.25}$$

Again these expressions are correct only to leading order in $|\mathbf{k}|$, and to that order

$$|\Phi(\mathbf{k}s,\mathbf{k}'s',-\mathbf{k}-\mathbf{k}'s'')|^2 \propto |\mathbf{k}|, \tag{17.26}$$

$$\Phi(\mathbf{k}s,-\mathbf{k}s,0s'') \propto |\mathbf{k}|, \tag{17.27}$$

for small $\mathbf{k}$ and s = acoustic.

Long-Wavelength Renormalized Acoustic Phonons

It is now possible to show that the phonon frequency shifts $\Delta_\kappa = \Delta(\mathbf{k}s)$ are proportional to $|\mathbf{k}|$ for small $\mathbf{k}$ and s = acoustic, for any lattice in any configuration at any temperature. The equation for $\Delta(\mathbf{k}s)$ is written out in (16.27), and all the potential energy coefficients appearing in that expression have already been shown to be proportional to $|\mathbf{k}|$, by (17.23), (17.26), and (17.27). The only other $\mathbf{k}$-dependence in (16.27) is contained in the square brackets in the cubic term; the complete term, which is a contribution to $\hbar\Delta_\kappa$, is

$$-\frac{18}{\hbar}\sum_{\kappa'\kappa''}|\Phi_{\kappa\kappa'\kappa''}|^2\left[\frac{(\bar{n}_{\kappa'}+\bar{n}_{\kappa''}+1)}{(\omega_\kappa+\omega_{\kappa'}+\omega_{\kappa''})_p}+\frac{(\bar{n}_{\kappa''}-\bar{n}_{\kappa'})}{(\omega_\kappa+\omega_{\kappa'}-\omega_{\kappa''})_p}\right.$$
$$\left.+\frac{(\bar{n}_{\kappa'}-\bar{n}_{\kappa''})}{(\omega_\kappa-\omega_{\kappa'}+\omega_{\kappa''})_p}-\frac{(\bar{n}_{\kappa'}+\bar{n}_{\kappa''}+1)}{(\omega_\kappa-\omega_{\kappa'}-\omega_{\kappa''})_p}\right]. \tag{17.28}$$

It is only necessary to show that to lowest order in $|\mathbf{k}|$ the expression in the square brackets is constant, i.e., of order $|\mathbf{k}|^0$. The sum on κ'' will be broken down into

$$\sum_{\kappa''}=\sum_{\mathbf{k}''s''}\rightarrow\sum_{s''} \quad \text{with } \mathbf{k}''=-\mathbf{k}-\mathbf{k}', \tag{17.29}$$

and it is expedient then to examine separately the cases $s''=s'$ and $s''\neq s'$.

Consider first $s''\neq s'$. In each term containing $\bar{n}_{\kappa''}$ in (17.28), interchange κ' and κ'' inside the $\Sigma_{\kappa'\kappa''}$ and note that $\Phi_{\kappa\kappa'\kappa''}=\Phi_{\kappa\kappa''\kappa'}$, to obtain

$$-(18/\hbar)\sum_{\kappa'\kappa''}|\Phi_{\kappa\kappa'\kappa''}|^2\{(2\bar{n}_{\kappa'}+1)[(\omega_\kappa+\omega_{\kappa'}+\omega_{\kappa''})_p^{-1}-(\omega_\kappa-\omega_{\kappa'}-\omega_{\kappa''})_p^{-1}]$$
$$+2\bar{n}_{\kappa'}[(\omega_\kappa-\omega_{\kappa'}+\omega_{\kappa''})_p^{-1}-(\omega_\kappa+\omega_{\kappa'}-\omega_{\kappa''})_p^{-1}]\}. \tag{17.30}$$

Now for small $\mathbf{k}$ we can write

$$\omega(\mathbf{k}''s'')=\omega(\mathbf{k}+\mathbf{k}'s'')=\omega(\mathbf{k}'s'')+O(|\mathbf{k}|), \tag{17.31}$$

and dropping all terms of order $|\mathbf{k}|$, the expression in braces in (17.30) is

$$2[2\bar{n}(\mathbf{k}'s')+1][\omega(\mathbf{k}'s')+\omega(\mathbf{k}'s'')]_p^{-1}-4\bar{n}(\mathbf{k}'s')[\omega(\mathbf{k}'s')-\omega(\mathbf{k}'s'')]_p^{-1}. \tag{17.32}$$

Inside the $\Sigma_{\kappa'\kappa''}$ this can be written

$$2[2\bar{n}(\mathbf{k}'s')+1]\{[\omega(\mathbf{k}'s')+\omega(\mathbf{k}'s'')]_p^{-1}-[\omega(\mathbf{k}'s')-\omega(\mathbf{k}'s'')]_p^{-1}\}, \tag{17.33}$$

since the added term in $[\omega(\mathbf{k}'s') - \omega(\mathbf{k}'s'')]_p^{-1}$ sums to zero. Hence for $s'' \neq s'$ the total contribution to (17.30) or (17.28) is

$$+(72/\hbar)\sum_{\mathbf{k}'s'}\sum_{s''}{}' |\Phi(\mathbf{k}s,\mathbf{k}'s',-\mathbf{k}-\mathbf{k}'s'')|^2 [2\bar{n}(\mathbf{k}'s')+1]\omega(\mathbf{k}'s'') \times \{[\omega(\mathbf{k}'s')]^2 - [\omega(\mathbf{k}'s'')]^2\}_p^{-1}. \tag{17.34}$$

Consider now the $s'' = s'$ contributions in (17.28). For the first and last terms in the square brackets again exchange κ'' and κ' in each function containing $\bar{n}_{\kappa''}$, to get the total contribution from these two terms,

$$-(18/\hbar)\sum_{\kappa'\kappa''} |\Phi_{\kappa\kappa'\kappa''}|^2 (2\bar{n}_{\kappa'}+1)[(\omega_\kappa+\omega_{\kappa'}+\omega_{\kappa''})_p^{-1} - (\omega_\kappa-\omega_{\kappa'}-\omega_{\kappa''})_p^{-1}]. \tag{17.35}$$

Then we can write

$$\omega(\mathbf{k}''s') = \omega(\mathbf{k}+\mathbf{k}'s') = \omega(\mathbf{k}'s') + O(|\mathbf{k}|), \tag{17.36}$$

and dropping all terms of order $|\mathbf{k}|$, (17.35) is

$$-(18/\hbar)\sum_{\mathbf{k}'s'} |\Phi(\mathbf{k}s,\mathbf{k}'s',-\mathbf{k}-\mathbf{k}'s')|^2 [2\bar{n}(\mathbf{k}'s')+1][\omega(\mathbf{k}'s')]^{-1}. \tag{17.37}$$

For the middle two terms in the square brackets in (17.28), both numerator and denominator are of leading order $|\mathbf{k}|$, so the contribution to the bracket is again of order $|\mathbf{k}|^0$. The total contribution from these two terms is

$$\begin{aligned}
-(18/\hbar)&\sum_{\kappa'\kappa''} |\Phi_{\kappa\kappa'\kappa''}|^2 (\bar{n}_{\kappa''}-\bar{n}_{\kappa'})[(\omega_\kappa+\omega_{\kappa'}-\omega_{\kappa''})_p^{-1} - (\omega_\kappa-\omega_{\kappa'}+\omega_{\kappa''})_p^{-1}] \\
&= -(36/\hbar)\sum_{\kappa'\kappa''} |\Phi_{\kappa\kappa'\kappa''}|^2 (\bar{n}_{\kappa''}-\bar{n}_{\kappa'})(\omega_{\kappa''}-\omega_{\kappa'})[\omega_\kappa^2 - (\omega_{\kappa''}-\omega_{\kappa'})^2]_p^{-1}.
\end{aligned} \tag{17.38}$$

In the final form of (17.38) the numerator and denominator are each of order $|\mathbf{k}|^2$. In order to study this contribution we note that $\omega(\mathbf{k}'s')$ is an analytic function of $\mathbf{k}'$, except at $\mathbf{k}' = 0$, and write an expansion

$$\omega(\mathbf{k}+\mathbf{k}'s') = \omega(\mathbf{k}'s') + \psi(\mathbf{k}'s',\hat{\mathbf{k}})\,|\mathbf{k}| + \cdots; \tag{17.39}$$

this expansion holds for small $|\mathbf{k}|$, and $\psi(\mathbf{k}'s',\hat{\mathbf{k}})$ is the directional derivative of $\omega(\mathbf{k}'s')$ with respect to $\mathbf{k}'$ in the direction $\hat{\mathbf{k}}$. Then since $\mathbf{k}'' = -\mathbf{k}-\mathbf{k}'$,

$$\omega(\mathbf{k}''s') - \omega(\mathbf{k}'s') = \psi(\mathbf{k}'s',\hat{\mathbf{k}})\,|\mathbf{k}|, \tag{17.40}$$

to order $|\mathbf{k}|$. In addition the phonon occupation numbers $\bar{n}(\mathbf{k}''s')$ may be expanded for small $\mathbf{k}$, and to order $|\mathbf{k}|$,

$$\bar{n}(\mathbf{k}''s') - \bar{n}(\mathbf{k}'s') = -\bar{n}(\mathbf{k}'s')[\bar{n}(\mathbf{k}'s')+1][\hbar\psi(\mathbf{k}'s',\hat{\mathbf{k}})/KT]\,|\mathbf{k}|. \tag{17.41}$$

Finally the harmonic $\omega(\mathbf{k}s)$ are expanded by the long-waves result (12.42),

$$\omega(\mathbf{k}s) = c(\hat{\mathbf{k}}s)\,|\mathbf{k}| + \cdots, \tag{17.42}$$

and to leading order in $|\mathbf{k}|$ the contribution (17.38) from the middle two terms in the square brackets in (17.28) when $s'' = s'$ is

$$(36/KT) \sum_{\mathbf{k}'s'} |\Phi(\mathbf{k}s,\mathbf{k}'s',-\mathbf{k}-\mathbf{k}'s')|^2\, \bar{n}(\mathbf{k}'s')[\bar{n}(\mathbf{k}'s')+1][\psi(\mathbf{k}'s',\hat{\mathbf{k}})]^2 \times \{[c(\hat{\mathbf{k}}s)]^2 - [\psi(\mathbf{k}'s',\hat{\mathbf{k}})]^2\}_p^{-1}. \tag{17.43}$$

The long-wavelength acoustic phonon frequency shifts $\Delta(\mathbf{k}s)$ can now be written from (16.27) for $\hbar\Delta(\mathbf{k}s)$, with the aid of (17.34), (17.37), and (17.43). For any lattice the explicit result is

$$\begin{aligned}\Delta(\mathbf{k}s) = {} & 12\hbar^{-1} \sum_{\mathbf{k}'s'} \Phi(\mathbf{k}s,-\mathbf{k}s,\mathbf{k}'s',-\mathbf{k}'s')[2\bar{n}(\mathbf{k}'s')+1] \\ & + 72\hbar^{-2} \sum_{\mathbf{k}'s'} \sum_{s''}{}' |\Phi(\mathbf{k}s,\mathbf{k}'s',-\mathbf{k}-\mathbf{k}'s'')|^2\, [2\bar{n}(\mathbf{k}'s')+1] \\ & \quad \times \omega(\mathbf{k}'s'')\{[\omega(\mathbf{k}'s')]^2 - [\omega(\mathbf{k}'s'')]^2\}_p^{-1} \\ & - 18\hbar^{-2} \sum_{\mathbf{k}'s'} |\Phi(\mathbf{k}s,\mathbf{k}'s',-\mathbf{k}-\mathbf{k}'s')|^2\, [2\bar{n}(\mathbf{k}'s')+1][\omega(\mathbf{k}'s')]^{-1} \\ & + (36/\hbar KT) \sum_{\mathbf{k}'s'} |\Phi(\mathbf{k}s,\mathbf{k}'s',-\mathbf{k}-\mathbf{k}'s')|^2\, \bar{n}(\mathbf{k}'s')[\bar{n}(\mathbf{k}'s')+1] \\ & \quad \times [\psi(\mathbf{k}'s',\hat{\mathbf{k}})]^2\{[c(\hat{\mathbf{k}}s)]^2 - [\psi(\mathbf{k}'s',\hat{\mathbf{k}})]^2\}_p^{-1} \\ & - 36\hbar^{-2} \sum_{\mathbf{k}'s'} \sum_{\substack{s'' \\ \text{(optic)}}} \Phi(\mathbf{k}s,-\mathbf{k}s,0s'')\Phi(\mathbf{k}'s',-\mathbf{k}'s',0s'') \\ & \quad \times [2\bar{n}(\mathbf{k}'s')+1][\omega(0s'')]^{-1}.\end{aligned} \tag{17.44}$$

This result is understood to be correct only to leading order in $|\mathbf{k}|$, and the long-wavelength potential energy coefficients are written out in (17.22), (17.24), and (17.25). Since these coefficients are all proportional to $|\mathbf{k}|$, we have shown

$$\Delta(\mathbf{k}s) \propto |\mathbf{k}|, \quad \text{for small } \mathbf{k} \text{ and } s = \text{acoustic}. \tag{17.45}$$

Thus the renormalized frequencies $\Omega(\mathbf{k}s)$ are still proportional to $|\mathbf{k}|$, with renormalized long-wavelength acoustic phonon velocities $C(\hat{\mathbf{k}}s)$:

$$\left.\begin{aligned}\Omega(\mathbf{k}s) &= C(\hat{\mathbf{k}}s)\,|\mathbf{k}|; \\ C(\hat{\mathbf{k}}s) &= c(\hat{\mathbf{k}}s) + |\mathbf{k}|^{-1}\Delta(\mathbf{k}s);\end{aligned}\right\} \quad \text{for small } \mathbf{k} \text{ and } s = \text{acoustic}. \tag{17.46}$$

Whereas $c(\hat{\mathbf{k}}s)$ depends only on the crystal configuration, the velocity shifts $|\mathbf{k}|^{-1}\Delta(\mathbf{k}s)$ depend on both the temperature and configuration.

At $T = 0$, $\Delta(\mathbf{k}s)$ is simplified by setting each $\bar{n}(\mathbf{k}'s')$, as well as $(KT)^{-1}\bar{n}(\mathbf{k}'s')$, equal to zero in (17.44). In addition, for a primitive lattice the last term in (17.44), involving $\Sigma_{s''}$ for $s'' =$ optic, is not present.

Strain Derivatives of the Phonon Frequencies

Since we are using the alternate phase factors $e^{i\mathbf{k}'\cdot\mathbf{R}(\mathrm{N})}$, the appropriate dynamical matrices are $\bar{\mathbf{D}}(\mathbf{k}')$; the elements of $\bar{\mathbf{D}}(\mathbf{k}')$ are written out in (11.64), and with surface effects eliminated that equation reads

$$\bar{D}_{ij}(\mu\nu,\mathbf{k}') = (M_\mu M_\nu)^{-1/2} \sum_{\mathrm{N}} \Phi_{ij}(0\mu,\mathrm{N}\nu) e^{i\mathbf{k}'\cdot\mathbf{R}(\mathrm{N})}. \tag{17.47}$$

Components of the normalized eigenvectors of $\bar{\mathbf{D}}(\mathbf{k}')$ are $\bar{w}_i(\mu,\mathbf{k}'s')$, and the eigenvalues are $[\omega(\mathbf{k}'s')]^2$. The only quantities in (17.47) which vary under a homogeneous deformation of the lattice are the potential coefficients $\Phi_{ij}(0\mu,\mathrm{N}\nu)$, and if the ion displacements are $\mathbf{U}(\mathrm{N}\nu)$ under a homogeneous deformation, the potential coefficients are expanded as [compare the first-order expansion (6.19)]

$$\begin{aligned} &\Phi_{ij}(0\mu,\mathrm{N}\nu) + \sum_{\mathrm{P}\pi} \sum_{k} \Phi_{ijk}(0\mu,\mathrm{N}\nu,\mathrm{P}\pi) U_k(\mathrm{P}\pi) \\ &\qquad + \tfrac{1}{2} \sum_{\mathrm{PQ}\pi\rho} \sum_{kl} \Phi_{ijkl}(0\mu,\mathrm{N}\nu,\mathrm{P}\pi,\mathrm{Q}\rho) U_k(\mathrm{P}\pi) U_l(\mathrm{Q}\rho). \end{aligned} \tag{17.48}$$

Now we want to calculate an explicit expression for the eigenvalues of $\bar{\mathbf{D}}(\mathbf{k}')$ to second order in the displacement gradients u_{mn}; it is therefore necessary to recognize that the ion displacements $\mathbf{U}(\mathrm{N}\nu)$ contain contributions of second order in the strains, namely the second-order sublattice displacements. In other words,

$$U_k(\mathrm{P}\pi) = U_k^{(1)}(\mathrm{P}\pi) + S_k^{(2)}(\pi); \tag{17.49}$$

the first-order ion displacements are given by (7.84),

$$U_k^{(1)}(\mathrm{P}\pi) = \sum_{mn} Y_{k,mn}(\mathrm{P}\pi) u_{mn}, \tag{17.50}$$

and the second-order sublattice displacements are given by (7.89), which we abbreviate as follows by defining the Z coefficients,

$$S_k^{(2)}(\pi) = \tfrac{1}{2} \sum_{mnpq} Z_{k,mnpq}(\pi) u_{mn} u_{pq}. \tag{17.51}$$

The dynamical matrix $\bar{\mathbf{D}}(\mathbf{k}')$ can be expanded to second order in the strains, as

$$\bar{D}_{ij}(\mu\nu,\mathbf{k}') + \sum_{mn} \bar{D}_{ij,mn}(\mu\nu,\mathbf{k}') u_{mn} + \tfrac{1}{2} \sum_{mnpq} \bar{D}_{ij,mnpq}(\mu\nu,\mathbf{k}') u_{mn} u_{pq}, \tag{17.52}$$

where the coefficients $\bar{D}_{ij,mn}(\mu\nu,\mathbf{k}')$ and $\bar{D}_{ij,mnpq}(\mu\nu,\mathbf{k}')$ can be written explicitly by using (17.47)–(17.51).

The eigenvalues of (17.52) can be calculated to second order in the strains by second-order perturbation theory; the zero-strain eigenvalue $[\omega(\mathbf{k}'s')]^2$ becomes the following expansion.

$$
\begin{aligned}
&[\omega(\mathbf{k}'s')]^2 + \sum_{\mu\nu}\sum_{ijmn} \bar{w}_i(\mu,-\mathbf{k}'s')\bar{D}_{ij,mn}(\mu\nu,\mathbf{k}')\bar{w}_j(\nu,\mathbf{k}'s')u_{mn} \\
&\quad + \tfrac{1}{2}\sum_{\mu\nu}\sum_{ijmnpq} \bar{w}_i(\mu,-\mathbf{k}'s')\bar{D}_{ij,mnpq}(\mu\nu,\mathbf{k}')\bar{w}_j(\nu,\mathbf{k}'s')u_{mn}u_{pq} \\
&\quad + \sum_{s''}{}' \left|\sum_{\mu\nu}\sum_{ijmn} \bar{w}_i(\mu,-\mathbf{k}'s'')\bar{D}_{ij,mn}(\mu\nu,\mathbf{k}')\bar{w}_j(\nu,\mathbf{k}'s')u_{mn}\right|^2 \\
&\quad \times \{[\omega(\mathbf{k}'s')]^2 - [\omega(\mathbf{k}'s'')]^2\}_P^{-1}.
\end{aligned} \tag{17.53}
$$

Finally we abbreviate the strain derivatives of the eigenvalues, evaluated at zero strain, by still another set of coefficients, namely the W coefficients, as follows.

$$\{\partial[\omega(\mathbf{k}'s')]^2/\partial u_{mn}\}_{u'} = W_{mn}(\mathbf{k}'s'); \tag{17.54}$$

$$\{\partial^2[\omega(\mathbf{k}'s')]^2/\partial u_{mn}\partial u_{pq}\}_{u'} = W_{mnpq}(\mathbf{k}'s'). \tag{17.55}$$

By starting with (17.47) and following through the above strain expansions, one finds the W coefficients to be given by

$$
\begin{aligned}
W_{mn}(\mathbf{k}'s') = &\sum_{\mathrm{NP}\mu\nu\pi}\sum_{ijk} \Phi_{ijk}(0\mu,\mathrm{N}\nu,\mathrm{P}\pi)(M_\mu M_\nu)^{-1/2} e^{i\mathbf{k}'\cdot\mathbf{R}(\mathrm{N})} \\
&\times \bar{w}_i(\mu,-\mathbf{k}'s')\bar{w}_j(\nu,\mathbf{k}'s')Y_{k,mn}(\mathrm{P}\pi);
\end{aligned} \tag{17.56}
$$

$$
\begin{aligned}
W_{mnpq}(\mathbf{k}'s') = &\sum_{\mathrm{NP}\mu\nu\pi}\sum_{ijk} \Phi_{ijk}(0\mu,\mathrm{N}\nu,\mathrm{P}\pi)(M_\mu M_\nu)^{-1/2} e^{i\mathbf{k}'\cdot\mathbf{R}(\mathrm{N})} \\
&\times \bar{w}_i(\mu,-\mathbf{k}'s')\bar{w}_j(\nu,\mathbf{k}'s')Z_{k,mnpq}(\pi) \\
&+ \sum_{\mathrm{NPQ}}\sum_{\mu\nu\pi\rho}\sum_{ijkl} \Phi_{ijkl}(0\mu,\mathrm{N}\nu,\mathrm{P}\pi,\mathrm{Q}\rho)(M_\mu M_\nu)^{-1/2} e^{i\mathbf{k}'\cdot\mathbf{R}(\mathrm{N})} \\
&\times \bar{w}_i(\mu,-\mathbf{k}'s')\bar{w}_j(\nu,\mathbf{k}'s')Y_{k,mn}(\mathrm{P}\pi)Y_{l,pq}(\mathrm{Q}\rho) \\
&+ 2\sum_{\mathrm{NN'PP'}}\sum_{\mu\mu'\nu\nu'\pi\pi'}\sum_{ii'jj'kk'} \Phi_{ijk}(0\mu,\mathrm{N}\nu,\mathrm{P}\pi)\Phi_{i'j'k'}(0\mu',\mathrm{N}'\nu',\mathrm{P}'\pi') \\
&\times (M_\mu M_{\mu'} M_\nu M_{\nu'})^{-1/2} e^{i\mathbf{k}'\cdot[\mathbf{R}(\mathrm{N})-\mathbf{R}(\mathrm{N}')]} \\
&\times \sum_{s''}{}' \{[\omega(\mathbf{k}'s')]^2 - [\omega(\mathbf{k}'s'')]^2\}_P^{-1} \bar{w}_i(\mu,-\mathbf{k}'s'')\bar{w}_{i'}(\mu',\mathbf{k}'s'') \\
&\times \bar{w}_j(\nu,\mathbf{k}'s')\bar{w}_{j'}(\nu',-\mathbf{k}'s')Y_{k,mn}(\mathrm{P}\pi)Y_{k',pq}(\mathrm{P}'\pi').
\end{aligned} \tag{17.57}
$$

In order to calculate the sound wave propagation coefficients A^T_{ijkl}, which are strain derivatives of the free energy, the strain derivatives of the harmonic phonon frequencies $\omega(\mathbf{k}'s')$ are required. Rewriting (17.54) gives

$$[\partial\omega(\mathbf{k}'s')/\partial u_{mn}]_{u'} = \tfrac{1}{2}[\omega(\mathbf{k}'s')]^{-1}W_{mn}(\mathbf{k}'s'), \tag{17.58}$$

and then (17.55) leads to

$$[\partial^2\omega(\mathbf{k}'s')/\partial u_{mn}\partial u_{pq}]_{u'} = \tfrac{1}{2}[\omega(\mathbf{k}'s')]^{-1}W_{mnpq}(\mathbf{k}'s') - \tfrac{1}{4}[\omega(\mathbf{k}'s')]^{-3}W_{mn}(\mathbf{k}'s')W_{pq}(\mathbf{k}'s'). \tag{17.59}$$

The strain derivatives (17.58) and (17.59) are evaluated at zero strain from the (arbitrary) initial configuration, and since for any configuration $\omega(\mathbf{k}'s') = \omega(-\mathbf{k}'s')$ must hold, the W coefficients must satisfy

$$W_{mn}(\mathbf{k}'s') = W_{mn}(-\mathbf{k}'s'); \qquad W_{mnpq}(\mathbf{k}'s') = W_{mnpq}(-\mathbf{k}'s'). \tag{17.60}$$

These relations can also be verified from the detailed expressions (17.56) and (17.57) for the W coefficients.

In order to calculate the phonon Grüneisen parameters, we transform the expansion of $[\omega(\mathbf{k}s)]^2$ in powers of the u_{ij} to an expansion in powers of the η_{ij}; this can be done since the $\omega(\mathbf{k}s)$ must be invariant under rotation of the crystal. The strain expansion of $[\omega(\mathbf{k}s)]^2$ is [see, e.g., (7.4) and (7.6), which relate the two strain expansions of Φ]

$$[\omega(\mathbf{k}s)]^2 + \sum_{ij} W_{ij}(\mathbf{k}s)u_{ij} + \tfrac{1}{2}\sum_{ijkl} W_{ijkl}(\mathbf{k}s)u_{ij}u_{kl} = [\omega(\mathbf{k}s)]^2 + \sum_{ij} W_{ij}(\mathbf{k}s)\eta_{ij} + \tfrac{1}{2}\sum_{ijkl}[W_{ijkl}(\mathbf{k}s) - W_{jl}(\mathbf{k}s)\delta_{ik}]\eta_{ij}\eta_{kl}, \tag{17.61}$$

where $[\omega(\mathbf{k}s)]^2$ is the value at zero strain. It follows that the coefficients $W_{ij}(\mathbf{k}s)$ and $W_{ijkl}(\mathbf{k}s) - W_{jl}(\mathbf{k}s)\delta_{ik}$ must have complete Voigt symmetry. The phonon Grüneisen parameters are defined in (16.57) and (16.58), and differentiation of (17.61) gives

$$\gamma_{K,ij} = \gamma_{ij}(\mathbf{k}s) = -\tfrac{1}{2}[\omega(\mathbf{k}s)]^{-2}W_{ij}(\mathbf{k}s); \tag{17.62}$$

$$\xi_{K,ijkl} = \xi_{ijkl}(\mathbf{k}s) = \tfrac{1}{2}[\omega(\mathbf{k}s)]^{-2}[W_{ijkl}(\mathbf{k}s) - W_{jl}(\mathbf{k}s)\delta_{ik}] - \tfrac{1}{4}[\omega(\mathbf{k}s)]^{-4}W_{ij}(\mathbf{k}s)W_{kl}(\mathbf{k}s). \tag{17.63}$$

These equations are useful for direct calculation of $\gamma_{ij}(\mathbf{k}s)$ and $\xi_{ijkl}(\mathbf{k}s)$ for a prescribed theoretical model.

Sound Waves at Zero Temperature

The propagation of long-wavelength adiabatic and isothermal sound waves (thermoelastic waves) was discussed in Section 3. The isothermal propagation coefficients A^T_{mnpq} are defined by

$$A^T_{mnpq} = V^{-1}(\partial^2 F/\partial u_{mn}\partial u_{pq})_{Tu'}, \tag{17.64}$$

and for any $\mathbf{k}$ the elements of the real symmetric 3×3 isothermal propagation matrix are given by

$$\sum_{nq} A^T_{mnpq} k_n k_q. \tag{17.65}$$

Note that the matrix elements (17.65) are $|\mathbf{k}|^2$ times the elements of the propagation matrix $\mathbf{L}(\hat{\mathbf{k}})$, as defined by (3.23). The eigenvalues of the propagation matrix (17.65) are $\rho(\omega_s^T)^2$, where ρ is the crystal density and ω_s^T for $s = 1,2,3$ are the three isothermal sound wave frequencies for the given $\mathbf{k}$.

Lattice-dynamical expressions for A^T_{mnpq} correct to fourth order are obtained from the free energy contributions

$$F = \Phi_0 + F_{\mathrm{H}}. \tag{17.66}$$

With this free energy a formal expansion of the isothermal propagation coefficients is

$$A^T_{mnpq} = \tilde{A}_{mnpq} + \delta A^T_{mnpq}, \tag{17.67}$$

where $\tilde{A}_{mnpq}$ are strain derivatives of Φ_0 and are discussed in Section 7, and

$$\delta A^T_{mnpq} = V^{-1}(\partial^2 F_{\mathrm{H}}/\partial u_{mn}\partial u_{pq})_{Tu'}. \tag{17.68}$$

The leading-order propagation matrix (potential approximation) has the matrix elements

$$\sum_{nq} \tilde{A}_{mnpq} k_n k_q = \sum_{nq} \hat{A}_{mnpq} k_n k_q, \tag{17.69}$$

and from the long-waves calculations of Section 12, particularly (12.36), the eigenvalues of (17.69) are $\rho[\omega(\mathbf{k}s)]^2$, where $\rho = M_C/V_C$ and $\omega(\mathbf{k}s)$ are the long-wavelength acoustic harmonic phonon frequencies, and the normalized eigenvectors of (17.69) are $\mathbf{y}(s)$. Therefore the eigenvalues $\rho(\omega_s^T)^2$ of the propagation matrix (17.65), to first order in δA^T_{mnpq}, are

$$\rho(\omega_s^T)^2 = \rho[\omega(\mathbf{k}s)]^2 + \sum_{mnpq} y_m(s)\delta A^T_{mnpq} k_n k_q y_p(s). \tag{17.70}$$

This result can be used to define a frequency shift $\delta\omega_s^T$ for the isothermal sound wave frequencies, according to

$$\omega_s^T = \omega(\mathbf{k}s) + \delta\omega_s^T, \tag{17.71}$$

where

$$\delta\omega_s^T = [2\omega(\mathbf{k}s)]^{-1}(V_C/M_C) \sum_{mnpq} y_m(s)\delta A^T_{mnpq} k_n k_q y_p(s). \tag{17.72}$$

Now (17.72) is valid for any lattice in any configuration at any temperature, where δA^T_{mnpq} is given by (17.68), and $\omega(\mathbf{k}s)$ is supposed to be $c(\hat{\mathbf{k}}s)\,|\mathbf{k}|$. We will evaluate (17.72) in detail at zero temperature, where

$$F_{\mathrm{H}} = \sum_{\mathbf{k}'s'} \tfrac{1}{2}\hbar\omega(\mathbf{k}'s'), \quad \text{for} \quad T = 0. \tag{17.73}$$

The second-strain derivatives of $\omega(\mathbf{k}'s')$ are given in (17.59), and with that expression the sound wave frequency shifts (17.72) at $T = 0$ are

$$\delta\omega_s^T = (\hbar/8M_C)[\omega(\mathbf{k}s)]^{-1}N_0^{-1}\sum_{\mathbf{k}'s'}\sum_{mnpq}\{[\omega(\mathbf{k}'s')]^{-1}W_{mnpq}(\mathbf{k}'s') - \tfrac{1}{2}[\omega(\mathbf{k}'s')]^{-3}W_{mn}(\mathbf{k}'s')W_{pq}(-\mathbf{k}'s')\}y_m(s)k_n y_p(s)k_q, \tag{17.74}$$

where we replaced $W_{pq}(\mathbf{k}'s')$ by $W_{pq}(-\mathbf{k}'s')$ for convenience. With the detailed equations (17.56) and (17.57) for the W coefficients, and with the factors $y_m(s)k_n$ replaced by H coefficients according to (17.21), the $\delta\omega_s^T$ at $T = 0$ can be transformed to an expression in terms of the long-wavelength potential energy coefficients (17.22), (17.24), and (17.25). The term in (17.57) involving the Z coefficients requires special attention, and will be worked out in detail.

The displacement gradients induced by a long-wavelength acoustic phonon $\mathbf{k}s$ are $y_m(s)\hat{k}_n$ according to (17.18), so

$$\sum_{mnpq} Z_{k,mnpq}(\pi)y_m(s)k_n y_p(s)k_q = |\mathbf{k}|^2 \sum_{mnpq} Z_{k,mnpq}(\pi)u_{mn}u_{pq} = 2\,|\mathbf{k}|^2\, S_k^{(2)}(\pi), \tag{17.75}$$

where the second-order sublattice displacements were introduced by (17.51). Then with the solution (7.89) for the $S_k^{(2)}(\pi)$, the quantity in (17.75) is

$$-|\mathbf{k}|^2 \sum_{\mathrm{NP}\mu\nu\rho}\sum_{ijl}\Gamma_{ki}(\pi\mu)\Phi_{ijl}(0\mu,\mathrm{N}\nu,\mathrm{P}\rho)U_j^{(1)}(\mathrm{N}\nu)U_l^{(1)}(\mathrm{P}\rho), \tag{17.76}$$

where the first-order ion displacements are given by (17.19). In addition, at $\mathbf{k}' = 0$, the dynamical matrix becomes

$$D_{ij}(\mu\pi,0) = \bar{D}_{ij}(\mu\pi,0) = (M_\mu M_\pi)^{-1/2}\sum_{\mathrm{P}}\Phi_{ij}(0\mu,\mathrm{P}\pi). \tag{17.77}$$

This matrix is singular, but if we strike out three rows and three columns we have left the nonsingular matrix whose eigenvalues are $[\omega(0s'')]^2$ for s'' = optic. Elements of the $(3n - 3) \times (3n - 3)$ inverse matrix are then [see, e.g., (10.59) for the general dynamical matrix inverse]

$$\sum_{s''=\mathrm{optic}} \bar{w}_i(\mu,0s'')[\omega(0s'')]^{-2}\bar{w}_j(\pi,0s'').$$

Now according to its definition (7.46), $\boldsymbol{\Gamma}$ is the inverse of the nonsingular part of the matrix $(M_\mu M_\pi)^{1/2}D_{ij}(\mu\pi,0)$; hence we can write for the nonzero elements of $\boldsymbol{\Gamma}$,

$$\Gamma_{ki}(\pi\mu) = (M_\pi M_\mu)^{-1/2}\sum_{s''=\mathrm{optic}} \bar{w}_k(\pi,0s'')[\omega(0s'')]^{-2}\bar{w}_i(\mu,0s''). \tag{17.78}$$

With this the quantity in (17.75) or (17.76) may be written

$$-|\mathbf{k}|^2 \sum_{\mathrm{NP}\mu\nu\rho} \sum_{ijl} \Phi_{ijl}(0\mu,\mathrm{N}\nu,\mathrm{P}\rho)(M_\pi M_\mu)^{-1/2} H_j(\mathrm{N}\nu,\hat{\mathbf{k}}s) H_l(\mathrm{P}\rho,\hat{\mathbf{k}}s)$$
$$\times \sum_{s''=\mathrm{optic}} [\omega(0s'')]^{-2} \bar{w}_k(\pi,0s'')\bar{w}_i(\mu,0s''). \tag{17.79}$$

Referring to (17.74) for $\delta\omega_s^T$, and (17.57) for $W_{mnpq}(\mathbf{k}'s')$, the contribution of the term containing $Z_{k,mnpq}(\pi)$ to $\delta\omega_s^T$ is written out with the aid of (17.79) as

$$-(\hbar/8M_C)N_0^{-1} \sum_{\mathbf{k}'s'} \sum_{s''=\mathrm{optic}} [\omega(\mathbf{k}s)\omega(\mathbf{k}'s')]^{-1}[\omega(0s'')]^{-2}$$
$$\times \sum_{\mathrm{NP}\mu\nu\pi} \sum_{ijk} \Phi_{ijk}(0\mu,\mathrm{N}\nu,\mathrm{P}\pi)(M_\mu M_\nu M_\pi)^{-1/2} e^{i\mathbf{k}'\cdot\mathbf{R}(\mathrm{N})}$$
$$\times \bar{w}_i(\mu,-\mathbf{k}'s')\bar{w}_j(\nu,\mathbf{k}'s')\bar{w}_k(\pi,0s'') \tag{17.80}$$
$$\times \sum_{\mathrm{N'P'}\mu'\nu'\pi'} \sum_{i'j'k'} \Phi_{i'j'k'}(0\mu',\mathrm{N}'\nu',\mathrm{P}'\pi')M_{\mu'}^{-1/2}\bar{w}_i(\mu',0s'')$$
$$\times |\mathbf{k}|^2 H_{j'}(\mathrm{N}'\nu',\hat{\mathbf{k}}s)H_{k'}(\mathrm{P}'\pi',\hat{\mathbf{k}}s).$$

Then with (17.25) for the long-wavelength coefficient $\Phi(\mathbf{k}s,-\mathbf{k}s,0s'')$, this is

$$-36\hbar^{-2}\sum_{\mathbf{k}'s'} \sum_{\substack{s''\\(\mathrm{optic})}} \Phi(\mathbf{k}s,-\mathbf{k}s,0s'')\Phi(\mathbf{k}'s',-\mathbf{k}'s',0s'')[\omega(0s'')]^{-1}. \tag{17.81}$$

In a similar way the remaining contributions to $\delta\omega_s^T$ in (17.74) are worked out to give the total sound wave frequency shift at $T = 0$:

$$\delta\omega_s^T = 12\hbar^{-1}\sum_{\mathbf{k}'s'}\Phi(\mathbf{k}s,-\mathbf{k}s,\mathbf{k}'s',-\mathbf{k}'s')$$
$$+ 72\hbar^{-2}\sum_{\mathbf{k}'s'}\sum_{s''}{}' |\Phi(\mathbf{k}s,\mathbf{k}'s',-\mathbf{k}-\mathbf{k}'s'')|^2\,\omega(\mathbf{k}'s'')$$
$$\times \{[\omega(\mathbf{k}'s')]^2 - [\omega(\mathbf{k}'s'')]^2\}_p^{-1} \tag{17.82}$$
$$- 18\hbar^{-2}\sum_{\mathbf{k}'s'}|\Phi(\mathbf{k}s,\mathbf{k}'s',-\mathbf{k}-\mathbf{k}'s')|^2\,[\omega(\mathbf{k}'s')]^{-1}$$
$$- 36\hbar^{-2}\sum_{\mathbf{k}'s'}\sum_{\substack{s''\\(\mathrm{optic})}}\Phi(\mathbf{k}s,-\mathbf{k}s,0s'')\Phi(\mathbf{k}'s',-\mathbf{k}'s',0s'')[\omega(0s'')]^{-1}.$$

Again this equation is understood to be correct to leading order in $|\mathbf{k}|$, and the expression is proportional to $|\mathbf{k}|$. Comparison of (17.82) with (17.44) for the renormalized acoustic phonon frequency shift $\Delta(\mathbf{k}s)$ shows at last

$$\Delta(\mathbf{k}s) = \delta\omega_s^T, \quad \text{to order } |\mathbf{k}| \text{ at } \quad T = 0. \tag{17.83}$$

The result (17.83) constitutes a detailed verification to second order in the anharmonicity of the theorem stated at the beginning of this section, that for any lattice and any configuration the sound wave frequencies and the long-wavelength renormalized acoustic phonon frequencies are the same at $T = 0$.

Sound Waves at Finite Temperatures

The isothermal sound wave frequency shifts $\delta\omega_s^T$ may be calculated from (17.72) at any temperature, by the same procedure as the above $T = 0$ calculation. At finite temperatures these shifts are no longer equal to the renormalized acoustic phonon frequency shifts $\Delta(\mathbf{k}s)$ given by (17.44); the difference is found to be

$$\Delta(\mathbf{k}s) - \delta\omega_s^T = (36/\hbar KT) \sum_{\mathbf{k}'s'} |\Phi(\mathbf{k}s,\mathbf{k}'s',-\mathbf{k}-\mathbf{k}'s')|^2 \times \bar{n}(\mathbf{k}'s')[\bar{n}(\mathbf{k}'s') + 1]\, [c(\hat{\mathbf{k}}s)]^2\{[c(\hat{\mathbf{k}}s)]^2 - [\psi(\mathbf{k}'s',\hat{\mathbf{k}})]^2\}_p^{-1}. \tag{17.84}$$

It will be recalled from (17.39) that $\psi(\mathbf{k}'s',\hat{\mathbf{k}})$ is the directional derivative of $\omega(\mathbf{k}'s')$ with respect to $\mathbf{k}'$ in the direction $\hat{\mathbf{k}}$; likewise the acoustic phonon velocity $c(\hat{\mathbf{k}}s)$ is the directional derivative of $\omega(\mathbf{k}s)$ with respect to $\mathbf{k}$ in the direction $\hat{\mathbf{k}}$, evaluated at $\mathbf{k} = 0$ and $s =$ acoustic. We expect in general that $\psi(\mathbf{k}'s',\hat{\mathbf{k}})$ will be smaller in magnitude than $c(\hat{\mathbf{k}}s)$ for most, but certainly not all, values of $\mathbf{k}'s'$ and $\hat{\mathbf{k}}s$. From this observation, it is then expected that (17.84) is positive, so that the long-wavelength acoustic renormalized phonon frequencies are likely to be greater than the isothermal sound wave frequencies;

$$\Omega(\mathbf{k}s) > \omega_s^T \quad \text{is expected at} \quad T \neq 0. \tag{17.85}$$

This is of course a qualitative prediction, and nature may well provide some crystals for which (17.85) is not always fulfilled.

The adiabatic sound wave frequencies may be calculated in a manner similar to the calculation of the isothermal frequencies, simply by using the adiabatic propagation coefficients A_{mnpq}^S in place of the isothermal A_{mnpq}^T. Since in leading order the adiabatic and isothermal propagation coefficients are the same, namely $\tilde{A}_{mnpq}$, we can write

$$A_{mnpq}^S = \tilde{A}_{mnpq} + \delta A_{mnpq}^S, \tag{17.86}$$

where the fourth-order contributions δA_{mnpq}^S may be calculated from

$$\delta A_{mnpq}^S - \delta A_{mnpq}^T = C_{mnpq}^S - C_{mnpq}^T. \tag{17.87}$$

Then the adiabatic analog of (17.70) is

$$\rho(\omega_s^S)^2 = \rho[\omega(\mathbf{k}s)]^2 + \sum_{mnpq} y_m(s)\delta A_{mnpq}^S k_n k_q y_p(s), \tag{17.88}$$

where ω_s^S for $s = 1,2,3$ are the three adiabatic sound wave frequencies for the given $\mathbf{k}$. Subtracting (17.70) from (17.88) gives

$$\rho(\omega_s^S)^2 - \rho(\omega_s^T)^2 = \sum_{mnpq} y_m(s)[\delta A_{mnpq}^S - \delta A_{mnpq}^T]k_n k_q y_p(s). \qquad (17.89)$$

Finally this can be transformed to give the adiabatic–isothermal frequency differences,

$$\omega_s^S - \omega_s^T = (2\rho\omega_s^T)^{-1} \sum_{mnpq} [C_{mnpq}^S - C_{mnpq}^T] y_m(s)k_n y_p(s)k_q. \qquad (17.90)$$

To evaluate (17.90) we can replace ω_s^T in the right-hand side by its leading contribution $\omega(\mathbf{k}s)$ for small $\mathbf{k}$ and $s =$ acoustic, and use (16.71) for $C_{mnpq}^S - C_{mnpq}^T$. It will be convenient to define $\varphi(\mathbf{k}'s',\hat{\mathbf{k}}s)$ as the total change in the harmonic phonon frequency $\omega(\mathbf{k}'s')$ due to the strain whose components are $y_m(s)\hat{k}_n$:

$$\varphi(\mathbf{k}'s',\hat{\mathbf{k}}s) = \sum_{mn} [\partial\omega(\mathbf{k}'s')/\partial u_{mn}]_{u'} y_m(s)\hat{k}_n. \qquad (17.91)$$

Then after a little algebra, (17.90) is

$$\omega_s^S - \omega_s^T = \tfrac{1}{2} |\mathbf{k}|^2 [M_C\omega(\mathbf{k}s)]^{-1}(TN_0/C_\eta) \times \left\{N_0^{-1} \sum_{\mathbf{k}'s'} \hbar\varphi(\mathbf{k}'s',\hat{\mathbf{k}}s)[\partial\bar{n}(\mathbf{k}'s')/\partial T]_\eta\right\}^2. \qquad (17.92)$$

Again this equation is understood to be correct to leading order in $|\mathbf{k}|$, and since $\omega(\mathbf{k}s) = c(\hat{\mathbf{k}}s)\,|\mathbf{k}|$, the right-hand side is proportional to $|\mathbf{k}|$. The important conclusion from (17.92) is that the adiabatic velocities are greater than the isothermal velocities at finite temperatures:

$$\omega_s^S > \omega_s^T, \quad \text{for each } \mathbf{k} \text{ and } s. \qquad (17.93)$$

Finally we should consider the question of whether or not the sound waves and long-wavelength acoustic phonons are in fact reasonably good, i.e., long-lived, modes of motion of a crystal at finite temperatures. To answer this question one needs to set up a hydrodynamical transport equation, and find the normal modes as solutions. Fortunately some useful conclusions can be established without a detailed consideration of this difficult problem. First of all it is well known experimentally that sound waves are good long-wavelength normal modes of motion in most crystals, since they generally have lifetimes that are long compared to their vibrational period. It is customarily *assumed* that sound propagates adiabatically in crystals; the validity of this assumption is a question worthy of further serious theoretical investigation. To the extent that sound waves are truly adiabatic, there is no reason to expect truly isothermal thermoelastic waves to survive in a crystal. Nevertheless the concept of isothermal waves is very useful in any theory dealing with equilibrium isothermal elastic processes, and such processes are, of course, observed in many experiments. Further, when the wave vector is sufficiently small so that the renormalized acoustic phonon frequency becomes as small as the half widths of most of the excited phonons in the crystal, then we are not able to guarantee that our renormalized acoustic phonons are long-lived. Nevertheless our free energy calculations are correct to fourth order in ϵ for any crystal and at all temperatures, even if some of the phonons are essentially nonexistent, since the free energy is calculated as a trace.

18. LOW-TEMPERATURE LIMIT

Absolute Zero

At zero temperature the free energy and the internal energy are equal, and are just the total ground-state energy $\mathscr{E}_G$:

$$F = U = \mathscr{E}_{\mathrm{G}} = \Phi_0 + \sum_{\kappa} \tfrac{1}{2}\hbar\omega_\kappa + G_4, \quad \text{for } T = 0, \tag{18.1}$$

where G_4 is written out in (13.92). Note that $G_2 = \Sigma_\kappa \frac{1}{2}\hbar\omega_\kappa$ and G_4 are the second- and fourth-order zero-point vibrational energies, respectively. The entropy and heat capacity are of course zero at $T = 0$, and the stresses and elastic constants are given by (16.68) and (16.69) evaluated at $T = 0$:

$$\tau_{ij}^0 = \tilde{C}_{ij} - V^{-1} \sum_{\kappa} \tfrac{1}{2}\hbar\omega_\kappa \gamma_{\kappa,ij}; \tag{18.2}$$

$$C_{ijkl}^0 = \tilde{C}_{ijkl} + V^{-1} \sum_{\kappa} \tfrac{1}{2}\hbar\omega_\kappa \xi_{\kappa,ijkl}. \tag{18.3}$$

The phonon frequency shifts Δ_κ have the $T = 0$ values θ_κ, according to (16.35), so that

$$\Omega_\kappa = \omega_\kappa + \theta_\kappa, \quad \text{for } T = 0. \tag{18.4}$$

The thermal stresses b_{ij} and thermal strains β_{ij} are also zero at $T = 0$. It may be noted that all the $T = 0$ values may be interpreted as limits as $T \to 0$, since the thermodynamic functions are analytic in T, except at phase transitions. We will now look at the explicit temperature-dependence of the thermodynamic functions in the low-temperature limit.

Renormalized Debye Temperature

Let us subtract the ground-state energy (18.1) from F to obtain the explicitly temperature-dependent part of F, evaluate this temperature-dependent part at low temperatures, and denote the result as F_{t}. Obviously, F_{t} has harmonic and anharmonic contributions,

$$F_{\mathrm{t}} = F_{\mathrm{Ht}} + F_{\mathrm{At}}, \tag{18.5}$$

and it will be possible to show that both F_{Ht} and F_{At} are proportional to T^4 at low temperature. From (16.13) for F_{H} and (16.36) for F_{A}, it follows that

$$F_{\mathrm{Ht}} = KT \sum_{\mathbf{k}s} \ln[1 - e^{-\hbar\omega(\mathbf{k}s)/KT}], \quad \text{at low } T; \tag{18.6}$$

$$F_{\mathrm{At}} = \sum_{\mathbf{k}s} \hbar\theta(\mathbf{k}s)\bar{n}(\mathbf{k}s), \quad \text{at low } T; \tag{18.7}$$

here the term in F_{A} of second order in the $\bar{n}(\mathbf{k}s)$ has been dropped since it goes as a higher power of T than T^4, i.e., it does not contribute to the *leading* temperature-dependence at low temperature.

At low temperatures only the long-wavelength acoustic phonons contribute to the $\Sigma_{\mathbf{k}s}$ in (18.6) and (18.7), since only for these modes are the energies as small as KT for sufficiently small T. Then in these sums $\omega(\mathbf{k}s)$ can be replaced by its long-wavelength limit $c(\hat{\mathbf{k}}s)\,|\mathbf{k}|$ for s = acoustic, and the sum over $\mathbf{k}$ can be replaced by an integral over all $\mathbf{k}$ space, i.e., with $|\mathbf{k}|$ going to infinity, since the integrand is cut off exponentially for values of $\hbar\omega(\mathbf{k}s) > KT$. The transformation from a sum to an integral is then accomplished by

$$\sum_{\mathbf{k}s} \rightarrow \frac{V}{(2\pi)^3} \sum_{\substack{s \\ \text{(acoustic)}}} \int_0^\infty d\mathbf{k}, \quad \text{at low } T, \tag{18.8}$$

since the density of allowed $\mathbf{k}$ vectors in inverse-lattice space is $V/(2\pi)^3$. Here $d\mathbf{k}$ is the three-dimensional differential

$$d\mathbf{k} = k^2\, dk\, d\Omega, \tag{18.9}$$

where k is the magnitude of $\mathbf{k}$ and Ω is the angle (direction) of $\mathbf{k}$. Also in (18.7), $\theta(\mathbf{k}s)$ may be replaced by its long-wavelength limit for s = acoustic. It is important to recognize that the leading temperature-dependence of F_{Ht} and F_{At} is determined by the $|\mathbf{k}|$-dependence of $\omega(\mathbf{k}s)$ and $\theta(\mathbf{k}s)$ at small $|\mathbf{k}|$ and s = acoustic.

To proceed with the evaluation of F_{Ht}, we write (18.6) as

$$F_{\mathrm{Ht}} = \frac{KTV}{(2\pi)^3} \sum_{\substack{s \\ \text{(acoustic)}}} \int_0^\infty \ln[1 - e^{-\hbar c(\hat{\mathbf{k}}s)|\mathbf{k}|/KT}]\, d\mathbf{k}. \tag{18.10}$$

The velocity $c(\hat{\mathbf{k}}s)$ depends only on the direction of $\mathbf{k}$, so the angular integration may be separated out by introducing the variable x,

$$x = \hbar c(\hat{\mathbf{k}}s)\,|\mathbf{k}|/KT, \qquad |\mathbf{k}| = xKT/\hbar c(\hat{\mathbf{k}}s); \tag{18.11}$$

so that

$$d\mathbf{k} = [KT/\hbar c(\hat{\mathbf{k}}s)]^3 x^2\, dx\, d\Omega. \tag{18.12}$$

Then (18.10) is

$$F_{\mathrm{Ht}} = \frac{KTV}{(2\pi)^3} \sum_{\substack{s \\ \text{(acoustic)}}} \int \left[\frac{KT}{\hbar c(\hat{\mathbf{k}}s)}\right]^3 d\Omega \int_0^\infty x^2 \ln(1 - e^{-x})\, dx. \tag{18.13}$$

The x integral may be evaluated by integration by parts to give

$$\int_0^\infty x^2 \ln(1 - e^{-x})\, dx = -\pi^4/45, \tag{18.14}$$

and F_{Ht} is

$$F_{Ht} = -\frac{\pi K^4 T^4 V}{360\hbar^3} \sum_{\substack{s \\ \text{(acoustic)}}} \int \frac{d\Omega}{[c(\hat{\mathbf{k}}s)]^3}. \tag{18.15}$$

Thus the harmonic free energy has a leading T^4-dependence at low temperature. Higher-order terms, proportional to $T^5, \cdots$, result from dispersion of the acoustic phonons, i.e., the terms of order $|\mathbf{k}|^2, \cdots$, in the expansion of $\omega(\mathbf{k}s)$ for small $\mathbf{k}$ and $s =$ acoustic. The optic modes may be neglected in the low-temperature expansion of F_H, since their contribution is exponentially small (see, e.g., the Einstein model results of Section 5).

It will be convenient to define the average $\langle [c(\hat{\mathbf{k}}s)]^{-3} \rangle$ of the velocities $c(\hat{\mathbf{k}}s)$, averaged over the three acoustic branches and over all angles of $\mathbf{k}$, as

$$\langle [c(\hat{\mathbf{k}}s)]^{-3} \rangle = \tfrac{1}{3} \sum_{\substack{s \\ \text{(acoustic)}}} \frac{1}{4\pi} \int \frac{d\Omega}{[c(\hat{\mathbf{k}}s)]^3}. \tag{18.16}$$

In addition we will define the $T = 0$ value of the harmonic Debye temperature as Θ_{H0}, given by the relation

$$(K\Theta_{H0})^{-3} = (V_A/6\pi^2\hbar^3)\langle [c(\hat{\mathbf{k}}s)]^{-3} \rangle, \tag{18.17}$$

where V_A is the volume per atom of the crystal,

$$V_A = V_C/n = V/nN_0. \tag{18.18}$$

Then (18.15) for F_{Ht} may be written

$$F_{Ht} = -nN_0(\pi^4/5)(KT)(T/\Theta_{H0})^3. \tag{18.19}$$

From this, the harmonic entropy S_H and the harmonic heat capacity at constant configuration $C_{\eta H}$ have the following values at low temperature.

$$S_H = nN_0 K(4\pi^4/5)(T/\Theta_{H0})^3; \tag{18.20}$$

$$C_{\eta H} = nN_0 K(12\pi^4/5)(T/\Theta_{H0})^3. \tag{18.21}$$

Thus with the definition (18.17) of the harmonic Debye temperature, the low-temperature harmonic heat capacity (18.21) is the same as the Debye formula (5.35), with the Debye temperature being Θ_{H0} and the number of atoms in the crystal nN_0.

It was shown in Section 17, particularly in (17.45), that all contributions to $\Delta(\mathbf{k}s)$, and hence also to $\theta(\mathbf{k}s)$, are proportional to $|\mathbf{k}|$ for small $\mathbf{k}$ and

s = acoustic. To evaluate (18.7) for F_{At}, let us write

$$\theta(\mathbf{k}s) = \delta c(\hat{\mathbf{k}}s)\,|\mathbf{k}|, \quad \text{for small } \mathbf{k} \text{ and } s = \text{acoustic}, \tag{18.22}$$

where $\delta c(\hat{\mathbf{k}}s)$ is obviously the long-wavelength acoustic phonon velocity shift at $T = 0$. Then F_{At} is

$$F_{\mathrm{At}} = \frac{V}{(2\pi)^3} \sum_{\substack{s \\ (\text{acoustic})}} \int_0^\infty \frac{\hbar\delta c(\hat{\mathbf{k}}s)\,|\mathbf{k}|}{e^{\hbar c(\hat{\mathbf{k}}s)|\mathbf{k}|/KT} - 1}\, d\mathbf{k}. \tag{18.23}$$

Again introducing the variable x by (18.11), the angular integration is separated, and the x integral becomes

$$\int_0^\infty x^3(e^x - 1)^{-1}\, dx = \pi^4/15. \tag{18.24}$$

The result for the temperature-dependent part of the anharmonic free energy at low temperature is then

$$F_{\mathrm{At}} = \frac{\pi K^4 T^4 V}{120\hbar^3} \sum_{\substack{s \\ (\text{acoustic})}} \int \frac{\delta c(\hat{\mathbf{k}}s)}{[c(\hat{\mathbf{k}}s)]^4}\, d\Omega. \tag{18.25}$$

This result shows that F_{At} also has a leading temperature-dependence of T^4, and further that this temperature-dependence follows directly from the fact that $\theta(\mathbf{k}s)$ is proportional to $|\mathbf{k}|$, as written in (18.22). If, for example, $\theta(\mathbf{k}s)$ were proportional to $|\mathbf{k}|^2$, then F_{At} would have a leading T^5-dependence.

We can now establish a particularly valuable result concerning the low-temperature thermodynamic functions. At $T = 0$ the long-wavelength renormalized acoustic phonon velocities are

$$C(\hat{\mathbf{k}}s) = c(\hat{\mathbf{k}}s) + \delta c(\hat{\mathbf{k}}s), \tag{18.26}$$

where by (18.22) [compare (17.46), which is valid for any temperature],

$$\delta c(\hat{\mathbf{k}}s) = |\mathbf{k}|^{-1}\theta(\mathbf{k}s), \quad \text{for small } \mathbf{k} \text{ and } s = \text{acoustic}. \tag{18.27}$$

The total temperature-dependent part of the free energy at low temperatures, correct to first order in the velocity shifts $\delta c(\hat{\mathbf{k}}s)$, is just the harmonic formula (18.15) with the harmonic velocities $c(\hat{\mathbf{k}}s)$ replaced by the renormalized velocities $C(\hat{\mathbf{k}}s)$:

$$F_{\mathrm{t}} = F_{\mathrm{Ht}} + F_{\mathrm{At}} = -\frac{\pi K^4 T^4 V}{360\hbar^3} \sum_{\substack{s \\ (\text{acoustic})}} \int \frac{d\Omega}{[C(\hat{\mathbf{k}}s)]^3}. \tag{18.28}$$

This may be verified by replacing $C(\hat{\mathbf{k}}s)$ by $c(\hat{\mathbf{k}}s) + \delta c(\hat{\mathbf{k}}s)$ and expanding to first order in $\delta c(\hat{\mathbf{k}}s)$, and noting that the zeroth-order term is just F_{Ht} of (18.15) and the first-order term is just F_{At} of (18.25). Now the low-temperature free energy, and hence all other thermodynamic functions, may be simply represented in terms of a renormalized $T = 0$ Debye temperature Θ_0, just as in the harmonic contributions given by (18.17)–(18.21). In particular, Θ_0 is defined by

$$(K\Theta_0)^{-3} = (V_A/6\pi^2\hbar^3)\langle[C(\hat{\mathbf{k}}s)]^{-3}\rangle, \tag{18.29}$$

where

$$\langle[C(\hat{\mathbf{k}}s)]^{-3}\rangle = \tfrac{1}{3} \sum_{\substack{s \\ \text{(acoustic)}}} \frac{1}{4\pi} \int \frac{d\Omega}{[C(\hat{\mathbf{k}}s)]^3}. \tag{18.30}$$

Then from (18.28),

$$F_t = -nN_0(\pi^4/5)(KT)(T/\Theta_0)^3, \tag{18.31}$$

and correct to leading order in anharmonicity (fourth order in ϵ) the low-temperature entropy S and heat capacity at constant configuration C_η are

$$S = nN_0K(4\pi^4/5)(T/\Theta_0)^3; \tag{18.32}$$

$$C_\eta = nN_0K(12\pi^4/5)(T/\Theta_0)^3. \tag{18.33}$$

The temperature-dependence of (18.31)–(18.33) is explicit, while the configuration-dependence is contained in Θ_0, which depends *only* on the crystal configuration.

In order to obtain explicit expressions for the anharmonic contributions to the low-temperature thermodynamic functions, let us separate Θ_0 into harmonic and anharmonic parts, as

$$\Theta_0 = \Theta_{H0} + \Theta_{A0}, \tag{18.34}$$

where Θ_{H0} is given by (18.17). An expansion of (18.29) for Θ_0, to first order in $\delta c(\hat{\mathbf{k}}s)$, then yields

$$\frac{\Theta_{A0}}{\Theta_{H0}} = \frac{\langle\delta c(\hat{\mathbf{k}}s)[c(\hat{\mathbf{k}}s)]^{-4}\rangle}{\langle[c(\hat{\mathbf{k}}s)]^{-3}\rangle}, \tag{18.35}$$

where as usual the average in the numerator is defined by

$$\langle\delta c(\hat{\mathbf{k}}s)[c(\hat{\mathbf{k}}s)]^{-4}\rangle = \tfrac{1}{3} \sum_{\text{(acoustic)}} \frac{1}{4\pi} \int \frac{\delta c(\hat{\mathbf{k}}s)}{[c(\hat{\mathbf{k}}s)]^4}\, d\Omega. \tag{18.36}$$

The anharmonic contributions to F_t, S, and C_η are then obtained by expanding (18.31)–(18.33) to give

$$F_{At} = -3(\Theta_{A0}/\Theta_{H0})F_{Ht}, \tag{18.37}$$

$$S_A = -3(\Theta_{A0}/\Theta_{H0})S_H, \tag{18.38}$$

$$C_{\eta A} = -3(\Theta_{A0}/\Theta_{H0})C_{\eta H}, \tag{18.39}$$

where the harmonic contributions are written in (18.19)–(18.21). In order to calculate Θ_{A0} from (18.35) for a given theoretical model, or to calculate F_{At} from (18.25), the velocity shifts may be calculated from (18.27), where $\theta(\mathbf{k}s)$ is the $T = 0$ value of (17.44) for $\Delta(\mathbf{k}s)$.

There are two factors which conspire to produce the result that the harmonic function F_{Ht} is correct also for the total F_t, when the harmonic velocities are replaced by the renormalized velocities. It is worthwhile to sketch an alternate calculation of F_t in order to clarify these factors. To begin, write the temperature-dependent part of F_H, with the harmonic frequencies $\omega(\mathbf{k}s)$ replaced by the renormalized frequencies $\Omega(\mathbf{k}s)$:

$$KT \sum_{\mathbf{k}s} \ln[1 - e^{-\hbar\Omega(\mathbf{k}s)/KT}]. \tag{18.40}$$

Since $\Omega(\mathbf{k}s) = \omega(\mathbf{k}s) + \Delta(\mathbf{k}s)$, the function (18.40) may be expanded for $\Delta(\mathbf{k}s)$ small compared to $\omega(\mathbf{k}s)$, and to first order in $\Delta(\mathbf{k}s)$ it is

$$KT \sum_{\mathbf{k}s} \ln[1 - e^{-\hbar\omega(\mathbf{k}s)/KT}] + \sum_{\mathbf{k}s} \hbar\Delta(\mathbf{k}s)\bar{n}(\mathbf{k}s). \tag{18.41}$$

But to obtain the leading low-temperature-dependence of the second term in (18.41), $\Delta(\mathbf{k}s)$ must be replaced by its $T = 0$ value $\theta(\mathbf{k}s)$, and then (18.41) is just (18.6) for F_{Ht} plus (18.7) for F_{At}, at low temperature. In other words, the leading low-temperature contribution to $F_{Ht} + F_{At}$ is obtained from the low-temperature expansion of (18.40). This is the first special factor, and it should be emphasized that such a simple result is not valid for F at arbitrary temperatures, or even at $T = 0$ when $F = \mathscr{E}_G$. The second factor is that $\Omega(\mathbf{k}s)$ is proportional to $|\mathbf{k}|$ for small $\mathbf{k}$ and s = acoustic, so when $\Omega(\mathbf{k}s)$ is replaced by $C(\hat{\mathbf{k}}s)\,|\mathbf{k}|$ in (18.40), the harmonic formulas are all obtained, only with $c(\hat{\mathbf{k}}s)$ replaced by $C(\hat{\mathbf{k}}s)$.

A further important point of interpretation is as follows. Since at $T = 0$ the sound wave velocities and acoustic phonon velocities are the same, then the renormalized Debye temperature given by (18.29) may be calculated from the measured sound velocities at $T = 0$, and this Debye temperature is the same as that obtained from the measured low-temperature heat capacity by means of equation (18.33). We consider this result to be established to leading order in the anharmonicity since, although by our theorem at the beginning of Section 17 the sound waves and long-wavelength acoustic phonons are identical to all orders at $T = 0$, the phonon statistical mechanics has been treated only to leading order in the anharmonicity, which is fourth order in ϵ. Therefore to the order we have carried the theory (and possibly to all orders),

$$\Theta_0(\text{propagation coefficients}) = \Theta_0(\text{heat capacity}). \tag{18.42}$$

The sound wave velocities can be calculated at all angles of $\mathbf{k}$ as eigenvalues of the propagation matrices, which in turn can be constructed from the measured set of independent propagation coefficients, as discussed in Section 3. Further, as it is shown below, the temperature-dependence of the sound wave velocities is small at low temperatures, namely it is T^4, so the $T = 0$ values are easily extrapolated from reasonably low-temperature measurements, say liquid helium measurements for most crystals. Finally, because our theory is valid for arbitrary initial equilibrium configuration of the crystal, the result (18.42) holds at any configuration, so the strain or volume derivatives of Θ_0(propagation coefficients) and of Θ_0(heat capacity) are also equal.

Another interesting and potentially useful observation is that the mass dependences of the harmonic and anharmonic contributions to the sound velocities at low temperature are different, so that a measurement of the low-temperature properties for different isotopes of the same crystal will allow in principle the experimental separation of harmonic and anharmonic contributions. Consider for simplicity a primitive lattice, or any lattice where all ions have the same mass. Then from the generalized-eigenvalue-problem equations (11.44)–(11.46), it is seen that the harmonic frequencies, and hence also the harmonic acoustic phonon velocities, have the dependence

$$\omega(\mathbf{k}s) \propto M_C^{-1/2}; \qquad c(\hat{\mathbf{k}}s) \propto M_C^{-1/2}. \tag{18.43}$$

On the other hand, for such a lattice, from (17.44) at $T = 0$, the frequency shifts and hence also the acoustic phonon velocity shifts have the dependence at $T = 0$

$$\Delta(\mathbf{k}s) \propto M_C^{-1}; \qquad \delta c(\hat{\mathbf{k}}s) \propto M_C^{-1}. \tag{18.44}$$

Then the harmonic and anharmonic contributions to the acoustic phonon velocities, and hence also to the sound velocities, are proportional to $M_C^{-1/2}$ and M_C^{-1}, respectively. Further, from (18.17) for Θ_{H0} and from (18.35) for Θ_{A0}/Θ_{H0}, it follows that

$$\Theta_{H0} \propto M_C^{-1/2}; \qquad \Theta_{A0} \propto M_C^{-1}. \tag{18.45}$$

Thus according to the above three equations, it is possible in principle to experimentally determine the harmonic and anharmonic contributions separately, on account of their different mass dependences, for the phonon frequencies, the sound velocities, and the Debye temperature, all at low temperatures. Considering the present-day accuracies of various experimental techniques, the easiest place to observe this isotope effect may be in the sound velocities.

Configuration Variations

In the low-temperature limit, the complete free energy is

$$F = \mathscr{E}_G - nN_0(\pi^4/5)(KT)(T/\Theta_0)^3, \tag{18.46}$$

where $\mathscr{E}_G$ is given to fourth order by (18.1). The stresses and isothermal elastic constants are written as strain derivatives of F in (2.27) and (2.34), respectively, and with (18.46) for F the stresses at low temperature are

$$\tau_{ij} = \tau_{ij}^0 + V_A^{-1}(3\pi^4/5)(KT)(T/\Theta_0)^3(\partial \ln \Theta_0/\partial \eta_{ij})_{\eta'}, \tag{18.47}$$

where τ_{ij}^0 is given to third order by (18.2). The isothermal elastic constants at low temperatures are

$$C_{ijkl}^T = C_{ijkl}^0 + V_A^{-1}(3\pi^4/5)(KT)(T/\Theta_0)^3[\Theta_0^{-1}(\partial^2\Theta_0/\partial\eta_{ij}\partial\eta_{kl})_{\eta'} - 4(\partial \ln \Theta_0/\partial\eta_{ij})_{\eta'}(\partial \ln \Theta_0/\partial\eta_{kl})_{\eta'}], \tag{18.48}$$

where C_{ijkl}^0 is given to fourth order by (18.3). According to (2.72), the thermal stresses are $b_{ij} = (\partial\tau_{ij}/\partial T)_\eta$, and from (18.47) the low-temperature evaluation is

$$b_{ij} = V_A^{-1}K(12\pi^4/5)(T/\Theta_0)^3(\partial \ln \Theta_0/\partial\eta_{ij})_{\eta'}. \tag{18.49}$$

From (2.78), the thermal strains are $\beta_{ij} = -\Sigma_{kl}\, S_{ijkl}^T b_{kl}$, and in the low-temperature limit the compliances may be replaced by their $T = 0$ values, to give

$$\beta_{ij} = -\sum_{kl} S_{ijkl}^0 b_{kl}. \tag{18.50}$$

Also from (2.102), the macroscopic Grüneisen parameter tensor is $\gamma_{ij} = -(V/C_\eta)b_{ij}$, and at low temperatures we use (18.33) for C_η and (18.49) for b_{ij}, to find

$$\gamma_{ij} = -(\partial \ln \Theta_0/\partial\eta_{ij})_{\eta'}, \quad \text{at} \quad T = 0. \tag{18.51}$$

Finally, according to (2.90) and (2.91), the difference between adiabatic and isothermal elastic constants is $(TV/C_\eta)b_{ij}b_{kl}$, and the low-temperature evaluation is

$$C_{ijkl}^S - C_{ijkl}^T = V_A^{-1}(12\pi^4/5)(KT)(T/\Theta_0)^3(\partial \ln \Theta_0/\partial\eta_{ij})_{\eta'}(\partial \ln \Theta_0/\partial\eta_{kl})_{\eta'}. \tag{18.52}$$

The above equations for various free energy strain derivatives are valid for arbitrary configuration. If we take the configuration to remain fixed, then the temperature-dependence of each of these functions is explicit, and the following temperature-dependences hold at constant configuration.

$$\tau_{ij},\ C_{ijkl}^T,\ A_{ijkl}^T,\ B_{ijkl}^T \ \text{go as}\ T^4. \tag{18.53}$$

In addition, since $C_{ijkl}^S - C_{ijkl}^T$ goes as T^4, according to (18.52), then

$$C_{ijkl}^S,\ A_{ijkl}^S,\ B_{ijkl}^S \ \text{go as}\ T^4. \tag{18.54}$$

Recall that at zero stress C_{ijkl}^T, A_{ijkl}^T, and B_{ijkl}^T are the same, and likewise the three adiabatic coefficients are the same. The thermal stresses and strains have the low-temperature-dependence

$$b_{ij}, \quad \beta_{ij} \quad \text{go as } T^3. \tag{18.55}$$

It is apparent from (18.53) that components of the stress required to hold the lattice configuration fixed go as T^4 at low temperature. Components of the thermal expansion tensor β_{ij} go as T^3 at constant configuration, and since $\beta_{ij} = (\partial\eta_{ij}/\partial T)_\tau$, the lattice strains η_{ij} measured from the $T = 0$ configuration go as T^4 when the stress is held constant:

$$\eta_{ij} = T\beta_{ij}, \quad \text{at low } T \text{ and constant stress.} \tag{18.56}$$

This last result allows us to determine the temperature-dependence at low temperatures of the thermodynamic functions under conditions of constant stress. At constant configuration, the free energy (18.46) goes as T^4, while at constant stress there is another T^4 contribution arising from the configuration-dependence of $\mathscr{E}_G$:

$$F = \mathscr{E}_G(\eta_{ij} = 0) + \sum_{ij} (\partial\mathscr{E}_G/\partial\eta_{ij})_{\eta'} T\beta_{ij} - nN_0(\pi^4/5)(KT)(T/\Theta_0)^3 + \cdots,$$

$$\text{at constant stress.} \tag{18.57}$$

Note that the term in $T\beta_{ij}$ goes as T^4. In the same way the elastic constants (adiabatic or isothermal) have a T^4 contribution arising from the variation of C^0_{ijkl} due to the thermal strains, in addition to the constant configuration T^4 contribution which is explicit in (18.48) for C^T_{ijkl}, and in (18.52) for $C^S_{ijkl} - C^T_{ijkl}$. It is important to recognize that a measurement of the T^4-dependence of the elastic constants, under conditions of constant stress, yields the sum of these two T^4 contributions; the same observation applies to the propagation coefficients, stress–strain coefficients, and to the compressibility and bulk modulus.

On the other hand, since there is no zero-point contribution to the entropy S of (18.32) and the heat capacity C_η of (18.33), the lowest-order temperature-dependence of these functions is T^3 as shown in those equations, and the thermal strains give rise to a T^7 contribution, which is negligible. [Note that S and C_η at constant stress are *not* obtained as temperature derivatives of (18.57) for F at constant stress.] Incidentally, the same conclusion is reached from (2.93), which is

$$C_\tau - C_\eta = -TV \sum_{ij} b_{ij}\beta_{ij},$$

and because of (18.55) this goes as T^7 at low temperature. The same dependence holds for $C_P - C_V$.

Detailed calculation of Θ_0 and its strain derivatives is straightforward. The value of Θ_0 is given by (18.29), in terms of the renormalized acoustic phonon velocities at $T = 0$, and differentiation of that equation with respect to strain leads to

$$\left(\frac{\partial \ln \Theta_0}{\partial \eta_{ij}}\right)_{\eta'} = -\frac{1}{3}\left(\frac{\partial \ln V_A}{\partial \eta_{ij}}\right)_{\eta'} + \frac{\langle [C(\hat{\mathbf{k}}s)]^{-3}[\partial \ln C(\hat{\mathbf{k}}s)/\partial \eta_{ij}]_{\eta'}\rangle}{\langle [C(\hat{\mathbf{k}}s)]^{-3}\rangle}. \tag{18.58}$$

Here the velocities and their strain derivatives are evaluated at $T = 0$. For a lattice-dynamical calculation the first approximation to be tested should generally be the harmonic approximation, where the acoustic phonon velocities and their strain derivatives are the bare velocities $c(\hat{\mathbf{k}}s)$, and the derivatives $\sigma_{ij}(\hat{\mathbf{k}}s)$ defined by (12.48). Thus in the harmonic approximation one has

$$\left(\frac{\partial \ln \Theta_{H0}}{\partial \eta_{ij}}\right)_{\eta'} = -\frac{1}{3}\left(\frac{\partial \ln V_A}{\partial \eta_{ij}}\right)_{\eta'} - \frac{\langle [c(\hat{\mathbf{k}}s)]^{-3}\sigma_{ij}(\hat{\mathbf{k}}s)\rangle}{\langle [c(\hat{\mathbf{k}}s)]^{-3}\rangle}. \tag{18.59}$$

The $\sigma_{ij}(\hat{\mathbf{k}}s)$ are easily evaluated from (12.50), and the angle averages may be carried out by straightforward numerical integration, as discussed in more detail in Appendix 1.

In order to calculate Θ_0 and its strain derivatives from the measured propagation coefficients and their strain derivatives, we begin by noting that the sound velocities v_s are given by [see e.g., (3.27)]

$$v_s^2 = \rho^{-1} \sum_{ik} w_{is} L_{ik} w_{ks}, \tag{18.60}$$

where ρ is the crystal density, and the v_s are understood to depend on the direction $\hat{\mathbf{k}}$. The propagation matrix elements are, from (3.23),

$$L_{ik} = \sum_{jl} A_{ijkl} \hat{k}_j \hat{k}_l. \tag{18.61}$$

Then with the eigenvalue derivative theorem (11.73), the strain derivatives of the velocities can be calculated at all angles $\hat{\mathbf{k}}$ from the measured set of strain derivatives of the propagation coefficients, according to

$$(\partial \ln v_s/\partial \eta_{mn})_{\eta'} = -\tfrac{1}{2}(\partial \ln \rho/\partial \eta_{mn})_{\eta'} + (2\rho v_s^2)^{-1} \sum_{ik} w_{is} (\partial L_{ik}/\partial \eta_{mn})_{\eta'} w_{ks}, \tag{18.62}$$

where

$$(\partial L_{ik}/\partial \eta_{mn})_{\eta'} = \sum_{jl} (\partial A_{ijkl}/\partial \eta_{mn})_{\eta'} \hat{k}_j \hat{k}_l. \tag{18.63}$$

Note that the direction $\hat{\mathbf{k}}$ has been taken to be fixed in the differentiation [see the discussion following (12.51)], and in order to calculate Θ_0 and its strain derivatives it is understood that (18.60)–(18.63) are to be evaluated from $T \approx 0$ measurements. Experimental determination of the propagation coefficients and their strain derivatives is discussed in more detail elsewhere.*

* D. C. Wallace, in *Solid State Physics*, edited by H. Ehrenreich, F. Seitz, and D. Turnbull, Academic Press Inc., New York, 1970, Vol. 25, p. 301.

Cubic Crystals

For the example of a cubic crystal under isotropic pressure there is a great deal of simplification of our equations. All the preceding equations of this section are of course still valid, but the stress and strain tensors all simplify to diagonal tensors with only one independent component in each, as shown in (2.115) and (2.116). The stress is $\tau_{ij} = -P\delta_{ij}$, and the pressure is given by

$$P = -(\partial F/\partial V)_T = P_0 - V_A^{-1}(3\pi^4/5)(KT)(T/\Theta_0)^3(d \ln \Theta_0/d \ln V), \quad (18.64)$$

where P_0 is the value at $T = 0$. From this it is seen that the pressure required to hold the volume fixed goes as T^4 at low temperatures. The independent component of b_{ij} is $-\beta B_T$, and it is given by

$$\beta B_T = -(\partial^2 F/\partial V \partial T)_{TV} = -V_A^{-1}K(12\pi^4/5)(T/\Theta_0)^3(d \ln \Theta_0/d \ln V). \quad (18.65)$$

The leading low-temperature contribution to the thermal expansion coefficient β may be written

$$\beta = \beta B_T/B_0, \quad (18.66)$$

where βB_T is given by (18.65) and B_0 is the bulk modulus at $T = 0$. Thus β goes as T^3 at low temperatures, and at constant pressure the crystal volume goes as T^4. The macroscopic Grüneisen parameter is $\gamma = V\beta B_T/C_V$, and it has the low-temperature limit

$$\gamma = -(d \ln \Theta_0/d \ln V), \quad \text{at } T = 0; \quad (18.67)$$

this is the same as the Debye result (5.38), where it may be noted that in the Debye approximation this result applies at all temperatures.

The bulk modulus can be calculated directly for a cubic crystal under isotropic pressure, because the only configuration variable is the volume. The isothermal bulk modulus is

$$B_T = V(\partial^2 F/\partial V^2)_T = B_0 + V_A^{-1}(3\pi^4/5)(KT)(T/\Theta_0)^3[(V^2/\Theta_0)(d^2\Theta_0/dV^2) - 4(d \ln \Theta_0/d \ln V)^2], \quad (18.68)$$

where

$$B_0 = V(d^2\mathscr{E}_G/dV^2). \quad (18.69)$$

A convenient form for $B_S - B_T$ is written in (16.84), and in the low-temperature limit it is

$$B_S - B_T = V_A^{-1}(12\pi^4/5)(KT)(T/\Theta_0)^3(d \ln \Theta_0/d \ln V)^2; \quad (18.70)$$

or with the aid of (18.68),

$$B_S = B_0 + V_A^{-1}(3\pi^4/5)(KT)(T/\Theta_0)^3(V^2/\Theta_0)(d^2\Theta_0/dV^2). \quad (18.71)$$

Again B_T and B_S go as T^4 at constant volume, according to (18.68) and (18.71), and at constant pressure there is an added T^4 term due to the volume variation of B_0. The volume variation of B_0 at constant pressure is calculated as follows, in the low-temperature limit.

$$\begin{aligned} B_0(V(T)) &= B_0(V_0) + (V - V_0)(dB_0/dV) \\ &= B_0(V_0) + (TV_0\beta)(-B_0/V_0)(dB_0/dP) \\ &= B_0(V_0) - T\beta B_T(dB_0/dP), \end{aligned} \tag{18.72}$$

where βB_T is given by (18.65). Thus B_S, for example, at constant pressure is to order T^4,

$$B_S(P = 0) = B_0(V_0) + V_A^{-1}(3\pi^4/5)(KT)(T/\Theta_0)^3[(V^2/\Theta_0)(d^2\Theta_0/dV^2) + 4(d \ln \Theta_0/d \ln V)(dB_0/dP)]. \tag{18.73}$$

The logarithmic volume derivative of Θ_0 is calculated from (18.29) for Θ_0 as

$$\frac{d \ln \Theta_0}{d \ln V} = \frac{\langle [C(\hat{\mathbf{k}}s)]^{-3}[-\frac{1}{3} + d \ln C(\hat{\mathbf{k}}s)/d \ln V]\rangle}{\langle [C(\hat{\mathbf{k}}s)]^{-3}\rangle}, \tag{18.74}$$

where the $-\frac{1}{3}$ arises from differentiation of the volume V_A in (18.29). In the harmonic approximation this is simply

$$\frac{d \ln \Theta_{H0}}{d \ln V} = -\frac{\langle [c(\hat{\mathbf{k}}s)]^{-3}[\frac{1}{3} + \sigma(\hat{\mathbf{k}}s)]\rangle}{\langle [c(\hat{\mathbf{k}}s)]^{-3}\rangle}, \tag{18.75}$$

where $\sigma(\hat{\mathbf{k}}s)$ is the logarithmic volume derivative of $c(\hat{\mathbf{k}}s)$, and may be calculated from (12.51).

For a cubic crystal under isotropic pressure, the propagation matrix elements are written out in (3.39) in terms of the three independent stress–strain coefficients B_{11}, B_{12}, and B_{44}; note that at zero pressure $B_{\alpha\beta} = C_{\alpha\beta}$. By a calculation similar to that leading to (18.62) for the strain derivatives of the sound velocities v_s, we find

$$d \ln v_s/d \ln V = \tfrac{1}{2} + (2\rho v_s^2)^{-1} \sum_{ij} w_{is}(V\, dL_{ij}/dV) w_{js}, \tag{18.76}$$

where the $\frac{1}{2}$ arises from differentiation of ρ^{-1}. It is convenient to transform the volume derivative of L_{ij} to pressure derivatives of the $B_{\alpha\beta}$, with the aid of the general relation (1.51), since the latter are generally reported as the results of experimental measurements; note that at zero pressure $dB_{\alpha\beta}/dP \neq dC_{\alpha\beta}/dP$. Then the experimental evaluation of the quantities $-\frac{1}{3} + d \ln C(\hat{\mathbf{k}}s)/d \ln V$ which appear in (18.74) may be accomplished as follows.

$$-\tfrac{1}{3} + d \ln v_s/d \ln V = \tfrac{1}{6} - (B_0/2\rho v_s^2) \sum_{ij} w_{is}(dL_{ij}/dP) w_{js}, \tag{18.77}$$

where

$$dL_{ii}/dP = (dB_{11}/dP)\hat{k}_i^2 + (dB_{44}/dP)(1 - \hat{k}_i^2),$$
$$dL_{ij}/dP = [(dB_{12}/dP) + (dB_{44}/dP)]\hat{k}_i\hat{k}_j, \quad i \neq j. \tag{18.78}$$

Again it is understood that for evaluation of Θ_0 and of $d \ln \Theta_0/d \ln V$ the measured $B_{\alpha\beta}$ and $dB_{\alpha\beta}/dP$ are at $T \approx 0$.

19. HIGH-TEMPERATURE LIMIT

Anharmonic Free Energy

In the high-temperature limit the anharmonic free energy (16.25) simplifies considerably. The high-temperature limit is here defined by the condition

$$KT > \hbar\omega_\kappa, \quad \text{for all } \kappa. \tag{19.1}$$

In this limit the statistical average phonon occupation number $\bar{n}_\kappa$ may be expanded in powers of $\hbar\omega_\kappa/KT$; from (16.16) for $\bar{n}_\kappa$ the result is

$$\bar{n}_\kappa = (KT/\hbar\omega_\kappa) - \tfrac{1}{2} + \tfrac{1}{12}(\hbar\omega_\kappa/KT) - \tfrac{1}{720}(\hbar\omega_\kappa/KT)^3 + \cdots. \tag{19.2}$$

It is seen that each $\bar{n}_\kappa$ increases as T at high temperatures. The expansion of F_A is straightforward but lengthy; the procedure will be illustrated by expanding the expression in square brackets in (16.25), which is the function

$$\frac{(\bar{n}_\kappa\bar{n}_{\kappa'} + \bar{n}_\kappa + \tfrac{1}{3})}{(\omega_\kappa + \omega_{\kappa'} + \omega_{\kappa''})_p} + \frac{(2\bar{n}_\kappa\bar{n}_{\kappa''} - \bar{n}_\kappa\bar{n}_{\kappa'} + \bar{n}_{\kappa''})}{(\omega_\kappa + \omega_{\kappa'} - \omega_{\kappa''})_p}. \tag{19.3}$$

This function appears inside $\Sigma_{\kappa\kappa'\kappa''}$, with the coefficient $-18\hbar^{-1}\,|\Phi_{\kappa\kappa'\kappa''}|^2$ being completely symmetric in κ,κ',κ''. Therefore the indices may be interchanged at will in (19.3), to produce the more convenient expression

$$\frac{(\bar{n}_\kappa + \tfrac{1}{2})(\bar{n}_{\kappa'} + \tfrac{1}{2}) + \tfrac{1}{12}}{(\omega_\kappa + \omega_{\kappa'} + \omega_{\kappa''})_p} + \frac{2(\bar{n}_{\kappa''} + \tfrac{1}{2})(\bar{n}_\kappa + \tfrac{1}{2}) - (\bar{n}_{\kappa'} + \tfrac{1}{2})(\bar{n}_\kappa + \tfrac{1}{2}) - \tfrac{1}{4}}{(\omega_\kappa + \omega_{\kappa'} - \omega_{\kappa''})_p}. \tag{19.4}$$

From (19.2), the high-temperature expansion of $\bar{n}_\kappa + \frac{1}{2}$ to order T^{-3} may be written

$$\bar{n}_\kappa + \tfrac{1}{2} = (KT/\hbar\omega_\kappa)[1 + \tfrac{1}{12}(\hbar\omega_\kappa/KT)^2 - \tfrac{1}{720}(\hbar\omega_\kappa/KT)^4], \tag{19.5}$$

and with this the first contribution in (19.4) is expanded as

$$\begin{aligned}(KT/\hbar)^2(\omega_\kappa\omega_{\kappa'})^{-1}&(\omega_\kappa + \omega_{\kappa'} + \omega_{\kappa''})_p^{-1} \\ &\times [1 + \tfrac{1}{12}(\hbar/KT)^2(\omega_\kappa^2 + \omega_{\kappa'}^2 + \omega_\kappa\omega_{\kappa'}) \\ &\qquad - \tfrac{1}{720}(\hbar/KT)^4(\omega_\kappa^4 + \omega_{\kappa'}^4 - 5\omega_\kappa^2\omega_{\kappa'}^2)].\end{aligned} \tag{19.6}$$

Now the trick is to put $\omega_{\kappa''}$ into both numerator and denominator of (19.6), and then since the coefficient $(\omega_\kappa\omega_{\kappa'}\omega_{\kappa''})^{-1}(\omega_\kappa + \omega_{\kappa'} + \omega_{\kappa''})_p^{-1}$ is symmetric in κ,κ',κ'', the indices may be interchanged at will in the numerator. After

$\omega_{\kappa''}$ is multiplied into the numerator of (19.6), the following interchanges simplify the function.

$$\begin{aligned} &\omega_{\kappa''} \to \tfrac{1}{3}(\omega_\kappa + \omega_{\kappa'} + \omega_{\kappa''}); \\ &\quad \omega_{\kappa''}(\omega_\kappa^2 + \omega_{\kappa'}^2 + \omega_\kappa\omega_{\kappa'}) \to \omega_\kappa\omega_{\kappa'}(\omega_\kappa + \omega_{\kappa'} + \omega_{\kappa''}); \\ &\quad \omega_{\kappa''}(\omega_\kappa^4 + \omega_{\kappa'}^4 - 5\omega_\kappa^2\omega_{\kappa'}^2) \\ &\qquad \to (2\omega_\kappa^3\omega_{\kappa'} - \omega_\kappa^2\omega_{\kappa'}^2 - 2\omega_\kappa^2\omega_{\kappa'}\omega_{\kappa''})(\omega_\kappa + \omega_{\kappa'} + \omega_{\kappa''}). \end{aligned} \tag{19.7}$$

Then since $(\omega_\kappa + \omega_{\kappa'} + \omega_{\kappa''})(\omega_\kappa + \omega_{\kappa'} + \omega_{\kappa''})_p^{-1} = 1$, the expansion (19.6) of the first contribution in (19.4) is

$$\begin{aligned} &(KT/\hbar)^2(\omega_\kappa\omega_{\kappa'}\omega_{\kappa''})^{-1}[\tfrac{1}{3} + \tfrac{1}{12}(\hbar/KT)^2\omega_\kappa\omega_{\kappa'} \\ &\qquad - \tfrac{1}{720}(\hbar/KT)^4(2\omega_\kappa^3\omega_{\kappa'} - \omega_\kappa^2\omega_{\kappa'}^2 - 2\omega_\kappa^2\omega_{\kappa'}\omega_{\kappa''})]. \end{aligned} \tag{19.8}$$

The second contribution in (19.4) is treated in a similar way, with $\omega_{\kappa'}$ or $\omega_{\kappa''}$ introduced in numerator and denominator, and indices interchanged in the numerator to produce the factor $\omega_\kappa + \omega_{\kappa'} - \omega_{\kappa''}$ in each term in the expansion. Then since $(\omega_\kappa + \omega_{\kappa'} - \omega_{\kappa''})(\omega_\kappa + \omega_{\kappa'} - \omega_{\kappa''})_p^{-1} = 1$, the second contribution in (19.4) becomes

$$\begin{aligned} &(KT/\hbar)^2(\omega_\kappa\omega_{\kappa'}\omega_{\kappa''})^{-1}[1 - \tfrac{1}{12}(\hbar/KT)^2\omega_\kappa\omega_{\kappa'} \\ &\qquad - \tfrac{1}{720}(\hbar/KT)^4(-2\omega_\kappa^3\omega_{\kappa'} - 3\omega_\kappa^2\omega_{\kappa'}^2 + 2\omega_\kappa^2\omega_{\kappa'}\omega_{\kappa''})]. \end{aligned} \tag{19.9}$$

The sum of (19.8) and (19.9) gives the complete expansion to order T^{-2} of (19.4):

$$(KT/\hbar)^2(\omega_\kappa\omega_{\kappa'}\omega_{\kappa''})^{-1}[\tfrac{4}{3} + \tfrac{4}{720}(\hbar/KT)^4\omega_\kappa^2\omega_{\kappa'}^2]. \tag{19.10}$$

With all the contributions in (16.25) expanded, the total high-temperature expansion of F_A correct to order T^{-2} is

$$\begin{aligned} F_A = {} & \frac{12(KT)^2}{\hbar^2}\sum_{\kappa\kappa'}\frac{\Phi_{\kappa,-\kappa,\kappa',-\kappa'}}{\omega_\kappa\omega_{\kappa'}}\left\{1 + \frac{1}{6}\left(\frac{\hbar\omega_\kappa}{KT}\right)^2\right. \\ & \left. - \frac{1}{720}\left[2\left(\frac{\hbar\omega_\kappa}{KT}\right)^4 - 5\left(\frac{\hbar\omega_\kappa}{KT}\right)^2\left(\frac{\hbar\omega_{\kappa'}}{KT}\right)^2\right]\right\} \\ & - \frac{24(KT)^2}{\hbar^3}\sum_{\kappa\kappa'\kappa''}\frac{|\Phi_{\kappa\kappa'\kappa''}|^2}{\omega_\kappa\omega_{\kappa'}\omega_{\kappa''}}\left[1 + \frac{1}{240}\left(\frac{\hbar\omega_\kappa}{KT}\right)^2\left(\frac{\hbar\omega_{\kappa'}}{KT}\right)^2\right] \\ & - \frac{36(KT)^2}{\hbar^3}\sum_{\kappa\kappa'\kappa''}\frac{\Phi_{\kappa,-\kappa,\kappa''}\Phi_{\kappa',-\kappa',-\kappa''}}{\omega_\kappa\omega_{\kappa'}\omega_{\kappa''}}\left\{1 + \frac{1}{6}\left(\frac{\hbar\omega_\kappa}{KT}\right)^2\right. \\ & \left. - \frac{1}{720}\left[2\left(\frac{\hbar\omega_\kappa}{KT}\right)^4 - 5\left(\frac{\hbar\omega_\kappa}{KT}\right)^2\left(\frac{\hbar\omega_{\kappa'}}{KT}\right)^2\right]\right\}. \end{aligned} \tag{19.11}$$

In accordance with the definition (16.9) of F_{A}, this expression includes the fourth-order ground-state energy G_4. It is convenient to abbreviate the expansion (19.11) by

$$F_{\mathrm{A}} = \mathscr{A}_2 T^2 + \mathscr{A}_0 + \mathscr{A}_{-2} T^{-2} + \cdots, \tag{19.12}$$

where the $\mathscr{A}$ coefficients depend only on the crystal configuration. Explicit expressions for the $\mathscr{A}$ coefficients may be written from (19.11); for example,

$$\begin{aligned}\mathscr{A}_2 = {} & 12K^2\hbar^{-2} \sum_{\kappa\kappa'} \Phi_{\kappa,-\kappa,\kappa',-\kappa'}(\omega_\kappa \omega_{\kappa'})^{-1} \\ & - 24K^2\hbar^{-3} \sum_{\kappa\kappa'\kappa''} [|\Phi_{\kappa\kappa'\kappa''}|^2 + \tfrac{3}{2}\Phi_{\kappa,-\kappa,\kappa''}\Phi_{\kappa',-\kappa',-\kappa''}](\omega_\kappa\omega_{\kappa'}\omega_{\kappa''})^{-1}.\end{aligned} \tag{19.13}$$

The condition that the expansion (19.11) be convergent when carried to all orders of $\hbar\omega_\kappa/KT$ is the condition (19.1); however in a detailed calculation the result for F_{A} will be accurate if the series (19.12) is found to converge well for the first few terms, and this should be the case as long as most of the frequencies ω_κ satisfy $\hbar\omega_\kappa < KT$. It is shown below that the harmonic free energy F_{H} goes as $T \ln T$ at high temperatures, and since F_{A} goes as T^2, F_{A} will become as large in magnitude as F_{H} at sufficiently high temperatures; the leading-order perturbation treatment of anharmonicity is clearly invalid at such high temperatures. It is not possible in general to determine the sign of $\mathscr{A}_2$ without a detailed evaluation of (19.13); in fact in model calculations which have been carried out the quartic term (involving $\Phi_{\kappa,-\kappa,\kappa',-\kappa'}$) has been found to be positive, while the cubic term involving $|\Phi_{\kappa\kappa'\kappa''}|^2$ is always negative, so it becomes a severe test of a theoretical calculation to determine the sign, let alone the magnitude, of $\mathscr{A}_2$.

There is a further possible simplification of the $\mathscr{A}$ coefficients, which is provided by introducing the matrices $\mathbf{E}(\mathbf{k})$ which are inverses to the dynamical matrices $\mathbf{D}(\mathbf{k})$. The algebra is illustrated by writing out in detail the quartic term in (19.11), with the definition (10.75) of $\Phi_{\kappa,-\kappa,\kappa',-\kappa'} = \Phi(\mathbf{k}s,-\mathbf{k}s,\mathbf{k}'s',-\mathbf{k}'s')$, as follows.

$$\begin{aligned}\tfrac{1}{8}(KT)^2 N_0^{-2} & \sum_{\mathrm{MNPQ}\mu\nu\pi\rho} \sum_{ijkl} \Phi_{ijkl}(\mathrm{M}\mu,\mathrm{N}\nu,\mathrm{P}\pi,\mathrm{Q}\rho)(M_\mu M_\nu M_\pi M_\rho)^{-1/2} \\ & \times \sum_{\mathbf{k}\mathbf{k}'ss'} e^{i\mathbf{k}\cdot[\mathbf{R}(\mathrm{M}\mu)-\mathbf{R}(\mathrm{N}\nu)]} e^{i\mathbf{k}'\cdot[\mathbf{R}(\mathrm{P}\pi)-\mathbf{R}(\mathrm{Q}\rho)]} \\ & \times w_i(\mu,\mathbf{k}s) w_j(\nu,-\mathbf{k}s) w_k(\pi,\mathbf{k}'s') w_l(\rho,-\mathbf{k}'s') \\ & \times \{[\omega(\mathbf{k}s)\omega(\mathbf{k}'s')]^{-2} + \tfrac{1}{6}(\hbar/KT)^2[\omega(\mathbf{k}'s')]^{-2} \\ & \quad - \tfrac{2}{720}(\hbar/KT)^4[\omega(\mathbf{k}s)]^2[\omega(\mathbf{k}'s')]^{-2} + \tfrac{5}{720}(\hbar/KT)^4\}.\end{aligned} \tag{19.14}$$

Now $\omega(\mathbf{k}s)$ appears in (19.14) in powers of -2, 0, $+2$, and for each term the sum over s can be carried out with the aid of (10.59) for $\mathbf{E}(\mathbf{k})$, or with the completeness relation (10.36), or with the expression (10.57) for **D(k)**. Likewise the sum over s' can be carried out, and (19.14) is transformed to

$$\tfrac{1}{8}(KT)^2N_0^{-2} \sum_{\mathrm{MNPQ}\mu\nu\pi\rho} \sum_{ijkl} \Phi_{ijkl}(\mathrm{M}\mu,\mathrm{N}\nu,\mathrm{P}\pi,\mathrm{Q}\rho)(M_\mu M_\nu M_\pi M_\rho)^{-1/2}$$
$$\times \sum_{\mathbf{kk}'} e^{i\mathbf{k}\cdot[\mathbf{R}(\mathrm{M}\mu)-\mathbf{R}(\mathrm{N}\nu)]}e^{i\mathbf{k}'\cdot[\mathbf{R}(\mathrm{P}\pi)-\mathbf{R}(\mathrm{Q}\rho)]} \tag{19.15}$$
$$\times \{E_{ij}(\mu\nu,\mathbf{k})E_{kl}(\pi\rho,\mathbf{k}') + \tfrac{1}{6}(\hbar/KT)^2\delta_{\mu\nu}\delta_{ij}E_{kl}(\pi\rho,\mathbf{k}')$$
$$- \tfrac{1}{720}(\hbar/KT)^4[2D_{ij}(\mu\nu,\mathbf{k})E_{kl}(\pi\rho,\mathbf{k}') - 5\delta_{\mu\nu}\delta_{\pi\rho}\delta_{ij}\delta_{kl}]\}.$$

The sums over **k** and **k**′ in (19.15) can now be carried out. By the lattice-vector translational invariance (6.57) of the second-order potential energy coefficients, it is possible to show that

$$N_0^{-1}\sum_{\mathbf{k}} D_{ij}(\mu\nu,\mathbf{k})e^{i\mathbf{k}\cdot[\mathbf{R}(\mathrm{M}\mu)-\mathbf{R}(\mathrm{N}\nu)]} = (M_\mu M_\nu)^{-1/2}\Phi_{ij}(\mathrm{M}\mu,\mathrm{N}\nu). \tag{19.16}$$

It is also convenient to define the Fourier-transformed inverse dynamical matrices by the function

$$J_{ij}(\mathrm{M}\mu,\mathrm{N}\nu) = N_0^{-1}\sum_{\mathbf{k}} E_{ij}(\mu\nu,\mathbf{k})e^{i\mathbf{k}\cdot[\mathbf{R}(\mathrm{M}\mu)-\mathbf{R}(\mathrm{N}\nu)]}. \tag{19.17}$$

Then the sums over **k** and **k**′ in (19.15) contain either (19.16) or (19.17), or else they may be reduced to deltas by observing that

$$N_0^{-1}\sum_{\mathbf{k}} e^{i\mathbf{k}\cdot[\mathbf{R}(\mathrm{M}\mu)-\mathbf{R}(\mathrm{N}\nu)]}\delta_{\mu\nu} = N_0^{-1}\sum_{\mathbf{k}} e^{i\mathbf{k}\cdot[\mathbf{R}(\mathrm{M})-\mathbf{R}(\mathrm{N})]}\delta_{\mu\nu} = \delta_{\mathrm{MN}}\delta_{\mu\nu}. \tag{19.18}$$

With (19.16)–(19.18) for the wave vector sums, (19.15) becomes

$$\tfrac{1}{8}(KT)^2 \sum_{\mathrm{MNPQ}\mu\nu\pi\rho} \sum_{ijkl} \Phi_{ijkl}(\mathrm{M}\mu,\mathrm{N}\nu,\mathrm{P}\pi,\mathrm{Q}\rho)(M_\mu M_\nu M_\pi M_\rho)^{-1/2}$$
$$\times \{J_{ij}(\mathrm{M}\mu,\mathrm{N}\nu)J_{kl}(\mathrm{P}\pi,\mathrm{Q}\rho) + \tfrac{1}{6}(\hbar/KT)^2\delta_{\mathrm{MN}}\delta_{\mu\nu}\delta_{ij}J_{kl}(\mathrm{P}\pi,\mathrm{Q}\rho)$$
$$- \tfrac{1}{720}(\hbar/KT)^4[2(M_\mu M_\nu)^{-1/2}\Phi_{ij}(\mathrm{M}\mu,\mathrm{N}\nu)J_{kl}(\mathrm{P}\pi,\mathrm{Q}\rho) \tag{19.19}$$
$$- 5\delta_{\mathrm{MN}}\delta_{\mathrm{PQ}}\delta_{\mu\nu}\delta_{\pi\rho}\delta_{ij}\delta_{kl}]\}.$$

The remaining (cubic) terms in (19.11) for F_{A} can be transformed in the same way as the quartic term above. The explicit result for $\mathscr{A}_2$ is

$$\mathscr{A}_2 = (K^2/8) \sum_{\mathrm{MNPQ}\mu\nu\pi\rho} \sum_{ijkl} \Phi_{ijkl}(\mathrm{M}\mu,\mathrm{N}\nu,\mathrm{P}\pi,\mathrm{Q}\rho)$$
$$\times (M_\mu M_\nu M_\pi M_\rho)^{-1/2}J_{ij}(\mathrm{M}\mu,\mathrm{N}\nu)J_{kl}(\mathrm{P}\pi,\mathrm{Q}\rho)$$
$$- (K^2/12) \sum_{\mathrm{MM'PP'QQ'}} \sum_{\mu\mu'\nu\nu'\pi\pi'} \sum_{ii'jj'kk'} \Phi_{ijk}(\mathrm{M}\mu,\mathrm{N}\nu,\mathrm{P}\pi) \tag{19.20}$$
$$\times \Phi_{i'j'k'}(\mathrm{M}'\mu',\mathrm{N}'\nu',\mathrm{P}'\pi')(M_\mu M_{\mu'} M_\nu M_{\nu'} M_\pi M_{\pi'})^{-1/2}$$
$$\times [J_{ii'}(\mathrm{M}\mu,\mathrm{M}'\mu')J_{jj'}(\mathrm{N}\nu,\mathrm{N}'\nu')J_{kk'}(\mathrm{P}\pi,\mathrm{P}'\pi')$$
$$+ \tfrac{3}{2}J_{ij}(\mathrm{M}\mu,\mathrm{N}\nu)J_{i'j'}(\mathrm{M}'\mu',\mathrm{N}'\nu')J_{kk'}(\mathrm{P}\pi,\mathrm{P}'\pi')].$$

The last term in (19.20), the one involving the factor $\frac{3}{2}$, does not appear for a primitive lattice, since it arises from the optic-mode contribution to (16.25). Equation (19.20) is probably the most convenient form for a theoretical calculation of the high-temperature anharmonic coefficient $\mathscr{A}_2$. Surface effects may be eliminated in each term by setting one of the unit cell indices, say M, equal to zero and replacing Σ_{M} by N_0; note this shows that each term in (19.20) is proportional to N_0. The sums over lattice points in (19.20) then converge rapidly in summing out from the origin. It should also be noted that the $\delta(\mathbf{k} + \mathbf{k}' + \mathbf{k}'')$ which appears in $\Phi(\mathbf{k}s,\mathbf{k}'s',\mathbf{k}''s'')$ is properly contained in the expression (19.20). Finally it may be noted that $D_{ij}(\mu\nu,\mathbf{k})$ is proportional to $(M_\mu M_\nu)^{-1/2}$, so that $E_{ij}(\mu\nu,\mathbf{k})$ and hence also $J_{ij}(\mathrm{M}\mu,\mathrm{N}\nu)$ is proportional to $(M_\mu M_\nu)^{1/2}$, and the mass dependence completely disappears from $\mathscr{A}_2$; this is to be expected since the high-temperature limit is a classical limit.

There is sufficient simplification for a primitive lattice to warrant rewriting some of our results explicitly for that case. The Fourier-transformed inverse dynamical matrix elements are

$$J_{ij}(\mathrm{M},\mathrm{N}) = N_0^{-1} \sum_{\mathbf{k}} E_{ij}(\mathbf{k}) e^{i\mathbf{k}\cdot[\mathbf{R}(\mathrm{M})-\mathbf{R}(\mathrm{N})]}, \tag{19.21}$$

and by lattice-vector translational invariance (in the interior) it follows that

$$J_{ij}(\mathrm{M},\mathrm{N}) = J_{ij}(0,\mathrm{N} - \mathrm{M}). \tag{19.22}$$

Further since $\mathbf{D}(\mathbf{k}) = \mathbf{D}(-\mathbf{k})$ for a primitive lattice, then $\mathbf{E}(\mathbf{k}) = \mathbf{E}(-\mathbf{k})$, and the J coefficients may be written

$$J_{ij}(0,\mathrm{N}) = N_0^{-1} \sum_{\mathbf{k}} E_{ij}(\mathbf{k}) \cos[\mathbf{k} \cdot \mathbf{R}(\mathrm{N})]. \tag{19.23}$$

The primitive lattice results for $\mathscr{A}_2$ and $\mathscr{A}_0$ are

$$\begin{aligned}\mathscr{A}_2 = {} & (K^2/8M_C^2) \sum_{\mathrm{MNPQ}} \sum_{ijkl} \Phi_{ijkl}(\mathrm{M},\mathrm{N},\mathrm{P},\mathrm{Q}) J_{ij}(\mathrm{M},\mathrm{N}) J_{kl}(\mathrm{P},\mathrm{Q}) \\ & - (K^2/12M_C^3) \sum_{\mathrm{MM'NN'PP'}} \sum_{ii'jj'kk'} \Phi_{ijk}(\mathrm{M},\mathrm{N},\mathrm{P}) \Phi_{i'j'k'}(\mathrm{M}',\mathrm{N}',\mathrm{P}') \\ & \times J_{ii'}(\mathrm{M},\mathrm{M}') J_{jj'}(\mathrm{N},\mathrm{N}') J_{kk'}(\mathrm{P},\mathrm{P}');\end{aligned} \tag{19.24}$$

$$\mathscr{A}_0 = (\hbar^2/48M_C^2) \sum_{\mathrm{MPQ}} \sum_{ikl} \Phi_{iikl}(\mathrm{M},\mathrm{M},\mathrm{P},\mathrm{Q}) J_{kl}(\mathrm{P},\mathrm{Q}). \tag{19.25}$$

Thermodynamic Functions

In the high-temperature limit the harmonic free energy (16.13) is expanded into

$$F_{\mathrm{H}} = -KT \sum_{\kappa} [\ln(KT/\hbar\omega_\kappa) - \tfrac{1}{24}(\hbar\omega_\kappa/KT)^2 + \tfrac{1}{2880}(\hbar\omega_\kappa/KT)^4 - \cdots]. \tag{19.26}$$

From this the high-temperature expansions of the harmonic contributions to the entropy and the heat capacity at constant configuration are

$$S_{\mathrm{H}} = K \sum_{\kappa} [\ln(KT/\hbar\omega_\kappa) + 1 + \tfrac{1}{24}(\hbar\omega_\kappa/KT)^2 - \tfrac{1}{960}(\hbar\omega_\kappa/KT)^4 + \cdots]; \tag{19.27}$$

$$C_{\eta\mathrm{H}} = K \sum_{\kappa} [1 - \tfrac{1}{12}(\hbar\omega_\kappa/KT)^2 + \tfrac{1}{240}(\hbar\omega_\kappa/KT)^4 - \cdots]. \tag{19.28}$$

The term $K\,\Sigma_\kappa\, 1$ which appears in S_{H} and $C_{\eta\mathrm{H}}$ is

$$K \sum_{\kappa} 1 = 3nN_0 K, \tag{19.29}$$

since there are $3nN_0$ terms in $\Sigma_\kappa = \Sigma_{\mathbf{k}s}$. At high temperatures S_{H} goes as $\ln T$, and the $-TS_{\mathrm{H}}$ term is the leading high-temperature contribution to F_{H}. The heat capacity $C_{\eta\mathrm{H}}$ approaches the classical limit $3nN_0K$ for a crystal with nN_0 atoms, at high temperatures.

The average harmonic phonon frequency squared is defined by [see also (10.62)]

$$\langle\omega_\kappa^2\rangle = (3nN_0)^{-1} \sum_{\kappa} \omega_\kappa^2, \tag{19.30}$$

and from (19.28), $C_{\eta\mathrm{H}}$ approaches the classical limit in the manner

$$C_{\eta\mathrm{H}} = 3nN_0K[1 - \tfrac{1}{12}(\hbar^2\langle\omega_\kappa^2\rangle/K^2T^2) + \cdots]. \tag{19.31}$$

If we define the high-temperature harmonic Debye temperature $\Theta_{\mathrm{H}\infty}$ by

$$K\Theta_{\mathrm{H}\infty} = [\tfrac{5}{3}\hbar^2\langle\omega_\kappa^2\rangle]^{1/2}, \tag{19.32}$$

then $C_{\eta\mathrm{H}}$ has the same form as the Debye heat capacity (5.37), namely

$$C_{\eta\mathrm{H}} = 3nN_0K[1 - \tfrac{1}{20}(\Theta_{\mathrm{H}\infty}/T)^2 + \cdots]. \tag{19.33}$$

Thus in the harmonic theory of lattice dynamics, the effective Debye temperature for the heat capacity is $\Theta_{\mathrm{H}\infty}$ of (19.32) at high temperatures, $\Theta_{\mathrm{H}0}$ of (18.17) at $T = 0$, and is a function of temperature at intermediate temperatures. Incidentally, the condition $T \geq \Theta_{\mathrm{H}\infty}$ serves as a useful qualitative measure of the high-temperature region.

It is convenient to discuss the anharmonic contributions to the thermodynamic functions in the high-temperature region in terms of the θ_κ, $\theta_{\kappa\kappa'}$ functions which contribute to the phonon frequency shifts Δ_κ. From (16.34) for Δ_κ, and the expansion (19.5) for $\bar{n}_{\kappa'} + \frac{1}{2}$, the frequency shifts are

$$\Delta_\kappa = \sum_{\kappa'} \theta_{\kappa\kappa'}[(KT/\hbar\omega_{\kappa'}) + \tfrac{1}{12}(\hbar\omega_{\kappa'}/KT) - \cdots]. \tag{19.34}$$

Thus at constant configuration the renormalized phonon frequencies $\Omega_\kappa = \omega_\kappa + \Delta_\kappa$ vary as T as high temperatures. The anharmonic free energy is written in terms of $\theta_\kappa, \theta_{\kappa\kappa'}$ in (16.36), and when that expression is expanded into

$$F_A = \mathscr{A}_2 T^2 + \mathscr{A}_0 + \mathscr{A}_{-2} T^{-2},$$

we find

$$\mathscr{A}_2 = (K^2/2\hbar) \sum_{\kappa\kappa'} (\theta_{\kappa\kappa'}/\omega_\kappa \omega_{\kappa'}); \tag{19.35}$$

$$\mathscr{A}_0 = G_4 - \tfrac{1}{4} \sum_{\kappa} \hbar\theta_\kappa + \tfrac{1}{12} \sum_{\kappa\kappa'} \hbar\theta_{\kappa\kappa'}(\omega_\kappa/\omega_{\kappa'}); \tag{19.36}$$

$$\mathscr{A}_{-2} = -(\hbar^3/1440K^2) \sum_{\kappa\kappa'} \theta_{\kappa\kappa'}[2(\omega_\kappa^3/\omega_{\kappa'}) - 5\omega_\kappa\omega_{\kappa'}]. \tag{19.37}$$

Then the anharmonic contributions to the entropy and heat capacity at high temperatures are

$$S_A = -2\mathscr{A}_2 T + 2\mathscr{A}_{-2} T^{-3} + \cdots; \tag{19.38}$$

$$C_{\eta A} = -2\mathscr{A}_2 T - 6\mathscr{A}_{-2} T^{-3} - \cdots. \tag{19.39}$$

The anharmonic contribution to the internal energy at high temperatures is $U_A = -\mathscr{A}_2 T^2 + \cdots$, so that U_A and $-TS_A$ both give the same order contribution to F_A at high temperatures.

The total lattice-dynamics entropy and heat capacity at high temperatures, correct to order T^{-4}, are

$$S = -2\mathscr{A}_2 T + K \sum_{\kappa} \ln(KT/\hbar\omega_\kappa) + 3nN_0K + (K/24) \sum_{\kappa} (\hbar\omega_\kappa/KT)^2 + 2\mathscr{A}_{-2} T^{-3} - (K/960) \sum_{\kappa} (\hbar\omega_\kappa/KT)^4 + \cdots; \tag{19.40}$$

$$C_\eta = -2\mathscr{A}_2 T + 3nN_0K - (K/12) \sum_{\kappa} (\hbar\omega_\kappa/KT)^2 - 6\mathscr{A}_{-2} T^{-3} + (K/240) \sum_{\kappa} (\hbar\omega_\kappa/KT)^4 - \cdots. \tag{19.41}$$

Because of the anharmonicity, the total theoretical heat capacity does not approach the Dulong–Petit limit, but continues to increase (or decrease) linearly with T at high temperatures. In order to compare the experimental heat capacity data with lattice-dynamics theory at high temperatures, we recommend the following procedure. The heat capacity at zero stress should be integrated to get $S(T)$ at zero stress, and this should then be corrected thermodynamically to a fixed configuration $\mathbf{X}_0$ with the equations of Section 4; the resulting $S(\mathbf{X}_0, T)$ may be fitted to the expansion (19.40) to obtain the lattice-dynamical coefficients $\mathscr{A}_2$, $\langle\omega_\kappa^2\rangle$, etc., evaluated at $\mathbf{X}_0$. Some examples of such analysis are provided in Chapter 7.

In writing the stresses and elastic constants at high temperatures, we will keep explicitly the contributions arising from Φ_0 and the leading high-temperature contributions arising from F_H and F_A. Thus we start with the free energy contributions

$$F = \Phi_0 - KT \sum_{\kappa} \ln(KT/\hbar\omega_\kappa) + \mathscr{A}_2 T^2, \quad \text{at high } T. \tag{19.42}$$

Then the stresses and isothermal elastic constants are

$$\tau_{ij} = \tilde{C}_{ij} - (KT/V) \sum_{\kappa} \gamma_{\kappa,ij} + (T^2/V)(\partial \mathscr{A}_2/\partial \eta_{ij})_{\eta'}; \tag{19.43}$$

$$C^T_{ijkl} = \tilde{C}_{ijkl} + (KT/V) \sum_{\kappa} (\xi_{\kappa,ijkl} - \gamma_{\kappa,ij}\gamma_{\kappa,kl}) + (T^2/V)(\partial^2 \mathscr{A}_2/\partial \eta_{ij}\partial \eta_{kl})_{\eta'}; \tag{19.44}$$

here the $\tilde{C}_{ij}$ and $\tilde{C}_{ijkl}$ are strain derivatives of Φ_0 as shown in (16.64) and (16.65), and the $\gamma_{\kappa,ij}$ and $\xi_{\kappa,ijkl}$ are strain derivatives of ω_κ as shown in (16.57) and (16.58). The thermal stresses are

$$b_{ij} = -(K/V) \sum_{\kappa} \gamma_{\kappa,ij} + (2T/V)(\partial \mathscr{A}_2/\partial \eta_{ij})_{\eta'}. \tag{19.45}$$

Note that we are now keeping terms in the stresses up to fifth order in ϵ, and in the elastic constants up to sixth order in ϵ. From the general results (2.90) and (2.91) for the adiabatic–isothermal differences, written in terms of the b_{ij}, the high-temperature evaluation up to sixth order in ϵ is

$$C^S_{ijkl} - C^T_{ijkl} = (K^2T/VC_\eta) \sum_{\kappa} \gamma_{\kappa,ij} \sum_{\kappa'} \gamma_{\kappa',kl} - (2KT^2/VC_\eta)\left[(\partial \mathscr{A}_2/\partial \eta_{ij})_{\eta'} \sum_{\kappa} \gamma_{\kappa,kl} + (\partial \mathscr{A}_2/\partial \eta_{kl})_{\eta'} \sum_{\kappa} \gamma_{\kappa,ij}\right]. \tag{19.46}$$

In order to see the explicit temperature-dependence of this equation, the high-temperature heat capacity may be written $C_\eta = 3nN_0K - 2\mathscr{A}_2 T$ from (19.41), and then (19.46) is to sixth order,

$$C^S_{ijkl} - C^T_{ijkl} = (KT/3nN_0V) \sum_{\kappa} \gamma_{\kappa,ij} \sum_{\kappa'} \gamma_{\kappa',kl} - (2T^2/3nN_0V)\left[(\partial \mathscr{A}_2/\partial \eta_{ij})_{\eta'} \sum_{\kappa} \gamma_{\kappa,kl} + (\partial \mathscr{A}_2/\partial \eta_{kl})_{\eta'} \sum_{\kappa} \gamma_{\kappa,ij} - (\mathscr{A}_2/3nN_0) \sum_{\kappa} \gamma_{\kappa,ij} \sum_{\kappa'} \gamma_{\kappa',kl}\right]. \tag{19.47}$$

It is interesting that the explicit temperature-dependence is the same in each order for both C^T_{ijkl} and C^S_{ijkl}, according to (19.44) and (19.47), namely it is T^0 in second order, T in fourth order, and T^2 in sixth order.

In the high-temperature region there is considerable difference between the temperature-dependence of thermodynamic functions evaluated at constant configuration and at constant stress. The stresses and elastic constants, as well as the thermal strains $\beta_{ij} = -\Sigma_{kl}\, S^T_{ijkl} b_{kl}$, may be calculated at constant configuration or at constant stress with the preceding equations. The constant-stress configuration may be determined theoretically by calculating τ_{ij} for various configurations at any desired temperature, and then picking out that configuration which corresponds to the required stresses at each temperature. In comparing theoretical calculations with experimental results, one may choose to study the theoretically simple b_{ij}, or the directly measurable β_{ij}, and one may choose to compare results at constant configuration or constant stress. In any case the comparison of constant-configuration calculations with zero-stress measurements is quite inappropriate—another point which has not been universally recognized.

As long as the leading-order anharmonic perturbation theory is valid, the equations of this section should be rapidly convergent series in the expansion parameter ϵ. In other words, for most crystals we should have

$$\Phi_0 \text{ contribution} \gg F_{\mathrm{H}} \text{ contribution} \gg F_{\mathrm{A}} \text{ contribution}, \qquad (19.48)$$

and this should hold up to quite high temperatures, where of course the first inequality does not apply for the temperature derivatives of F, namely S, C_η, b_{ij}, since the Φ_0 contribution is zero for these functions. We therefore expect that the high-temperature lattice-dynamics equations should be applicable for most crystals for temperatures from about $\Theta_{\mathrm{H}\infty}$ to two or three times $\Theta_{\mathrm{H}\infty}$.

Cubic Crystals

Again for cubic crystals under isotropic pressure there is a great deal of simplification of our general formulas. If we take the free energy contributions written out in (19.42), the pressure and the isothermal bulk modulus are

$$\mathrm{P} = \tilde{P} + (KT/V) \sum_\kappa \gamma_\kappa - T^2(d\mathscr{A}_2/dV); \qquad (19.49)$$

$$B_T = \tilde{B} + (KT/V) \sum_\kappa (\xi_\kappa - \gamma_\kappa^2) + T^2 V(d^2\mathscr{A}_2/dV^2); \qquad (19.50)$$

where $\tilde{P}$ and $\tilde{B}$ are the volume derivatives of Φ_0 defined by (16.78) and (16.79), and γ_κ and ξ_κ are the volume derivatives of ω_κ defined by (16.72) and (16.73). The quantity βB_T is

$$\beta B_T = (K/V) \sum_\kappa \gamma_\kappa - 2T(d\mathscr{A}_2/dV), \qquad (19.51)$$

and to sixth order in ϵ the difference $B_S - B_T$ is written from (16.84),

$$B_S - B_T = (K^2 T/VC_V)\left[\sum_\kappa \gamma_\kappa\right]^2 - (4KT^2/C_V)(d\mathscr{A}_2/dV) \sum_\kappa \gamma_\kappa. \qquad (19.52)$$

Because $\beta B_T = (\partial P/\partial T)_V$, according to (1.29), it is possible to put the condition of zero pressure into the expression for βB_T, and hence evaluate βB_T at $P = 0$. From (19.49), the $P = 0$ condition may be written

$$(K/V) \sum_{\kappa} \gamma_{\kappa} = T^{-1}[(d\Phi_0/dV) + T^2(d\mathscr{A}_2/dV)], \qquad (19.53)$$

since $\tilde{P} = -(d\Phi_0/dV)$, and with this (19.51) is

$$\beta B_T = T^{-1}[(d\Phi_0/dV) - T^2(d\mathscr{A}_2/dV)], \quad \text{at } P = 0. \qquad (19.54)$$

If one neglects the variation of $(d\Phi_0/dV)$ and $(d\mathscr{A}_2/dV)$ with the volume, corresponding to the change in volume with temperature at $P = 0$, then the $P = 0$ value of βB_T is proportional to T^{-1} with a small correction due to the $(d\mathscr{A}_2/dV)$ term in (19.54). For a more accurate analysis, it is more convenient to study βB_T at fixed volume, rather than to account for the volume-dependence of the expression in square brackets in (19.54) at $P = 0$.

We will now illustrate the analysis of high-temperature thermodynamic data in terms of lattice-dynamics theory, in more detail for the example of a cubic crystal under isotropic pressure. Let us denote an average over all phonons κ by brackets $\langle\ \rangle$, as

$$\langle f_{\kappa} \rangle = (3nN_0)^{-1} \sum_{\kappa} f_{\kappa}, \qquad (19.55)$$

where f_{κ} is any function, as for example, ω_{κ}, γ_{κ}, etc. From (19.40) and (19.41), the high-temperature entropy and heat capacity may be written

$$S = 3nN_0K\{\langle \ln(KT/\hbar\omega_{\kappa})\rangle + 1 + \tfrac{1}{24}\langle(\hbar\omega_{\kappa}/KT)^2\rangle - \tfrac{1}{960}\langle(\hbar\omega_{\kappa}/KT)^4\rangle + \cdots$$
$$+ (3nN_0K)^{-1}[-2\mathscr{A}_2T + 2\mathscr{A}_{-2}T^{-3} + \cdots]\}; \qquad (19.56)$$

$$C_V = 3nN_0K\{1 - \tfrac{1}{12}\langle(\hbar\omega_{\kappa}/KT)^2\rangle + \tfrac{1}{240}\langle(\hbar\omega_{\kappa}/KT)^4\rangle - \cdots$$
$$+ (3nN_0K)^{-1}[-2\mathscr{A}_2T - 6\mathscr{A}_{-2}T^{-3} - \cdots]\}. \qquad (19.57)$$

Keeping terms to order T^{-4}, the quantity βB_T is conveniently written as

$$V\beta B_T = 3nN_0K\{\langle\gamma_{\kappa}\rangle - \tfrac{1}{12}\langle\gamma_{\kappa}(\hbar\omega_{\kappa}/KT)^2\rangle + \tfrac{1}{240}\langle\gamma_{\kappa}(\hbar\omega_{\kappa}/KT)^4\rangle - \cdots$$
$$+ (3nN_0K)^{-1}[-2V(d\mathscr{A}_2/dV)T + 2V(d\mathscr{A}_{-2}/dV)T^{-3} + \cdots]\}. \qquad (19.58)$$

In (19.56)–(19.58) the functions are arranged in two series in decreasing powers of T, namely the series of harmonic contributions and the series of anharmonic contributions. The dominant term in each of these equations is the leading high-temperature harmonic term, and all other terms are considered small compared to the dominant term at high temperatures. Note that the anharmonic contributions appear in ratios such as $\mathscr{A}_2T^2/3nN_0KT$, which is the ratio of anharmonic-to-harmonic energies at high temperatures. In the spirit of such expansions, the heat capacity (19.57) may be inverted, and keeping all terms to order T^{-4} and to leading order in anharmonicity (fourth order in ϵ), the result is

$$C_V^{-1} = (3nN_0K)^{-1}\{1 + \tfrac{1}{12}\langle(\hbar\omega_\kappa/KT)^2\rangle - \tfrac{1}{240}\langle(\hbar\omega_\kappa/KT)^4\rangle$$
$$+ \tfrac{1}{144}\langle(\hbar\omega_\kappa/KT)^2\rangle^2 + \cdots \tag{19.59}$$
$$+ (3nN_0K)^{-1}[2\mathscr{A}_2T + 6\mathscr{A}_{-2}T^{-3} + \cdots$$
$$+ \tfrac{1}{3}\mathscr{A}_2T\langle(\hbar\omega_\kappa/KT)^2\rangle - \tfrac{1}{60}\mathscr{A}_2T\langle(\hbar\omega_\kappa/KT)^4\rangle + \cdots]\}.$$

The bulk moduli B_T and B_S are given by (19.50) and (19.52); the leading high-temperature harmonic and anharmonic contributions to their constant volume–temperature derivatives may be written

$$V(\partial B_T/\partial T)_V = 3nN_0K[\langle\xi_\kappa - \gamma_\kappa^2\rangle + (2TV^2/3nN_0K)(d^2\mathscr{A}_2/dV^2)]; \tag{19.60}$$

$$V(\partial B_S/\partial T)_V = 3nN_0K\{\langle\xi_\kappa - \gamma_\kappa^2\rangle + \langle\gamma_\kappa\rangle^2 + (2T/3nN_0K)[V^2(d^2\mathscr{A}_2/dV^2)$$
$$- 4\langle\gamma_\kappa\rangle V(d\mathscr{A}_2/dV) + 2\langle\gamma_\kappa\rangle^2\mathscr{A}_2]\}. \tag{19.61}$$

In deriving (19.61), we used the leading terms in the expansion (19.59) of C_V^{-1} in the expression (19.52) for $B_S - B_T$. The macroscopic Grüneisen parameter is $\gamma = V\beta B_T/C_V$, and with (19.58) and (19.59) the high-temperature expansion is

$$\gamma = \langle\gamma_\kappa\rangle\{1 + \tfrac{1}{12}[\langle(\hbar\omega_\kappa/KT)^2\rangle - \langle\gamma_\kappa\rangle^{-1}\langle\gamma_\kappa(\hbar\omega_\kappa/KT)^2\rangle] + \cdots$$
$$+ (3nN_0K)^{-1}[2\mathscr{A}_2T - 2\langle\gamma_\kappa\rangle^{-1}V(d\mathscr{A}_2/dV)T + \cdots]\}. \tag{19.62}$$

In a similar way the high-temperature expansion of β may be calculated from the ratio of the expansions for $V\beta B_T$ and VB_T.

Now by assembling the experimental data on S, β, B_T, and B_S, and by correcting such functions as S, $V\beta B_T$, and VB_T or VB_S to fixed volume and studying the temperature-dependence in the high-temperature region, one can determine a great deal of detailed experimental information about the lattice-dynamical properties of a given crystal. The first test to be made is whether or not the temperature-dependence at constant volume agrees with that predicted by the appropriate equations above; if it does, then the coefficients in the expansions may be extracted to within limits placed by the experimental uncertainties. These coefficients include $\langle\omega_\kappa^2\rangle$, $\langle\omega_\kappa^4\rangle, \cdots$, and $\mathscr{A}_2$, $\mathscr{A}_{-2}, \cdots$, from the entropy; $\langle\gamma_\kappa\rangle$, $\langle\gamma_\kappa\omega_\kappa^2\rangle, \cdots$, and $(d\mathscr{A}_2/dV)$, $(d\mathscr{A}_{-2}/dV), \cdots$, from $V\beta B_T$; and $\langle\xi_\kappa - \gamma_\kappa^2\rangle$ and $(d^2\mathscr{A}_2/dV^2)$ from VB_T or VB_S.

In practice the entropy analysis can be done quite accurately with presently available experimental results, and the available data concerning $V\beta B_T$ is generally sufficient to determine the leading coefficients with respectable accuracy. It is difficult to find data to correct B_T or B_S to a fixed volume over a range of temperatures; however if a measurement of say $(\partial B_S/\partial P)_T$ is available at room temperature, then $(\partial B_S/\partial T)_V$ may be calculated at the room temperature volume (not V_0) from the general relation (1.52). We note that the heat capacity gives the same information as the entropy, so the entropy should be used for analysis since the correction of S to fixed volume is much simpler than the correction of C_V to fixed volume [see (4.29) and (4.30)]. We also note that in the high-temperature expansion (19.62) of γ, the harmonic temperature-dependent terms tend to cancel (we cannot say the same about the anharmonic terms in general), and while the correction of γ to fixed volume is quite difficult, a plot of γ *vs* T may be expected to reach the constant limit $\langle \gamma_\kappa \rangle$ at $T < \Theta_{H\infty}$, perhaps at sufficiently low T that the volume correction is unimportant.

20. CALCULATIONS BASED ON MEASURED PHONON FREQUENCIES

Errors in the Use of Harmonic Formulas

The frequencies which are measured by inelastic neutron-scattering experiments are the renormalized frequencies $\Omega_\kappa = \omega_\kappa + \Delta_\kappa$; these frequencies contain anharmonic contributions, and they are explicitly dependent on both the crystal configuration and the temperature. The thermodynamic functions are written in Section 16 in terms of the harmonic and anharmonic contributions to Ω_κ, namely ω_κ and Δ_κ, respectively, and the point of the present section is that if the thermodynamic functions are calculated from the harmonic formulas, with the ω_κ replaced by Ω_κ, then the anharmonic contributions are not given correctly in general. In order to calculate the thermodynamic functions from the measured Ω_κ, one may separate experimentally the contributions to Ω_κ, or one may derive appropriate thermodynamic formulas. These ideas will be discussed in particular for the entropy and the heat capacity; the extension to other thermodynamic functions is straightforward.

Before proceeding to the general discussion, we note that there is one region where the use of harmonic formulas with measured frequencies is valid, and that is in calculating the temperature-dependent part of the thermodynamic functions at low temperatures. Thus to leading order in the anharmonicity, as shown in Section 18, the $T = 0$ Debye temperature Θ_0 is given by the harmonic formula with the harmonic velocities replaced by the renormalized velocities for the long-wavelength acoustic phonons, and this holds also for all strain derivatives of Θ_0. The same replacement does *not* hold for the ground-state energy: The harmonic ground-state energy is $G_2 = \Sigma_\kappa \frac{1}{2}\hbar\omega_\kappa$, and the harmonic plus anharmonic ground-state energy is

$G_2 + G_4$, but this is not the same as $\Sigma_\kappa \frac{1}{2}\hbar\Omega_\kappa$ since obviously

$$\sum_\kappa \tfrac{1}{2}\hbar\Delta_\kappa \neq G_4. \tag{20.1}$$

The lattice-dynamical expressions for the entropy $S = S_{\mathrm{H}} + S_{\mathrm{A}}$ and for the heat capacity $C_\eta = C_{\eta\mathrm{H}} + C_{\eta\mathrm{A}}$ were derived in Section 16, and are rewritten here for convenience:

$$S_{\mathrm{H}} = S_{\mathrm{H}}(\omega_\kappa) = K \sum_\kappa [(\bar{n}_\kappa + 1)\ln(\bar{n}_\kappa + 1) - \bar{n}_\kappa \ln \bar{n}_\kappa]; \tag{20.2}$$

$$S_{\mathrm{A}} = -\sum_\kappa \hbar\Delta_\kappa(\partial\bar{n}_\kappa/\partial T)_\eta; \tag{20.3}$$

$$C_{\eta\mathrm{H}} = C_{\eta\mathrm{H}}(\omega_\kappa) = \sum_\kappa \hbar\omega_\kappa(\partial\bar{n}_\kappa/\partial T)_\eta; \tag{20.4}$$

$$C_{\eta\mathrm{A}} = -T\sum_\kappa \hbar\Delta_\kappa(\partial^2\bar{n}_\kappa/\partial T^2)_\eta - T\sum_{\kappa\kappa'} \hbar\theta_{\kappa\kappa'}(\partial\bar{n}_\kappa/\partial T)_\eta(\partial\bar{n}_{\kappa'}/\partial T)_\eta; \tag{20.5}$$

where $(\partial\bar{n}_\kappa/\partial T)_\eta$ and $(\partial^2\bar{n}_\kappa/\partial T^2)_\eta$ are given by (16.40) and (16.41), respectively, and where

$$\bar{n}_\kappa = \bar{n}_\kappa(\omega_\kappa) = [e^{\hbar\omega_\kappa/KT} - 1]^{-1}. \tag{20.6}$$

Let us use a hat to denote a harmonic function with the frequencies ω_κ replaced by Ω_κ, as for example,

$$\hat{n}_\kappa = [e^{\hbar\Omega_\kappa/KT} - 1]^{-1}. \tag{20.7}$$

To leading order, the anharmonic contributions to such functions are obtained by expanding them to leading order in the frequency shifts Δ_κ; some preliminary results are as follows.

$$\begin{aligned} \Omega_\kappa &= \omega_\kappa + \Delta_\kappa; \\ \hat{n}_\kappa &= \bar{n}_\kappa[1 - (\hbar\Delta_\kappa/KT)(\bar{n}_\kappa + 1)]; \\ \hat{n}_\kappa + 1 &= (\bar{n}_\kappa + 1)[1 - (\hbar\Delta_\kappa/KT)\bar{n}_\kappa]. \end{aligned} \tag{20.8}$$

Now we can expand $\hat{S}_{\mathrm{H}}$ to leading order in Δ_κ, to see what we would in fact be calculating if we used the harmonic formula with the measured frequencies; the expansion of (20.2) for S_{H} gives

$$\hat{S}_{\mathrm{H}} = S_{\mathrm{H}} - \sum_\kappa \hbar\Delta_\kappa(\partial\bar{n}_\kappa/\partial T)_\eta. \tag{20.9}$$

By comparing this with (20.3) for S_{A}, it is seen that to leading order in the anharmonicity the entropy is given correctly by the harmonic formula with the measured frequencies:

$$\hat{S}_{\mathrm{H}} = S_{\mathrm{H}} + S_{\mathrm{A}}. \tag{20.10}$$

This holds for all temperatures and all configurations, but it must be recognized that the Δ_κ are functions of both temperature and configuration, so that S is given by $\hat{S}_{\mathrm{H}}$ in terms of the measured Ω_κ *only at the temperature and configuration* at which the Ω_κ are measured.

A similar expansion of the heat capacity gives the result

$$\hat{C}_{\eta\mathrm{H}} = C_{\eta\mathrm{H}} - T \sum_\kappa \hbar\Delta_\kappa(\partial^2\bar{n}_\kappa/\partial T^2)_\eta. \tag{20.11}$$

By comparing this with (20.5) for $C_{\eta\mathrm{A}}$, it is seen that the anharmonic contribution to the heat capacity is not generally included correctly by the harmonic formula with the measured frequencies. The same conclusion is found to hold for the free energy and the internal energy,

$$\hat{F}_{\mathrm{H}} \neq F_{\mathrm{H}} + F_{\mathrm{A}}, \qquad \hat{U}_{\mathrm{H}} \neq U_{\mathrm{H}} + U_{\mathrm{A}}; \tag{20.12}$$

and these inequalities still hold if the ground-state energy contributions are omitted from consideration. The error in calculating the heat capacity by using the harmonic formula with the measured frequencies is, from (20.5) and (20.11),

$$C_\eta - \hat{C}_{\eta\mathrm{H}} = -T \sum_{\kappa\kappa'} \hbar\theta_{\kappa\kappa'}(\partial\bar{n}_\kappa/\partial T)_\eta(\partial\bar{n}_{\kappa'}/\partial T)_\eta. \tag{20.13}$$

This function is of lowest-order quadratic in the $\bar{n}_\kappa$, and hence does not contribute to the leading T^3 temperature-dependence of C_η at low temperatures. This confirms the result for the heat capacity that in the low-temperature limit the harmonic formula with renormalized frequencies is correct. On the other hand, the error (20.13) is quite significant at high temperatures. Expansion of (20.11) shows that the leading high-temperature anharmonic contribution to $\hat{C}_{\eta\mathrm{H}}$ goes as T^{-1}:

$$\hat{C}_{\eta\mathrm{H}} = C_{\eta\mathrm{H}} - (\hbar/6T) \sum_{\kappa\kappa'} \theta_{\kappa\kappa'}(\omega_\kappa/\omega_{\kappa'}), \quad \text{at high } T; \tag{20.14}$$

note that the high-temperature expansion (19.34) of Δ_κ was used in obtaining this result. Thus the function $\hat{C}_{\eta\mathrm{H}}$ completely misses the leading anharmonic contribution to C_η at high temperature, which is $-2\mathscr{A}_2T$ according to (19.39). Indeed this contribution is the leading term in the high-temperature expansion of the error (20.13); that is,

$$\begin{aligned} C_\eta - \hat{C}_{\eta\mathrm{H}} &= -(K^2T/\hbar) \sum_{\kappa\kappa'} (\theta_{\kappa\kappa'}/\omega_\kappa\omega_{\kappa'}), \quad \text{at high } T, \\ &= -2\mathscr{A}_2T. \end{aligned} \tag{20.15}$$

The errors in calculating thermodynamic functions from harmonic formulas, with the harmonic frequencies ω_κ replaced by the measured frequencies Ω_κ, follow from two sources, namely (a) double counting of certain contributions, as mentioned in the discussion preceding (16.36), and

(b) the incorrect use of temperature-dependent energies in standard statistical mechanical equations. It should be obvious from the preceding equations that it is in fact quite difficult to use the Ω_κ measured at some temperature and configuration to calculate thermodynamic functions; thus our philosophy is to consider all thermodynamic functions, including also the Ω_κ, as sources of information for, or as tests of, theoretical calculations.

Corrections for the Errors

For simplicity we will now restrict our calculations to the case of cubic crystals with P–V–T variables; extension to noncubic crystals is straightforward. The frequency shifts are leading anharmonic quantities, i.e., $\hbar\Delta_\kappa \sim \epsilon^4$, and their volume derivatives are of higher order, so to leading order in the anharmonicity the shifts may be considered independent of volume. Then since the harmonic frequencies ω_κ depend only on the volume, the explicit volume- and temperature-dependence of the Ω_κ may be taken to be

$$\Omega_\kappa(V,T) = \omega_\kappa(V) + \Delta_\kappa(T). \tag{20.16}$$

It is possible in principle to partially separate the harmonic and anharmonic contributions to Ω_κ, by measuring both the temperature- and pressure-dependence of Ω_κ. Let V be the volume at any temperature and zero pressure, and V_0 be the volume at zero temperature and pressure. Then from (20.16),

$$\Omega_\kappa(V_0,T) - \Omega_\kappa(V_0,0) = \Delta_\kappa(T) - \Delta_\kappa(0), \tag{20.17}$$

and also

$$\Omega_\kappa(V,T) - \Omega_\kappa(V_0,T) = (V - V_0)(d\omega_\kappa/dV). \tag{20.18}$$

Adding these two equations gives

$$\Omega_\kappa(V,T) - \Omega_\kappa(V_0,0) = (V - V_0)(d\omega_\kappa/dV) + \Delta_\kappa(T) - \Delta_\kappa(0). \tag{20.19}$$

Since $\hbar\Delta_\kappa$ is of order ϵ^4, so is $\hbar\Delta_\kappa(T) - \hbar\Delta_\kappa(0)$; also $\hbar(d\omega_\kappa/dV) \sim \epsilon^3$ and $V - V_0 \sim \epsilon$, so the quantities on the right of (20.19) are both of the same order. Strictly speaking these quantities are to be evaluated at V_0, according to their definitions in (20.17) and (20.18), but their volume dependences are of higher order, and will be neglected here.

In addition, because of the volume- and temperature-dependence of Ω_κ shown in (20.16), the shifts Δ_κ do not contribute to the pressure derivative, and with the aid of the general relation (1.51) we can write

$$(\partial\Omega_\kappa/\partial P)_T = -(V/B_T)(d\omega_\kappa/dV). \tag{20.20}$$

Both (20.19) for the variation of Ω_κ with temperature at zero pressure, and (20.20) for the variation of Ω_κ with pressure at constant temperature, are amenable to experimental determination. A combination of the two equations leads to

$$\Delta_\kappa(T) - \Delta_\kappa(0) = [\Omega_\kappa(V,T) - \Omega_\kappa(V_0,0)] + [(V - V_0)/V]B_T(\partial\Omega_\kappa/\partial P)_T. \tag{20.21}$$

With (16.35) for the shifts Δ_κ, the total renormalized phonon frequencies may be written

$$\Omega_\kappa = \omega_\kappa + \theta_\kappa + \sum_{\kappa'} \theta_{\kappa\kappa'}\bar{n}_{\kappa'}; \tag{20.22}$$

thus if the phonon frequencies are measured at $T = 0$, and also if the temperature variations $\Delta_\kappa(T) - \Delta_\kappa(0)$ are determined from (20.21), then the contributions to Ω_κ can be separated into the two terms

$$\omega_\kappa + \theta_\kappa = \Omega_\kappa(T = 0); \tag{20.23}$$

$$\sum_{\kappa'} \theta_{\kappa\kappa'}\bar{n}_{\kappa'} = \Delta_\kappa(T) - \Delta_\kappa(0). \tag{20.24}$$

It may be noted that, except for the ground-state energy contributions, the thermodynamic functions can be properly constructed from the quantities in (20.23) and (20.24). The free energy, for example, may be written

$$F = \mathscr{E}_G + KT \sum_\kappa \ln[1 - e^{-\hbar(\omega_\kappa+\theta_\kappa)/KT}] + \tfrac{1}{2} \sum_{\kappa\kappa'} \hbar\theta_{\kappa\kappa'}\bar{n}_\kappa\bar{n}_{\kappa'}, \tag{20.25}$$

correct to leading order in the anharmonicity. However, the pressure derivatives (20.20) and the temperature variations (20.19) or (20.21) are of more direct theoretical interest than are the sums involved in thermodynamic functions.

We will now illustrate, for the example of the heat capacity, the procedure for correcting the errors involved in calculating thermodynamic functions from harmonic formulas with measured frequencies. For a cubic crystal the appropriate heat capacity is C_V, and the object is to derive a formula by which the error $C_V - \hat{C}_{VH}$, given in general by (20.13), may be evaluated from experimental data. Referring again to the volume- and temperature-dependence of Ω_κ shown in (20.16), the temperature derivative at constant pressure may be written

$$(\partial\Omega_\kappa/\partial T)_P = (d\omega_\kappa/dV)(\partial V/\partial T)_P + (d\Delta_\kappa/dT). \tag{20.26}$$

But from (16.35) for Δ_κ,

$$d\Delta_\kappa/dT = \sum_{\kappa'} \theta_{\kappa\kappa'}(\partial\bar{n}_{\kappa'}/\partial T)_V, \tag{20.27}$$

and since $(\partial V/\partial T)_P = V\beta$ according to the definition (1.28) of β, then (20.26) is

$$(\partial\Omega_\kappa/\partial T)_P = V\beta(d\omega_\kappa/dV) + \sum_{\kappa'} \theta_{\kappa\kappa'}(\partial\bar{n}_{\kappa'}/\partial T)_V. \tag{20.28}$$

Multiplying this by $\hbar T(\partial \bar{n}_\kappa/\partial T)_V$, summing over κ, and solving for the term involving $\theta_{\kappa\kappa'}$, gives

$$-T \sum_{\kappa\kappa'} \hbar\theta_{\kappa\kappa'}(\partial \bar{n}_\kappa/\partial T)_V(\partial \bar{n}_{\kappa'}/\partial T)_V$$
$$= TV\beta \sum_\kappa \hbar(d\omega_\kappa/dV)(\partial \bar{n}_\kappa/\partial T)_V - T \sum_\kappa \hbar(\partial\Omega_\kappa/\partial T)_P(\partial \bar{n}_\kappa/\partial T)_V. \tag{20.29}$$

Now from (20.13) the left-hand side of this equation is just $C_V - \hat{C}_{V\mathrm{H}}$, and from (16.83) for βB_T for a cubic crystal the first term on the right-hand side is just $-TV\beta^2 B_T$, so we have the simple result

$$C_V - \hat{C}_{V\mathrm{H}} = -TV\beta^2 B_T - T\sum_\kappa \hbar(\partial\Omega_\kappa/\partial T)_P(\partial \bar{n}_\kappa/\partial T)_V. \tag{20.30}$$

Thus if it is desired to calculate the heat capacity C_V, one may calculate $\hat{C}_{V\mathrm{H}}$, which is the harmonic formula for C_V with the ω_κ replaced by Ω_κ, and then add the correction $C_V - \hat{C}_{V\mathrm{H}}$ given by (20.30). Both sides of (20.30) are evaluated at the same volume and temperature, and each term on the right-hand side is of fourth order. The quantities $(\partial \bar{n}_\kappa/\partial T)_V$ are as expressed by (16.40), and to leading order in anharmonicity these may be evaluated with ω_κ replaced by Ω_κ. The temperature derivatives $(\partial\Omega_\kappa/\partial T)_P$ are reasonably accessible to experimental determination and we would ordinarily expect them to be negative, but it is not generally possible to determine the sign of $C_V - \hat{C}_{V\mathrm{H}}$ without a detailed evaluation of (20.30).

We noted in (20.15) that in the high-temperature limit the correction $C_V - \hat{C}_{V\mathrm{H}}$ is just the high-temperature anharmonic heat capacity $-2\mathscr{A}_2 T$. Therefore evaluation of (20.30) at high temperature, where $(\partial \bar{n}_\kappa/\partial T)_V = K/\hbar\omega_\kappa$, leads to an expression for $\mathscr{A}_2$,

$$\mathscr{A}_2 = \tfrac{1}{2}V\beta^2 B_T + \tfrac{1}{2}K\sum_\kappa (\partial\Omega_\kappa/\partial T)_P\omega_\kappa^{-1}, \quad \text{at high } T. \tag{20.31}$$

Again both sides of (20.31) are evaluated at the same volume and temperature, and the equation should be valid at $T \gtrsim \Theta_{\mathrm{H}\infty}$. Also to the order to which (20.31) is valid, ω_κ^{-1} may be replaced by Ω_κ^{-1}.

5

BAND-STRUCTURE THEORY

In principle the total lattice potential Φ may be calculated for any crystal from the results of an electronic band-structure calculation, since Φ is the total energy of the band electrons, plus the interactions of the ion cores. The general expressions for Φ, and for the dynamical matrices, are derived here in the framework of the one-electron approximation for band-structure calculations; in practice these expressions can be evaluated by rather extensive computation. The additional contributions to the thermodynamic functions of metals, due to thermal excitation of conduction electrons, are discussed, and the electron–phonon interaction effects are calculated in leading-order perturbation.

21. GENERAL FORMULATION FOR NONMETALS

Adiabatic Approximation

The crystal is composed of "ions" and "band electrons," and subscripts I and E will be used to denote the collection of ions and the collection of band electrons, respectively. Each ion is composed of the nucleus plus the core electrons; the core electrons are presumed to look essentially the same as they do in free atoms, so they may be adequately represented by atomic wave functions. The remaining electrons are called collectively the band electrons, and these may comprise one or more filled valence bands in the case of nonmetals. The band-electron wave functions are to be found by a band-structure calculation; we note that the number of electrons from each atom which are assigned to the bands is arbitrary.

The total Hamiltonian of the crystal may be written

$$\mathscr{H} = KE_{\text{I}} + KE_{\text{E}} + \Omega_{\text{I}} + \Omega_{\text{E}} + \Omega_{\text{IE}}. \tag{21.1}$$

Here KE_{I} and KE_{E} are the total kinetic energies of the ions and the band electrons, respectively; Ω_{I} and Ω_{E} represent the interactions among the ions and the interactions among the band electrons, respectively; and Ω_{IE} is the total interaction between the ions and the band electrons. An important point of this description of a crystal is as follows. If the ion cores are reasonably small, i.e., if all the outer electrons have been put into the bands, then the only significant contribution to Ω_{I} is the Coulomb interaction among point ions, since the overlaps of the cores, and their polarizations, will be negligible. Thus all the complications arising from the deformations of the ions in the usual ionic models, as for example, representing NaCl as a collection of Na^+ and Cl^- ions, are eliminated here; more precisely, the important effects resulting from deformations of the outer electrons are absorbed into the band-electron calculations.

To begin, we separate out the Hamiltonian $\mathscr{H}_{\text{E}}$ for the band electrons:

$$\mathscr{H}_{\text{E}} = KE_{\text{E}} + \Omega_{\text{IE}} + \Omega_{\text{E}}. \tag{21.2}$$

This is to be solved for the states of the band electrons, with the ions in arbitrary positions, but not moving. Let us denote the ion positions by $\tilde{\mathbf{R}}(\text{N}\nu)$, and write

$$\tilde{\mathbf{R}}(\text{N}\nu) = \mathbf{R}(\text{N}\nu) + \mathbf{U}(\text{N}\nu), \tag{21.3}$$

where as usual $\mathbf{R}(\text{N}\nu)$ are the equilibrium positions, and $\mathbf{U}(\text{N}\nu)$ the displacements from equilibrium. The wave equation for the band electrons is

$$\mathscr{H}_{\text{E}}\Psi_{\Lambda} = E_{\Lambda}\Psi_{\Lambda}, \tag{21.4}$$

where E_{Λ} and Ψ_{Λ} are the total energy and total wave function, respectively, of the total state labeled Λ. Now Ω_{IE} depends explicitly on the positions $\tilde{\mathbf{R}}(\text{N}\nu)$ of the ions, and as a result Ω_{E} also depends implicitly, through the electronic wave function, on the $\tilde{\mathbf{R}}(\text{N}\nu)$. The band electrons are labeled $\alpha, \beta, \cdots$, their position coordinates are $\mathbf{r}(\alpha)$, and the dependence of E_{Λ} and Ψ_{Λ} on all the position variables is given by

$$E_{\Lambda} = E_{\Lambda}(\tilde{\mathbf{R}}(\text{N}\nu)); \qquad \Psi_{\Lambda} = \Psi_{\Lambda}(\tilde{\mathbf{R}}(\text{N}\nu), \mathbf{r}(\alpha)). \tag{21.5}$$

The wave functions Ψ_{Λ} form a complete set, complete in the space of the electron coordinates $\mathbf{r}(\alpha)$, and they are orthonormal in this space. The total crystal wave function Ξ_{Π}, corresponding to a total crystal state labeled Π, may be expanded in the set of electronic wave functions as

$$\Xi_{\Pi} = \sum_{\Lambda} \chi_{\Pi\Lambda}\Psi_{\Lambda}, \tag{21.6}$$

where the coefficients $\chi_{\Pi\Lambda}$ depend only on the ion positions $\tilde{\mathbf{R}}(N\nu)$. The crystal wave equation is then

$$\mathscr{H}\Xi_{\Pi} = \mathscr{E}_{\Pi}\Xi_{\Pi}, \tag{21.7}$$

where $\mathscr{E}_{\Pi}$ is the total crystal energy. With $\mathscr{H}$ given by (21.1) and (21.2), and Ξ_{Π} given by (21.6), the crystal wave equation may be written

$$(KE_{\mathrm{I}} + \Omega_{\mathrm{I}} + \mathscr{H}_{\mathrm{E}} - \mathscr{E}_{\Pi}) \sum_{\Lambda'} \chi_{\Pi\Lambda'}\Psi_{\Lambda'} = 0. \tag{21.8}$$

Now $\mathscr{H}_{\mathrm{E}}$ operates only on the $\Psi_{\Lambda'}$, to give $E_{\Lambda'}\Psi_{\Lambda'}$, while KE_{I} operates on both the $\chi_{\Pi\Lambda'}$ and $\Psi_{\Lambda'}$; with this in mind, (21.8) may be written

$$\sum_{\Lambda'} \Psi_{\Lambda'}(KE_{\mathrm{I}} + \Omega_{\mathrm{I}} + E_{\Lambda'} - \mathscr{E}_{\Pi})\chi_{\Pi\Lambda'} + \sum_{\Lambda'} (KE_{\mathrm{I}}\Psi_{\Lambda'} - \Psi_{\Lambda'}KE_{\mathrm{I}})\chi_{\Pi\Lambda'} = 0, \tag{21.9}$$

where KE_{I} operates only on the function to the right. Multiplying (21.9) by Ψ_{Λ} and integrating over all electronic coordinates gives

$$(KE_{\mathrm{I}} + \Omega_{\mathrm{I}} + E_{\Lambda} - \mathscr{E}_{\Pi})\chi_{\Pi\Lambda} = -\sum_{\Lambda'} \int \Psi_{\Lambda}(KE_{\mathrm{I}}\Psi_{\Lambda'} - \Psi_{\Lambda'}KE_{\mathrm{I}})\chi_{\Pi\Lambda'}\, d\{\mathbf{r}(\alpha)\}, \tag{21.10}$$

where the integral on the right is over the volume of the crystal for all $\mathbf{r}(\alpha)$.

The terms on the right of (21.10) are the nonadiabatic terms. These terms give a small contribution to the total energy in all crystals, namely a contribution of the order of the ratio of electron to nuclear masses, relative to the electronic energy E_{Λ}. We will therefore neglect the nonadiabatic terms for nonmetals. The nonadiabatic effects are included for metals, in Section 25, since these effects modify the ordinary electronic thermodynamic functions.

In the adiabatic approximation, we are neglecting the effect of the ion kinetic energy KE_{I} operating on the electronic wave functions Ψ_{Λ}, and the total energy $\mathscr{E}_{\Pi}$ is given as a solution to (21.10) with the right-hand side set equal to zero:

$$(KE_{\mathrm{I}} + \Omega_{\mathrm{I}} + E_{\Lambda})\chi_{\Pi\Lambda} = \mathscr{E}_{\Pi}\chi_{\Pi\Lambda}. \tag{21.11}$$

The total crystal wave function is a product $\chi_{\Pi\Lambda}\Psi_{\Lambda}$, since if we write

$$\mathscr{H}\chi_{\Pi\Lambda}\Psi_{\Pi\Lambda} = (KE_{\mathrm{I}} + \Omega_{\mathrm{I}} + \mathscr{H}_{\mathrm{E}})\chi_{\Pi\Lambda}\Psi_{\Lambda} = \mathscr{E}_{\Pi}\chi_{\Pi\Lambda}\Psi_{\Lambda} \tag{21.12}$$

and neglect the result of KE_{I} operating on Ψ_{Λ}, then (21.12) is the same as (21.11). Thus the different states of the band electrons are not mixed in the adiabatic approximation, and of course for nonmetals the appropriate electronic state is the ground state corresponding to filled valence bands. The electronic ground-state energy and wave function are denoted by E_{G} and Ψ_{G}, respectively, and from (21.11) the adiabatic Hamiltonian for the crystal is

$$\mathscr{H} = KE_{\mathrm{I}} + \Omega_{\mathrm{I}} + E_{\mathrm{G}}. \tag{21.13}$$

This is just the lattice-dynamics Hamiltonian, with the total adiabatic potential for the ions given by

$$\Phi = \Omega_{\rm I} + E_{\rm G}, \tag{21.14}$$

and where both $\Omega_{\rm I}$ and $E_{\rm G}$ depend on the ion positions $\tilde{\mathbf{R}}(\mathrm{N}\nu)$, and the energies $\mathscr{E}_{\Pi}$ which are eigenvalues of $\mathscr{H}$ are the total crystal energy levels.

One-Electron Approximation

Before proceeding with the calculation of the energy of the band electrons, it is useful to introduce the Fourier transforms appropriate for functions defined everywhere in the crystal. The position variable within the volume V of the crystal is $\mathbf{r}$, and the plane waves which form a complete and orthonormal set are

$$|\mathbf{q}\rangle = V^{-1/2} e^{i\mathbf{q}\cdot\mathbf{r}}, \tag{21.15}$$

where $\mathbf{q}$ are wave vectors quantized in V, i.e., the $\mathbf{q}$ satisfy the same periodic boundary condition as (10.15) for the phonon $\mathbf{k}$ vectors. The $\mathbf{q}$ are not restricted to a Brillouin zone or an inverse-lattice cell, but take on all values allowed by the periodic boundary condition, hence an infinite number of values. The orthonormality and completeness relations are

$$V^{-1} \int e^{i(\mathbf{q}-\mathbf{q}')\cdot\mathbf{r}}\, d\mathbf{r} = \delta_{\mathbf{q}\mathbf{q}'}, \qquad \text{orthonormality}; \tag{21.16}$$

$$V^{-1} \sum_{\mathbf{q}} e^{i\mathbf{q}\cdot(\mathbf{r}-\mathbf{r}')} = \delta(\mathbf{r}-\mathbf{r}'), \quad \text{completeness}. \tag{21.17}$$

Here $\int d\mathbf{r}$ is over the entire volume V. A general function $f(\mathbf{r})$ may be Fourier transformed as

$$f(\mathbf{r}) = \sum_{\mathbf{q}} f(\mathbf{q}) e^{i\mathbf{q}\cdot\mathbf{r}}; \tag{21.18}$$

with the orthonormality relation (21.16) this transform may be inverted to give

$$f(\mathbf{q}) = V^{-1} \int f(\mathbf{r}) e^{-i\mathbf{q}\cdot\mathbf{r}}\, d\mathbf{r}. \tag{21.19}$$

Also the volume integral of the product of two functions $f(\mathbf{r})$ and $g(\mathbf{r})$ may be simply expressed as a Fourier sum:

$$\int f(\mathbf{r}) g(\mathbf{r})\, d\mathbf{r} = \sum_{\mathbf{q}\mathbf{q}'} f(\mathbf{q}) g(\mathbf{q}') \int e^{i(\mathbf{q}+\mathbf{q}')\cdot\mathbf{r}}\, d\mathbf{r} = V \sum_{\mathbf{q}} f(\mathbf{q}) g(-\mathbf{q}). \tag{21.20}$$

Since the density of $\mathbf{q}$ vectors in inverse-lattice space is $V/(2\pi)^3$, the same as for the phonon $\mathbf{k}$ vectors, a $\Sigma_{\mathbf{q}}$ can be transformed to an $\int d\mathbf{q}$ according to

$$\sum_{\mathbf{q}} \to [V/(2\pi)^3] \int d\mathbf{q}, \tag{21.21}$$

where $\int d\mathbf{q}$ is over all space for $\mathbf{q}$.

The total Hamiltonian $\mathscr{H}_{\mathrm{E}}$ for the band electrons is written in (21.2), where the interaction Ω_{E} among the band electrons is just the Coulomb interaction

$$\Omega_{\mathrm{E}} = \tfrac{1}{2}e^2 \sum_{\alpha\beta}' |\mathbf{r}(\alpha) - \mathbf{r}(\beta)|^{-1}, \tag{21.22}$$

and e is the magnitude of the electronic charge. In keeping with the philosophy of band-structure calculations, the band-electron problem will be represented as a one-electron problem. Obviously the kinetic energy KE_{E} is a sum of single-electron kinetic energies, and Ω_{IE} is the same potential for each band electron. If Ω_{E} is replaced by a self-consistent field which is the same for each band electron, then the total electronic Hamiltonian can be written as a sum of single-electron Hamiltonians,

$$\mathscr{H}_{\mathrm{E}} \to \sum_{\alpha} h(\mathbf{r}(\alpha)). \tag{21.23}$$

Here the one-electron Hamiltonian $h(\mathbf{r})$ is the same for each and every band electron, and may be written

$$h(\mathbf{r}) = ke(\mathbf{r}) + \Omega_{\mathrm{B}}(\mathbf{r}) + \Omega_{\mathrm{S}}(\mathbf{r}), \tag{21.24}$$

where ke is the kinetic energy operator for one electron, Ω_{B} is the potential for each band electron due to all the ions (subscript B for bare potential), and Ω_{S} is the self-consistent field for each band electron (subscript S for screening potential). The total one-electron potential is Ω, given by

$$\Omega = \Omega_{\mathrm{B}} + \Omega_{\mathrm{S}}, \tag{21.25}$$

and this is effectively a screened ion potential.

With the introduction of the one-electron approximation, exchange effects arise, as a result of the antisymmetrization of the total electronic wave function. We make the further approximation that all exchange effects can be adequately represented in terms of local one-electron potentials, i.e., potentials which are functions of $\mathbf{r}$, not operators. The potential Ω_{B} due to the ions, which includes effects of exchange between the core and band electrons, is then a local potential, and in fact may be taken as a sum of potentials due to single ions:

$$\Omega_{\mathrm{B}}(\mathbf{r}) = \sum_{N\nu} v_{\nu}(\mathbf{r} - \tilde{\mathbf{R}}(N\nu)). \tag{21.26}$$

Here the form of the bare ion potential v_ν may be different for each type of ion, i.e., for each ν, and since the deformation of ion cores is presumed negligible, each potential v_ν is taken to move rigidly with its ion. To illustrate our calculations, the Hartree approximation will be used for the self-consistent field Ω_S, and a density-dependent approximation for exchange among band electrons will be included in the following section.

The one-electron states which are solutions to $h(\mathbf{r})$ are enumerated by the indices $\lambda, \lambda', \cdots$, where λ stands for the spin, the wave vector or momentum, and the band index. The one-electron energies and wave functions are ϵ_λ and ψ_λ, respectively, and they are solutions of

$$h\psi_\lambda = \epsilon_\lambda \psi_\lambda. \tag{21.27}$$

The ψ_λ are orthonormal in the volume of the crystal:

$$\int \psi_\lambda(\mathbf{r})^* \psi_{\lambda'}(\mathbf{r})\, d\mathbf{r} = \delta_{\lambda\lambda'}, \tag{21.28}$$

and in addition the ψ_λ form a complete set of functions in the space $\mathbf{r}$. It is convenient to define the ground-state occupation numbers g_λ, where $g_\lambda = 1$ for the occupied states, and $g_\lambda = 0$ for the unoccupied states which lie above the filled bands. Then the total density of band electrons at any position $\mathbf{r}$, denoted by $\rho(\mathbf{r})$, may be written

$$\rho(\mathbf{r}) = \sum_\lambda g_\lambda \psi_\lambda(\mathbf{r})^* \psi_\lambda(\mathbf{r}), \tag{21.29}$$

where here and in the following the Σ_λ goes over all λ in the complete set. Note that $\rho(\mathbf{r})$ is an *electron* density, not a *charge* density. In terms of the electronic density, the Hartree potential is

$$\Omega_S(\mathbf{r}) = e^2 \int \rho(\mathbf{r}')\, |\mathbf{r} - \mathbf{r}'|^{-1}\, d\mathbf{r}', \tag{21.30}$$

and with (21.29) this is

$$\Omega_S(\mathbf{r}) = e^2 \sum_\lambda g_\lambda \int \psi_\lambda(\mathbf{r}')^*\, |\mathbf{r} - \mathbf{r}'|^{-1}\, \psi_\lambda(\mathbf{r}')\, d\mathbf{r}'. \tag{21.31}$$

Fourier components of the Hartree potential are written

$$\begin{aligned} \Omega_S(\mathbf{q}) &= V^{-1} \int \Omega_S(\mathbf{r}) e^{-i\mathbf{q}\cdot\mathbf{r}}\, d\mathbf{r} \\ &= e^2 V^{-1} \iint \rho(\mathbf{r}')\, |\mathbf{r} - \mathbf{r}'|^{-1}\, e^{-i\mathbf{q}\cdot\mathbf{r}}\, d\mathbf{r}\, d\mathbf{r}' \\ &= e^2 V^{-1} \iint |\mathbf{r} - \mathbf{r}'|^{-1}\, e^{-i\mathbf{q}\cdot(\mathbf{r}-\mathbf{r}')}\, d\mathbf{r} \rho(\mathbf{r}') e^{-i\mathbf{q}\cdot\mathbf{r}'}\, d\mathbf{r}', \end{aligned} \tag{21.32}$$

where we used (21.30) for $\Omega_S(\mathbf{r})$. In the last line of (21.32), the $\int d\mathbf{r}$ is independent of $\mathbf{r}'$ (neglecting surface effects), and the $\int d\mathbf{r}'$ is just $\rho(\mathbf{q})$, so that

$$\Omega_S(\mathbf{q}) = e^2\rho(\mathbf{q})\int |\mathbf{r}|^{-1}\, e^{-i\mathbf{q}\cdot\mathbf{r}}\, d\mathbf{r}.$$

Here the integral is evaluated by introducing a convergence factor ζ,

$$\int |\mathbf{r}|^{-1}\, e^{-i\mathbf{q}\cdot\mathbf{r}} e^{-\zeta|\mathbf{r}|}\, d\mathbf{r} = 4\pi/(q^2 + \zeta^2), \tag{21.33}$$

and taking the limit $\zeta \to 0$ then gives

$$\Omega_S(\mathbf{q}) = (4\pi e^2/q^2)\rho(\mathbf{q}), \quad \mathbf{q} \neq 0. \tag{21.34}$$

Except for the uniform component of the band-electron density, which is $\rho(\mathbf{q} = 0)$, the Hartree field satisfies identically Poisson's equation

$$\nabla^2\Omega_S(\mathbf{r}) = -4\pi e^2\rho(\mathbf{r}); \tag{21.35}$$

this is shown by the following calculation.

$$\begin{aligned} \nabla^2\Omega_S(\mathbf{r}) &= \nabla^2 \sum_{\mathbf{q}} \Omega_S(\mathbf{q})e^{i\mathbf{q}\cdot\mathbf{r}} \\ &= -\sum_{\mathbf{q}}{}' q^2\Omega_S(\mathbf{q})e^{i\mathbf{q}\cdot\mathbf{r}} \\ &= -4\pi e^2 \sum_{\mathbf{q}}{}' \rho(\mathbf{q})e^{i\mathbf{q}\cdot\mathbf{r}}, \end{aligned} \tag{21.36}$$

where we used (21.34). The uniform band-electron charge density is exactly canceled by the uniform component of the ion charge density, so the net uniform charge density of the crystal is zero.

The total Hartree wave function for the band electrons in the ground state is

$$\Psi_G = \Pi_\lambda \psi_\lambda, \tag{21.37}$$

where the product is over all occupied states. The total ground-state energy is

$$E_G = \int \Psi_G \mathscr{H}_E \Psi_G\, d\{\mathbf{r}(\alpha)\}, \tag{21.38}$$

and the integral is over the crystal volume for all the electron coordinates $\mathbf{r}(\alpha)$. The $\mathscr{H}_E$ is given by (21.2) together with (21.22) for Ω_E, and if each electron is put into one of the occupied states ψ_λ, the integral in (21.38) becomes

$$\begin{aligned} E_G = &\sum_\lambda g_\lambda \int \psi_\lambda(\mathbf{r})^*[ke(\mathbf{r}) + \Omega_B(\mathbf{r})]\psi_\lambda(\mathbf{r})\, d\mathbf{r} \\ &+ \tfrac{1}{2}e^2 \sum_{\lambda\lambda'}{}' g_\lambda g_{\lambda'} \iint \psi_\lambda(\mathbf{r})^*\psi_{\lambda'}(\mathbf{r}')^*\, |\mathbf{r} - \mathbf{r}'|^{-1}\, \psi_{\lambda'}(\mathbf{r}')\psi_\lambda(\mathbf{r})\, d\mathbf{r}\, d\mathbf{r}'. \end{aligned} \tag{21.39}$$

Now the $\Sigma'_{\lambda\lambda'}$ can be replaced by $\Sigma_{\lambda\lambda'}$ with negligible error, i.e., with relative error of order N_0^{-1} in crystals, and with (21.31) for $\Omega_S(\mathbf{r})$ the expression (21.39) becomes

$$E_G = \sum_\lambda g_\lambda \int \psi_\lambda(\mathbf{r})^*[ke(\mathbf{r}) + \Omega_B(\mathbf{r}) + \tfrac{1}{2}\Omega_S(\mathbf{r})]\psi_\lambda(\mathbf{r})\, d\mathbf{r}. \qquad (21.40)$$

Adding and subtracting $\frac{1}{2}\Omega_S(\mathbf{r})$ in the integrand, and using the one-electron Hamiltonian (21.24) and one-electron wave equation (21.27), then gives

$$\begin{aligned} E_G &= \sum_\lambda g_\lambda \int \psi_\lambda(\mathbf{r})^*[h(\mathbf{r}) - \tfrac{1}{2}\Omega_S(\mathbf{r})]\psi_\lambda(\mathbf{r})\, d\mathbf{r} \\ &= \sum_\lambda g_\lambda \epsilon_\lambda - \tfrac{1}{2}\sum_\lambda g_\lambda \int \psi_\lambda(\mathbf{r})^*\Omega_S(\mathbf{r})\psi_\lambda(\mathbf{r})\, d\mathbf{r}. \end{aligned} \qquad (21.41)$$

Thus the total ground-state energy is the sum of occupied one-electron energies, i.e., $\Sigma_\lambda g_\lambda \epsilon_\lambda$, minus one half the total Hartree field contribution to $\Sigma_\lambda g_\lambda \epsilon_\lambda$. In other words, in summing ϵ_λ over all occupied states, the total Coulomb energy Ω_E is counted twice, so a double counting correction must be subtracted, as written in (21.41). This is *exactly the same type of double counting correction* as discussed in Section 13, where the total lattice-dynamics energy levels were calculated, and from that discussion we could write down (21.41) for E_G directly.

Expansion in Ion Displacements

We will now treat the displacements $\mathbf{U}(\mathrm{N}\nu)$ of the ions from their equilibrium positions $\mathbf{R}(\mathrm{N}\nu)$ as a perturbation, and calculate E_G to second order in the displacements. In the total one-electron potential Ω of (21.25), Ω_B depends explicitly, and Ω_S implicitly, on the ion displacements. The total potential may be formally expanded in powers of the $\mathbf{U}(\mathrm{N}\nu)$ as

$$\Omega = \Omega^{(0)} + \Omega^{(1)} + \Omega^{(2)} + \cdots, \qquad (21.42)$$

where

$$\Omega^{(0)} = \Omega(\mathbf{R}(\mathrm{N}\nu)); \qquad (21.43)$$

$$\Omega^{(1)} = \sum_{\mathrm{N}\nu}\sum_i [\partial\Omega/\partial U_i(\mathrm{N}\nu)]U_i(\mathrm{N}\nu); \qquad (21.44)$$

$$\Omega^{(2)} = \tfrac{1}{2}\sum_{\mathrm{NP}\nu\pi}\sum_{ij} [\partial^2\Omega/\partial U_i(\mathrm{N}\nu)\partial U_j(\mathrm{P}\pi)]U_i(\mathrm{N}\nu)U_j(\mathrm{P}\pi). \qquad (21.45)$$

The derivatives are evaluated at the initial equilibrium configuration, which is at all $\mathbf{U}(\mathrm{N}\nu) = 0$. Similarly the contributions Ω_B and Ω_S may each be expanded,

$$\Omega_B = \Omega_B^{(0)} + \Omega_B^{(1)} + \Omega_B^{(2)} + \cdots, \qquad (21.46)$$

$$\Omega_S = \Omega_S^{(0)} + \Omega_S^{(1)} + \Omega_S^{(2)} + \cdots, \qquad (21.47)$$

where the various terms are defined as in (21.43)–(21.45) for the expansion of Ω. The band-electron energies and wave functions may be formally expanded in powers of the displacements, and calculated in perturbation theory:

$$\epsilon_\lambda = \epsilon_\lambda^{(0)} + \epsilon_\lambda^{(1)} + \epsilon_\lambda^{(2)} + \cdots, \tag{21.48}$$

$$\psi_\lambda = \psi_\lambda^{(0)} + \psi_\lambda^{(1)} + \psi_\lambda^{(2)} + \cdots. \tag{21.49}$$

Let us introduce the shorthand notation for matrix elements

$$\langle\psi_\lambda|\,\Omega\,|\psi_{\lambda'}\rangle = \int \psi_\lambda(\mathbf{r})^*\Omega(\mathbf{r})\psi_{\lambda'}(\mathbf{r})\,d\mathbf{r}, \tag{21.50}$$

where it may be noted that this matrix element, as well as all other matrix elements we shall encounter, vanishes unless the two spins σ and σ' are the same. From (21.41), the zeroth-order ground state energy is

$$E_{\mathrm{G}}^{(0)} = \sum_\lambda g_\lambda \epsilon_\lambda^{(0)} - \tfrac{1}{2}\sum_\lambda g_\lambda \langle\psi_\lambda^{(0)}|\,\Omega_{\mathrm{S}}^{(0)}\,|\psi_\lambda^{(0)}\rangle, \tag{21.51}$$

and this is the energy of the band electrons when all the ions are at their equilibrium positions. The first-order ground-state energy may be written

$$E_{\mathrm{G}}^{(1)} = \sum_\lambda g_\lambda \epsilon_\lambda^{(1)} - \tfrac{1}{2}\sum_\lambda g_\lambda \langle\psi_\lambda|\,\Omega_{\mathrm{S}}\,|\psi_\lambda\rangle^{(1)}, \tag{21.52}$$

where the first-order one-electron energies are

$$\epsilon_\lambda^{(1)} = \langle\psi_\lambda^{(0)}|\,\Omega^{(1)}\,|\psi_\lambda^{(0)}\rangle, \tag{21.53}$$

and the last term in (21.52) is meant to include all the first-order terms in the matrix elements, according to

$$\langle\psi_\lambda|\,\Omega_{\mathrm{S}}\,|\psi_\lambda\rangle^{(1)} = \langle\psi_\lambda^{(1)}|\,\Omega_{\mathrm{S}}^{(0)}\,|\psi_\lambda^{(0)}\rangle + \langle\psi_\lambda^{(0)}|\,\Omega_{\mathrm{S}}^{(1)}\,|\psi_\lambda^{(0)}\rangle + \langle\psi_\lambda^{(0)}|\,\Omega_{\mathrm{S}}^{(0)}\,|\psi_\lambda^{(1)}\rangle. \tag{21.54}$$

The contributions to $E_{\mathrm{G}}^{(1)}$ of the matrix elements in (21.54) may be rewritten and combined to simplify (21.52) for $E_{\mathrm{G}}^{(1)}$, as follows. From (21.31) for $\Omega_{\mathrm{S}}(\mathbf{r})$, we can write

$$\sum_\lambda g_\lambda[\langle\psi_\lambda^{(1)}|\,\Omega_{\mathrm{S}}^{(0)}\,|\psi_\lambda^{(0)}\rangle + \langle\psi_\lambda^{(0)}|\,\Omega_{\mathrm{S}}^{(0)}\,|\psi_\lambda^{(1)}\rangle]$$

$$= e^2\sum_{\lambda\lambda'} g_\lambda g_{\lambda'}\iint [\psi_\lambda^{(1)}(\mathbf{r})^*\psi_{\lambda'}^{(0)}(\mathbf{r}')^*\,|\mathbf{r}-\mathbf{r}'|^{-1}\,\psi_{\lambda'}^{(0)}(\mathbf{r}')\psi_\lambda^{(0)}(\mathbf{r}) \tag{21.55}$$

$$+ \psi_\lambda^{(0)}(\mathbf{r})^*\psi_{\lambda'}^{(0)}(\mathbf{r}')^*\,|\mathbf{r}-\mathbf{r}'|^{-1}\,\psi_{\lambda'}^{(0)}(\mathbf{r}')\psi_\lambda^{(1)}(\mathbf{r})]\,d\mathbf{r}\,d\mathbf{r}'.$$

In the same way the sum of matrix elements of $\Omega_S^{(1)}$ can be written

$$\sum_\lambda g_\lambda \langle \psi_\lambda^{(0)} | \Omega_S^{(1)} | \psi_\lambda^{(0)} \rangle$$
$$= e^2 \sum_{\lambda\lambda'} g_\lambda g_{\lambda'} \iint [\psi_\lambda^{(0)}(\mathbf{r})^* \psi_{\lambda'}^{(1)}(\mathbf{r}')^* |\mathbf{r} - \mathbf{r}'|^{-1} \psi_{\lambda'}^{(0)}(\mathbf{r}')\psi_\lambda^{(0)}(\mathbf{r}) \quad (21.56)$$
$$+ \psi_\lambda^{(0)}(\mathbf{r})^* \psi_{\lambda'}^{(0)}(\mathbf{r}')^* |\mathbf{r} - \mathbf{r}'|^{-1} \psi_{\lambda'}^{(1)}(\mathbf{r}')\psi_\lambda^{(0)}(\mathbf{r})] \, d\mathbf{r} \, d\mathbf{r}'.$$

Now by interchanging λ and λ', and $\mathbf{r}$ and $\mathbf{r}'$, in say (21.55), it is the same as (21.56), and this means

$$\sum_\lambda g_\lambda \langle \psi_\lambda^{(0)} | \Omega_S^{(1)} | \psi_\lambda^{(0)} \rangle = \sum_\lambda g_\lambda [\langle \psi_\lambda^{(1)} | \Omega_S^{(0)} | \psi_\lambda^{(0)} \rangle + \langle \psi_\lambda^{(0)} | \Omega_S^{(0)} | \psi_\lambda^{(1)} \rangle]. \quad (21.57)$$

Then with (21.54) for $\langle \psi_\lambda | \Omega_S | \psi_\lambda \rangle^{(1)}$, it follows that

$$-\tfrac{1}{2} \sum_\lambda g_\lambda \langle \psi_\lambda | \Omega_S | \psi_\lambda \rangle^{(1)} = -\sum_\lambda g_\lambda \langle \psi_\lambda^{(0)} | \Omega_S^{(1)} | \psi_\lambda^{(0)} \rangle. \quad (21.58)$$

Finally since $\epsilon_\lambda^{(1)}$ is given by (21.53), in which $\Omega^{(1)} = \Omega_B^{(1)} + \Omega_S^{(1)}$, then (21.52) for $E_G^{(1)}$ becomes simply

$$E_G^{(1)} = \sum_\lambda g_\lambda \langle \psi_\lambda^{(0)} | \Omega_B^{(1)} | \psi_\lambda^{(0)} \rangle. \quad (21.59)$$

Thus $E_G^{(1)}$ contains only diagonal matrix elements of the bare ion potential $\Omega_B^{(1)}$, and the screening field $\Omega_S^{(1)}$ has been eliminated.

The first-order corrections to the one-electron wave functions are given by

$$\psi_\lambda^{(1)} = \sum_{\lambda'}{}' \frac{\langle \psi_{\lambda'}^{(0)} | \Omega^{(1)} | \psi_\lambda^{(0)} \rangle}{[\epsilon_\lambda^{(0)} - \epsilon_{\lambda'}^{(0)}]_p} \psi_{\lambda'}^{(0)}, \quad (21.60)$$

and the second-order contributions to ϵ_λ may be written

$$\epsilon_\lambda^{(2)} = \tfrac{1}{2}[\langle \psi_\lambda^{(1)} | \Omega^{(1)} | \psi_\lambda^{(0)} \rangle + \langle \psi_\lambda^{(0)} | \Omega^{(1)} | \psi_\lambda^{(1)} \rangle] + \langle \psi_\lambda^{(0)} | \Omega^{(2)} | \psi_\lambda^{(0)} \rangle. \quad (21.61)$$

Again from (21.41) for E_G, the second-order ground-state energy is

$$E_G^{(2)} = \sum_\lambda g_\lambda \epsilon_\lambda^{(2)} - \tfrac{1}{2} \sum_\lambda g_\lambda \langle \psi_\lambda | \Omega_S | \psi_\lambda \rangle^{(2)}. \quad (21.62)$$

Here the second-order matrix elements $\langle \psi_\lambda | \Omega_S | \psi_\lambda \rangle^{(2)}$ contain six contributions, as follows.

$$\langle \psi_\lambda | \Omega_S | \psi_\lambda \rangle^{(2)} = \langle \psi_\lambda^{(2)} | \Omega_S^{(0)} | \psi_\lambda^{(0)} \rangle + \langle \psi_\lambda^{(0)} | \Omega_S^{(2)} | \psi_\lambda^{(0)} \rangle + \langle \psi_\lambda^{(0)} | \Omega_S^{(0)} | \psi_\lambda^{(2)} \rangle$$
$$+ \langle \psi_\lambda^{(1)} | \Omega_S^{(0)} | \psi_\lambda^{(1)} \rangle + \langle \psi_\lambda^{(1)} | \Omega_S^{(1)} | \psi_\lambda^{(0)} \rangle + \langle \psi_\lambda^{(0)} | \Omega_S^{(1)} | \psi_\lambda^{(1)} \rangle. \quad (21.63)$$

The contributions of these matrix elements to $E_G^{(2)}$ can be written out, as in (21.55) and (21.56) for the first-order screening contributions, and again by interchanging λ and λ' inside sums and $\mathbf{r}$ and $\mathbf{r}'$ inside integrals, it is possible to show

$$\sum_\lambda g_\lambda \langle \psi_\lambda^{(0)} | \Omega_S^{(2)} | \psi_\lambda^{(0)} \rangle = \sum_\lambda g_\lambda [\langle \psi_\lambda^{(2)} | \Omega_S^{(0)} | \psi_\lambda^{(0)} \rangle + \langle \psi_\lambda^{(0)} | \Omega_S^{(0)} | \psi_\lambda^{(2)} \rangle + \langle \psi_\lambda^{(1)} | \Omega_S^{(0)} | \psi_\lambda^{(1)} \rangle]. \quad (21.64)$$

Then the total double counting correction in (21.62) for $E_G^{(2)}$ simplifies to

$$-\tfrac{1}{2} \sum_\lambda g_\lambda \langle \psi_\lambda | \Omega_S | \psi_\lambda \rangle^{(2)} = -\tfrac{1}{2} \sum_\lambda g_\lambda [\langle \psi_\lambda^{(1)} | \Omega_S^{(1)} | \psi_\lambda^{(0)} \rangle + \langle \psi_\lambda^{(0)} | \Omega_S^{(1)} | \psi_\lambda^{(1)} \rangle] - \sum_\lambda g_\lambda \langle \psi_\lambda^{(0)} | \Omega_S^{(2)} | \psi_\lambda^{(0)} \rangle. \quad (21.65)$$

Finally with (21.61) for $\epsilon_\lambda^{(2)}$, and with (21.65) for the double counting correction, $E_G^{(2)}$ may be written

$$E_G^{(2)} = \tfrac{1}{2} \sum_\lambda g_\lambda [\langle \psi_\lambda^{(1)} | \Omega_B^{(1)} | \psi_\lambda^{(0)} \rangle + \langle \psi_\lambda^{(0)} | \Omega_B^{(1)} | \psi_\lambda^{(1)} \rangle] + \sum_\lambda g_\lambda \langle \psi_\lambda^{(0)} | \Omega_B^{(2)} | \psi_\lambda^{(0)} \rangle. \quad (21.66)$$

Thus $E_G^{(2)}$ is written here in terms of matrix elements of the bare ion potentials $\Omega_B^{(1)}$ and $\Omega_B^{(2)}$. The second-order screening field $\Omega_S^{(2)}$ has been eliminated from $E_G^{(2)}$, but $\Omega_S^{(1)}$ has not been eliminated since it appears in $\psi_\lambda^{(1)}$ according to (21.60). For computational purposes it is convenient to eliminate $\Omega_S^{(1)}$, so that all computations can be based on the zeroth-order wave functions $\psi_\lambda^{(0)}$ and the bare ion potential Ω_B; this elimination is carried out in terms of the dielectric function.

Dielectric Function

The local potentials we are dealing with can be Fourier transformed according to (21.18), and a useful result for any local potential Ω is

$$\langle \psi_{\lambda'}^{(0)} | \Omega | \psi_\lambda^{(0)} \rangle = \int \psi_{\lambda'}^{(0)}(\mathbf{r})^* \Omega(\mathbf{r}) \psi_\lambda^{(0)}(\mathbf{r}) \, d\mathbf{r} = \sum_{\mathbf{q}} \Omega(\mathbf{q}) \langle \psi_{\lambda'}^{(0)} | e^{i\mathbf{q}\cdot\mathbf{r}} | \psi_\lambda^{(0)} \rangle. \quad (21.67)$$

This general result will be used below with different local potentials in place of Ω.

The band-electronic density $\rho(\mathbf{r})$ may be expanded in powers of the ion displacements as

$$\rho(\mathbf{r}) = \rho^{(0)}(\mathbf{r}) + \rho^{(1)}(\mathbf{r}) + \cdots, \quad (21.68)$$

and from (21.29) for $\rho(\mathbf{r})$ it follows that

$$\rho^{(1)}(\mathbf{r}) = \sum_{\lambda} g_\lambda[\psi_\lambda^{(1)}(\mathbf{r})\psi_\lambda^{(0)}(\mathbf{r})^* + \text{complex conjugate}]. \tag{21.69}$$

With (21.60) for $\psi_\lambda^{(1)}$, this is written out as

$$\rho^{(1)}(\mathbf{r}) = \sum_{\lambda} g_\lambda \sum_{\lambda'}{}' \left[\frac{\langle\psi_{\lambda'}^{(0)}|\, \Omega^{(1)}\, |\psi_\lambda^{(0)}\rangle}{[\epsilon_\lambda^{(0)} - \epsilon_{\lambda'}^{(0)}]_p} \psi_{\lambda'}^{(0)}(\mathbf{r})\psi_\lambda^{(0)}(\mathbf{r})^* + \text{complex conjugate}\right]. \tag{21.70}$$

By interchanging λ and λ' in the complex conjugate term this becomes

$$\rho^{(1)}(\mathbf{r}) = \sum_{\lambda\lambda'}{}' (g_\lambda - g_{\lambda'}) \frac{\langle\psi_{\lambda'}^{(0)}|\, \Omega^{(1)}\, |\psi_\lambda^{(0)}\rangle}{[\epsilon_\lambda^{(0)} - \epsilon_{\lambda'}^{(0)}]_p} \psi_\lambda^{(0)}(\mathbf{r})^* \psi_{\lambda'}^{(0)}(\mathbf{r}). \tag{21.71}$$

Then the Fourier components

$$\rho^{(1)}(\mathbf{q}) = V^{-1}\int \rho^{(1)}(\mathbf{r})e^{-i\mathbf{q}\cdot\mathbf{r}}\, d\mathbf{r} \tag{21.72}$$

are calculated from (21.71), with the aid of (21.67) for the matrix elements of $\Omega^{(1)}$, to obtain

$$\rho^{(1)}(\mathbf{q}) = \sum_{q'} a(\mathbf{q},\mathbf{q}')\Omega^{(1)}(\mathbf{q}'), \tag{21.73}$$

where

$$a(\mathbf{q},\mathbf{q}') = V^{-1}\sum_{\lambda\lambda'}{}' (g_\lambda - g_{\lambda'}) \frac{\langle\psi_\lambda^{(0)}|\, e^{-i\mathbf{q}\cdot\mathbf{r}}\, |\psi_{\lambda'}^{(0)}\rangle\langle\psi_{\lambda'}^{(0)}|\, e^{i\mathbf{q}'\cdot\mathbf{r}}\, |\psi_\lambda^{(0)}\rangle}{[\epsilon_\lambda^{(0)} - \epsilon_{\lambda'}^{(0)}]_p}. \tag{21.74}$$

By interchanging λ and λ' inside the sum, it is seen that the coefficients $a(\mathbf{q},\mathbf{q}')$ satisfy the symmetry

$$a(\mathbf{q},\mathbf{q}') = a(-\mathbf{q}',-\mathbf{q}), \tag{21.75}$$

and the complex conjugate of (21.74) gives

$$a(\mathbf{q},\mathbf{q}')^* = a(\mathbf{q}',\mathbf{q}). \tag{21.76}$$

We also note that $a(\mathbf{q},\mathbf{q}')$ vanishes when either $\mathbf{q}$ or $\mathbf{q}'$ is zero.

The Hartree field $\Omega_S(\mathbf{r})$ satisfies Poisson's equation by virtue of (21.34), which is

$$\Omega_S(\mathbf{q}) = (4\pi e^2/q^2)\rho(\mathbf{q}).$$

Since this holds for arbitrary positions of the ions, it must hold in each order of the displacements of the ions from equilibrium, so that

$$\Omega_S^{(1)}(\mathbf{q}) = (4\pi e^2/q^2)\rho^{(1)}(\mathbf{q}). \tag{21.77}$$

Eliminating $\rho^{(1)}(\mathbf{q})$ by (21.73), and writing $\Omega^{(1)}(\mathbf{q}') = \Omega_B^{(1)}(\mathbf{q}') + \Omega_S^{(1)}(\mathbf{q}')$, then transforms (21.77) to

$$\Omega_S^{(1)}(\mathbf{q}) = (4\pi e^2/q^2) \sum_{\mathbf{q}'} a(\mathbf{q},\mathbf{q}')[\Omega_B^{(1)}(\mathbf{q}') + \Omega_S^{(1)}(\mathbf{q}')], \tag{21.78}$$

or

$$\sum_{\mathbf{q}'} [\delta_{\mathbf{qq}'} - (4\pi e^2/q^2)a(\mathbf{q},\mathbf{q}')]\Omega_S^{(1)}(\mathbf{q}') = (4\pi e^2/q^2) \sum_{\mathbf{q}'} a(\mathbf{q},\mathbf{q}')\Omega_B^{(1)}(\mathbf{q}'). \tag{21.79}$$

This leads us to the dielectric function $d(\mathbf{q},\mathbf{q}')$,

$$d(\mathbf{q},\mathbf{q}') = \delta_{\mathbf{qq}'} - (4\pi e^2/q^2)a(\mathbf{q},\mathbf{q}'), \tag{21.80}$$

and in terms of the dielectric function, (21.79) may be written

$$\sum_{\mathbf{q}'} d(\mathbf{q},\mathbf{q}')\Omega_S^{(1)}(\mathbf{q}') = \sum_{\mathbf{q}'} [\delta_{\mathbf{qq}'} - d(\mathbf{q},\mathbf{q}')]\Omega_B^{(1)}(\mathbf{q}'). \tag{21.81}$$

In order to solve for the screening field $\Omega_S^{(1)}$, we define the inverse dielectric function $b(\mathbf{q},\mathbf{q}')$ by the relation

$$\sum_{\mathbf{q}'} b(\mathbf{q},\mathbf{q}')d(\mathbf{q}',\mathbf{q}'') = \sum_{\mathbf{q}'} d(\mathbf{q},\mathbf{q}')b(\mathbf{q}',\mathbf{q}'') = \delta_{\mathbf{qq}''}. \tag{21.82}$$

Then multiplying (21.81) by $b(\mathbf{q}'',\mathbf{q})$, summing over $\mathbf{q}$, and relabeling indices gives

$$\Omega_S^{(1)}(\mathbf{q}) = \sum_{\mathbf{q}'} [b(\mathbf{q},\mathbf{q}') - \delta_{\mathbf{qq}'}]\Omega_B^{(1)}(\mathbf{q}'). \tag{21.83}$$

Finally it is convenient to define the screening potential–bare ion potential correlation function $c(\mathbf{q},\mathbf{q}')$,

$$c(\mathbf{q},\mathbf{q}') = (q^2 V_C/4\pi e^2)[b(\mathbf{q},\mathbf{q}') - \delta_{\mathbf{qq}'}], \tag{21.84}$$

where as usual V_C is the volume per unit cell. Then from (21.83), $\Omega_S^{(1)}(\mathbf{q})$ is expressed as

$$\Omega_S^{(1)}(\mathbf{q}) = (4\pi e^2/q^2 V_C) \sum_{\mathbf{q}'} c(\mathbf{q},\mathbf{q}')\Omega_B^{(1)}(\mathbf{q}'), \tag{21.85}$$

and this result will be used to eliminate the screening field components $\Omega_S^{(1)}(\mathbf{q})$ in the following calculations. Alternatively one can add $\Omega_B^{(1)}(\mathbf{q})$ to both sides of (21.83) to obtain

$$\Omega^{(1)}(\mathbf{q}) = \sum_{\mathbf{q}'} b(\mathbf{q},\mathbf{q}')\Omega_B^{(1)}(\mathbf{q}'), \tag{21.86}$$

which relates the total potential to the bare ion potential through the inverse dielectric function $b(\mathbf{q},\mathbf{q}')$.

The dielectric function $d(\mathbf{q},\mathbf{q}')$, and the correlation function $c(\mathbf{q},\mathbf{q}')$, are expressed by the above equations in a form which is amenable to computer calculation. We do not claim the calculations to be easy, but they should not be essentially more difficult than the work which is presently being carried out in determining band structures. The zeroth-order one-electron wave functions are evaluated with all ions located at their equilibrium positions, and hence they satisfy the Bloch theorem, which may be written

$$\psi_\lambda^{(0)}(\mathbf{r}) = e^{i\mathbf{p}\cdot\mathbf{R}(N)}\psi_\lambda^{(0)}(\mathbf{r} - \mathbf{R}(N)), \tag{21.87}$$

where $\mathbf{p}$ is the wave vector of the state λ. Because of this property, the matrix elements $\langle\psi_\lambda^{(0)}|\, e^{i\mathbf{q}\cdot\mathbf{r}}\, |\psi_{\lambda'}^{(0)}\rangle$ can be represented as a sum of integrals over all the unit cells in the crystal, and the result is

$$\langle\psi_\lambda^{(0)}|\, e^{i\mathbf{q}\cdot\mathbf{r}}\, |\psi_{\lambda'}^{(0)}\rangle = \delta(\mathbf{q} - \mathbf{p} + \mathbf{p}')N_0\int_C \psi_\lambda^{(0)}(\mathbf{r})^* e^{i\mathbf{q}\cdot\mathbf{r}}\psi_{\lambda'}^{(0)}(\mathbf{r})\, d\mathbf{r}, \tag{21.88}$$

where $\delta(\mathbf{q} - \mathbf{p} + \mathbf{p}')$ requires $\mathbf{q} - \mathbf{p} + \mathbf{p}' = \mathbf{Q}$ with $\mathbf{Q}$ any reciprocal lattice vector, and where the $\int_C d\mathbf{r}$ is over one unit cell. Note that $N_0\int_C d\mathbf{r}$ in (21.88) is of order 1, not N_0, since the $\psi_\lambda^{(0)}$ are normalized to the volume $V = N_0V_C$. The result (21.88) then leads to the restriction that the dielectric function vanishes unless $\mathbf{q}' = \mathbf{q} + \mathbf{Q}$, for any inverse-lattice vector $\mathbf{Q}$:

$$a(\mathbf{q},\mathbf{q}') = \delta_{\mathbf{q}',\mathbf{q}+\mathbf{Q}}\, a(\mathbf{q},\mathbf{q} + \mathbf{Q}). \tag{21.89}$$

We also note that when either $\mathbf{q}$ or $\mathbf{q}'$ is large, say as large as several inverse unit cell vectors, then $a(\mathbf{q},\mathbf{q}')$ becomes negligibly small, since the matrix elements such as $\langle\psi_\lambda^{(0)}|\, e^{i\mathbf{q}\cdot\mathbf{r}}\, |\psi_{\lambda'}^{(0)}\rangle$ become small. These results simplify the computation of the dielectric function.

For $\mathbf{q} \neq \mathbf{q}'$, we expect $a(\mathbf{q},\mathbf{q}')$ to have simple zeros at $\mathbf{q} = 0$ and at $\mathbf{q}' = 0$; hence $d(\mathbf{q},\mathbf{q}')$ and $b(\mathbf{q},\mathbf{q}')$ should each have a simple pole at $\mathbf{q} = 0$ and a simple zero at $\mathbf{q}' = 0$, and $c(\mathbf{q},\mathbf{q}')$ should have simple zeros at $\mathbf{q} = 0$ and at $\mathbf{q}' = 0$. In the present calculations the $\mathbf{q} = 0$ and $\mathbf{q}' = 0$ terms are to be interpreted in the sense of limits as $\mathbf{q} \to 0$ and as $\mathbf{q}' \to 0$, respectively, and those terms generally depend on the direction along which the vector approaches zero. Alternately, the $\mathbf{q} = 0$ and $\mathbf{q}' = 0$ terms may be explicitly omitted in the above calculations, and in those of the following two sections, with a corresponding redefinition of $b(\mathbf{q},\mathbf{q}')$.

22. APPROXIMATION FOR BAND-ELECTRON EXCHANGE

Density-Dependent Exchange

In order to provide a theory of lattice dynamics of crystals which can be evaluated with the results of current band-structure calculations, we consider exchange effects to be approximated by a local one-electron potential, rather than proceed to a Hartree–Fock approximation. Here the calculation of E_G

is extended to include an approximation for exchange which depends only on the density $\rho(\mathbf{r})$ of the band electrons; alternate forms of a local exchange potential may easily be included by the same method. The exchange interactions are limited to electrons with the same spin, but since all our calculations are spin-independent, the density of band electrons with each spin is the same function of $\mathbf{r}$, namely $\frac{1}{2}\rho(\mathbf{r})$. The total exchange energy of the band electrons in their ground state is presumed to be

$$\sum_\lambda g_\lambda \int \psi_\lambda(\mathbf{r})^* X(\mathbf{r}) \psi_\lambda(\mathbf{r})\, d\mathbf{r}, \tag{22.1}$$

where $X(\mathbf{r})$ represents exchange, and is a function only of $\rho(\mathbf{r})$:

$$X(\mathbf{r}) = X(\rho(\mathbf{r})). \tag{22.2}$$

The Slater type of expression for $X(\mathbf{r})$ is discussed at the end of this section.

It is instructive to proceed with a variational calculation, which is designed to determine the best one-electron wave equation by minimizing the total ground-state energy E_{G} with respect to variations $\delta\psi_\lambda$ of the one-electron wave functions ψ_λ. The variation of E_{G} is set equal to zero,

$$\delta E_{\mathrm{G}} = \sum_\lambda g_\lambda \int \{[\partial E_{\mathrm{G}}/\partial \psi_\lambda(\mathbf{r})^*]\delta\psi_\lambda(\mathbf{r})^* + [\partial E_{\mathrm{G}}/\partial \psi_\lambda(\mathbf{r})]\delta\psi_\lambda(\mathbf{r})\}\, d\mathbf{r} = 0, \tag{22.3}$$

subject to the condition that the wave functions remain normalized,

$$\delta \int \psi_\lambda(\mathbf{r})^* \psi_\lambda(\mathbf{r})\, d\mathbf{r} = \int [\delta\psi_\lambda(\mathbf{r})^* \psi_\lambda(\mathbf{r}) + \psi_\lambda(\mathbf{r})^* \delta\psi_\lambda(\mathbf{r})]\, d\mathbf{r} = 0. \tag{22.4}$$

We multiply (22.4) by the Lagrange multipliers ϵ_λ and sum over states occupied in the ground state; then subtract the result from (22.3) to obtain

$$\sum_\lambda g_\lambda \int \{[\partial E_{\mathrm{G}}/\partial \psi_\lambda(\mathbf{r})^*] - \epsilon_\lambda \psi_\lambda(\mathbf{r})\}\delta\psi_\lambda(\mathbf{r})^*\, d\mathbf{r} + \text{complex conjugate} = 0. \tag{22.5}$$

Now each $\delta\psi_\lambda(\mathbf{r})$ is an arbitrary function, so in the integrand of (22.5) the coefficients of the real and imaginary parts of each $\delta\psi_\lambda(\mathbf{r})$ must vanish everywhere, and this gives the one-electron wave equations

$$[\partial E_{\mathrm{G}}/\partial \psi_\lambda(\mathbf{r})^*] - \epsilon_\lambda \psi_\lambda(\mathbf{r}) = 0. \tag{22.6}$$

The Hartree ground-state energy is written in (21.40), and with the approximate exchange energy (22.1) included, the total ground-state energy is

$$E_{\mathrm{G}} = \sum_\lambda g_\lambda \int \psi_\lambda(\mathbf{r})^* [ke(\mathbf{r}) + \Omega_{\mathrm{B}}(\mathbf{r}) + \tfrac{1}{2}\Omega_{\mathrm{S}}(\mathbf{r}) + X(\mathbf{r})]\psi_\lambda(\mathbf{r})\, d\mathbf{r}. \tag{22.7}$$

With the aid of (21.29) for $\rho(\mathbf{r})$ and (21.30) for $\Omega_{\mathrm{S}}(\mathbf{r})$, the contributions in

(22.7) arising from Coulomb and exchange interactions among the band electrons may be written as follows.

$$\tfrac{1}{2}\sum_{\lambda} g_\lambda \int \psi_\lambda(\mathbf{r})^* \Omega_S(\mathbf{r})\psi_\lambda(\mathbf{r})\, d\mathbf{r} = \tfrac{1}{2}e^2 \iint \rho(\mathbf{r})\rho(\mathbf{r}')\, |\mathbf{r}-\mathbf{r}'|^{-1}\, d\mathbf{r}\, d\mathbf{r}'; \tag{22.8}$$

$$\sum_{\lambda} g_\lambda \int \psi_\lambda(\mathbf{r})^* X(\mathbf{r})\psi_\lambda(\mathbf{r})\, d\mathbf{r} = \int \rho(\mathbf{r})X(\mathbf{r})\, d\mathbf{r}. \tag{22.9}$$

Also from (21.29) for $\rho(\mathbf{r})$, it follows that

$$\partial\rho(\mathbf{r})/\partial\psi_\lambda(\mathbf{r})^* = \psi_\lambda(\mathbf{r}), \tag{22.10}$$

so that differentiation of (22.8) and (22.9) with respect to $\psi_\lambda(\mathbf{r})^*$ gives the following two contributions to $\partial E_G/\partial\psi_\lambda(\mathbf{r})^*$, respectively:

$$e^2 \int \psi_\lambda(\mathbf{r})\rho(\mathbf{r}')\, |\mathbf{r}-\mathbf{r}'|^{-1}\, d\mathbf{r}' = \Omega_S(\mathbf{r})\psi_\lambda(\mathbf{r});$$

$$[\partial(\rho X)/\partial\rho]\psi_\lambda(\mathbf{r}).$$

Then with these two contributions, and noting that ke and Ω_B in (22.7) do not depend on the $\psi_\lambda(\mathbf{r})$, the one-electron wave equation (22.6) becomes

$$(ke + \Omega)\psi_\lambda = \epsilon_\lambda\psi_\lambda, \tag{22.11}$$

where the total one-electron potential Ω is given by

$$\Omega = \Omega_B + \Omega_S + \Omega_X; \tag{22.12}$$

$$\Omega_X = \partial(\rho X)/\partial\rho. \tag{22.13}$$

Thus the one-electron Hamiltonian is just the Hartree Hamiltonian (21.24), plus the one-electron exchange potential Ω_X.

The total ground-state energy may again be written as a sum of one-electron energies ϵ_λ, with double counting corrections subtracted. By adding and subtracting $\tfrac{1}{2}\Omega_S + \Omega_X$ in the integrand of (22.7), and using the one-electron wave equation (22.11), the ground-state energy becomes

$$E_G = \sum_\lambda g_\lambda\epsilon_\lambda - \tfrac{1}{2}\sum_\lambda g_\lambda\langle\psi_\lambda|\, \Omega_S\, |\psi_\lambda\rangle + \sum_\lambda g_\lambda\langle\psi_\lambda|\, X - \Omega_X\, |\psi_\lambda\rangle. \tag{22.14}$$

Here the Coulomb double counting correction is of course the same as in (21.41) for the Hartree ground-state energy, and the last term in (22.14) is the "double counting" correction for the band-electron exchange energy. It is convenient to rewrite the exchange double counting correction as follows. From (22.13), Ω_X may be written

$$\Omega_X = X + \rho(\partial X/\partial\rho), \tag{22.15}$$

and then the last term in (22.14) is

$$\sum_\lambda g_\lambda \langle \psi_\lambda | \, X - \Omega_X \, | \psi_\lambda \rangle = -\sum_\lambda g_\lambda \langle \psi_\lambda | \, \rho(\partial X/\partial \rho) \, | \psi_\lambda \rangle = -\int \rho^2 (\partial X/\partial \rho) \, d\mathbf{r}, \tag{22.16}$$

where the last form on the right is obtained from (21.29) for ρ.

The next step is to expand the ground-state energy in powers of the displacements of the ions from equilibrium. We abbreviate the derivatives of X with respect to ρ, evaluated at the equilibrium configuration, by the notation

$$X' = (\partial X/\partial \rho)^{(0)}; \quad X'' = (\partial^2 X/\partial \rho^2)^{(0)}; \quad \text{etc.} \tag{22.17}$$

Then the expansion of $X(\rho)$, correct to second order, is

$$X(\rho) = X^{(0)} + X'\rho^{(1)} + \tfrac{1}{2}X''[\rho^{(1)}]^2 + X'\rho^{(2)} + \cdots. \tag{22.18}$$

Similarly, since $\rho = \rho^{(0)} + \rho^{(1)} + \rho^{(2)} + \cdots$, we can write

$$\rho(\partial X/\partial \rho) = \rho^{(0)}X' + [\rho^{(0)}X'' + X']\rho^{(1)} + [\tfrac{1}{2}\rho^{(0)}X''' + X''][\rho^{(1)}]^2 + [\rho^{(0)}X'' + X']\rho^{(2)} + \cdots. \tag{22.19}$$

Also, according to (22.15), the expansion of Ω_X is the sum of (22.18) and (22.19), and the expansion of $\rho^2(\partial X/\partial \rho)$ is just ρ times (22.19).

The zeroth-order term in (22.19) is $\rho^{(0)}X'$, and with this in the second expression of (22.16) the zeroth-order terms in (22.14) for E_G may be written

$$E_G^{(0)} = \sum_\lambda g_\lambda [\epsilon_\lambda^{(0)} - \langle \psi_\lambda^{(0)} | \, \tfrac{1}{2}\Omega_S^{(0)} + \rho^{(0)}X' \, | \psi_\lambda^{(0)} \rangle]. \tag{22.20}$$

It is convenient to use the third expression of (22.16) to express $E_G^{(1)}$ as

$$E_G^{(1)} = \sum_\lambda g_\lambda \epsilon_\lambda^{(1)} - \tfrac{1}{2}\sum_\lambda g_\lambda \langle \psi_\lambda | \, \Omega_S \, | \psi_\lambda \rangle^{(1)} - \int [\rho^2(\partial X/\partial \rho)]^{(1)} \, d\mathbf{r}. \tag{22.21}$$

Now of course $\epsilon_\lambda^{(1)}$ and $-\tfrac{1}{2}\Sigma_\lambda \, g_\lambda \langle \psi_\lambda | \, \Omega_S \, | \psi_\lambda \rangle^{(1)}$ are still given by (21.53) and (21.58), respectively, and the last term in (22.21) is evaluated by the calculation

$$\sum_\lambda g_\lambda \langle \psi_\lambda^{(0)} | \, \Omega_X^{(1)} \, | \psi_\lambda^{(0)} \rangle = \int \rho^{(0)} \Omega_X^{(1)} \, d\mathbf{r} = \int \rho^{(0)} [\rho^{(0)}X'' + 2X']\rho^{(1)} \, d\mathbf{r} = \int [\rho^2(\partial X/\partial \rho)]^{(1)} \, d\mathbf{r}. \tag{22.22}$$

Therefore (22.21) for $E_G^{(1)}$ is simply

$$E_G^{(1)} = \sum_\lambda g_\lambda \langle \psi_\lambda^{(0)} | \Omega_B^{(1)} | \psi_\lambda^{(0)} \rangle, \tag{22.23}$$

and this is exactly the result obtained previously, in (21.59), without exchange. The wave functions $\psi_\lambda^{(0)}$ are now different, however, since they satisfy the one-electron wave equation (22.11) which includes the exchange potential. Finally the second-order contribution to the exchange double counting correction (22.16) is evaluated by the calculation

$$\begin{aligned} \tfrac{1}{2} \sum_\lambda g_\lambda [\langle \psi_\lambda^{(1)} | \Omega_X^{(1)} | \psi_\lambda^{(0)} \rangle + \langle \psi_\lambda^{(0)} | \Omega_X^{(1)} | \psi_\lambda^{(1)} \rangle] &+ \sum_\lambda g_\lambda \langle \psi_\lambda^{(0)} | \Omega_X^{(2)} | \psi_\lambda^{(0)} \rangle \\ &= \int [\tfrac{1}{2} \rho^{(1)} \Omega_X^{(1)} + \rho^{(0)} \Omega_X^{(2)}] \, d\mathbf{r} \\ &= \int [\rho^2 (\partial X / \partial \rho)]^{(2)} \, d\mathbf{r} \end{aligned} \tag{22.24}$$

Combining this result with (21.61) for $\epsilon_\lambda^{(2)}$ and (21.65) for $-\frac{1}{2}\Sigma_\lambda g_\lambda \langle \psi_\lambda | \Omega_S | \psi_\lambda \rangle^{(2)}$, the sum of second-order terms in (22.14) for E_G is

$$\begin{aligned} E_G^{(2)} = \tfrac{1}{2} \sum_\lambda g_\lambda [\langle \psi_\lambda^{(1)} | \Omega_B^{(1)} | \psi_\lambda^{(0)} \rangle &+ \langle \psi_\lambda^{(0)} | \Omega_B^{(1)} | \psi_\lambda^{(1)} \rangle] \\ &+ \sum_\lambda g_\lambda \langle \psi_\lambda^{(0)} | \Omega_B^{(2)} | \psi_\lambda^{(0)} \rangle. \end{aligned} \tag{22.25}$$

Again this is the result obtained in (21.66) without exchange, and $\psi_\lambda^{(1)}$ is still given by (21.60), where now

$$\Omega^{(1)} = \Omega_B^{(1)} + \Omega_S^{(1)} + \Omega_X^{(1)}. \tag{22.26}$$

Modified Dielectric Function

As we have seen, the Hartree potential $\Omega_S^{(1)}$ can be eliminated in favor of the bare ion potential $\Omega_B^{(1)}$, by means of the dielectric function (21.80). In the same way the exchange potential $\Omega_X^{(1)}$, or any other local potential, can also be eliminated by a suitable modification of the dielectric function. The first-order term in the expansion of Ω_X is the sum of first-order terms in (22.18) and (22.19):

$$\begin{aligned} \Omega_X^{(1)} &= [\rho^{(0)} X'' + 2X'] \rho^{(1)} \\ &= [\partial^2 (\rho X) / \partial \rho^2]^{(0)} \rho^{(1)}. \end{aligned} \tag{22.27}$$

With $\rho^{(1)}(\mathbf{r})$ given by (21.71), the Fourier components of $\Omega_X^{(1)}$ are

$$\Omega_X^{(1)}(\mathbf{q}) = \sum_{\mathbf{q}'} a_X(\mathbf{q},\mathbf{q}') \Omega^{(1)}(\mathbf{q}'), \tag{22.28}$$

where

$$a_X(\mathbf{q},\mathbf{q}') = V^{-1} \sum_{\lambda\lambda'}{}' (g_\lambda - g_{\lambda'}) \langle \psi_\lambda^{(0)} | [\partial^2(\rho X)/\partial\rho^2]^{(0)} e^{-i\mathbf{q}'\cdot\mathbf{r}} | \psi_{\lambda'}^{(0)} \rangle$$

$$\times \frac{\langle \psi_{\lambda'}^{(0)} | e^{i\mathbf{q}'\cdot\mathbf{r}} | \psi_\lambda^{(0)} \rangle}{[\epsilon_\lambda^{(0)} - \epsilon_{\lambda'}^{(0)}]_p}. \tag{22.29}$$

Let us abbreviate $\Omega_S + \Omega_X$ by Ω_{SX}, so that

$$\Omega_{SX}^{(1)}(\mathbf{q}) = \Omega_S^{(1)}(\mathbf{q}) + \Omega_X^{(1)}(\mathbf{q}), \tag{22.30}$$

and components of the total first-order potential are written

$$\Omega^{(1)}(\mathbf{q}) = \Omega_B^{(1)}(\mathbf{q}) + \Omega_{SX}^{(1)}(\mathbf{q}). \tag{22.31}$$

The expression (21.78) for $\Omega_S^{(1)}(\mathbf{q})$ in terms of the *total potential* components still holds, i.e.,

$$\Omega_S^{(1)}(\mathbf{q}) = (4\pi e^2/q^2) \sum_{\mathbf{q}'} a(\mathbf{q},\mathbf{q}')\Omega^{(1)}(\mathbf{q}'), \tag{22.32}$$

and adding $\Omega_X^{(1)}(\mathbf{q})$ according to (22.28) gives

$$\Omega_{SX}^{(1)}(\mathbf{q}) = \sum_{\mathbf{q}'} [(4\pi e^2/q^2)a(\mathbf{q},\mathbf{q}') + a_X(\mathbf{q},\mathbf{q}')]\Omega^{(1)}(\mathbf{q}'). \tag{22.33}$$

From this point the components $\Omega_{SX}^{(1)}(\mathbf{q})$ can be eliminated in favor of $\Omega_B^{(1)}(\mathbf{q})$, just as in (21.85) or (21.86) which serve to eliminate $\Omega_S^{(1)}(\mathbf{q})$, where now instead of (21.80) the dielectric function is given by

$$\hat{d}(\mathbf{q},\mathbf{q}') = \delta_{\mathbf{q}\mathbf{q}'} - (4\pi e^2/q^2)a(\mathbf{q},\mathbf{q}') - a_X(\mathbf{q},\mathbf{q}'). \tag{22.34}$$

The hat is used here to indicate inclusion of exchange in the dielectric function.

In particular the correlation function is now defined by

$$\hat{c}(\mathbf{q},\mathbf{q}') = (q^2 V_C/4\pi e^2)[\hat{b}(\mathbf{q},\mathbf{q}') - \delta_{\mathbf{q}\mathbf{q}'}], \tag{22.35}$$

where $\hat{b}(\mathbf{q},\mathbf{q}')$ are elements of the tensor inverse to $\hat{d}(\mathbf{q},\mathbf{q}')$. Then (21.85) becomes

$$\Omega_{SX}^{(1)}(\mathbf{q}) = (4\pi e^2/q^2 V_C) \sum_{\mathbf{q}'} \hat{c}(\mathbf{q},\mathbf{q}')\Omega_B^{(1)}(\mathbf{q}'), \tag{22.36}$$

and (21.86) becomes

$$\Omega^{(1)}(\mathbf{q}) = \sum_{\mathbf{q}'} \hat{b}(\mathbf{q},\mathbf{q}')\Omega_B^{(1)}(\mathbf{q}'). \tag{22.37}$$

This last result allows us to find the new relation between $\Omega_S^{(1)}$ and $\Omega_B^{(1)}$. By using (22.37) in the right-hand side of (22.32), we have

$$\Omega_S^{(1)}(\mathbf{q}) = (4\pi e^2/q^2) \sum_{\mathbf{q}'\mathbf{q}''} a(\mathbf{q},\mathbf{q}')\hat{b}(\mathbf{q}',\mathbf{q}'')\Omega_B^{(1)}(\mathbf{q}''). \tag{22.38}$$

Then solving (22.34) for $a(\mathbf{q},\mathbf{q}')$ gives

$$a(\mathbf{q},\mathbf{q}') = (q^2/4\pi e^2)[\delta_{\mathbf{q}\mathbf{q}'} - \hat{d}(\mathbf{q},\mathbf{q}') - a_X(\mathbf{q},\mathbf{q}')], \tag{22.39}$$

and the $\Sigma_{\mathbf{q}'}$ in (22.38) becomes

$$\sum_{\mathbf{q}'} a(\mathbf{q},\mathbf{q}')\hat{b}(\mathbf{q}',\mathbf{q}'') = (q^2/4\pi e^2)\Big[\hat{b}(\mathbf{q},\mathbf{q}'') - \delta_{\mathbf{q}\mathbf{q}''} - \sum_{\mathbf{q}'} a_{\mathrm{X}}(\mathbf{q},\mathbf{q}')\hat{b}(\mathbf{q}',\mathbf{q}'')\Big]. \tag{22.40}$$

Now if we define the function

$$c_{\mathrm{X}}(\mathbf{q},\mathbf{q}'') = (q^2 V_C/4\pi e^2) \sum_{\mathbf{q}'} a_{\mathrm{X}}(\mathbf{q},\mathbf{q}')\hat{b}(\mathbf{q}',\mathbf{q}''), \tag{22.41}$$

then (22.40) is

$$\sum_{\mathbf{q}'} a(\mathbf{q},\mathbf{q}')\hat{b}(\mathbf{q}',\mathbf{q}'') = V_C^{-1}\tilde{c}(\mathbf{q},\mathbf{q}''), \tag{22.42}$$

where

$$\tilde{c}(\mathbf{q},\mathbf{q}') = \hat{c}(\mathbf{q},\mathbf{q}') - c_{\mathrm{X}}(\mathbf{q},\mathbf{q}'), \tag{22.43}$$

and $\hat{c}(\mathbf{q},\mathbf{q}')$ is given by (22.35). The desired result is now obtained by using (22.42) in (22.38):

$$\Omega_{\mathrm{S}}^{(1)}(\mathbf{q}) = (4\pi e^2/q^2 V_C) \sum_{\mathbf{q}'} \tilde{c}(\mathbf{q},\mathbf{q}')\Omega_{\mathrm{B}}^{(1)}(\mathbf{q}'). \tag{22.44}$$

The preceding methods of transforming and eliminating the exchange potential, and of modifying the dielectric function, can easily be carried over to alternate forms of one-electron potentials.

Further Approximations

The expression (22.29) for $a_{\mathrm{X}}(\mathbf{q},\mathbf{q}')$ differs from (21.74) for $a(\mathbf{q},\mathbf{q}')$ only in the presence of the extra factor $[\partial^2(\rho X)/\partial\rho^2]^{(0)}$ in the matrix elements in $a_{\mathrm{X}}(\mathbf{q},\mathbf{q}')$. This observation suggests a most useful general approximation, in which the exchange potential $\Omega_{\mathrm{X}}^{(1)}$ is related to the Hartree potential $\Omega_{\mathrm{S}}^{(1)}$, as for example by the expression

$$\Omega_{\mathrm{SX}}^{(1)}(\mathbf{q}) = \Omega_{\mathrm{S}}^{(1)}(\mathbf{q})[1 - Y(\mathbf{q})], \tag{22.45}$$

where the function $Y(\mathbf{q})$ may be chosen as desired. In view of (22.33) for $\Omega_{\mathrm{SX}}^{(1)}(\mathbf{q})$ and (22.32) for $\Omega_{\mathrm{S}}^{(1)}(\mathbf{q})$, the approximation (22.45) is equivalent to

$$(4\pi e^2/q^2)a(\mathbf{q},\mathbf{q}') + a_{\mathrm{X}}(\mathbf{q},\mathbf{q}') = (4\pi e^2/q^2)[1 - Y(\mathbf{q})]a(\mathbf{q},\mathbf{q}'),$$

or

$$a_{\mathrm{X}}(\mathbf{q},\mathbf{q}') = -(4\pi e^2/q^2)Y(\mathbf{q})a(\mathbf{q},\mathbf{q}'). \tag{22.46}$$

Then the modified dielectric function (22.34) is obviously

$$\hat{d}(\mathbf{q},\mathbf{q}') = \delta_{\mathbf{q}\mathbf{q}'} - (4\pi e^2/q^2)[1 - Y(\mathbf{q})]a(\mathbf{q},\mathbf{q}'), \tag{22.47}$$

and in addition it follows that

$$\tilde{c}(\mathbf{q},\mathbf{q}') = [1 - Y(\mathbf{q})]^{-1}\hat{c}(\mathbf{q},\mathbf{q}'), \tag{22.48}$$

where $\tilde{c}(\mathbf{q},\mathbf{q}')$ is given by (22.43) and $\hat{c}(\mathbf{q},\mathbf{q}')$ is given by (22.35).

Assume, for example, that the function $[\partial^2(\rho X)/\partial\rho^2]^{(0)}$ is independent of the position $\mathbf{r}$. Note that this approximation is implied if one assumes the electronic density $\rho^{(0)}(\mathbf{r})$ is constant, since X depends only on ρ, but it does not necessarily require $\rho^{(0)}(\mathbf{r})$ to be constant. Then by taking $[\partial^2(\rho X)/\partial\rho^2]^{(0)}$ outside the matrix elements in (22.29), we can write

$$a_X(\mathbf{q},\mathbf{q}') \to [\partial^2(\rho X)/\partial\rho^2]^{(0)}a(\mathbf{q},\mathbf{q}'), \tag{22.49}$$

and by comparison with (22.46),

$$Y(\mathbf{q}) \to -(q^2/4\pi e^2)[\partial^2(\rho X)/\partial\rho^2]^{(0)}. \tag{22.50}$$

The Slater type of approximation for exchange takes the function $X(\mathbf{r})$ to be proportional to $[\rho(\mathbf{r})]^{1/3}$:

$$X(\mathbf{r}) = \alpha_X[\rho(\mathbf{r})]^{1/3}, \tag{22.51}$$

where α_X is a parameter which may be chosen to minimize the total energy, or to give the correct exchange energy for a uniform electron gas. It is well known* that the exchange energy per electron for an electron gas of constant density ρ_0 is $-(3e^2/4)(3\rho_0/\pi)^{1/3}$. Also, evaluation of (22.1) with $X(\mathbf{r})$ given by (22.51) and with $\rho(\mathbf{r}) = \rho_0$ gives $\alpha_X\rho_0^{1/3}$ for the exchange energy per electron. Thus for a uniform electron gas,

$$\alpha_X = -(3e^2/4)(3/\pi)^{1/3}, \quad \text{for constant density;} \tag{22.52}$$

this value of α_X is known as the Kohn–Sham value.† Corresponding to the Slater type of exchange (22.51), the one-electron exchange potential (22.13) is

$$\Omega_X(\mathbf{r}) = \tfrac{4}{3}\alpha_X[\rho(\mathbf{r})]^{1/3}, \tag{22.53}$$

and $a_X(\mathbf{q},\mathbf{q}')$ is given by (22.29) along with

$$[\partial^2(\rho X)/\partial\rho^2]^{(0)} = \tfrac{4}{9}\alpha_X[\rho^{(0)}(\mathbf{r})]^{-2/3}. \tag{22.54}$$

Now the $\rho^{1/3}$ approximation for exchange, with α_X given by (22.52), is correct in the limit of slowly varying band-electron density, so we might expect this approximation to give a reasonable representation of $Y(\mathbf{q})$ in the limit of small $\mathbf{q}$. Evaluation of (22.50) for $Y(\mathbf{q})$, with the aid of (22.52) and (22.54), gives

$$Y(\mathbf{q}) = \frac{q^2}{12\pi(\pi\rho_0^2/3)^{1/3}}, \quad \text{for small } \mathbf{q}. \tag{22.55}$$

* See, e.g., C. Kittel, *Quantum Theory of Solids*, John Wiley & Sons, Inc., New York, 1963.

† W. Kohn and L. J. Sham, *Phys. Rev.* **140,** A1133 (1965).

If now we follow the argument of Hubbard* that the effect of exchange should be to cancel half of the direct Coulomb contribution to the dielectric function at large $\mathbf{q}$, then $1 - Y(\mathbf{q}) \to \frac{1}{2}$ at large $\mathbf{q}$, and (22.55) may be replaced by the interpolation formula

$$Y(\mathbf{q}) = \frac{q^2}{2[q^2 + 6\pi(\pi\rho_0^2/3)^{1/3}]}. \tag{22.56}$$

It should be recognized that this approximation is not consistent with the Slater type of exchange at large $\mathbf{q}$.

23. TOTAL ADIABATIC POTENTIAL

Electronic Ground-State Energy

The total adiabatic potential is $\Phi = \Omega_{\rm I} + E_{\rm G}$, and this potential is the basis for lattice-statics calculations, as in Chapter 2, and lattice-dynamics calculations, as in Chapter 3. The band-electron ground-state energy $E_{\rm G}$ was derived in the last two sections, and in preparation for lattice-dynamics calculations we will derive the contributions of $E_{\rm G}$ to the potential energy coefficients evaluated in the interior of the crystal. It is assumed that effects of exchange among band electrons are included in a local potential approximation, such that the results (22.23) for $E_{\rm G}^{(1)}$ and (22.25) for $E_{\rm G}^{(2)}$ hold, and the total potential $\Omega^{(1)}$ is related to the bare ion potential $\Omega_{\rm B}^{(1)}$ by means of (22.37).

The first-order electronic ground-state energy (22.23) is

$$E_{\rm G}^{(1)} = \sum_{\lambda} g_\lambda \langle \psi_\lambda^{(0)} | \, \Omega_{\rm B}^{(1)} \, | \psi_\lambda^{(0)} \rangle,$$

and by introducing the Fourier transform of $\Omega_{\rm B}^{(1)}$ this may be written

$$E_{\rm G}^{(1)} = N_0 \sum_{\mathbf{q}} c(\mathbf{q}) \Omega_{\rm B}^{(1)}(\mathbf{q}), \tag{23.1}$$

where

$$c(\mathbf{q}) = N_0^{-1} \sum_{\lambda} g_\lambda \langle \psi_\lambda^{(0)} | \, e^{i\mathbf{q}\cdot\mathbf{r}} \, | \psi_\lambda^{(0)} \rangle = V_C \rho^{(0)}(-\mathbf{q}). \tag{23.2}$$

The second-order electronic ground-state energy is given by (22.25), and with (21.60) for the first-order electronic wave functions, the result is

$$E_{\rm G}^{(2)} = \tfrac{1}{2} \sum_{\lambda\lambda'}{}' (g_\lambda - g_{\lambda'}) \frac{\langle \psi_\lambda^{(0)} | \, \Omega_{\rm B}^{(1)} \, | \psi_{\lambda'}^{(0)} \rangle \langle \psi_{\lambda'}^{(0)} | \, \Omega^{(1)} \, | \psi_\lambda^{(0)} \rangle}{[\epsilon_\lambda^{(0)} - \epsilon_{\lambda'}^{(0)}]_p} + \sum_{\lambda} g_\lambda \langle \psi_\lambda^{(0)} | \, \Omega_{\rm B}^{(2)} \, | \psi_\lambda^{(0)} \rangle. \tag{23.3}$$

* J. Hubbard, *Proc. Roy. Soc. (London)* **A243,** 336 (1958).

Again by introducing the Fourier transforms of the potentials, as illustrated by the general relation (21.67), it follows that

$$E_G^{(2)} = \tfrac{1}{2}V \sum_{\mathbf{q}\mathbf{q}'} \Omega_B^{(1)}(-\mathbf{q})\Omega^{(1)}(\mathbf{q}')a(\mathbf{q},\mathbf{q}') + N_0 \sum_{\mathbf{q}} c(\mathbf{q})\Omega_B^{(2)}(\mathbf{q}), \tag{23.4}$$

where $a(\mathbf{q},\mathbf{q}')$ is given by (21.74). To evaluate the double sum in (23.4), we eliminate $\Omega^{(1)}(\mathbf{q}')$ in favor of $\Omega_B^{(1)}(\mathbf{q}')$ by means of (22.37) to obtain the expression

$$\tfrac{1}{2}V \sum_{\mathbf{q}\mathbf{q}'\mathbf{q}''} \Omega_B^{(1)}(-\mathbf{q})\Omega_B^{(1)}(\mathbf{q}'')a(\mathbf{q},\mathbf{q}')\hat{b}(\mathbf{q}',\mathbf{q}''),$$

and then evaluate the $\Sigma_{\mathbf{q}'}$ here by means of (22.42), which gives

$$\tfrac{1}{2}N_0 \sum_{\mathbf{q}\mathbf{q}'} \Omega_B^{(1)}(-\mathbf{q})\Omega_B^{(1)}(\mathbf{q}')\tilde{c}(\mathbf{q},\mathbf{q}').$$

Then the second-order ground-state energy (23.4) is written entirely in terms of the bare ion potential, as

$$E_G^{(2)} = \tfrac{1}{2}N_0 \sum_{\mathbf{q}\mathbf{q}'} \tilde{c}(\mathbf{q},\mathbf{q}')\Omega_B^{(1)}(-\mathbf{q})\Omega_B^{(1)}(\mathbf{q}') + N_0 \sum_{\mathbf{q}} c(\mathbf{q})\Omega_B^{(2)}(\mathbf{q}). \tag{23.5}$$

The total bare ion potential $\Omega_B(\mathbf{r})$ is represented as a sum of single ion potentials in (21.26), and the corresponding Fourier components are

$$\begin{aligned}\Omega_B(\mathbf{q}) &= V^{-1}\int \Omega_B(\mathbf{r})e^{-i\mathbf{q}\cdot\mathbf{r}}\,d\mathbf{r} \\ &= V^{-1}\sum_{N\nu} e^{-i\mathbf{q}\cdot\tilde{\mathbf{R}}(N\nu)} \int v_\nu(\mathbf{r} - \tilde{\mathbf{R}}(N\nu))e^{-i\mathbf{q}\cdot[\mathbf{r}-\tilde{\mathbf{R}}(N\nu)]}\,d\mathbf{r}.\end{aligned} \tag{23.6}$$

Here the $\int d\mathbf{r}$ does not depend on the location $\tilde{\mathbf{R}}(N\nu)$ of ion $N\nu$, and in fact depends on ν only through the form of the potential $v_\nu(\mathbf{r})$, so we define the Fourier components of the single ion potential as

$$v_\nu(\mathbf{q}) = V_C^{-1}\int v_\nu(\mathbf{r})e^{-i\mathbf{q}\cdot\mathbf{r}}\,d\mathbf{r}. \tag{23.7}$$

The integral in (23.7) is still over the entire crystal volume V, but $v_\nu(\mathbf{q})$ has been "normalized" to V_C rather than V because $v_\nu(\mathbf{r})$ is appreciable only over a volume of the order V_C. With (23.7), the total $\Omega_B(\mathbf{q})$ of (23.6) is

$$\Omega_B(\mathbf{q}) = N_0^{-1}\sum_{N\nu} e^{-i\mathbf{q}\cdot\tilde{\mathbf{R}}(N\nu)}\,v_\nu(\mathbf{q}). \tag{23.8}$$

For the perfect crystal, where all ions are located at their equilibrium positions $\mathbf{R}(N\nu) = \mathbf{R}(N) + \mathbf{R}(\nu)$, (23.8) simplifies to

$$\Omega_B^{(0)}(\mathbf{q}) = N_0^{-1}\sum_{N} e^{-i\mathbf{q}\cdot\mathbf{R}(N)} \sum_{\nu} e^{-i\mathbf{q}\cdot\mathbf{R}(\nu)}v_\nu(\mathbf{q}); \tag{23.9}$$

further since the $N_0^{-1}\,\Sigma_{\mathrm{N}}$ here is a $\delta(\mathbf{q})$ according to (10.22), then

$$\Omega_{\mathrm{B}}^{(0)}(\mathbf{q}) = \delta_{\mathbf{qQ}} \sum_{\nu} e^{-i\mathbf{Q}\cdot\mathbf{R}(\nu)} v_\nu(\mathbf{Q}), \tag{23.10}$$

where $\mathbf{Q}$ is any inverse-lattice vector.

The derivatives of $\Omega_{\mathrm{B}}(\mathbf{q})$ with respect to displacements of the ions from equilibrium, evaluated at the equilibrium configuration, are easily calculated from (23.8) as

$$\partial\Omega_{\mathrm{B}}(\mathbf{q})/\partial U_i(\mathrm{M}\mu) = -iN_0^{-1} q_i e^{-i\mathbf{q}\cdot\mathbf{R}(\mathrm{M}\mu)}\, v_\mu(\mathbf{q}); \tag{23.11}$$

$$\partial^2\Omega_{\mathrm{B}}(\mathbf{q})/\partial U_i(\mathrm{M}\mu)\partial U_j(\mathrm{N}\nu) = -\delta_{\mathrm{MN}}\delta_{\mu\nu} N_0^{-1} q_i q_j e^{-i\mathbf{q}\cdot\mathbf{R}(\mathrm{M}\mu)}\, v_\mu(\mathbf{q}). \tag{23.12}$$

Now $\Omega_{\mathrm{B}}^{(1)}(\mathbf{q})$ and $\Omega_{\mathrm{B}}^{(2)}(\mathbf{q})$ are the sums of first- and second-order terms, respectively, in the expansion of $\Omega_{\mathrm{B}}(\mathbf{q})$ in powers of the ion displacements [compare (21.44) for $\Omega^{(1)}(\mathbf{r})$ and (21.45) for $\Omega^{(2)}(\mathbf{r})$], and with the derivatives (23.11) and (23.12) we have

$$\Omega_{\mathrm{B}}^{(1)}(\mathbf{q}) = -iN_0^{-1} \sum_{\mathrm{M}\mu} \sum_{i} q_i e^{-i\mathbf{q}\cdot\mathbf{R}(\mathrm{M}\mu)} v_\mu(\mathbf{q}) U_i(\mathrm{M}\mu); \tag{23.13}$$

$$\Omega_{\mathrm{B}}^{(2)}(\mathbf{q}) = -\tfrac{1}{2}N_0^{-1} \sum_{\mathrm{M}\mu} \sum_{ij} q_i q_j e^{-i\mathbf{q}\cdot\mathbf{R}(\mathrm{M}\mu)}\, v_\mu(\mathbf{q}) U_i(\mathrm{M}\mu) U_j(\mathrm{M}\mu). \tag{23.14}$$

Then with these results, $E_{\mathrm{G}}^{(1)}$ and $E_{\mathrm{G}}^{(2)}$, given by (23.1) and (23.5), respectively, may be written in terms of the ion displacements as follows.

$$E_{\mathrm{G}}^{(1)} = \sum_{\mathrm{M}\mu} \sum_{i} \Phi_i(\mathrm{M}\mu)_{\mathrm{E}} U_i(\mathrm{M}\mu), \tag{23.15}$$

$$E_{\mathrm{G}}^{(2)} = \tfrac{1}{2} \sum_{\mathrm{MN}\mu\nu} \sum_{ij} \Phi_{ij}(\mathrm{M}\mu,\mathrm{N}\nu)_{\mathrm{E}} U_i(\mathrm{M}\mu) U_j(\mathrm{N}\nu), \tag{23.16}$$

where

$$\Phi_i(\mathrm{M}\mu)_{\mathrm{E}} = -i \sum_{\mathbf{q}} c(\mathbf{q}) q_i e^{-i\mathbf{q}\cdot\mathbf{R}(\mathrm{M}\mu)} v_\mu(\mathbf{q}), \tag{23.17}$$

$$\begin{aligned}\Phi_{ij}(\mathrm{M}\mu,\mathrm{N}\nu)_{\mathrm{E}} = {} & N_0^{-1} \sum_{\mathbf{qq}'} \tilde{c}(\mathbf{q},\mathbf{q}') q_i q_j' e^{i[\mathbf{q}\cdot\mathbf{R}(\mathrm{M}\mu) - \mathbf{q}'\cdot\mathbf{R}(\mathrm{N}\nu)]} v_\mu(-\mathbf{q}) v_\nu(\mathbf{q}') \\ & - \delta_{\mathrm{MN}}\delta_{\mu\nu} \sum_{\mathbf{q}} c(\mathbf{q}) q_i q_j e^{-i\mathbf{q}\cdot\mathbf{R}(\mathrm{M}\mu)} v_\mu(\mathbf{q}).\end{aligned} \tag{23.18}$$

The subscript E is to remind us that $\Phi_i(\mathrm{M}\mu)_{\mathrm{E}}$ and $\Phi_{ij}(\mathrm{M}\mu,\mathrm{N}\nu)_{\mathrm{E}}$ are just the electronic ground-state energy contributions to the potential energy coefficients.

According to (6.25), the translational invariance condition involving the second-order potential coefficients is

$$\sum_{\mathrm{N}\nu} \Phi_{ij}(\mathrm{M}\mu,\mathrm{N}\nu) = 0. \tag{23.19}$$

This must hold, of course, for the *total* potential coefficients, and it is quite useful to show that translational invariance is satisfied separately by the electronic contribution $\Phi_{ij}(\mathrm{M}\mu,\mathrm{N}\nu)_{\mathrm{E}}$; that is,

$$\sum_{\mathrm{N}\nu} \Phi_{ij}(\mathrm{M}\mu,\mathrm{N}\nu)_{\mathrm{E}} = 0. \tag{23.20}$$

To begin the exercise, consider the equilibrium configuration of the crystal in which the Hartree potential is $\Omega_{\mathrm{S}}^{(0)}(\mathbf{r})$; then let the ions undergo a uniform infinitesimal translation $\boldsymbol{\epsilon}$ so that the Hartree potential at $\mathbf{r}$ becomes $\Omega_{\mathrm{S}}^{(0)}(\mathbf{r}) + \Omega_{\mathrm{S}}^{(1)}(\mathbf{r})$, to first order in $\boldsymbol{\epsilon}$. Then

$$\Omega_{\mathrm{S}}^{(1)}(\mathbf{r}) = \sum_{\mathbf{q}} \Omega_{\mathrm{S}}^{(1)}(\mathbf{q}) e^{i\mathbf{q}\cdot\mathbf{r}}, \tag{23.21}$$

and from (22.44)

$$\Omega_{\mathrm{S}}^{(1)}(\mathbf{q}) = (4\pi e^2/q^2 V_C) \sum_{\mathbf{q}'} \tilde{c}(\mathbf{q},\mathbf{q}')\Omega_{\mathrm{B}}^{(1)}(\mathbf{q}'), \tag{23.22}$$

and from (23.13) for $\Omega_{\mathrm{B}}^{(1)}(\mathbf{q}')$ with the ion displacements $U_i(\mathrm{M}\mu)$ set equal to ϵ_i,

$$\Omega_{\mathrm{S}}^{(1)}(\mathbf{q}) = -i(4\pi e^2/q^2 V_C) \sum_{\mathbf{q}'} \tilde{c}(\mathbf{q},\mathbf{q}') N_0^{-1} \sum_{\mathrm{M}\mu}\sum_{i} q_i' e^{-i\mathbf{q}'\cdot\mathbf{R}(\mathrm{M}\mu)} v_\mu(\mathbf{q}')\epsilon_i. \tag{23.23}$$

Now because the screening electrons move rigidly with the ions under a uniform translation of the ions, the translated potential $\Omega_{\mathrm{S}}^{(0)}(\mathbf{r}) + \Omega_{\mathrm{S}}^{(1)}(\mathbf{r})$ must be the same as that at the position $\mathbf{r} - \boldsymbol{\epsilon}$ in the original equilibrium configuration of the crystal. That potential is, to first order in $\boldsymbol{\epsilon}$,

$$\begin{aligned} \Omega_{\mathrm{S}}^{(0)}(\mathbf{r} - \boldsymbol{\epsilon}) &= \sum_{\mathbf{q}} \Omega_{\mathrm{S}}^{(0)}(\mathbf{q}) e^{i\mathbf{q}\cdot(\mathbf{r}-\boldsymbol{\epsilon})} \\ &= \sum_{\mathbf{q}} \Omega_{\mathrm{S}}^{(0)}(\mathbf{q}) e^{i\mathbf{q}\cdot\mathbf{r}}[1 - i\mathbf{q}\cdot\boldsymbol{\epsilon}] \\ &= \Omega_{\mathrm{S}}^{(0)}(\mathbf{r}) - i \sum_{\mathbf{q}} \Omega_{\mathrm{S}}^{(0)}(\mathbf{q}) e^{i\mathbf{q}\cdot\mathbf{r}} \mathbf{q}\cdot\boldsymbol{\epsilon}. \end{aligned} \tag{23.24}$$

Equating this to $\Omega_{\mathrm{S}}^{(0)}(\mathbf{r}) + \Omega_{\mathrm{S}}^{(1)}(\mathbf{r})$ gives

$$\Omega_{\mathrm{S}}^{(1)}(\mathbf{r}) = -i \sum_{\mathbf{q}} \Omega_{\mathrm{S}}^{(0)}(\mathbf{q}) e^{i\mathbf{q}\cdot\mathbf{r}} \mathbf{q}\cdot\boldsymbol{\epsilon}, \tag{23.25}$$

or

$$\Omega_{\mathrm{S}}^{(1)}(\mathbf{q}) = -i \sum_{i} \Omega_{\mathrm{S}}^{(0)}(\mathbf{q}) q_i \epsilon_i. \tag{23.26}$$

Since the components ϵ_i of the translation are arbitrary, the coefficients of each ϵ_i in the two expressions (23.23) and (23.26) for $\Omega_{\mathrm{S}}^{(1)}(\mathbf{q})$ must be equal, so that

$$\Omega_{\mathrm{S}}^{(0)}(\mathbf{q}) q_i = (4\pi e^2/q^2 V_C) \sum_{\mathbf{q}'} \tilde{c}(\mathbf{q},\mathbf{q}') q_i' \, N_0^{-1} \sum_{\mathrm{M}\mu} e^{-i\mathbf{q}'\cdot\mathbf{R}(\mathrm{M}\mu)} v_\mu(\mathbf{q}'),$$

or

$$N_0^{-1} \sum_{\mathrm{N}\nu}\sum_{\mathbf{q}'} \tilde{c}(\mathbf{q},\mathbf{q}') q_j' e^{-i\mathbf{q}'\cdot\mathbf{R}(\mathrm{N}\nu)} v_\nu(\mathbf{q}') = (q^2 V_C/4\pi e^2)\Omega_{\mathrm{S}}^{(0)}(\mathbf{q}) q_j. \tag{23.27}$$

When (23.18) is summed over Nν, the term involving $\tilde{c}(\mathbf{q},\mathbf{q}')$ is simplified by (23.27), and the result is

$$\sum_{N\nu} \Phi_{ij}(M\mu,N\nu)_E = \sum_{\mathbf{q}} (q^2 V_C/4\pi e^2)\Omega_S^{(0)}(\mathbf{q}) q_i q_j e^{i\mathbf{q}\cdot\mathbf{R}(M\mu)} v_\mu(-\mathbf{q})$$
$$- \sum_{\mathbf{q}} c(\mathbf{q}) q_i q_j e^{-i\mathbf{q}\cdot\mathbf{R}(M\mu)} v_\mu(\mathbf{q}). \qquad (23.28)$$

Now by combining (23.2) for $c(\mathbf{q})$ and (21.34) evaluated for $\rho^{(0)}(-\mathbf{q})$, it follows that

$$c(\mathbf{q}) = (q^2 V_C/4\pi e^2)\Omega_S^{(0)}(-\mathbf{q}), \qquad (23.29)$$

and then by changing the sign of $\mathbf{q}$ inside $\Sigma_{\mathbf{q}}$ in one of the two contributions on the right of (23.28), it is seen that these two contributions cancel and the translational invariance condition (23.20) is verified.

Ion–Ion Coulomb Interactions

The ion cores are assumed to be sufficiently small and nonpolarizable so that their interactions are adequately represented by the Coulomb interactions among a set of point charges. If an ion of type μ has charge $z_\mu e$, the potential $\varphi_\mu(R)$ at a distance R from the ion is

$$\varphi_\mu(R) = z_\mu e/R, \qquad (23.30)$$

where we are taking the convention $e > 0$. It is convenient to represent $\varphi_\mu(R)$ as the sum of two contributions:

$$\varphi_\mu(R) = \varphi_{\mu 1}(R) + \varphi_{\mu 2}(R), \qquad (23.31)$$

where

$$\varphi_{\mu 1}(R) = \frac{z_\mu e}{R} \frac{2}{\sqrt{\pi}} \int_0^{\eta R} e^{-x^2}\, dx, \qquad (23.32)$$

$$\varphi_{\mu 2}(R) = \frac{z_\mu e}{R} \frac{2}{\sqrt{\pi}} \int_{\eta R}^{\infty} e^{-x^2}\, dx. \qquad (23.33)$$

The two expressions (23.30) and (23.31) for $\varphi_\mu(R)$ are the same since

$$\frac{2}{\sqrt{\pi}} \int_0^{\infty} e^{-x^2}\, dx = 1. \qquad (23.34)$$

The reason for splitting $\varphi_\mu(R)$ into these two contributions is to avoid the slowly converging Coulomb sums over the crystal lattice. At large R, the main contribution to $\varphi_\mu(R)$ comes from $\varphi_{\mu 1}(R)$, so this term will be Fourier transformed. Then when the formal calculations are done, the parameter η, which has the dimension of R^{-1}, may be chosen to obtain optimum convergence of both direct- and inverse-lattice sums in any particular computation.

The Fourier components of $\varphi_{\mu 1}(R)$ are calculated as follows.

$$\begin{aligned}\varphi_{\mu 1}(\mathbf{q}) &= V^{-1}\int \varphi_{\mu 1}(R)e^{-i\mathbf{q}\cdot\mathbf{R}}\,d\mathbf{R} \\ &= (4\pi/qV)\int_0^\infty \sin(qR)\varphi_{\mu 1}(R)R\,dR \\ &= (8\sqrt{\pi}z_\mu e/qV)\int_0^\infty \sin(qR)\int_0^{\eta R} e^{-x^2}\,dx\,dR.\end{aligned} \tag{23.35}$$

The double integral may be evaluated by parts, with the result

$$\int_0^\infty \sin(qR)\int_0^{\eta R} e^{-x^2}\,dx\,dR = (\sqrt{\pi}/2q)e^{-q^2/4\eta^2}. \tag{23.36}$$

Then (23.35) becomes

$$\varphi_{\mu 1}(\mathbf{q}) = (4\pi z_\mu e/q^2 V)e^{-q^2/4\eta^2}, \tag{23.37}$$

and in terms of the Fourier components, $\varphi_{\mu 1}(R)$ is

$$\varphi_{\mu 1}(R) = \sum_{\mathbf{q}} \varphi_{\mu 1}(\mathbf{q})e^{i\mathbf{q}\cdot\mathbf{R}}. \tag{23.38}$$

Obviously the right-hand side of (23.38) depends only on the magnitude R, since $\varphi_{\mu 1}(\mathbf{q})$ of (23.37) depends only on the magnitude q. A useful limit is $\varphi_{\mu 1}(R)$ as $R \to 0$, and is evaluated from (23.32) as follows.

$$\begin{aligned}\lim_{R\to 0}\varphi_{\mu 1}(R) &= \lim_{R\to 0}\frac{2z_\mu e}{R\sqrt{\pi}}\int_0^{\eta R}(1 - x^2 + \cdots)\,dx \\ &= 2z_\mu e\eta/\sqrt{\pi}.\end{aligned} \tag{23.39}$$

The second contribution $\varphi_{\mu 2}(R)$ is conveniently written, from (23.33),

$$\varphi_{\mu 2}(R) = (z_\mu e/R)\operatorname{erfc}(\eta R), \tag{23.40}$$

where the complementary error function (erfc) is defined by

$$\operatorname{erfc}(y) = \frac{2}{\sqrt{\pi}}\int_y^\infty e^{-x^2}\,dx. \tag{23.41}$$

Then the total Coulomb potential at a distance R from the point charge $z_\mu e$ is

$$\varphi_\mu(R) = (4\pi z_\mu e/V)\sum_{\mathbf{q}} q^{-2}e^{-q^2/4\eta^2}e^{i\mathbf{q}\cdot\mathbf{R}} + (z_\mu e/R)\operatorname{erfc}(\eta R). \tag{23.42}$$

With the ion positions $\tilde{\mathbf{R}}(\mathrm{M}\mu)$ arbitrary in the volume V of the crystal, the total energy $\Omega_{\rm I}$ of the Coulomb interactions among the ions is

$$\Omega_{\rm I} = \tfrac{1}{2} \sum_{{\rm MN}\mu\nu}' \frac{z_\mu z_\nu e^2}{|\tilde{\mathbf{R}}({\rm M}\mu) - \tilde{\mathbf{R}}({\rm N}\nu)|}. \tag{23.43}$$

This can be written in terms of the single-ion Coulomb potentials $\varphi_\mu(R)$ as

$$\Omega_{\rm I} = \tfrac{1}{2} \sum_{{\rm MN}\mu\nu}' z_\nu e \varphi_\mu(|\tilde{\mathbf{R}}({\rm M}\mu) - \tilde{\mathbf{R}}({\rm N}\nu)|), \tag{23.44}$$

and with (23.42) for $\varphi_\mu(R)$ it is

$$\begin{aligned} \Omega_{\rm I} = {} & (2\pi e^2/V) \sum_{\mathbf{q}} q^{-2} e^{-q^2/4\eta^2} \sum_{{\rm MN}\mu\nu}' z_\mu z_\nu e^{i\mathbf{q}\cdot[\tilde{\mathbf{R}}({\rm M}\mu) - \tilde{\mathbf{R}}({\rm N}\nu)]} \\ & + \tfrac{1}{2} e^2 \sum_{{\rm MN}\mu\nu}' z_\mu z_\nu \, |\tilde{\mathbf{R}}({\rm M}\mu) - \tilde{\mathbf{R}}({\rm N}\nu)|^{-1} \operatorname{erfc}[\eta \, |\tilde{\mathbf{R}}({\rm M}\mu) - \tilde{\mathbf{R}}({\rm N}\nu)|]. \end{aligned} \tag{23.45}$$

In the first term of $\Omega_{\rm I}$, the $\Sigma_{\mathbf{q}}$ term in (23.45), we can add the terms with ${\rm M}\mu = {\rm N}\nu$ by removing the prime from $\Sigma'_{{\rm MN}\mu\nu}$, and subtract the same terms the form

$$\lim_{R\to 0} \tfrac{1}{2} \sum_{{\rm M}\mu} z_\mu e \varphi_{\mu 1}(R) = \frac{N_0 e^2 \eta}{\sqrt{\pi}} \sum_\mu z_\mu^2, \tag{23.46}$$

where (23.39) was used in this evaluation. Then the Fourier sum in (23.45) for $\Omega_{\rm I}$ becomes

$$(2\pi e^2/V) \sum_{\mathbf{q}} q^{-2} e^{-q^2/4\eta^2} \sum_{{\rm MN}\mu\nu} z_\mu z_\nu e^{i\mathbf{q}\cdot[\tilde{\mathbf{R}}({\rm M}\mu) - \tilde{\mathbf{R}}({\rm N}\nu)]} - N_0(e^2\eta/\sqrt{\pi}) \sum_\mu z_\mu^2. \tag{23.47}$$

Now the $\mathbf{q} = 0$ term in (23.47), which arises from the self-energy of a uniform positive charge distribution, is divergent. However, such a term should not appear in the *total* potential $\Phi = \Omega_{\rm I} + E_{\rm G}$, since the net uniform component ($\mathbf{q} = 0$ component) of the total charge distribution vanishes. In other words, the two divergent $\mathbf{q} = 0$ terms arising from the ion–ion interactions and the electron–electron interactions are just canceled by the $\mathbf{q} = 0$ term arising from the ion–electron interactions. However, we cannot simply drop the $\mathbf{q} = 0$ term from (23.47), since as a result of splitting the ion–ion interactions into two contributions, (23.47) contains only a part of the ion–ion Coulomb energy. The correct elimination of the divergent $\mathbf{q} = 0$ terms may be accomplished as follows.

The total charge density $e\rho_{\rm I}(\mathbf{r})$ of the ions may be written

$$e\rho_{\rm I}(\mathbf{r}) = e \sum_{{\rm M}\mu} z_\mu \delta(\mathbf{r} - \tilde{\mathbf{R}}({\rm M}\mu)), \tag{23.48}$$

and the Fourier components are simply

$$e\rho_{\rm I}(\mathbf{q}) = eV^{-1} \sum_{{\rm M}\mu} z_\mu e^{-i\mathbf{q}\cdot\tilde{\mathbf{R}}({\rm M}\mu)}. \tag{23.49}$$

Since the electrostatic potential due to the ionic charges must satisfy Poisson's equation, components of this electrostatic potential are $(4\pi e^2/q^2)\rho_I(\mathbf{q})$, and the total ion–ion energy is

$$\Omega_I = \tfrac{1}{2}V \sum_{\mathbf{q}} \rho_I(\mathbf{q})(4\pi e^2/q^2)\rho_I(-\mathbf{q}), \tag{23.50}$$

where the factor $\tfrac{1}{2}$ is to avoid double counting, the same as the factor $\tfrac{1}{2}$ in (23.43). In the same way the total electron–electron and ion–electron electrostatic energies may be written, respectively,

$$\tfrac{1}{2}V \sum_{\mathbf{q}} \rho(\mathbf{q})(4\pi e^2/q^2)\rho(-\mathbf{q}); \tag{23.51}$$

$$-\tfrac{1}{2}V \sum_{\mathbf{q}} (4\pi e^2/q^2)[\rho(\mathbf{q})\rho_I(-\mathbf{q}) + \rho_I(\mathbf{q})\rho(-\mathbf{q})]. \tag{23.52}$$

In keeping with our previous notation, $\rho(\mathbf{q})$ are components of the band-electron density; the corresponding charge density components are $-e\rho(\mathbf{q})$. Now as $\mathbf{q} \to 0$, $\rho(\mathbf{q}) \to \rho_I(\mathbf{q})$, and the divergent terms in (23.50)–(23.52) sum to zero. However, the Fourier sum in (23.47) does not include the total Ω_I of (23.50), but only the fraction $e^{-q^2/4\eta^2}$ of each Fourier component; this is easily verified by writing out (23.50) for Ω_I with the aid of (23.49) for $\rho_I(\mathbf{q})$, and comparing the result with (23.47). Therefore in collecting all the divergent terms in the total potential Φ, we have to evaluate

$$\lim_{\mathbf{q}\to 0} \frac{2\pi e^2 V}{q^2} \rho_I(\mathbf{q})\rho_I(-\mathbf{q})[e^{-q^2/4\eta^2} + 1 - 2] = -\frac{\pi e^2 V}{2\eta^2}\left(\frac{\sum_{\mu} z_{\mu}}{V_C}\right)^2, \tag{23.53}$$

where we used

$$\rho_I(\mathbf{q} = 0) = V_C^{-1} \sum_{\mu} z_{\mu}. \tag{23.54}$$

Thus all the electrostatic self-energies cancel, and the use of the convergence factor η in calculating the ion–ion Coulomb energy leaves the contribution (23.53) when the $\mathbf{q} = 0$ term is omitted from the Fourier sum (23.47). Note that (23.53) $\to 0$ when $\eta \to \infty$. Now the energy of an electron at $\mathbf{r}$ in the presence of the ion $M\mu$ at $\tilde{\mathbf{R}}(M\mu)$ is the bare ion potential $v_{\mu}(\mathbf{r} - \tilde{\mathbf{R}}(M\mu))$, and this potential presumably contains a point-charge Coulomb contribution $-z_{\mu}e^2\,|\mathbf{r} - \tilde{\mathbf{R}}(M\mu)|^{-1}$, plus a remaining contribution which is localized to the core region. Therefore $\Omega_B(\mathbf{r})$ may be represented as

$$\Omega_B(\mathbf{r}) = -e \sum_{M\mu} \varphi_{\mu}(|\mathbf{r} - \tilde{\mathbf{R}}(M\mu)|) + \text{core contributions}, \tag{23.55}$$

where φ_{μ} is given by (23.30). In the process of canceling out the divergent electrostatic self-energies, all $\mathbf{q} = 0$ components of the point-ion Coulomb part of Ω_{B}, as written out explicitly in (23.55), as well as all $\mathbf{q} = 0$ components of the Hartree potential Ω_{S}, are also canceled from the band-electron ground-state energy E_{G}. This applies, of course, to each order in the expansion of E_{G} in powers of the ion displacements, namely to $E_{\mathrm{G}}^{(0)}, E_{\mathrm{G}}^{(1)}, E_{\mathrm{G}}^{(2)}, \cdots$. It should be noted that in including *all* the electrostatic $\mathbf{q} = 0$ components, we are transforming Ω_{I} from the total Coulomb energy of a collection of point ions to the total Coulomb energy of these point ions together with a uniform compensating negative background charge. As a result, at normal crystal densities Ω_{I} is dominated by the attractive interaction between the ions and the background charge, and is negative.

Our final expression for Ω_{I} is obtained from (23.45), with the Fourier sum transformed to (23.47) and the $\mathbf{q} = 0$ term omitted by adding (23.53):

$$\begin{aligned}\Omega_{\mathrm{I}} = {} & (2\pi e^2/V) \sum_{\mathbf{q}}{}' q^{-2} e^{-q^2/4\eta^2} \sum_{\mathrm{MN}\mu\nu} z_\mu z_\nu e^{i\mathbf{q}\cdot[\tilde{\mathbf{R}}(\mathrm{M}\mu) - \tilde{\mathbf{R}}(\mathrm{N}\nu)]} \\ & + \tfrac{1}{2} e^2 \sum_{\mathrm{MN}\mu\nu}{}' z_\mu z_\nu |\tilde{\mathbf{R}}(\mathrm{M}\mu) - \tilde{\mathbf{R}}(\mathrm{N}\nu)|^{-1} \operatorname{erfc}[\eta\, |\tilde{\mathbf{R}}(\mathrm{M}\mu) - \tilde{\mathbf{R}}(\mathrm{N}\nu)|] \qquad (23.56) \\ & - N_0 e^2 \left[(\eta/\sqrt{\pi}) \sum_\mu z_\mu^2 + (\pi/2\eta^2 V_C) \left(\sum_\mu z_\mu \right)^2 \right].\end{aligned}$$

When all the ions are located at their equilibrium positions $\mathbf{R}(\mathrm{M}\mu)$, the value of Ω_{I} is denoted $\Omega_{\mathrm{I}}^{(0)}$, and (23.56) simplifies because

$$\underset{\mathrm{MN}\mu\nu}{} z_\mu z_\nu e^{i\mathbf{q}\cdot[\mathbf{R}(\mathrm{M}\mu) - \mathbf{R}(\mathrm{N}\nu)]} = N_0^2 \delta(\mathbf{q}) \sum_{\mu\nu} z_\mu z_\nu e^{i\mathbf{q}\cdot[\mathbf{R}(\mu) - \mathbf{R}(\nu)]}. \qquad (23.57)$$

Here we used (10.22) for $\delta(\mathbf{q})$, and noted that $e^{-i\mathbf{Q}\cdot\mathbf{R}(\mathrm{N})} = 1$. When (23.57) is used in the Fourier sum in (23.56), the $\Sigma_{\mathbf{q}}'$ is reduced to a $\Sigma_{\mathbf{Q}}'$, and because the coefficient $Q^{-2} e^{-Q^2/4\eta^2}$ is the same for $\pm\mathbf{Q}$, then $e^{i\mathbf{Q}\cdot[\mathbf{R}(\mu) - \mathbf{R}(\nu)]}$ may be replaced by $\cos[\mathbf{Q} \cdot [\mathbf{R}(\mu) - \mathbf{R}(\nu)]]$. In addition, surface effects may be eliminated from the lattice sum in (23.56), and $\Omega_{\mathrm{I}}^{(0)}$ becomes

$$\begin{aligned}\Omega_{\mathrm{I}}^{(0)} = {} & N_0 (2\pi e^2/V_C) \sum_{\mathbf{Q}}{}' Q^{-2} e^{-Q^2/4\eta^2} \sum_{\mu\nu} z_\mu z_\nu \cos[\mathbf{Q} \cdot [\mathbf{R}(\mu) - \mathbf{R}(\nu)]] \\ & + \tfrac{1}{2} N_0 e^2 \sum_{\mathrm{N}\mu\nu}{}' z_\mu z_\nu |\mathbf{R}(\mu) - \mathbf{R}(\mathrm{N}\nu)|^{-1} \operatorname{erfc}[\eta\, |\mathbf{R}(\mu) - \mathbf{R}(\mathrm{N}\nu)|] \qquad (23.58) \\ & - N_0 e^2 \left[(\eta/\sqrt{\pi}) \sum_\mu z_\mu^2 + (\pi/2\eta^2 V_C) \left(\sum_\mu z_\mu \right)^2 \right].\end{aligned}$$

This expression is independent of η, and if η is properly chosen both the direct- and inverse-lattice sums converge very rapidly.

Now in calculating the contribution of E_{G} to the potential energy coefficients, equations (23.17) and (23.18), all $\mathbf{q} = 0$ terms were retained, so the $\mathbf{q} = 0$ term must also be retained in calculating the contribution of Ω_{I} to the potential coefficients. This procedure will allow us to demonstrate the explicit cancellation of these terms in the final expressions for the dynamical matrices. The Fourier sum in (23.47) is easily expanded in powers of the displacements $\mathbf{U}(\mathrm{M}\mu)$ of the ions from their equilibrium positions $\mathbf{R}(\mathrm{M}\mu)$, and the expansion coefficients are found to be

$$\Phi_i(\mathrm{M}\mu)_{\mathrm{IF}} = -(4\pi e^2/V_C)z_\mu \sum_\nu z_\nu \sum_{\mathbf{Q}} Q^{-2} e^{-Q^2/4\eta^2} Q_i \sin[\mathbf{Q}\cdot[\mathbf{R}(\mu) - \mathbf{R}(\nu)]]; \tag{23.59}$$

$$\begin{aligned}\Phi_{ij}(\mathrm{M}\mu,\mathrm{N}\nu)_{\mathrm{IF}} = {} & (4\pi e^2/V)z_\mu z_\nu \sum_{\mathbf{q}} q^{-2} e^{-q^2/4\eta^2} q_i q_j \cos[\mathbf{q}\cdot[\mathbf{R}(\mathrm{M}\mu) - \mathbf{R}(\mathrm{N}\nu)]] \\ & - \delta_{\mathrm{MN}}\delta_{\mu\nu}(4\pi e^2/V_C)z_\mu \sum_\pi z_\pi \sum_{\mathbf{Q}} Q^{-2} e^{-Q^2/4\eta^2} Q_i Q_j \\ & \times \cos[\mathbf{Q}\cdot[\mathbf{R}(\mu) - \mathbf{R}(\pi)]].\end{aligned} \tag{23.60}$$

Here the subscripts IF are to remind us that these contributions to the potential energy coefficients arise from the Fourier sum contribution to Ω_{I}. It is obvious that the contribution (23.60) to the second-order potential coefficients satisfies the translational invariance condition:

$$\sum_{\mathrm{N}\nu} \Phi_{ij}(\mathrm{M}\mu,\mathrm{N}\nu)_{\mathrm{IF}} = 0. \tag{23.61}$$

The lattice sum contribution to Ω_{I}, as shown in (23.45), is a sum over all pairs of ions of a central potential of the form

$$\varphi_{\mu\nu}(R) = z_\mu z_\nu e^2 R^{-1}\,\mathrm{erfc}(\eta R), \tag{23.62}$$

and the corresponding contributions to the potential energy coefficients are given by the results of Section 9. From (9.9), the first-order coefficients are found to be

$$\begin{aligned}\Phi_i(\mathrm{M}\mu)_{\mathrm{IL}} = {} & -e^2 z_\mu \sum_{\mathrm{N}\nu}{}' z_\nu \{(2\eta/\sqrt{\pi})\,|\mathbf{R}(\mathrm{M}\mu,\mathrm{N}\nu)|^{-2}\, e^{-\eta^2|\mathbf{R}(\mathrm{M}\mu,\mathrm{N}\nu)|^2} \\ & + |\mathbf{R}(\mathrm{M}\mu,\mathrm{N}\nu)|^{-3}\,\mathrm{erfc}[\eta\,|\mathbf{R}(\mathrm{M}\mu,\mathrm{N}\nu)|]\} R_i(\mathrm{M}\mu,\mathrm{N}\nu),\end{aligned} \tag{23.63}$$

where we are again using the abbreviation (9.7):

$$\mathbf{R}(\mathrm{M}\mu,\mathrm{N}\nu) = \mathbf{R}(\mathrm{M}\mu) - \mathbf{R}(\mathrm{N}\nu).$$

From (9.10), the second-order coefficients for $\mathrm{M}\mu \neq \mathrm{N}\nu$ are found to be

$$\begin{aligned}& \Phi_{ij}(\mathrm{M}\mu,\mathrm{N}\nu)_{\mathrm{IL}} \\ & = -e^2 z_\mu z_\nu \{(2\eta/\sqrt{\pi})[3\,|\mathbf{R}(\mathrm{M}\mu,\mathrm{N}\nu)|^{-4} + 2\eta^2\,|\mathbf{R}(\mathrm{M}\mu,\mathrm{N}\nu)|^{-2}] e^{-\eta^2|\mathbf{R}(\mathrm{M}\mu,\mathrm{N}\nu)|^2} \\ & \quad + 3\,|\mathbf{R}(\mathrm{M}\mu,\mathrm{N}\nu)|^{-5}\,\mathrm{erfc}[\eta\,|\mathbf{R}(\mathrm{M}\mu,\mathrm{N}\nu)|]\} R_i(\mathrm{M}\mu,\mathrm{N}\nu) R_j(\mathrm{M}\mu,\mathrm{N}\nu) \\ & \quad + \delta_{ij} e^2 z_\mu z_\nu \{(2\eta/\sqrt{\pi})\,|\mathbf{R}(\mathrm{M}\mu,\mathrm{N}\nu)|^{-2}\, e^{-\eta^2|\mathbf{R}(\mathrm{M}\mu,\mathrm{N}\nu)|^2} \\ & \quad + |\mathbf{R}(\mathrm{M}\mu,\mathrm{N}\nu)|^{-3}\,\mathrm{erfc}[\eta\,|\mathbf{R}(\mathrm{M}\mu,\mathrm{N}\nu)|]\}.\end{aligned} \tag{23.64}$$

The subscripts IL indicate contributions to the potential coefficients arising from the lattice sum contribution to Ω_{I}.

Now the total point-ion Coulomb energy must be invariant with respect to a uniform translation of the lattice, so that the contributions from Ω_{I} to the potential energy coefficients must separately satisfy translational invariance conditions. Since the Fourier sum contributions satisfy translational invariance, according to (23.61), then the lattice sum contributions must also satisfy the same condition, and this allows us to determine the lattice sum contribution to the self-coupling coefficient $\Phi_{ij}(\mathrm{M}\mu,\mathrm{M}\mu)_{\mathrm{IL}}$, by the equation

$$\Phi_{ij}(\mathrm{M}\mu, \mathrm{M}\mu)_{\mathrm{IL}} = -\sum_{\mathrm{N}\nu}{}' \Phi_{ij}(\mathrm{M}\mu,\mathrm{N}\nu)_{\mathrm{IL}}. \tag{23.65}$$

The coefficients on the right of (23.65) are given by (23.64).

Lattice Statics and Dynamics

In Section 21, especially (21.13) and (21.14), the total Hamiltonian in the adiabatic approximation is shown to be

$$\mathscr{H} = KE_{\mathrm{I}} + \Phi, \tag{23.66}$$

where Φ is the total adiabatic potential $\Omega_{\mathrm{I}} + E_{\mathrm{G}}$. According to the theory of Section 10, this Hamiltonian represents a set of weakly interacting phonons, and with the neglect of surface effects it may be written

$$\mathscr{H} = \Phi_0 + \mathscr{H}_{\mathrm{p}} + \mathscr{H}_{\mathrm{pp}}, \tag{23.67}$$

where $\mathscr{H}_{\mathrm{p}}$ is the set of harmonic phonons

$$\mathscr{H}_p = \sum_{\kappa} \hbar\omega_{\kappa}(A_{\kappa}^{\dagger}A_{\kappa} + \tfrac{1}{2}), \tag{23.68}$$

and $\mathscr{H}_{\mathrm{pp}}$ is the phonon–phonon interaction contribution

$$\mathscr{H}_{\mathrm{pp}} = \Phi_3 + \Phi_4 + \cdots. \tag{23.69}$$

Now Φ_0 is the total crystal energy when all ions are located at their equilibrium positions and the band electrons are in their ground state,

$$\Phi_0 = \Omega_{\mathrm{I}}^{(0)} + E_{\mathrm{G}}^{(0)}, \tag{23.70}$$

and Φ_0 depends only on the macroscopic configuration of the crystal. The harmonic phonon Hamiltonian $\mathscr{H}_{\mathrm{p}}$ is based on the potential energy coefficients evaluated in the interior of the crystal, and we shall presently discuss the equilibrium conditions and the dynamical matrices. The phonon–phonon interactions require evaluation of third- and fourth-order potential energy coefficients, and the band-electron ground-state energy contributions to these coefficients have not been discussed in the present formulation. A straightforward approach to the anharmonic terms $E_G^{(3)}$ and $E_G^{(4)}$ requires extension of the self-consistent screening theory of Sections 21 and 22 to higher orders. Alternate approaches are possible, and there is a need for more work on this complicated problem.

The total potential Φ_0 is most easily calculated from (23.58) for $\Omega_I^{(0)}$, plus the total $E_G^{(0)}$ as obtained from a band-structure calculation [represented by (22.20) for the Hartree plus exchange model, for example]. Surface effects have been eliminated from the expression for $\Omega_I^{(0)}$, and further the strain and volume derivatives of $\Omega_I^{(0)}$ are easily calculated by direct differentiation of (23.58). Likewise, surface effects are eliminated from $E_G^{(0)}$ in band-structure calculations by working with electron wave functions in the space of one unit cell, or an equivalent polyhedron, and it is practical to calculate also the strain and volume derivatives of $E_G^{(0)}$. The potential contributions to the stresses and elastic constants, which arise from strain derivatives of Φ_0, are discussed in Section 16. For example, the pressure is $\tilde{P} = -d\Phi_0/dV$ in the potential approximation, and we note that even if one requires $\tilde{P} = 0$ as an approximate zero pressure condition, the volume derivatives of $\Omega_I^{(0)}$ and $E_G^{(0)}$ are not each zero. Indeed the attractive interaction between the ions and the background charge tend to contract the crystal, giving rise to $d\Omega_I^{(0)}/dV$ large and positive. On the other hand, the band electron energy $E_G^{(0)}$ is dominated by the electronic kinetic energy, which corresponds to $dE_G^{(0)}/dV$ large and negative. Similar remarks apply to contributions from $\Omega_I^{(0)}$ and $E_G^{(0)}$ to components of the stress tensor and the elastic constants. Incidentally, to set the zero of energy to correspond to the free atoms at infinite separation, the total ionization energy of the collection of ions and electrons is presumably contained in $E_G^{(0)}$.

For the crystal in the presence of arbitrary externally applied stress, but excluding externally applied fields which penetrate the crystal, the equilibrium condition for ions in the interior is (7.1), which requires $\Phi_i(M\mu)$ to vanish for all ions in the interior, or

$$\Phi_i(0\mu) = 0, \quad \text{for all } \mu, i. \tag{23.71}$$

These coefficients may be calculated from the three contributions written out in (23.17), (23.59), and (23.63):

$$\Phi_i(0\mu) = \Phi_i(0\mu)_E + \Phi_i(0\mu)_{IF} + \Phi_i(0\mu)_{IL}. \tag{23.72}$$

The equilibrium condition must be satisfied for any prescribed model in order to ensure the validity of lattice-statics or lattice-dynamics calculations. Further, the equilibrium condition is just the condition which determines the sublattice displacements under a homogeneous deformation of the lattice from the initial equilibrium configuration, as worked out in Section 7. In sufficiently simple crystals it is possible to show that the equilibrium condition is satisfied by symmetry, and that the sublattice displacements are zero; in more complicated cases it is necessary to solve for the sublattice displacements in order to calculate strain derivatives of Φ_0.

The dynamical matrices are given by the general formula (10.29):

$$D_{ij}(\mu\nu,\mathbf{k}) = (M_\mu M_\nu)^{-1/2} N_0^{-1} \sum_{\mathrm{MN}} \Phi_{ij}(\mathrm{M}\mu,\mathrm{N}\nu) e^{-i\mathbf{k}\cdot[\mathbf{R}(\mathrm{M}\mu)-\mathbf{R}(\mathrm{N}\nu)]}. \tag{23.73}$$

We will work out this formula for each of the contributions to the potential coefficients $\Phi_{ij}(\mathrm{M}\mu,\mathrm{N}\nu)$, and for $\mathbf{k} \neq 0$. In the process we use the δ function (10.23), which may be written

$$N_0^{-1} \sum_{\mathrm{M}} e^{i(\mathbf{q}-\mathbf{k})\cdot\mathbf{R}(\mathrm{M}\mu)} = \delta(\mathbf{q}-\mathbf{k}) e^{i(\mathbf{q}-\mathbf{k})\cdot\mathbf{R}(\mu)}; \tag{23.74}$$

this implies $\mathbf{q} - \mathbf{k} = \mathbf{Q}$, and can be used to reduce a $\Sigma_{\mathbf{q}}$ to a $\Sigma_{\mathbf{Q}}$. The band-electron ground-state potential coefficients $\Phi_{ij}(\mathrm{M}\mu,\mathrm{N}\nu)_{\mathrm{E}}$ are written in (23.18), and their contribution to the dynamical matrix is

$$\begin{aligned} D_{ij}(\mu\nu,\mathbf{k})_{\mathrm{E}} = {} & (M_\mu M_\nu)^{-1/2} \sum_{\mathbf{QQ}'} \tilde{c}(\mathbf{Q}+\mathbf{k},\mathbf{Q}'+\mathbf{k})(\mathbf{Q}+\mathbf{k})_i(\mathbf{Q}'+\mathbf{k})_j \\ & \times v_\mu(-\mathbf{Q}-\mathbf{k}) v_\nu(\mathbf{Q}'+\mathbf{k}) e^{i[\mathbf{Q}\cdot\mathbf{R}(\mu)-\mathbf{Q}'\cdot\mathbf{R}(\nu)]} \\ & - \delta_{\mu\nu} M_\mu^{-1} \sum_{\mathbf{Q}} c(\mathbf{Q}) Q_i Q_j v_\mu(\mathbf{Q}) e^{-i\mathbf{Q}\cdot\mathbf{R}(\mu)}. \end{aligned} \tag{23.75}$$

For the Fourier sum part of the point ion Coulomb energy Ω_{I}, the second-order potential coefficients are written in (23.60) and the corresponding contribution to the dynamical matrix is

$$\begin{aligned} D_{ij}(\mu\nu,\mathbf{k})_{\mathrm{IF}} = {} & (M_\mu M_\nu)^{-1/2} (4\pi e^2/V_C) z_\mu z_\nu \sum_{\mathbf{Q}} e^{-|\mathbf{Q}+\mathbf{k}|^2/4\eta^2} |\mathbf{Q}+\mathbf{k}|^{-2} \\ & \times (\mathbf{Q}+\mathbf{k})_i (\mathbf{Q}+\mathbf{k})_j e^{i\mathbf{Q}\cdot[\mathbf{R}(\mu)-\mathbf{R}(\nu)]} \\ & - \delta_{\mu\nu} M_\mu^{-1} (4\pi e^2/V_C) z_\mu \sum_{\pi} z_\pi \sum_{\mathbf{Q}} e^{-Q^2/4\eta^2} Q^{-2} Q_i Q_j \\ & \times \cos[\mathbf{Q}\cdot[\mathbf{R}(\mu)-\mathbf{R}(\pi)]]. \end{aligned} \tag{23.76}$$

Finally for the lattice sum part of the point-ion Coulomb energy, the dynamical matrix with surface effects eliminated and with the self-coupling potential coefficients explicitly eliminated is

$$\begin{aligned} D_{ij}(\mu\nu,\mathbf{k})_{\mathrm{IL}} = (M_\mu M_\nu)^{-1/2} \Big\{ & \sum_{\mathrm{N}}{}'' \Phi_{ij}(0\mu,\mathrm{N}\nu)_{\mathrm{IL}} e^{i\mathbf{k}\cdot[\mathbf{R}(\mathrm{N}\nu)-\mathbf{R}(\mu)]} \\ & - \delta_{\mu\nu} \sum_{\mathrm{P}\pi}{}' \Phi_{ij}(0\mu,\mathrm{P}\pi)_{\mathrm{IL}} \Big\}. \end{aligned} \tag{23.77}$$

Here the notation Σ''_N means to sum over all N when $\mu \neq \nu$, and to omit the term $\mathrm{N} = 0$ when $\mu = \nu$; the $\Sigma'_{\mathrm{P}\pi}$ means to omit the term $\mathrm{P}\pi = 0\mu$, as usual. The coefficients $\Phi_{ij}(0\mu,\mathrm{N}\nu)_{\mathrm{IL}}$ appearing in (23.77) are given by (23.64).

The total dynamical matrix is, of course, the sum of (23.75)–(23.77):

$$D_{ij}(\mu\nu,\mathbf{k}) = D_{ij}(\mu\nu,\mathbf{k})_{\mathrm{E}} + D_{ij}(\mu\nu,\mathbf{k})_{\mathrm{IF}} + D_{ij}(\mu\nu,\mathbf{k})_{\mathrm{IL}}. \tag{23.78}$$

Further, since our equations are valid for arbitrary initial equilibrium configuration of the lattice, the strain derivatives of $D_{ij}(\mu\nu,\mathbf{k})$ may be calculated from the contributions (23.75)–(23.77), and the strain derivatives of the phonon frequencies may then be calculated from (11.76).

Now $D_{ij}(\mu\nu,\mathbf{k})_{\mathrm{E}}$ and $D_{ij}(\mu\nu,\mathbf{k})_{\mathrm{IF}}$ each contain a $\Sigma_{\mathbf{Q}}$ which is independent of $\mathbf{k}$, and the $\mathbf{Q} = 0$ terms cancel between these two sums. To show this, evaluate the $\mathbf{Q} = 0$ terms as a limit as $\mathbf{q} \to 0$, with $\mathbf{q}$ going to zero in the *same direction* in each term, and use the relations

$$c(\mathbf{q}) \to c(0) = \sum_{\nu} z_{\nu}, \qquad \text{as } \mathbf{q} \to 0; \tag{23.79}$$

$$v_{\mu}(\mathbf{q}) \to -4\pi e^2 z_{\mu}/q^2 V_C, \quad \text{as } \mathbf{q} \to 0. \tag{23.80}$$

To evaluate (23.79) we used (23.2), and (23.80) represents the fact that $v_{\mu}(\mathbf{r})$ is a Coulomb potential at large $\mathbf{r}$.

It is also instructive to examine the role of translational invariance in the dynamical matrix in the limit $\mathbf{k} \to 0$. Since $\tilde{c}(\mathbf{q},\mathbf{q}')$ contains a $\delta(\mathbf{q} - \mathbf{q}')$, it can be written in the form $\tilde{c}(\mathbf{Q} + \mathbf{k},\mathbf{Q}' + \mathbf{k})$, and (23.18) may be expressed as

$$\begin{aligned}\Phi_{ij}(\mathrm{M}\mu,\mathrm{N}\nu)_{\mathrm{E}} = N_0^{-1} &\sum_{\mathbf{k}} \sum_{\mathbf{QQ}'} \tilde{c}(\mathbf{Q} + \mathbf{k},\mathbf{Q}' + \mathbf{k})(\mathbf{Q} + \mathbf{k})_i(\mathbf{Q}' + \mathbf{k})_j \\ &\times v_{\mu}(-\mathbf{Q} - \mathbf{k})v_{\nu}(\mathbf{Q}' + \mathbf{k})e^{i[(\mathbf{Q}+\mathbf{k}\cdot\mathbf{R}(\mathrm{M}\mu)-(\mathbf{Q}'+\mathbf{k})\mathbf{R}\cdot(\mathrm{N}\nu)]} \\ &- \delta_{\mathrm{MN}}\delta_{\mu\nu} \sum_{\mathbf{k}} \sum_{\mathbf{Q}} c(\mathbf{Q} + \mathbf{k})(\mathbf{Q} + \mathbf{k})_i(\mathbf{Q} + \mathbf{k})_j \\ &\times v_{\mu}(\mathbf{Q} + \mathbf{k})e^{-i(\mathbf{Q}+\mathbf{k})\cdot\mathbf{R}(\mathrm{M}\mu)},\end{aligned} \tag{23.81}$$

where $\Sigma_{\mathbf{k}}$ is over the Brillouin zone. According to the translational invariance condition (23.20), the sum over $\mathrm{N}\nu$ of (23.81) must vanish; when the sum is carried out the result is

$$\begin{aligned}\lim_{\mathbf{k}\to 0} \Big\{ &\sum_{\nu} \sum_{\mathbf{QQ}'} \tilde{c}(\mathbf{Q} + \mathbf{k},\mathbf{Q}' + \mathbf{k})(\mathbf{Q} + \mathbf{k})_i(\mathbf{Q}' + \mathbf{k})_j \\ &\times v_{\mu}(-\mathbf{Q} - \mathbf{k})v_{\nu}(\mathbf{Q}' + \mathbf{k})e^{i[\mathbf{Q}\cdot\mathbf{R}(\mu)-\mathbf{Q}'\cdot\mathbf{R}(\nu)]} \\ &- \sum_{\mathbf{Q}} c(\mathbf{Q} + \mathbf{k})(\mathbf{Q} + \mathbf{k})_i(\mathbf{Q} + \mathbf{k})_j v_{\mu}(\mathbf{Q} + \mathbf{k})e^{-i\mathbf{Q}\cdot\mathbf{R}(\mu)}\Big\} = 0.\end{aligned} \tag{23.82}$$

Here we have recognized that $c(\mathbf{q})$ contains a $\delta(\mathbf{q})$, and that $\delta(\mathbf{k})$ implies $\mathbf{k} = 0$ since $\mathbf{k}$ lies in the Brillouin zone, and have replaced $\mathbf{k} = 0$ by the limit $\mathbf{k} \to 0$ with the understanding that $c(\mathbf{Q} + \mathbf{k}) \to c(\mathbf{Q})$ as $\mathbf{k} \to 0$. But in (23.82) there are terms which depend on the direction of $\mathbf{k}$ as $\mathbf{k} \to 0$, and terms which do not, and since (23.82) must hold for *any* direction of $\mathbf{k}$, it contains in fact two separate conditions. The terms with $\mathbf{Q} \neq 0$ and $\mathbf{Q}' \neq 0$ do not depend on the direction of $\mathbf{k}$, and letting $\mathbf{k} = 0$ in these terms gives the condition

$$\sum_{\nu} \sum_{\mathbf{Q}}{}' \sum_{\mathbf{Q}'}{}' \tilde{c}(\mathbf{Q},\mathbf{Q}') Q_i Q_j' v_\mu(-\mathbf{Q}) v_\nu(\mathbf{Q}') e^{i[\mathbf{Q}\cdot\mathbf{R}(\mu) - \mathbf{Q}'\cdot\mathbf{R}(\nu)]} = \sum_{\mathbf{Q}}{}' c(\mathbf{Q}) Q_i Q_j v_\mu(\mathbf{Q}) e^{-i\mathbf{Q}\cdot\mathbf{R}(\mu)}. \qquad (23.83)$$

The remaining terms in (23.82) may be simplified with the aid of (23.79) and (23.80) to give the condition

$$\lim_{\mathbf{k}\to 0} \sum_{\nu} \Big\{ [(4\pi e^2/k^2 V_C)\tilde{c}(\mathbf{k},\mathbf{k}) + 1] k^{-2} k_i k_j z_\mu z_\nu - \sum_{\mathbf{Q}}{}' \tilde{c}(\mathbf{Q}+\mathbf{k},\mathbf{k}) k^{-2} Q_i k_j z_\nu v_\mu(-\mathbf{Q}) e^{i\mathbf{Q}\cdot\mathbf{R}(\mu)} - \sum_{\mathbf{Q}}{}' \tilde{c}(\mathbf{k},\mathbf{Q}+\mathbf{k}) k^{-2} k_i Q_j z_\mu v_\nu(\mathbf{Q}) e^{-i\mathbf{Q}\cdot\mathbf{R}(\nu)} \Big\} = 0. \qquad (23.84)$$

These last two equations must hold for each μ. If $\tilde{c}(\mathbf{q},\mathbf{q}')$ is diagonal, i.e. contains a $\delta_{\mathbf{qq}'}$, then (23.84) reduces to

$$\lim_{\mathbf{k}\to 0} \tilde{c}(\mathbf{k},\mathbf{k}) = -k^2 V_C / 4\pi e^2, \quad \text{for } \tilde{c}(\mathbf{q},\mathbf{q}') \text{ diagonal.} \qquad (23.85)$$

It can now be seen that the translational invariance condition (23.82), or the two conditions (23.83) and (23.84), leads to

$$\lim_{\mathbf{k}\to 0} \sum_{\nu} (M_\mu M_\nu)^{1/2} D_{ij}(\mu\nu,\mathbf{k})_{\mathrm{E}} = 0, \quad \text{for each } \mu, \qquad (23.86)$$

where the last $\Sigma_{\mathbf{Q}}$ of (23.75) for $D_{ij}(\mu\nu,\mathbf{k})_{\mathrm{E}}$ is evaluated by replacing $\mathbf{Q}$ by $\mathbf{Q} + \mathbf{k}$ with $\mathbf{k} \to 0$. The limit (23.86) holds also for the contributions $D_{ij}(\mu\nu,\mathbf{k})_{\mathrm{IF}}$ and $D_{ij}(\mu\nu,\mathbf{k})_{\mathrm{IL}}$, so the total dynamical matrix obeys the condition

$$\lim_{\mathbf{k}\to 0} \sum_{\nu} (M_\mu M_\nu)^{1/2} D_{ij}(\mu\nu,\mathbf{k}) = 0, \quad \text{for each } \mu. \qquad (23.87)$$

According to the long-waves theory of Section 12, this is just the condition which requires the acoustic phonon frequencies to go to zero as $\mathbf{k} \to 0$.

24. METALS

Koopmans' Theorem

A metal crystal is described as a collection of ions and band electrons, just as the description of nonmetals in Section 21, except that for metals there is a partially filled conduction band, as well as the filled valence bands. In addition to the ordinary lattice-dynamical thermodynamics discussed in Chapter 4, there is a contribution to the free energy, and hence to each thermodynamic function, arising from the thermal excitation of conduction electrons in metals. This contribution arises because there are one-electron states in the conduction band which are unoccupied in the ground state of the band electrons, and which lie in an energy range KT above the occupied states, for any $KT > 0$. In nonmetals, the energy gap is presumed to be large compared to KT, for all temperatures of interest to us. The excitation of conduction electrons represents a small contribution to the total free energy of a metal, but this contribution is nevertheless observable in measurements of the heat capacity and other properties. By the same token, nonadiabatic effects still give a very small contribution to the total free energy of a metal, but these effects must be included in the theory since they modify the electronic excitation contributions in a significant way. Our description of nonadiabatic effects in metals will take the form of electron–phonon interactions.

The first step in the treatment of metals is to sort out the double counting corrections required in calculating the total crystal energy when excited electronic states are included; we will do this for the Hartree plus exchange approximation of Section 22. The one-electron occupation numbers will be denoted by $f_\lambda = 0,1$ for the band electrons in an arbitrary state in the crystal, and the total energy of the band electrons is $E_\Lambda = E_\Lambda(\cdots f_\lambda \cdots)$.* The one-electron wave equation (22.11) may be written

$$h\psi_\lambda = \epsilon_\lambda \psi_\lambda, \tag{24.1}$$

where

$$h = ke + \Omega_B + \Omega_S + \Omega_X. \tag{24.2}$$

With the band electrons in an arbitrary state, not necessarily the ground state, the band-electron density is, instead of (21.29),

$$\rho(\mathbf{r}) = \sum_\lambda f_\lambda \psi_\lambda(\mathbf{r})^* \psi_\lambda(\mathbf{r}). \tag{24.3}$$

* The f_λ are the same as the n_λ of Section 13; we use a different symbol here to avoid any possible confusion with the phonon occupation numbers n_κ.

Further, the Hartree self-consistent potential is, instead of (21.31),

$$\Omega_S(\mathbf{r}) = e^2 \sum_\lambda f_\lambda \int \psi_\lambda(\mathbf{r}')^* \, |\mathbf{r} - \mathbf{r}'|^{-1} \, \psi_\lambda(\mathbf{r}') \, d\mathbf{r}'. \tag{24.4}$$

Finally the total energy is of the same form as (22.7) for E_G, only with g_λ replaced by f_λ and of course with $\rho(\mathbf{r})$ and $\Omega_S(\mathbf{r})$ evaluated from (24.3) and (24.4); the result may be written

$$\begin{aligned} E_\Lambda = {} & \sum_\lambda f_\lambda \int \psi_\lambda(\mathbf{r})^*[ke(\mathbf{r}) + \Omega_B(\mathbf{r})]\psi_\lambda(\mathbf{r}) \, d\mathbf{r} \\ & + \tfrac{1}{2}e^2 \sum_{\lambda\lambda'}{}' f_\lambda f_{\lambda'} \iint \psi_\lambda(\mathbf{r})^* \psi_{\lambda'}(\mathbf{r}')^* \, |\mathbf{r} - \mathbf{r}'|^{-1} \, \psi_{\lambda'}(\mathbf{r}')\psi_\lambda(\mathbf{r}) \, d\mathbf{r} \, d\mathbf{r}' \\ & + \sum_\lambda f_\lambda \int \psi_\lambda(\mathbf{r})^* X(\mathbf{r}) \psi_\lambda(\mathbf{r}) \, d\mathbf{r}. \end{aligned} \tag{24.5}$$

We now make the argument which is central to Koopmans' theorem.* First of all the electron states which are of importance to us are states in which only a small fraction of the band electrons are excited. The number of band electrons is of order N_0, the number of unit cells in the crystal, and when one electron is excited from the ground state the Hartree and exchange potentials Ω_S and Ω_X, which appear in the one-electron Hamiltonian (24.2), vary by an amount of order N_0^{-1}. Therefore each self-consistent wave function ψ_λ, which is a solution to the one-electron Hamiltonian, will vary in order N_0^{-1}. But the total energy is stationary with respect to variations of the ψ_λ, according to the variational calculation of Section 22, so the change in E_Λ due to the variation of each ψ_λ is of relative order N_0^{-2}. Hence the total variation in E_Λ, due to the self-consistent variation of all the ψ_λ when a single electron is excited, is of relative order N_0^{-1} and is negligible. Extending the argument to include excitation of several electrons, we conclude that the variations of the ψ_λ may be neglected in calculating E_Λ, as long as the number of excited electrons is small compared to the total number of band electrons. This means that in calculating E_Λ by (24.5), the wave functions ψ_λ may be taken to be those determined as solutions of the one-electron Hamiltonian, *where the potentials* Ω_S *and* Ω_X *are evaluated with all electrons in the ground state;* these are just the one-electron wave functions which are obtained in principle in band-structure calculations.

Let us assume that the number of excited electrons is small, and write f_λ as

$$f_\lambda = g_\lambda + (f_\lambda - g_\lambda); \tag{24.6}$$

* See, e.g., F. Seitz, *The Modern Theory of Solids*, McGraw-Hill Book Co., Inc., New York, 1940.

then each $\Sigma_\lambda f_\lambda$ can be expanded by treating $\Sigma_\lambda (f_\lambda - g_\lambda)$ as small compared to $\Sigma_\lambda g_\lambda$. Since E_{G} is just (24.5) with each f_λ replaced by g_λ, the expansion of (24.5) to first order in the $f_\lambda - g_\lambda$ gives

$$E_\Lambda = E_{\mathrm{G}} + \sum_\lambda (f_\lambda - g_\lambda) \int \psi_\lambda(\mathbf{r})^*[ke(\mathbf{r}) + \Omega_{\mathrm{B}}(\mathbf{r})]\psi_\lambda(\mathbf{r})\, d\mathbf{r}$$
$$+ e^2 \sum_{\lambda\lambda'}{}' (f_\lambda - g_\lambda) g_{\lambda'} \iint \psi_\lambda(\mathbf{r})^* \psi_{\lambda'}(\mathbf{r}')^* \, |\mathbf{r} - \mathbf{r}'|^{-1} \, \psi_{\lambda'}(\mathbf{r}')\psi_\lambda(\mathbf{r})\, d\mathbf{r}\, d\mathbf{r}' \quad (24.7)$$
$$+ \sum_\lambda (f_\lambda - g_\lambda) \int \psi_\lambda(\mathbf{r})^*[\partial(\rho X)/\partial\rho]\psi_\lambda(\mathbf{r})\, d\mathbf{r}.$$

Note that the expansion of the exchange term in (24.5) is accomplished by writing $\Sigma_\lambda f_\lambda \int \psi_\lambda{}^* X \psi_\lambda \, d\mathbf{r}$ as $\int \rho X \, d\mathbf{r}$, and expanding the integrand to first order in the change in the band-electron density, which is $\Sigma_\lambda (f_\lambda - g_\lambda)\psi_\lambda{}^*\psi_\lambda$. In the Coulomb term in (24.7), the $\Sigma'_{\lambda\lambda'}$ may be replaced by $\Sigma_{\lambda\lambda'}$ with error of relative order N_0^{-1} in crystals, and from (21.31) for the Hartree potential, the Coulomb term is

$$\sum_\lambda (f_\lambda - g_\lambda) \int \psi_\lambda(\mathbf{r})^* \Omega_{\mathrm{S}}(\mathbf{r}) \psi_\lambda(\mathbf{r})\, d\mathbf{r}.$$

Also from (22.13) for Ω_X, the exchange term in (24.7) is

$$\sum_\lambda (f_\lambda - g_\lambda) \int \psi_\lambda(\mathbf{r})^* \Omega_{\mathrm{X}}(\mathbf{r}) \psi_\lambda(\mathbf{r})\, d\mathbf{r}.$$

Then the total energy (24.7) may be written

$$E_\Lambda = E_{\mathrm{G}} + \sum_\lambda (f_\lambda - g_\lambda) \int \psi_\lambda(\mathbf{r})^* h(\mathbf{r}) \psi_\lambda(\mathbf{r})\, d\mathbf{r}, \quad (24.8)$$

where $h(\mathbf{r})$ is given by (24.2). Finally, because of the one-electron wave equation (24.1), we have

$$E_\Lambda = E_{\mathrm{G}} + \sum_\lambda (f_\lambda - g_\lambda)\epsilon_\lambda. \quad (24.9)$$

This important result is Koopmans' theorem, and it tells us that the total excitation energy of the band electrons is just the sum of single electron excitation energies, with no double counting corrections required. It should be remembered, however, that for the ground-state energy expressed as a sum of one-electron energies, there *are* double counting corrections, as expressed in (22.14). It will be convenient to denote the ground-state double counting corrections as E_{GC}, so that (22.14) for E_{G} is written

$$E_{\mathrm{G}} = \sum_\lambda g_\lambda \epsilon_\lambda - E_{\mathrm{GC}}, \quad (24.10)$$

where

$$E_{\mathrm{GC}} = \tfrac{1}{2} \sum_{\lambda} g_\lambda \langle \psi_\lambda | \, \Omega_{\mathrm{S}} \, | \psi_\lambda \rangle + \sum_{\lambda} g_\lambda \langle \psi_\lambda | \, \Omega_{\mathrm{X}} - X \, | \psi_\lambda \rangle. \tag{24.11}$$

Then the total energy (24.9) may be expressed as

$$E_\Lambda = \sum_{\lambda} f_\lambda \epsilon_\lambda - E_{\mathrm{GC}}. \tag{24.12}$$

We note that E_{GC} is a function of the ion positions $\tilde{\mathbf{R}}(\mathrm{N}\nu)$.

Electrons and Phonons

Consider now the total system of ions and band electrons representing a metal crystal. We want to construct for this system a Hamiltonian which has the correct total energy levels, and which corresponds to a set of weakly interacting phonons and electrons. The total Hamiltonian is still given by (21.1), and the first step is to add and subtract E_{G}, to obtain

$$\mathscr{H} = (KE_{\mathrm{I}} + \Omega_{\mathrm{I}} + E_{\mathrm{G}}) + (\mathscr{H}_{\mathrm{E}} - E_{\mathrm{G}}), \tag{24.13}$$

where $\mathscr{H}_{\mathrm{E}}$ is given by (21.2). The next step is to replace the band-electron Hamiltonian $\mathscr{H}_{\mathrm{E}}$ by a set of independent one-electron Hamiltonians $h(\mathbf{r}(\alpha))$,

$$\mathscr{H}_{\mathrm{E}} \to \sum_{\alpha} h(\mathbf{r}(\alpha)), \tag{24.14}$$

where $h(\mathbf{r})$ may be the Hartree approximation (21.24), or the Hartree plus exchange approximation (22.11), or any other suitable approximation. Now the only part of $\mathscr{H}$ which operates on electron coordinates is $\mathscr{H}_{\mathrm{E}}$, and its eigenvalues are E_Λ; however the eigenvalues of the one-electron approximation (24.14) are $\Sigma_\lambda f_\lambda \epsilon_\lambda$, and according to (24.12) this is $E_\Lambda + E_{\mathrm{GC}}$. Hence in order to retain the correct energy levels when the one-electron approximation (24.14) is introduced, we must subtract E_{GC} from $\mathscr{H}$:

$$\mathscr{H} = (KE_{\mathrm{I}} + \Omega_{\mathrm{I}} + E_{\mathrm{G}}) + \left[\sum_{\alpha} h(\mathbf{r}(\alpha)) - E_{\mathrm{G}} - E_{\mathrm{GC}} \right]. \tag{24.15}$$

Subtracting E_{GC} is just the band-electron double counting correction, evaluated in the ground state because of Koopmans' theorem; it matters not that E_{G} has already been corrected for double counting since E_{G} is added and subtracted in (24.15). Finally $E_{\mathrm{G}} + E_{\mathrm{GC}}$ may be replaced by $\Sigma_\lambda g_\lambda \epsilon_\lambda$, according to (24.10), and the total Hamiltonian is then

$$\mathscr{H} = (KE_{\mathrm{I}} + \Omega_{\mathrm{I}} + E_{\mathrm{G}}) + \left[\sum_{\alpha} h(\mathbf{r}(\alpha)) - \sum_{\lambda} g_\lambda \epsilon_\lambda \right]. \tag{24.16}$$

We will now transform $\mathscr{H}$ to a description in terms of phonons and electrons. The first three terms (in parentheses) in (24.16) constitute the total lattice-dynamics Hamiltonian, based on the total adiabatic potential $\Phi = \Omega_I + E_G$, as discussed in Section 23. Following (23.67), this may be written

$$KE_I + \Omega_I + E_G = \Phi_0 + \mathscr{H}_p + \mathscr{H}_{pp}. \tag{24.17}$$

The adiabatic phonons based on the potential Φ are the best phonons we know how to construct for a metal, and since E_G is the ground-state energy of the electrons, the adiabatic phonons are independent of the electronic excitations. On the other hand, the electronic part of (24.16) still depends on the ion positions. Each one-electron Hamiltonian is

$$h = ke + \Omega, \tag{24.18}$$

and this depends on the ion positions through the potential Ω. In order to make the zeroth-order electrons independent of the phonons, we expand each quantity in the square bracket in (24.16) in powers of the ion displacements, as

$$\sum_\alpha h(\mathbf{r}(\alpha)) - \sum_\lambda g_\lambda \epsilon_\lambda = \sum_\alpha h^{(0)}(\mathbf{r}(\alpha)) - \sum_\lambda g_\lambda \epsilon_\lambda^{(0)}$$
$$+ \left\{\Omega^{(1)} + \Omega^{(2)} + \cdots - \sum_\lambda g_\lambda [\epsilon_\lambda^{(1)} + \epsilon_\lambda^{(2)} + \cdots]\right\}. \tag{24.19}$$

The energy levels of $\Sigma_\alpha h^{(0)}(\mathbf{r}(\alpha))$ are $\Sigma_\lambda f_\lambda \epsilon_\lambda^{(0)}$, so the zeroth-order terms in (24.19) represent a set of independent electronic excitations above the ground state, with energy levels $\Sigma_\lambda (f_\lambda - g_\lambda)\epsilon_\lambda^{(0)}$. Thus the zeroth-order terms in (24.19) may be replaced by the electronic excitation Hamiltonian $\mathscr{H}_e$, which is

$$\mathscr{H}_e = \sum_\lambda \epsilon_\lambda^{(0)}(C_\lambda^\dagger C_\lambda - g_\lambda). \tag{24.20}$$

Here $C_\lambda^\dagger$ is the fermion creation operator which places an electron in the state λ, and the eigenvalue of $C_\lambda^\dagger C_\lambda$ is f_λ. The one-electron energies $\epsilon_\lambda^{(0)}$ are based on the band-structure calculation with all ions located at their equilibrium positions, and hence the zeroth-order electrons are independent of the phonons.

The remaining terms in the total Hamiltonian, namely the last line of (24.19), should give a small contribution to the total energy of the crystal, and hence should be adequately treated by perturbation theory, for the following reason. If the vibrations of the ions are neglected, so that the ions are held fixed at the positions $\mathbf{R}(N\nu) + \mathbf{U}(N\nu)$, and the band electrons are in their ground state, then the total contribution to the crystal energy due to the perturbation $\Omega^{(1)}$ is just $\Sigma_\lambda g_\lambda \epsilon_\lambda^{(1)}$ in first order, and the total contribution to the crystal energy due to $\Omega^{(1)} + \Omega^{(2)}$ is just $\Sigma_\lambda g_\lambda [\epsilon_\lambda^{(1)} + \epsilon_\lambda^{(2)}]$ to second order in the ion displacements. Note that this statement simply reflects the perturbation calculation of the band-electron ground-state energy in Sections 21 and 22, without double counting corrections included. Therefore when vibration of the ions and excitation of the electrons are neglected, the expression in braces in (24.19) gives a contribution to the crystal energy which vanishes in each order of perturbation. The last line of (24.19) is denoted $\mathscr{H}_{\mathrm{ep}}$, which stands for electron–phonon interactions:

$$\mathscr{H}_{\mathrm{ep}} = \Omega^{(1)} + \Omega^{(2)} + \cdots - \sum_{\lambda} g_{\lambda}[\epsilon_{\lambda}^{(1)} + \epsilon_{\lambda}^{(2)} + \cdots]. \tag{24.21}$$

The total Hamiltonian is composed of (24.17), (24.20), and (24.21),

$$\mathscr{H} = \Phi_0 + \mathscr{H}_{\mathrm{p}} + \mathscr{H}_{\mathrm{e}} + (\mathscr{H}_{\mathrm{pp}} + \mathscr{H}_{\mathrm{ep}}), \tag{24.22}$$

where $\mathscr{H}_{\mathrm{pp}}$ and $\mathscr{H}_{\mathrm{ep}}$ are to be treated as perturbations. The phonon–phonon interactions were treated in detail in Chapters 3 and 4, and the electron–phonon interactions will be studied in the following section. Two important points about these interaction terms are as follows. First, when $\mathscr{H}_{\mathrm{pp}}$ and $\mathscr{H}_{\mathrm{ep}}$ are each treated to the leading order in which they contribute to the total energy, which is to second order in perturbation theory, these two perturbations do not interfere, that is, there are no cross terms, and the two perturbations may be worked out separately and each adds a separate contribution to the free energy. Second, the contribution of $\mathscr{H}_{\mathrm{ep}}$ to the free energy can be regarded as a simple modification of the contribution of $\mathscr{H}_{\mathrm{e}}$, and this modification can easily be included in the ordinary contributions to the thermodynamic functions arising from $\mathscr{H}_{\mathrm{e}}$. We also note that there are no electron–electron interaction terms in $\mathscr{H}$ because these have been replaced by a self-consistent field in the one-electron approximation. Along this line, it may be recalled that terms of order $(f_{\lambda} - g_{\lambda})^2$ were dropped in deriving Koopmans' theorem (24.9); such terms, if retained, would appear as electron–electron interactions.

Normal Fermion Statistics

In preparation for calculating the electronic excitation contribution to the crystal free energy, we sketch a derivation of the asymptotic "low-temperature" expansion of the Helmholtz free energy F_{e} for a normal system of independent electrons (fermions). Here the low-temperature condition means

$$KT \ll \mu, \tag{24.23}$$

where μ is the chemical potential; hence all temperatures of practical interest, say up to the melting temperature, are included for most metals. A normal fermion system means that bound-state effects, such as superconductivity, are unimportant. The ordinary derivation of the free energy, through the grand canonical partition function, is well known, so an alternate derivation is presented here.

The total number of electrons in the system is N_e, and this must equal the sum of all the occupation numbers f_λ in any total electronic state:

$$N_e = \sum_\lambda f_\lambda. \tag{24.24}$$

The operator form of N_e is $\Sigma_\lambda C_\lambda^\dagger C_\lambda$, and with $\mathscr{H}_e$ given by (24.20) it follows that

$$\mathscr{H}_e - \mu N_e = \sum_\lambda (\epsilon_\lambda - \mu) C_\lambda^\dagger C_\lambda - \sum_\lambda g_\lambda \epsilon_\lambda, \tag{24.25}$$

where here and in the following $\epsilon_\lambda^{(0)}$ is written simply ϵ_λ. Since the electron operators satisfy the anticommutators

$$\{C_\lambda, C_{\lambda'}^\dagger\} = \delta_{\lambda\lambda'}, \qquad \{C_\lambda, C_{\lambda'}\} = 0, \tag{24.26}$$

then from (24.25) we have

$$[\mathscr{H}_e - \mu N_e, C_\lambda^\dagger] = (\epsilon_\lambda - \mu) C_\lambda^\dagger. \tag{24.27}$$

The statistical average of an electron number operator $C_\lambda^\dagger C_\lambda$ is

$$\langle C_\lambda^\dagger C_\lambda \rangle = \text{Tr}\, C_\lambda^\dagger C_\lambda e^{-(\mathscr{H}_e - \mu N_e)/KT}; \tag{24.28}$$

in view of (24.27) this average is given by the zeroth-order statistical perturbation equation (14.15), with the appropriate energy being $\epsilon_\lambda - \mu$:

$$\langle C_\lambda^\dagger C_\lambda \rangle = [e^{(\epsilon_\lambda - \mu)/KT} + 1]^{-1} \langle \{C_\lambda, C_\lambda^\dagger\} \rangle, \tag{24.29}$$

or

$$\langle C_\lambda^\dagger C_\lambda \rangle = \bar{f}_\lambda, \tag{24.30}$$

where

$$\bar{f}_\lambda = [e^{(\epsilon_\lambda - \mu)/KT} + 1]^{-1}. \tag{24.31}$$

The canonical partition function is the sum over all the energy levels E of the electronic system,

$$Z = \sum e^{-E/KT}, \tag{24.32}$$

and the electronic Helmholtz free energy is

$$F_e = -KT \ln Z. \tag{24.33}$$

The electronic energy levels depend on the set of occupation numbers f_λ, and from (24.20) for $\mathscr{H}_e$, the total energies are

$$E(\cdots f_\lambda \cdots) = \sum_\lambda \epsilon_\lambda (f_\lambda - g_\lambda). \tag{24.34}$$

It is convenient to add and subtract $\mu N_e = \Sigma_\lambda \mu f_\lambda$ in (24.34), to obtain

$$E(\cdots f_\lambda \cdots) = \mu N_e - \sum_\lambda g_\lambda \epsilon_\lambda + \sum_\lambda f_\lambda (\epsilon_\lambda - \mu). \tag{24.35}$$

Then since each $f_\lambda = 0$ or 1, the partition function (24.32) is evaluated as follows.

$$\begin{aligned} Z &= e^{-(\mu N_e - \Sigma_\lambda g_\lambda \epsilon_\lambda)/KT} \sum_{(\cdots f_\lambda \cdots)} e^{-\Sigma_\lambda f_\lambda (\epsilon_\lambda - \mu)/KT} \\ &= e^{-(\mu N_e - \Sigma_\lambda g_\lambda \epsilon_\lambda)/KT} \sum_{f_\lambda = 0}^{1} \prod_\lambda e^{-f_\lambda(\epsilon_\lambda - \mu)/KT} \\ &= e^{-(\mu N_e - \Sigma_\lambda g_\lambda \epsilon_\lambda)/KT} \prod_\lambda [1 + e^{-(\epsilon_\lambda - \mu)/KT}]. \end{aligned} \tag{24.36}$$

The electronic free energy (24.33) is therefore

$$F_e = \mu N_e - \sum_\lambda g_\lambda \epsilon_\lambda - KT \sum_\lambda \ln[1 + e^{-(\epsilon_\lambda - \mu)/KT}]. \tag{24.37}$$

In order to find the explicit temperature-dependence of F_e at low temperatures ($KT \ll \mu$), it is convenient to transform the Σ_λ to an integral, and we will now discuss the general procedure for such a transformation. The continuous variable representing the one-electron energies is ϵ, and the total number of one-electron states with energies in the range ϵ to $\epsilon + d\epsilon$ is $n(\epsilon)d\epsilon$. At any energy, the total density of states $n(\epsilon)$ may represent the sum of contributions from several bands, and includes both spins; the density of states function is generally computed in band-structure calculations. The statistical average electron occupation number $\bar{f}_\lambda$ is given by (24.31), and as a function of ϵ this is the Fermi distribution function $f(\epsilon)$,

$$f(\epsilon) = [e^{(\epsilon - \mu)/KT} + 1]^{-1}. \tag{24.38}$$

For any function G_λ of ϵ_λ, a sum such as $\Sigma_\lambda \bar{f}_\lambda G_\lambda$ may be written

$$\sum_\lambda \bar{f}_\lambda G_\lambda = \int_0^\infty n(\epsilon) f(\epsilon) G(\epsilon)\, d\epsilon. \tag{24.39}$$

If now $G(\epsilon)$ is a slowly varying function of ϵ, i.e., as long as $G(\epsilon)$ varies only slightly when ϵ varies over a range KT, then the Sommerfeld low-temperature expansion of the general integral (24.39) is*

$$\int_0^\infty n(\epsilon) f(\epsilon) G(\epsilon)\, d\epsilon = \int_0^\mu n(\epsilon) G(\epsilon)\, d\epsilon + \tfrac{1}{6}\pi^2 (KT)^2 \{d[n(\epsilon)G(\epsilon)]/d\epsilon\}_{\epsilon=\mu} + \cdots . \tag{24.40}$$

Here the $+ \cdots$ represents terms of order $(KT)^4$ and higher.

The variation of μ with temperature is obtained from the requirement (24.24) for the total number of electrons N_e; transforming the statistical

* See, e.g., A. H. Wilson, *The Theory of Metals*, Cambridge University Press, Cambridge, England, 1954, Second Edition, Appendix A4.

average of N_e to an integral and making the low-temperature expansion leads to

$$\begin{aligned} N_e &= \int_0^\infty n(\epsilon) f(\epsilon)\, d\epsilon \\ &= \int_0^\mu n(\epsilon)\, d\epsilon + \tfrac{1}{6}\pi^2 (KT)^2 (dn/d\epsilon)_\mu + \cdots . \end{aligned} \tag{24.41}$$

At $T = 0$, the chemical potential is μ_0, and (24.41) becomes

$$N_e = \int_0^{\mu_0} n(\epsilon)\, d\epsilon. \tag{24.42}$$

At finite T we can expand the integral $\int_0^\mu$ about μ_0 in (24.41), and keep terms to first order in $\mu - \mu_0$ and in $(KT)^2$, to write

$$N_e = \int_0^{\mu_0} n(\epsilon)\, d\epsilon + (\mu - \mu_0) n(\mu_0) + \tfrac{1}{6}\pi^2 (KT)^2 (dn/d\epsilon)_{\mu_0}. \tag{24.43}$$

Then eliminating the integral $\int_0^{\mu_0}$ by means of (24.42), the temperature-dependence of μ is found to be

$$\mu = \mu_0 - \tfrac{1}{6}\pi^2 (KT)^2 [d \ln n(\epsilon)/d\epsilon]_{\mu_0}. \tag{24.44}$$

It is also useful to separate out the explicit temperature-dependence of the general expansion (24.40), which contains the arbitrary function $G(\epsilon)$. The right-hand side of (24.40) is expanded into

$$\int_0^{\mu_0} n(\epsilon) G(\epsilon)\, d\epsilon + (\mu - \mu_0) n(\mu_0) G(\mu_0) + \tfrac{1}{6}\pi^2 (KT)^2 \{d[n(\epsilon)G(\epsilon)]/d\epsilon\}_{\mu_0}, \tag{24.45}$$

and with (24.44) for $\mu - \mu_0$ this is

$$\int_0^{\mu_0} n(\epsilon) G(\epsilon)\, d\epsilon + \tfrac{1}{6}\pi^2 (KT)^2 \{-G(\mu_0)[dn/d\epsilon]_{\mu_0} + [dnG/d\epsilon]_{\mu_0}\}. \tag{24.46}$$

The expression in braces is simply $n(\mu_0)[dG/d\epsilon]_{\mu_0}$, so the general relation (24.40) is, to order $(KT)^2$,

$$\int_0^\infty n(\epsilon) f(\epsilon) G(\epsilon)\, d\epsilon = \int_0^{\mu_0} n(\epsilon) G(\epsilon)\, d\epsilon + \tfrac{1}{6}\pi^2 (KT)^2 n(\mu_0) [dG(\epsilon)/d\epsilon]_{\mu_0}. \tag{24.47}$$

We now return to the low-temperature evaluation of the electronic free energy (24.37). The ground-state evaluation of $f(\epsilon)$ is $g(\epsilon)$, which is

$$\begin{aligned} g(\epsilon) &= 1, \quad 0 \le \epsilon < \mu_0; \\ g(\epsilon) &= 0, \quad \epsilon > \mu_0. \end{aligned} \tag{24.48}$$

Therefore, a sum such as $\Sigma_\lambda\, g_\lambda G_\lambda$ may be written

$$\sum_\lambda g_\lambda G_\lambda = \int_0^{\mu_0} n(\epsilon)G(\epsilon)\, d\epsilon, \tag{24.49}$$

which is of course the $T = 0$ value of (24.47). The integral form of (24.37) for F_e is then

$$F_e = \mu N_e - \int_0^{\mu_0} n(\epsilon)\epsilon\, d\epsilon - KT\int_0^{\infty} n(\epsilon)\ln[1 + e^{(\mu-\epsilon)/KT}]\, d\epsilon. \tag{24.50}$$

Here the integral $\int_0^\infty$ may be transformed by integration by parts, to

$$KT\int_0^{\infty} n(\epsilon)\ln[1 + e^{(\mu-\epsilon)/KT}]\, d\epsilon = \int_0^{\infty} f(\epsilon)t(\epsilon)\, d\epsilon, \tag{24.51}$$

where

$$t(\epsilon) = \int_0^{\epsilon} n(\epsilon')\, d\epsilon'. \tag{24.52}$$

Then the integral on the right of (24.51) may be expanded by the low-temperature expansion (24.40), with $n(\epsilon)G(\epsilon)$ in (24.40) replaced by $t(\epsilon)$, to obtain

$$\begin{aligned} KT\int_0^{\infty} n(\epsilon)\ln[1 + e^{(\mu-\epsilon)/KT}]\, d\epsilon \\ &= \int_0^{\mu_0} t(\epsilon)\, d\epsilon + (\mu - \mu_0)t(\mu_0) + \tfrac{1}{6}\pi^2(KT)^2[dt(\epsilon)/d\epsilon]_{\mu_0} \\ &= \mu_0 N_e - \int_0^{\mu_0} n(\epsilon)\epsilon\, d\epsilon + (\mu - \mu_0)N_e + \tfrac{1}{6}\pi^2(KT)^2 n(\mu_0). \end{aligned} \tag{24.53}$$

The total F_e of (24.50) is then simply

$$F_e = -\tfrac{1}{6}\pi^2(KT)^2 n(\mu_0), \tag{24.54}$$

to order $(KT)^2$.

Thermodynamic Functions

With the neglect of electron–phonon interactions, the total free energy of a normal metal is

$$F = \Phi_0 + F_H + F_e + F_A. \tag{24.55}$$

The ordinary lattice contributions $\Phi_0 + F_H + F_A$ were determined in Chapter 4. Because we started with the electronic *excitation* Hamiltonian $\mathscr{H}_e$, the electronic free energy F_e contains only the *excitation* contribution (24.54), the electronic ground-state energy having been absorbed into the lattice free energy. It is convenient to separate the explicit temperature-dependence and the configuration-dependence of F_e by writing

$$F_e = -\tfrac{1}{2}\Gamma T^2, \tag{24.56}$$

where Γ is a function only of the macroscopic configuration of the crystal, and from (24.54),

$$\Gamma = \tfrac{1}{3}\pi^2 K^2 n(\mu_0). \tag{24.57}$$

For all thermodynamic functions which are simply derivatives of the free energy, and not ratios of such derivatives, the electronic contribution may be calculated from F_e, and is to be added to the lattice contribution. For example, the electronic entropy S_e and heat capacity at constant configuration $C_{\eta e}$ are

$$S_e = C_{\eta e} = \Gamma T. \tag{24.58}$$

The electronic contributions to the stresses and isothermal elastic constants are given by

$$\tau_{ij}(\text{electronic}) = -\tfrac{1}{2}V^{-1}(\partial\Gamma/\partial\eta_{ij})_{\eta'}T^2; \tag{24.59}$$

$$C^T_{ijkl}(\text{electronic}) = -\tfrac{1}{2}V^{-1}(\partial^2\Gamma/\partial\eta_{ij}\partial\eta_{kl})_{\eta'}T^2. \tag{24.60}$$

In addition, the electronic contributions to the thermal stresses are

$$b_{ij}(\text{electronic}) = -V^{-1}(\partial\Gamma/\partial\eta_{ij})_{\eta'}T, \tag{24.61}$$

and the total thermal stresses are, of course,

$$b_{ij} = b_{ij}(\text{lattice}) + b_{ij}(\text{electronic}). \tag{24.62}$$

The separation of electronic contributions to other thermodynamic functions is not so straightforward. Consider the thermal strains β_{ij}; according to (2.78) these are given by

$$\beta_{ij} = -\sum_{kl} S^T_{ijkl} b_{kl}. \tag{24.63}$$

The total b_{kl} is the sum of lattice and electronic contributions, according to (24.62); however the lattice and electronic contributions to the isothermal compliances S^T_{ijkl} are rather well mixed together, and hence the same is true of the β_{ij}. It is convenient, however, to *define* the electronic thermal strains by the equations

$$\beta_{ij}(\text{electronic}) = -\sum_{kl} S^T_{ijkl} b_{kl}(\text{electronic}), \tag{24.64}$$

where the compliances appearing here are the *total* compliances. Then with the lattice contributions defined by

$$\beta_{ij}(\text{lattice}) = -\sum_{kl} S^T_{ijkl} b_{kl}(\text{lattice}), \tag{24.65}$$

the total β_{ij} are the sums of lattice and electronic contributions:

$$\beta_{ij} = \beta_{ij}(\text{lattice}) + \beta_{ij}(\text{electronic}). \tag{24.66}$$

It should always be recognized that "the electronic contribution to the thermal strains" is not a well-defined concept, without some accompanying definition such as (24.64).

The macroscopic Grüneisen tensor γ_{ij} is given by (2.102) as

$$\gamma_{ij} = -(V/C_\eta)b_{ij}. \tag{24.67}$$

If we *define* an electronic Grüneisen tensor by

$$\gamma_{ij}(\text{electronic}) = -(V/C_{\eta e})b_{ij}(\text{electronic}), \tag{24.68}$$

then from (24.58) for $C_{\eta e}$ and (24.61) for b_{ij}(electronic), it follows that

$$\gamma_{ij}(\text{electronic}) = (\partial \ln \Gamma/\partial \eta_{ij})_{\eta'}. \tag{24.69}$$

Note that with the definition (24.68) of γ_{ij}(electronic), the total Grüneisen tensor is *not* simply the sum of lattice and electronic contributions, but is given by

$$\gamma_{ij} = -(V/C_\eta)b_{ij}(\text{lattice}) + (C_{\eta e}/C_\eta)\gamma_{ij}(\text{electronic}), \tag{24.70}$$

where C_η is the total (lattice plus electronic) heat capacity. Once again we note that the quantities which are simplest from the theoretical point of view are the entropy and heat capacity, the stresses and isothermal elastic constants, and the thermal stresses b_{ij}.

The electronic contributions to S, C_η, and b_{ij} are all proportional to T, while the electronic contributions to τ_{ij} and C_{ijkl}^T are proportional to T^2. (These temperature-dependences are slightly modified, by electron–phonon interactions, at intermediate temperatures only, say from a few degrees to the Debye temperature; see Section 25.) Thus at sufficiently low temperatures, i.e., $T \ll \Theta_0$, the electronic contribution dominates the temperature-dependent part of each thermodynamic function, since the lattice-dynamical contribution is of order T^2 times the electronic contribution. It is therefore possible to determine experimentally the electronic contributions to the thermodynamic functions, from measurements of the temperature-dependence at very low temperatures. In addition, since b_{ij}(electronic) goes as T, and since for $T \ll \Theta_0$ the elastic compliances are constant to lowest order in T, then from (24.64)

$$\beta_{ij}(\text{electronic}) \propto T, \quad \text{at low } T. \tag{24.71}$$

This means that the lattice strains η_{ij} measured from the $T = 0$ configuration go as T^2 when the stress is held constant, since $\eta_{ij} = \beta_{ij}T$ in the low-temperature limit and at constant stress. Thus the variation of the crystal configuration with temperature at constant stress leads to additional T^3 contributions to the electronic parts of the functions S, C_η, and b_{ij}, and leads also to additional T^2 contributions to the lattice parts of the functions τ_{ij} and C_{ijkl}^T, at low temperatures.

In the high-temperature region $T > \Theta_{H\infty}$, but of course with $KT \ll \mu_0$, the anharmonic free energy has a leading temperature-dependence of T^2, and is given by (19.12) as

$$F_A = \mathscr{A}_2 T^2 + \text{lower orders of } T. \tag{24.72}$$

Therefore all the thermodynamic calculations of Section 19 for the high-temperature region will also include the electronic contributions if $\mathscr{A}_2$ is replaced by $\mathscr{A}_2 - \frac{1}{2}\Gamma$:

$$\mathscr{A}_2 \to \mathscr{A}_2 - \tfrac{1}{2}\Gamma, \quad \text{includes electronic contributions at high } T. \tag{24.73}$$

Let us consider in more detail the electronic contribution to the thermal expansion coefficient β for cubic crystals. The electronic contributions to the pressure P and to the isothermal bulk modulus B_T are

$$P_e = \tfrac{1}{2}(d\Gamma/dV)T^2; \tag{24.74}$$

$$B_{Te} = -\tfrac{1}{2}V(d^2\Gamma/dV^2)T^2. \tag{24.75}$$

Also the electronic contribution to βB_T is

$$(\beta B_T)_e = (d\Gamma/dV)T. \tag{24.76}$$

At low temperatures ($T \ll \Theta_0$), the lattice-dynamical contribution to βB_T is given by (18.65), and the total βB_T may be written

$$\beta B_T = (d\Gamma/dV)T + aT^3, \quad \text{at low } T, \tag{24.77}$$

where

$$a = -V_A^{-1}K(12\pi^4/5)\Theta_0^{-3}(d \ln \Theta_0/d \ln V). \tag{24.78}$$

Also the lattice contribution to B_T is $B_0 + O(T^4)$ at low temperatures, where B_0 is the value at $T = 0$, so with (24.75) the bulk modulus to lowest order in T is

$$B_T = B_0 - \tfrac{1}{2}V(d^2\Gamma/dV^2)T^2, \quad \text{at low } T. \tag{24.79}$$

Now β may be calculated by dividing (24.77) by (24.79), and to order T^3 the result is

$$\beta = B_0^{-1}(d\Gamma/dV)T + B_0^{-1}[a + \tfrac{1}{2}VB_0^{-1}(d\Gamma/dV)(d^2\Gamma/dV^2)]T^3. \tag{24.80}$$

This equation shows how the lattice-dynamical contribution, represented by the function a, and the electronic contribution, represented by the volume derivatives of Γ, mix in the T^3 term in β at low temperatures. Of course the leading contribution to β at low temperatures is linear in T, and measurement of this contribution gives a measurement of the coefficient $B_0^{-1}(d\Gamma/dV)$. At zero pressure, there is an added T^2 contribution to the lattice part of B_T, and a T^3 contribution to the electronic part of βB_T, leading to additional

T^3 contributions to β, again due to the electronic contribution to the variation of the crystal volume.

Finally it should be noted that the electronic contributions to the adiabatic elastic constants and the adiabatic bulk modulus are not simply separated from the lattice contributions. This is because $C^S_{ijkl} - C^T_{ijkl}$ contains the factor $b_{ij}b_{kl}$, and each b_{ij} is a sum of lattice plus electronic parts. Similarly, $B_S - B_T$ contains the factor $(\beta B_T)^2$, and βB_T is a sum of lattice plus electronic parts.

25. ELECTRON–PHONON INTERACTIONS

The Interaction Hamiltonian

The electron–phonon interaction Hamiltonian $\mathscr{H}_{\text{ep}}$ is written in (24.21), and we will consider here only the terms up to second order in the ion displacements. The one-electron potentials $\Omega^{(1)}$ and $\Omega^{(2)}$ depend on both the ion and the electron coordinates, while the one-electron energies $\epsilon_\lambda^{(1)}$ and $\epsilon_\lambda^{(2)}$ depend only on the ion coordinates. An important point is that the contribution of $\mathscr{H}_{\text{ep}}$ to the total crystal energy will depend on the solution which has been obtained for the band electrons. In other words, if various representations of the band electrons are taken for a given metal, ranging from say free electrons to an accurate band-structure calculation, then the major contributions to the total Hamiltonian written out in (24.22), namely Φ_0, $\mathscr{H}_{\text{p}}$, and $\mathscr{H}_{\text{e}}$, will be different for these different representations; likewise the contribution $\mathscr{H}_{\text{ep}}$ will be different for the different cases. One should therefore be careful to avoid the conceptual error of evaluating $\mathscr{H}_{\text{ep}}$ for a given electronic model, and then concluding from the results of this evaluation that electron–phonon interactions give a large (or small) contribution to the free energy of the metal under consideration.

We will treat the electron–phonon interactions only for the very simple case of a primitive lattice with all the band electrons in a single conduction band. The band-electron index λ then stands for a wave vector $\mathbf{p}$ and spin σ, with

$$\lambda = \mathbf{p}\sigma; \qquad \sum_\lambda = \sum_{\mathbf{p}\sigma} = 2\sum_{\mathbf{p}}. \tag{25.1}$$

Here since all the required matrix elements are independent of spin, the sum over spins Σ_σ can always be replaced by 2, and the sum over wave vectors $\Sigma_{\mathbf{p}}$ is unrestricted, i.e., $\mathbf{p}$ goes over all values in the complete Fourier set. The phonons are still labeled by κ, where for a primitive lattice

$$\kappa = \mathbf{k}s, \quad s = 1,2,3, \tag{25.2}$$

and where all the independent phonon wave vectors are contained in one Brillouin zone.

To calculate displacement derivatives of the one-electron potential Ω, we make the approximation that Ω can be represented as a sum of single ion potentials which move rigidly with their ions:

$$\Omega(\mathbf{r}) = \sum_{\mathrm{M}} u(\mathbf{r} - \tilde{\mathbf{R}}(\mathrm{M})), \tag{25.3}$$

where as usual the ion positions $\tilde{\mathbf{R}}(\mathrm{M})$ are arbitrary. Then the displacement derivatives, evaluated at the equilibrium configuration of the ions, are

$$\begin{aligned} \partial\Omega/\partial U_i(\mathrm{M}) &= [\partial u(\mathbf{r} - \mathbf{R})/\partial R_i]_{\mathbf{R}=\mathbf{R}(\mathrm{M})} \\ &\equiv u_i(\mathbf{r} - \mathbf{R}(\mathrm{M})); \end{aligned} \tag{25.4}$$

$$\begin{aligned} \partial^2\Omega/\partial U_i(\mathrm{M})\partial U_j(\mathrm{N}) &= \delta_{\mathrm{MN}}[\partial^2 u(\mathbf{r} - \mathbf{R})/\partial R_i \partial R_j]_{\mathbf{R}=\mathbf{R}(\mathrm{M})} \\ &\equiv \delta_{\mathrm{MN}} u_{ij}(\mathbf{r} - \mathbf{R}(\mathrm{M})). \end{aligned} \tag{25.5}$$

Since the bare ion potential $\Omega_{\mathrm{B}}(\mathbf{r})$ is a sum of rigid ion potentials, according to (21.26), then (25.3) is valid if there is a linear relation between components $\Omega(\mathbf{q})$ of the total potential and $\Omega_{\mathrm{B}}(\mathbf{q})$ of the bare potential, the linear relation being independent of the positions of the ions. There is such a relation in first order, namely (22.37) which relates $\Omega^{(1)}(\mathbf{q})$ and $\Omega_{\mathrm{B}}^{(1)}(\mathbf{q}')$ through the inverse dielectric function $\hat{b}(\mathbf{q},\mathbf{q}')$, and with that relation it is easy to write $u_i(\mathbf{r} - \mathbf{R}(\mathrm{M}))$ in terms of the bare ion potentials $v(\mathbf{r} - \mathbf{R}(\mathrm{M}))$. However the second-order derivative (25.5) has to be regarded as an approximation. We note that the pseudopotential perturbation approximation of Chapter 6 satisfies linear screening in all orders of the ion displacements, and further there is no *essential* complication in the following calculations if the approximation (25.5) is not followed.

In order to transform $\mathscr{H}_{\mathrm{ep}}$ to a representation in terms of electron and phonon operators, we need to calculate matrix elements such as $\langle\psi_{\lambda'}^{(0)}|\,\Omega^{(1)}\,|\psi_{\lambda}^{(0)}\rangle$. For a primitive lattice, $\Omega^{(1)}$ is [see, e.g., (21.44)]

$$\Omega^{(1)} = \sum_{\mathrm{M}}\sum_{i} [\partial\Omega/\partial U_i(\mathrm{M})] U_i(\mathrm{M}), \tag{25.6}$$

and then with (25.4) the matrix elements are

$$\langle\psi_{\lambda'}^{(0)}|\,\Omega^{(1)}\,|\psi_{\lambda}^{(0)}\rangle = \sum_{\mathrm{M}}\sum_{i} U_i(\mathrm{M}) \int \psi_{\lambda'}^{(0)}(\mathbf{r})^* u_i(\mathbf{r} - \mathbf{R}(\mathrm{M}))\psi_{\lambda}^{(0)}(\mathbf{r})\, d\mathbf{r}. \tag{25.7}$$

Since the $\psi_{\lambda}^{(0)}$ are Bloch functions, and satisfy the Bloch theorem (21.87), then (25.7) may be written

$$\langle\psi_{\lambda'}^{(0)}|\,\Omega^{(1)}\,|\psi_{\lambda}^{(0)}\rangle = N_0^{-1} \sum_{\mathrm{M}}\sum_{i} U_i(\mathrm{M}) e^{i(\mathbf{p}-\mathbf{p}')\cdot\mathbf{R}(\mathrm{M})} \theta_i(\lambda'\lambda), \tag{25.8}$$

where

$$\theta_i(\lambda'\lambda) = N_0 \int \psi_{\lambda'}^{(0)}(\mathbf{r} - \mathbf{R}(\mathrm{M}))^* u_i(\mathbf{r} - \mathbf{R}(\mathrm{M}))\psi_{\lambda}^{(0)}(\mathbf{r} - \mathbf{R}(\mathrm{M}))\, d\mathbf{r}. \quad (25.9)$$

The integral $\theta_i(\lambda'\lambda)$ is of order 1, not of order N_0, since the $\psi_\lambda^{(0)}(\mathbf{r})$ are normalized to the volume V of the crystal, while $u_i(\mathbf{r} - \mathbf{R}(\mathrm{M}))$ is appreciable only over a volume of order V_C. Obviously $\theta_i(\lambda'\lambda)$ is independent of $\mathbf{R}(\mathrm{M})$, and also satisfies

$$\begin{aligned} &\theta_i(\lambda'\lambda)^* = \theta_i(\lambda\lambda'); \\ &\theta_i(\lambda'\lambda) \quad \text{contains a } \delta_{\sigma\sigma'}. \end{aligned} \quad (25.10)$$

The expansion of ion displacements in phonon coordinates is described in Section 10; by combining (10.40) and (10.66) we can write, for a primitive lattice,

$$U_i(\mathrm{M}) = \sum_\kappa{}' (\hbar/2N_0M_C\omega_\kappa)^{1/2}(A_\kappa + A_{-\kappa}^\dagger)e^{i\mathbf{k}\cdot\mathbf{R}(\mathrm{M})}w_i(\kappa). \quad (25.11)$$

Here the $\mathbf{k}$ is understood to correspond to $\kappa = \mathbf{k}s$, $w_i(\kappa)$ is the i component of the phonon eigenvector, and $\mathbf{k} = 0$ is explicitly omitted from the sum on κ in (25.11) since $\mathbf{k} = 0$ represents a uniform translation for a primitive lattice. With (25.11) for the ion displacements, the matrix element (25.8) is

$$\begin{aligned} &\langle\psi_{\lambda'}^{(0)}|\,\Omega^{(1)}\,|\psi_\lambda^{(0)}\rangle \\ &= \sum_\kappa{}' \delta(\mathbf{k} + \mathbf{p} - \mathbf{p}')(\hbar/2N_0M_C\omega_\kappa)^{1/2}(A_\kappa + A_{-\kappa}^\dagger)\sum_i w_i(\kappa)\theta_i(\lambda'\lambda), \end{aligned} \quad (25.12)$$

where the δ function arises from a sum over lattice vectors, according to (10.22). The phonon operators are explicitly contained in (25.12), and the operator equivalent of $\Omega^{(1)}$ may therefore be constructed from the expression

$$\Omega^{(1)} = \sum_{\lambda\lambda'} \langle\psi_{\lambda'}^{(0)}|\,\Omega^{(1)}\,|\psi_\lambda^{(0)}\rangle C_{\lambda'}^\dagger C_\lambda. \quad (25.13)$$

Now $\lambda' = \mathbf{p}'\sigma'$, and since (25.10) requires $\sigma' = \sigma$, and (25.12) requires $\mathbf{p}' = \mathbf{p} + \mathbf{k} + \mathbf{Q}$ where $\mathbf{Q}$ is any inverse-lattice vector, then $\Sigma_{\lambda'}$ can be eliminated from (25.13) in favor of a $\Sigma_\mathbf{Q}$, and the operator form of $\Omega^{(1)}$ may be written

$$\Omega^{(1)} = \sum_\kappa{}' \sum_\mathbf{Q} \sum_\lambda V_{\kappa+\mathbf{Q},\lambda}(A_\kappa + A_{-\kappa}^\dagger)C_{\lambda+\kappa+\mathbf{Q}}^\dagger C_\lambda. \quad (25.14)$$

Here we are using the abbreviated notation

$$\begin{aligned} \kappa + \mathbf{Q} &= \mathbf{k} + \mathbf{Q}, s, \\ \lambda + \kappa + \mathbf{Q} &= \mathbf{p} + \mathbf{k} + \mathbf{Q}, \sigma, \end{aligned} \quad (25.15)$$

and the electron–phonon interaction coefficient is

$$V_{\kappa+\mathbf{Q},\lambda} = (\hbar/2N_0M_C\omega_\kappa)^{1/2}\sum_i w_i(\kappa)\theta_i(\lambda + \kappa + \mathbf{Q}, \lambda). \quad (25.16)$$

The operator $\Omega^{(1)}$ is easily shown to be Hermitian, with the aid of (25.10); note also that $\Omega^{(1)}$ is nondiagonal in phonon coordinates, and also in electron coordinates because $\kappa = 0$ is omitted from the sum in (25.14). In the $\Sigma_{\mathbf{Q}}$ in $\Omega^{(1)}$, the terms with $\mathbf{Q} \neq 0$ are the electron–phonon Umklapp interactions.

For the second-order operator $\Omega^{(2)}$ it is only necessary to calculate the diagonal matrix elements in order to obtain the total energy contributions correct to second order in the ion displacements. Proceeding as in the calculation of (25.14) for $\Omega^{(1)}$, the terms in $\Omega^{(2)}$ which are diagonal in both phonons and electrons are found to be

$$\Omega^{(2)}(\text{diagonal}) = \sum_{\kappa}\sum_{\lambda} W_{\kappa\lambda}(A_\kappa^\dagger A_\kappa + \tfrac{1}{2})C_\lambda^\dagger C_\lambda, \tag{25.17}$$

where

$$W_{\kappa\lambda} = (\hbar/2N_0 M_C \omega_\kappa) \sum_{ij} w_i(\kappa) w_j(\kappa) \theta_{ij}(\lambda); \tag{25.18}$$

$$\theta_{ij}(\lambda) = N_0 \int \psi_\lambda^{(0)}(\mathbf{r} - \mathbf{R}(\mathrm{M}))^* u_{ij}(\mathbf{r} - \mathbf{R}(\mathrm{M})) \psi_\lambda^{(0)}(\mathbf{r} - \mathbf{R}(\mathrm{M}))\, d\mathbf{r}. \tag{25.19}$$

Again $\theta_{ij}(\lambda)$ is of order 1, not N_0, and is independent of $\mathbf{R}(\mathrm{M})$; the potential derivatives $u_{ij}(\mathbf{r} - \mathbf{R}(\mathrm{M}))$ are defined in (25.5).

Let us now turn to the evaluation of the one-electron energy contributions $\epsilon_\lambda^{(1)}$ and $\epsilon_\lambda^{(2)}$ in (24.21) for $\mathscr{H}_{\text{ep}}$. According to (21.53), which holds also when $\Omega^{(1)}$ includes exchange,

$$\epsilon_\lambda^{(1)} = \langle \psi_\lambda^{(0)} | \, \Omega^{(1)} \, | \psi_\lambda^{(0)} \rangle.$$

But in view of (25.14), $\Omega^{(1)}$ has no diagonal elements, so for a primitive lattice,

$$\epsilon_\lambda^{(1)} = 0. \tag{25.20}$$

Also $\epsilon_\lambda^{(2)}$ is given by (21.61), and $\psi_\lambda^{(1)}$ is given by (21.60); with those results we can write

$$\sum_\lambda g_\lambda \epsilon_\lambda^{(2)} = \sum_{\lambda\lambda'}{}' \frac{g_\lambda \langle \psi_\lambda^{(0)} | \, \Omega^{(1)} \, | \psi_{\lambda'}^{(0)} \rangle \langle \psi_{\lambda'}^{(0)} | \, \Omega^{(1)} \, | \psi_\lambda^{(0)} \rangle}{[\epsilon_\lambda^{(0)} - \epsilon_{\lambda'}^{(0)}]_p} + \sum_\lambda g_\lambda \langle \psi_\lambda^{(0)} | \, \Omega^{(2)} \, | \psi_\lambda^{(0)} \rangle. \tag{25.21}$$

This is a function only of the phonon coordinates, and only those terms diagonal in phonon operators are required in calculating the total energy correct to second order in the ion displacements. With the aid of (25.14) for $\Omega^{(1)}$ and (25.17) for $\Omega^{(2)}$, the diagonal operator form of (25.21) is found to be

$$\begin{aligned} \sum_\lambda g_\lambda \epsilon_\lambda^{(2)}(\text{diagonal}) &= \sum_{\kappa}{}' \sum_{\mathbf{Q}} \sum_{\lambda} |V_{\kappa+\mathbf{Q},\lambda}|^2 \, g_\lambda (1 - g_{\lambda+\kappa+\mathbf{Q}}) [\epsilon_\lambda^{(0)} - \epsilon_{\lambda+\kappa+\mathbf{Q}}^{(0)}]_p^{-1} \\ &\quad \times (A_\kappa^\dagger A_\kappa + A_{-\kappa}^\dagger A_{-\kappa} + 1) + \sum_\kappa \sum_\lambda W_{\kappa\lambda} g_\lambda (A_\kappa^\dagger A_\kappa + \tfrac{1}{2}). \end{aligned} \tag{25.22}$$

Renormalized Electrons and Phonons

The contribution of $\mathscr{H}_{\text{ep}}$ to the total system energy levels could be calculated by ordinary perturbation theory. However, it is instructive to see how the "bare" electrons and phonons are renormalized due to the first-order interactions between them, and then to calculate the renormalized particle energies to second order. For this calculation it is convenient to simplify the notation in several ways. The total Hamiltonian is written in (24.22) and we will drop the phonon–phonon interactions $\mathscr{H}_{\text{pp}}$, as well as the constant term Φ_0, and the constant term $-\Sigma_\lambda g_\lambda \epsilon_\lambda^{(0)}$ which appears in $\mathscr{H}_{\text{e}}$. The remaining terms in $\mathscr{H}$ are written

$$\mathscr{H}^{(0)} + \mathscr{H}_{\text{ep}}^{(1)} + \mathscr{H}_{\text{ep}}^{(2)}, \tag{25.23}$$

where

$$\mathscr{H}^{(0)} = \sum_\kappa \hbar\omega_\kappa(A_\kappa^\dagger A_\kappa + \tfrac{1}{2}) + \sum_\lambda \epsilon_\lambda C_\lambda^\dagger C_\lambda, \tag{25.24}$$

$$\mathscr{H}_{\text{ep}}^{(1)} = \sum_{\kappa\lambda}{}' V_{\kappa\lambda}(A_\kappa + A_{-\kappa}^\dagger)C_{\lambda+\kappa}^\dagger C_\lambda, \tag{25.25}$$

$$\mathscr{H}_{\text{ep}}^{(2)} = -\sum_{\kappa\lambda}{}' |V_{\kappa\lambda}|^2 g_\lambda(1 - g_{\lambda+\kappa})(\epsilon_\lambda - \epsilon_{\lambda+\kappa})_p^{-1}(A_\kappa^\dagger A_\kappa + A_{-\kappa}^\dagger A_{-\kappa} + 1) + \sum_{\kappa\lambda} W_{\kappa\lambda}(A_\kappa^\dagger A_\kappa + \tfrac{1}{2})(C_\lambda^\dagger C_\lambda - g_\lambda). \tag{25.26}$$

Here $\mathscr{H}^{(0)}$ is just the harmonic phonon Hamiltonian $\mathscr{H}_{\text{p}}$, as written out in (23.68), for example, plus the electronic Hamiltonian $\mathscr{H}_{\text{e}}$ of (24.20), with $-\Sigma_\lambda g_\lambda \epsilon_\lambda^{(0)}$ omitted; $\mathscr{H}_{\text{ep}}^{(1)}$ is $\Omega^{(1)}$, given by (25.14), and $\mathscr{H}_{\text{ep}}^{(2)}$ is $\Omega^{(2)} - \Sigma_\lambda g_\lambda \epsilon_\lambda^{(2)}$, given by (25.17) and (25.22). In (25.24) and (25.26), and in the following calculations, $\epsilon_\lambda^{(0)}$ is written simply as ϵ_λ. The explicit representation of Umklapp terms in $\mathscr{H}_{\text{ep}}^{(1)}$ and $\mathscr{H}_{\text{ep}}^{(2)}$, given by sums over $\mathbf{Q}$ with $\mathbf{Q} \neq 0$, has been dropped for abbreviation; the Umklapp terms may be restored explicitly at any point in the following calculations by the replacement

$$V_{\kappa\lambda}G_\kappa \to \sum_{\mathbf{Q}} V_{\kappa+\mathbf{Q},\lambda}G_{\kappa+\mathbf{Q}}, \tag{25.27}$$

where G_κ is the appropriate function coupled with $V_{\kappa\lambda}$. The condition that $\mathscr{H}_{\text{ep}}^{(1)}$ is Hermitian is

$$V_{\kappa\lambda}^* = V_{-\kappa,\lambda+\kappa}, \tag{25.28}$$

and this is satisfied by virtue of (25.10). Finally, $\Sigma'_{\kappa\lambda}$ means omit the $\kappa = 0$ term.

The phonon and electron operators satisfy the following commutators and anticommutators.

$$[A_\kappa, A_{\kappa'}^\dagger] = \delta_{\kappa\kappa'}, \qquad [A_\kappa, A_{\kappa'}] = 0; \tag{25.29}$$

$$\{C_\lambda, C_{\lambda'}^\dagger\} = \delta_{\lambda\lambda'}, \qquad \{C_\lambda, C_{\lambda'}\} = 0; \tag{25.30}$$

$$[A_\kappa, C_\lambda] = [A_\kappa, C_\lambda^\dagger] = 0. \tag{25.31}$$

From these relations, and from (25.24) for $\mathscr{H}^{(0)}$, it is seen that the zeroth-order Hamiltonian commutator equations are satisfied:

$$[\mathscr{H}^{(0)},A_\kappa^\dagger] = \hbar\omega_\kappa A_\kappa^\dagger; \tag{25.32}$$

$$[\mathscr{H}^{(0)},C_\lambda^\dagger] = \epsilon_\lambda C_\lambda^\dagger. \tag{25.33}$$

The first step in the operator renormalization process is to calculate the commutators of $\mathscr{H}_{\text{ep}}^{(1)}$ with the zeroth-order phonon and electron operators; from (25.25) for $\mathscr{H}_{\text{ep}}^{(1)}$ the results are

$$[\mathscr{H}_{\text{ep}}^{(1)},A_\kappa^\dagger] = \sum_\lambda V_{\kappa\lambda} C_{\lambda+\kappa}^\dagger C_\lambda; \tag{25.34}$$

$$[\mathscr{H}_{\text{ep}}^{(1)},C_\lambda^\dagger] = \sum_\kappa{}' V_{\kappa\lambda}(A_\kappa + A_{-\kappa}^\dagger)C_{\lambda+\kappa}^\dagger. \tag{25.35}$$

The next step is to calculate the first-order single-particle energy shifts from the general equations (13.36) for bosons and (13.38) for fermions. Since (25.34) is diagonal in phonon coordinates, the matrix element (13.36) in first order vanishes, and

$$\hbar\omega_{1\kappa} = 0. \tag{25.36}$$

Also since (25.35) is nondiagonal in phonon coordinates, the matrix element (13.38) in first order vanishes, and

$$\epsilon_{1\lambda} = 0. \tag{25.37}$$

Now from the form of the right-hand side of (25.34), a general form for the first-order correction $B_\kappa^\dagger$ for the renormalized phonon creation operators is written

$$B_\kappa^\dagger = \sum_\lambda V_{\kappa\lambda}\alpha_{\kappa\lambda} C_{\lambda+\kappa}^\dagger C_\lambda, \tag{25.38}$$

and from the form of the right-hand side of (25.35), a general form for the first-order correction $D_\lambda^\dagger$ for the renormalized electron creation operator is

$$D_\lambda^\dagger = \sum_\kappa{}' V_{\kappa\lambda}(\eta_{\kappa\lambda}A_\kappa + \zeta_{\kappa\lambda}A_{-\kappa}^\dagger)C_{\lambda+\kappa}^\dagger, \tag{25.39}$$

where $\alpha_{\kappa\lambda}$, $\eta_{\kappa\lambda}$, and $\zeta_{\kappa\lambda}$ are numerical coefficients. These coefficients are then determined by requiring the renormalized phonon operators $A_\kappa^\dagger + B_\kappa^\dagger$ and electron operators $C_\lambda^\dagger + D_\lambda^\dagger$ to be creation operators of the total Hamiltonian to first order; that is, we require the following first-order Hamiltonian commutator equations to be satisfied.

$$[\mathscr{H}^{(0)},B_\kappa^\dagger] + [\mathscr{H}_{\text{ep}}^{(1)},A_\kappa^\dagger] = (\hbar\omega_\kappa + i\gamma)B_\kappa^\dagger; \tag{25.40}$$

$$[\mathscr{H}^{(0)},D_\lambda^\dagger] + [\mathscr{H}_{\text{ep}}^{(1)},C_\lambda^\dagger] = (\epsilon_\lambda + i\gamma)D_\lambda^\dagger. \tag{25.41}$$

Here γ is a real positive infinitesimal number, introduced to avoid divergencies just as in the phonon renormalization of Section 13, and the first-order terms $\hbar\omega_{1\kappa}A_\kappa^\dagger$ and $\epsilon_{1\lambda}C_\lambda^\dagger$ do not appear on the right sides of (25.40) and (25.41), respectively, since the first-order energies vanish. The commutators of $\mathscr{H}^{(0)}$ with the first-order creation operators are evaluated to give

$$[\mathscr{H}^{(0)},B_\kappa^\dagger] = \sum_\lambda V_{\kappa\lambda}\alpha_{\kappa\lambda}(\epsilon_{\lambda+\kappa} - \epsilon_\lambda)C_{\lambda+\kappa}^\dagger C_\lambda; \tag{25.42}$$

$$[\mathscr{H}^{(0)},D_\lambda^\dagger] = \sum_\kappa{}' V_{\kappa\lambda}[(\epsilon_{\lambda+\kappa} - \hbar\omega_\kappa)\eta_{\kappa\lambda}A_\kappa + (\epsilon_{\lambda+\kappa} + \hbar\omega_\kappa)\zeta_{\kappa\lambda}A_{-\kappa}^\dagger]C_{\lambda+\kappa}^\dagger. \tag{25.43}$$

Then with (25.38) for $B_\kappa^\dagger$, and the commutators (25.34) and (25.42), the first-order equation (25.40) determines the $\alpha_{\kappa\lambda}$ as

$$\alpha_{\kappa\lambda} = (\epsilon_\lambda - \epsilon_{\lambda+\kappa} + \hbar\omega_\kappa + i\gamma)^{-1}. \tag{25.44}$$

Similarly with (25.39) for $D_\lambda^\dagger$, and the commutators (25.35) and (25.43), the first-order equation (25.41) determines the $\eta_{\kappa\lambda}$ and $\zeta_{\kappa\lambda}$ as

$$\eta_{\kappa\lambda} = (\epsilon_\lambda - \epsilon_{\lambda+\kappa} + \hbar\omega_\kappa + i\gamma)^{-1} = \alpha_{\kappa\lambda}; \tag{25.45}$$

$$\zeta_{\kappa\lambda} = (\epsilon_\lambda - \epsilon_{\lambda+\kappa} - \hbar\omega_\kappa + i\gamma)^{-1}. \tag{25.46}$$

The last step in the second-order renormalization procedure is to calculate the commutators of $\mathscr{H}_{\text{ep}}^{(2)}$ with the zeroth-order operators $A_\kappa^\dagger$ and $C_\lambda^\dagger$, and the commutators of $\mathscr{H}_{\text{ep}}^{(1)}$ with the first-order operators $B_\kappa^\dagger$ and $D_\lambda^\dagger$, and then from these commutators to calculate the second-order phonon and electron energies, according to the general equations (13.36) and (13.38). Such calculations are illustrated in some detail in Section 13 for the interacting phonon case; the present calculations are even simpler than those of Section 13, and the results are

$$\begin{aligned}\hbar\omega_{2\kappa} = {} & \sum_\lambda |V_{\kappa\lambda}|^2\,\alpha_{\kappa\lambda}(f_\lambda - f_{\lambda+\kappa}) + \sum_\lambda W_{\kappa\lambda}(f_\lambda - g_\lambda) \\ & - \sum_\lambda [|V_{\kappa\lambda}|^2\,g_\lambda(1 - g_{\lambda+\kappa})(\epsilon_\lambda - \epsilon_{\lambda+\kappa})_p^{-1} \\ & \qquad + |V_{-\kappa\lambda}|^2\,g_\lambda(1 - g_{\lambda-\kappa})(\epsilon_\lambda - \epsilon_{\lambda-\kappa})_p^{-1}];\end{aligned} \tag{25.47}$$

$$\begin{aligned}\epsilon_{2\lambda} = {} & \sum_\kappa{}' |V_{\kappa\lambda}|^2\,[\eta_{\kappa\lambda}(n_\kappa + f_{\lambda+\kappa}) + \zeta_{\kappa\lambda}(n_{-\kappa} + 1 - f_{\lambda+\kappa})] \\ & + \sum_\kappa W_{\kappa\lambda}(n_\kappa + \tfrac{1}{2}).\end{aligned} \tag{25.48}$$

In deriving these two equations we used (25.28) to write $V_{\kappa\lambda}V_{-\kappa,\lambda+\kappa}$ as $|V_{\kappa\lambda}|^2$. Recall that n_κ is the number of phonons in state κ, f_λ is the number of electrons in state λ, and g_λ is the number of electrons in state λ when the electrons are in their ground state. It is possible to simplify $\hbar\omega_{2\kappa}$ by making index interchanges inside the last Σ_λ in (25.47). In the final term, which contains $|V_{-\kappa\lambda}|^2$, take $\lambda \to \lambda + \kappa$, so that $|V_{-\kappa\lambda}|^2 \to |V_{-\kappa,\lambda+\kappa}|^2 = |V_{\kappa\lambda}|^2$, and the last Σ_λ in (25.47) becomes

$$-\sum_\lambda |V_{\kappa\lambda}|^2 [g_\lambda(1 - g_{\lambda+\kappa})(\epsilon_\lambda - \epsilon_{\lambda+\kappa})_p^{-1} + g_{\lambda+\kappa}(1 - g_\lambda)(\epsilon_{\lambda+\kappa} - \epsilon_\lambda)_p^{-1}]$$
$$= -\sum_\lambda |V_{\kappa\lambda}|^2 (g_\lambda - g_{\lambda+\kappa})(\epsilon_\lambda - \epsilon_{\lambda+\kappa})_p^{-1}. \quad (25.49)$$

Then the total $\hbar\omega_{2\kappa}$ of (25.47) simplifies to

$$\hbar\omega_{2\kappa} = \sum_\lambda |V_{\kappa\lambda}|^2 [(f_\lambda - f_{\lambda+\kappa})\alpha_{\kappa\lambda} - (g_\lambda - g_{\lambda+\kappa})(\epsilon_\lambda - \epsilon_{\lambda+\kappa})_p^{-1}]$$
$$+ \sum_\lambda W_{\kappa\lambda}(f_\lambda - g_\lambda). \quad (25.50)$$

From the above calculations, we have the following picture of the electrons and phonons in the presence of electron–phonon interactions. The renormalized phonons have creation operators $A_\kappa^\dagger + B_\kappa^\dagger$. According to (25.38) for $B_\kappa^\dagger$, a dressed phonon is accompanied by a set of electron-hole pairs, and in each pair the electron-hole momentum difference (wave vector difference) is $\mathbf{p} + \mathbf{k} - \mathbf{p} = \mathbf{k}$, the same as the wave vector of the bare phonon. The renormalized electrons have creation operators $C_\lambda^\dagger + D_\lambda^\dagger$, and according to (25.39) for $D_\lambda^\dagger$ a dressed electron is accompanied by two sets of particle pairs. One type of pair consists of an electron plus a phonon, and the total wave vector for these pairs is $\mathbf{p} + \mathbf{k} - \mathbf{k} = \mathbf{p}$, the same as that of the bare electron. The second type of pair consists of an electron plus a removed phonon, and the wave vector difference for these pairs is $\mathbf{p} + \mathbf{k} - \mathbf{k} = \mathbf{p}$, again the same as that of the bare electron. The renormalization cloud of particle pairs which is associated with each phonon and electron is designed to screen out the interactions between renormalized electrons and phonons in first order. In addition, if the system is in a given quantum state corresponding to a set of occupation numbers n_κ and f_λ, the energy required to create a renormalized phonon is $\hbar\omega_\kappa + \hbar\omega_{2\kappa}$, and to create a renormalized electron it is $\epsilon_\lambda + \epsilon_{2\lambda}$, correct to second order.

The zeroth-order statistical average of $\hbar\omega_{2\kappa}$ is obtained by replacing each f_λ in (25.50) by the function $\bar{f}_\lambda$, given by (24.31). Then taking the limit $\gamma \to 0$, the real and imaginary parts of $\langle\hbar\omega_{2\kappa}\rangle$ are obtained according to (16.19), and the result may be written

$$\langle\hbar\omega_{2\kappa}\rangle = \mathscr{R}\langle\hbar\omega_{2\kappa}\rangle - i\mathscr{I}\langle\hbar\omega_{2\kappa}\rangle, \quad (25.51)$$

where

$$\mathscr{R}\langle\hbar\omega_{2\kappa}\rangle = \sum_\lambda |V_{\kappa\lambda}|^2 [(\bar{f}_\lambda - \bar{f}_{\lambda+\kappa})(\epsilon_\lambda - \epsilon_{\lambda+\kappa} + \hbar\omega_\kappa)_p^{-1}$$
$$- (g_\lambda - g_{\lambda+\kappa})(\epsilon_\lambda - \epsilon_{\lambda+\kappa})_p^{-1}] \quad (25.52)$$
$$+ \sum_\lambda W_{\kappa\lambda}(\bar{f}_\lambda - g_\lambda);$$

$$\mathscr{I}\langle\hbar\omega_{2\kappa}\rangle = \pi \sum_\lambda |V_{\kappa\lambda}|^2 (\bar{f}_\lambda - \bar{f}_{\lambda+\kappa})\delta(\epsilon_\lambda - \epsilon_{\lambda+\kappa} + \hbar\omega_\kappa). \quad (25.53)$$

Thus because of the electron–phonon interactions, there is an additional contribution $\mathscr{R}\langle\hbar\omega_{2\kappa}\rangle$ to the energy of a phonon which is observed in inelastic neutron-scattering experiments, and an additional contribution $\mathscr{I}\langle\hbar\omega_{2\kappa}\rangle$ to the half-width of the neutron line.

The zeroth-order statistical average of the electron energy renormalization is written

$$\langle\epsilon_{2\lambda}\rangle = \mathscr{R}\langle\epsilon_{2\lambda}\rangle - i\mathscr{I}\langle\epsilon_{2\lambda}\rangle, \tag{25.54}$$

and is obtained by replacing each f_λ by $\bar{f}_\lambda$ and each n_κ by $\bar{n}_\kappa$ in (25.48), and then taking the limit $\gamma \to 0$; note that $\bar{n}_\kappa$ is given by (16.16). The result is

$$\begin{aligned}\mathscr{R}\langle\epsilon_{2\lambda}\rangle = \sum_\kappa{}' |V_{\kappa\lambda}|^2 [(\bar{n}_\kappa + \bar{f}_{\lambda+\kappa})(\epsilon_\lambda - \epsilon_{\lambda+\kappa} + \hbar\omega_\kappa)_p^{-1} \\ + (\bar{n}_\kappa + 1 - \bar{f}_{\lambda+\kappa})(\epsilon_\lambda - \epsilon_{\lambda+\kappa} - \hbar\omega_\kappa)_p^{-1}] \\ + \sum_\kappa W_{\kappa\lambda}(\bar{n}_\kappa + \tfrac{1}{2});\end{aligned} \tag{25.55}$$

$$\begin{aligned}\mathscr{I}\langle\epsilon_{2\lambda}\rangle = \pi \sum_\kappa{}' |V_{\kappa\lambda}|^2 [(\bar{n}_\kappa + \bar{f}_{\lambda+\kappa})\delta(\epsilon_\lambda - \epsilon_{\lambda+\kappa} + \hbar\omega_\kappa) \\ + (\bar{n}_\kappa + 1 - \bar{f}_{\lambda+\kappa})\delta(\epsilon_\lambda - \epsilon_{\lambda+\kappa} - \hbar\omega_\kappa)].\end{aligned} \tag{25.56}$$

Thus because of interactions with the phonons, each electron has an added contribution $\mathscr{R}\langle\epsilon_{2\lambda}\rangle$ to its statistical average energy, and a contribution $\hbar/\mathscr{I}\langle\epsilon_{2\lambda}\rangle$ to its statistical average lifetime. Note that these energy contributions apply to all the electrons in the conduction band. Also, the Umklapp terms are explicitly restored to (25.52)–(25.56) by the replacement

$$|V_{\kappa\lambda}|^2 G_\kappa \to \sum_{\mathbf{Q}} |V_{\kappa+\mathbf{Q},\lambda}|^2 G_{\kappa+\mathbf{Q}}, \tag{25.57}$$

where G_κ is the function coupled with $V_{\kappa\lambda}$ in each expression. In the case of a nonprimitive lattice, there is an additional contribution to $\epsilon_{2\lambda}$ which arises from interaction of the conduction electrons with the $\mathbf{k} = 0$ optic phonons.*

Total System Energy Levels

The process of summing the renormalized electron and phonon energies to get the total system energy is somewhat intricate, but quite instructive. The energy renormalization arising from electron–phonon interactions applies to each and every conduction electron and phonon; therefore to get the double counting corrections right, we start with the vacuum (no particle state) and add the phonons and electrons one at a time to obtain the desired final state. The total energy is then the energy of the vacuum plus the sum of all the electron and phonon energies, with each two-particle energy contribution counted only half. The zeroth-order energies are of course just $\Sigma_\kappa n_\kappa \hbar\omega_\kappa$

* D. C. Wallace, *Phys. Rev.* **152**, 247 (1966).

for the phonons and $\Sigma_\lambda f_\lambda \epsilon_\lambda$ for the electrons; henceforth we will consider only the total electron–phonon interaction energy.

The perturbation Hamiltonian $\mathscr{H}_{\text{ep}}^{(1)} + \mathscr{H}_{\text{ep}}^{(2)}$ is given by (25.25) and (25.26), and obviously $\mathscr{H}_{\text{ep}}^{(1)}$ gives no first- *or* second-order contribution to the energy of the vacuum, since C_λ operates on the vacuum to give zero. On the other hand, $\mathscr{H}_{\text{ep}}^{(2)}$ obviously has the vacuum value (diagonal matrix element)

$$-\sum_{\kappa\lambda}{}' |V_{\kappa\lambda}|^2 g_\lambda(1 - g_{\lambda+\kappa})(\epsilon_\lambda - \epsilon_{\lambda+\kappa})_p^{-1} - \tfrac{1}{2}\sum_{\kappa\lambda} W_{\kappa\lambda} g_\lambda. \tag{25.58}$$

Note that $C_\lambda^\dagger C_\lambda$ is zero in the vacuum, but of course the g_λ in $\mathscr{H}_{\text{ep}}^{(2)}$ are not to be set equal to zero since they are numbers, not operators. In the first contribution of (25.58), let $\kappa \to -\kappa$ followed by $\lambda \to \lambda + \kappa$ to show that the term involving $g_\lambda g_{\lambda+\kappa}$ vanishes, and further that (25.58) may be rewritten as

$$-\tfrac{1}{2}\sum_{\kappa\lambda}{}' |V_{\kappa\lambda}|^2 (g_\lambda - g_{\lambda+\kappa})(\epsilon_\lambda - \epsilon_{\lambda+\kappa})_p^{-1} - \tfrac{1}{2}\sum_{\kappa\lambda} W_{\kappa\lambda} g_\lambda. \tag{25.59}$$

The phonon energy renormalization is written in (25.50); from this the total energy contribution arising from the $\hbar\omega_{2\kappa}$ is

$$\sum_{\kappa\lambda}{}' |V_{\kappa\lambda}|^2 [\tfrac{1}{2}(f_\lambda - f_{\lambda+\kappa})\alpha_{\kappa\lambda} - (g_\lambda - g_{\lambda+\kappa})(\epsilon_\lambda - \epsilon_{\lambda+\kappa})_p^{-1}] n_\kappa + \sum_{\kappa\lambda} W_{\kappa\lambda}(\tfrac{1}{2}f_\lambda - g_\lambda) n_\kappa. \tag{25.60}$$

Here the electron–phonon terms $f_\lambda n_\kappa$ have been counted one-half, but the terms $g_\lambda n_\kappa$ have been fully counted, since again the g_λ arise from numbers in the Hamiltonian, not operators. Finally the electron energy renormalization is written in (25.48), and the corresponding total energy contribution is

$$\sum_{\kappa\lambda}{}' |V_{\kappa\lambda}|^2 [\eta_{\kappa\lambda}(\tfrac{1}{2}n_\kappa + \tfrac{1}{2}f_{\lambda+\kappa}) + \zeta_{\kappa\lambda}(\tfrac{1}{2}n_{-\kappa} + 1 - \tfrac{1}{2}f_{\lambda+\kappa})] f_\lambda + \sum_{\kappa\lambda} W_{\kappa\lambda}(\tfrac{1}{2}n_\kappa + \tfrac{1}{2}) f_\lambda. \tag{25.61}$$

Then the total system energy due to electron–phonon interactions is the sum of (25.59)–(25.61), and the real part of the statistical average of this total energy is denoted $\langle \mathscr{E}_{\text{ep}} \rangle$. When $\alpha_{\kappa\lambda}$, $\eta_{\kappa\lambda}$, and $\zeta_{\kappa\lambda}$ are replaced by their definitions (25.44)–(25.46), and the statistical average is taken by replacing n_κ, f_λ by $\bar{n}_\kappa, \bar{f}_\lambda$, and the limit $\gamma \to 0$ is taken, the result is

$$\begin{aligned}
\langle \mathscr{E}_{\text{ep}} \rangle = \tfrac{1}{2}\sum_{\kappa\lambda}{}' |V_{\kappa\lambda}|^2 \{ & (\epsilon_\lambda - \epsilon_{\lambda+\kappa} - \hbar\omega_\kappa)_p^{-1}(\bar{n}_\kappa + 2 - \bar{f}_{\lambda+\kappa})\bar{f}_\lambda \\
& + (\epsilon_\lambda - \epsilon_{\lambda+\kappa} + \hbar\omega_\kappa)_p^{-1}[(\bar{f}_\lambda - \bar{f}_{\lambda+\kappa})\bar{n}_\kappa + (\bar{n}_\kappa + \bar{f}_{\lambda+\kappa})\bar{f}_\lambda] \\
& - (\epsilon_\lambda - \epsilon_{\lambda+\kappa})_p^{-1}(g_\lambda - g_{\lambda+\kappa})(2\bar{n}_\kappa + 1)\} \\
& + \sum_{\kappa\lambda} W_{\kappa\lambda}(\bar{f}_\lambda - g_\lambda)(\bar{n}_\kappa + \tfrac{1}{2}).
\end{aligned} \tag{25.62}$$

To simplify the total interaction energy, let us regroup the terms in (25.62) as follows.

$$\langle \mathscr{E}_{\text{ep}} \rangle = \tfrac{1}{2} \sum_{\kappa\lambda}{}' |V_{\kappa\lambda}|^2 \{(\epsilon_\lambda - \epsilon_{\lambda+\kappa} - \hbar\omega_\kappa)_p^{-1}[\bar{f}_\lambda(\bar{n}_\kappa + 1) - \bar{f}_\lambda(\bar{f}_{\lambda+\kappa} - 1)]$$
$$+ (\epsilon_\lambda - \epsilon_{\lambda+\kappa} + \hbar\omega_\kappa)_p^{-1}[(\bar{f}_\lambda - \bar{f}_{\lambda+\kappa})\bar{n}_\kappa + \bar{f}_\lambda(\bar{n}_\kappa + 1) + \bar{f}_\lambda(\bar{f}_{\lambda+\kappa} - 1)]$$
$$- (\epsilon_\lambda - \epsilon_{\lambda+\kappa})_p^{-1}(g_\lambda - g_{\lambda+\kappa})(2\bar{n}_\kappa + 1)\}$$
$$+ \sum_{\kappa\lambda} W_{\kappa\lambda}(\bar{f}_\lambda - g_\lambda)(\bar{n}_\kappa + \tfrac{1}{2}). \tag{25.63}$$

Now under the interchange $\kappa \to -\kappa$, followed by $\lambda \to \lambda + \kappa$ inside the $\Sigma'_{\kappa\lambda}$, we have

$$\bar{f}_\lambda \to \bar{f}_{\lambda+\kappa}, \qquad \bar{f}_{\lambda+\kappa} \to \bar{f}_\lambda;$$
$$(\epsilon_\lambda - \epsilon_{\lambda+\kappa} - \hbar\omega_\kappa)_p^{-1} \to -(\epsilon_\lambda - \epsilon_{\lambda+\kappa} + \hbar\omega_\kappa)_p^{-1}; \tag{25.64}$$
$$|V_{\kappa\lambda}|^2 \to |V_{-\kappa,\lambda+\kappa}|^2 = |V_{\kappa\lambda}|^2.$$

In the term in (25.63) which contains $(\epsilon_\lambda - \epsilon_{\lambda+\kappa} - \hbar\omega_\kappa)_p^{-1}$, let us make the above interchange for just the part $\bar{f}_\lambda(\bar{n}_\kappa + 1)$, and rewrite (25.63) as

$$\langle \mathscr{E}_{\text{ep}} \rangle = \sum_{\kappa\lambda}{}' |V_{\kappa\lambda}|^2 [(\epsilon_\lambda - \epsilon_{\lambda+\kappa} + \hbar\omega_\kappa)_p^{-1}(\bar{f}_\lambda - \bar{f}_{\lambda+\kappa})(\bar{n}_\kappa + \tfrac{1}{2})$$
$$- (\epsilon_\lambda - \epsilon_{\lambda+\kappa})_p^{-1}(g_\lambda - g_{\lambda+\kappa})(\bar{n}_\kappa + \tfrac{1}{2})] + \sum_{\kappa\lambda} W_{\kappa\lambda}(\bar{f}_\lambda - g_\lambda)(\bar{n}_\kappa + \tfrac{1}{2})$$
$$+ \tfrac{1}{2} \sum_{\kappa\lambda}{}' |V_{\kappa\lambda}|^2 \bar{f}_\lambda(\bar{f}_{\lambda+\kappa} - 1) \tag{25.65}$$
$$\times [(\epsilon_\lambda - \epsilon_{\lambda+\kappa} + \hbar\omega_\kappa)_p^{-1} - (\epsilon_\lambda - \epsilon_{\lambda+\kappa} - \hbar\omega_\kappa)_p^{-1}].$$

With the introduction of a slight approximation, $\langle \mathscr{E}_{\text{ep}} \rangle$ can now be simplified. In the terms containing $(\bar{f}_\lambda - \bar{f}_{\lambda+\kappa})(\epsilon_\lambda - \epsilon_{\lambda+\kappa} + \hbar\omega_\kappa)_p^{-1}$, there is no divergence problem as $\hbar\omega_\kappa \to 0$, so such terms can be expanded for $\hbar\omega_\kappa \ll \epsilon_\lambda - \epsilon_{\lambda+\kappa}$. Further since the ϵ_λ vary over a range of order μ_0, and the $\hbar\omega_\kappa$ vary over a range of order $K\Theta_0$, then this expansion is in powers of $K\Theta_0/\mu_0$; in other words, inside the $\Sigma'_{\kappa\lambda}$ we can write

$$(\bar{f}_\lambda - \bar{f}_{\lambda+\kappa})(\epsilon_\lambda - \epsilon_{\lambda+\kappa} + \hbar\omega_\kappa)_p^{-1} \to (\bar{f}_\lambda - \bar{f}_{\lambda+\kappa})(\epsilon_\lambda - \epsilon_{\lambda+\kappa})_p^{-1} + O(K\Theta_0/\mu_0). \tag{25.66}$$

Then neglecting the terms of order $K\Theta_0/\mu_0$, the first two lines of (25.65) may be written

$$\sum_{\kappa\lambda}{}' |V_{\kappa\lambda}|^2 (\epsilon_\lambda - \epsilon_{\lambda+\kappa})_p^{-1}(\bar{n}_\kappa + \tfrac{1}{2})[(\bar{f}_\lambda - \bar{f}_{\lambda+\kappa}) - (g_\lambda - g_{\lambda+\kappa})]$$
$$+ \sum_{\kappa\lambda} W_{\kappa\lambda}(\bar{n}_\kappa + \tfrac{1}{2})(\bar{f}_\lambda - g_\lambda). \tag{25.67}$$

Now in the terms containing $\bar{f}_{\lambda+\kappa}$ and $g_{\lambda+\kappa}$, the interchange $\kappa \to -\kappa$ followed by $\lambda \to \lambda + \kappa$ can be made, to transform (25.67) to

$$\sum_{\kappa\lambda}' |V_{\kappa\lambda}|^2 (\epsilon_\lambda - \epsilon_{\lambda+\kappa})_p^{-1}(2\bar{n}_\kappa + 1)(\bar{f}_\lambda - g_\lambda) + \sum_{\kappa\lambda} W_{\kappa\lambda}(\bar{n}_\kappa + \tfrac{1}{2})(\bar{f}_\lambda - g_\lambda). \tag{25.68}$$

This last expression has a simple interpretation in terms of the second-order one-electron energies $\epsilon_\lambda^{(2)}$. The $\epsilon_\lambda^{(2)}$ are given by (21.61), along with (21.60) for the first-order electron wave functions $\psi_\lambda^{(1)}$, and with the expressions (25.14) for $\Omega^{(1)}$ and (25.17) for $\Omega^{(2)}$, it follows that the statistical average (a phonon average) of $\epsilon_\lambda^{(2)}$ is

$$\langle \epsilon_\lambda^{(2)} \rangle = \sum_{\kappa}' |V_{\kappa\lambda}|^2 (\epsilon_\lambda - \epsilon_{\lambda+\kappa})_p^{-1}(2\bar{n}_\kappa + 1) + \sum_{\kappa} W_{\kappa\lambda}(\bar{n}_\kappa + \tfrac{1}{2}). \tag{25.69}$$

Therefore (25.68) is simply

$$\sum_{\lambda} (\bar{f}_\lambda - g_\lambda)\langle \epsilon_\lambda^{(2)} \rangle. \tag{25.70}$$

Finally the total interaction energy (25.65) may be expressed as

$$\langle \mathscr{E}_{\text{ep}} \rangle = \Delta_{\text{ep}} + \sum_{\lambda} (\bar{f}_\lambda - g_\lambda)\langle \epsilon_\lambda^{(2)} \rangle, \tag{25.71}$$

where

$$\Delta_{\text{ep}} = \tfrac{1}{2} \sum_{\kappa\lambda}' |V_{\kappa\lambda}|^2 \bar{f}_\lambda(\bar{f}_{\lambda+\kappa} - 1) \times [(\epsilon_\lambda - \epsilon_{\lambda+\kappa} + \hbar\omega_\kappa)_p^{-1} - (\epsilon_\lambda - \epsilon_{\lambda+\kappa} - \hbar\omega_\kappa)_p^{-1}]. \tag{25.72}$$

Again the Umklapp terms are explicitly restored in (25.69) and (25.72) by the replacement (25.57), and in (25.72) the $\Sigma_{\kappa\lambda}'$ is $\Sigma_\kappa' \, \Sigma_\lambda$.

The Interaction Free Energy

By following an argument which is essentially the same as the derivation of the phonon–phonon interaction contribution to the free energy in Section 16, the electron–phonon interaction contribution F_{ep}, in leading-order perturbation, is obviously just the statistical average interaction energy $\langle \mathscr{E}_{\text{ep}} \rangle$ given by (25.71):

$$F_{\text{ep}} = \Delta_{\text{ep}} + \sum_{\lambda} (\bar{f}_\lambda - g_\lambda)\langle \epsilon_\lambda^{(2)} \rangle. \tag{25.73}$$

Let us compare this free energy contribution to the zeroth-order electronic excitation contribution $F_{\text{e}} = -\tfrac{1}{2}\Gamma T^2$. It is easily shown by the methods of Section 24 that the statistical average of the zeroth-order excitation energy of the band electrons is

$$\sum_{\lambda} (\bar{f}_{\lambda} - g_{\lambda})\epsilon_{\lambda}^{(0)} = \tfrac{1}{2}\Gamma T^2 = -F_{\mathrm{e}}. \tag{25.74}$$

But $\epsilon_{\lambda}^{(2)}$ is the second-order correction to the one-electron energies, due to the vibration of the ions, and we expect $\langle \epsilon_{\lambda}^{(2)} \rangle$ to be small compared to $\epsilon_{\lambda}^{(0)}$ for temperatures up to several times the Debye temperature. It is therefore a reasonable approximation to neglect the term $\Sigma_{\lambda}\ (\bar{f}_{\lambda} - g_{\lambda})\langle \epsilon_{\lambda}^{(2)} \rangle$ in the interaction free energy (25.73), as being small compared to the magnitude of F_{e}. It is of course not necessary to neglect this term, and its evaluation for a given model is not difficult. Since $\bar{n}_{\kappa}$ goes as T at high T, this term has the following temperature-dependence:

$$\sum_{\lambda} (\bar{f}_{\lambda} - g_{\lambda})\langle \epsilon_{\lambda}^{(2)} \rangle \propto \begin{cases} T^2, & \text{for} \quad T \ll \Theta_0; \\ T^3, & \text{for} \quad T > \Theta_0. \end{cases} \tag{25.75}$$

Some model evaluations of the low- and high-temperature coefficients are given in Chapter 8.

It is more difficult to estimate the magnitude of Δ_{ep}, as compared to F_{e}; however Δ_{ep} can be transformed to find the explicit temperature-dependence at temperatures well below, and above, the Debye temperature. To proceed with this calculation, we make one further approximation, namely that the electron–phonon interaction coefficient $V_{\kappa\lambda}$ is independent of λ:

$$V_{\kappa\lambda} = V_{\kappa} = V(\mathbf{k}s). \tag{25.76}$$

Recall that the present treatment of electron–phonon interactions is limited to a primitive lattice with a single conduction band, for which $\lambda = \mathbf{p}\sigma$, and $V_{\kappa\lambda}$ is obviously independent of σ. Further the dependence on λ of $V_{\kappa\lambda}$ is contained in $\theta_i(\lambda + \kappa, \lambda)$, and from (25.9) it is seen that $\theta_i(\lambda + \kappa, \lambda)$ is independent of λ if each electron wave function is a single plane wave, or an orthogonalized plane wave.

With the Umklapp terms restored, the Σ'_{κ} in Δ_{ep} is $\Sigma_s \Sigma'_{\mathbf{k}} \Sigma_{\mathbf{Q}} = \Sigma_s \Sigma'_{\mathbf{p}} = \tfrac{1}{2}\Sigma_s \Sigma'_{\lambda}$ since the Σ_{σ} contained in Σ_{λ} is just 2. Then replacing $\lambda + \kappa + \mathbf{Q}$ by λ', and using the approximation (25.76), the expression (25.72) for Δ_{ep} may be written

$$\Delta_{\mathrm{ep}} = \tfrac{1}{4} \sum_{s} \sum_{\lambda\lambda'}{}' |V(\mathbf{p}' - \mathbf{p}, s)|^2 \bar{f}_{\lambda}(\bar{f}_{\lambda'} - 1)$$
$$\times \{[\epsilon_{\lambda} - \epsilon_{\lambda'} + \hbar\omega(\mathbf{p}' - \mathbf{p}, s)]_p^{-1} - [\epsilon_{\lambda} - \epsilon_{\lambda'} - \hbar\omega(\mathbf{p}' - \mathbf{p}, s)]_p^{-1}\}. \tag{25.77}$$

In order to transform the $\Sigma'_{\lambda\lambda'}$ to a double integral over energies, we define $G(\epsilon_{\lambda}, \epsilon_{\lambda'})$ as the angle average of the summand in (25.77) over the surface of constant $\epsilon_\lambda = \epsilon_{\mathbf{p}}$ and the surface of constant $\epsilon_{\lambda'} = \epsilon_{\mathbf{p}'}$:

$$G(\epsilon_\lambda,\epsilon_{\lambda'}) = \tfrac{1}{2}\sum_s \int \frac{d\Omega_{\mathbf{p}}}{4\pi}\int \frac{d\Omega_{\mathbf{p}'}}{4\pi} \times \left\{\frac{|V(\mathbf{p}'-\mathbf{p},s)|^2}{[\epsilon_{\mathbf{p}} - \epsilon_{\mathbf{p}'} + \hbar\omega(\mathbf{p}'-\mathbf{p},s)]_p} - \frac{|V(\mathbf{p}'-\mathbf{p},s)|^2}{[\epsilon_{\mathbf{p}} - \epsilon_{\mathbf{p}'} - \hbar\omega(\mathbf{p}'-\mathbf{p},s)]_p}\right\}, \tag{25.78}$$

where $\int d\Omega_{\mathbf{p}}$ is over all angles of $\mathbf{p}$, etc. Then (25.77) is transformed to an integral representation by means of the general formula (24.39), to obtain

$$\Delta_{\text{ep}} = \tfrac{1}{2}\int_0^\infty n(\epsilon)f(\epsilon)\int_0^\infty n(\epsilon')[f(\epsilon') - 1]G(\epsilon,\epsilon')\,d\epsilon'\,d\epsilon. \tag{25.79}$$

Let us now examine some of the properties of $G(\epsilon_\lambda,\epsilon_{\lambda'})$. In keeping with the approximation (25.76), it follows that $V(\mathbf{k}s)^* = V(-\mathbf{k}s)$, and since $\omega(\mathbf{k}s) = \omega(-\mathbf{k}s)$, then

$$G(\epsilon_\lambda,\epsilon_{\lambda'}) = G(\epsilon_{\lambda'},\epsilon_\lambda), \tag{25.80}$$

and also

$$\partial G(\epsilon_\lambda,\epsilon_{\lambda'})/\partial\epsilon_\lambda = -\partial G(\epsilon_\lambda,\epsilon_{\lambda'})/\partial\epsilon_{\lambda'}. \tag{25.81}$$

Now the electron energies ϵ_λ vary over a range of order μ_0, while the phonon energies $\hbar\omega(\mathbf{k}s)$ vary over a range of order $K\Theta_0$, and of course $K\Theta_0 \ll \mu_0$. Because of the energy denominators in (25.78), $G(\epsilon_\lambda,\epsilon_{\lambda'})$ is appreciable only when $|\epsilon_\lambda - \epsilon_{\lambda'}| \lesssim K\Theta_0$, and in fact a crude representation of $G(\epsilon_\lambda,\epsilon_{\lambda'})$ may be expressed as

$$\begin{aligned} G(\epsilon_\lambda,\epsilon_{\lambda'}) &\approx G(\epsilon_\lambda,\epsilon_\lambda), &&\text{for}\quad |\epsilon_\lambda - \epsilon_{\lambda'}| < K\Theta_0;\\ G(\epsilon_\lambda,\epsilon_{\lambda'}) &\approx (K\Theta_0/\mu_0)G(\epsilon_\lambda,\epsilon_\lambda), &&\text{for}\quad |\epsilon_\lambda - \epsilon_{\lambda'}| > K\Theta_0. \end{aligned} \tag{25.82}$$

The important point is that $G(\epsilon_\lambda,\epsilon_{\lambda'})$ varies significantly when ϵ_λ or $\epsilon_{\lambda'}$ varies over a range of order $K\Theta_0$.

For very low temperatures, i.e., $T \ll \Theta_0$, $G(\epsilon_\lambda,\epsilon_{\lambda'})$ does not vary significantly over energy ranges of order KT, and the Sommerfeld expansion may be applied to (25.79) for Δ_{ep}. The low-temperature evaluation of the double integral in (25.79) makes use of the relations (25.80) and (25.81), and is a recommended exercise; the result is

$$\Delta_{\text{ep}} = \text{constant} - \tfrac{1}{6}\pi^2(KT)^2[n(\mu_0)]^2G(\mu_0,\mu_0), \quad \text{for}\quad T \ll \Theta_0. \tag{25.83}$$

At high temperatures, i.e., $T > \Theta_0$, we use the approximation (25.82) to replace ϵ' by ϵ and $\int_0^\infty d\epsilon'$ by $2K\Theta_0$ in (25.79); evaluation of $\int_0^\infty d\epsilon$ then gives a power series in KT/μ_0 (essentially a Sommerfeld expansion), and the leading term is

$$\Delta_{ep} = -(K\Theta_0)(KT)[n(\mu_0)]^2 G(\mu_0,\mu_0), \quad \text{for} \quad T > \Theta_0. \tag{25.84}$$

A constant term in Δ_{ep} may be neglected since it is extremely small compared to the static lattice potential Φ_0, and the temperature-dependent part of Δ_{ep} may be incorporated as a correction to the zeroth-order electronic excitation free energy F_e, as follows:

$$F_e + \Delta_{ep} = -\tfrac{1}{2}\Gamma(1 + \Delta_\Gamma)T^2. \tag{25.85}$$

The electron–phonon correction Δ_Γ is presumably small compared to 1, is temperature-dependent, and has the low- and high-temperature limits

$$\begin{aligned} \Delta_\Gamma &= n(\mu_0)G(\mu_0,\mu_0), && \text{for} \quad T \ll \Theta_0; \\ \Delta_\Gamma &\approx (\Theta_0/T)n(\mu_0)G(\mu_0,\mu_0), && \text{for} \quad T > \Theta_0. \end{aligned} \tag{25.86}$$

Finally from (25.78), the function $G(\mu_0,\mu_0)$ is simply

$$G(\mu_0,\mu_0) = \sum_s \int \frac{d\Omega_{\mathbf{p}}}{4\pi} \int \frac{d\Omega_{\mathbf{p}'}}{4\pi} \frac{|V(\mathbf{p}' - \mathbf{p},s)|^2}{\hbar\omega(\mathbf{p}' - \mathbf{p},s)}, \tag{25.87}$$

where the angle integrals are over the $T = 0$ Fermi surface. From this equation it is seen that $G(\mu_0,\mu_0)$ is always positive, and hence

$$\Delta_\Gamma > 0, \quad \text{for} \quad T \ll \Theta_0, \quad \text{and for} \quad T > \Theta_0. \tag{25.88}$$

Since Δ_Γ depends explicitly on the temperature, the electron–phonon interaction contributions to thermodynamic functions are not included simply by replacing Γ with $\Gamma(1 + \Delta_\Gamma)$ in the zeroth-order electronic excitation contributions of Section 24. In particular, with the inclusion of Δ_{ep} in the free energy, according to (25.85), the total electronic plus electron–phonon contributions to the entropy and the heat capacity at constant configuration are

$$S_e + S_{ep} = C_{\eta e} + C_{\eta ep} = \Gamma(1 + \Delta_\Gamma)T, \quad \text{for} \quad T \ll \Theta_0; \tag{25.89}$$

$$\begin{aligned} S_e + S_{ep} &= \Gamma T + \text{constant} && \text{for} \quad T > \Theta_0, \\ C_{\eta e} + C_{\eta ep} &= C_{\eta e} = \Gamma T && \text{for} \quad T > \Theta_0. \end{aligned} \tag{25.90}$$

An unfortunate property of the electron–phonon contribution to the entropy and heat capacity at low temperatures is that this contribution, like the electronic excitation contribution, has no dependence on the mass of the ions. According to (25.16) the matrix element $V(\mathbf{p}' - \mathbf{p},s)$ is proportional to

$[M_C\omega(\mathbf{p}' - \mathbf{p},s)]^{-1/2}$, so the integrand in (25.87) for $G(\mu_0,\mu_0)$ is proportional to $\{M_C[\omega(\mathbf{p}' - \mathbf{p},s)]^2\}^{-1}$, and since the phonon frequencies $\omega(\mathbf{k}s)$ have the dependence $M_C^{-1/2}$, it follows for both low- and high-temperature limits that

$$\Gamma \quad \text{and} \quad \Delta_\Gamma \quad \text{are independent of } M_C. \tag{25.91}$$

Thus it is not possible in principle to experimentally separate the electronic excitation contribution and the electron–phonon contribution, by carrying out measurements on different isotopes of the same metal (in low- or high-temperature limits).

6

PSEUDOPOTENTIAL PERTURBATION THEORY

The Fermi energy and the ground-state energy of the conduction electrons are calculated to second order in the local pseudopotential, with approximate inclusion of exchange and correlation effects, and for the ions in arbitrary positions. Illustrative calculations of the elastic constants in the potential approximation, and of the dynamical matrices, are carried out. Because the pseudopotential is treated as a perturbation, the theory is not internally consistent in each order of the ion displacements; this shows up as a difference between homogeneous deformation and long-waves calculations. The inconsistency is removed by expansions of the dielectric function and the dynamical matrices in powers of Umklapp terms in the pseudopotential. The electronic and electron–phonon interaction contributions to the free energy are also derived.

26. LOCAL PSEUDOPOTENTIALS

The Perturbation Approximation

The pseudopotential perturbation theory is based on the same principles as the band-structure theory of the preceding chapter. However, by means of approximate solutions for the conduction electrons in metals, the lattice-statics and lattice-dynamics calculations are greatly simplified, but at the same time are less faithful to the theoretical principles. Nevertheless the theory is quite useful, since within the approximations made, it is possible to carry out calculations for liquids as well as crystals, for defects and impurities in crystals, and for alloys. Here we limit our discussion to crystals with all atoms the same, but arbitrary crystal structures; the extension to alloys is straightforward.

The metal crystal is now described as a collection of closed shell ions, with all the electrons outside the last closed shell placed in a single conduction band. Thus each conduction electron is characterized by a wave vector **p** and a spin σ, and the spin index will be suppressed since all our calculations are independent of spin. The one-electron wave equation is

$$h\psi_{\mathbf{p}} = \epsilon_{\mathbf{p}}\psi_{\mathbf{p}}, \tag{26.1}$$

where

$$h = ke + \Omega, \tag{26.2}$$

and the one-electron potential Ω is composed of bare ion and screening contributions. The orthonormality of the conduction electrons is expressed by

$$\langle \psi_{\mathbf{p}} \mid \psi_{\mathbf{p}'} \rangle = \delta_{\mathbf{p}\mathbf{p}'}. \tag{26.3}$$

The electrons in the ion cores are denoted by the index α, where α enumerates all the core electrons in all the ions, and a single core-electron wave function is represented by $|\alpha\rangle$. As Herring* has pointed out, the $\psi_{\mathbf{p}}$ must be orthogonal to the $|\alpha\rangle$, that is,

$$\langle \alpha \mid \psi_{\mathbf{p}} \rangle = 0, \tag{26.4}$$

and this can be ensured by writing $\psi_{\mathbf{p}}$ in the form

$$\psi_{\mathbf{p}} = (1 - P)\varphi_{\mathbf{p}}, \tag{26.5}$$

where

$$P = \sum_{\alpha} |\alpha\rangle\langle\alpha| \tag{26.6}$$

and $\varphi_{\mathbf{p}}$ is arbitrary at the moment. Since the core states satisfy $\langle \alpha \mid \alpha' \rangle = \delta_{\alpha\alpha'}$, it is obvious from (26.5) and (26.6) that (26.4) is satisfied, and it also follows that the projection operator P satisfies

$$P^{\dagger} = P; \qquad P^2 = P. \tag{26.7}$$

It should also be noted that since the $\psi_{\mathbf{p}}$ are orthogonal, according to (26.3), then the $\varphi_{\mathbf{p}}$ cannot be orthogonal.

It is now possible to write a pseudo-wave equation which has as solutions the pseudo-wave functions $\varphi_{\mathbf{p}}$ and the correct conduction electron energies $\epsilon_{\mathbf{p}}$. From (26.1) and (26.5),

$$h(1 - P)\varphi_{\mathbf{p}} = \epsilon_{\mathbf{p}}(1 - P)\varphi_{\mathbf{p}}, \tag{26.8}$$

* C. Herring, *Phys. Rev.* **57**, 1169 (1940).

or

$$[h(1 - P) + \epsilon_{\mathbf{p}} P]\varphi_{\mathbf{p}} = \epsilon_{\mathbf{p}}\varphi_{\mathbf{p}}. \tag{26.9}$$

By introducing the one-electron pseudopotential W, this last equation is

$$(ke + W)\varphi_{\mathbf{p}} = \epsilon_{\mathbf{p}}\varphi_{\mathbf{p}}, \tag{26.10}$$

where

$$W\varphi_{\mathbf{p}} = [\Omega + (\epsilon_{\mathbf{p}} - h)P]\varphi_{\mathbf{p}}. \tag{26.11}$$

At this point W is an operator, and a very unhandy operator at that. In the first place, any combination of core functions can be added to $\varphi_{\mathbf{p}}$ and the pseudo-wave equation (26.10) is still satisfied with the same $\epsilon_{\mathbf{p}}$, and also $\psi_{\mathbf{p}}$ of (26.5) is unchanged. This arbitrariness in $\varphi_{\mathbf{p}}$ leads to arbitrariness in W. In the second place, (26.11) does not define W, but only the result of W operating on $\varphi_{\mathbf{p}}$, and since that result contains the energy $\epsilon_{\mathbf{p}}$, then W is a non-Hermitian operator.* On the other hand, arguments can be made that $W\varphi_{\mathbf{p}}$ ought to be small compared to $ke\varphi_{\mathbf{p}}$,† and hence that W may be treated as a perturbation. The only contribution to W in the region outside the ion cores is Ω, and this is composed of Coulomb potentials arising from the ions and from the other electrons (plus exchange corrections). Qualitatively this total Ω may be assumed small compared to the electronic kinetic energy; the weakness of this assumption is clarified in (26.87) below. Inside the ion cores Ω is large and negative, due to the unscreened nuclear charge, while $(\epsilon_{\mathbf{p}} - h)P$ is large and positive, because h operators on P to produce core electron energies ϵ_α and the band-electron energies $\epsilon_{\mathbf{p}}$ lie well above the core energies. Hence it is at least plausible that W is small inside the cores, and we will therefore treat $W\varphi_{\mathbf{p}}$ as a perturbation.

Ordinary perturbation theory can be carried through, even though W is a non-Hermitian operator. The zeroth-order Hamiltonian in (26.10) is just ke, and solutions are the complete set of orthonormal plane waves $|\mathbf{p}\rangle$:

$$ke\,|\mathbf{p}\rangle = e_{\mathbf{p}}\,|\mathbf{p}\rangle, \qquad e_{\mathbf{p}} = \hbar^2 p^2/2m; \tag{26.12}$$

where m is the electron mass. The $|\mathbf{p}\rangle$ are the same as the $|\mathbf{q}\rangle$ defined in (21.15),

$$|\mathbf{p}\rangle = V^{-1/2} e^{i\mathbf{p}\cdot\mathbf{r}}, \tag{26.13}$$

where the $\mathbf{p}$ also satisfy the periodic boundary conditions in the crystal volume V, and the orthonormality (21.16) and completeness (21.17) may be written

* B. J. Austin, V. Heine, and L. J. Sham, *Phys. Rev.* **127,** 276 (1962).

† J. C. Phillips and L. Kleinman, *Phys. Rev.* **116,** 287 (1959); E. Antoncik, *J. Phys. Chem. Solids* **10,** 314 (1959).

in the form

$$\langle \mathbf{p} \mid \mathbf{p}' \rangle = \delta_{\mathbf{p}\mathbf{p}'}, \quad \text{orthonormality;} \tag{26.14}$$

$$\sum_{\mathbf{p}} |\mathbf{p}\rangle\langle\mathbf{p}| = 1, \quad \text{completeness.} \tag{26.15}$$

The pseudo-wave functions correct to first order are then

$$\varphi_{\mathbf{p}} = |\mathbf{p}\rangle + \sum_{\mathbf{q}} A(\mathbf{q},\mathbf{p})\,|\mathbf{p}+\mathbf{q}\rangle, \tag{26.16}$$

where

$$A(\mathbf{q},\mathbf{p}) = \frac{\langle \mathbf{p}+\mathbf{q}|\, W\, |\mathbf{p}\rangle}{e_{\mathbf{p}} - e_{\mathbf{p}+\mathbf{q}}}, \quad \mathbf{q} \neq 0; \tag{26.17}$$

the coefficient $A(0,\mathbf{p})$ is determined in (26.23) below from a normalization condition. Finally the electron energies $\epsilon_{\mathbf{p}}$ correct to second order in W are given by

$$\epsilon_{\mathbf{p}} = e_{\mathbf{p}} + \langle \mathbf{p}|\, W\, |\mathbf{p}\rangle + \sum_{\mathbf{q}}{}' A(\mathbf{q},\mathbf{p})\langle\mathbf{p}|\, W\, |\mathbf{p}+\mathbf{q}\rangle. \tag{26.18}$$

The condition (26.3) for orthonormality of the $\psi_{\mathbf{p}}$ can be written in terms of the $\varphi_{\mathbf{p}}$ by means of (26.5),

$$\delta_{\mathbf{p}\mathbf{p}'} = \langle \psi_{\mathbf{p}} \mid \psi_{\mathbf{p}'}\rangle = \langle\varphi_{\mathbf{p}}|\,(1-P)(1-P)\,|\varphi_{\mathbf{p}'}\rangle = \langle\varphi_{\mathbf{p}}|\, 1-P\, |\varphi_{\mathbf{p}'}\rangle, \tag{26.19}$$

since $P^2 = P$, and with the expansion (26.16) of $\varphi_{\mathbf{p}}$ this leads to

$$\langle\mathbf{p}|\, P\, |\mathbf{p}'\rangle = \sum_{\mathbf{q}} [A(\mathbf{q},\mathbf{p})^* \langle\mathbf{p}+\mathbf{q}|\, 1-P\, |\mathbf{p}'\rangle + A(\mathbf{q},\mathbf{p}')\langle\mathbf{p}|\, 1-P\, |\mathbf{p}'+\mathbf{q}\rangle], \tag{26.20}$$

to first order in the $A(\mathbf{q},\mathbf{p})$ coefficients. But this implies $\langle\mathbf{p}|\, P\, |\mathbf{p}'\rangle$ is of first order, since the right-hand side is of lowest order first order; hence the operator P can be dropped from the right-hand side to get the first-order equation

$$\langle\mathbf{p}|\, P\, |\mathbf{p}'\rangle = A(\mathbf{p}'-\mathbf{p},\mathbf{p})^* + A(\mathbf{p}-\mathbf{p}',\mathbf{p}'). \tag{26.21}$$

Then by relabeling the wave vectors, and using the result (26.17) for $A(\mathbf{q},\mathbf{p})$ and noting that $\langle\mathbf{p}|\, W\, |\mathbf{p}+\mathbf{q}\rangle^* = \langle\mathbf{p}+\mathbf{q}|\, W^\dagger\, |\mathbf{p}\rangle$, (26.21) may be written

$$\langle\mathbf{p}+\mathbf{q}|\, P\, |\mathbf{p}\rangle = \frac{\langle\mathbf{p}+\mathbf{q}|\, W - W^\dagger\, |\mathbf{p}\rangle}{e_{\mathbf{p}} - e_{\mathbf{p}+\mathbf{q}}}, \quad \mathbf{q} \neq 0; \tag{26.22}$$

$$\langle\mathbf{p}|\, P\, |\mathbf{p}\rangle = A(0,\mathbf{p})^* + A(0,\mathbf{p}). \tag{26.23}$$

Although it will not concern us in the subsequent calculations, it should be noted that since matrix elements of P are first-order quantities, there is some difficulty in maintaining the condition $P^2 = P$ within a consistent perturbation approach.

THE LOCAL APPROXIMATION

The theoretical development has been continued up to the calculation of the total energy of the conduction electrons, with W treated as a non-Hermitian operator.* However, in the interest of providing a theory which is quite simple, even including anharmonic lattice-dynamics calculations, we make the local approximation at this point; that is, we assume W is a real function of the position variable $\mathbf{r}$, not an operator:

$$W = W^\dagger = W(\mathbf{r}), \quad \text{local approximation.} \tag{26.24}$$

It is instructive, and somewhat surprising, to look into the significance of this approximation. First, the matrix elements $\langle \mathbf{p}+\mathbf{q}| W |\mathbf{p}\rangle$ are independent of $\mathbf{p}$, since

$$\begin{aligned}\langle \mathbf{p}+\mathbf{q}| W |\mathbf{p}\rangle &= V^{-1}\int e^{-i(\mathbf{p}+\mathbf{q})\cdot\mathbf{r}} W(\mathbf{r}) e^{i\mathbf{p}\cdot\mathbf{r}}\, d\mathbf{r} \\ &= V^{-1}\int W(\mathbf{r}) e^{-i\mathbf{q}\cdot\mathbf{r}}\, d\mathbf{r} \\ &= W(\mathbf{q}).\end{aligned}$$

This leads to the following simplifications:

$$\langle \mathbf{p}+\mathbf{q}| W |\mathbf{p}\rangle = W(\mathbf{q}), \qquad W(\mathbf{q})^* = W(-\mathbf{q}); \tag{26.25}$$

$$A(\mathbf{q},\mathbf{p}) = W(\mathbf{q})/(e_{\mathbf{p}} - e_{\mathbf{p}+\mathbf{q}}), \qquad A(\mathbf{q},\mathbf{p})^* = A(-\mathbf{q},-\mathbf{p}) = -A(-\mathbf{q},\mathbf{p}+\mathbf{q}). \tag{26.26}$$

Second, from (26.22), the local approximation leads to $\langle \mathbf{p}+\mathbf{q}| P |\mathbf{p}\rangle = 0$ for $\mathbf{q} \neq 0$ and to first order in W; this implies P = constant, and then $P^2 \neq P$, again to first order in W. Then in view of (26.5) it follows that $\psi_{\mathbf{p}}$ and $\varphi_{\mathbf{p}}$ are equal up to first order in W, and up to a normalization constant $1 - P$. This means we can write

$$(ke + W)\psi_{\mathbf{p}} = \epsilon_{\mathbf{p}}\psi_{\mathbf{p}}, \tag{26.27}$$

where the orthonormalized $\psi_{\mathbf{p}}$ are obtained from (26.16) for $\varphi_{\mathbf{p}}$,

$$\psi_{\mathbf{p}} = |\mathbf{p}\rangle + \sum_{\mathbf{q}}{}' A(\mathbf{q},\mathbf{p})\, |\mathbf{p}+\mathbf{q}\rangle. \tag{26.28}$$

The orthonormality of these $\psi_{\mathbf{p}}$ is easily verified; to first order in the $A(\mathbf{q},\mathbf{p})$ we have from (26.28)

$$\langle \psi_{\mathbf{p}} \mid \psi_{\mathbf{p}'}\rangle = \langle \mathbf{p} \mid \mathbf{p}'\rangle + \sum_{\mathbf{q}}{}' [A(\mathbf{q},\mathbf{p})^*\langle \mathbf{p}+\mathbf{q} \mid \mathbf{p}'\rangle + A(\mathbf{q},\mathbf{p}')\langle \mathbf{p} \mid \mathbf{p}'+\mathbf{q}\rangle]. \tag{26.29}$$

* W. A. Harrison, *Pseudopotentials in the Theory of Metals*, W. A. Benjamin Inc., New York, 1966; R. Pick and G. Sarma, *Phys. Rev.* **135**, A1363 (1964).

For $\mathbf{p} = \mathbf{p}'$ this is obviously

$$\langle \psi_{\mathbf{p}} \mid \psi_{\mathbf{p}} \rangle = \langle \mathbf{p} \mid \mathbf{p} \rangle = 1, \tag{26.30}$$

while for $\mathbf{p} \neq \mathbf{p}'$, (26.26) may be used to show

$$\langle \psi_{\mathbf{p}} \mid \psi_{\mathbf{p}'} \rangle = A(\mathbf{p}' - \mathbf{p},\mathbf{p})^* + A(\mathbf{p} - \mathbf{p}',\mathbf{p}') = 0. \tag{26.31}$$

It is interesting to note that the relation $\psi_{\mathbf{p}} = (1 - P)\varphi_{\mathbf{p}}$ now implies $P = A(0,\mathbf{p})$ to first order, by comparison of (26.16) and (26.28); this is not the same as (26.23) because now $P^2 \neq P$.

It is not particularly instructive to discuss further the subtle complications which are associated with abandoning the proper conditions $W \neq W^\dagger$ and $P^2 = P$. Instead we simply look upon (26.27) as an approximate description of the conduction-electron problem; the band-structure theory of Chapter 5 now applies, with the added simplification that the pseudopotential may be treated as a perturbation. The electronic wave functions are given to first order by (26.28), and the electronic energies are given to second order by the local potential evaluation of (26.18):

$$\epsilon_{\mathbf{p}} = e_{\mathbf{p}} + W(\mathbf{q} = 0) + \sum_{\mathbf{q}}{}' \, |W(\mathbf{q})|^2 \, (e_{\mathbf{p}} - e_{\mathbf{p}+\mathbf{q}})_P^{-1}. \tag{26.32}$$

Since the experimentally observed Fermi surfaces for many metals show only small deviations from the free electron Fermi surfaces, the electronic energies may be fairly accurately represented by a small local $W(\mathbf{r})$; however the electronic wave functions will not be represented accurately, especially inside the ion cores.

Let us now turn to the construction of $W(\mathbf{r})$. From the theory of Sections 21 and 22, the total one-electron potential should be the sum of the bare ion potential W_{B}, the Hartree potential W_{S}, and any desired exchange contribution W_{X}:

$$W = W_{\mathrm{B}} + W_{\mathrm{S}} + W_{\mathrm{X}}. \tag{26.33}$$

The potential $W_{\mathrm{S}} + W_{\mathrm{X}}$ is the total screening field, and will be discussed below. The total W_{B} is a sum of single ion potentials w_{B}, each of which is assumed to move rigidly with its ion:

$$W_{\mathrm{B}}(\mathbf{r}) = \sum_{\mathrm{N}\nu} w_{\mathrm{B}}(\mathbf{r} - \tilde{\mathbf{R}}(\mathrm{N}\nu)). \tag{26.34}$$

Here the positions $\tilde{\mathbf{R}}(\mathrm{N}\nu)$ of the ions are arbitrary in the volume V of the crystal; since all ions are the same, each single ion function w_{B} is the same and the index ν serves only to distinguish the different ions in a unit cell, as for example in the hcp lattice. Recall that there are N_0 unit cells in the crystal and n ions in each unit cell. The Fourier components of W_{B} are written and simplified as follows.

$$W_B(\mathbf{q}) = V^{-1}\int W_B(\mathbf{r})e^{-i\mathbf{q}\cdot\mathbf{r}}\,d\mathbf{r}$$

$$= V^{-1}\sum_{N\nu} e^{-i\mathbf{q}\cdot\tilde{\mathbf{R}}(N\nu)}\int w_B(\mathbf{r}-\tilde{\mathbf{R}}(N\nu))e^{-i\mathbf{q}\cdot[\mathbf{r}-\tilde{\mathbf{R}}(N\nu)]}\,d\mathbf{r} \tag{26.35}$$

$$= S(\mathbf{q})w_B(\mathbf{q}),$$

where the structure factor $S(\mathbf{q})$ is given by

$$S(\mathbf{q}) = (nN_0)^{-1}\sum_{N\nu} e^{-i\mathbf{q}\cdot\tilde{\mathbf{R}}(N\nu)}, \tag{26.36}$$

and the Fourier component $w_B(\mathbf{q})$ of the single ion potential is given by

$$w_B(\mathbf{q}) = V_A^{-1}\int w_B(\mathbf{r}-\tilde{\mathbf{R}}(N\nu))e^{-i\mathbf{q}\cdot[\mathbf{r}-\tilde{\mathbf{R}}(N\nu)]}\,d\mathbf{r}. \tag{26.37}$$

In keeping with the most commonly used definition, $S(\mathbf{q})$ is normalized to the total number of ions nN_0, so that $w_B(\mathbf{q})$ is then normalized to the volume per atom V_A.

Now to the extent that $w_B(\mathbf{r} - \tilde{\mathbf{R}}(N\nu))$ is carried rigidly with its ion, it ought to be a central potential $w_B(r)$ depending only on the distance r from the ion center. In other words, the potential due to a single isolated closed shell ion is presumably central, when exchange effects are approximated by a potential instead of an operator, and it must remain central in the crystal if it is considered to move undeformed with the ion. Then the Fourier components $w_B(\mathbf{q})$ depend only on the magnitude $q = |\mathbf{q}|$, and may be written

$$w_B(q) = V_A^{-1}\int w_B(r)e^{-i\mathbf{q}\cdot\mathbf{r}}\,d\mathbf{r}. \tag{26.38}$$

The structure factor $S(\mathbf{q})$, of course, depends on both the direction and magnitude of $\mathbf{q}$.

It is useful to discuss some well-known central potential models for the single ion potential. If the ion is located at $r = 0$, the potential energy of an electron at a distance r, due to its interaction with the ion, is $w_B(r)$. The charge of each ion is taken to be $+ze$, so that outside the ion core $w_B(r)$ is $-ze^2/r$. It is convenient to express the total $w_B(r)$ as the sum of this Coulomb term and a contribution $w_C(r)$ which is localized to the core region, and which tends to cancel the Coulomb term inside the core:

$$w_B(r) = -ze^2/r + w_C(r). \tag{26.39}$$

For example, the Harrison model represents $w_C(r)$ by a function of the form of a 1s electron density, namely $e^{-r/\hat{\rho}}$ where $\hat{\rho}$ is a parameter. The Fourier components for this model are then calculated as follows. For the Coulomb term, from (21.33),

$$V_A^{-1}\int(-ze^2/r)e^{-i\mathbf{q}\cdot\mathbf{r}}\,d\mathbf{r} = -4\pi ze^2/q^2V_A, \quad q \neq 0. \tag{26.40}$$

For the core part we can write

$$V_A^{-1}\int e^{-r/\hat{\rho}}e^{-i\mathbf{q}\cdot\mathbf{r}}\,d\mathbf{r} \propto V_A^{-1}(1 + q^2\hat{\rho}^2)^{-2};$$

then multiplying this by another arbitrary parameter $\hat{\beta}$, the total $w_B(q)$ for the Harrison model is

$$w_B(q) = V_A^{-1}[-(4\pi ze^2/q^2) + \hat{\beta}(1 + q^2\hat{\rho}^2)^{-2}], \tag{26.41}$$

where $\hat{\beta}$ and $\hat{\rho}$ are positive. Obviously for this model,

$$w_C(q = 0) = \hat{\beta}/V_A. \tag{26.42}$$

The Heine–Abarenkov local model takes $w_B(r)$ to be a square well inside the core region, and the ion Coulomb potential outside the core:

$$w_B(r) = -\bar{\beta}, \quad r \leq \bar{\rho}; \qquad w_B(r) = -ze^2/r, \quad r > \bar{\rho}. \tag{26.43}$$

The Fourier transform of this potential is

$$w_B(q) = -(4\pi/q^2V_A)[(ze^2 - \bar{\beta}\bar{\rho})\cos(q\bar{\rho}) + (\bar{\beta}/q)\sin(q\bar{\rho})]. \tag{26.44}$$

When the ion Coulomb term (26.40) is subtracted from this expression, the remaining part is $w_C(q)$, and the $q = 0$ limit of this is found to be

$$w_C(q = 0) = (2\pi\bar{\rho}^2/V_A)(ze^2 - \tfrac{2}{3}\bar{\beta}\bar{\rho}). \tag{26.45}$$

Finally the special case of the Heine–Abarenkov model obtained when the well depth $\bar{\beta}$ is set equal to zero is the Ashcroft model, and for this potential there is only one parameter, namely the core radius $\bar{\rho}$.

Before proceeding with the general theory, it should be pointed out that the rigid ion approximation is generally less valid in the present pseudopotential formulation than in the band-structure formulation. This is because the band electrons are here taken as just those outside the last closed shell, and this leaves rather large cores for the atoms on the left side of the periodic table, particularly the alkali metals. The deformations of these large cores as the ions vibrate in the crystal introduce a departure from the rigid ion potential approximation, expressed by (26.34), and also introduce direct ion–ion interactions in addition to the point-ion Coulomb interactions discussed in Section 23.

Self-Consistent Fermi Surface

The chemical potential at $T = 0$ is μ_0, and this will also be referred to here as the Fermi energy ϵ_F. In zeroth order in the pseudopotential, the electronic energy is just the kinetic energy $e_{\mathbf{p}}$, and the Fermi surface is a sphere of constant $e_{\mathbf{p}}$ located at $|\mathbf{p}| = |\mathbf{f}|$, where $\mathbf{f}$ is the zeroth-order Fermi wave vector. Just as in Section 24, the Fermi energy is determined by requiring that the total number of conduction electrons znN_0 is equal to the number of electron states with energy equal to or less than the Fermi energy; in zeroth order this requirement is simply

$$znN_0 = (2V/8\pi^3)\int_0^{\mathbf{f}} d\mathbf{p} = V\,|\mathbf{f}|^3/3\pi^2, \tag{26.46}$$

where the factor 2 preceding the integral is to account for spins. Since $V = nN_0V_A$, the result for the magnitude $f = |\mathbf{f}|$ is

$$f^3 = 3\pi^2 z/V_A = 3\pi^2\rho_0, \tag{26.47}$$

where ρ_0 is the average conduction electron density, i.e., the $\mathbf{q} = 0$ component of $\rho(\mathbf{q})$.

Now let the Fermi wave vector, correct to second order in the pseudopotential, be $\mathbf{F}$. The Fermi energy ϵ_F correct to second order is then just (26.32) for $\epsilon_{\mathbf{p}}$ evaluated at $\mathbf{p} = \mathbf{F}$:

$$\epsilon_F = (\hbar^2/2m)\,|\mathbf{F}|^2 + W(\mathbf{q} = 0) + \sum_{\mathbf{q}}{}'\,|W(\mathbf{q})|^2\,(e_{\mathbf{F}} - e_{\mathbf{F}+\mathbf{q}})_p^{-1}. \tag{26.48}$$

Since $W(\mathbf{q} = 0)$ is a constant, it is irrelevant to the present discussion, and may be dropped from (26.48). Also in the second-order term in (26.48), $\mathbf{F}$ may be replaced by the zeroth-order value $\mathbf{f}$; then solving for $|\mathbf{F}|^2$ gives

$$|\mathbf{F}|^2 = (2m\epsilon_F/\hbar^2)\left[1 - \epsilon_F^{-1}\sum_{\mathbf{q}}{}'\,|W(\mathbf{q})|^2\,(e_{\mathbf{f}} - e_{\mathbf{f}+\mathbf{q}})_p^{-1}\right], \tag{26.49}$$

where here and in the following equations, $\mathbf{F}$ and $\mathbf{f}$ are parallel. Again in the second-order term in (26.49), ϵ_F may be replaced by the zeroth-order value $\hbar^2 f^2/2m$, and the square root of (26.49) correct to second order is

$$|\mathbf{F}| = (2m\epsilon_F/\hbar^2)^{1/2} - (m/\hbar^2 f)\sum_{\mathbf{q}}{}'\,|W(\mathbf{q})|^2\,(e_{\mathbf{f}} - e_{\mathbf{f}+\mathbf{q}})_p^{-1}. \tag{26.50}$$

To determine ϵ_F to second order, we again require that the number of states with energy equal to or less than ϵ_F is znN_0:

$$znN_0 = (2V/8\pi^3)\int_0^{\mathbf{F}} d\mathbf{p}, \tag{26.51}$$

where $|\mathbf{F}|$ is a function of angles. The integral can be written as $\int_0^{\mathbf{f}} d\mathbf{p} + \int_{\mathbf{f}}^{\mathbf{F}} d\mathbf{p}$, and the first integral is already equal to znN_0, according to (26.46), so the second integral must vanish:

$$\int_{\mathbf{f}}^{\mathbf{F}} d\mathbf{p} = 0. \tag{26.52}$$

The integrand is $d\Omega p^2 dp$, where $d\Omega$ is over all angles, and since $dp = |\mathbf{F}| - |\mathbf{f}|$ is a second-order quantity, p^2 may be replaced by the zeroth-order value f^2 and taken outside the integral. The remaining angle integral must then vanish:

$$\int (|\mathbf{F}| - |\mathbf{f}|)\, d\Omega = 0. \tag{26.53}$$

With (26.50) for $|\mathbf{F}|$, and noting that ϵ_F and $|\mathbf{f}| = f$ are independent of angles, (26.53) may be written

$$4\pi[(2m\epsilon_F/\hbar^2)^{1/2} - f] - (m/\hbar^2 f) \sum_{\mathbf{q}}{}' |W(\mathbf{q})|^2 \int (e_{\mathbf{f}} - e_{\mathbf{f+q}})_p^{-1}\, d\Omega = 0, \tag{26.54}$$

where $\int d\Omega$ is over angles of $\mathbf{f}$. The principal part evaluation of the integral in (26.54) is as follows.

$$\int \frac{d\Omega}{(e_{\mathbf{f}} - e_{\mathbf{f+q}})_p} = -\frac{2m}{\hbar^2} \int \frac{d\Omega}{(q^2 + 2\mathbf{q}\cdot\mathbf{f})_p} = -\frac{2\pi m}{\hbar^2 q f} \ln \left| \frac{1 + (q/2f)}{1 - (q/2f)} \right|. \tag{26.55}$$

Finally (26.54) may be solved for ϵ_F, and with (26.55) for the integral, the result correct to second order is

$$\epsilon_F = \frac{\hbar^2 f^2}{2m} - \frac{m}{2\hbar^2 f} \sum_{\mathbf{q}}{}' \frac{|W(\mathbf{q})|^2}{q} \ln \left| \frac{1 + (q/2f)}{1 - (q/2f)} \right|. \tag{26.56}$$

It may be noted that $W(\mathbf{q})$ contains the factor $S(\mathbf{q})$ [see, e.g., (26.85) below], and for the equilibrium configuration of the crystal $S(\mathbf{q})$ contains a $\delta(\mathbf{q})$, as can be seen from (26.36), so the $\Sigma'_{\mathbf{q}}$ becomes a $\Sigma'_{\mathbf{Q}}$ in (26.56) when the ions are at their equilibrium positions.

We can now put the solution (26.56) for ϵ_F back into (26.50), and find the following solution for the Fermi wave vector $\mathbf{F}$, correct to second order.

$$|\mathbf{F}| = f - \frac{m}{\hbar^2 f} \sum_{\mathbf{q}}{}' |W(\mathbf{q})|^2 \left[\frac{1}{(e_{\mathbf{f}} - e_{\mathbf{f+q}})_p} + \frac{m}{2\hbar^2 q f} \ln \left| \frac{1 + (q/2f)}{1 - (q/2f)} \right| \right]. \tag{26.57}$$

Here the dependence of $|\mathbf{F}|$ on the direction of $\mathbf{F}$ is contained in the dependence of the right-hand side on the direction of $\mathbf{f}$, through the energy denominators $(e_{\mathbf{f}} - e_{\mathbf{f+q}})_p^{-1}$; recall that $\mathbf{F}$ and $\mathbf{f}$ lie in the same direction. The result (26.57) is most interesting; it gives the Fermi surface correct to second order in the pseudopotential, for arbitrary positions of the ions, and hence can be used to calculate the Fermi surface for a liquid, for example, and also to calculate fluctuations in the Fermi surface due to the motions of the ions.

Again for the equilibrium configuration of a crystal, the $\Sigma'_{\mathbf{q}}$ in (26.57) is reduced to a $\Sigma'_{\mathbf{Q}}$.

It is useful to calculate the total kinetic energy of the conduction electrons in their ground state, correct to second order in the pseudopotential. The ground-state occupation numbers are $g_{\mathbf{p}}$, and the total ground-state kinetic energy is

$$\sum_{\mathbf{p}} g_{\mathbf{p}} e_{\mathbf{p}} = \frac{2V}{8\pi^3} \int_0^{\mathbf{F}} \frac{\hbar^2 p^2}{2m} \, d\mathbf{p}, \tag{26.58}$$

where the factor 2 for spin is implicit in the $\Sigma_{\mathbf{p}}$ on the left, and explicit in the numerator on the right. Again the integral can be written $\int_0^{\mathbf{f}} + \int_{\mathbf{f}}^{\mathbf{F}}$, and since the second integral is of second order, it can be reduced to an angle integral by setting $p^2 dp = f^2(|\mathbf{F}| - |\mathbf{f}|)$, and (26.58) becomes

$$\sum_{\mathbf{p}} g_{\mathbf{p}} e_{\mathbf{p}} = \frac{2V}{8\pi^3} \int_0^{\mathbf{f}} \frac{\hbar^2 p^2}{2m} \, d\mathbf{p} + \frac{2V}{8\pi^3} \frac{\hbar^2 f^4}{2m} \int (|\mathbf{F}| - |\mathbf{f}|) \, d\Omega. \tag{26.59}$$

But the angle integral vanishes by (26.53), so the second-order contribution to the total ground-state kinetic energy is zero, and this holds for arbitrary positions of the ions. The first integral in (26.59) is evaluated, and with (26.47) it is reduced to the usual form of the free electron ground-state kinetic energy:

$$\sum_{\mathbf{p}} g_{\mathbf{p}} e_{\mathbf{p}} = \tfrac{3}{5} z n N_0 e_{\mathrm{F}}; \qquad e_{\mathrm{F}} = \hbar^2 f^2/2m. \tag{26.60}$$

Screening

The total ground-state density of conduction electrons is denoted as $\rho(\mathbf{r})$, and may be written [see, e.g., (21.29); note we use the electron wave functions $\psi_{\mathbf{p}}$, and not the pseudo-wave functions $\varphi_{\mathbf{p}}$]

$$\rho(\mathbf{r}) = \sum_{\mathbf{p}} g_{\mathbf{p}} \psi_{\mathbf{p}}(\mathbf{r})^* \psi_{\mathbf{p}}(\mathbf{r}). \tag{26.61}$$

This is to be evaluated correct to first order. With the expansion (26.28) of $\psi_{\mathbf{p}}$, and the explicit representation (26.13) of the plane waves, $\rho(\mathbf{r})$ is

$$\rho(\mathbf{r}) = V^{-1} \sum_{\mathbf{p}} g_{\mathbf{p}} \left\{ 1 + \sum_{\mathbf{q}}{}' \, [A(\mathbf{q},\mathbf{p})^* e^{-i\mathbf{q}\cdot\mathbf{r}} + A(\mathbf{q},\mathbf{p}) e^{i\mathbf{q}\cdot\mathbf{r}}] \right\}. \tag{26.62}$$

In the first-order terms in (26.62) the $\Sigma_{\mathbf{p}}$ can be taken over the zeroth-order ground state, which is the sphere $|\mathbf{p}| \leq f$. Then $g_{\mathbf{p}} = g_{-\mathbf{p}}$, and by taking $\mathbf{p} \to -\mathbf{p}$ and $\mathbf{q} \to -\mathbf{q}$ inside the sums, it follows with (26.26) that $A(\mathbf{q},\mathbf{p})^* e^{-i\mathbf{q}\cdot\mathbf{r}} \to A(\mathbf{q},\mathbf{p}) e^{i\mathbf{q}\cdot\mathbf{r}}$, and $\rho(\mathbf{r})$ becomes

$$\rho(\mathbf{r}) = V^{-1} \sum_{\mathbf{p}} g_{\mathbf{p}} \left[1 + 2 \sum_{\mathbf{q}}{}' A(\mathbf{q},\mathbf{p}) e^{i\mathbf{q}\cdot\mathbf{r}} \right]. \tag{26.63}$$

The Fourier components $\rho(\mathbf{q})$ are then written down by inspection of (26.63):

$$\rho_0 = \rho(\mathbf{q} = 0) = V^{-1} \sum_{\mathbf{p}} g_{\mathbf{p}} = znN_0/V = z/V_A; \tag{26.64}$$

$$\rho(\mathbf{q}) = 2V^{-1} \sum_{\mathbf{p}} g_{\mathbf{p}} A(\mathbf{q},\mathbf{p}), \quad \mathbf{q} \neq 0. \tag{26.65}$$

In order to evaluate (26.65) it is useful first to find the zeroth-order ground-state sum of energy denominators, defined by

$$\sum_{\mathbf{p}} g_{\mathbf{p}}(e_{\mathbf{p}} - e_{\mathbf{p}+\mathbf{q}})_{\mathscr{P}}^{-1} = (2V/8\pi^3) \int_0^{\mathbf{f}} (e_{\mathbf{p}} - e_{\mathbf{p}+\mathbf{q}})_{\mathscr{P}}^{-1} \, d\mathbf{p}. \tag{26.66}$$

The angle integral is carried out in (26.55), and the remaining integral on p is evaluated to obtain

$$\sum_{\mathbf{p}} g_{\mathbf{p}}(e_{\mathbf{p}} - e_{\mathbf{p}+\mathbf{q}})_{\mathscr{P}}^{-1} = (q^2 V/8\pi e^2)[1 - H(q)], \tag{26.67}$$

where $H(q)$ is the static Hartree dielectric function, given by

$$H(q) - 1 = \frac{2me^2 f}{\pi \hbar^2 q^2} \left[\frac{1 - (q/2f)^2}{2(q/2f)} \ln \left| \frac{1 + (q/2f)}{1 - (q/2f)} \right| + 1 \right]. \tag{26.68}$$

Now with $A(\mathbf{q},\mathbf{p})$ given by (26.26), the zeroth-order evaluation of (26.65) for $\rho(\mathbf{q})$ is expressed by means of (26.67) as

$$\rho(\mathbf{q}) = (q^2/4\pi e^2) W(\mathbf{q})[1 - H(q)], \quad \mathbf{q} \neq 0. \tag{26.69}$$

Components $W_S(\mathbf{q})$ of the Hartree potential may now be written down by means of Poisson's equation, which requires [see, e.g., (21.34)]

$$W_S(\mathbf{q}) = (4\pi e^2/q^2)\rho(\mathbf{q}), \quad \mathbf{q} \neq 0; \tag{26.70}$$

then eliminating $\rho(\mathbf{q})$ by (26.69) gives

$$W_S(\mathbf{q}) = W(\mathbf{q})[1 - H(q)]. \tag{26.71}$$

If now the total W is just $W_B + W_S$, then (26.71) may be solved for $W(\mathbf{q})$ in terms of $W_B(\mathbf{q})$ to find $W(\mathbf{q}) = W_B(\mathbf{q})/H(q)$, and comparison of this with the general result (21.86) shows that $H(q)$ plays the role of a dielectric function.

The remarkable simplicity of the local pseudopotential perturbation theory is contained essentially in (26.71). The two important points are that the dielectric function $d(\mathbf{q},\mathbf{q}')$ is here approximated by a diagonal function $H(q)\delta_{\mathbf{q}\mathbf{q}'}$, and that the linear relation between $W(\mathbf{q})$ and $W_S(\mathbf{q})$ is independent of the positions of the ions. Since W_B is a sum of rigid single ion potentials, (26.71) implies that $W_B + W_S$ is also a sum of screened single ion potentials.

A characteristic property of metals is that the dielectric function diverges as $\mathbf{q} \rightarrow 0$; in fact the small q expansion of (26.68) correct to order q^0 is

$$H(q) - 1 = \frac{4me^2 f}{\pi\hbar^2 q^2} - \frac{me^2}{3\pi\hbar^2 f} + \cdots . \tag{26.72}$$

Also, because just the electrons in the zeroth-order Fermi sphere contribute to the screening, the function $H(q)$ has a singular behavior at $q = 2f$. The singular part of (26.68) is

$$[1 - (q/2f)][1 + (q/2f)] \ln |[1 + (q/2f)]/[1 - (q/2f)]|, \tag{26.73}$$

and this function vanishes at $q = 2f$, but its derivatives with respect to q have singularities at $q = 2f$. The leading term in the large q expansion of (26.68) is given by

$$H(q) - 1 = (16me^2 f^3/3\pi\hbar^2 q^4) + \cdots , \tag{26.74}$$

so $H(q) - 1$ converges as q^{-4} at large q.

Up to this point we have considered interactions between the conduction electrons and the ions, and Coulomb interactions among the electrons in a Hartree approximation. We now assume that all other effective interactions among the electrons, namely exchange and correlation effects, are included in a local one-electron potential W_X, and that this potential is adequately represented by an approximation of the type shown in (22.45), so that

$$W_S(\mathbf{q}) + W_X(\mathbf{q}) = W_{SX}(\mathbf{q}) = W_S(\mathbf{q})[1 - Y(q)], \tag{26.75}$$

where now this equation holds for arbitrary positions of the ions, and $Y(q)$ is independent of the positions of the ions. In addition, $Y(q)$ will be expressed by an interpolation approximation of the form of (22.56),

$$Y(q) = \frac{q^2}{2(q^2 + \xi f^2)}, \tag{26.76}$$

where ξ is a parameter to be determined so that $Y(q)$ has the correct behavior as $q \rightarrow 0$. This $q = 0$ limit is obtained from the appropriate energy of a uniform electron gas; in fact if the electron density ρ is a constant ρ_0, then (22.50) holds where now $X(\rho_0)$ is the total exchange and correlation energy per electron. Thus from (22.50) and (26.76),

$$\frac{\partial^2[\rho_0 X(\rho_0)]}{\partial \rho_0^2} = \lim_{q\rightarrow 0} -\frac{4\pi e^2}{q^2} Y(q) = -\frac{2\pi e^2}{\xi f^2}, \tag{26.77}$$

and this equation determines ξ.

For an electron gas of constant density ρ_0, the exchange energy per electron e_X is [see the text following (22.51)]

$$e_X = -\frac{3e^2}{4}\left(\frac{3\rho_0}{\pi}\right)^{1/3} = -\frac{3e^2 f}{4\pi} = -\frac{0.916}{r_s}\,\text{Ry}, \tag{26.78}$$

where we used the relation (26.47) between ρ_0 and f^3, and where r_s is the radius of the sphere whose volume is the volume per electron:

$$\tfrac{4}{3}\pi r_s^{\,3} = V_A/z = \rho_0^{-1}. \tag{26.79}$$

In (26.78), r_s is in units of the Bohr radius $a_0 = \hbar^2/me^2$. For a uniform electron gas, the correlation energy per electron e_{C} may be approximated by the Pines–Nozières formula*

$$e_{\mathrm{C}} = -(0.115 - 0.031 \ln r_s)\ \mathrm{Ry}, \tag{26.80}$$

with r_s in a_0. Now if we include only the exchange energy, then $X(\rho_0) = e_{\mathrm{X}}$, and (26.77) leads to

$$\xi = 2, \quad \text{exchange only.} \tag{26.81}$$

Incidentally, with $\xi = 2$, then (26.76) for $Y(q)$ is the same as (22.56), as it should be. With exchange and correlation both included, we take $X(\rho_0) = e_{\mathrm{X}} + e_{\mathrm{C}}$, and use (26.77) to obtain

$$\xi = 0.916/(0.458 + 0.012 r_s), \quad \text{exchange plus correlation.} \tag{26.82}$$

By combining (26.71) and (26.75), the total screening potential may be written

$$W_{\mathrm{SX}}(\mathbf{q}) = W(\mathbf{q})[1 - H(q)][1 - Y(q)], \tag{26.83}$$

and since $W(\mathbf{q}) = W_{\mathrm{B}}(\mathbf{q}) + W_{\mathrm{SX}}(\mathbf{q})$, (26.83) can be solved for $W(\mathbf{q})$ in terms of $W_{\mathrm{B}}(\mathbf{q})$ to obtain

$$W(\mathbf{q}) = W_{\mathrm{B}}(\mathbf{q})\{1 + [H(q) - 1][1 - Y(q)]\}^{-1}. \tag{26.84}$$

The modified dielectric function is therefore $1 + [H(q) - 1][1 - Y(q)]$. Further, with (26.35) for the components $W_{\mathrm{B}}(\mathbf{q})$, (26.84) becomes

$$W(\mathbf{q}) = S(\mathbf{q})w(q), \tag{26.85}$$

where the screened ion form factor $w(q)$ is given in terms of the bare ion form factor $w_{\mathrm{B}}(q)$ by

$$w(q) = w_{\mathrm{B}}(q)\{1 + [H(q) - 1][1 - Y(q)]\}^{-1}. \tag{26.86}$$

An important property of the $w(q)$, which follows from the fact that the bare ion potential $w_{\mathrm{B}}(r)$ is Coulombic at large distances r, is

$$\lim_{q\to 0} w(q) = -\tfrac{2}{3}e_{\mathrm{F}}, \tag{26.87}$$

where $e_{\mathrm{F}} = \hbar^2 f^2/2m$. This limit is easily calculated by noting $Y(q) \to 0$ as $q \to 0$, and by using (26.40) for the Coulomb part of $w_{\mathrm{B}}(q)$ and (26.72) for $H(q)$ at small q.

* D. Pines and P. Nozières, *The Theory of Quantum Liquids*, W. A. Benjamin Inc., New York, 1966, Volume I.

Now (26.87) reveals the major weakness in treating the pseudopotential as a perturbation; obviously at small $\mathbf{q}$, $w(q)$ is not small compared to the electronic kinetic energy, which is of the order of e_F. For certain applications, as for example, in calculating the static lattice potential, components of the pseudopotential are required only at inverse-lattice vectors $\mathbf{Q} \neq 0$, where $w(q)$ is indeed small. On the other hand, lattice-dynamics calculations depend on $w(q)$ at all values of q.

Finally we note that $w(q)$ is independent of the positions $\tilde{\mathbf{R}}(N\nu)$ of the ions, and the dependence of $W(\mathbf{q})$ on the $\tilde{\mathbf{R}}(N\nu)$ is contained entirely in the structure factor $S(\mathbf{q})$ in (26.85). In fact the total screened potential $W(\mathbf{r})$ is still a sum of screened single ion potentials, each moving rigidly with its ion; that is,

$$W(\mathbf{r}) = \sum_{N\nu} w(\mathbf{r} - \tilde{\mathbf{R}}(N\nu)), \tag{26.88}$$

where from (26.85) it follows that

$$w(\mathbf{r} - \tilde{\mathbf{R}}(N\nu)) = (nN_0)^{-1} \sum_{\mathbf{q}} w(q) e^{i\mathbf{q}\cdot[\mathbf{r} - \tilde{\mathbf{R}}(N\nu)]}. \tag{26.89}$$

27. TOTAL ADIABATIC POTENTIAL

Conduction-Electron Ground-State Energy

Here we conclude the calculation of the total ground-state energy E_G of the conduction electrons, correct to second order in the pseudopotential. The one-electron energy $\epsilon_{\mathbf{p}}$ is written in (26.32), and the ground-state sum of $\epsilon_{\mathbf{p}}$ is

$$\sum_{\mathbf{p}} g_{\mathbf{p}} \epsilon_{\mathbf{p}} = \sum_{\mathbf{p}} g_{\mathbf{p}} e_{\mathbf{p}} + znN_0 W(\mathbf{q} = 0) + \sum_{\mathbf{q}}{}' W(\mathbf{q}) W(-\mathbf{q}) \sum_{\mathbf{p}} g_{\mathbf{p}} (e_{\mathbf{p}} - e_{\mathbf{p}+\mathbf{q}})_P^{-1}, \tag{27.1}$$

where znN_0 is the total number of conduction electrons, and $|W(\mathbf{q})|^2$ has been written $W(\mathbf{q})W(-\mathbf{q})$. The first term on the right of (27.1) is just the total electronic kinetic energy, and according to (26.60) it is $\frac{3}{5} znN_0 e_F$, correct to second order.

The corrections for double counting, and the proper removal of divergent $\mathbf{q} = 0$ terms, were discussed in Sections 21–23, and we briefly outline those procedures for the pseudopotential perturbation theory. The total electron–electron Coulomb energy, which has been double counted in (27.1), may be written

$$\tfrac{1}{2} \int \rho(\mathbf{r}) W_S(\mathbf{r})\, d\mathbf{r} = \tfrac{1}{2} V \sum_{\mathbf{q}} \rho(\mathbf{q}) W_S(-\mathbf{q}) \rightarrow \tfrac{1}{2} V \sum_{\mathbf{q}}{}' \rho(\mathbf{q}) W_S(-\mathbf{q}), \tag{27.2}$$

where the $\mathbf{q} = 0$ term has been dropped since it is one of the divergent electrostatic terms which sum to zero. The total double counting correction for the exchange and correlation energies is given by (22.16), which may be written in the form

$$\int \rho(X - W_X)\, d\mathbf{r} = -\int \rho^2(\partial X/\partial \rho)\, d\mathbf{r}, \tag{27.3}$$

where $W_X = \partial(\rho X)/\partial \rho$ according to (22.13), and where X and W_X are functions only of ρ. Now the electronic density $\rho(\mathbf{r})$ may be expanded in powers of the pseudopotential, as

$$\rho = \rho_0 + \rho_1 + \rho_2, \tag{27.4}$$

where ρ_0 is the constant given by (26.64), the Fourier components of ρ_1 are given by (26.65), and the explicit expression for ρ_2 will not be required here. Then the functions X, W_X, and $\rho^2(\partial X/\partial \rho)$ may be expanded in powers of the pseudopotential, through the expansion (27.4) of ρ, just as in the expansions of these functions in powers of the displacements of the ions in Section 22. It is convenient to express (27.3) as a Fourier sum, and from the left-hand side it is obvious that the $\mathbf{q} = 0$ component is $V\rho_0[X(\mathbf{q} = 0) - W_X(\mathbf{q} = 0)]$; note that this expression is correct to all orders in the pseudopotential expansion of the functions X and W_X. The $\mathbf{q} \neq 0$ components of the right-hand side of (27.3) may be written as $-\frac{1}{2}\int \rho_1 W_{X1}\, d\mathbf{r}$, just as in (22.24), so that (27.3) correct to second order in the pseudopotential is

$$V\rho_0[X(\mathbf{q} = 0) - W_X(\mathbf{q} = 0)] - \tfrac{1}{2}V \sum_{\mathbf{q}}{}' \rho(\mathbf{q}) W_X(-\mathbf{q}). \tag{27.5}$$

Note that this result is valid for an arbitrary form of the exchange and correlation potential $W_X(\rho)$. The total double counting correction, which is to be added to (27.1), is (27.5) minus (27.2):

$$V\rho_0[X(\mathbf{q} = 0) - W_X(\mathbf{q} = 0)] - \tfrac{1}{2}V \sum_{\mathbf{q}}{}' \rho(\mathbf{q}) W_{SX}(-\mathbf{q}). \tag{27.6}$$

Let us now sum the $\mathbf{q} = 0$ contributions to the total energy. In (27.1), $W(\mathbf{q} = 0)$ is $W_B(\mathbf{q} = 0) + W_S(\mathbf{q} = 0) + W_X(\mathbf{q} = 0)$, and the ion Coulomb part of $W_B(\mathbf{q} = 0)$, as well as $W_S(\mathbf{q} = 0)$, are the divergent electrostatic terms to be omitted; this leaves $W_C(\mathbf{q} = 0) + W_X(\mathbf{q} = 0)$, where W_C is the core part defined by (26.39). Then adding the $\mathbf{q} = 0$ terms from (27.6), and noting $V\rho_0 = znN_0$, the sum of $\mathbf{q} = 0$ terms is

$$znN_0[W_C(\mathbf{q} = 0) + X(\mathbf{q} = 0)]. \tag{27.7}$$

The result $W_B(\mathbf{q}) = S(\mathbf{q})w_B(\mathbf{q})$, equation (26.35), holds separately for each additive part of $W_B(\mathbf{q})$, so that $W_C(\mathbf{q}) = S(\mathbf{q})w_C(q)$, and since $S(\mathbf{q} = 0) = 1$, then $W_C(\mathbf{q} = 0) = w_C(q = 0)$. Also $X(\mathbf{q} = 0)$ is just $X(\rho_0)$, the total exchange and correlation energy per electron for a uniform electron gas. Hence (27.7) may be rewritten

$$znN_0[w_C(q = 0) + e_X + e_C]. \tag{27.8}$$

The $\mathbf{q} \neq 0$ contributions to the total double counting correction are given by the Fourier sum in (27.6), and with (26.65) for $\rho(\mathbf{q})$ and (26.26) for $A(\mathbf{q},\mathbf{p})$, that Fourier sum is

$$-\sum_{\mathbf{q}}' W(\mathbf{q})W_{SX}(-\mathbf{q}) \sum_{\mathbf{p}} g_{\mathbf{p}}(e_{\mathbf{p}} - e_{\mathbf{p+q}})_P^{-1}. \tag{27.9}$$

When this is added to the Fourier sum in (27.1), the result is

$$\sum_{\mathbf{q}}' W(\mathbf{q})W_B(-\mathbf{q}) \sum_{\mathbf{p}} g_{\mathbf{p}}(e_{\mathbf{p}} + e_{\mathbf{p+q}})_P^{-1}, \tag{27.10}$$

since $W(-\mathbf{q}) = W_B(-\mathbf{q}) + W_{SX}(-\mathbf{q})$. Since (27.10) is second order in the pseudopotential, the $\Sigma_{\mathbf{p}}$ may be replaced by its zeroth-order evaluation (26.67), and (27.10) is

$$(V/8\pi e^2) \sum_{\mathbf{q}}' q^2 W(\mathbf{q})W_B(-\mathbf{q})[1 - H(q)]. \tag{27.11}$$

Finally we can use (26.84) to replace $W(\mathbf{q})$ in terms of $W_B(\mathbf{q})$, and (26.35) to write $W_B(\mathbf{q}) = S(\mathbf{q})w_B(q)$, and we can express (27.11) in the form

$$nN_0 \sum_{\mathbf{q}}' S(\mathbf{q})S(-\mathbf{q})F(q), \tag{27.12}$$

where $F(q)$ is the energy wave-number characteristic, given by

$$F(q) = \frac{q^2 V_A}{8\pi e^2} \frac{[w_B(q)]^2[1 - H(q)]}{1 + [H(q) - 1][1 - Y(q)]}. \tag{27.13}$$

The quantity (27.12) is called the band-structure energy, and represents the second-order pseudopotential correction to the energy of the conduction electrons. We note that $F(q)$ depends only on the magnitude q, as a consequence of assuming central bare ion potentials, and that $F(q) < 0$ since $H(q) - 1 > 0$.

The total conduction-electron energy is the sum of the kinetic energy in (27.1), the $\mathbf{q} = 0$ terms in (27.8), and the band-structure energy:

$$E_G = znN_0[\tfrac{3}{5}e_F + e_X + e_C + w_C(q = 0)] + nN_0 \sum_{\mathbf{q}}' S(\mathbf{q})S(-\mathbf{q})F(q). \tag{27.14}$$

This result holds for arbitrary positions of the ions, and hence may be applied to liquids as well as crystals; the dependence of E_G on the ion positions is contained in the structure factor $S(\mathbf{q})$.

As it might be expected, the band-structure energy can be expressed as a sum of central potentials among the ions, plus a volume-dependent term. With $S(\mathbf{q})$ given by (26.36), it is immediately possible to rewrite (27.12) in the form

$$\tfrac{1}{2} \sum_{\mathrm{MN}\mu\nu}{}' v(|\tilde{\mathbf{R}}(\mathrm{M}\mu) - \tilde{\mathbf{R}}(\mathrm{N}\nu)|) + \sum_{\mathbf{q}}{}' F(q), \tag{27.15}$$

where

$$v(R) = 2(nN_0)^{-1} \sum_{\mathbf{q}}{}' F(q) e^{i\mathbf{q}\cdot\mathbf{R}} = 2(nN_0)^{-1} \sum_{\mathbf{q}}{}' F(q)(qR)^{-1}\sin(qR). \tag{27.16}$$

There are two points which should be noted in connection with representing the band-structure energy in terms of central potentials. First, $v(R)$ is a long-range potential, having the Friedel oscillatory behavior $R^{-3}\cos(qR)$ at large R, and because of the slow convergence of the lattice sum in (27.15), it is not at all convenient to use the central potential formulation in lattice-statics and lattice-dynamics calculations. Second, in addition to the sum of central potentials, the band-structure energy (27.15) contains the term $\Sigma_{\mathbf{q}}' F(q)$, which depends on the volume of the crystal.

Ion–Ion Interactions

The Coulomb interactions in the system composed of point ions in a uniform compensating background charge were discussed in Section 23. In the theory of the present chapter, all ions are the same and each has charge $+ze$, so the general expression (23.56) for Ω_{I} is simplified to

$$\begin{aligned} \Omega_{\mathrm{I}}(\text{point ions}) = {} & nN_0(2\pi z^2 e^2/V_A) \sum_{\mathbf{q}}{}' S(\mathbf{q})S(-\mathbf{q})q^{-2}e^{-q^2 4\eta^2} \\ & + \tfrac{1}{2}z^2e^2 \sum_{\mathrm{MN}\mu\nu}{}' |\tilde{\mathbf{R}}(\mathrm{M}\mu) - \tilde{\mathbf{R}}(\mathrm{N}\nu)|^{-1}\,\mathrm{erfc}[\eta\,|\tilde{\mathbf{R}}(\mathrm{M}\mu) - \tilde{\mathbf{R}}(\mathrm{N}\nu)|] \qquad (27.17) \\ & - nN_0 z^2 e^2[(\eta/\sqrt{\pi}) + (\pi/2\eta^2 V_A)]. \end{aligned}$$

In transforming (23.56) to (27.17), we used (26.36) for $S(\mathbf{q})$, and noted $V = nN_0V_A$ and $V_{\mathrm{C}} = nV_A$. The result (27.17) is valid for arbitrary positions $\tilde{\mathbf{R}}(\mathrm{N}\nu)$ of the ions, and is independent of the parameter η. For the equilibrium configuration of the crystal, $S(\mathbf{q})$ contains a $\delta(\mathbf{q})$ and the $\Sigma_{\mathbf{q}}'$ in (27.17) reduces to a $\Sigma_{\mathbf{Q}}'$. It is useful to calculate Ω_{I}(point ions) for the equilibrium configuration of a crystal, and to tabulate the results as a function of the macroscopic configuration. Let us write Ω_{I}(point ions) in the form

$$(nN_0)^{-1}\Omega^{\mathrm{I}}(\text{point ions}) = -\alpha_{\mathrm{I}}(z^2/r_A)\ \mathrm{Ry}, \tag{27.18}$$

Table 3. Values of α_{I}, defined by (27.18), for simple lattices. For the hcp lattice, α_{I} is maximum at or very near to the ideal c/a ratio of $\sqrt{8/3}$.

Lattice	α_{I}
fcc	1.79172
bcc	1.79186
hcp, c/a = 1.4	1.78664
1.5	1.78997
1.6	1.79156
$\sqrt{8/3}$	1.791676
1.7	1.79128
1.8	1.78909

where r_A is the radius of the sphere whose volume is the volume per atom,

$$\tfrac{4}{3}\pi r_A^3 = V_A, \tag{27.19}$$

and (27.18) is the point-ion Coulomb energy per ion, with r_A in units of a_0. Then α_{I} is a constant for each primitive cubic lattice, and α_{I} depends only on the c/a ratio for the hcp lattice. Some values of α_{I} are listed in Table 3.

As we noted in the text following (26.45), the ion cores are large for the closed shell ions on the left side of the periodic table, and these cores may overlap one another to a significant degree as the ions vibrate in the crystal. Let us represent this effect by including a Born–Mayer repulsive central potential between the ions, of the form $\alpha_{\mathrm{B}} e^{-\gamma_{\mathrm{B}} R}$ when two ions are separated by a distance R, so that the total ion–ion interaction energy also contains the contribution

$$\Omega_{\mathrm{I}}(\text{Born–Mayer}) = \tfrac{1}{2}\alpha_{\mathrm{B}} \sum_{MN\mu\nu}{}' e^{-\gamma_{\mathrm{B}}|\tilde{\mathbf{R}}(M\mu)-\tilde{\mathbf{R}}(N\nu)|}. \tag{27.20}$$

Here α_{B} and γ_{B} are positive quantities, and they may be considered as adjustable parameters, or they may be determined by some appropriate calculation.

In all our calculations we have not considered explicitly the zero of energy for the system of interacting ions and electrons; in thermodynamics calculations this zero is customarily taken to be the energy of the free atoms all infinitely separated. Now in the explicit pseudopotential models discussed in Section 26, the local bare ion potential $w_{\mathrm{B}}(r)$ was taken to be the energy of an electron in the presence of the ion, and the zero of $w_{\mathrm{B}}(r)$ corresponds to the ion and electron at infinite separation. Hence the zero of energy for the total system corresponds to all the ions and electrons at infinite separation, and this energy lies above the energy when all atoms are at infinite separation by the total ionization energy of the system. If the ionization energy of each ion is I_z, the energy required to remove the z outer electrons to infinity, then we need to add the total ionization energy nN_0I_z so that our system energy is measured relative to the free atoms at infinite separation. Thus the total adiabatic potential in the present formulation is

$$\Phi = nN_0I_z + E_G + \Omega_I(\text{point ions}) + \Omega_I(\text{Born–Mayer}). \quad (27.21)$$

In the band-structure theory of Chapter 5, the total ionization energy is considered part of the bare ion potential Ω_B; the explicit appearance of nN_0I_z in (27.21) corresponds to the customary definition of w_B as the energy of an electron in the presence of a free ion.

When all the ions are at their equilibrium positions $\mathbf{R}(N\nu)$, the structure factor (26.36) reduces to

$$S(\mathbf{q}) = (nN_0)^{-1}\sum_{N\nu} e^{-i\mathbf{q}\cdot\mathbf{R}(N\nu)} = \delta(\mathbf{q})n^{-1}\sum_{\nu} e^{-i\mathbf{q}\cdot\mathbf{R}(\nu)}, \quad (27.22)$$

and this requires $\mathbf{q} = \mathbf{Q}$ for any inverse-lattice vector $\mathbf{Q}$. By gathering up the contributions to (27.21), the total adiabatic potential for the equilibrium configuration is conveniently written

$$\begin{aligned}\Phi_0 = nN_0\{I_z + z[\tfrac{3}{5}e_F + e_X + e_C + w_C(q=0)]\} \\ + nN_0\sum_{\mathbf{Q}}{}' \left| n^{-1}\sum_{\nu} e^{-i\mathbf{Q}\cdot\mathbf{R}(\nu)}\right|^2 F(Q) - nN_0\alpha_I(z^2/r_A) \quad (27.23) \\ + \tfrac{1}{2}N_0\alpha_B\sum_{N\mu\nu}{}' e^{-\gamma_B|\mathbf{R}(\mu)-\mathbf{R}(N\nu)|}.\end{aligned}$$

Here surface effects have been eliminated, and the strain and volume derivatives of Φ_0 are easily calculated.

For arbitrary positions of the ions, it is convenient to regroup the terms in the total adiabatic potential, and express Φ as a Fourier sum, a sum of central potentials, and a term which depends only on the volume of the crystal. The energy E_G is given by (27.14), and the two contributions to Ω_I are given by (27.17) and (27.20), so that the total Φ may be written

$$\begin{aligned}\Phi = nN_0\sum_{\mathbf{q}}{}' S(\mathbf{q})S(-\mathbf{q})G(q) + \tfrac{1}{2}\sum_{MN\mu\nu}{}' \varphi(|\tilde{\mathbf{R}}(M\mu) - \tilde{\mathbf{R}}(N\nu)|) \quad (27.24) \\ + nN_0\{I_z + z[\tfrac{3}{5}e_F + e_X + e_C + w_C(q=0)] - z^2e^2[(\eta/\sqrt{\pi}) + (\pi/2\eta^2V_A)]\}.\end{aligned}$$

Here $G(q)$ contains contributions from the band-structure energy and from the Fourier sum in Ω_I(point ions),

$$G(q) = F(q) + (2\pi z^2e^2/V_A)q^{-2}e^{-q^2/4\eta^2}; \quad (27.25)$$

the central potential contains contributions from the point-ion and Born–Mayer parts of Ω_{I},

$$\varphi(R) = z^2e^2R^{-1}\,\mathrm{erfc}(\eta R) + \alpha_{\mathrm{B}}e^{-\gamma_{\mathrm{B}}R}; \tag{27.26}$$

and the term in braces in (27.24) depends only on the volume of the crystal.

Homogeneous Deformation

It is instructive to illustrate some of the procedures involved in calculating the elastic constants. We consider a cubic crystal in the presence of isotropic pressure P, and introduce the set of six nonsymmetric deformation parameters of Fuchs,* ϵ_x, ϵ_y, ϵ_z, γ_{xy}, γ_{yz}, and γ_{zx}. These are infinitesimal deformation parameters, and they correspond to the displacement gradients u_{ij} of (2.6) having the following values.

$$\begin{gathered} u_{xx} = \epsilon_x, \qquad u_{yy} = \epsilon_y, \qquad u_{zz} = \epsilon_z; \\ u_{xy} = \gamma_{xy}, \qquad u_{yz} = \gamma_{yz}, \qquad u_{zx} = \gamma_{zx}; \\ u_{yx} = u_{zy} = u_{xz} = 0. \end{gathered} \tag{27.27}$$

To second order in the displacement gradients, the free energy expansion (3.15) is

$$F(\mathbf{x},T) = F(\mathbf{X},T) + V\sum_{ij}\tau_{ij}u_{ij} + \tfrac{1}{2}V\sum_{ijkl}A^T_{ijkl}u_{ij}u_{kl}. \tag{27.28}$$

For isotropic pressure the stress is

$$\tau_{ij} = -P\delta_{ij}, \tag{27.29}$$

and the isothermal wave-propagation coefficients are given by (3.30),

$$A^T_{ijkl} = -P\delta_{jl}\delta_{ik} + C^T_{ijkl}. \tag{27.30}$$

Now for a cubic crystal there are only three independent second-order isothermal elastic constants, namely C^T_{11}, C^T_{12}, and C^T_{44}, in Voigt notation. With these three independent elastic constants, there are only four independent A^T_{ijkl}, and it is convenient to express these A^T_{ijkl} in terms of the stress–strain coefficients B^T_{11}, B^T_{12}, and B^T_{44}, with the aid of (3.38):

$$\begin{aligned} A^T_{iiii} &= A^T_{1111} = C^T_{11} - P = B^T_{11}, &&\text{for all } i; \\ A^T_{iijj} &= A^T_{1122} = C^T_{12} = B^T_{12} - P, &&\text{for } i \neq j; \\ A^T_{ijij} &= A^T_{1212} = C^T_{44} - P = B^T_{44}, &&\text{for } i \neq j; \\ A^T_{ijji} &= A^T_{1221} = C^T_{44} = B^T_{44} + P, &&\text{for } i \neq j. \end{aligned} \tag{27.31}$$

* K. Fuchs, *Proc. Roy. Soc. (London)* **A153,** 622 (1936); **A157,** 444 (1936).

All other A^T_{ijkl} vanish, and with the coefficients (27.29) and (27.31) and the strains (27.27), the free energy expansion (27.28) simplifies to

$$
\begin{aligned}
F(\mathbf{x},T) = F(\mathbf{X},T) &- VP(\epsilon_x + \epsilon_y + \epsilon_z) \\
&+ \tfrac{1}{2}V[A^T_{1111}(\epsilon_x^2 + \epsilon_y^2 + \epsilon_z^2) + 2A^T_{1122}(\epsilon_x\epsilon_y + \epsilon_y\epsilon_z + \epsilon_z\epsilon_x) \\
&+ A^T_{1212}(\gamma_{xy}^2 + \gamma_{yz}^2 + \gamma_{zx}^2)].
\end{aligned} \tag{27.32}
$$

To calculate the three independent $C^T_{\alpha\beta}$, or the $B^T_{\alpha\beta}$, we follow Fuchs and consider three particular deformations, as described below.

1. Expansion of the crystal by the relative amount ϵ_x in the x direction, with a sufficient compression ϵ_y in the y direction so as to conserve volume: In this case the deformation parameters are written

$$
(1 + \epsilon_y) = 1/(1 + \epsilon_x), \quad \epsilon_y = -\epsilon_x + \epsilon_x^2; \qquad \epsilon_z = \gamma_{xy} = \gamma_{yz} = \gamma_{zx} = 0. \tag{27.33}
$$

Then the free energy expansion (27.32) becomes

$$
F(\mathbf{x},T) = F(\mathbf{X},T) - VP\epsilon_x^2 + V(A^T_{1111} - A^T_{1122})\epsilon_x^2, \tag{27.34}
$$

and it follows that

$$
\tfrac{1}{2}V^{-1}(\partial^2 F/\partial\epsilon_x^2)_T = -P + A^T_{1111} - A^T_{1122} = B^T_{11} - B^T_{12}. \tag{27.35}
$$

2. A shear by the amount γ_{xy} in the xy plane, conserving volume: In this case the deformation parameters are all zero except for γ_{xy}, and the free energy expansion (27.32) is

$$
F(\mathbf{x},T) = F(\mathbf{X},T) + \tfrac{1}{2}VA^T_{1212}\gamma_{xy}^2, \tag{27.36}
$$

so that

$$
V^{-1}(\partial^2 F/\partial\gamma_{xy}^2)_T = A^T_{1212} = B^T_{44}. \tag{27.37}
$$

3. For the third independent strain we take simply a uniform volume change, and use the results of Sections 1 and 2 to write

$$
P = -(\partial F/\partial V)_T, \tag{27.38}
$$

$$
B_T = V(\partial^2 F/\partial V^2)_T = \tfrac{1}{3}(B^T_{11} + 2B^T_{12}). \tag{27.39}
$$

Let us now turn to the evaluation of the pressure and the stress–strain coefficients in the pseudopotential theory. This calculation will be restricted to a primitive cubic lattice, and to the potential approximation $\Phi_0 \approx F$. A convenient way to include homogeneous deformations of the primitive cubic lattice is to write down the total static potential Φ_0 for an arbitrary primitive lattice; then since $S(\mathbf{q}) = \delta(\mathbf{q})$ for *any* primitive lattice, we have from (27.24)

$$\Phi_0 = N_0 \sum_{\mathbf{Q}}' G(Q) + \tfrac{1}{2}N_0 \sum_{\mathrm{N}}' \varphi(0,\mathrm{N}) + N_0\{I_z + z[\tfrac{3}{5}e_{\mathrm{F}} + e_{\mathrm{X}} + e_{\mathrm{C}} + w_{\mathrm{C}}(q=0)] - z^2e^2[(\eta/\sqrt{\pi}) + (\pi/2\eta^2 V_A)]\}. \tag{27.40}$$

In this expression the lattice vectors $\mathbf{R}(\mathrm{N})$ and corresponding inverse-lattice vectors $\mathbf{Q}$ represent an arbitrary primitive lattice, and $\varphi(0,\mathrm{N}) = \varphi(|\mathbf{R}(\mathrm{N})|)$, the same as our notation of Sections 9 and 15. An important point of the present illustration is that since $G(Q)$ depends only on the magnitude $Q = |\mathbf{Q}|$, the simplest way to evaluate and compute strain derivatives of $G(Q)$ is to use chain-rule differentiation to express the results in terms of derivatives of $G(Q)$ with respect to Q or Q^2. This is, of course, the procedure we have always followed in treating the central potentials $\varphi(R)$. Another important point is that for any macroscopic configuration (27.40) is independent of η, and η can be taken as strictly constant, or η can be taken to vary with the configuration, whichever is most convenient. Here η is taken to be constant.

The derivatives of $G(Q)$ and $\varphi(R)$ are conveniently abbreviated as

$$G'(Q) = \partial G/\partial Q^2, \qquad G''(Q) = \partial^2 G/\partial(Q^2)^2;$$
$$\varphi'(R) = d\varphi/dR^2, \qquad \varphi''(R) = d^2\varphi/d(R^2)^2. \tag{27.41}$$

The partial derivatives in the case of $G(Q)$ are used to indicate conditions of fixed volume, since $G(Q)$ contains V_A and the volume-dependent Fermi wave vector f. Now according to (27.33), when the primitive cubic lattice with lattice vectors $\mathbf{R}(\mathrm{N})$ and inverse-lattice vectors $\mathbf{Q}$ undergoes strain 1 above, the deformed lattice corresponds to $\bar{\mathbf{R}}(\mathrm{N})$ and $\bar{\mathbf{Q}}$ given by

$$\bar{R}_x(\mathrm{N}) = (1+\epsilon_x)R_x(\mathrm{N}), \qquad \bar{R}_y(\mathrm{N}) = (1+\epsilon_x)^{-1}R_y(\mathrm{N}), \qquad \bar{R}_z(\mathrm{N}) = R_z(\mathrm{N}); \tag{27.42}$$

$$\bar{Q}_x = (1+\epsilon_x)^{-1}Q_x, \qquad \bar{Q}_y = (1+\epsilon_x)Q_y, \qquad \bar{Q}_z = Q_z. \tag{27.43}$$

The derivatives of $|\bar{\mathbf{R}}(\mathrm{N})|^2$ and $|\bar{\mathbf{Q}}|^2$ with respect to ϵ_x are now calculated from these expressions; for example,

$$d\bar{Q}^2/d\epsilon_x = -2Q_x^2 + 2Q_y^2,$$
$$d^2\bar{Q}^2/d\epsilon_x^2 = 6Q_x^2 + 2Q_y^2, \tag{27.44}$$

evaluated at $\epsilon_x = 0$. Then with our usual notation $\tilde{B}_{\alpha\beta}$ for coefficients calculated in the potential approximation, (27.35) is written

$$\tilde{B}_{11} - \tilde{B}_{12} = \tfrac{1}{2}V^{-1}(d^2\Phi_0/d\epsilon_x^2), \tag{27.45}$$

and the differentiation of (27.40) for Φ_0 is carried out to give

$$\tilde{B}_{11} - \tilde{B}_{12} = 4V_A^{-1} \sum_{\mathbf{Q}}' [\tfrac{1}{3}Q^2G'(Q) + (Q_x^4 - Q_x^2Q_y^2)G''(Q)]$$
$$+ 2V_A^{-1} \sum_{\mathrm{N}}' \{\tfrac{1}{3}R(\mathrm{N})^2\varphi'(0,\mathrm{N}) + [R_x(\mathrm{N})^4 - R_x(\mathrm{N})^2R_y(\mathrm{N})^2]\varphi''(0,\mathrm{N})\}. \tag{27.46}$$

Since the shear strain 1 corresponds to fixed volume, the term in braces in (27.40) does not contribute to $d^2\Phi_0/d\epsilon_x^2$.

Under strain 2 above, the lattice and inverse-lattice vectors become

$$\bar{R}_x(\mathrm{N}) = R_x(\mathrm{N}) + \gamma_{xy}R_y(\mathrm{N}), \qquad \bar{R}_y(\mathrm{N}) = R_y(\mathrm{N}), \qquad \bar{R}_z(\mathrm{N}) = R_z(\mathrm{N}); \tag{27.47}$$

$$\bar{Q}_x = Q_x, \qquad \bar{Q}_y = Q_y - \gamma_{xy}Q_x, \qquad \bar{Q}_z = Q_z. \tag{27.48}$$

Then from (27.37), in the potential approximation,

$$\tilde{B}_{44} = V^{-1}(d^2\Phi_0/d\gamma_{xy}^2), \tag{27.49}$$

and the differentiation of (27.40) for Φ_0 is carried out to give

$$\tilde{B}_{44} = 2V_A^{-1} \sum_{\mathbf{Q}}' [\tfrac{1}{3}Q^2G'(Q) + 2Q_x^2Q_y^2G''(Q)]$$
$$+ V_A^{-1} \sum_{\mathrm{N}}' [\tfrac{1}{3}R(\mathrm{N})^2\varphi'(0,\mathrm{N}) + 2R_x(\mathrm{N})^2R_y(\mathrm{N})^2\varphi''(0,\mathrm{N})]. \tag{27.50}$$

For the volume derivatives of Φ_0, it is generally more convenient to use the form (27.23) of Φ_0, for detailed computations. However, we continue with the form (27.40) for the present illustration, and note that the term in braces contributes to the volume derivatives. The exchange and correlation energies per electron are written in (26.78) and (26.80), respectively, and in the same units we can write $\frac{3}{5}e_F = 2.21/r_s^2$. Further, because of the normalization of the electronic wave functions, $w_C(q = 0) = \hat{\beta}/V_A$ with $\hat{\beta}$ a constant. Then the explicit volume-dependence of the expression in braces in (27.40) is shown by writing

$$\tfrac{3}{5}e_F + e_X + e_C + w_C(q = 0)$$
$$= (2.21/r_s^2) - (0.916/r_s) \doteq 0.115 + 0.031 \ln r_s + (\hat{\beta}/V_A). \tag{27.51}$$

According to (27.25) for $G(q)$, together with (27.13) for $F(q)$, the quantity

$G(Q)$ depends on the volume through Q, through V_A, and through f which is contained in $F(q)$; note that each term in $G(Q)$ is proportional to V_A^{-1} since $w_B(Q)$ in $F(Q)$ is proportional to V_A^{-1}. With these explicit volume-dependences, and those shown in (27.51), the volume derivatives of Φ_0 are as follows.

$$\begin{aligned}\tilde{P} &= -d\Phi_0/dV \\ &= V_A^{-1}\Big\{\sum_{\mathbf{Q}}{}' \,[G(Q) + \tfrac{2}{3}Q^2G'(Q) + \Delta_1(Q)] - \tfrac{1}{3}\sum_{\mathrm{N}}{}' \,R(\mathrm{N})^2\varphi'(0,\mathrm{N}) \\ &\quad + z[\tfrac{2}{5}e_{\mathrm{F}} + \tfrac{1}{3}e_{\mathrm{X}} - 0.010\ \mathrm{Ry} + w_{\mathrm{C}}(q=0)] - z^2e^2(\pi/2\eta^2V_A)\Big\};\end{aligned} \tag{27.52}$$

$$\begin{aligned}\tilde{B} &= \tfrac{1}{3}(\tilde{B}_{11} + 2\tilde{B}_{12}) = V(d^2\Phi_0/dV^2) \\ &= V_A^{-1}\Big\{\sum_{\mathbf{Q}}{}'[2G(Q) + \tfrac{22}{9}Q^2G'(Q) + \tfrac{4}{9}Q^4G''(Q) + \Delta_2(Q)] \\ &\quad + \tfrac{1}{9}\sum_{\mathrm{N}}{}' \,[-R(\mathrm{N})^2\varphi'(0,\mathrm{N}) + 2R(\mathrm{N})^4\varphi''(0,\mathrm{N})] \\ &\quad + z[\tfrac{2}{3}e_{\mathrm{F}} + \tfrac{4}{9}e_{\mathrm{X}} - 0.010\ \mathrm{Ry} + 2w_{\mathrm{C}}(q=0)] - z^2e^2(\pi/\eta^2V_A)\Big\}.\end{aligned} \tag{27.53}$$

Here the terms $\Delta_1(Q)$ in $\tilde{P}$ and $\Delta_2(Q)$ in $\tilde{B}$ arise from the dependence of $F(Q)$ on the Fermi wave vector magnitude f, and since f is proportional to $V_A^{-1/3}$ according to (26.47), these terms are found to be

$$\Delta_1(Q) = \tfrac{1}{3}f\,[\partial F(Q)/\partial f], \tag{27.54}$$

$$\Delta_2(Q) = \tfrac{1}{9}\{10f\,[\partial F(Q)/\partial f] + f^2[\partial^2F(Q)/\partial f^2] + 2Qf\,[\partial^2F(Q)/\partial Q\partial f]\}. \tag{27.55}$$

In principle the volume-dependence of the parameter ξ of (26.82) is to be included in $\Delta_1(Q)$ and $\Delta_2(Q)$ as an f-dependence; in practice the volume-dependence of ξ can be neglected entirely.

The pressure and the three independent stress–strain coefficients, or the elastic constants, can easily be computed for a primitive cubic lattice from (27.46), (27.50), (27.52), and (27.53), and the extension to higher-order elastic constants is straightforward.

Lattice Dynamics

The expression (27.24) for Φ can be applied to any crystal composed of a single type of atom, and since the positions of the ions are arbitrary in (27.24), the potential energy coefficients of all orders may be calculated. Under conditions of constant crystal volume, the term in braces in (27.24) does not contribute to the potential coefficients, and one has to consider only the Fourier sum and the central potentials. We will illustrate the calculation of potential coefficients for the example of a primitive lattice.

The structure factor for a primitive lattice is

$$S(\mathbf{q}) = N_0^{-1} \sum_{\mathrm{M}} e^{-i\mathbf{q}\cdot[\mathbf{R}(\mathrm{M})+\mathbf{U}(\mathrm{M})]}, \tag{27.56}$$

where as usual $\mathbf{U}(\mathrm{M})$ are displacements of the ions from their equilibrium positions $\mathbf{R}(\mathrm{M})$. If the factor $S(\mathbf{q})S(-\mathbf{q})$ is expanded in powers of the $\mathbf{U}(\mathrm{M})$, and the resulting expansion of the Fourier sum in (27.24) is compared to the general expansion (6.1) of Φ, the Fourier sum contributions to the potential coefficients for a primitive lattice are found to be as follows, for $\mathrm{M} \neq \mathrm{N}$.

$$\Phi_i(\mathrm{M})_{\mathrm{F}} = 2i \sum_{\mathbf{Q}}{}' G(Q)Q_i = 0; \tag{27.57}$$

$$\Phi_{ij}(\mathrm{M,N})_{\mathrm{F}} = 2N_0^{-1} \sum_{\mathbf{q}}{}' G(q)q_i q_j \cos[\mathbf{q} \cdot [\mathbf{R}(\mathrm{M}) - \mathbf{R}(\mathrm{N})]]; \tag{27.58}$$

$$\Phi_{ijk}(\mathrm{M,M,N})_{\mathrm{F}} = -2N_0^{-1} \sum_{\mathbf{q}}{}' G(q)q_i q_j q_k \sin[q \cdot [\mathbf{R}(\mathrm{M}) - \mathbf{R}(\mathrm{N})]]; \tag{27.59}$$

$$\Phi_{ijkl}(\mathrm{M,M,M,N})_{\mathrm{F}} = -2N_0^{-1} \sum_{\mathbf{q}}{}' G(q)q_i q_j q_k q_l \cos[\mathbf{q} \cdot [\mathbf{R}(\mathrm{M}) - \mathbf{R}(\mathrm{N})]]. \tag{27.60}$$

We note that (27.57) vanishes because the inverse lattice is primitive, and for every $\mathbf{Q}$ there is a $-\mathbf{Q}$. In addition, the self-coupling coefficients also appear in the expansion of the Fourier sum, and these are found to be

$$\Phi_{ij}(\mathrm{M,M})_{\mathrm{F}} = 2 \sum_{\mathbf{q}}{}' G(q)q_i q_j [N_0^{-1} - \delta(\mathbf{q})]; \tag{27.61}$$

$$\Phi_{ijk}(\mathrm{M,M,M})_{\mathrm{F}} = 2i \sum_{\mathbf{q}}{}' G(q)q_i q_j q_k [3N_0^{-1} - \delta(\mathbf{q})] = 0; \tag{27.62}$$

$$\Phi_{ijkl}(\mathrm{M,M,M,M})_{\mathrm{F}} = -2 \sum_{\mathbf{q}}{}' G(q)q_i q_j q_k q_l [N_0^{-1} - \delta(\mathbf{q})]. \tag{27.63}$$

It is a straightforward calculation to verify that these contributions to the potential coefficients satisfy the translational and rotational invariance conditions. In fact, since the Fourier sum in (27.24) can be represented as a sum of central potentials plus a volume-dependent term, the coefficients (27.57)–(27.63) have the same properties as the potential coefficients corresponding to central potentials, and this guarantees their proper invariance. Finally, the remaining third- and fourth-order coefficients can be obtained from the ones given above by the relations [see (9.22) and (9.23)]

$$\Phi_{ijk}(\mathrm{M,M,N})_{\mathrm{F}} = -\Phi_{ijk}(\mathrm{M,N,N})_{\mathrm{F}}; \tag{27.64}$$

$$\Phi_{ijkl}(\mathrm{M,M,M,N})_{\mathrm{F}} = -\Phi_{ijkl}(\mathrm{M,M,N,N})_{\mathrm{F}} = \Phi_{ijkl}(\mathrm{M,N,N,N})_{\mathrm{F}}. \tag{27.65}$$

The contributions of the lattice sum in (27.24) to the potential coefficients may be written directly from the central potential results of Section 9; the first- and second-order coefficients evaluated in the interior of the crystal are

$$\Phi_i(0)_{\mathrm{L}} = -2 \sum_{\mathrm{N}}' \varphi'(0,\mathrm{N}) R_i(\mathrm{N}) = 0; \tag{27.66}$$

$$\Phi_{ij}(0,\mathrm{N})_{\mathrm{L}} = -2[\varphi'(0,\mathrm{N})\delta_{ij} + 2\varphi''(0,\mathrm{N}) R_i(\mathrm{N}) R_j(\mathrm{N})]. \tag{27.67}$$

Again (27.66) vanishes by symmetry, because for every $\mathbf{R}(\mathrm{N})$ there is a $-\mathbf{R}(\mathrm{N})$, and (27.67) holds for $\mathrm{N} \neq 0$, with the self-coupling coefficient given by the translational invariance condition:

$$\Phi_{ij}(0,0)_{\mathrm{L}} = -\sum_{\mathrm{N}}' \Phi_{ij}(0,\mathrm{N})_{\mathrm{L}}. \tag{27.68}$$

For a primitive lattice the general expression (10.39) for elements of the dynamical matrices, with surface effects eliminated, is

$$D_{ij}(\mathbf{k}) = M_{\mathrm{C}}^{-1} \sum_{\mathrm{N}} \Phi_{ij}(0,\mathrm{N}) e^{i\mathbf{k}\cdot\mathbf{R}(\mathrm{N})}. \tag{27.69}$$

From (27.58) and (27.61), we can write an equation for $\Phi_{ij}(0,\mathrm{N})_{\mathrm{F}}$ which includes $\mathrm{N} = 0$ as well as $\mathrm{N} \neq 0$, namely,

$$\Phi_{ij}(0,\mathrm{N})_{\mathrm{F}} = 2 \sum_{\mathbf{q}}' G(q) q_i q_j [N_0^{-1} e^{-i\mathbf{q}\cdot\mathbf{R}(\mathrm{N})} - \delta(\mathbf{q})\delta(\mathbf{R}(\mathrm{N}))]. \tag{27.70}$$

With this result, the Fourier sum contribution to (27.69) is seen to be

$$D_{ij}(\mathbf{k})_{\mathrm{F}} = 2M_{\mathrm{C}}^{-1} \sum_{\mathbf{q}}' G(q) q_i q_j [\delta(\mathbf{k} - \mathbf{q}) - \delta(\mathbf{q})]. \tag{27.71}$$

Now since $\mathbf{k}$ lies in the first Brillouin zone, $\mathbf{k}$ is never equal to an inverse-lattice vector $\mathbf{Q}$ except when $\mathbf{k} = 0$; therefore for $\mathbf{k} \neq 0$, the $\delta(\mathbf{k} - \mathbf{q})$ requires $\mathbf{q} = \mathbf{k} + \mathbf{Q}$ and this is never zero, so in the term containing $\delta(\mathbf{k} - \mathbf{q})$ in (27.71) the $\Sigma_{\mathbf{q}}'$ becomes a $\Sigma_{\mathbf{Q}}$. Then (27.71) is

$$D_{ij}(\mathbf{k})_{\mathrm{F}} = 2M_C^{-1} \sum_{\mathbf{Q}} G(|\mathbf{Q} + \mathbf{k}|)(\mathbf{Q} + \mathbf{k})_i (\mathbf{Q} + \mathbf{k})_j - 2M_C^{-1} \sum_{\mathbf{Q}}' G(Q) Q_i Q_j. \tag{27.72}$$

For $\mathbf{k} = 0$, the $\Sigma_{\mathbf{Q}}$ in (27.72) must be replaced by $\Sigma_{\mathbf{Q}}'$, and then $D_{ij}(0)_{\mathrm{F}}$ vanishes. In addition, with (27.25) for $G(Q)$ together with (27.13) for $F(Q)$, it is a straightforward calculation to show that $D_{ij}(\mathbf{k})_{\mathrm{F}}$ is proportional to $|\mathbf{k}|^2$ in the limit of small $|\mathbf{k}|$.

The lattice sum contributions to the dynamical matrices may be written from the general result (15.1) for a primitive lattice with central potentials:

$$D_{ij}(\mathbf{k})_{\mathrm{L}} = M_C^{-1} \sum_{\mathrm{N}}' [2\varphi'(0,\mathrm{N})\delta_{ij} + 4\varphi''(0,\mathrm{N}) R_i(\mathrm{N}) R_j(\mathrm{N})]\{1 - \cos[\mathbf{k}\cdot\mathbf{R}(\mathrm{N})]\}. \tag{27.73}$$

The appropriate central potential $\varphi(R)$ is (27.26), and the following derivatives are useful in evaluating (27.73) for any primitive lattice.

$$\varphi'(R) = -z^2e^2[(1/\sqrt{\pi})\eta R^{-2}e^{-\eta^2R^2} + \tfrac{1}{2}R^{-3}\operatorname{erfc}(\eta R)] - \tfrac{1}{2}\alpha_B\gamma_B R^{-1}e^{-\gamma_B R}; \tag{27.74}$$

$$\varphi''(R) = z^2e^2[(1/\sqrt{\pi})(\eta^3R^{-2} + \tfrac{3}{2}\eta R^{-4})e^{-\eta^2R^2} + \tfrac{3}{4}R^{-5}\operatorname{erfc}(\eta R)] + \tfrac{1}{4}\alpha_B\gamma_B(\gamma_B R^{-2} + R^{-3})e^{-\gamma_B R}. \tag{27.75}$$

The total dynamical matrix is, of course, the sum of (27.72) and (27.73):

$$D_{ij}(\mathbf{k}) = D_{ij}(\mathbf{k})_F + D_{ij}(\mathbf{k})_L. \tag{27.76}$$

There is some further simplification of the dynamical matrix for a primitive cubic lattice, where

$$\left.\begin{aligned} \sum_{\mathbf{Q}}{}' G(Q)Q_iQ_j &= \tfrac{1}{3}\delta_{ij}\sum_{\mathbf{Q}}{}' G(Q)Q^2, \\ \sum_{\mathrm{N}}{}' \varphi''(0,\mathrm{N})R_i(\mathrm{N})R_j(\mathrm{N}) &= \tfrac{1}{3}\delta_{ij}\sum_{\mathrm{N}}{}' \varphi''(0,\mathrm{N})R(\mathrm{N})^2. \end{aligned}\right\} \quad \text{cubic lattice} \tag{27.77}$$

Then the ion–ion interaction contributions to these sums, which appear in (27.72) and (27.73), can be simplified by the theta function transformation of (28.6) below.

For the case of two like atoms per unit cell, the dynamical matrix can be transformed to the real symmetric 6×6 form shown in (15.29). This 6×6 matrix is constructed from three real symmetrix 3×3 matrices, namely $\mathbf{f}(\mathbf{k})$, $\mathscr{R}\mathbf{g}(\mathbf{k})$, and $\mathscr{I}\mathbf{g}(\mathbf{k})$, and in the present pseudopotential theory the lattice sum contributions to these three matrices are given by the general results (15.19), (15.21), and (15.22) for central potentials. The Fourier sum contributions to these matrices are as follows.

$$f_{ij}(\mathbf{k})_F = \sum_{\mathbf{Q}} G(|\mathbf{Q}+\mathbf{k}|)(\mathbf{Q}+\mathbf{k})_i(\mathbf{Q}+\mathbf{k})_j - \sum_{\mathbf{Q}}{}' G(Q)Q_iQ_j\{1 + \cos[\mathbf{Q}\cdot\mathbf{R}(1)]\}; \tag{27.78}$$

$$\mathscr{R}g_{ij}(\mathbf{k})_F = \sum_{\mathbf{Q}} G(|\mathbf{Q}+\mathbf{k}|)(\mathbf{Q}+\mathbf{k})_i(\mathbf{Q}+\mathbf{k})_j\cos[\mathbf{Q}\cdot\mathbf{R}(1)]; \tag{27.79}$$

$$\mathscr{I}g_{ij}(\mathbf{k})_F = -\sum_{\mathbf{Q}} G(|\mathbf{Q}+\mathbf{k}|)(\mathbf{Q}+\mathbf{k})_i(\mathbf{Q}+\mathbf{k})_j\sin[\mathbf{Q}\cdot\mathbf{R}(1)]. \tag{27.80}$$

28. ADDITIONAL TOPICS

Properties of Harmonic Phonons

In order to make preliminary comparisons of theory and experiment, or to develop theoretical estimates of vibrational contributions to thermodynamic functions, it is useful to calculate the average phonon frequency squared $\langle[\omega(\mathbf{k}s)]^2\rangle$, averaged over the Brillouin zone. In preparation for evaluating this average for the pseudopotential theory, we will first derive the theta function transformation. Let the set of ion positions $\tilde{\mathbf{R}}(\mathrm{N}\nu)$ be arbitrary in the volume V of the crystal, and define the function $T(\mathbf{r},\eta)$ as

$$T(\mathbf{r},\eta) = (2/\sqrt{\pi}) \sum_{\mathrm{N}\nu} e^{-|\tilde{\mathbf{R}}(\mathrm{N}\nu)-\mathbf{r}|^2\eta^2}. \tag{28.1}$$

The Fourier components of $T(\mathbf{r},\eta)$ are written down and simplified as follows.

$$\begin{aligned} T(\mathbf{q}) &= V^{-1}\int T(\mathbf{r},\eta)e^{-i\mathbf{q}\cdot\mathbf{r}}\, d\mathbf{r} \\ &= (2/\sqrt{\pi}V) \sum_{\mathrm{N}\nu} e^{-i\mathbf{q}\cdot\tilde{\mathbf{R}}(\mathrm{N}\nu)} \int e^{-|\tilde{\mathbf{R}}(\mathrm{N}\nu)-\mathbf{r}|^2\eta^2} e^{-i\mathbf{q}\cdot[\mathbf{r}-\tilde{\mathbf{R}}(\mathrm{N}\nu)]}\, d\mathbf{r} \\ &= (2/\sqrt{\pi}V_A)S(\mathbf{q})\int e^{-|\mathbf{r}|^2\eta^2}e^{-i\mathbf{q}\cdot\mathbf{r}}\, d\mathbf{r}, \end{aligned} \tag{28.2}$$

where we used (26.36) for $S(\mathbf{q})$, with $V = nN_0V_A$, and noted that the integral in the second line of (28.2) is independent of $\tilde{\mathbf{R}}(\mathrm{N}\nu)$. The integral in the last line of (28.2) is evaluated according to

$$\int e^{-|\mathbf{r}|^2\eta^2}e^{-i\mathbf{q}\cdot\mathbf{r}}\, d\mathbf{r} = (\sqrt{\pi}/\eta)^3 e^{-q^2/4\eta^2}, \tag{28.3}$$

so that (28.2) becomes

$$T(\mathbf{q}) = (2\pi/\eta^3V_A)S(\mathbf{q})e^{-q^2/4\eta^2}. \tag{28.4}$$

Now we simply equate (28.1) for $T(\mathbf{r},\eta)$ to its Fourier sum representation $\Sigma_{\mathbf{q}}\, T(\mathbf{q})e^{i\mathbf{q}\cdot\mathbf{r}}$, to obtain the generalized theta function transformation:

$$(2/\sqrt{\pi}) \sum_{\mathrm{N}\nu} e^{-|\tilde{\mathbf{R}}(\mathrm{N}\nu)-\mathbf{r}|^2\eta^2} = (2\pi\ \eta^3V_A) \sum_{\mathbf{q}} S(\mathbf{q})e^{-q^2/4\eta^2}e^{i\mathbf{q}\cdot\mathbf{r}}. \tag{28.5}$$

If the ion positions are taken to be the equilibrium positions $\mathbf{R}(\mathrm{N})$ for a primitive lattice, then $S(\mathbf{q}) = \delta(\mathbf{q})$, and (28.5) reduces to the ordinary theta function transformation

$$(2/\sqrt{\pi}) \sum_{\mathrm{N}} e^{-|\mathbf{R}(\mathrm{N})-\mathbf{r}|^2\eta^2} = (2\pi/\eta^3V_A) \sum_{\mathbf{Q}} e^{-Q^2/4\eta^2}\, e^{i\mathbf{Q}\cdot\mathbf{r}}. \tag{28.6}$$

We now proceed to the detailed calculation of $\langle[\omega(\mathbf{k}s)]^2\rangle$ for a primitive lattice; this may be written from (10.62),

$$\langle[\omega(\mathbf{k}s)]^2\rangle = \tfrac{1}{3}M_C^{-1} \sum_i \Phi_{ii}(0,0). \tag{28.7}$$

The Fourier sum contribution to (28.7) is obtained from (27.61) for

$\Phi_{ij}(0,0)_F$, as

$$\langle[\omega(\mathbf{k}s)]^2\rangle_F = \tfrac{2}{3}M_C^{-1}\sum_{\mathbf{q}}{}' G(q)q^2[N_0^{-1} - \delta(\mathbf{q})], \tag{28.8}$$

and the lattice sum contribution to (28.7) is obtained from (27.67) and (27.68) for $\Phi_{ij}(0,0)_L$, as

$$\langle[\omega(\mathbf{k}s)]^2\rangle_L = \tfrac{2}{3}M_C^{-1}\sum_{\mathrm{N}}{}' [3\varphi'(0,\mathrm{N}) + 2\varphi''(0,\mathrm{N})R(\mathrm{N})^2]. \tag{28.9}$$

It is convenient to separate out the contribution to $\langle[\omega(\mathbf{k}s)]^2\rangle$ which arises from the Coulomb interactions among the point ions in a uniform compensating background charge, that is, the contribution arising from Ω_{I}(point ions) of (27.17). According to (27.25) and (27.26), Ω_{I}(point ions) gives the contribution $(2\pi z^2e^2/V_A)q^{-2}e^{-q^2/4\eta^2}$ to $G(q)$, and the contributions to $\varphi'(R)$ and $\varphi''(R)$ which correspond to the term $z^2e^2R^{-1}\operatorname{erfc}(\eta R)$ in $\varphi(R)$. From such a term in $\varphi(R)$, the corresponding value of $3\varphi'(R) + 2R^2\varphi''(R)$ is just $(2z^2e^2\eta^3/\sqrt{\pi})e^{-\eta^2R^2}$, so the total ion–ion interaction contribution to $\langle[\omega(\mathbf{k}s)]^2\rangle$ is

$$\begin{aligned}\langle[\omega(\mathbf{k}s)]^2\rangle_{\mathrm{I}} = (4\pi z^2e^2/3V_AM_C)\sum_{\mathbf{q}}{}' e^{-q^2/4\eta^2}[N_0^{-1} - \delta(\mathbf{q})]\\ + (4z^2e^2\eta^3/3\sqrt{\pi}\,M_C)\sum_{\mathrm{N}}{}' e^{-\eta^2|\mathbf{R}(\mathrm{N})|^2}.\end{aligned} \tag{28.10}$$

The term in (28.10) involving $N_0^{-1}\Sigma_{\mathbf{q}}'$ is an unrestricted sum over $\mathbf{q}$, and $\Sigma_{\mathbf{q}}'$ can be replaced by $\Sigma_{\mathbf{q}}$ with an error of relative order N_0^{-1}, and the sum can be transformed to an integral and evaluated as follows.

$$\begin{aligned}\frac{4\pi z^2e^2}{3N_0V_AM_C}\sum_{\mathbf{q}} e^{-q^2/4\eta^2} &= \frac{4\pi z^2e^2}{3N_0V_AM_C}\left(\frac{V}{8\pi^3}\right)\int e^{-q^2/4\eta^2}\,d\mathbf{q}\\ &= \frac{z^2e^2}{6\pi^2M_C}4\pi\int_0^\infty q^2e^{-q^2/4\eta^2}\,dq\\ &= \frac{4z^2e^2\eta^3}{3\sqrt{\pi}\,M_C}.\end{aligned} \tag{28.11}$$

Now this term is obviously just the term omitted in the Σ_{N}' in (28.10), namely the $\mathbf{R}(\mathrm{N}) = 0$ term. Also, in the remaining $\Sigma_{\mathbf{q}}'$ in (28.10), which contains $\delta(\mathbf{q})$, we can add and subtract the $\mathbf{q} = 0$ term, so that the total (28.10) becomes

$$\begin{aligned}&\langle[\omega(\mathbf{k}s)]^2\rangle_{\mathrm{I}}\\ &= (4\pi z^2e^2/3V_AM_C)\left\{1 - \sum_{\mathbf{Q}} e^{-Q^2/4\eta^2}\right\} + (4z^2e^2\eta^3/3\sqrt{\pi}\,M_C)\sum_{\mathrm{N}} e^{-\eta^2|\mathbf{R}(\mathrm{N})|^2}.\end{aligned} \tag{28.12}$$

Then the Σ_Q and Σ_N in (28.12) cancel, by virtue of the theta function transformation (28.6) evaluated at $\mathbf{r} = 0$, and the net result is

$$\langle [\omega(\mathbf{k}s)]^2 \rangle_I = 4\pi z^2 e^2 / 3 V_A M_C. \tag{28.13}$$

According to (27.25) and (27.26), the remaining contribution to $G(q)$ is the energy wave-number characteristic $F(q)$, and the remaining contributions to $\varphi'(R)$ and $\varphi''(R)$ arise from the Born–Mayer repulsion $\alpha_B e^{-\gamma_B R}$ in $\varphi(R)$. With these contributions in (28.8) and (28.9), plus (28.13), the total $\langle [\omega(\mathbf{k}s)]^2 \rangle$ is

$$\begin{aligned} \langle [\omega(\mathbf{k}s)]^2 \rangle &= \tfrac{2}{3} M_C^{-1} \sum_{\mathbf{q}}{}' F(q) q^2 [N_0^{-1} - \delta(\mathbf{q})] \\ &+ \tfrac{1}{3} \alpha_B \gamma_B M_C^{-1} \sum_{\mathrm{N}}{}' [\gamma_B - 2R(\mathrm{N})^{-1}] e^{-\gamma_B R(\mathrm{N})} + (4\pi z^2 e^2 / 3 V_A M_C). \end{aligned} \tag{28.14}$$

Again in the term $N_0^{-1} \Sigma'_{\mathbf{q}}$, the $\Sigma'_{\mathbf{q}}$ can be replaced by $\Sigma_{\mathbf{q}}$ with an error of relative order N_0^{-1}, and in the term containing $\Sigma'_{\mathbf{q}}\, \delta(\mathbf{q})$ we can add and subtract the $\mathbf{q} = 0$ term with the aid of the following limit:

$$\lim_{q \to 0} q^2 F(q) = - \frac{2\pi z^2 e^2}{V_A}. \tag{28.15}$$

We note that the essential contribution to (28.15) comes from the ion Coulomb part (26.40) of $w_B(q)$. Our final result for the average frequency squared is then

$$\begin{aligned} &\langle [\omega(\mathbf{k}s)]^2 \rangle \\ &= \tfrac{2}{3} M_C^{-1} \sum_{\mathbf{q}} F(q) q^2 [N_0^{-1} - \delta(\mathbf{q})] + \tfrac{1}{3} \alpha_B \gamma_B M_C^{-1} \sum_{\mathrm{N}}{}' [\gamma_B - 2R(\mathrm{N})^{-1}] e^{-\gamma_B R(\mathrm{N})} \end{aligned} \tag{28.16}$$

Here the term involving $N_0^{-1} \Sigma_{\mathbf{q}}$ may be computed as an integral, and the term involving $\delta(\mathbf{q})$ reduces to a Σ_Q.

It should be noted that the Coulomb interactions among a lattice of point ions will contribute nothing to $\langle [\omega(\mathbf{k}s)]^2 \rangle$, and that the contribution (28.13) arises because Ω_I(point ions) also includes the Coulomb interactions of the point ions with the compensating background charge. In fact (28.13) is independent of η, and is naturally returned to the band-structure contribution in (28.16).

Another property of the harmonic phonons is the presence of Kohn anomalies in the curves of $\omega(\mathbf{k}s)$ *vs* $\mathbf{k}$. These anomalies are small wiggles in the curves at certain points in $\mathbf{k}$ space; they arise from the band-structure contribution to the dynamical matrices and are due to the singular behavior of the Hartree dielectric function at $q = 2f$. The phenomenon of Kohn anomalies is not restricted to a simple pseudopotential theory, but is quite general, since the real dielectric function of a real metal will show some sort of "singular" behavior associated with the Fermi surface.

Consider in particular a primitive lattice, for which the **k**-dependent part of the band-structure contribution to $D_{ij}(\mathbf{k})$ is obtained from (27.72) as

$$D_{ij}(\mathbf{k})_{\mathrm{BS}} = 2M_C^{-1} \sum_{\mathbf{Q}} F(|\mathbf{Q} + \mathbf{k}|)(\mathbf{Q} + \mathbf{k})_i(\mathbf{Q} + \mathbf{k}_j). \tag{28.17}$$

A Kohn anomaly appears whenever $|\mathbf{Q} + \mathbf{k}|$ passes through $2f$ as **k** varies, and some information about the magnitude of such an anomaly is obtained from the diagonalization condition (10.34),

$$\sum_{ij} w_i(\mathbf{k}s)^* D_{ij}(\mathbf{k}) w_j(\mathbf{k}s) = [\omega(\mathbf{k}s)]^2, \tag{28.18}$$

where $\mathbf{w}(\mathbf{k}s)$ are the phonon eigenvectors. The contribution of (28.17) to $[\omega(\mathbf{k}s)]^2$ is therefore

$$[\omega(\mathbf{k}s)]^2_{\mathrm{BS}} = 2M_C^{-1} \sum_{\mathbf{Q}} F(|\mathbf{Q} + \mathbf{k}|)[(\mathbf{Q} + \mathbf{k}) \cdot \mathbf{w}(\mathbf{k}s)]^2. \tag{28.19}$$

We might estimate the *relative* magnitudes of the different Kohn anomalies in a given metal by the function

$$K(\mathbf{k}s) = \sum_{\mathbf{Q}}{}'' \, [(\mathbf{Q} + \mathbf{k}) \cdot \mathbf{w}(\mathbf{k}s)]^2, \tag{28.20}$$

where $\Sigma''_{\mathbf{Q}}$ is over all **Q** for which $|\mathbf{Q} + \mathbf{k}| = 2f$ for the given **k**. An important observation is that the Kohn anomalies should be more pronounced in curves of the phonon Grüneisen parameters $\gamma(\mathbf{k}s)$, since the singularity in the volume derivative of $F(q)$ is more pronounced than in $F(q)$ itself. The relative magnitudes of Kohn anomalies in the $\gamma(\mathbf{k}s)$ curves for a given metal might be expected to be given qualitatively by the function $K(\mathbf{k}s)[\omega(\mathbf{k}s)]^{-2}$.

The volume and strain derivatives of the dynamical matrices are easily calculated by direct differentiation of the dynamical matrix equations of Section 27, and in this differentiation the parameter η may be taken to be constant or to vary with the configuration. For a primitive lattice, the equation for the ordinary phonon Grüneisen parameters $\gamma(\mathbf{k}s)$ is (15.67), and for the generalized phonon Grüneisen parameters $\gamma_{ij}(\mathbf{k}s)$ it is (15.65). Results for some model calculations of $\gamma(\mathbf{k}s)$ are shown in Chapter 8.

A final important point to be discussed is a failure of the pseudopotential perturbation theory which has been presented in this chapter; recognition of this failure clarifies the relation between the pseudopotential perturbation procedure and the band-structure theory of Chapter 5. The essential difference between the two methods is that the present pseudopotential theory treats the pseudopotential as a perturbation, while the band-structure theory treats the ion displacements as a perturbation. We have previously noted that the results of homogeneous-deformation calculations and of long-waves calculations are not the same for the pseudopotential perturbation theory.* In particular if one calculates the stress–strain coefficients $\tilde{B}_{\alpha\beta}$ for a primitive cubic lattice by the method of long waves, then (27.46) for $\tilde{B}_{11} - \tilde{B}_{12}$ and (27.50) for $\tilde{B}_{44}$ are recovered exactly, and (27.53) for $\tilde{B} = \frac{1}{3}(\tilde{B}_{11} + 2\tilde{B}_{12})$ is recovered except for the term $\Delta_2(Q)$ appearing in the Σ'_Q. This $\Delta_2(Q)$ term arises in the method of homogeneous deformation entirely from the volume variation of the screening parts of $F(Q)$ for $Q \neq 0$, and only through the dependence of the Fermi wave vector magnitude f (and the parameter ξ) on the volume. This information would be contained in the method of long-waves calculation if we included the local variation of f in the presence of a long-wavelength compressional (longitudinal) wave; such a local variation of f is given by (26.50) as a term of second order in the pseudopotential, and since the dynamical matrices are already second order in the pseudopotential it was appropriate to use the constant zeroth-order value of f. In fact our consistent calculation to second order in the pseudopotential would be a good approximation if the pseudopotential were always small, but it is just in the long-wavelength region where the phonon frequencies depend strongly on $w(q)$ as $q \to 0$, and since according to (26.87) $w(q \to 0)$ is not small, the long-waves–homogeneous-deformation difference can be significant.

The difference between long-waves and homogeneous-deformation results for the pseudopotential perturbation theory does not represent a failure of the general equivalence of these two methods shown in Section 12, because there we treated all potential contributions which are of second order in displacements of the ions from equilibrium. Indeed in the pseudopotential perturbation theory, there are contributions to the total adiabatic potential which are of second order in the ion displacements, and which appear in every order of the pseudopotential. The key to pseudopotential perturbation theory is, of course, that $W(\mathbf{q})$ is small for $\mathbf{q}$ of the order or larger than the first inverse-lattice vector, and the static-lattice potential Φ_0 of say (27.23) should be reasonably accurate since it contains only $W(Q)$ for $Q \neq 0$. This gives us the tip that the dynamical matrices can be calculated to the same accuracy from the formulation of Chapter 5, with the dielectric function calculated to second order in Umklapp terms, that is to second order in $W(Q)$ for $Q \neq 0$. We will now outline this calculation.

Expansion in Umklapp Processes

The total dynamical matrix is given by (23.78) as the sum of contributions from the electronic ground-state energy and the ion–ion interaction energy. We consider only primitive lattices, and write these contributions to $D_{ij}(\mathbf{k})$

* D. C. Wallace, *Phys. Rev.* **182,** 778 (1969); *Phys. Rev.* **B1,** 942 (1970).

in the language of pseudopotential theory:

$$D_{ij}(\mathbf{k}) = D_{ij}(\mathbf{k})_{\mathrm{E}} + D_{ij}(\mathbf{k})_{\mathrm{IF}} + D_{ij}(\mathbf{k})_{\mathrm{IL}}, \tag{28.21}$$

where from (23.75)

$$D_{ij}(\mathbf{k})_{\mathrm{E}} = M_C^{-1} \sum_{\mathbf{QQ}'} \tilde{c}(\mathbf{Q}+\mathbf{k},\mathbf{Q}'+\mathbf{k})(\mathbf{Q}+\mathbf{k})_i(\mathbf{Q}'+\mathbf{k})_j w_{\mathrm{B}}(|\mathbf{Q}+\mathbf{k}|) \times w_{\mathrm{B}}(|\mathbf{Q}'+\mathbf{k}|) - M_C^{-1} \sum_{\mathbf{Q}}' c(\mathbf{Q}) Q_i Q_j w_{\mathrm{B}}(Q), \tag{28.22}$$

and from (23.76), or from (27.72) with $G(q)$ taken to be just the ion–ion Coulomb part $(2\pi z^2 e^2/V_A) q^{-2} e^{-q^2/4\eta^2}$,

$$D_{ij}(\mathbf{k})_{\mathrm{IF}} = (4\pi z^2 e^2/V_A M_C) \sum_{\mathbf{Q}} |\mathbf{Q}+\mathbf{k}|^{-2} e^{-|\mathbf{Q}+\mathbf{k}|^2/4\eta^2} (\mathbf{Q}+\mathbf{k})_i (\mathbf{Q}+\mathbf{k})_j - (4\pi z^2 e^2/V_A M_C) \sum_{\mathbf{Q}}' Q^{-2} e^{-Q^2/4\eta^2} Q_i Q_j, \tag{28.23}$$

and the remaining term $D_{ij}(\mathbf{k})_{\mathrm{IL}}$ in (28.21) is the same as $D_{ij}(\mathbf{k})_{\mathrm{L}}$, which is given by (27.73) and includes the Born–Mayer repulsion. Note the $\mathbf{Q} = 0$ terms in the $\mathbf{k}$-independent sums have been cancelled between (28.22) and (28.23). The pseudopotential contribution is contained entirely in $D_{ij}(\mathbf{k})_{\mathrm{E}}$, and it will be necessary to work out the correlation functions in the framework of the pseudopotential theory.

The correlation functions in (28.22) are evaluated at the equilibrium configuration, and for a primitive lattice $S(\mathbf{Q}) = 1$ so that $W(\mathbf{Q}) = w(Q)$. According to (23.2), $c(\mathbf{Q}) = V_A \rho^{(0)}(-\mathbf{Q})$, where $V_A = V_C$ for a primitive lattice, and from (26.69) for $\rho(\mathbf{q})$, it follows that

$$c(\mathbf{Q}) = c(Q) = (Q^2 V_A/4\pi e^2) w(Q)[1 - H(Q)]. \tag{28.24}$$

Then the second contribution in (28.22), the $\Sigma'_{\mathbf{Q}}$, becomes

$$-M_C^{-1} \sum_{\mathbf{Q}}' (Q^2 V_A/4\pi e^2) Q_i Q_j w(Q) w_{\mathrm{B}}(Q)[1 - H(Q)] = -2M_C^{-1} \sum_{\mathbf{Q}}' F(Q) Q_i Q_j, \tag{28.25}$$

where we used (27.13) for $F(Q)$. The function (28.25) is of second order in Umklapp terms.

Consider now the $\Sigma_{\mathbf{QQ}'}$ term in (28.22). A formal expansion of $\tilde{c}(\mathbf{q},\mathbf{q}')$ in powers of Umklapp processes may be written

$$\tilde{c}(\mathbf{q},\mathbf{q}') = \tilde{c}_0(\mathbf{q},\mathbf{q}') + \tilde{c}_1(\mathbf{q},\mathbf{q}') + \tilde{c}_2(\mathbf{q},\mathbf{q}') + \cdots, \tag{28.26}$$

and it is shown below that the $\tilde{c}_1(\mathbf{q},\mathbf{q}) = 0$. Then the terms which should be kept in the $\Sigma_{\mathbf{QQ}'}$ in (28.22) are as follows.

$$\begin{aligned} &M_C^{-1}\sum_{\mathbf{QQ}'}\tilde{c}_0(\mathbf{Q}+\mathbf{k},\mathbf{Q}'+\mathbf{k})(\mathbf{Q}+\mathbf{k})_i(\mathbf{Q}'+\mathbf{k})_j w_B(|\mathbf{Q}+\mathbf{k}|)w_B(|\mathbf{Q}'+\mathbf{k}|) \\ &\quad + M_C^{-1}\sum_{\mathbf{Q}}{}'\,[\tilde{c}_1(\mathbf{Q}+\mathbf{k},\mathbf{k})(\mathbf{Q}+\mathbf{k})_i k_j + \tilde{c}_1(\mathbf{k},\mathbf{Q}+\mathbf{k})k_i(\mathbf{Q}+\mathbf{k})_j] \qquad (28.27) \\ &\quad \times w_B(|\mathbf{Q}+\mathbf{k}|)w_B(k) + M_C^{-1}\tilde{c}_2(\mathbf{k},\mathbf{k})k_i k_j[w_B(k)]^2. \end{aligned}$$

In the limit $\mathbf{k}\to 0$, (28.27) gives all the contributions to the $\Sigma_{\mathbf{QQ}'}$ up to second order in Umklapp terms, that is up to second order in $w_B(Q\neq 0)$. Further since $|\mathbf{k}|$ is at most of order $\frac{1}{2}|\mathbf{Q}|$ for the smallest $\mathbf{Q}$, then (28.27) should be a reasonably accurate representation of the $\Sigma_{\mathbf{QQ}'}$ for all $\mathbf{k}$, although (28.27) is not strictly correct to second order in $w_B(Q\neq 0)$.

To calculate the expansion (28.26) of $\tilde{c}(\mathbf{q},\mathbf{q}')$, we begin with $a(\mathbf{q},\mathbf{q}')$ given by (21.74), put in the pseudopotential expansions (26.28) of the wave functions $\psi_{\mathbf{p}}$ and (26.32) of the energies $\epsilon_{\mathbf{p}}$, and write out an expansion of $a(\mathbf{q},\mathbf{q}')$ in powers of the pseudopotential. Note that in this procedure one needs to include explicitly the second-order pseudopotential contributions to the $\psi_{\mathbf{p}}$, but these contributions can be eliminated in favor of first-order contributions by the wave function normalization requirements. The contribution to $a(\mathbf{q},\mathbf{q}')$ which arises from the zeroth-order wave functions $|\mathbf{p}\rangle$ is

$$\delta_{\mathbf{qq}'}V^{-1}\sum_{\mathbf{p}}(g_{\mathbf{p}}-g_{\mathbf{p}+\mathbf{q}})(\epsilon_{\mathbf{p}}-\epsilon_{\mathbf{p}+\mathbf{q}})_P^{-1};$$

this sum gives two second-order contributions to $a(\mathbf{q},\mathbf{q})$, one arising from the second-order contribution to the energy denominators, and the other arising from carrying the $\Sigma_{\mathbf{p}}g_{\mathbf{p}}$ up to the Fermi surface correct to second order, as defined by the Fermi wave vector $|\mathbf{F}|$ of (26.50). The results for the expansion of $a(\mathbf{q},\mathbf{q}')$ are as follows.

$$a_0(\mathbf{q},\mathbf{q}') = \delta_{\mathbf{qq}'}(q^2/4\pi e^2)[1-H(q)]; \qquad (28.28)$$

$$\begin{aligned} a_1(\mathbf{q},\mathbf{q}) &= 0, \\ a_1(\mathbf{q},\mathbf{q}') &= V^{-1}\sum_{\mathbf{p}}\{(g_{\mathbf{p}}-g_{\mathbf{p}+\mathbf{q}})(e_{\mathbf{p}}-e_{\mathbf{p}+\mathbf{q}})_P^{-1}[A(\mathbf{q}-\mathbf{q}',\mathbf{p})+A(\mathbf{q}'-\mathbf{q},\mathbf{p}+\mathbf{q})] \\ &\quad + (g_{\mathbf{p}}-g_{\mathbf{p}+\mathbf{q}'})(e_{\mathbf{p}}-e_{\mathbf{p}+\mathbf{q}'})_P^{-1}[A(\mathbf{q}'-\mathbf{q},\mathbf{p})+A(\mathbf{q}-\mathbf{q}',\mathbf{p}+\mathbf{q}')]\}; \\ &\qquad (28.29) \end{aligned}$$

$$\begin{aligned} a_2(\mathbf{q},\mathbf{q}) &= (f^2/2\pi^3)\int(|\mathbf{F}|-|\mathbf{f}|)(e_{\mathbf{f}}-e_{\mathbf{f}+\mathbf{q}})_P^{-1}\,d\Omega_{\mathbf{f}} \\ &\quad - V^{-1}\sum_{\mathbf{p}}(g_{\mathbf{p}}-g_{\mathbf{p}+\mathbf{q}})\Big\{(e_{\mathbf{p}}-e_{\mathbf{p}+\mathbf{q}})_P^{-1}\sum_{\mathbf{Q}}{}'\,[A(\mathbf{Q},\mathbf{p})-A(\mathbf{Q},\mathbf{p}+\mathbf{q})]^2 \\ &\quad + (e_{\mathbf{p}}-e_{\mathbf{p}+\mathbf{q}})_P^{-2}\sum_{\mathbf{Q}}{}'\,w(Q)[A(\mathbf{Q},\mathbf{p})-A(\mathbf{Q},\mathbf{p}+\mathbf{q})]\Big\} \qquad (28.30) \\ &\quad + V^{-1}\sum_{\mathbf{p}}\sum_{\mathbf{Q}}{}'\,(g_{\mathbf{p}}-g_{\mathbf{p}+\mathbf{q}+\mathbf{Q}})(e_{\mathbf{p}}-e_{\mathbf{p}+\mathbf{q}+\mathbf{Q}})_P^{-1} \\ &\quad \times [A(\mathbf{Q},\mathbf{p})-A(\mathbf{Q},\mathbf{p}+\mathbf{q})]^2. \end{aligned}$$

Here we have used $A(\mathbf{q},\mathbf{p}) = A(\mathbf{q},\mathbf{p})^*$ for the equilibrium configuration of the lattice. Each $\Sigma_{\mathbf{p}}$ in (28.29) and (28.30) contains a factor 2 for spin, and is to be evaluated to zeroth order. We note that in second order, only the diagonal elements $a_2(\mathbf{q},\mathbf{q})$ are required.

With the approximation (26.75) for exchange and correlation effects, the dielectric function $\hat{d}(\mathbf{q},\mathbf{q}')$ is given by (22.47) and $\tilde{c}(\mathbf{q},\mathbf{q}')$ is given by (22.48). After a short calculation, the expansion of $\tilde{c}(\mathbf{q},\mathbf{q}')$ is found to be given by the following equations.

$$\tilde{c}_0(\mathbf{q},\mathbf{q}') = \delta_{\mathbf{q}\mathbf{q}'}(q^2V_A/4\pi e^2)Z(q)[1 - H(q)]; \tag{28.31}$$

$$\begin{aligned} \tilde{c}_1(\mathbf{q},\mathbf{q}) &= 0, \\ \tilde{c}_1(\mathbf{q},\mathbf{q}') &= V_A Z(q)Z(q')a_1(\mathbf{q},\mathbf{q}'); \end{aligned} \tag{28.32}$$

$$\begin{aligned} \tilde{c}_2(\mathbf{q},\mathbf{q}) = V_A[Z(q)]^2\Big\{a_2(\mathbf{q},\mathbf{q}) &+ \sum_{\mathbf{q}'}(4\pi e^2/q'^2)Z(q') \\ &\times [1 - Y(q')]a_1(\mathbf{q},\mathbf{q}')a_1(\mathbf{q}',\mathbf{q})\Big\}. \end{aligned} \tag{28.33}$$

Here the function $Z(q)$ is the abbreviation

$$Z(q) = \{1 + [H(q) - 1][1 - Y(q)]\}^{-1}. \tag{28.34}$$

Returning to the expansion (28.27) of the $\Sigma_{\mathbf{Q}\mathbf{Q}'}$ in (28.22), the terms $\tilde{c}_0(\mathbf{Q} + \mathbf{k},\mathbf{Q}' + \mathbf{k})$ simplify to $\delta_{\mathbf{Q}\mathbf{Q}'}\tilde{c}_0(\mathbf{Q} + \mathbf{k},\mathbf{Q} + \mathbf{k})$ since $\tilde{c}_0(\mathbf{q},\mathbf{q}')$ is diagonal; these terms further simplify because $\tilde{c}_0(\mathbf{q},\mathbf{q})[w_{\mathrm{B}}(q)]^2 = 2F(q)$, from (27.13) for $F(q)$. Also since $a_1(\mathbf{q},\mathbf{q}') = a_1(\mathbf{q}',\mathbf{q})$ according to (28.29), then $\tilde{c}_1(\mathbf{q},\mathbf{q}') = \tilde{c}_1(\mathbf{q}',\mathbf{q})$ according to (28.32), and the terms in (28.27) containing $\tilde{c}_1(\mathbf{Q} + \mathbf{k},\mathbf{k})$ and $\tilde{c}_1(\mathbf{k},\mathbf{Q} + \mathbf{k})$ can be combined. With these simplifications, and including the contribution (28.25), the total pseudopotential contribution to the dynamical matrix is written

$$\begin{aligned} D_{ij}(\mathbf{k})_{\mathrm{E}} = {} & 2M_C^{-1}\sum_{\mathbf{Q}} F(|\mathbf{Q} + \mathbf{k}|)(\mathbf{Q} + \mathbf{k})_i(\mathbf{Q} + \mathbf{k})_j - 2M_C^{-1}\sum_{\mathbf{Q}}{}' F(Q)Q_iQ_j \\ & + M_C^{-1}\sum_{\mathbf{Q}}{}' \tilde{c}_1(\mathbf{Q} + \mathbf{k},\mathbf{k})[(\mathbf{Q} + \mathbf{k})_i k_j + k_i(\mathbf{Q} + \mathbf{k})_j]w_{\mathrm{B}}(|\mathbf{Q} + \mathbf{k}|)w_{\mathrm{B}}(k) \\ & + M_C^{-1}\tilde{c}_2(\mathbf{k},\mathbf{k})k_ik_j[w_{\mathrm{B}}(k)]^2. \end{aligned} \tag{28.35}$$

The first two sums, which contain $F(|\mathbf{Q} + \mathbf{k}|)$ and $F(Q)$, are just the ordinary band-structure contribution to the dynamical matrix for a primitive lattice, as contained in (27.72). The sum containing $\tilde{c}_1(\mathbf{Q} + \mathbf{k},\mathbf{k})$ is formally of third order, and the term containing $\tilde{c}_2(\mathbf{k},\mathbf{k})$ is formally of fourth order, in the pseudopotential.

Electronic Free Energy

Here we carry out the pseudopotential calculation of the electronic and electron–phonon interaction contributions to the Helmholtz free energy, for the example of a primitive lattice. The general theory of these effects is presented in Sections 24 and 25, and the local pseudopotential perturbation theory satisfies the approximations of Section 25, namely equations (25.3) and (25.76). The calculations here will be done explicitly to second order in the pseudopotential.

The first step is to calculate the zeroth-order electronic density of states at the Fermi energy. The general expression for the density of states $n(\epsilon)$ is

$$n(\epsilon) = (2V/8\pi^3)\int_\epsilon |\nabla\epsilon|^{-1}\, dS, \tag{28.36}$$

where the factor 2 in the numerator accounts for spin, the integral is over the surface of constant energy at the value ϵ, and $\nabla\epsilon$ is the gradient of ϵ evaluated on the surface. For free electrons, which is zeroth order in the pseudopotential, the electronic energies are $e_{\mathbf{p}} = \hbar^2 p^2/2m$ and the Fermi surface is at $e_F = \hbar^2 f^2/2m$, so that (28.36) gives

$$n(e_F) = \frac{mVf}{\pi^2\hbar^2} = \frac{3zN_0}{2e_F}, \quad \text{primitive lattice,} \tag{28.37}$$

where we used (26.47) to eliminate V_A in favor of f. Now the electronic energy $\epsilon_{\mathbf{p}}$ correct to second order in the pseudopotential is given by (26.32), and from this one can calculate $\nabla\epsilon$ and then $n(\epsilon)$ correct to second order. An alternate approach, however, is somewhat simpler.

Recall that in the present pseudopotential theory there is only one electronic band, the conduction band, for which the electron index λ is simply $\mathbf{p}\sigma$, and the spin σ is unimportant except for counting. The electronic Hamiltonian (24.20) is $\Sigma_\lambda\, \epsilon_\lambda^{(0)}(C_\lambda^\dagger C_\lambda - g_\lambda)$, and here we can write $\epsilon_\lambda^{(0)} = e_{\mathbf{p}} + \delta\epsilon_\lambda^{(0)}$, where $\delta\epsilon_\lambda^{(0)}$ is the second-order pseudopotential contribution to $\epsilon_{\mathbf{p}}$ in (26.32), evaluated at equilibrium. Note that the first-order term $W(\mathbf{q} = 0)$ in (26.32) does not contribute to the electronic Hamiltonian since it is a number, and $\Sigma_\lambda\, C_\lambda^\dagger C_\lambda = \Sigma_\lambda\, g_\lambda$. Then treating $\delta\epsilon_\lambda^{(0)}$ as a perturbation, the electronic Hamiltonian is a system of free electrons plus the perturbation $\Sigma_\lambda\, \delta\epsilon_\lambda^{(0)}(C_\lambda^\dagger C_\lambda - g_\lambda)$, and this perturbation gives a leading-order contribution $\Sigma_\lambda\, \delta\epsilon_\lambda^{(0)}(\bar{f}_\lambda - g_\lambda)$ to the free energy. This free energy contribution may be added to (25.73) for F_{ep} to get a total free energy contribution, correct to second order in the pseudopotential, given by

$$F(\text{pseudopotential perturbation}) = \Delta_{\text{ep}} + \sum_\lambda (\bar{f}_\lambda - g_\lambda)\langle \delta\epsilon_\lambda^{(0)} + \delta\epsilon_\lambda^{(2)}\rangle. \tag{28.38}$$

Here we used the fact that $\epsilon_\lambda^{(2)}$ in (25.73) is second order in the ion displacements, and only the pseudopotential part $\delta\epsilon_\lambda$ of ϵ_λ depends on the positions of the ions, so $\epsilon_\lambda^{(2)} = \delta\epsilon_\lambda^{(2)}$. The brackets $\langle\ \rangle$ in (28.38) represent a phonon statistical average.

To evaluate the Σ_λ in (28.38), we write $\delta\epsilon_\lambda$ from (26.32), average this over angles with the aid of (26.55), then make the Sommerfeld low-temperature expansion (24.47), and evaluate the resulting expression to leading order (second order) in the pseudopotential by replacing $n(\mu_0)$ with $n(e_F)$. Since $W(\mathbf{q}) = S(\mathbf{q})w(q)$, and only $S(\mathbf{q})$ depends on the ion positions, then only $S(\mathbf{q})$ is involved in the phonon statistical average, and we obtain the result

$$\sum_\lambda (\bar{f}_\lambda - g_\lambda)\langle\delta\epsilon_\lambda\rangle = -\tfrac{1}{6}\pi^2(KT)^2 n(e_F)\Delta_W, \tag{28.39}$$

where

$$\Delta_W = \sum_{\mathbf{q}}{}' L(q)\langle S(\mathbf{q})S(-\mathbf{q})\rangle, \tag{28.40}$$

$$L(q) = \frac{\pi}{2e^2 f e_F}\frac{(q/2f)^2}{1-(q/2f)^2}[w(q)]^2[1-H(q)]. \tag{28.41}$$

Now $S(\mathbf{q})S(-\mathbf{q})$ is expanded in powers of displacements of the ions from equilibrium; since the zeroth-order term is $\delta(\mathbf{q})$ for a primitive lattice, (28.40) leads to

$$\Delta_W^{(0)} = \sum_{\mathbf{Q}}{}' L(Q). \tag{28.42}$$

This term is just the band-structure enhancement of the density of states at the Fermi energy.

The first-order term in the expansion of $S(\mathbf{q})S(-\mathbf{q})$ is linear in the ion displacements, and hence averages to zero; the second-order term is worked out for a primitive lattice to give

$$\begin{aligned}\Delta_W^{(2)} = {} & N_0^{-1}\sum_{\mathbf{k}s}{}' [\hbar/M_C\omega(\mathbf{k}s)][\bar{n}(\mathbf{k}s)+\tfrac{1}{2}] \\ & \times\left\{\sum_{\mathbf{Q}} [(\mathbf{Q}+\mathbf{k})\cdot\mathbf{w}(\mathbf{k}s)]^2 L(|\mathbf{Q}+\mathbf{k}|) - \sum_{\mathbf{Q}}{}' [\mathbf{Q}\cdot\mathbf{w}(\mathbf{k}s)]^2 L(Q)\right\}.\end{aligned} \tag{28.43}$$

This is the very small term $\Sigma_\lambda\,(\bar{f}_\lambda - g_\lambda)\langle\epsilon_\lambda^{(2)}\rangle$ discussed in the text preceding (25.75), and we may approximate (28.43) by replacing each phonon frequency $\omega(\mathbf{k}s)$ by the average over the zone $\langle\omega(\mathbf{k}s)\rangle$, and then by replacing $\Sigma_s\,[(\mathbf{Q}+\mathbf{k})\cdot\mathbf{w}(\mathbf{k}s)]^2$ by $|\mathbf{Q}+\mathbf{k}|^2$, and $\Sigma_s\,[\mathbf{Q}\cdot\mathbf{w}(\mathbf{k}s)]^2$ by Q^2, to obtain thc order of magnitude approximate result

$$\begin{aligned}\Delta_W^{(2)} &\approx \frac{\tfrac{1}{2}\hbar\langle\omega(\mathbf{k}s)\rangle}{M_C\langle\omega(\mathbf{k}s)\rangle^2}\sum_{\mathbf{q}}{}' q^2L(q)[N_0^{-1}-\delta(\mathbf{q})], \quad \text{at low } T; \\ \Delta_W^{(2)} &\approx \frac{KT}{M_C\langle\omega(\mathbf{k}s)\rangle^2}\sum_{\mathbf{q}}{}' q^2L(q)[N_0^{-1}-\delta(\mathbf{q})], \quad \text{at high } T.\end{aligned} \tag{28.44}$$

Here of course low temperature is $T \ll \Theta_0$, and high temperature is $T > \Theta_0$.

We now turn to the pseudopotential evaluation of Δ_{ep}, the major electron–phonon interaction contribution to the free energy. The first step is to evaluate the electron–phonon interaction integral $\theta_i(\lambda'\lambda)$ given by (25.9). As we showed in (26.88), the total screened pseudopotential is the sum of screened single ion pseudopotentials $w(\mathbf{r} - \tilde{\mathbf{R}}(N))$, and this function takes the place of the total single ion potential $u(\mathbf{r} - \tilde{\mathbf{R}}(N))$ defined by (25.3). Further, to zeroth order in the pseudopotential the electronic wave functions are $\psi_\lambda(\mathbf{r}) = V^{-1/2}e^{i\mathbf{p}\cdot\mathbf{r}}$, and with these results $\theta_i(\lambda'\lambda)$ is written

$$\theta_i(\lambda'\lambda) = \theta_i(\mathbf{p}'\mathbf{p}) = (N_0/V)\int e^{i(\mathbf{p}-\mathbf{p}')\cdot[\mathbf{r}-\mathbf{R}(N)]} w_i(\mathbf{r} - \mathbf{R}(N))\, d\mathbf{r}, \quad (28.45)$$

where

$$w_i(\mathbf{r} - \mathbf{R}(N)) = [\partial w(\mathbf{r} - \mathbf{R})/\partial R_i]_{\mathbf{R}=\mathbf{R}(N)}. \quad (28.46)$$

But the integral $\int e^{i(\mathbf{p}-\mathbf{p}')\cdot(\mathbf{r}-\mathbf{R})} w(\mathbf{r} - \mathbf{R})\, d\mathbf{r}$ is obviously independent of $\mathbf{R}$, so its derivative with respect to a component R_i is zero, and this means that the derivative of $w(\mathbf{r} - \mathbf{R}(N))$ in (28.45) can be transferred to a derivative of $e^{i(\mathbf{p}-\mathbf{p}')\cdot[\mathbf{r}-\mathbf{R}(N)]}$ with a change of sign, so that (28.45) is

$$\begin{aligned}\theta_i(\mathbf{p}'\mathbf{p}) &= V_A^{-1}\, [i(\mathbf{p} - \mathbf{p}')_i] \int e^{i(\mathbf{p}-\mathbf{p}')\cdot[\mathbf{r}-\mathbf{R}(N)]} w(\mathbf{r} - \mathbf{R}(N))\, d\mathbf{r} \\ &= i(\mathbf{p} - \mathbf{p}')_i w(|\mathbf{p}' - \mathbf{p}|),\end{aligned} \quad (28.47)$$

where the Fourier components $w(q)$ are defined by (26.86). With (28.47), the electron–phonon matrix element $V_{\kappa\lambda}$ defined by (25.16) is found to be independent of λ, and may be written

$$V(\mathbf{k}s) = -i[\hbar/2N_0 M_C w(\mathbf{k}s)]^{1/2}[\mathbf{k} \cdot \mathbf{w}(\mathbf{k}s)]w(k), \quad (28.48)$$

where there should be no confusion between the phonon eigenvector $\mathbf{w}(\mathbf{k}s)$ and the screened single ion form factor $w(k)$.

The function $G(\mu_0,\mu_0)$ is given by (25.87) as a double angle integral over the Fermi surface, and when this is evaluated over the zeroth-order Fermi sphere the integrals can be transformed to a single volume integral. Consider the double integral

$$\int d\Omega_{\mathbf{p}} \int d\Omega_{\mathbf{p}'} J(\mathbf{p} - \mathbf{p}'), \quad (28.49)$$

where $J(\mathbf{p} - \mathbf{p}')$ is any function of the vector $\mathbf{p} - \mathbf{p}' = \mathbf{q}$, and $\mathbf{p}$ and $\mathbf{p}'$ both have magnitude f. Let $\mathbf{p}'$ define a polar axis, so that θ and φ are the angles of $\mathbf{p}$ with respect to $\mathbf{p}'$; then with θ and φ fixed the integral $\int d\Omega_{\mathbf{p}'}$ is the same as $\int d\Omega_{\mathbf{q}}$. Also $q^2 = 2f^2(1 - \cos\theta)$, so that $d\Omega_{\mathbf{p}} = \sin\theta d\theta d\varphi$ is the same as

$(qdq/f^2)d\varphi$, and (28.49) can be simplified as follows.

$$\int d\Omega_{\mathbf{p}} \int d\Omega_{\mathbf{p}'} J(\mathbf{p}-\mathbf{p}') = \int d\Omega_{\mathbf{p}} \int d\Omega_{\mathbf{q}} J(\mathbf{q})$$

$$= 2\pi \int_0^{2f} (q/f^2)\, dq \int d\Omega_{\mathbf{q}} J(\mathbf{q}) \tag{28.50}$$

$$= (2\pi/f^2) \int_0^{2\mathrm{f}} q^{-1} J(\mathbf{q})\, d\mathbf{q}.$$

With this simplification, and with (28.48) for $V(\mathbf{k}s)$, the function $n(\mu_0)G(\mu_0,\mu_0)$ correct to second order in the pseudopotential is simply

$$n(\mu_0)G(\mu_0,\mu_0) = \frac{3zm}{16\pi M_C \hbar^2 f^4} \int_0^{2\mathrm{f}} \frac{[w(q)]^2}{q} \sum_s \frac{[\mathbf{q}\cdot\mathbf{w}(\mathbf{q}s)]^2}{[\omega(\mathbf{q}s)]^2}\, d\mathbf{q}. \tag{28.51}$$

The high- and low-temperature limits of Δ_{ep} are given in terms of $n(\mu_0)G(\mu_0,\mu_0)$ by (25.83) and (25.84).

Finally the total electronic plus electron–phonon interaction free energy is just the sum of the pseudopotential part (28.38), plus the free electron part $-\frac{1}{6}\pi^2(KT)^2 n(e_{\mathrm{F}})$. We can write this total in the form

$$F_{\mathrm{e}} + F_{\mathrm{ep}} = -\tfrac{1}{6}\pi^2(KT)^2 n(e_{\mathrm{F}})[1 + \Delta_W^{(0)} + \Delta_W^{(2)} + \Delta_\Gamma], \tag{28.52}$$

where each function Δ is of second order in the pseudopotential, and (28.52) is correct to second order. The quantities in (28.52) are related to the general theory of Sections 24 and 25 as follows: $F_{\mathrm{e}} = -\frac{1}{2}\Gamma T^2$, according to (24.56), where now

$$\Gamma = \tfrac{1}{3}\pi^2 K^2 n(e_{\mathrm{F}})[1 + \Delta_W^{(0)}]. \tag{28.53}$$

According to (25.73), $F_{\mathrm{ep}} = \Delta_{\mathrm{ep}} + \Sigma_\lambda\, (\bar{f}_\lambda - g_\lambda)\langle \epsilon_\lambda^{(2)} \rangle$; the Σ_λ is just the term in (28.52) which contains $\Delta_W^{(2)}$, while Δ_{ep} is the term in (28.52) which contains Δ_Γ. High- and low-temperature limits of $\Delta_W^{(2)}$ are approximated in (28.44), and the high- and low-temperature limits of Δ_Γ are given by (25.86) in terms of $n(\mu_0)G(\mu_0,\mu_0)$.

Comments on Pseudopotentials

Because of the simplicity of the pseudopotential perturbation theory, it is possible to carry lattice-statics and lattice-dynamics calculations nearly to completion by analytic methods, leaving only minor numerical computations to be performed for final evaluation. In addition this theory represents a respectable approximation for the metals in the upper left corner of the periodic table; consequently for these metals the theory of lattice dynamics is now advanced farther than for any other class of materials except for the rare gas crystals, namely to the point of calculating harmonic and anharmonic thermodynamic functions on the basis of physically meaningful potentials. Extension of the calculations to alloys will undoubtedly be forthcoming.

There are two areas in which significant advances can be made by extension of the present pseudopotential perturbation theory. In the first place the pseudopotential should not be treated as a perturbation in the small $\mathbf{q}$ region, and the way around this difficulty was outlined above. The pseudopotential contribution to the dynamical matrix for a primitive lattice, carried to second order in the large $\mathbf{q}$ components of the pseudopotential, was written in (28.35), and that equation is an important improvement over the ordinary pseudopotential perturbation result (27.72). In the second place, as soon as one gets down to the noble metals and the transition metals in the periodic table, it is not a good approximation to represent all the closed shell electrons as a rigid ion core, and it becomes necessary to include the outer closed shell as another band of electrons. Unfortunately these closed shell bands are not like free electrons, and it may be more appropriate to describe them in terms of tight binding wave functions. A reasonable approach for the $3d$ metals would be a two-band model, which in zeroth order has the s electrons in a free electron band and the d electrons in a tight binding or APW band, and in which the departure from this zeroth-order description is represented by a local potential which is to be treated as a perturbation.

In order to go beyond such simple considerations as those just mentioned, one should proceed by the band-structure theory of Chapter 5.

7

ANALYSIS OF EXPERIMENTAL DATA

The qualitative temperature- and volume-dependences of thermodynamic functions are studied for the low-temperature, intermediate-temperature, and high-temperature regions. The results of detailed analyses of experimental data are presented; these analyses give the lattice and electronic thermodynamic functions at low temperatures, and the harmonic and anharmonic lattice contributions at high temperatures. Approximate relations among the experimentally determined lattice-dynamical functions are discussed. Because of limited available experimental data, the cubic-crystal theory is used for the present analyses; this procedure is approximate for the noncubic crystals studied. All numbered references to experimental data are listed in Appendix 2.

29. LOW TEMPERATURES

Temperature-Dependence of Experimental Data

In the limit of low temperatures, the measured thermodynamic functions of crystals exhibit the temperature-dependences corresponding to lattice-dynamical plus electronic contributions for metals, and lattice-dynamical contributions alone for insulators. Therefore, by fitting measured results to the appropriate equations, information is obtained about the renormalized

Debye temperature Θ_0, and the total electronic excitation plus electron–phonon coefficient $\hat{\Gamma}$, where

$$\hat{\Gamma} = \Gamma(1 + \Delta_\Gamma). \tag{29.1}$$

Note that we are following (25.85) and writing $F_e + \Delta_{ep} = -\frac{1}{2}\hat{\Gamma}T^2$, and we are neglecting the very small electron–phonon term (25.75).

For the heat capacity at constant configuration, for example, the lattice contribution is given by (18.33) and the electronic plus electron–phonon contribution is given by (25.89), so the total low-temperature limit of C_η is

$$C_\eta = \hat{\Gamma}T + nN_0K(12\pi^4/5)(T/\Theta_0)^3. \tag{29.2}$$

If this equation is evaluated at zero pressure, one must keep in mind that the crystal configuration varies with the temperature, and this variation gives additional explicit temperature-dependence to C_η due to the configuration-dependence of $\hat{\Gamma}$ and of Θ_0. For nonmetals $\hat{\Gamma} = 0$, and the configuration-dependence of Θ_0 introduce a negligible T^7 contribution to C_η. However for metals, the configuration-dependence of $\hat{\Gamma}$ introduces another T^3 contribution to C_η, and it is necessary in principle to include this contribution with the lattice term in (29.2).

To analyze the zero-pressure heat capacity measurements, we consider polycrystalline samples for which the only configuration variable is the volume, so that $C_\eta = C_V$. The leading low-temperature contribution to the thermal expansion coefficient β is $B_0^{-1}(d\hat{\Gamma}/dV)T$, where B_0 is the bulk modulus at $T = 0$ [see, e.g., (24.80) for β for a cubic crystal]. Then from (29.2), the explicit temperature-dependence of C_V at low temperatures and at zero pressure is

$$C_V = \hat{\Gamma}(V_0)T + (1 + x)nN_0K(12\pi^4/5)(T/\Theta_0)^3, \quad \text{at} \quad P = 0, \tag{29.3}$$

where V_0 is the volume at $T = 0$ and $P = 0$, and where

$$x = \frac{\hat{\Gamma}^2(d \ln \hat{\Gamma}/d \ln V)^2\Theta_0^3}{nN_0K(12\pi^4/5)V_0B_0}. \tag{29.4}$$

The quantity x was calculated from experimental data for several metals, and is listed in Table 4; since $x \lesssim 10^{-5}$ for these metals it is negligible in (29.3), and it is safe to assume that the same result holds for all metals. In addition, $C_P - C_V$ may be written from (1.40) as $TV\beta^2B_T$, and for nonmetals this goes as T^7 at low temperatures and hence is negligible compared to C_V. For metals the leading low-temperature contribution to βB_T is $(d\hat{\Gamma}/dV)T$ [see, e.g., (24.76) for a cubic crystal], and then from (29.3) it follows that

$$C_P = \hat{\Gamma}(V_0)T + (1 + 2x)nN_0K(12\pi^4/5)(T/\Theta_0)^3, \quad \text{at} \quad P = 0. \tag{29.5}$$

Table 4. Values of x defined by (29.4). Note that $d \ln \hat{\Gamma}/d \ln V$ for the hexagonal metals was measured for polycrystalline samples.

Metal	x
Cu	5×10^{-9}
Al	3×10^{-7}
Pb	2×10^{-8}
Cd	1×10^{-9}
Ti	2×10^{-5}

Finally since x is negligible compared to 1, we can analyze the measured C_P at $P = 0$ in the low-temperature limit by means of the equation

$$C_P \approx \hat{\Gamma} T + nN_0 K(12\pi^4/5)(T/\Theta_0)^3, \quad \text{at} \quad P = 0. \tag{29.6}$$

The measured values of $\hat{\Gamma}$ and Θ_0 for several materials are listed in Table 5; these were obtained by fitting C_P to (29.6) in the very low temperature region, for T of order $\Theta_0/100$. The zero-temperature and -pressure value of the molar volume and the bulk modulus for each material is also listed in Table 5. The V_0 were obtained from room temperature x-ray lattice parameters, together with the integrated thermal expansion from $T = 0$ to room temperature, and the B_0 were obtained from elastic constants or from pressure–volume measurements; references for these data are listed in Appendix 2.

Again for polycrystalline samples, the low-temperature thermal expansion may be analyzed by means of the equations derived for cubic crystals; from (18.65) and (18.66) the lattice contribution is

$$\beta(\text{lattice}) = -(B_0 V_A)^{-1} K(12\pi^4/5)(T/\Theta_0)^3 (d \ln \Theta_0/d \ln V). \tag{29.7}$$

This is the entire contribution to β for nonmetals, and again the variation of volume at $P = 0$ introduces a negligible T^7 contribution, so that (29.7) may be compared with the β measured at $P = 0$ for nonmetals. For metals the additional electronic plus electron–phonon contribution, evaluated at fixed volume, may be written from (24.80) as

$$\beta(\text{electronic}) = B_0^{-1}(d\hat{\Gamma}/dV)T + \tfrac{1}{2} V B_0^{-2}(d\hat{\Gamma}/dV)(d^2\hat{\Gamma}/dV^2)T^3. \tag{29.8}$$

Thus in order to analyze the measured β for metals, we have the same problem as in the case of the heat capacity C_P, namely the presence of a direct electronic T^3 term at fixed volume, shown in (29.8), and another T^3

Table 5. Low-temperature data for several materials. The Debye temperatures Θ_0 and the electronic coefficients $\hat{\Gamma}$ were obtained from heat capacity measurements.

Material	V_0 (cm^3/mole)	B_0 (10^{10} dyne/cm^2)	Θ_0 (°K)	$\hat{\Gamma}$ (10^{-4} cal/mole °K^2)	Reference for Θ_0, $\hat{\Gamma}$
Na	22.804	7.41	152.5	3.30	1
K	43.31	3.67	90.6	4.97	1
Cu	7.0421	142.0	345	1.67	2,3,4
Ag	10.143	108.7	226	1.54	2,4,5
Au	10.112	180.3	162	1.74	2,6,7
Al	9.8709	79.4	428	3.23	8
Pb	17.873	48.8	105	7.17	9
Th	19.643	58.1	170	11.2	10
Zn	8.9793	66.1	336	1.57	7
Cd	12.712	52.5	204	1.64	11
Ti	10.574	110.1	430	8.50	12
Zr	13.970	97.2	310	7.25	12
NaCl	26.391	26.6	321	0	13
KCl	36.690	19.7	235	0	13
KBr	42.231	17.6	174	0	13
KI	51.653	12.8	132	0	13
Ar	22.557	2.68	92.0	0	14
Kr	27.097	3.45	71.9	0	14

term at $P = 0$ arising from the volume-dependence of $B_0^{-1}(d\hat{\Gamma}/dV)$. For the metals we are studying, there are sufficient experimental data to evaluate these terms for the case of Pb; here the electronic T^3 term in (29.8) is 3×10^{-6} times the lattice T^3 term given by (29.7), and the volume variation of $B_0^{-1}(d\hat{\Gamma}/dV)$ leads to a T^3 term at $P = 0$ which is 6×10^{-6} times the lattice T^3 term. Hence for Pb, and presumably for all metals, these electronic contributions are negligible, and the measured β at $P = 0$ in the low-temperature limit may be analyzed by means of the equation

$$\begin{aligned}\beta &\approx B_0^{-1}(d\hat{\Gamma}/dV)T \\ &\quad - (B_0V_A)^{-1}K(12\pi^4/5)(d\ln\Theta_0/d\ln V)(T/\Theta_0)^3, \quad \text{at} \quad P = 0.\end{aligned} \tag{29.9}$$

Measured values of $d\ln\hat{\Gamma}/d\ln V$ and $d\ln\Theta_0/d\ln V$ for several materials are listed in Table 6; these were obtained by fitting β to (29.9) for temperatures $T \leqslant \Theta_0/20$. In a similar way the temperature-dependence at low temperatures of single-crystal β_{ij} can be analyzed by the equations of Sections 18 and 24, to find experimental values of the strain derivatives $d\hat{\Gamma}/d\eta_{ij}$ and

Table 6. Values of the logarithmic volume derivatives of Θ_0 and $\hat{\Gamma}$, obtained from low-temperature thermal expansion measurements on polycrystalline samples.

Material	$-d \ln \Theta_0/d \ln V$	$d \ln \hat{\Gamma}/d \ln V$	Reference
Cu	1.69	0.7	15
Ag	2.2		15
Al	2.65	1.8	15
Pb	2.7	1.7	15
Cd	2	0.5	16
Ti	5	6	16
NaCl	0.93	0	15
KCl	0.32	0	15
KBr	0.30	0	15
KI	0.28	0	15

$d\Theta_0/d\eta_{ij}$. We might expect such strain derivatives to be highly anisotropic, especially for anisotropic crystals.

The temperature-dependence of single-crystal elastic coefficients may also be analyzed with the aid of the lattice and electronic contributions derived in Sections 18 and 24, respectively. For nonmetals these coefficients have a lowest-order temperature-dependence of T^4 at constant configuration and also at constant stress; however, such T^4 contributions are very small and have not yet been experimentally identified. For cubic or polycrystalline metals, the lowest-order temperature-dependence of B_T is given by (24.79) as

$$B_T = B_0 - \tfrac{1}{2}V(d^2\hat{\Gamma}/dV^2)T^2. \tag{29.10}$$

Also from (1.41)–(1.43) it follows that $B_S - B_T = TV(\beta B_T)^2/C_V$, and the leading electronic plus electron–phonon contribution to this at low temperatures is $(V/\hat{\Gamma})(d\hat{\Gamma}/dV)^2T^2$, so that

$$B_S = B_0 + [(V/\hat{\Gamma})(d\hat{\Gamma}/dV)^2 - \tfrac{1}{2}V(d^2\hat{\Gamma}/dV^2)]T^2. \tag{29.11}$$

Finally we bring out the explicit temperature-dependence of B_S at $P = 0$ by accounting for the low-temperature thermal expansion and the corresponding change of B_0 with volume, to find

$$\begin{aligned} B_S = B_0(V_0) + [(V/\hat{\Gamma})(d\hat{\Gamma}/dV)^2 - \tfrac{1}{2}V(d^2\hat{\Gamma}/dV^2) \\ - (dB_0/dP)(d\hat{\Gamma}/dV)]T^2, \quad \text{at} \quad P = 0. \end{aligned} \tag{29.12}$$

The T^2 contribution to B_S at $P = 0$ was observed for several metals by Alers and Waldorf,* who carried out ultrasonic measurements in the temperature range 1–4°K. Their results for Pb are given by $B_S = B_0[1 - 23.5\ (10^{-7}\ {}^\circ K^{-2})T^2]$; from this together with our other tabulated data for Pb we find the first and last terms in the square brackets in (29.12) are negligible compared to the middle term, and $(V_0/B_0)(d^2\hat{\Gamma}/dV^2) = 47\ (10^{-7}\ {}^\circ K^{-2})$ for Pb. Measurement of this quantity for additional metals would be useful.

Calculations from Wave-Propagation Coefficients

According to the theory of Section 18, the $T = 0$ renormalized Debye temperature Θ_0 and its strain derivatives may be determined either from the temperature-dependences of thermodynamic functions at low temperatures, or from the $T = 0$ wave-propagation coefficients and their strain derivatives. Here we compare the results of these two methods of determination for several cubic crystals. The $T = 0$ extrapolations of the three independent stress–strain coefficients B_{11}, B_{12}, and B_{44} are listed in Table 7. From these coefficients we computed the propagation matrices, written out in (3.39) for cubic crystals, diagonalized these matrices to obtain the sound velocities as

Table 7. Experimental values of $B_{\alpha\beta}$, extrapolated to $T = 0$ from measurements down to liquid helium temperatures, for several cubic crystals. The Debye temperature calculated from these $B_{\alpha\beta}$ is denoted $\Theta_0(B_{\alpha\beta})$, and that obtained from heat capacity measurements is $\Theta_0(C_V)$.

Material	$B_{\alpha\beta}$ (10^{10} dyne/cm^2)			Reference for $B_{\alpha\beta}$	$\Theta_0(B_{\alpha\beta})$ (°K)	$\Theta_0(C_V)$ (°K)
	B_{11}	B_{12}	B_{44}			
K	4.17	3.41	2.86	17	90.9	90.6
Cu	176.2	124.9	81.8	18	344.5	345
Ag	131.5	97.3	51.1	19	226.5	226
Au	201.6	169.7	45.4	19	161.6	162
Al	114.3	61.92	31.62	20	430.5	428
Pb	55.54	45.42	19.42	21	105.3	105
NaCl	57.33	11.23	13.31	22	321.0	321
KCl	48.3	5.4	6.6	23	236	235
KBr	41.8	5.6	5.2	24	172	174
KI	33.8	2.2	3.7	23	131	132
Ar	4.39	1.83	1.64	25	90.5	92.0

* G. A. Alers and D. L. Waldorf, *Phys. Rev. Letters* **6**, 677 (1961).

Table 8. Experimental values of $(\partial B_{\alpha\beta}/\partial P)_T$ for several cubic crystals; the $B_{\alpha\beta}$ are adiabatic in all cases except for Al, where $(\partial B^T_{\alpha\beta}/\partial P)_T$ were reported. The calculation of $d \ln \Theta_0/d \ln V$ from these pressure derivatives is approximate, since that calculation should be based on $dB_{\alpha\beta}/dP$ measured at low temperature.

Material	T (°K)	$\left(\frac{\partial B_{11}}{\partial P}\right)_T$	$\left(\frac{\partial B_{12}}{\partial P}\right)_T$	$\left(\frac{\partial B_{44}}{\partial P}\right)_T$	Reference for $(\partial B_{\alpha\beta}/\partial P)_T$	$-d \ln \Theta_0/d \ln V$ from $(\partial B_{\alpha\beta}/\partial P)_T$	$-d \ln \Theta_0/d \ln V$ from β at $T \approx 0$
K	300	4.30	3.80	1.62	26	1.11	
Cu	300	6.4	5.5	2.5	27	1.64	1.69
Ag	300	7.03	5.75	2.31	28	2.14	2.2
Au	300	7.01	6.14	1.79	28	2.91	
Al	300	7.34	4.05	2.39	29	2.62	
Al	200	6.99	3.80	2.30	29	2.52	
Al	77.4	6.94	3.64	2.25	29	2.51	2.65
Pb	296	5.93	5.33	2.06	30	1.94	
Pb	195	5.82	5.26	1.97	30	1.83	2.7
NaCl	195	11.5	1.95	0.32	31	1.09	0.93
KCl	195	12.8	1.61	−0.42	31	0.36	0.32
KBr	300	13.0	1.59	−0.33	32	0.46	0.30
KI	300	14.0	2.42	−0.24	32	0.43	0.28

Recent work by G. K. White gives 1.04 for NaCl for the last column in Table 8.

functions of the propagation direction, and carried out the angle averages indicated in (18.29) to calculate Θ_0. These results are compared in Table 7 with Θ_0 determined from the heat capacity; in all cases the two values of the Debye temperature are in excellent agreement, and are within reasonable limits of experimental uncertainty.

Most presently available results for the pressure derivatives of wave-propagation coefficients are at room temperature. The values of $(\partial B_{\alpha\beta}/\partial P)_T$ for most of the cubic crystals we are studying, and the measurement temperatures, are listed in Table 8. From these results, we computed $d \ln \Theta_0/d \ln V$ with the aid of (18.74), (18.77), and (18.78); these calculated $d \ln \Theta_0/d \ln V$ are compared with those determined from low-temperature thermal expansion measurements in Table 8. We should of course use low-temperature measurements of $dB_{\alpha\beta}/dP$, and our failure to do this is the source of most of the discrepancies between the values of $d \ln \Theta_0/d \ln V$ for a given material, as listed in Table 8. The case of Pb suggests that one of the measurements, either the low-temperature β or the high-temperature $(\partial B_{\alpha\beta}/\partial P)_T$, is in error. The results for Al are interesting; they show a convergence of $-d \ln \Theta_0/d \ln V$ as calculated from the measured $(\partial B_{\alpha\beta}/\partial P)_T$ to a low-temperature value of about 2.50, with very little temperature-dependence for T below $\frac{1}{2}\Theta_0$.

30. QUALITATIVE TEMPERATURE- AND VOLUME-DEPENDENCES

Theoretical Expectations

There are two good reasons for studying the qualitative temperature- and volume-dependences of measured thermodynamic functions for a large number of materials. First, such a study serves as preparation for accurate comparisons of theoretical and experimental results for thermodynamic properties; one finds, for example, that it is quite inappropriate to compare a calculated curve of $\beta(V_0)$ *vs* T with a measured $\beta(V)$ *vs* T at high temperatures. Second, the qualitative T- and V-dependences provide a basis for extrapolating available experimental data to other regions of T and V; it is inaccurate, for example, to assume that C_V is represented by the classical limit $3nN_0K$ for all temperatures well above the Debye temperature.

We have collected experimental results on the thermal expansion coefficient, the heat capacity at constant pressure, and the bulk modulus for each of the 18 materials listed in Table 5. The measured β and C_P are generally for polycrystalline samples, while in most cases the bulk modulus and its pressure derivative were calculated from wave-propagation coefficients and their pressure derivatives. For the cubic crystals, these data allow us to find the explicit temperature-dependences of the thermodynamic functions at fixed volume, with the aid of the volume corrections derived in Section 4. We should like to be able to calculate the detailed configuration-dependences for the noncubic crystals, but the required experimental data are not generally available at present. The noncubic crystals Zn, Cd, Ti, and Zr were therefore treated by the same procedure as the cubic crystals; this analysis is subject to an error since the single-crystal bulk modulus and its pressure derivative are not the same as those for an isotropic polycrystalline specimen. Nevertheless, the error should be small, and the results of our analysis should be qualitatively reliable. There is much work to be done in experimental and theoretical analysis of the configuration-dependences of thermodynamic functions for noncubic crystals.

The basic experimental data are tabulated in Appendix 2; from these data we computed $C_P - C_V$ and $B_S - B_T$, as well as the various quantities required to correct the thermodynamic functions to fixed volume. The pressure derivative of the bulk modulus was computed from measured pressure derivatives of the adiabatic wave-propagation coefficients where such data were available, and $(\partial B_S/\partial P)_T$ was converted to $(\partial B_T/\partial P)_T$ by means of (1.59). For Th, Zn, and Zr we used Bridgman's measurements of $(\partial B_T/\partial P)_T$, and applied the appropriate calibration corrections. For Ar and Kr we used results obtained by fitting Stewart's *V vs P* data. Results for $(\partial B_T/\partial P)_T$, at near room temperature except for Ar and Kr, are collected in Table 9. For Cu, Al, Pb, NaCl, and KCl the experimental $(\partial B_S/\partial P)_T$ was also provided at temperatures significantly below room temperature in the references cited in Table 9; we were therefore able to include an estimate of the temperature-dependence of $(\partial B_T/\partial P)_T$ in our calculations for these materials. For all other materials, $(\partial B_T/\partial P)_T$ was taken as constant.

The equations required for correcting the thermodynamic functions to the fixed volume V_0 are developed as power series in $a_0 = (V - V_0)/V_0$, where V is the volume at any T and $P = 0$, and V_0 is the volume at $T = 0$ and $P = 0$. Of particular interest here are (4.34) for $B_T(V_0)$, (4.36) for $\beta B_T(V_0)$, (4.37) for $S(V_0)$, (4.41) for $C_V(V_0)$, and (4.44) or (4.45) for $\beta(V_0)$. In the present calculations the first volume corrections, i.e., the terms linear in a_0, are quite accurate for β and S. Because of near cancellation of two terms, the first volume correction for βB_T and the second volume correction for S are only qualitatively reliable. The first volume correction for B_T is only qualitatively reliable in cases where the temperature-dependence of $(\partial B_T/\partial P)_T$ is not known. The first volume correction for C_V and the second volume correction for β are felt to be quite unreliable, and in some cases they may even have the wrong sign. These qualitative conclusions regarding the accuracy of various volume corrections reflect the present-day availability

Table 9. Bulk modulus pressure derivatives, measured near room temperature except for Ar and Kr. These results are based on pressure derivatives of the adiabatic wave-propagation coefficients for all materials except Th, Zn, Zr, Ar, and Kr; at present there is no measurement for Ti.

Material	$(\partial B_S/\partial P)_T$	$(\partial B_T/\partial P)_T$	Reference
Na	3.60	3.71	33
K	3.97	3.94	26
Cu	5.8	5.8	27
Ag	6.18	6.20	28
Au	6.43	6.52	28
Al		5.15	29
Pb	5.53	5.70	30
Th		6.2	34
Zn		4.9	34
Cd	6.65	6.63	35
Zr		9.9	34
NaCl	5.27	5.28	31
KCl	5.35	5.51	31
KBr	5.38	5.62	32
KI	6.28	6.16	32
Ar (65°K)		8.0	36
Kr (77°K)		9.0	36

and accuracy of experimental data, and the same conclusions are more or less applicable to all materials.

As we saw in Section 29, the volume corrections are quantitatively negligible in the low-temperature region. At intermediate temperatures, say for $T \lessdot \Theta_0$, the volume corrections are small, but at high temperatures they can be quite significant. Let us examine *very approximately* the temperature- and volume-dependences of the thermodynamic functions at high T, as predicted by the lattice-dynamics theory of Section 19; note that the effect of the electronic excitation is included by replacing $\mathscr{A}_2$ by $\mathscr{A}_2 - \frac{1}{2}\Gamma$, according to (24.73), and the electron–phonon interaction effects may be neglected at sufficiently high T, according to (25.86). The orders of magnitude of the lattice potential, the harmonic vibrational energy, and the anharmonic plus electronic excitation energy are given for most materials by the following very approximate relations.

$$\Phi_0 \text{ and } V^2(d^2\Phi_0/dV^2) \sim 10^4 \text{ cal/mole;} \tag{30.1}$$

$$3nN_0K \sim \text{cal/mole °K;} \tag{30.2}$$

$$\mathscr{A}_2 - \tfrac{1}{2}\Gamma \sim 10^{-4} \text{ cal/mole °K}^2. \tag{30.3}$$

The leading high-temperature contributions to the heat capacity at constant volume are written from (19.57) as

$$C_V = 3nN_0K\{1 - [(2\mathscr{A}_2 - \Gamma)/3nN_0K]T\}; \tag{30.4}$$

the order of magnitude theoretical estimate of C_V at high T is then

$$C_V \sim 3nN_0K[1 + 0(10^{-4}T)], \quad T \text{ in } {}^\circ\text{K}, \tag{30.5}$$

where $0(10^{-4}T)$ means a term of the order of $10^{-4}T$. Hence the leading term of C_V is $3nN_0K$, which is strictly constant, so that C_V should be only weakly dependent on V and T in the high-T region. An important point to recognize is that only the temperature-dependent term in (30.4) for C_V depends on the volume, so when one corrects C_V to fixed volume one is actually correcting only the anharmonic plus electronic contribution to fixed volume, at high T. In other words, at high T,

$$(\partial C_V/\partial V)_T = -[d(2\mathscr{A}_2 - \Gamma)/dV]T; \tag{30.6}$$

it is no wonder the thermodynamic correction of C_V to fixed volume is difficult to calculate accurately from available experimental data!

The high-temperature expansion of $V\beta B_T$ is given by (19.58), and the leading high-temperature contributions may be written

$$V\beta B_T = 3nN_0K\langle\gamma_\kappa\rangle\left\{1 - \frac{1}{\langle\gamma_\kappa\rangle}\frac{d\ln(\mathscr{A}_2 - \frac{1}{2}\Gamma)}{d\ln V}\frac{(2\mathscr{A}_2 - \Gamma)}{3nN_0K}T\right\}. \tag{30.7}$$

Now $\langle\gamma_\kappa\rangle$ is of order 1, and we further expect $d\ln(\mathscr{A}_2 - \frac{1}{2}\Gamma)/d\ln V$ to be of order 1, so the order of magnitude expression for $V\beta B_T$ at high T is

$$V\beta B_T \sim 3nN_0K\langle\gamma_\kappa\rangle[1 + 0(10^{-4}T)], \quad T \text{ in } {}^\circ\text{K}. \tag{30.8}$$

Here, in contrast to the heat capacity expression (30.5), the leading term depends on the volume, so we expect $V\beta B_T$ to have a noticeable volume-dependence, while the temperature-dependence at fixed volume should be small, in the high-T region. Actually the volume-dependence of $\langle\gamma_\kappa\rangle$ appears to be rather small for most materials, so that the volume-dependence of $V\beta B_T$ turns out to be small for the materials we have studied. According to (19.62) the leading contribution to the macroscopic Grüneisen parameter γ is $\langle\gamma_\kappa\rangle$ in the high-T region; hence an estimate of $d\ln\langle\gamma_\kappa\rangle/d\ln V$ is given by $(\partial\ln\gamma/\partial\ln V)_T$ evaluated for $T \gtrsim \Theta_0$. We have calculated $(\partial\ln\gamma/\partial\ln V)_T$ from the available experimental data by means of (1.62); the results are listed in Table 10, and must be regarded as order of magnitude estimates—in some cases our result may even have the wrong sign. Nevertheless we can conclude that $(\partial\ln\gamma/\partial\ln V)_T$ is of order 1, and hence the volume-dependence of $V\beta B_T$ should be of the same order as the volume itself, that is,

Table 10. Estimated room temperature values of $(\partial \ln \gamma/\partial \ln V)_T$. The results in which we have the least confidence are enclosed in parentheses.

Material	$(\partial \ln \gamma/\partial \ln V)_T$	Material	$(\partial \ln \gamma/\partial \ln V)_T$
Na	1.4	Zn	(1.3)
K	(−0.2)	Cd	0.4
Cu	0.9	Zr	(−2.5)
Ag	0.7	NaCl	0.6
Au	1.7	KCl	2.2
Al	1.3	KBr	2.8
Pb	1.8	KI	−0.7
Th	(−3.4)		

the relative change in $V\beta B_T$ in going from V to V_0 should be of order $(V - V_0)/V_0$.

The high-temperature expansion (19.50) of B_T is conveniently written in the form

$$VB_T = V^2(d^2\Phi_0/dV^2) + 3nN_0K\langle \xi_\kappa - \gamma_\kappa^2\rangle T + V^2[d^2(\mathscr{A}_2 - \tfrac{1}{2}\Gamma)/dV^2]T^2, \quad (30.9)$$

where as usual $\mathscr{A}_2$ has been replaced by $\mathscr{A}_2 - \frac{1}{2}\Gamma$ to include electronic effects. It is reasonable to assume that $\langle \xi_\kappa \rangle$ is of order 1 and $V^2[d^2(\mathscr{A}_2 - \frac{1}{2}\Gamma)/dV^2]$ is of order $\mathscr{A}_2 - \frac{1}{2}\Gamma$; then the order of magnitude representation of (30.9) is

$$VB_T \sim V^2(d^2\Phi_0/dV^2)[1 + 0(10^{-4}T) + 0(10^{-8}T^2)], \quad T \text{ in } {}^\circ\text{K}. \quad (30.10)$$

Again the leading term depends on the volume, and we note from Table 9 that $(\partial B_T/\partial P)_T = -(\partial \ln B_T/\partial \ln V)_T$ is of order 5, so the relative volume-dependence of VB_T at high T should be considerably larger than that of $V\beta B_T$. On the other hand, at fixed volume the temperature-dependence of B_T should again be small at high T.

Dividing (30.8) for $V\beta B_T$ by (30.10) for VB_T gives the order of magnitude expression for β at high T:

$$\beta \approx \frac{3nN_0K\langle \gamma_\kappa \rangle}{V^2(d^2\Phi_0/dV^2)}[1 + 0(10^{-4}T) + 0(10^{-8}T^2)]. \quad (30.11)$$

From this we expect the relative volume-dependence of β to be about the same as that of B_T at high T, and larger than that of $V\beta B_T$, while at fixed volume the temperature-dependence of β should be small at high T.

We may note that the same qualitative arguments apply to the tensor properties of noncubic crystals, and the qualitative temperature- and strain-dependences of C_η, b_{ij}, and C^T_{ijkl} should be similar to those of C_V, βB_T, and B_T, respectively.

Experimental Results

We have corrected $C_V(V)$ to $C_V(V_0)$ for all the 18 materials listed in Table 5, except Ti. In all cases there is a large cancellation among the three terms in the square bracket in (4.41), and the volume correction is quite small at high temperature. The temperature-dependences of both $C_V(V)$ and $C_V(V_0)$ are small in the high-temperature region. The example of KCl is shown in Figure 7. The heat capacity volume correction for KCl is rather larger than for most of the other materials; also $C_V(V) - C_V(V_0)$ is negative for KCl at high T, so from (30.6) we can conclude

$$d(\mathscr{A}_2 - \tfrac{1}{2}\Gamma)/dV > 0 \quad \text{for KCl.} \tag{30.12}$$

Such an estimate of the sign of an anharmonic function can be of value in testing model calculations and in extrapolating thermodynamic functions to volumes and temperatures beyond where they are measured.

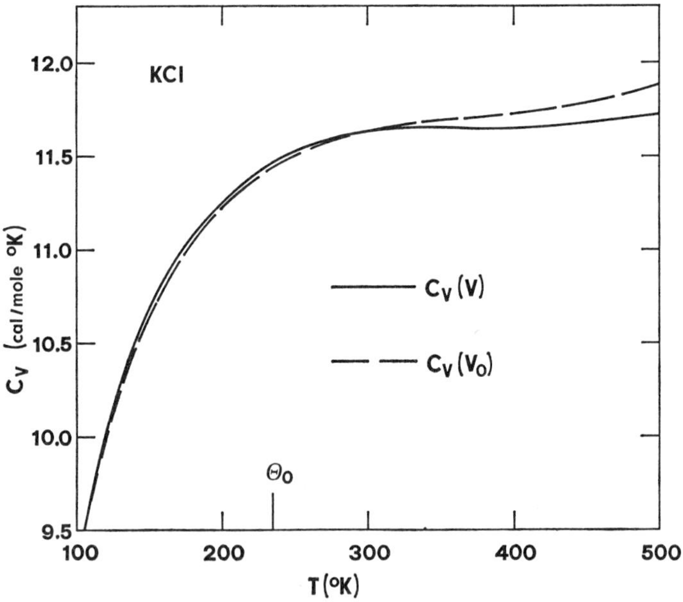

Figure 7. The experimental curves of $C_V(V)$ and $C_V(V_0)$ for KCl.

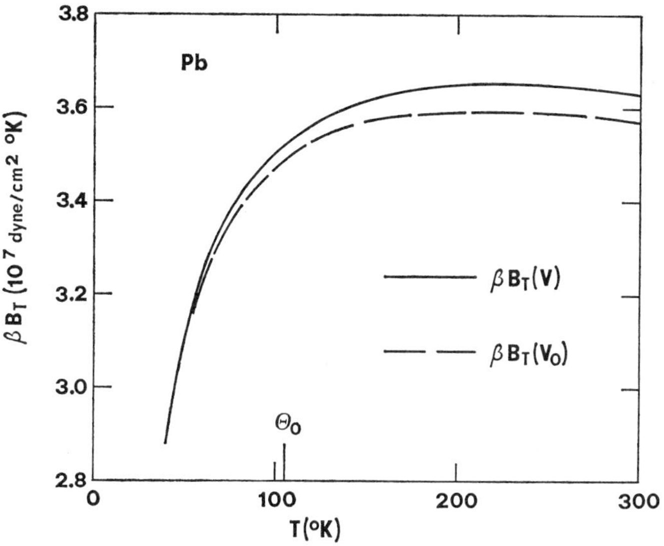

Figure 8. The experimental curves of $\beta B_T(V)$ and $\beta B_T(V_0)$ for Pb.

We have also corrected βB_T from V to V_0 for all the 18 materials in Table 5 except Ti, and again the volume corrections are quite small and $\beta B_T(V)$ and $\beta B_T(V_0)$ both have small temperature-dependence at high T. The two terms in the square bracket in (4.36) arise from the separate volume corrections for β and B_T, and while each term is fairly large, they tend to cancel and leave a net small volume correction for βB_T. The example of Pb is shown in Figure 8, where it is seen that $\beta B_T(V) - \beta B_T(V_0)$ is positive at high T. Then from (30.8) we expect that

$$d \ln\langle\gamma_\kappa\rangle / d \ln V > 1 \quad \text{for Pb,} \tag{30.13}$$

and since this is approximately equal to $(\partial \ln \gamma/\partial \ln V)_T$ at high T, the result (30.13) is in agreement with the value $(\partial \ln \gamma/\partial \ln V)_T = 1.8$ listed in Table 10 for Pb.

Volume corrections for the isothermal bulk modulus B_T are shown for several examples in Figures 9–11. From these examples, and the others not shown, we can conclude that $B_T(V_0)$ is more constant in temperature than is $B_T(V)$. At the present time, the volume corrections are not accurate enough to give us any reliable information about the temperature derivatives of B_T at the fixed volume V_0. We note that such temperature derivatives can be obtained with reasonable accuracy at the volume and temperature at which

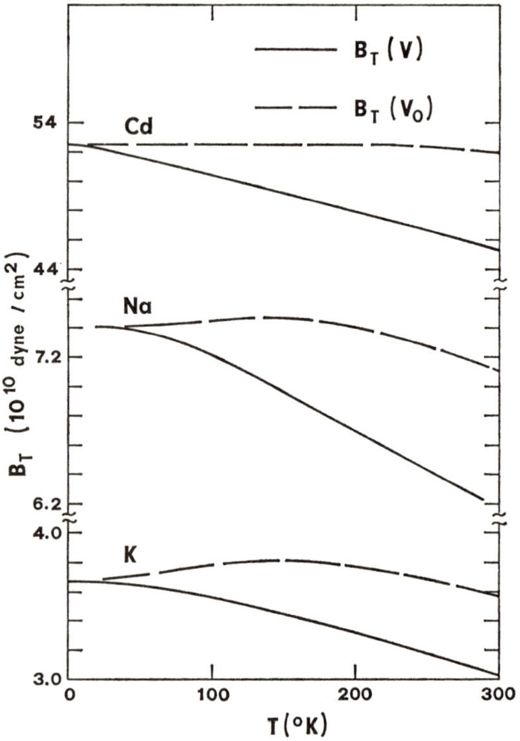

Figure 9. The experimental curves of $B_T(V)$ and $B_T(V_0)$ for Cd, Na, and K. Note the changes in scale for B_T.

$(\partial B_S/\partial P)_T$ or $(\partial B_T/\partial P)_T$ is measured; such calculations for room temperature and zero pressure are illustrated in Section 32. In some cases, particularly for Na, K, and KI, $B_T(V_0)$ increases at first as T increases from $T = 0$. Although this increase may be a real experimental result, in all fairness we have to admit that the combined errors in our $B_T(V_0)$ curves are at least as large as the apparent increases, and that we have no basis for trying to interpret the observed detailed temperature-dependences of $B_T(V_0)$. In addition we note that there is no increase of $B_T(V_0)$ as T increases for NaCl, KCl, and KBr, in contrast to the result for KI.

For the thermal expansion coefficient β at high temperatures, the volume correction is large, and $\beta(V_0)$ is nearly constant with T; these results hold for all the materials we have studied for temperatures in the range $\Theta_0 \leq T \leq 1.5\Theta_0$. For those materials for which we have good high-temperature data, $\beta(V_0)$ remains very weakly temperature-dependent for T up to 2 or 3 times

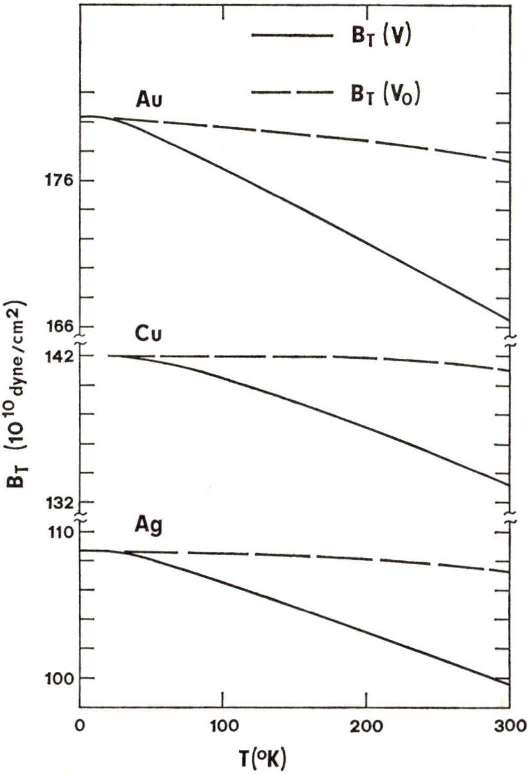

Figure 10. The experimental curves of $B_T(V)$ and $B_T(V_0)$ for Au, Cu, and Ag. Note the breaks in scale for B_T.

Θ_0. This qualitative behavior of β is in agreement with the theoretical estimate given by (30.11). Illustrative experimental curves of $\beta(V)$ and $\beta(V_0)$ are shown in Figures 12–15.

In the discussion of the qualitative behavior of thermodynamic functions, the macroscopic Grüneisen parameter $\gamma = V\beta B_T/C_V$ is useful because it depends only weakly on the temperature, except at low temperature, and on the volume. For the materials we have studied, $\gamma(V)$ is fairly constant in the temperature range $0.7\Theta_0 \leq T \leq 1.5\Theta_0$, and in some cases to considerably higher temperatures. Some illustrative curves of $\gamma(V)$ are shown in Figures 16 and 17. The fact that $\gamma(V)$ is constant down to temperatures well below the Debye temperature is expected from the high-temperature expansion (19.62) of γ, where the two harmonic lattice-dynamics terms of order T^{-2} should tend to cancel. The continual variation of $\gamma(V)$ for Zn from 100 to

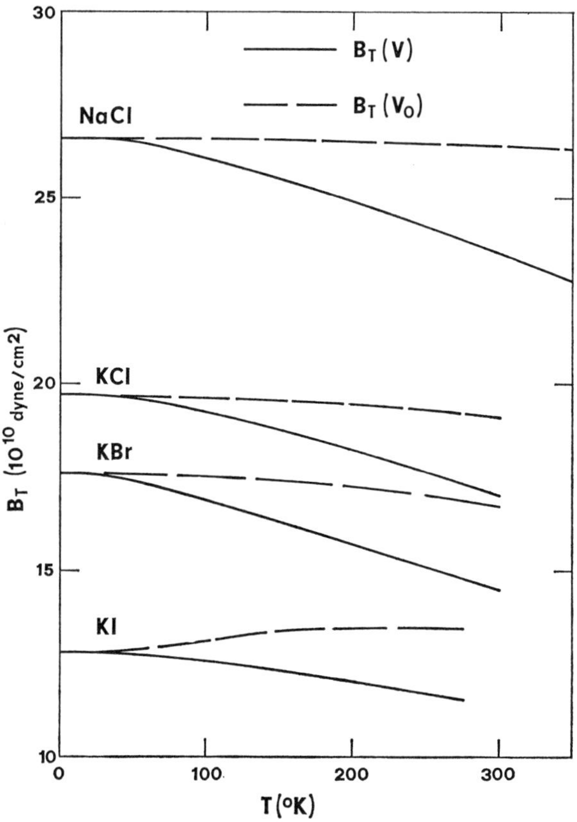

Figure 11. The experimental curves of $B_T(V)$ and $B_T(V_0)$ for NaCl, KCl, KBr, and KI.

500°K may be the result of the large anisotropy of this crystal and the corresponding continual variation of the configuration, say c/a, as the temperature is varied; Cd shows similar behavior.

We calculated $\gamma(V_0)$ from the functions $\beta B_T(V_0)$ and $C_V(V_0)$; hence there is considerable accumulated uncertainty in our results for $\gamma(V) - \gamma(V_0)$. Two cases where we have some faith in the volume corrections for γ are shown in Figure 18; for both Kr and KCl it appears that $\gamma(V_0)$ is more strongly temperature-dependent than is $\gamma(V)$ in the high-temperature region. For both cases shown in Figure 18, the decrease in $\gamma(V_0)$ with increasing T at high T is due both to the corresponding decrease in $\beta B_T(V_0)$ and the increase in $C_V(V_0)$. Hence from the theoretical point of view, the temperature variation of $\gamma(V_0)$ at high temperatures must be considered to contain

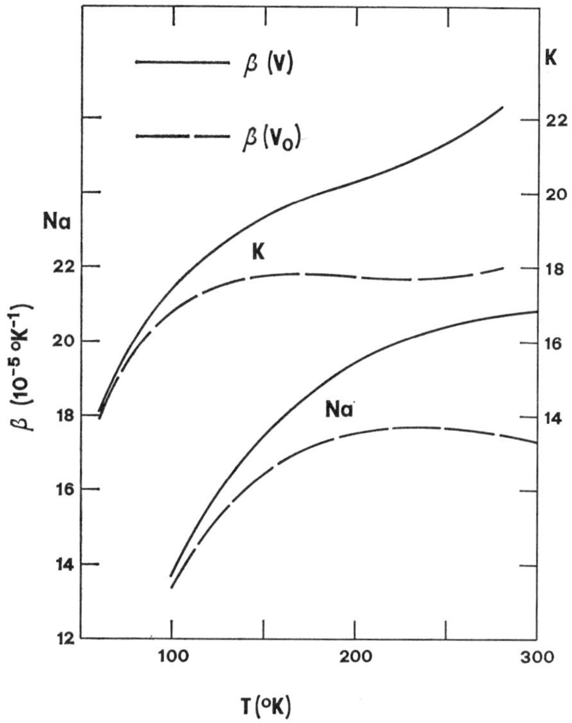

Figure 12. The experimental curves of $\beta(V)$ and $\beta(V_0)$ for K (right scale) and Na (left scale).

significant contributions from *both* the anharmonic terms written out in (19.62).

Elastic Coefficients

We have not made a detailed study of the qualitative temperature- and volume-dependences of the various elastic coefficients, but by analogy with the preceding discussion of the bulk modulus, it is reasonable to expect in general that the strain derivatives of the free energy are more nearly constant in temperature at fixed crystal configuration than at zero stress. There may well be exceptions, of course. For a cubic crystal under isotropic pressure there are three independent elastic constants $C_{\alpha\beta}$ and three independent stress–strain coefficients $B_{\alpha\beta}$; the $C_{\alpha\beta}$ and $B_{\alpha\beta}$ are related according to (3.38). The three $B_{\alpha\beta}$ may be considered to be one pure volume constant and two pure shear constants; the volume constant is of course $B = \frac{1}{3}(B_{11} + 2B_{12})$, and the two shear constants are B_{44} and $\frac{1}{2}(B_{11} - B_{12})$. Now from (2.120), the adiabatic and isothermal shear constants are the same, and then with the

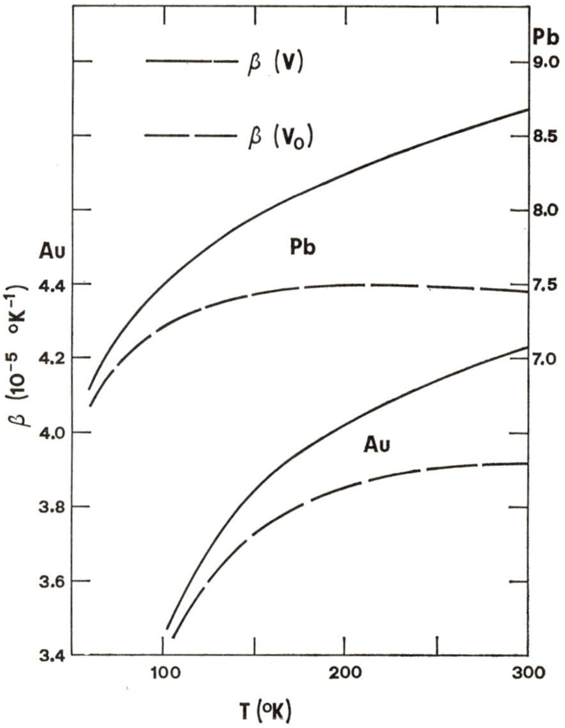

Figure 13. The experimental curves of $\beta(V)$ and $\beta(V_0)$ for Pb (right scale) and Au (left scale).

aid of (3.38) we can write for adiabatic or isothermal cases

$$C_{44} = B_{44} + P,$$
$$\tfrac{1}{2}(C_{11} - C_{12}) = \tfrac{1}{2}(B_{11} - B_{12}) + P. \tag{30.14}$$

At $P = 0$ the $C_{\alpha\beta}$ and $B_{\alpha\beta}$ are the same, but at the fixed volume V_0 they are not the same, except at $T = 0$. When (30.14) is evaluated at V_0, the appropriate pressure is $P_0(T)$, the pressure required to reduce the volume to V_0 at a given T. At room temperature, P_0 is generally quite small compared to the bulk modulus, but not so small compared to the shear constants, and the temperature-dependence of P_0 is a significant part of the temperature-dependence of the $B_{\alpha\beta}(V_0)$. Values of P_0 at 300°K, as evaluated from (4.12), are listed in Table 11.

As it was pointed out in Section 3, particularly in (3.32), when the stress is isotropic pressure the measured wave propagation coefficients can always be interpreted as stress–strain coefficients, and hence measurement of the

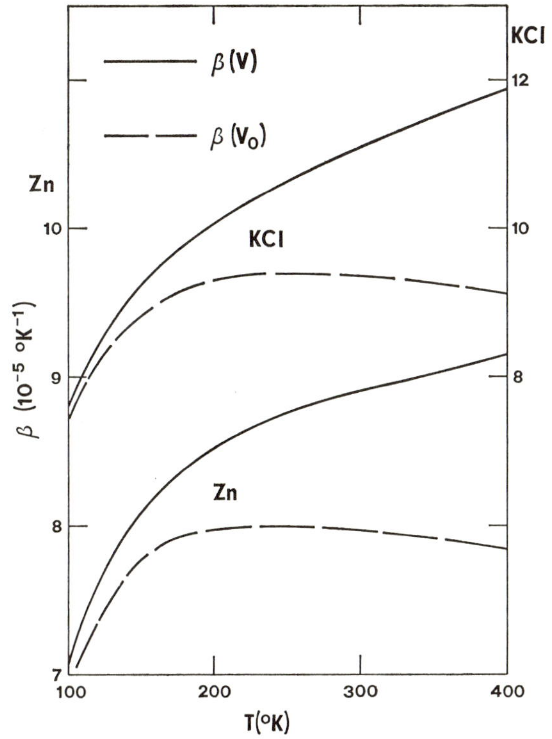

Figure 14. The experimental curves of $\beta(V)$ and $\beta(V_0)$ for KCl (right scale) and Zn (left scale).

pressure derivatives of sound velocities yields directly $(\partial B_{\alpha\beta}/\partial P)_T$. For cubic crystals these experimental results can be used to correct the $P = 0$ values of shear combinations of $B_{\alpha\beta}$ to the fixed volume V_0, according to the following equations which are correct to first order in a_0.

$$B_{44}(V_0) = B_{44}(V) + a_0 B_T(\partial B_{44}/\partial P)_T + \cdots ;$$
$$(B_{11} - B_{12})(V_0) = (B_{11} - B_{12})(V) + a_0 B_T[\partial(B_{11} - B_{12})/\partial P]_T + \cdots . \tag{30.15}$$

We have corrected the shear combinations of $B_{\alpha\beta}$ to fixed volume by means of (30.15) for the example of Pb, and we have also calculated the shear combinations of $C_{\alpha\beta}(V_0)$ by means of (30.14) with $P = P_0$. The results are shown in Figure 19, and perhaps the most important observation is the large differences between $B_{44}(V)$, $B_{44}(V_0)$, and $C_{44}(V_0)$, and the similar differences for the other shear coefficient. These differences must be recognized when

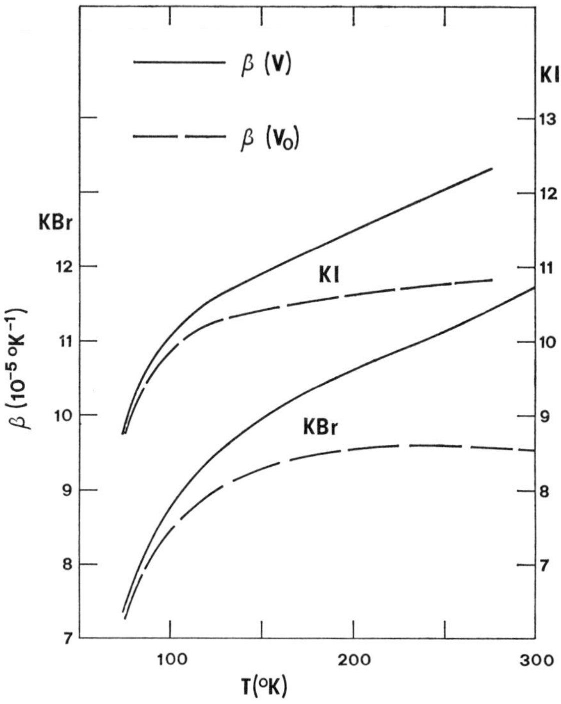

Figure 15. The experimental curves of $\beta(V)$ and $\beta(V_0)$ for KI (right scale) and KBr (left scale).

Table 11. Values of P_0 at 300°K, calculated by (4.12).

Material	P_0 (10^{10} dyne/cm^2)	Material	P_0 (10^{10} dyne/cm^2)
Na	0.290	Zn	1.245
K	0.169	Cd	1.106
Cu	1.384	Ti	0.50
Ag	1.316	Zr	0.322
Au	1.717	NaCl	0.599
Al	0.973	KCl	0.418
Pb	0.965	KBr	0.397
Th	0.445	KI	0.36

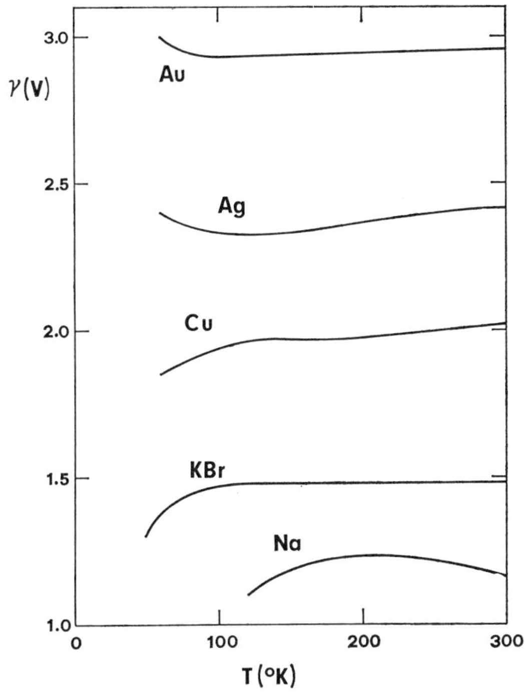

Figure 16. The experimental curves of $\gamma(V)$ for Au, Ag, Cu, KBr, and Na.

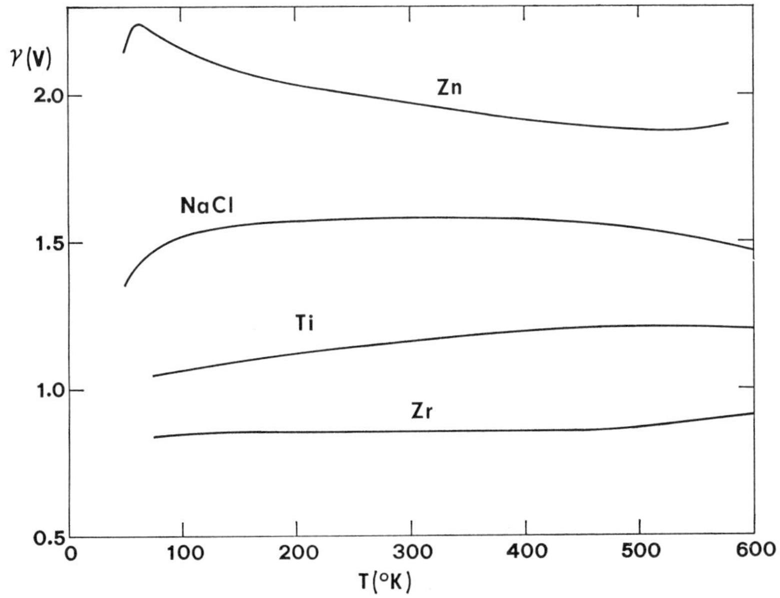

Figure 17. The experimental curves of $\gamma(V)$ for Zn, NaCl, Ti, and Zr.

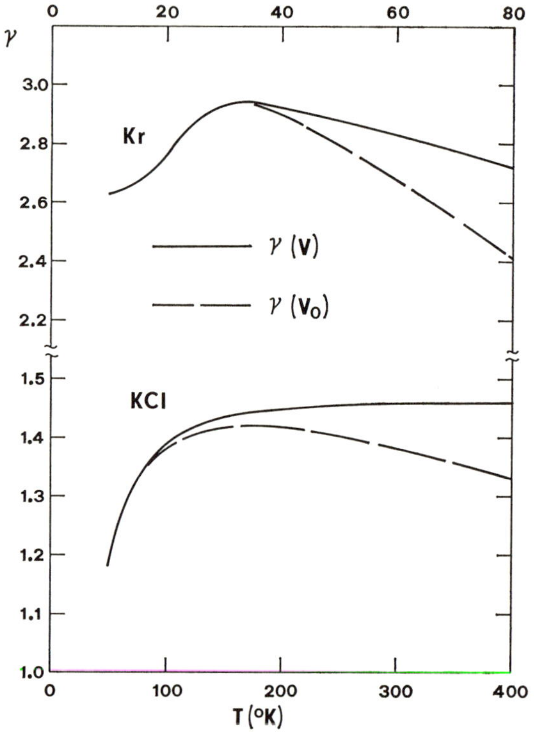

Figure 18. The experimental curves of $\gamma(V)$ and $\gamma(V_0)$ for Kr (temperature scale at the top) and KCl (temperature scale at the bottom). Note the change in scale for γ.

one compares theoretical lattice-dynamics calculations with experimental results—a requirement which has generally been overlooked.

31. HIGH-TEMPERATURE ANALYSIS

Procedures

As it was pointed out in Section 19, a quantitative analysis of experimental data in the intermediate- to high-temperature region can give a great deal of information about the harmonic and anharmonic lattice-dynamical properties of crystals. If the experimental data are sufficiently accurate, one can correct the measured thermodynamic functions to fixed volume, or to fixed crystal configuration in the case of noncubic crystals, and use a computer to

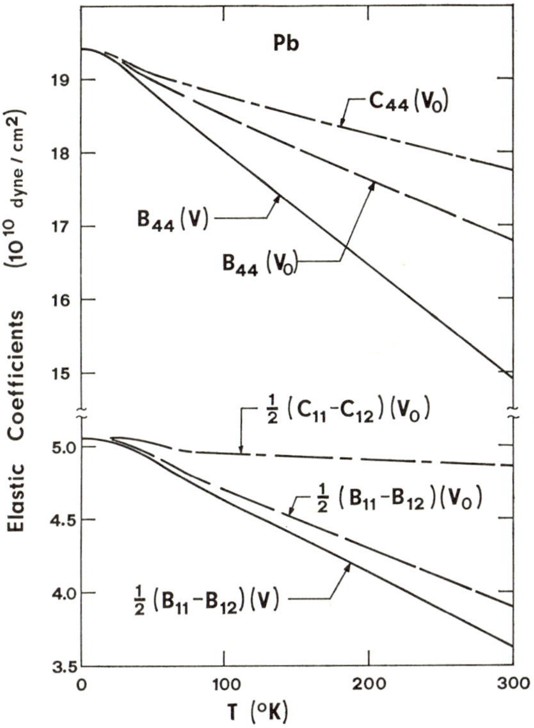

Figure 19. The experimental curves of shear combinations of $B_{\alpha\beta}$ at V and at V_0, and of shear combinations of $C_{\alpha\beta}$ at V_0, for Pb. Note the change in scale for the elastic coefficients.

fit the temperature-dependence to the theoretical expressions of Section 19. Consider, for example, the high-temperature expansion (19.40) for the entropy. This expansion is valid for any crystal configuration, and contains the series of terms of order T, ln T, constant, T^{-2}, T^{-3}, T^{-4}, $\cdots$. It would be extremely difficult to use this expansion to analyze the temperature-dependence of the entropy at $P = 0$, since each coefficient in the expansion depends on the crystal configuration, and hence each coefficient also varies with T at $P = 0$. However, the thermodynamic correction of the entropy to a condition of constant configuration can be made quite accurately, and then a computer analysis will show quantitatively (a) whether the theoretical expansion (19.40) fits the experimental data, and (b) how accurately each of the coefficients may be determined from the data.

In the entropy analysis, the lattice-dynamical terms of order T^{-2}, T^{-3}, T^{-4}, $\cdots$, are presumably important in the temperature range from say

$\frac{1}{2}\Theta_{H\infty}$ to $\Theta_{H\infty}$, and should be determined from analysis throughout this temperature range. At slightly higher temperatures, say from $\Theta_{H\infty}$ to $\frac{3}{2}\Theta_{H\infty}$, we might expect the leading terms of order T and $\ln T$ to be most important in the temperature-dependence of the entropy at fixed configuration.

For metals there is the added contribution from electronic excitations and electron–phonon interactions; at high temperature the corresponding entropy contribution is $\Gamma T +$ constant, and the constant, which arises from electron–phonon interactions, is negligible compared to the lattice-dynamical constant term $3nN_0K$. Therefore for $T \gtrsim \Theta_{H\infty}$ the electron–phonon interactions may be quantitatively neglected in the entropy, and the electronic excitation contribution is included by replacing $\mathscr{A}_2$ by $\mathscr{A}_2 - \frac{1}{2}\Gamma$, as noted in (24.73). By similar arguments the electron–phonon interactions may be quantitatively neglected in the high-temperature expansions of the heat capacity C_η, the thermal stresses b_{ij}, and the function $V\beta B_T$. We will also neglect electron–phonon contributions to thermodynamic functions at temperatures below $\Theta_{H\infty}$, where such contributions are small but not entirely negligible in general.

As a start in the analysis of the thermodynamic functions at high temperatures, we have carried out a graphical analysis to see if the temperature-dependences do indeed agree with the theoretical expectations. Again all the materials were treated by the equations appropriate for cubic crystals (or isotropic materials), and this procedure is only approximate for the hexagonal crystals we analyzed. The high-temperature expansion of the entropy is written in (19.56), and with electronic effects included this is

$$S = 3nN_0K[\langle\ln(KT/\hbar\omega_\kappa)\rangle + 1 + \tfrac{1}{24}\langle(\hbar\omega_\kappa/KT)^2\rangle - \tfrac{1}{960}\langle(\hbar\omega_\kappa/KT)^4\rangle + \cdots] - 2(\mathscr{A}_2 - \tfrac{1}{2}\Gamma)T + 2\mathscr{A}_{-2}T^{-3} + \cdots. \quad (31.1)$$

It is convenient to introduce a constant temperature T_0, and rewrite the log term in (31.1) as

$$\langle\ln(KT/\hbar\omega_\kappa)\rangle = \ln(T/T_0) + \langle\ln(KT_0/\hbar\omega_\kappa)\rangle. \quad (31.2)$$

We then define the function Q by

$$Q = S - 3nN_0K\ln(T/T_0), \quad (31.3)$$

and from (31.1) and (31.2) it follows that

$$Q = 3nN_0K[1 + \langle\ln(KT_0/\hbar\omega_\kappa)\rangle] + 3nN_0K[\tfrac{1}{24}\langle(\hbar\omega_\kappa/KT)^2\rangle - \tfrac{1}{960}\langle(\hbar\omega_\kappa/KT)^4\rangle + \cdots] - 2(\mathscr{A}_2 - \tfrac{1}{2}\Gamma)T + 2\mathscr{A}_{-2}T^{-3} + \cdots. \quad (31.4)$$

Finally the temperature derivative of Q at constant volume, arranged in a series of decreasing powers of T, is given by

$$(\partial Q/\partial T)_V = (\Gamma - 2\mathscr{A}_2) - (3nN_0K\hbar^2\langle\omega_\kappa^2\rangle/12K^2)T^{-3} - 6\mathscr{A}_{-2}T^{-4} + (3nN_0K\hbar^4\langle\omega_\kappa^4\rangle/240K^4)T^{-5} + \cdots. \tag{31.5}$$

Thus if $(\partial Q/\partial T)_V$ is evaluated at fixed volume and plotted as a function of T^{-3}, the graph should be a straight line at high temperatures, with intercept $\Gamma - 2\mathscr{A}_2$ and slope $-(3nN_0K\hbar^2\langle\omega_\kappa^2\rangle/12K^2)$. At lower temperatures the graph should depart from a straight line, due to the T^{-4} and T^{-5} terms in (31.5); of these two terms the harmonic T^{-5} term should be dominant, and since this is always positive we expect the function $(\partial Q/\partial T)_V$ to rise above the high-temperature straight line as the temperature is lowered. Note that the temperature T_0 drops out of $(\partial Q/\partial T)_V$, and can be chosen as desired for convenience in the analysis.

We have corrected S to fixed volume for the 18 materials listed in Table 5, calculated Q and then $(\partial Q/\partial T)_V$ evaluated at V_0, and found the intercept and slope of the straight line fitting $(\partial Q/\partial T)_V$ *vs* T^{-3} in the high-temperature region. The correction of $S(V)$ to $S(V_0)$ was calculated by means of (4.37); the first volume correction $a_0V_0\beta B_T$ was obtained quite accurately for all the materials we studied, and the second volume correction [the a_0^2 term in (4.37)] was obtained with qualitative reliability. The first volume correction was important in this entropy analysis, while the second volume correction was found to make no significant differences in the intercepts and slopes for all the 18 materials we studied.

Presently available thermal expansion measurements are also of sufficient accuracy to allow their quantitative analysis at high temperatures. The high-temperature expansion of $V\beta B_T$ for a cubic crystal is written in (19.58), and with inclusion of electronic effects this becomes

$$V\beta B_T/3nN_0K = \langle\gamma_\kappa\rangle - \tfrac{1}{12}\langle\gamma_\kappa(\hbar\omega_\kappa/KT)^2\rangle + O(T^{-4}) + (3nN_0K)^{-1}[Vd(\Gamma - 2\mathscr{A}_2)/dV]T + O(T^{-3}). \tag{31.6}$$

The temperature derivative of this function, evaluated at constant volume and plotted as a function of T^{-3}, should again be a straight line in the high-temperature region, with intercept $(3nN_0K)^{-1}[Vd(\Gamma - 2\mathscr{A}_2)/dV]$ and slope $\hbar^2\langle\gamma_\kappa\omega_\kappa^2\rangle/6K^2$. Unfortunately the experimental accuracy for $V\beta B_T(V_0)$ is not sufficient to allow us to analyze the temperature derivative in this manner, so instead we simply plotted $V\beta B_T(V_0)/3nN_0K$ as a function of T^{-2}. For this graph, if the electronic plus anharmonic term of order T in (31.6) is negligible, we should find a straight line in the high-temperature region, with intercept $\langle\gamma_\kappa\rangle$ and slope $-\hbar^2\langle\gamma_\kappa\omega_\kappa^2\rangle/12K^2$. Such high-temperature dependence of $V\beta B_T(V_0)$ was found for many, but not all, of the 18 materials we studied.

In connection with the analysis of thermodynamic data at high temperatures, there are two procedures which we wish to caution against. For the entropy or heat capacity, for example, the electronic contribution at high T is ΓT. It is not fair, however, to take a constant value of Γ and subtract ΓT from $P = 0$ values of S or C_V to get a "lattice" contribution to S or C_V at $P = 0$, since Γ varies with the volume, and hence also varies with T at $P = 0$. Alternatively, if one *must* try to remove the electronic contribution to S or C_V at $P = 0$, then the volume variation of Γ should be taken into account.

A second caution regards the method of correcting thermodynamic functions to fixed volume. A procedure which has been used for the entropy and heat capacity is to find the effective Debye temperature $\Theta(V)$ corresponding to the experimental results at zero pressure, and then correct this Debye temperature to a fixed volume, i.e., to $\Theta(V_0)$, by the Debye theory relation (5.39), which may be written

$$\Theta_D(V_0)/\Theta_D(V) = (V/V_0)^\gamma, \tag{31.7}$$

where γ is the macroscopic Grüneisen parameter. But the Debye theory is a harmonic theory, and the correction (31.7) therefore does not include any effect

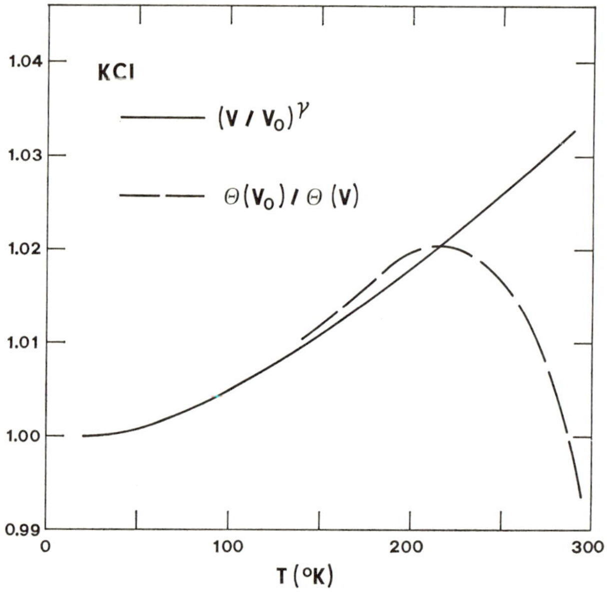

Figure 20. Comparison of the experimentally determined ratio $\Theta(V_0)/\Theta(V)$, for the heat capacity Debye temperatures Θ, with the Debye approximation $(V/V_0)^\gamma$ for this ratio, for KCl. The Debye approximation fails at high temperatures.

of anharmonicity, while it is just the anharmonic terms which one is generally seeking in the high-temperature analysis. To clarify this point further, we note from (30.6) that the volume variation of C_V at high temperature arises entirely from the volume variation of $2\mathscr{A}_2 - \Gamma$, while the volume variation of Θ given by (31.7) has nothing to do with $2\mathscr{A}_2 - \Gamma$, and hence in the case of the heat capacity the volume variation (31.7) is completely spurious at high temperatures.

We have calculated $C_V(V_0)$ from the experimental $C_V(V)$ for KCl, where we have some faith in the accuracy of the thermodynamic volume correction, and then found the corresponding heat capacity Debye temperatures $\Theta(V_0)$ and $\Theta(V)$. In Figure 20 the ratio $\Theta(V_0)/\Theta(V)$ is compared to $(V/V_0)^\gamma$ for KCl. At low temperatures, where the volume variation of C_V is due mainly to the harmonic contribution to C_V, the two curves are nearly the same. In the high-temperature region, above about 240°K, the volume variation of $2\mathscr{A}_2 - \Gamma$ dominates the volume variation of C_V, and causes $\Theta(V_0)/\Theta(V)$ to diverge markedly from the Debye expression (31.7). In fact $\Theta(V_0)/\Theta(V)$ becomes less than 1 at about 287°K for KCl.

Results

Our results of the high-temperature entropy analysis are illustrated in Figures 21–24. In all cases the plot of $(\partial Q/\partial T)_V$ evaluated at V_0, as a function of T^{-3}, exhibits a straight-line portion over a considerable temperature range; the average straight-line temperature range for the 18 materials studied is $0.7\Theta_0 \leq T \leq 1.3\Theta_0$. Also, in all cases the curve rises above the straight line at lower temperatures, presumably because of the harmonic T^{-5} term in (31.5). It is mildly surprising that the hexagonal crystals also follow a straight-line behavior, as shown by the graph for Cd in Figure 23 and those for Ti and Zr in Figure 24. In several cases the experimental data are reliable for temperatures well above $1.3\Theta_0$, and an experimentally meaningful departure from the straight-line behavior is observed at higher temperatures. We attribute such departures to higher-order anharmonic effects; these effects represent very small contributions to the total entropy at the temperatures considered here, say up to $2\Theta_0$, but they are exaggerated and made quite noticeable in the graphs of $(\partial Q/\partial T)_V$ *vs* T^{-3}. Particularly noticeable are the high-temperature departures from the straight lines shown for Kr in Figure 21, for Na and K in Figure 22, and for NaCl in Figure 24.

The values of $\Gamma - 2\mathscr{A}_2$ and $\langle\omega_\kappa^2\rangle$ obtained in the high-temperature entropy analysis, from the intercepts and slopes of the straight lines interpreted according to (31.5), are listed in Table 12. The error limits given for $\Gamma - 2\mathscr{A}_2$ have nothing to do with the accuracy of the primary experimental data, but simply reflect the scatter of the points in our graphs. The sign of

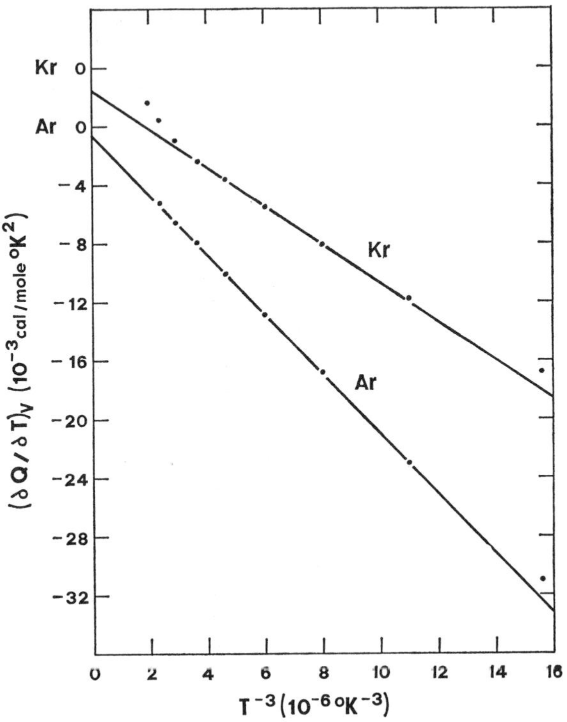

Figure 21. The high-temperature analysis of $S(V_0)$, according to (31.5), for Kr and Ar. The dots are experimental data points, and the straight lines are graphically fitted to these points. The vertical scale is labeled for Ar; for Kr the scale is the same but is displaced upward, and only the zero is labeled.

the high-temperature deviation of $(\partial Q/\partial T)_V$ from the straight line is also listed in Table 12, for all cases in which this deviation appears to be experimentally meaningful. It should be noted that the values obtained for $\Gamma - 2\mathscr{A}_2$ and $\langle \omega_\kappa^2 \rangle$ are appropriate for the volume V_0, and that these quantities are dependent on the volume.

It is of interest to compare our results for $\langle \omega_\kappa^2 \rangle$ with values obtained from direct measurements of the phonon frequencies by inelastic neutron-scattering experiments. Neutron-scattering experiments measure the renormalized frequencies $\Omega_\kappa = \omega_\kappa + \Delta_\kappa$, according to (16.29), and the Ω_κ depend on both the crystal volume and the temperature. It is customary to use interpolation procedures, to estimate Ω_κ at points throughout the Brillouin zone from the values measured along symmetry directions, and then to compute $\langle \Omega_\kappa^2 \rangle$

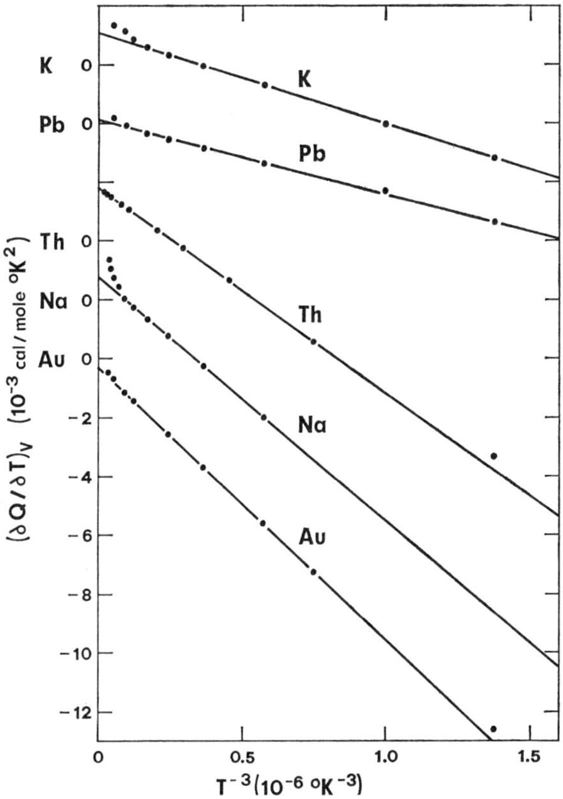

Figure 22. The high-temperature analysis of $S(V_0)$, according to (31.5), for K, Pb, Th, Na, and Au. The straight lines are graphically fitted to the experimental points (the dots), and the vertical scales are all the same but are displaced from one another.

with these interpolated values. Comparison of our $\langle\omega_\kappa^2\rangle(V_0)$ with the $\langle\Omega_\kappa^2\rangle(V,T)$ determined from neutron-scattering experiments is shown in Table 13; for the crystals listed there and the temperatures of the neutron-scattering measurements, we expect $\langle\omega_\kappa^2\rangle$ and $\langle\Omega_\kappa^2\rangle$ to differ by no more than 1 or 2%. The agreement between the two sets of results is quite satisfactory.

With regard to the thermal expansion analysis, our graphs of the temperature derivatives $[\partial(V\beta B_T/3nN_0K)/\partial V]_{V=V_0}$ as functions of T^{-3} exhibited so much scatter of the experimental points that we could not clearly observe straight-line regions at high temperatures. It should be noted that to find $[\partial(V\beta B_T/3nN_0K)/\partial T]_{V=V_0}$ accurately as a function of temperature, one

Table 12. Results of graphical analysis of $S(V_0)$ at high temperatures by means of (31.5).

Material	$\Gamma - 2\mathscr{A}_2$ (10^{-4} cal/mole °K^2)	$\langle\omega_\kappa^2\rangle$ (10^{26} sec^{-2})	Deviation at high T
Na	8.0 ± 1	2.87	+
K	11.0 ± 1	1.07	+
Cu	1.4 ± 1	10.4	
Ag	0.5 ± 0.5	4.56	
Au	−3.0 ± 1	3.19	
Al	3.0 ± 1	17.1	
Pb	1.0 ± 1	0.91	
Th	18.0 ± 0.5	2.41	
Zn	0.0 ± 1	6.2	+
Cd	−1.0 ± 0.5	3.07	
Ti	11.0 ± 1	12.7	+
Zr	10.0 ± 1	6.7	+
NaCl	2.1 ± 1	8.53	+
KCl	5.0 ± 1	5.52	
KBr	8.0 ± 1	3.40	
KI	3.5 ± 0.5	2.49	
Ar	−7.0 ± 1	0.70	
Kr	−16 ± 5	0.46	+

Table 13. Values of $\langle\Omega_\kappa^2\rangle$ determined from inelastic neutron-scattering experiments, compared with $\langle\omega_\kappa^2\rangle$ determined from the high-temperature entropy analysis. The temperatures of the neutron-scattering experiments are also listed.

Material	$\langle\Omega_\kappa^2\rangle$ (10^{26} sec^{-2})	T (°K)	Reference for $\langle\Omega_\kappa^2\rangle$	$\langle\omega_\kappa^2\rangle$ (10^{26} sec^{-2})
Na	2.87	90	37	2.87
K	1.07	9	38	1.07
Cu	10.1	300	39	10.4
Cu	10.3	49	40	10.4
Al	17.08	80	41	17.1
KBr	3.46	90	42	3.40

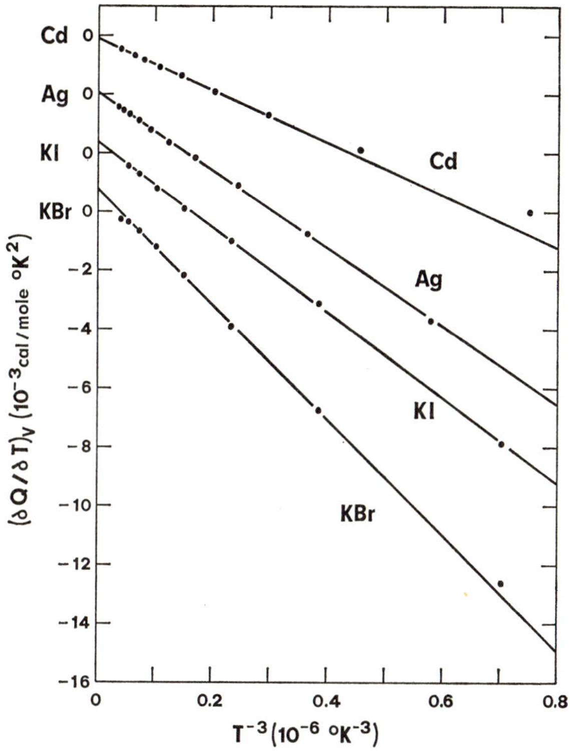

Figure 23. The high-temperature analysis of $S(V_0)$, according to (31.5), for Cd, Ag, KI, and KBr. The straight lines are graphically fitted to the experimental points (the dots), and the vertical scales are all the same but are displaced from one another.

needs extremely accurate results for β and B_T *vs* T, and also a reasonably accurate volume correction for βB_T. The graph for Pb, which is one of the better examples, is shown in Figure 25. From such graphs it appears that the intercept $(3nN_0K)^{-1}[Vd(\Gamma - 2\mathscr{A}_2)/dV]$ is quite small, of the order of 10^{-4} °K^{-1}, for Cu, Ag, Au, Al, Pb, Cd, Zr, NaCl, KCl, KBr, and KI. In units of 10^{-3} °K^{-1}, the intercept appears to be about 0.5 for K, -0.5 for Zn, $+1$ for Na and Th, and -8 for Ar and Kr. Although these values must be regarded as very approximate, they indicate nonnegligible anharmonic plus electronic contributions to $V\beta B_T$ for these last six materials in the high-temperature region. More extensive and accurate experimental measurements, and more detailed analyses, are in order for these materials.

Graphs of $V\beta B_T(V_0)/3nN_0K$ as functions of T^{-2} are shown in Figures

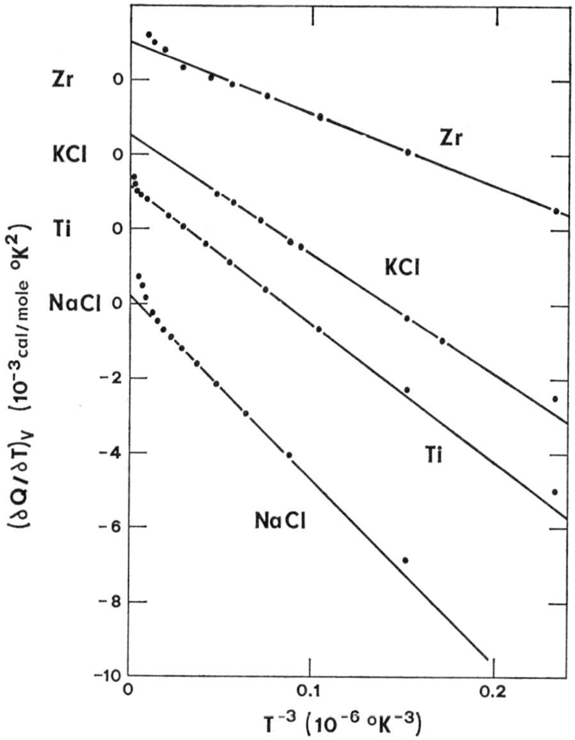

Figure 24. The high-temperature analysis of $S(V_0)$, according to (31.5), for Zr, KCl, Ti, and NaCl. The straight lines are graphically fitted to the experimental points (the dots), and the vertical scales are all the same but are displaced from one another.

26–28 for all cases where a reasonable straight-line behavior is observed at high temperatures; Ti is not included because we were unable to make the volume correction for $V\beta B_T$. For the materials shown the average straight-line temperature range is $0.6\Theta_0 \leq T \leq 1.2\Theta_0$. Again it is somewhat surprising that the hexagonal crystals follow straight-line behavior, as may be seen for Cd in Figure 26, and for Zn and Zr in Figure 27. Several of the materials show a departure from the straight-line behavior at the highest temperatures; this departure is rather abrupt for Cd, KBr, Zn, KCl, Zr, and NaCl. The experimental data are not sufficiently accurate to determine whether this departure is due to the anharmonic plus electronic term in (31.6), or to higher-order anharmonic effects, or both.

The values of $\langle\gamma_\kappa\rangle$ and $\langle\gamma_\kappa\omega_\kappa^2\rangle$ obtained in the analysis of $V\beta B_T(V_0)$, from the intercepts and slopes of the straight lines drawn in Figures 26–28, are

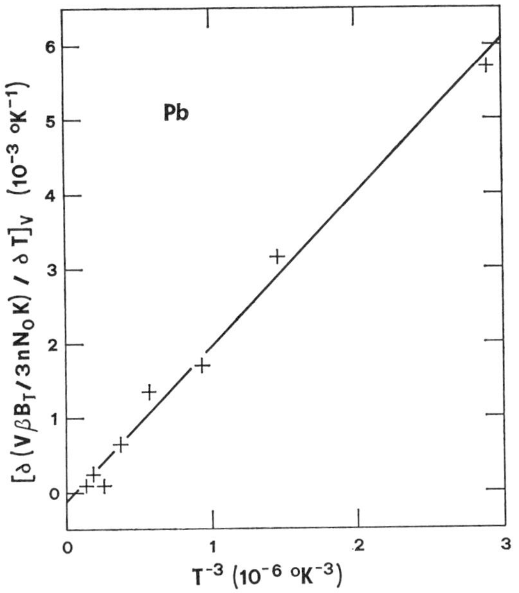

Figure 25. The temperature derivative

$$[\partial(V\beta B_T/3nN_0K)/\partial T]_V$$

evaluated at V_0, as a function of T^{-3} for Pb. The crosses are experimental data points, and the straight line is an approximate graphical fit to these points.

listed in Table 14. Estimates of $\langle\gamma_\kappa\rangle$ for the other materials are also listed in Table 14; these estimates are based on the curves of $V\beta B_T(V_0)/3nN_0K$, with approximate account of the term linear in T in (31.6). All the quantities listed in Table 14 are appropriate for the volume V_0. For each of the materials shown in Figures 26–28, the value of $\langle\gamma_\kappa\rangle$ is close to the value of the macroscopic Grüneisen parameter $\gamma(V)$ or $\gamma(V_0)$ at temperatures in the vicinity of the Debye temperature. The small differences in these various quantities are significant, however; for KCl, for example, it is seen in Figure 18 that $\gamma(V)$ levels off at about 1.46, $\gamma(V_0)$ does not level off at all, and the intercept in Figure 27 gives $\langle\gamma_\kappa\rangle = 1.42$.

The graphical analyses we have presented are quite encouraging, since the experimental data generally follow the theoretical temperature-dependences and yield accurate values of the lattice-dynamical quantities. It is hoped that the present results will stimulate more detailed analysis of those thermodynamic functions which are obtained accurately at fixed volume (S and $V\beta B_T$), and elementary analysis similar to the present graphical procedures

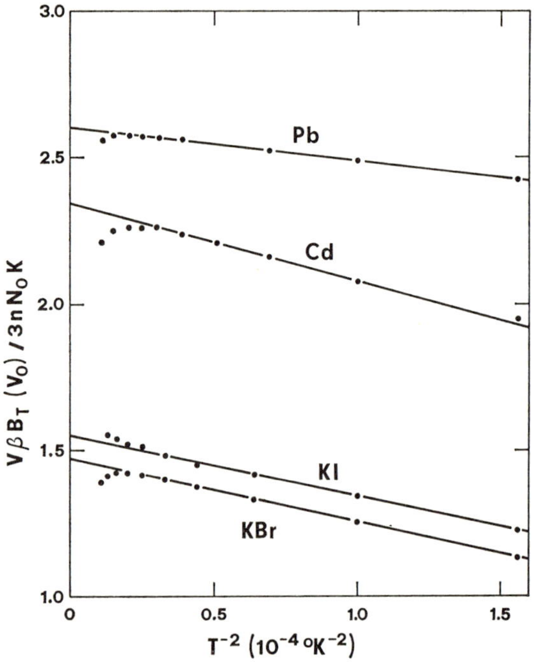

Figure 26. The high-temperature analysis of $\beta B_T(V_0)$, according to (31.6), for Pb, Cd, KI, and KBr. The dots are experimental data points, and the straight lines are graphically fitted to these points.

for those functions which are obtained less accurately at fixed volume (B_T, B_S, and the elastic coefficients). For noncubic crystals analysis of the entropy, the thermal stresses, and the elastic coefficients, corrected to conditions of fixed crystal configuration, will be most useful; a start in this direction is provided by the recent work of T. H. K. Barron.

32. APPROXIMATIONS AND CORRELATIONS

Average Phonon Parameters

The high-temperature harmonic Debye temperature $\Theta_{H\infty}$ is defined in (19.32), by the condition that the harmonic heat capacity have the same high-temperature behavior as the Debye heat capacity. The $\Theta_{H\infty}$ is simply related to the average phonon frequency squared $\langle\omega_\kappa^2\rangle$, and in Table 15 we

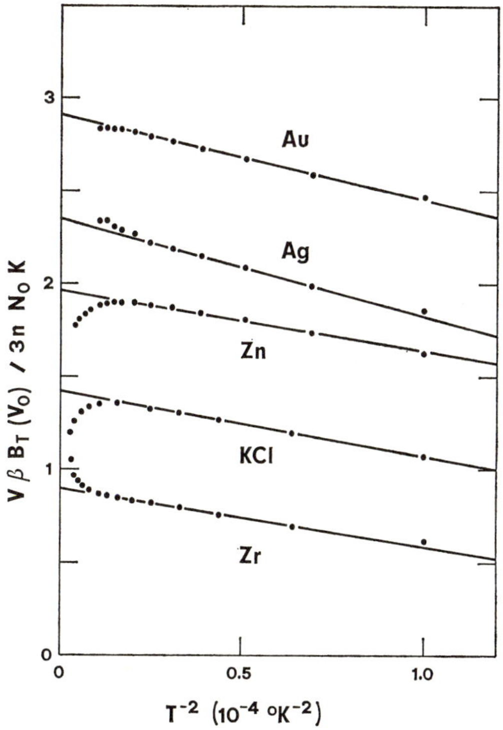

Figure 27. The high-temperature analysis of $\beta B_T(V_0)$, according to (31.6), for Au, Ag, Zn, KCl, and Zr. The straight lines are graphically fitted to the experimental points. For Zn, the experimental points are not as well fitted by a straight line as for the other materials shown.

list the values of $\Theta_{H\infty}$ which correspond to the $\langle\omega_\kappa^2\rangle$ given in Table 12. In the Debye theory, of course, Θ_D is independent of the temperature, and from this we might expect $\Theta_{H\infty}$ and Θ_{H0} to be roughly the same for the simplest and most isotropic crystals, say primitive cubic lattices. For non-primitive lattices the optic modes contribute to $\Theta_{H\infty}$ but not to Θ_{H0}, and for any lattice a marked anisotropy should lead to a difference between $\Theta_{H\infty}$ and Θ_{H0}.

The experimentally available Θ_0 contains anharmonic plus harmonic contributions, according to Section 18, and it will be interesting to separate out the harmonic part Θ_{H0} when the appropriate data become available. In the absence of such information we can only guess that the anharmonic contribution to Θ_0 is at most a few percent for the materials listed in Table 15.

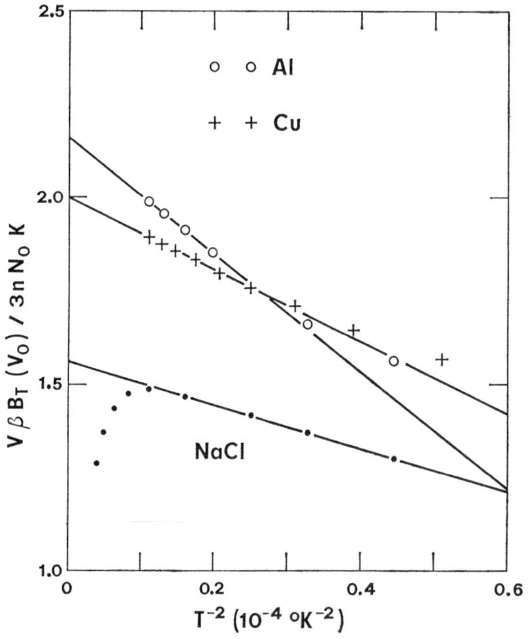

Figure 28. The high-temperature analysis of $\beta B_T(V_0)$, according to (31.6), for Al, Cu, and NaCl. The straight lines are graphically fitted to the experimental points.

The ratio $\Theta_{H\infty}/\Theta_0$ is listed for each material in Table 15; this ratio varies from 0.90 to 1.13 for the primitive cubic lattices, and is not significantly different for the four alkali halides. The four hexagonal crystals are noticeably different, having $\Theta_{H\infty}/\Theta_0$ in the range 0.73–0.85.

The phonon Grüneisen parameters γ_κ generally do not vary greatly over most of the Brillouin zone, although they may show large variations for both optic and acoustic branches near the zone center. Therefore in any Brillouin zone average which contains the γ_κ, it is a reasonable approximation to take these parameters out as the average $\langle\gamma_\kappa\rangle$. In other words, for example, we expect $\langle\gamma_\kappa\omega_\kappa^2\rangle$ to be approximately equal to $\langle\gamma_\kappa\rangle\langle\omega_\kappa^2\rangle$. The ratio $\langle\gamma_\kappa\rangle\langle\omega_\kappa^2\rangle/\langle\gamma_\kappa\omega_\kappa^2\rangle$ is listed in Table 15 for each of the materials for which our high-temperature analysis of $V\beta B_T$ appeared to give a reliable result for $\langle\gamma_\kappa\omega_\kappa^2\rangle$; this ratio ranges from 0.90 to 1.14 for all the materials except the highly anisotropic metals Zn and Cd, where it is 1.9 and 1.3, respectively. This may represent a real difference for Zn and Cd, or it may be that our isotropic treatment of these metals has given some significantly incorrect results.

Table 14. Results of graphical analysis of $V\beta B_T/3nN_0K$ corrected to the fixed volume V_0; the analysis follows (31.6).

Material	$\langle\gamma_\kappa\rangle$	$\langle\gamma_\kappa\omega_\kappa^2\rangle$ (10^{26} sec^{-2})
Na	~1.1	
K	~1.1	
Cu	2.00	19.9
Ag	2.35	10.8
Au	2.91	9.5
Al	2.16	32.5
Pb	2.60	2.28
Th	~1.2	
Zn	~2.0	~6.6
Cd	2.34	5.4
Ti	~1.2	
Zr	0.90	6.4
NaCl	1.56	12.1
KCl	1.42	7.1
KBr	1.47	4.42
KI	1.55	4.28
Ar	~2.7	
Kr	~3.0	

According to the cubic-crystal equations of Section 19, the constant-volume temperature derivatives of the adiabatic and isothermal bulk moduli, evaluated in the high-temperature region, can give values for the Brillouin zone averages of the phonon Grüneisen parameters. In particular the leading (harmonic) contributions to these derivatives may be written from (19.60) and (19.61) as

$$(V/3nN_0K)(\partial B_T/\partial T)_V = \langle \xi_\kappa - \gamma_\kappa^2 \rangle; \tag{32.1}$$

$$(V/3nN_0K)(\partial B_S/\partial T)_V = \langle \xi_\kappa - \gamma_\kappa^2 \rangle + \langle \gamma_\kappa \rangle^2. \tag{32.2}$$

Defining $\Delta B = B_S - B_T$, it follows that

$$(V/3nN_0K)(\partial \Delta B/\partial T)_V = \langle \gamma_\kappa \rangle^2. \tag{32.3}$$

We have tested this last equation in the following way. The general relation (1.52) can be used to write

$$(\partial \Delta B/\partial T)_V = (\partial \Delta B/\partial T)_P + \beta B_T (\partial \Delta B/\partial P)_T. \tag{32.4}$$

Table 15. The high-temperature harmonic Debye temperature $\Theta_{H\infty}$, the ratio $\Theta_{H\infty}/\Theta_0$, and the ratio $\langle\gamma_\kappa\rangle\langle\omega_\kappa^2\rangle/\langle\gamma_\kappa\omega_\kappa^2\rangle$ which should be approximately 1.

Material	$\Theta_{H\infty}$ (°K)	$\Theta_{H\infty}/\Theta_0$	$\langle\gamma_\kappa\rangle\langle\omega_\kappa^2\rangle/\langle\gamma_\kappa\omega_\kappa^2\rangle$
Na	167	1.10	
K	102	1.13	
Cu	318	0.92	1.05
Ag	211	0.93	0.99
Au	176	1.09	0.98
Al	408	0.95	1.14
Pb	94	0.90	1.04
Th	153	0.90	
Zn	245	0.73	1.9
Cd	173	0.85	1.3
Ti	351	0.82	
Zr	255	0.82	0.94
NaCl	288	0.90	1.10
KCl	232	0.99	1.10
KBr	182	1.05	1.13
KI	156	1.18	0.90
Ar	83	0.90	
Kr	67	0.93	

We calculated ΔB from experimental data as a function of T at $P = 0$, and from this we obtained $(\partial\Delta B/\partial T)_{P=0}$ at 300°K. We also calculated $(\partial\Delta B/\partial P)_T$ at $P = 0$ and 300°K from (1.59) or (1.61), and then used (32.4) to obtain $(\partial\Delta B/\partial T)_V$ at $P = 0$ and $T = 300$°K. For those materials for which we could get a reliable estimate of $(\partial\Delta B/\partial T)_V$ in this way, the results are given in Table 16 in the form of the ratio $(V/3nN_0K)(\partial\Delta B/\partial T)_V/\langle\gamma_\kappa\rangle^2$, where $\langle\gamma_\kappa\rangle$ was taken to be $\langle\gamma_\kappa\rangle(V_0)$ as determined in Section 31. According to (32.3), this ratio should be 1 under the following three approximations. First, the temperature of evaluation, namely 300°K, should be of the order or greater than the Debye temperature. Second, the explicit anharmonic and electronic contributions to $(\partial B_T/\partial T)_V$ and $(\partial B_S/\partial T)_V$ should be small, since these contributions were dropped in (32.1)–(32.3). Finally, the difference between $\langle\gamma_\kappa\rangle(V)$ and $\langle\gamma_\kappa\rangle(V_0)$ should be negligible; this is probably true since $d \ln \langle\gamma_\kappa\rangle/d \ln V$ is of order 1. For all cases listed in Table 16 the ratio is 1 to within the accuracy of our estimate of $(\partial\Delta B/\partial T)_V$, and we take this as evidence that the anharmonic and electronic contribution to $(\partial\Delta B/\partial T)_V$ is not large.

Another useful approximation is obtained from (32.2); since $\langle \gamma_\kappa^2 \rangle$ and $\langle \gamma_\kappa \rangle^2$ should be about the same, a reasonable approximation for $\langle \xi_\kappa \rangle$ is

$$(V/3nN_0K)(\partial B_S/\partial T)_V \approx \langle \xi_\kappa \rangle, \quad \text{at high } T. \tag{32.5}$$

The left-hand side may be evaluated from

$$(\partial B_S/\partial T)_V = (\partial B_S/\partial T)_P + \beta B_T(\partial B_S/\partial P)_T. \tag{32.6}$$

The approximate $\langle \xi_\kappa \rangle$ evaluated from (32.5) is given in Table 16 for those few materials for which we have reliable estimates of $(\partial B_S/\partial T)_V$ at $P = 0$ and 300°K, by means of (32.6), and for which the Debye temperature is well below 300°K. The tabulated $\langle \xi_\kappa \rangle$ are appropriate for the volume $V(P = 0, T = 300°\text{K})$, and the volume-dependence should be small. There appears to be some correlation between $\langle \xi_\kappa \rangle$ and $\langle \gamma_\kappa \rangle$; this bears further study when more extensive and accurate estimates of $\langle \xi_\kappa \rangle$ are made.

Additional Correlations

Let us begin by finding an approximate relation between two harmonic quantities, namely the average phonon frequency squared and the bulk modulus. The lattice-dynamical expression for $\langle \omega_\kappa^2 \rangle$ is (10.62):

$$\langle \omega_\kappa^2 \rangle = (3n)^{-1} \sum_{\nu i} M_\nu^{-1} \Phi_{ii}(0\nu,0\nu). \tag{32.7}$$

Table 16. The function $(V/3nN_0K)(\partial \Delta B/\partial T)_V$ divided by $\langle \gamma_\kappa \rangle^2$, which should be approximately 1, and the function $(V/3nN_0K)(\partial B_S/\partial T)_V$, which should be approximately $\langle \xi_\kappa \rangle$. These approximations hold for $T \gtrsim \Theta_{H\infty}$. The derivatives $(\partial \Delta B/\partial T)_V$ and $(\partial B_S/\partial T)_V$ were evaluated at 300°K and $P = 0$.

Material	$\dfrac{(V/3nN_0K)(\partial \Delta B/\partial T)_V}{\langle \gamma_\kappa \rangle^2}$	$(V/3nN_0K)(\partial B_S/\partial T)_V$ (approximately $\langle \xi_\kappa \rangle$)
Cu	1.1	
Ag	1.2	6.4
Au	1.0	6.9
Al	1.1	
Pb	0.9	3.9
Zn	0.8	
Cd	0.7	
Zr	1.0	
NaCl	1.1	
KCl	1.0	0.4
KBr	1.0	0.1
KI	1.1	

We approximate this by replacing each inverse mass M_ν^{-1} by the average of M_ν^{-1}, averaged over all the ions in one unit cell, so that (32.7) is

$$\langle \omega_\kappa^2 \rangle \approx \langle M_\nu^{-1} \rangle (3n)^{-1} \sum_{\nu i} \Phi_{ii}(0\nu,0\nu), \tag{32.8}$$

where

$$\langle M_\nu^{-1} \rangle = n^{-1} \sum_\nu M_\nu^{-1}. \tag{32.9}$$

Writing out the coefficients $\Phi_{ii}(0\nu,0\nu)$ as derivatives of the total lattice potential Φ then gives

$$\langle M_\nu^{-1} \rangle^{-1} \langle \omega_\kappa^2 \rangle \approx (3n)^{-1} \sum_{\nu i} \partial^2\Phi/\partial U_i(0\nu)\partial U_i(0\nu). \tag{32.10}$$

The right of (32.10) is the average over all ions in a unit cell of the second displacement derivative of the total potential; for a uniform volume change this ought to be proportional to a second derivative of Φ with respect to a distance such as the nearest-neighbor distance R_1, so that

$$\langle M_\nu^{-1} \rangle^{-1} \langle \omega_\kappa^2 \rangle \propto d^2\Phi/dR_1^2. \tag{32.11}$$

Actually the major approximation we will make in applying (32.11) is that the proportionality constant is the same for all materials.

Now R_1^3 is proportional to the crystal volume V, so under a uniform volume change (32.11) leads to

$$\begin{aligned} \langle M_\nu^{-1} \rangle^{-1} R_1^2 \langle \omega_\kappa^2 \rangle &\propto R_1^2(d^2\Phi/dR_1^2) \\ &\propto V^2(d^2\Phi/dV^2) + \tfrac{2}{3}V(d\Phi/dV). \end{aligned} \tag{32.12}$$

The volume derivatives of Φ give the leading (potential approximation) contributions to the pressure $\tilde{P}$ and the bulk modulus $\tilde{B}$, and (32.12) may be written

$$\langle M_\nu^{-1} \rangle^{-1} R_1^2 \langle \omega_\kappa^2 \rangle \propto V\tilde{B} - \tfrac{2}{3}V\tilde{P}. \tag{32.13}$$

At zero temperature and pressure, $\tilde{B}$ should be fairly close to the total bulk modulus B_0, and (32.13) may be expressed as

$$V_0 B_0 \propto nN_0 \langle M_\nu^{-1} \rangle^{-1} R_1^2 \langle \omega_\kappa^2 \rangle, \quad \text{at} \quad P = 0. \tag{32.14}$$

The right-hand side has been multiplied by nN_0 so that both sides of (32.14) may be expressed in calories per mole. A test of this approximation is shown in Figure 29, where it is seen to be qualitatively satisfied by the 18 materials we have studied.

It is of interest to test the approximation (32.13) further by differentiating with respect to volume. Again taking $R_1^3 \propto V$, the logarithmic volume derivative of (32.13) evaluated at $\tilde{P} = 0$ is

$$\tfrac{2}{3} + d\ln\langle \omega_\kappa^2 \rangle / d\ln V = \tfrac{5}{3} + d\ln \tilde{B}/d\ln V, \tag{32.15}$$

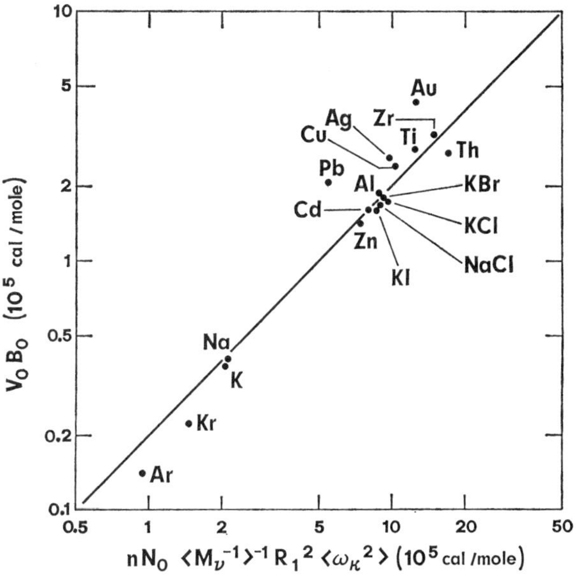

Figure 29. The correlation between the bulk modulus and $\langle\omega_\kappa^2\rangle$, following (32.14). The straight line is merely a graphical fit to the data points.

where we used the potential approximation $Vd\tilde{P}/dV = -\tilde{B}$. Now since $\gamma_\kappa = -d\ln\omega_\kappa/d\ln V = -\frac{1}{2}\,d\ln\omega_\kappa^2/d\ln V$, it is reasonable to approximate the left side of (32.15) by means of

$$d\ln\langle\omega_\kappa^2\rangle/d\ln V \approx -2\langle\gamma_\kappa\rangle. \tag{32.16}$$

Further, on the right of (32.15) we can use $\tilde{B} \approx B_T$, and then $d\ln\tilde{B}/d\ln V \approx -(\partial B_T/\partial P)_T$, so that (32.15) gives

$$(\partial B_T/\partial P)_T \approx 2\langle\gamma_\kappa\rangle + 1. \tag{32.17}$$

This type of approximation, relating a bulk modulus pressure derivative to a Grüneisen parameter, was obtained long ago by Slater and others. Figure 30 shows a plot of the room temperature values of $(\partial B_T/\partial P)_T$ *vs* the values of $\langle\gamma_\kappa\rangle$ which were obtained in our analysis of $V\beta B_T$ curves in Section 31. The data are in poor agreement with the straight line $2\langle\gamma_\kappa\rangle + 1$, and although a line at about $2\langle\gamma_\kappa\rangle + 2$ would fit the data better, the correlation between $(\partial B_T/\partial P)_T$ and $\langle\gamma_\kappa\rangle$ is quite poor in any case. Incidentally, in connection with Figure 30, our analysis of experimental data has already led us to suspect the value of $(\partial B_T/\partial P)_T$ for Zr.

The approximate relations between B_T and $\langle\omega_\kappa^2\rangle$, and between their volume derivatives, are essentially dimensional analysis relations. We do

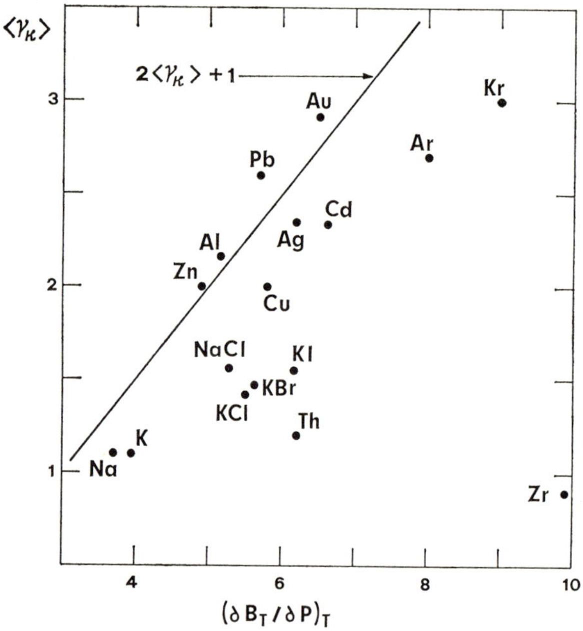

Figure 30. The correlation between the bulk modulus pressure derivative (room temperature values) and $\langle\gamma_\kappa\rangle$, following (32.17). For the data presented here, there is almost no correlation.

not advocate their use in estimating unknown quantities from available ones, and have presented Figures 29 and 30 simply to test the relations. We have tried without success to find a correlation between the anharmonic quantity $\langle\gamma_\kappa\rangle$ and some more easily determined harmonic quantity. An interesting graph of $\langle\gamma_\kappa\rangle$ *vs* $\Theta_{H\infty}$ is shown in Figure 31; the lack of correlation is obvious.

Finally we wish to correlate the high-temperature anharmonic coefficient $\mathscr{A}_2$ with simpler thermodynamic functions. The values of $\Gamma - 2\mathscr{A}_2$ were found in the high-temperature entropy analysis, and are listed in Table 12. Unfortunately we do not have values for Γ, but have only the low-temperature coefficients $\hat{\Gamma}$; note $\Gamma \lessdot \hat{\Gamma}$. Rather than make guesses for Γ, we used the approximation $\Gamma \approx \hat{\Gamma}$ and calculated $\mathscr{A}_2$ from

$$\mathscr{A}_2 \approx \tfrac{1}{2}[\hat{\Gamma} - (\Gamma - 2\mathscr{A}_2)]. \tag{32.18}$$

This approximation is satisfactory for the present purpose of seeking a qualitative correlation involving $\mathscr{A}_2$; for any quantitative calculation one should estimate Γ more carefully.

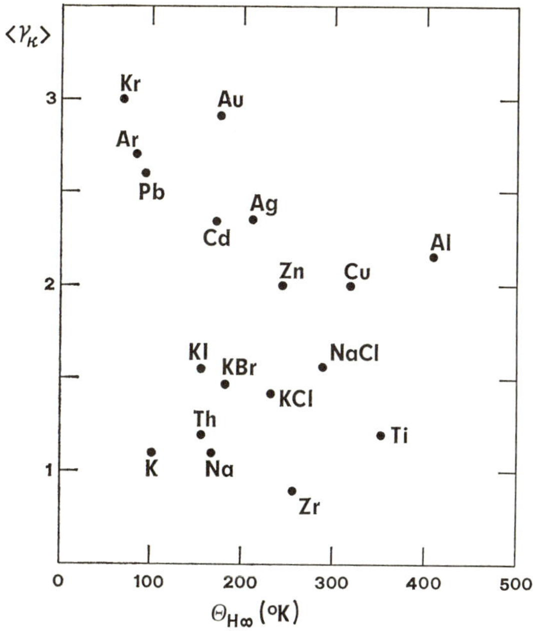

Figure 31. Illustration of the lack of correlation between $\langle\gamma_\kappa\rangle$ and $\Theta_{H\infty}$.

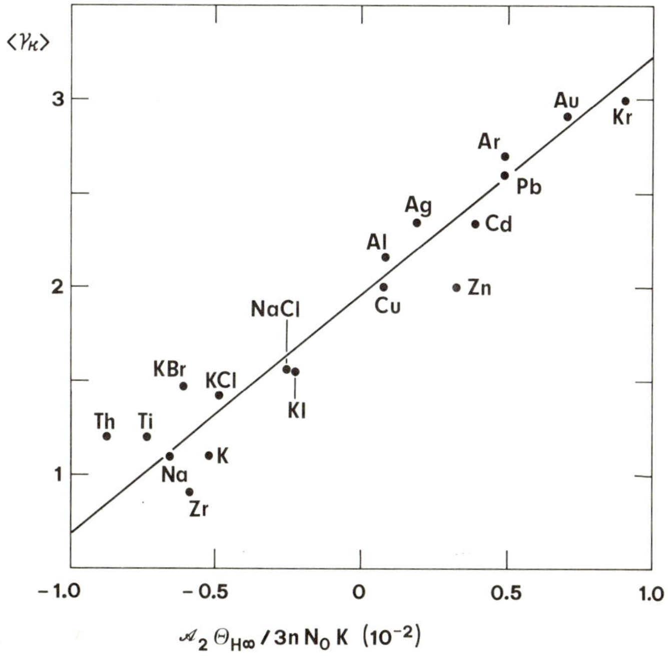

Figure 32. The correlation between $\mathscr{A}_2\Theta_{H\infty}/3nN_0K$ and $\langle\gamma_\kappa\rangle$. The straight line is (32.19).

The lattice-dynamical expressions for $\mathscr{A}_2$ were derived in Section 19, and from (19.13), for example, we might expect $\mathscr{A}_2$ to be approximated by a function of $\langle\gamma_\kappa\rangle$ and $\langle\xi_\kappa\rangle$ divided by a function of $\langle\omega_\kappa^2\rangle$. A correlation of this type does hold, as shown in Figure 32, which is a plot of $\mathscr{A}_2\Theta_{H\infty}/3nN_0K$ *vs* $\langle\gamma_\kappa\rangle$. The experimental points in Figure 32 are reasonably well fitted by the straight line drawn there, which is given by

$$\mathscr{A}_2\Theta_{H\infty}/3nN_0K = 0.0078\langle\gamma_\kappa\rangle - 0.0154. \tag{32.19}$$

The departure of experimental points from this straight line appears to be within reasonable estimates of our errors except for Zn and KBr. This correlation is probably the most valuable result of the present section since it allows an estimation of the anharmonic contribution to the entropy and heat capacity at high temperatures, if one knows $\langle\gamma_\kappa\rangle$. Incidentally, the relation stated in (32.19) is not faithful to the theoretical principles, in that $\mathscr{A}_2$ was shown in Section 19 to depend on the potential energy coefficients, but not on the ion masses, whereas $\Theta_{H\infty}$ depends on the masses as well as the potential coefficients. Hopefully, more accurate data and more extensive analysis will remove this difficulty.

8

MODEL CALCULATIONS

The lattice-dynamical perturbation theory of thermodynamic functions of crystals is illustrated by several calculations based on simple model potentials. Harmonic and anharmonic quantities are calculated for solid Ar and Kr, on the basis of a Lennard–Jones central potential, and lattice and electronic quantities are calculated for the simple metals Na, K, and Al, on the basis of local pseudopotentials. Comparisons of theory and experiment allow us to determine inadequacies in the simple model potentials, and directions for improving the theory are discussed.

33. RARE GAS CRYSTALS

The Model Potential

In order to illustrate the general quality of the theoretical calculations, we will treat a Lennard–Jones central potential model, and compare the results with experiment for fcc Ar and Kr. According to this model, the potential between two atoms separated by a distance R is

$$\varphi(R) = aR^{-\alpha} - bR^{-\beta}, \tag{33.1}$$

where a, b, α, and β are positive parameters and $\alpha > \beta$. The potential $\varphi(R)$ is composed of a short-range repulsive term $aR^{-\alpha}$ and a longer-range attractive term $-bR^{-\beta}$, and we will further restrict the present calculations to the case $\alpha = 12$, $\beta = 6$. The parameters a and b will be determined so as to obtain agreement between the theory, in the potential approximation, and the experimental results for the crystal volume and bulk modulus at $T = 0$ and $P = 0$. Then with the potential parameters so determined, several thermodynamic functions will be calculated and compared with experiment.

It should be noted that the perturbation expansion of the free energy is quite justified for the rare gas crystals, except solid helium; this expansion is given by (16.7) in the form

$$F = \Phi_0 + F_{\mathrm{H}} + F_{\mathrm{A}}. \tag{33.2}$$

For solid Ar with the Lennard–Jones 12,6 potential model, the relative contributions in percentage units of the three terms on the right of (33.2) are about -111.0, $+10.6$, $+0.4$ at $T = 0$, and they are about -92, -10, $+2$ at the normal melting temperature. The series converges somewhat more slowly for solid Ne, and more rapidly for solid Kr and Xe. The potential approximation $F \approx \Phi_0$ is used in our $T = 0$ calculations for simplicity; inclusion of the zero-point vibrational contribution F_{H} at $T = 0$ does not significantly alter any of the theoretical results for the present model, but it greatly complicates the equations. The magnitudes of the zero-point contributions are discussed at the end of this section.

With the origin of coordinates at a lattice site $\mathbf{R}(0) = 0$ in the interior of the crystal, and neglecting surface effects, the static-lattice potential Φ_0 for N_0 atoms may be written

$$\Phi_0 = \tfrac{1}{2}N_0\left\{a \sum_{\mathrm{N}}{}' R(\mathrm{N})^{-\alpha} - b \sum_{\mathrm{N}}{}' R(\mathrm{N})^{-\beta}\right\}. \tag{33.3}$$

It is convenient to introduce the nearest-neighbor distance R_1, and define lattice sums such as

$$S_\alpha = \sum_{\mathrm{N}}{}' [R_1/R(\mathrm{N})]^\alpha, \tag{33.4}$$

so that (33.3) is simply

$$\Phi_0 = \tfrac{1}{2}N_0[aS_\alpha R_1^{-\alpha} - bS_\beta R_1^{-\beta}]. \tag{33.5}$$

Since the crystal volume V is proportional to R_1^3, it follows that

$$V(d\Phi_0/dV) = \tfrac{1}{3}R_1(d\Phi_0/dR_1); \tag{33.6}$$

$$V^2(d^2\Phi_0/dV^2) = \tfrac{1}{9}[R_1^2(d^2\Phi_0/dR_1^2) - 2R_1(d\Phi_0/dR_1)]. \tag{33.7}$$

Then in the potential approximation the pressure $\tilde{P}$ is given by

$$V\tilde{P} = -\tfrac{1}{3}R_1(d\Phi_0/dR_1), \tag{33.8}$$

and from (33.5) for Φ_0 the condition $\tilde{P} = 0$ gives

$$b = (\alpha S_\alpha/\beta S_\beta)aR_1^{\beta-\alpha}, \quad \text{at} \quad \tilde{P} = 0. \tag{33.9}$$

Table 17. Experimental results for R_1 and V_0B_0 at $T = 0$ and $P = 0$, and the values of a and b required to fit these results with the Lennard–Jones 12,6 potential in the potential approximation, for Ar and Kr.

Quantity	Ar	Kr
R_1 (Å)	3.7555	3.9923
V_0B_0 (10^{-12} erg/atom)	1.004	1.551
a (10^{-7} erg Å^{12})	1.629	5.241
b (10^{-10} erg Å^{6})	0.975	2.173

The bulk modulus $\tilde{B}$ in the potential approximation is given by

$$V\tilde{B} = V^2(d^2\Phi_0/dV^2), \tag{33.10}$$

and at $\tilde{P} = 0$ the result for the present model may be written

$$V_0\tilde{B} = \tfrac{1}{18}N_0\alpha(\alpha - \beta)aS_\alpha R_1^{-\alpha}, \quad \text{at} \quad \tilde{P} = 0. \tag{33.11}$$

For $\alpha = 12$, $\beta = 6$, the required lattice sums for the fcc lattice are*

$$S_{12} = 12.1318802, \qquad S_6 = 14.4539211. \tag{33.12}$$

The parameter a was determined so that (33.11) agrees with the measured value V_0B_0 at $T = 0$ and $P = 0$, and then b was calculated from (33.9). In this determination R_1 was appropriately taken as the measured value at $T = 0$ and $P = 0$. The results for Ar and Kr are listed in Table 17.

Anharmonic Free Energy at High Temperature

The high-temperature expansion (19.12) for the anharmonic free energy is

$$F_A = \mathscr{A}_2T^2 + \mathscr{A}_0 + \mathscr{A}_{-2}T^{-2} + \cdots, \tag{33.13}$$

and the coefficient $\mathscr{A}_2$ is simplified for a primitive lattice in (19.24). Central potentials are two-body potentials, and for them the expression for $\mathscr{A}_2$ can be further simplified because the potential coefficients $\Phi_{ijk}(\mathrm{M,N,P})$ and $\Phi_{ijkl}(\mathrm{M,N,P,Q})$ couple only two different ions at most (see Section 9 for the potential coefficients corresponding to central potentials).

From (19.24) the quartic contribution to $\mathscr{A}_2$, i.e., the contribution which

* Tables of accurate lattice sums S_α, and other related sums, are listed for several simple lattices by D. C. Wallace and J. L. Patrick, *Phys. Rev.* **137**, A152 (1965).

contains fourth-order potential coefficients, contains the sum

$$\sum_{\mathrm{MNPQ}} \sum_{ijkl} \Phi_{ijkl}(\mathrm{M,N,P,Q}) J_{ij}(\mathrm{M,N}) J_{kl}(\mathrm{P,Q}), \tag{33.14}$$

and by eliminating surface effects this sum is reduced to

$$N_0 \sum_{\mathrm{NPQ}} \sum_{ijkl} \Phi_{ijkl}(0,\mathrm{N,P,Q}) J_{ij}(0,\mathrm{N}) J_{kl}(\mathrm{P,Q}). \tag{33.15}$$

But for two-body potentials, $\Phi_{ijkl}(0,\mathrm{N,P,Q})$ vanishes unless the index set NPQ corresponds to one of the eight sets 000, 00N, 0N0, N00, 0NN, N0N, NN0, or NNN, for any N $\neq$ 0. Thus the Σ_{NPQ} in (33.15) gives rise to eight terms; these may be written out and grouped together by making use of the symmetries (9.15), (9.21), and (9.23):

$$\Phi_{ijkl}(0,0,0,0) = -\sum_{\mathrm{N}}{}' \Phi_{ijkl}(0,0,0,\mathrm{N}), \tag{33.16}$$

$$\begin{aligned} \Phi_{ijkl}(0,0,0,\mathrm{N}) &= \Phi_{ijkl}(0,0,\mathrm{N},0) = \Phi_{ijkl}(0,\mathrm{N},0,0) \\ &= -\Phi_{ijkl}(0,0,\mathrm{N,N}) = -\Phi_{ijkl}(0,\mathrm{N},0,\mathrm{N}) \\ &= -\Phi_{ijkl}(0,\mathrm{N,N},0) = \Phi_{ijkl}(0,\mathrm{N,N,N}); \end{aligned} \tag{33.17}$$

and also by making use of the symmetries which follow from (19.22) and (19.23):

$$J_{ij}(0,\mathrm{N}) = J_{ij}(\mathrm{N},0); \tag{33.18}$$

$$J_{ij}(0,0) = J_{ij}(\mathrm{N,N}). \tag{33.19}$$

Simplification of (33.15) then proceeds as follows.

$$\begin{aligned} &N_0 \sum_{\mathrm{NPQ}} \sum_{ijki} \Phi_{ijkl}(0,\mathrm{N,P,Q}) J_{ij}(0,\mathrm{N}) J_{kl}(\mathrm{P,Q}) \\ &= 2N_0 \sum_{\mathrm{N}}{}' \sum_{ijkl} \Phi_{ijkl}(0,0,\mathrm{N,N})[J_{ij}(0,0)J_{kl}(0,0) - J_{ij}(0,0)J_{kl}(0,\mathrm{N}) \\ &\qquad\qquad - J_{ij}(0,\mathrm{N})J_{kl}(0,0) + J_{ij}(0,\mathrm{N})J_{kl}(0,\mathrm{N})] \\ &= 2N_0 \sum_{\mathrm{N}}{}' \sum_{ijkl} \Phi_{ijkl}(0,0,\mathrm{N,N})[J_{ij}(0,0) - J_{ij}(0,\mathrm{N})][J_{kl}(0,0) - J_{kl}(0,\mathrm{N})]. \end{aligned} \tag{33.20}$$

The cubic contribution to (19.24) for $\mathscr{A}_2$, i.e., the contribution which contains third-order potential coefficients, contains the sum

$$\begin{aligned} \sum_{\mathrm{MM'NN'PP'}} \sum_{ii'jj'kk'} &\Phi_{ijk}(\mathrm{M,N,P}) \Phi_{i'j'k'}(\mathrm{M',N',P'}) \\ &\times J_{ii'}(\mathrm{M,M'}) J_{jj'}(\mathrm{N,N'}) J_{kk'}(\mathrm{P,P'}). \end{aligned} \tag{33.21}$$

Eliminating surface effects from (33.21) gives

$$N_0 \sum_{\mathrm{M'NN'PP'}} \sum_{ii'jj'kk'} \Phi_{ijk}(0,\mathrm{N,P}) \Phi_{i'j'k'}(\mathrm{M',N',P'}) J_{ii'}(0,\mathrm{M'}) J_{jj'}(\mathrm{N,N'}) J_{kk'}(\mathrm{P,P'});$$

then using the translational invariance condition $\Phi_{i'j'k'}(\mathrm{M}',\mathrm{N}',\mathrm{P}') = \Phi_{i'j'k'}(0,\mathrm{N}'-\mathrm{M}',\mathrm{P}'-\mathrm{M}')$ and replacing N′ by N′ + M′ and P′ by P′ + M′ inside the sum gives

$$N_0 \sum_{\mathrm{M}'\mathrm{N}\mathrm{N}'\mathrm{P}\mathrm{P}'} \sum_{ii'jj'kk'} \Phi_{ijk}(0,\mathrm{N},\mathrm{P})\Phi_{i'j'k'}(0,\mathrm{N}',\mathrm{P}') \times J_{ii'}(0,\mathrm{M}')J_{jj'}(\mathrm{N},\mathrm{N}'+\mathrm{M}')J_{kk'}(\mathrm{P},\mathrm{P}'+\mathrm{M}');$$

and finally using the translational invariance condition $J_{jj'}(\mathrm{N},\mathrm{N}'+\mathrm{M}') = J_{jj'}(0,\mathrm{N}'+\mathrm{M}'-\mathrm{N})$ gives

$$N_0 \sum_{\mathrm{M}'\mathrm{N}\mathrm{N}'\mathrm{P}\mathrm{P}'} \sum_{ii'jj'kk'} \Phi_{ijk}(0,\mathrm{N},\mathrm{P})\Phi_{i'j'k'}(0,\mathrm{N}',\mathrm{P}') \times J_{ii'}(0,\mathrm{M}')J_{jj'}(0,\mathrm{N}'+\mathrm{M}'-\mathrm{N})J_{kk'}(0,\mathrm{P}'+\mathrm{M}'-\mathrm{P}). \tag{33.22}$$

For two-body potentials $\Phi_{ijk}(0,\mathrm{N},\mathrm{P})$ vanishes unless the index pair NP corresponds to one of the four sets 00, 0N, N0, or NN, for any N ≠ 0. Thus the sums Σ_{NP} and $\Sigma_{\mathrm{N}'\mathrm{P}'}$ in (33.22) each give rise to four terms; the 16 terms in $\Sigma_{\mathrm{NN}'\mathrm{PP}'}$ may be written out and factored by making use of the symmetries (9.14), (9.20), and (9.22):

$$\Phi_{ijk}(0,0,0) = -\sum_{\mathrm{N}}{}' \Phi_{ijk}(0,0,\mathrm{N}); \tag{33.23}$$

$$\Phi_{ijk}(0,0,\mathrm{N}) = \Phi_{ijk}(0,\mathrm{N},0) = -\Phi_{ijk}(0,\mathrm{N},\mathrm{N}). \tag{33.24}$$

After (33.22) is simplified, the complete expression for $\mathscr{A}_2$ for a primitive lattice crystal with two-body potentials is found to be

$$\begin{aligned}
\mathscr{A}_2 = {} & (N_0K^2/4M_C^2) \sum_{\mathrm{N}}{}' \sum_{ijkl} \Phi_{ijkl}(0,0,\mathrm{N},\mathrm{N}) \\
& \times [J_{ij}(0,0) - J_{ij}(0,\mathrm{N})][J_{kl}(0,0) - J_{kl}(0,\mathrm{N})] \\
& - (N_0K^2/12M_C^3) \sum_{\mathrm{M}} \sum_{\mathrm{N}}{}' \sum_{\mathrm{N}'}{}' \sum_{ii'jj'kk'} \Phi_{ijk}(0,0,\mathrm{N})\Phi_{i'j'k'}(0,0,\mathrm{N}')J_{ii'}(0,\mathrm{M}) \\
& \times [J_{jj'}(0,\mathrm{M}) - J_{jj'}(0,\mathrm{M}-\mathrm{N}) - J_{jj'}(0,\mathrm{M}+\mathrm{N}') \\
& + J_{jj'}(0,\mathrm{M}-\mathrm{N}+\mathrm{N}')] \\
& \times [J_{kk'}(0,\mathrm{M}) - J_{kk'}(0,\mathrm{M}-\mathrm{N}) - J_{kk'}(0,\mathrm{M}+\mathrm{N}') \\
& + J_{kk'}(0,\mathrm{M}-\mathrm{N}+\mathrm{N}')].
\end{aligned} \tag{33.25}$$

For central potentials, the potential coefficients appearing in (33.25) are given by (9.11) and (9.12).

We have previously evaluated (33.25) for several Lennard–Jones potentials for the fcc lattice;* for the 12,6 potential under the condition $\tilde{P} = 0$, the cubic and quartic contributions to $\mathscr{A}_2$ are conveniently expressed in terms of

* D. C. Wallace, *Phys. Rev.* **131,** 2046 (1963).

the total static-lattice potential as follows.

$$\mathscr{A}_2(\text{cubic}) = 2.09 N_0^2 K^2/\Phi_0,$$
$$\mathscr{A}_2(\text{quartic}) = -3.75 N_0^2 K^2/\Phi_0, \qquad (33.26)$$
$$\mathscr{A}_2(\text{total}) = -1.66 N_0^2 K^2/\Phi_0.$$

Note that Φ_0 is negative, so $\mathscr{A}_2$ is positive. In all our calculations of $\mathscr{A}_2$, the sums over lattice points in (33.25) converge rapidly when summing out from the origin, so that accurate results may be obtained by summing over nearest neighbors only [note that the Σ_{M} in the cubic part of (33.25) also includes the point $\mathrm{M} = 0$]. For the 12,6 potential, for example, if the lattice point sums are restricted to nearest neighbors only, the error in $\mathscr{A}_2$(cubic) is 0.4% and the error in $\mathscr{A}_2$(quartic) is <0.1%.

We have also used a crude approximation for $\mathscr{A}_2$ to estimate its volume derivative;* the result for the present model is

$$d \ln \mathscr{A}_2/d \ln V \approx 6.3. \qquad (33.27)$$

This volume derivative is required to calculate the anharmonic contribution to the thermal expansion coefficient at high temperature.

Referring again to the high-temperature expansion (33.13) of F_{A}, it is seen that the term $\mathscr{A}_0$ does not contribute to thermodynamic functions which are temperature derivatives of F, such as the heat capacity C_V and the function βB_T. An estimate of the magnitude of the term $\mathscr{A}_{-2}T^{-2}$ may be obtained from (19.11) for F_{A}; by replacing each ω_κ^2 inside the sums in that equation by the average $\langle\omega_\kappa^2\rangle$, we have the approximation

$$|\mathscr{A}_{-2}|\, T^{-2} \approx \tfrac{1}{240}[\hbar^2\langle\omega_\kappa^2\rangle/K^2T^2]^2\, |\mathscr{A}_2|\, T^2,$$

or with the definition (19.32) of $\Theta_{\mathrm{H}\infty}$,

$$|\mathscr{A}_{-2}|\, T^{-2} \approx \tfrac{1}{240}(\Theta_{\mathrm{H}\infty}/T)^4\, |\mathscr{A}_2|\, T^2. \qquad (33.28)$$

Thus the term $\mathscr{A}_{-2}T^{-2}$ may be expected to be about the same magnitude as $\mathscr{A}_2T^2$ at a temperature $T \approx \frac{1}{4}\Theta_{\mathrm{H}\infty}$, and it may be expected to be negligible compared to $\mathscr{A}_2T^2$ for $T \gtrsim \frac{1}{2}\Theta_{\mathrm{H}\infty}$. As a first approximation, then, we will calculate the anharmonic contributions to C_V and to βB_T in the intermediate- to high-temperature region from the leading high-temperature term $F_{\mathrm{A}} = \mathscr{A}_2T^2$.

Comparison of Theory and Experiment

With the condition (33.9) for $\tilde{P} = 0$, the total static-lattice potential of (33.5) becomes

$$\Phi_0 = -\tfrac{1}{2}N_0[(\alpha - \beta)/\beta]aS_\alpha R_1^{-\alpha}. \qquad (33.29)$$

* D. C. Wallace, *Phys. Rev.* **139**, A877 (1965).

Table 18. Comparison of lattice-dynamical calculations and measured quantities for Ar and Kr.

Quantity		Ar		Kr	
Theory	Experiment	Theory	Experiment	Theory	Experiment
Φ_0	F_0 (kcal/mole)	-1.806	-1.846	-2.790	-2.666
$d\tilde{B}/dP$	$(\partial B_T/\partial P)_T$	8	8.0 (65°K)	8	9.0 (77°K)
$\tilde{C}_{11}$	C_{11} (10^{10} dyne/cm^2)	3.75	4.39 (0°K)		
$\tilde{C}_{12}$	C_{12} (10^{10} dyne/cm^2)	2.15	1.83 (0°K)		
$\tilde{C}_{44}$	C_{44} (10^{10} dyne/cm^2)	2.15	1.64 (0°K)		
Θ_{H0}	Θ_0 (°K)	87.9	92.0	71.0	71.9
$\langle\omega_\kappa^2\rangle$	$\langle\omega_\kappa^2\rangle$ (10^{26} sec^{-2})	0.698	0.70	0.455	0.46
$\langle\gamma_\kappa\rangle$	$\langle\gamma_\kappa\rangle$	2.964	2.7	2.964	3.0
$\mathscr{A}_2$	$\mathscr{A}_2$ (10^{-4} cal/mole °K^2)	36	3.5	23	8

The calculated values of Φ_0 are compared with $F_0 = F(T = 0)$ in Table 18; note that F_0 is the negative of the crystal binding energy at $T = 0$. The differences between theory and experiment for the binding energies are small, but they are outside the limits of experimental error. To calculate pressure derivatives of the bulk modulus in the potential approximation, we begin with the general formula

$$\tilde{B} = V(d^2\Phi_0/dV^2), \tag{33.30}$$

which is valid for any volume, and transform volume derivatives to pressure derivatives according to

$$d\tilde{B}/dP = (d\tilde{B}/dV)(dV/dP) = -(V/\tilde{B})(d\tilde{B}/dV). \tag{33.31}$$

Final results for the Lennard–Jones potential, evaluated at $\tilde{P} = 0$, are

$$d\tilde{B}/dP = \tfrac{1}{3}(\alpha + \beta + 6), \tag{33.32}$$

$$\tilde{B}(d^2\tilde{B}/dP^2) = -\tfrac{1}{9}(\alpha\beta + 3\alpha + 3\beta + 9). \tag{33.33}$$

For the 12,6 potential these results are simply

$$d\tilde{B}/dP = 8, \qquad \tilde{B}(d^2\tilde{B}/dP^2) = -15. \tag{33.34}$$

Experimental values of $(\partial B_T/\partial P)_T$ for Ar and Kr at finite temperatures are listed in Table 18, and these agree well with the potential approximation value of 8.

For central potentials the second-order elastic constants $\tilde{C}_{ijkl}$ in the potential approximation are given by (9.39). That expression is simplified for a primitive lattice with Lennard–Jones potentials, and for $\tilde{P} = 0$, to

$$\tilde{C}_{ijkl} = [9\tilde{B}/(\alpha - \beta)][(\alpha + 2)S_\alpha^{-1}S_{ijkl}^\alpha - (\beta + 2)S_\beta^{-1}S_{ijkl}^\beta]. \tag{33.35}$$

The lattice sums introduced in (33.35) are defined by

$$S_{ijkl}^{\alpha} = \sum_{\mathrm{N}}' [R_1/R(\mathrm{N})]^{\alpha}[R_i(\mathrm{N})R_j(\mathrm{N})R_k(\mathrm{N})R_l(\mathrm{N})/R(\mathrm{N})^4]. \tag{33.36}$$

The S_{ijkl}^{α} vanish for a primitive cubic lattice unless the Cartesian indices are equal in pairs, and the two independent S_{ijkl}^{α} are S_{xxxx}^{α} and S_{xxyy}^{α}. These two sums are obviously related to S_{α} by

$$3S_{xxxx}^{\alpha} + 6S_{xxyy}^{\alpha} = S_{\alpha}. \tag{33.37}$$

For $\alpha = 12$, $\beta = 6$, the required lattice sums for the fcc lattice are

$$S_{xxxx}^{12} = 2.037770, \qquad S_{xxxx}^{6} = 2.563592. \tag{33.38}$$

The theoretical elastic constants $\tilde{C}_{11}$ and $\tilde{C}_{12} = \tilde{C}_{44}$ (Voigt notation) are compared with the $T = 0$ experimental values for Ar in Table 18; here also the differences between theory and experiment are larger than experimental errors.

We have also calculated the harmonic Debye temperature $\Theta_{\mathrm{H}0}$ at $T = 0$, the Brillouin zone averages $\langle \omega_{\kappa}^2 \rangle$ and $\langle \gamma_{\kappa} \rangle$, and the high-temperature anharmonic coefficient $\mathscr{A}_2$, for the present central potential models for Ar and Kr. These calculations are compared with experimental results in Table 18, where the experimental values of $\langle \omega_{\kappa}^2 \rangle$, $\langle \gamma_{\kappa} \rangle$, and $\mathscr{A}_2$ are taken from the high-temperature thermodynamic analysis of Section 31. The agreement between theory and experiment is good for the harmonic quantity $\langle \omega_{\kappa}^2 \rangle$; this agreement may be expected since the potential was determined to fit the experimental value of the harmonic quantity $V_0 B_0$. Further, the differences between the calculated $\Theta_{\mathrm{H}0}$ and the measured Θ_0 are of the same sign and order of magnitude as expected from theoretical estimates of the anharmonic contribution to Θ_0. The difference between calculated and measured $\langle \gamma_{\kappa} \rangle$ is small for Kr, but larger than experimental error for Ar. Finally the theoretical $\mathscr{A}_2$ is much too large, being larger than experiment by a factor of 10 for Ar and a factor of 3 for Kr. These large discrepancies for $\mathscr{A}_2$ provide clear evidence that the Lennard–Jones 12,6 potential is inaccurate for Ar and Kr.

We have calculated the constant-volume heat capacity C_V from the measured C_P, and then corrected C_V to the fixed volume V_0 as a function of T for Ar and Kr. The corresponding experimental heat capacity Debye temperatures $\Theta(V_0)$, which of course include anharmonicity, are plotted in Figure 33. As T increases from $T = 0$, the initial decrease and the subsequent increase of $\Theta(V_0)$ is due to the phonon dispersion, and the continued increase of $\Theta(V_0)$ at temperatures above about 20°K is due to the anharmonic contribution to $C_V(V_0)$. The harmonic heat capacity $C_{V\mathrm{H}}(V_0)$ was calculated from (16.54) for Ar and Kr, and the corresponding Debye temperatures $\Theta_{\mathrm{H}}(V_0)$

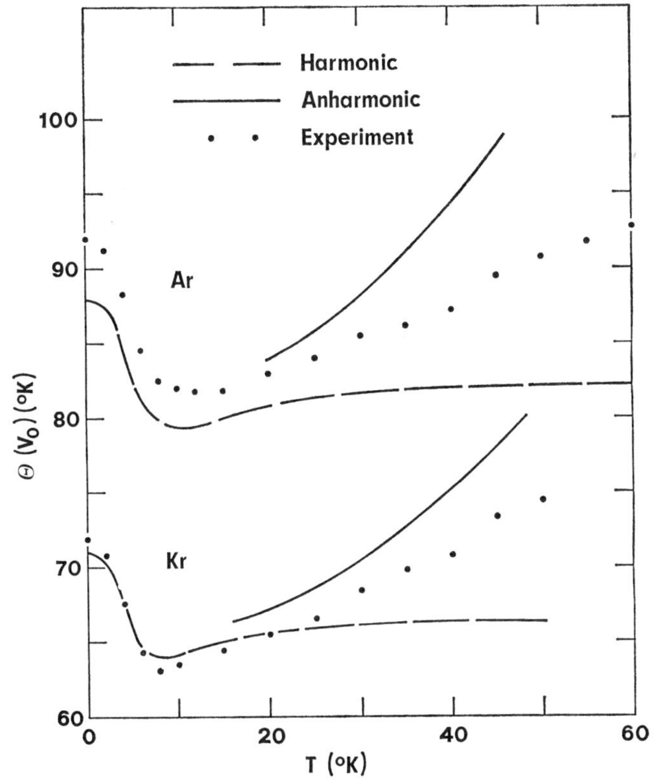

Figure 33. Comparison of theoretical and experimental Debye temperatures, evaluated from $C_V(V_0)$, for Ar and Kr.

are shown by the dashed lines in Figure 33. For both Ar and Kr, the effects of phonon dispersion are qualitatively reproduced in the model calculations of $\Theta_{\mathrm{H}}(V_0)$. At higher temperatures the total theoretical heat capacity $C_V(V_0)$ was approximated by including the leading anharmonic term, according to

$$C_V(V_0) = C_{V\mathrm{H}}(V_0) - 2\mathscr{A}_2 T, \tag{33.39}$$

where $\mathscr{A}_2$ is the theoretical value listed in Table 18. The corresponding calculated Debye temperatures $\Theta(V_0)$ are shown by the solid lines in Figure 33, and it is again seen that the calculated anharmonic contribution is much larger than experiment for both Ar and Kr.

The measured results for βB_T were corrected to the fixed volume V_0, and the experimental curves are shown in the form of the dimensionless quantity $V\beta B_T(V_0)/3N_0K$ for Ar in Figure 34 and for Kr in Figure 35. The harmonic approximation $(\beta B_T)_{\mathrm{H}}$ was calculated from (16.83), and the results are the

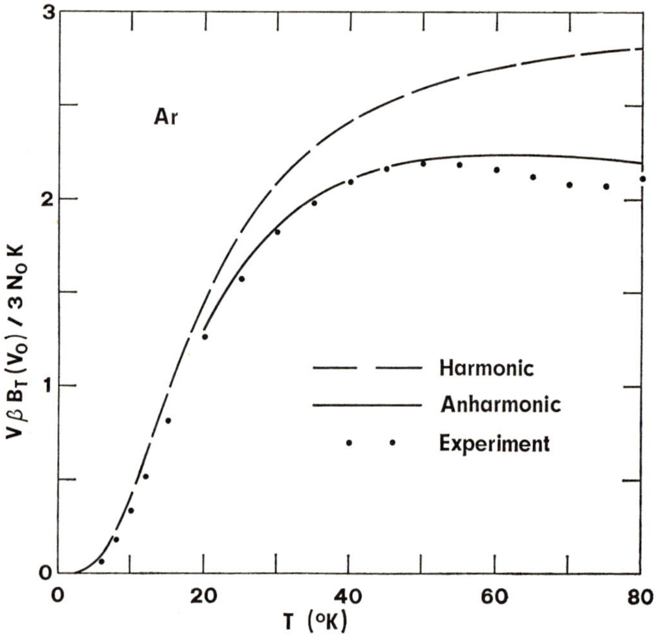

Figure 34. Comparison of theory and experiment for the function $\beta B_T(V_0)$ for Ar.

dashed lines in Figures 34 and 35. Again at higher temperatures the total βB_T was approximated by including the leading anharmonic term, according to

$$V\beta B_T = V(\beta B_T)_{\rm H} - 2\mathscr{A}_2 T(d \ln \mathscr{A}_2/d \ln V), \tag{33.40}$$

where the approximation (33.27) was used for $d \ln \mathscr{A}_2/d \ln V$. Results of this high-temperature approximation are shown by the solid lines in Figures 34 and 35. From these figures it is seen that the present theory, including anharmonicity approximately at intermediate to high temperatures, gives a qualitatively correct description of the measured temperature-dependence of βB_T for Ar and Kr.

Comments on Improvement of the Theory

The calculations for Ar and Kr presented here are based on an over-simplified model for the potential of interaction among the atoms in the crystal; these calculations are designed primarily to illustrate the procedures for comparing theory and experiment for the thermodynamic properties of crystals. In order to provide a significantly improved theory for the rare

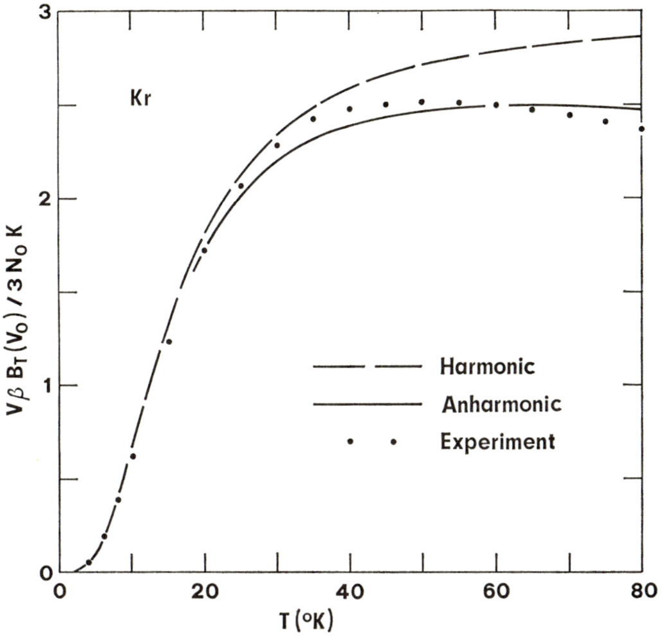

Figure 35. Comparison of theory and experiment for the function $\beta B_T(V_0)$ for Kr.

gas crystals, it is necessary to begin with a physically more realistic interatomic potential, and to eliminate some of the approximations made in the present lattice-dynamical calculations.

Consider first the interatomic potential. The simple Lennard–Jones 12,6 form is based on a van der Waals attraction proportional to R^{-6} between two rare gas atoms separated by a distance R, and some sort of short-range repulsion between two rare gas atoms when they are close enough so that their electrons begin to overlap. But the complete van der Waals interaction between two rare gas atoms contains additional terms in inverse powers of R, and may be written

$$\varphi(R)(\text{van der Waals}) = -bR^{-6} - cR^{-8} - \cdots . \qquad (33.41)$$

The coefficients b and c are positive, and may be calculated theoretically. One weakness of the present 12,6 potential is that the coefficient b, determined by fitting the measured crystal volume and bulk modulus at $T = 0$ and $P = 0$, is much larger than the theoretical value of b; this weakness is characteristic of all similar models in which the potential is determined by fitting crystal data. In particular, the present model values of b, in units of 10^{-10} erg Å^6, are 0.975 for Ar and 2.173 for Kr, while the corresponding theoretical values for van der Waals interactions are 0.626 and 1.253, respectively.* In addition, the van der Waals R^{-8} term in the two-body potential is not negligible; for the nearest neighbors in Ar, for example, cR^{-8} is about 15% of bR^{-6}. Thus for a physically realistic model, the attractive two-body van der Waals interaction should be taken from (33.41), with the theoretical values of the coefficients b and c.

The short-range repulsive potential is taken proportional to R^{-12} in the Lennard–Jones 12,6 model. Although an accurate theoretical calculation of the repulsion between two overlapping rare gas atoms is difficult, experiments have been carried out on the short-range scattering between rare gas atoms.† These experiments may be interpreted in terms of a short-range exponential potential, so it would seem more reasonable to use the form

$$\varphi(R)(\text{repulsive}) = ae^{-\alpha R}, \tag{33.42}$$

with the positive parameters a and α determined from scattering experiments. A repulsive potential of this form might lead to a considerable reduction in the magnitude of the theoretical value of $\mathscr{A}_2$.

Another contribution to the interatomic potential, which is not included in the Lennard–Jones model, arises from noncentral many-body van der Waals interactions. Such interactions can be expressed as a series containing three-body potentials, four-body potentials, and so on. The form of the three-body van der Waals potential was derived by Axilrod and Teller;‡ this potential is basically repulsive and gives rise to a nonnegligible positive contribution to the static-lattice potential Φ_0. For example, the contribution amounts to $0.09\,|\Phi_0|$ for Ar and Kr. Again a physically realistic model should include the three-body van der Waals potential.

For Ar and Kr, we have constructed potentials composed of the two-body van der Waals attraction (33.41), the short-range repulsion (33.42), and the three-body van der Waals interactions. The coefficients were all taken from theoretical calculations, except for the repulsion, where the parameters were determined from scattering data. Because of the significantly smaller values of the van der Waals coefficient b, these potentials are not deep enough to account for the experimentally observed binding energies. In other words, at $\tilde{P} = 0$, the calculated values of $|\Phi_0|$ are slightly less than the measured binding energies. This suggests the presence of additional *attractive* forces in the rare gas crystals, which might arise from short-range electronic exchange among near neighbors. Such an exchange effect may also be pictured as a property of the electronic band structure of the crystal, and gives rise to a

* A. E. Kingston, *Phys. Rev.* **135,** A1018 (1964).

† See, e.g., the graphs of A. A. Abrahamson, *Phys. Rev.* **130,** 693 (1963).

‡ B. M. Axilrod and E. Teller, *J. Chem. Phys.* **11,** 299 (1943).

noncentral many-body potential. In fact the crystal energy is presumably lowered when the atomic electrons become band electrons, and the corresponding lowering of the total potential Φ_0 is obtainable from a band-structure calculation. On the other hand, the entire lattice-dynamics calculation for the rare gas crystals may be carried out from the band-structure theory of Chapter 5; such a procedure includes in principle all the van der Waals, the repulsive, and the exchange interactions.

Let us now consider some of the approximations made in the present lattice-dynamics calculations. First of all the measurements on rare gas crystals are generally carried out with the crystals in the presence of their normal vapor pressure, and although the effect is small it should be recognized, since the vapor pressure depends on the temperature. In addition, at sufficiently high temperatures, above the Debye temperature for Ar and Kr, the thermal production of vacancies can contribute significantly to the thermodynamic functions. The experimental separation of vacancy contributions is quite difficult, and for a serious test of a lattice-dynamics model it would seem appropriate at first to consider only temperatures at which vacancies are unimportant.

Further the harmonic vibrational free energy F_{H} should be included at all temperatures in calculations of the free energy and its volume or strain derivatives. At $T = 0$ the harmonic free energy is

$$F_{\mathrm{H}0} = \tfrac{3}{2} N_0 \hbar \langle \omega_\kappa \rangle, \tag{33.43}$$

and this should be included in the calculation of the binding energy, the pressure (which is presumably zero), and the bulk modulus and elastic constants. For the present Lennard–Jones 12,6 model, at $T = 0$ and $\tilde{P} = 0$,

$$\Phi_0 + F_{\mathrm{H}0} = 0.8990\Phi_0 \quad \text{for Ar;} \qquad \Phi_0 + F_{\mathrm{H}0} = 0.9472\Phi_0 \quad \text{for Kr.} \tag{33.44}$$

The harmonic contribution to the bulk modulus at $T = 0$ is

$$B_{\mathrm{H}0} = \tfrac{3}{2} N_0 \hbar V (d^2 \langle \omega_\kappa \rangle / dV^2), \tag{33.45}$$

and for the present model at $\tilde{P} = 0$,

$$\tilde{B} + B_{\mathrm{H}0} = 1.123\tilde{B} \quad \text{for Ar;} \qquad \tilde{B} + B_{\mathrm{H}0} = 1.064\tilde{B} \quad \text{for Kr.} \tag{33.46}$$

These harmonic contributions are not negligible for Ar and Kr. We have in fact included the harmonic free energy at $T = 0$, and redetermined the potential parameters a and b so as to fit the measured crystal volume and bulk modulus at $T = 0$ and $P = 0$, for Ar and Kr. This improved model calculation leads to slight changes in all the theoretical thermodynamic functions, but no major changes are found.

Finally in the calculation of temperature derivatives of the free energy, namely C_V and βB_T, the present theory includes the anharmonic free energy accurately at high temperature, and approximately at intermediate temperature. For a more accurate lattice-dynamics calculation it would be most useful to include an accurate calculation of the anharmonic contribution to Θ_0 and its volume derivative, and a more respectable estimate of the anharmonic contributions to C_V and βB_T at intermediate temperatures.

34. SIMPLE METALS

Sodium and Potassium: The Pseudopotential Model

We have applied the local pseudopotential perturbation theory of Chapter 6 to a detailed lattice-dynamical calculation for Na and K.* The calculations were done for the bcc lattice with one conduction electron per ion ($z = 1$), and experimental data for Na were restricted to the bcc phase, with no consideration of the hcp phase (below 35°K). The Harrison model was taken for the local pseudopotential $w_{\mathrm{B}}(r)$; the corresponding Fourier components $w_{\mathrm{B}}(q)$ are given by (26.41) in terms of the two parameters $\hat{\beta}$ and $\hat{\rho}$. A Born–Mayer repulsion of the form $\alpha_{\mathrm{B}} e^{-\gamma_{\mathrm{B}} R}$ between two ions separated by a distance R was also included. The parameters were then determined by fitting the potential approximation theory to the experimental results for the energy, the crystal volume, and the bulk modulus at $T = 0$ and $P = 0$. Of these three experimental quantities, the repulsive potential contributes significantly only to the bulk modulus, so the quality of the theoretical fits remains essentially unchanged under correlated changes in α_{B} and γ_{B}, e.g., a simultaneous increase in both. We therefore fixed γ_{B} at the value found by Tosi† in a study of the potentials for alkali halides:

$$1/\gamma_{\mathrm{B}} = 0.339(10^{-8}\ \mathrm{cm}). \tag{34.1}$$

The total static-lattice potential is written in (27.23), and for a primitive lattice it is convenient to separate this into several contributions, as follows.

$$\Phi_0/N_0 = I_z + E_{\mathrm{EG}} + E_{\mathrm{BS}} + E_{\mathrm{ES}} + E_{\mathrm{R}}, \tag{34.2}$$

where the ionization energy per atom is I_z, the uniform electron gas energy per atom is

$$E_{\mathrm{EG}} = z(\tfrac{3}{5}e_{\mathrm{F}} + e_{\mathrm{X}} + e_{\mathrm{C}} + \hat{\beta}/V_A), \tag{34.3}$$

* D. C. Wallace, *Phys. Rev.* **176,** 832 (1968).

† M. P. Tosi, in *Solid State Physics*, edited by F. Seitz and D. Turnbull, Academic Press Inc., New York, 1964, Vol. 16, p. 1.

the band-structure energy per atom is

$$E_{\mathrm{BS}} = \sum_{\mathbf{Q}}' F(Q), \tag{34.4}$$

the electrostatic energy per atom is

$$E_{\mathrm{ES}} = -\alpha_{\mathrm{I}}(z^2/r_A), \tag{34.5}$$

and the Born–Mayer repulsive energy per atom is

$$E_{\mathrm{R}} = \tfrac{1}{2} \sum_{\mathrm{N}}' \alpha_{\mathrm{B}} e^{-\gamma_{\mathrm{B}} R(\mathrm{N})}. \tag{34.6}$$

In (34.3) the free-electron Fermi energy, exchange energy per electron, and correlation energy per electron are e_{F}, e_{X}, and e_{C}, respectively, and $\hat{\beta}/V_A$ is $w_{\mathrm{C}}(q = 0)$ for the Harrison model, according to (26.42). In (34.5), r_A is the radius of the sphere whose volume is V_A, according to (27.19), and from Table 3 the value of α_{I} for the bcc lattice is 1.79186. The free energy at $T = 0$ is F_0, the negative of the crystal binding energy, and in the potential approximation the value per atom is

$$N_0^{-1} F_0 \approx N_0^{-1} \Phi_0. \tag{34.7}$$

Also in the potential approximation the pressure is $\tilde{P}$ and the bulk modulus is $\tilde{B}$, and these are conveniently calculated from the expressions

$$V_A \tilde{P} = -V_A[d(\Phi_0/N_0)/dV_A], \tag{34.8}$$

$$V_A \tilde{B} = V_A^2[d^2(\Phi_0/N_0)/dV_A^2]. \tag{34.9}$$

With the parameter γ_{B} fixed according to (34.1), we found that it is possible to fit the measured binding energy and bulk modulus, and to satisfy the condition $\tilde{P} = 0$ at the measured atomic volume V_A, for a range of values of the remaining parameters $\hat{\beta}$, $\hat{\rho}$, and α_{B}. We therefore required the additional condition that theory and experiment should agree for the Brillouin zone average $\langle \omega_\kappa^2 \rangle$; this average was calculated from (28.16), and the experimental value was taken from the high-temperature entropy analysis of Section 31. This additional condition served to fix uniquely the potential parameters. The experimental parameters used in the calculations for Na and K are listed in Table 19. Here r_s is the radius of the sphere whose volume is the volume per electron, and $r_s = r_A$ since $z = 1$. The effects of electronic exchange and correlation on the screening are included by the interpolation approximation (26.76), in terms of the parameter ξ defined by (26.82). The potential parameters which are determined by fitting theory to experiment, as described above, are also listed in Table 19.

The various calculated contributions to the static-lattice potential per atom, and its volume derivatives, are listed separately in Table 20. Several

Table 19. Experimental quantities and the corresponding fitted potential parameters for Na and K.

Quantity	Na	K
Experimental quantities		
z	1	1
V_A (a_0^3)	255.5	485.3
$r_s = r_A$ (a_0)	3.937	4.875
ξ	1.81	1.77
$N_0^{-1}F_0$ (Ry)	−0.08	−0.07
I_z (Ry)	0.38	0.32
$V_A B_0$ (Ry)	0.129	0.121
Potential parameters		
α_B (Ry)	10.5	124
$\hat{\beta}$ (Ry a_0^3)	37	66
$\hat{\rho}$ (a_0)	0.50	0.69

interesting results for Na and K may be observed from this table. Consider first the energy, listed in Table 20 in the form $(\Phi_0/N_0) - I_z$. Although the four contributions to the uniform electron gas energy E_{EG} are not small, they nearly cancel, and the total binding energy of the crystal is due almost entirely to the electrostatic energy E_{ES} of the lattice of point ions in a uniform negative background charge. On the other hand, for the pressure in the form $V_A\tilde{P}$, the contributions from the volume derivatives of E_{EG} and of E_{ES} are both large in magnitude, but are opposite in sign and nearly cancel. The

Table 20. Theoretical contributions to the static-lattice potential and its volume derivatives, evaluated at $V = V_0$. All quantities are in Rydbergs.

Potential contribution	Na			K		
	$(\Phi_0/N_0) - I_z$	$V_A\tilde{P}$	$V_A\tilde{B}$	$(\Phi_0/N_0) - I_z$	$V_A\tilde{P}$	$V_A\tilde{B}$
$\frac{3}{5}e_{\mathrm{F}}$	0.14	0.095	0.159	0.09	0.062	0.103
e_{X}	−0.23	−0.078	−0.103	−0.19	−0.063	−0.084
e_{C}	−0.08	−0.010	−0.010	−0.07	−0.010	−0.010
$\hat{\beta}/V_A$	0.15	0.145	0.290	0.14	0.136	0.272
E_{EG}	−0.02	0.152	0.336	−0.03	0.125	0.281
E_{BS}	−0.00	−0.004	−0.021	−0.00	−0.005	−0.017
E_{ES}	−0.45	−0.152	−0.202	−0.37	−0.123	−0.163
E_{R}	+0.00	0.004	0.016	+0.00	0.004	0.020
Total	−0.47	0.000	0.129	−0.40	0.001	0.121
Experiment	−0.46	0.000	0.129	−0.39	0.000	0.121

Table 21. Comparison of theory and experiment for the elastic stiffness coefficients and their pressure derivatives for Na and K. The measured pressure derivatives are at room temperature.

Quantity		Na		K	
Theory	Experiment	Theory	Experiment	Theory	Experiment
$V_A\tilde{B}_{11}$	V_AB_{11} (Ry)	0.139	0.145	0.131	0.138
$V_A\tilde{B}_{12}$	V_AB_{12} (Ry)	0.124	0.121	0.116	0.113
$V_A\tilde{B}_{44}$	V_AB_{44} (Ry)	0.094	0.104	0.087	0.094
$V_A\tilde{B}$	V_AB (Ry)	0.129	0.129	0.121	0.121
$d\tilde{B}_{11}/dP$	$(\partial B_{11}^S/\partial P)_T$	3.8	3.90	4.0	4.30
$d\tilde{B}_{12}/dP$	$(\partial B_{12}^S/\partial P)_T$	3.5	3.45	3.7	3.80
$d\tilde{B}_{44}/dP$	$(\partial B_{44}^S/\partial P)_T$	1.5	1.63	1.6	1.62
$d\tilde{B}/dP$	$(\partial B_S/\partial P)_T$	3.6	3.60	3.8	3.97

smallness of the band-structure energy contribution to the potential and its volume derivatives is in agreement with the nearly free-electron characteristics of the alkali metals. The smallness of the Born–Mayer repulsion contribution to the potential and its volume derivatives indicates that the ion cores do not overlap appreciably. In fact from the contributions to $V_A\tilde{P}$, it is obviously the outward pressure of the uniform electron gas, and not the repulsion between ion cores, which keeps the crystal from collapsing inward at $P = 0$. Finally, we note that the band-structure and Born–Mayer energies are of increasing relative importance in higher-volume derivatives of the potential; this indicates the importance of these energies in the anharmonic properties.

It is much more justified to neglect the harmonic vibrational free energy at $T = 0$ for the alkali metals than for the rare gas crystals. For the present pseudopotential models for Na and K, the zero-point contributions to F_0/N_0 and to V_AP are about 0.001 Ry.

Sodium and Potassium: Comparison with Experiment

The elastic coefficients $\tilde{B}_{\alpha\beta}$ were calculated by the method of homogeneous deformation, as outlined in Section 27, for the present pseudopotential models for Na and K. The calculated results are compared with $T = 0$ experimental values in Table 21. The experimental data for K are taken from Table 7, while for Na we used the low-temperature extrapolation of the data of Diederich and Trivisonno* for bcc Na. The agreement between theory

* M. E. Diederich and J. Trivisonno, *J. Phys. Chem. Solids* **27**, 637 (1966).

and experiment for the elastic stiffness coefficients is reasonably good, but not within experimental error. Note that the Cauchy relation $\tilde{B}_{12} = \tilde{B}_{44}$ at $\tilde{P} = 0$ does not hold for the pseudopotential theory. The zero-point vibrational contributions to $V_A B_{\alpha\beta}$, which of course are not included in our calculated $V_A \tilde{B}_{\alpha\beta}$, are estimated to be about 0.001 Ry in magnitude for Na and K.

We also calculated the volume derivatives of the $\tilde{B}_{\alpha\beta}$ by evaluating them at several different volumes, and then converted the volume derivatives to pressure derivatives by the equation

$$d\tilde{B}_{\alpha\beta}/dP = -(V/\tilde{B})(d\tilde{B}_{\alpha\beta}/dV). \tag{34.10}$$

These calculations are compared with room temperature measurements of $(\partial B^S_{\alpha\beta}/\partial P)_T$ in Table 21. The experimental data for K are taken from Table 8, while for Na we used the results of Daniels.* The calculations, of course, correspond to conditions of $T = 0$ and $P = 0$; the agreement with measured pressure derivatives is again reasonably good.

As it was noted in Section 28, following (28.20), the coefficients $\tilde{B}_{\alpha\beta}$ calculated from the pseudopotential perturbation theory by the method of long waves do not agree with those calculated by the method of homogeneous deformation. Let us write this difference for the bulk modulus $\tilde{B}$ as

$$V_A\tilde{B}(\text{long waves}) = V_A\tilde{B}(\text{homogeneous deformation}) - \Delta; \tag{34.11}$$

the same difference holds for $V_A\tilde{B}_{11}$ and $V_A\tilde{B}_{12}$, while for $V_A\tilde{B}_{44}$ there is no difference between the two methods of calculation. For the present pseudopotential models for Na and K, Δ is -0.008 and -0.010 Ry, respectively. Also for the pressure derivatives we can write

$$d\tilde{B}/dP(\text{long waves}) = d\tilde{B}/dP(\text{homogeneous deformation}) - \Delta'; \tag{34.12}$$

again the same difference holds for $d\tilde{B}_{11}/dP$ and $d\tilde{B}_{12}/dP$, and there is no difference for $d\tilde{B}_{44}/dP$. For the present models, Δ' has the value -0.3 for Na and K. These long-waves–homogeneous-deformation differences are quite large, and demonstrate the importance of the improved pseudopotential lattice-dynamics theory of Section 28.

The phonon frequencies calculated from the dynamical matrix (27.76) for the present pseudopotential models are in good agreement with experiment, as shown in Figures 36 and 37 for Na and K, respectively. In these figures the theoretical curves are the harmonic phonon frequencies and the experimental points are the frequencies measured by inelastic neutron scattering at fairly low temperatures; the frequencies are plotted in terms of ν_κ, where

$$\nu_\kappa = \omega_\kappa/2\pi. \tag{34.13}$$

* W. B. Daniels, *Phys. Rev.* **119**, 1246 (1960).

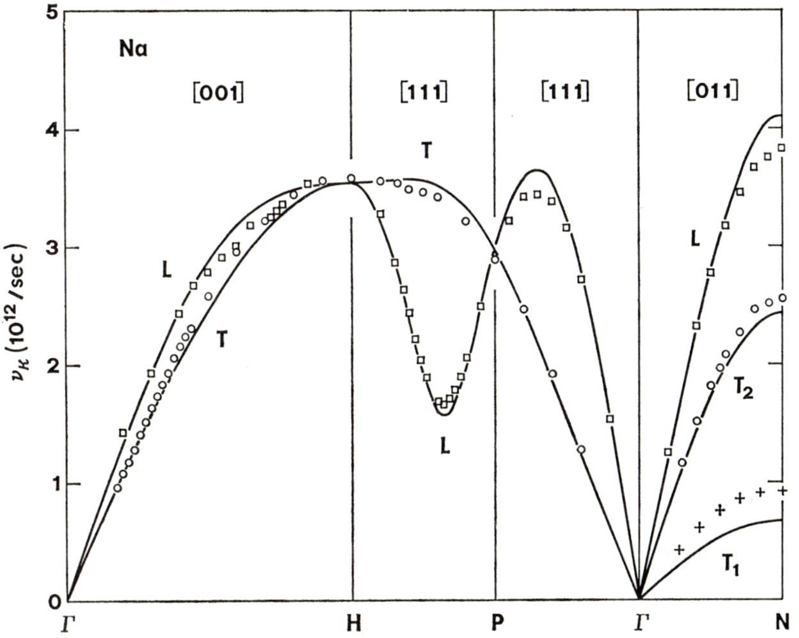

Figure 36. Calculated phonon-dispersion curves (solid lines) for bcc Na, compared with the experimental results (reference 37, Appendix 2) at 90°K.

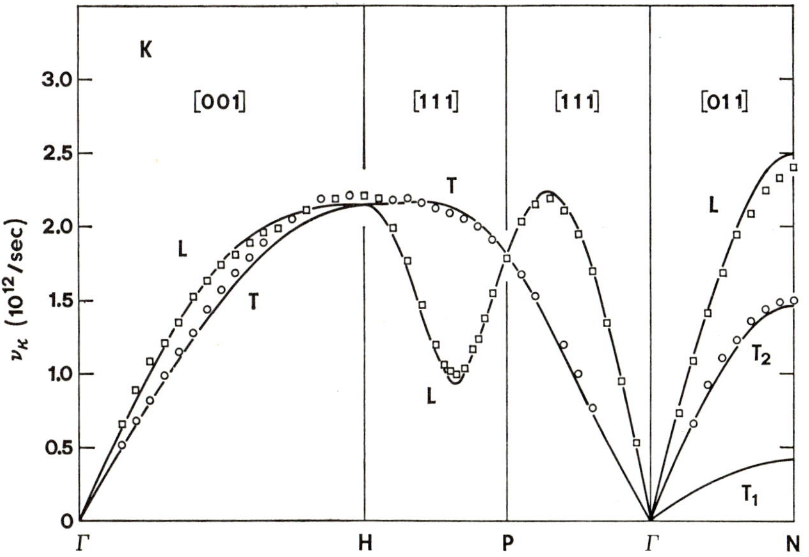

Figure 37. Calculated phonon-dispersion curves (solid lines) for bcc K, compared with the experimental results (reference 38, Appendix 2) at 9°K.

At the temperatures for which the experiments were carried out, namely 90°K for Na and 9°K for K, the anharmonic contributions to the frequencies should be quite small, say $\leqslant 1\%$. By slight variations of the potential parameters, the theoretical frequencies can be brought into even better agreement with the measured frequencies, generally within experimental error, except that the present simple model does not show the observed dip in the longitudinal phonon branch for **k** along [001]. We have no explanation for this observed dip in the phonon-dispersion curve; it appears to be much too large to be a Kohn anomaly or an anharmonic effect, but it might be accounted for by the additional terms in the expansion of the dynamical matrix in Umklapp processes in (28.35).

The theoretical phonon Grüneisen parameters γ_κ for **k** along symmetry directions are shown in Figures 38 and 39 for Na and K, respectively. These parameters are all positive, and show large variations over the Brillouin zone. All the calculated γ_κ lie in the range ~0.9–1.8 for Na, and ~0.95–1.7 for K. At the present time these calculations can be compared with experiment only through the measured thermal expansion coefficient, as shown below; we hope to see direct neutron-scattering measurements of the pressure-dependence of the phonon frequencies in the near future.

For Na and K, the heat capacity C_V was calculated from the measured C_P and then corrected to conditions of fixed volume V_0, and the electronic

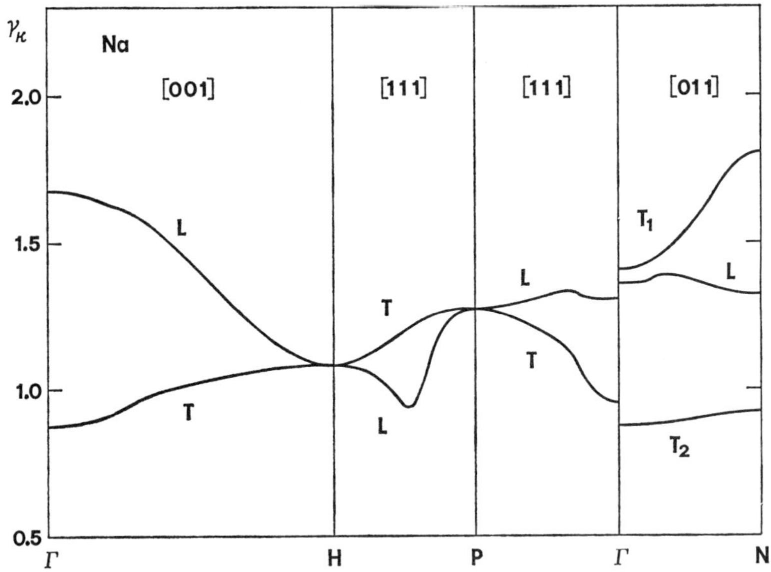

Figure 38. Calculated phonon Grüneisen parameters for bcc Na.

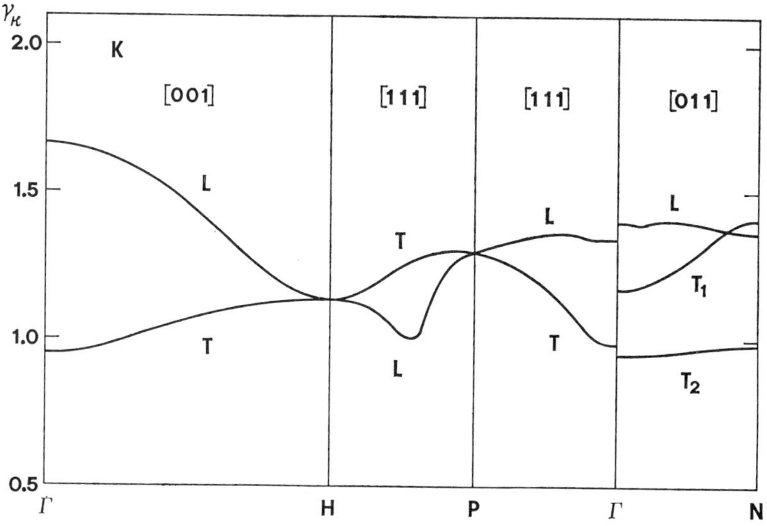

Figure 39. Calculated phonon Grüneisen parameters for bcc K.

contribution was subtracted to obtain the lattice heat capacity at V_0. The corresponding experimental Debye temperatures $\Theta(V_0)$ are shown in Figures 40 and 41 for Na and K, respectively. As the temperature increases from $T = 0$, the initial decrease and subsequent increase of $\Theta(V_0)$ is due to the dispersion of the phonons; at higher temperatures $\Theta(V_0)$ begins to decrease again, as shown clearly in Figure 40 for $T > 60°K$, because of the anharmonic contribution to $C_V(V_0)$. The theoretical curves of the harmonic Debye temperature $\Theta_H(V_0)$ are also shown in Figures 40 and 41; these curves approach the limits $\Theta_{H\infty}(V_0)$ at high temperatures, and do not show the decrease in $\Theta(V_0)$ due to anharmonicity. The theory and experiment agree for $\Theta_{H\infty}(V_0)$, since the theoretical potential parameters were fitted to the experimental $\langle\omega_\kappa^2\rangle$. Incidentally, it is clear from Figures 40 and 41 that it is quite difficult to determine the harmonic limit $\Theta_{H\infty}(V_0)$ from a graph of the experimental $\Theta(V_0)$ *vs* T.

At $T = 0$ the theoretical Debye temperature is much too low for Na and K. In particular the theoretical and experimental values are 132.4 and 152.5, respectively, for Na, and 80.3 and 90.6, respectively, for K. Note that the theoretical values here correspond to calculations by the method of long waves, appropriate to the theoretical phonon-dispersion curves, and are different from the values which would be calculated from the theoretical homogeneous-deformation elastic constants listed in Table 21. For K, for example, the homogeneous-deformation elastic constants of Table 21 lead to $\Theta_{H0} = 78.3°K$. Hence the discrepancies between the calculated Θ_{H0} and

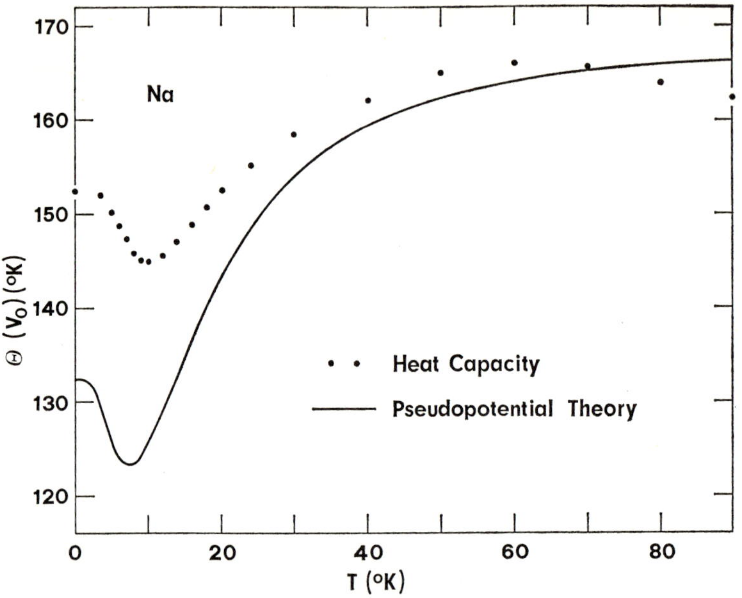

Figure 40. Comparison of the theoretical harmonic Debye temperature $\Theta_H(V_0)$ with the experimental lattice Debye temperature $\Theta(V_0)$, for Na.

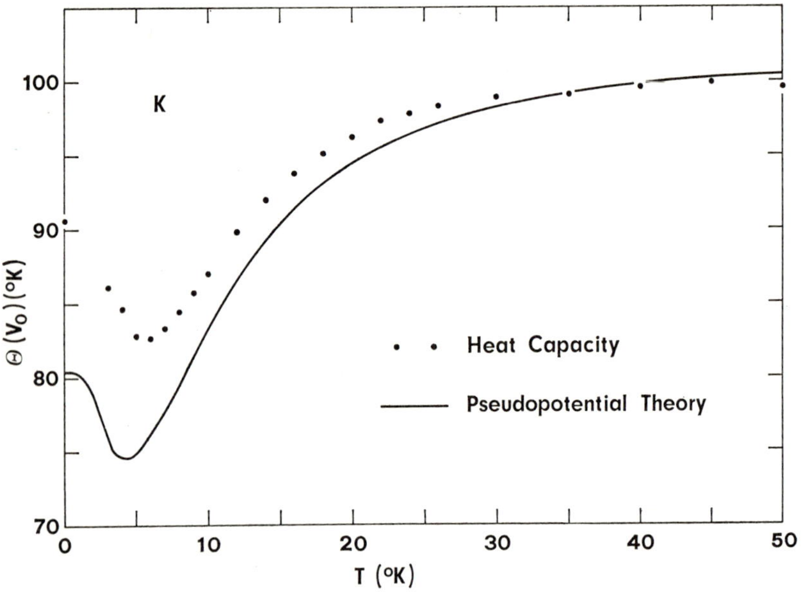

Figure 41. Comparison of the theoretical harmonic Debye temperature $\Theta_H(V_0)$ with the experimental lattice Debye temperature $\Theta(V_0)$, for K.

measured Θ_0 are not due simply to the long-waves–homogeneous-deformation differences. Furthermore, the discrepancies are much too large to be due to anharmonicity, i.e., to true differences between Θ_{H0} and Θ_0 for Na and K. It would be of interest to carry out an improved calculation for Na and K, including the additional terms in (28.35) for the electronic contribution to the dynamical matrices and allowing the potential parameters to be redetermined, to see if the present agreement between theory and experiment for the phonon frequencies in general can be retained while the agreement for the $T = 0$ elastic constants and Debye temperature can be improved.

It is interesting that the local pseudopotential theory does reproduce reasonably well the magnitudes and positions of the minima in the Debye temperature curves for Na and K, according to Figures 40 and 41. In connection with the temperature-dependence of the Debye temperature, we note that for $T \leqslant 20°K$ the frequencies which contribute to the heat capacity correspond to $\nu_\kappa \leqslant 0.4 \times 10^{12}$/sec, and these frequencies are below the lowest measured frequencies shown in Figures 36 and 37 for Na and K. Hence the experimental elastic constants and the temperature-dependence of $\Theta(V_0)$ at low temperatures provide important information about the phonon frequencies, in *addition to* the information provided by presently available neutron-scattering experiments.

The theory and experiment for the quantity $V\beta B_T(V_0)/3N_0K$ are compared in Figures 42 and 43 for Na and K, respectively. The theoretical curves are

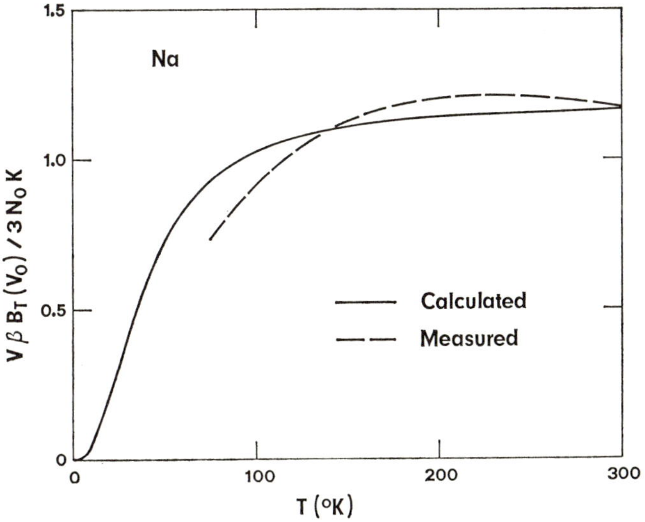

Figure 42. Comparison of theoretical (harmonic approximation) and experimental curves of $V\beta B_T/3N_0K$, evaluated at the fixed volume V_0, for Na.

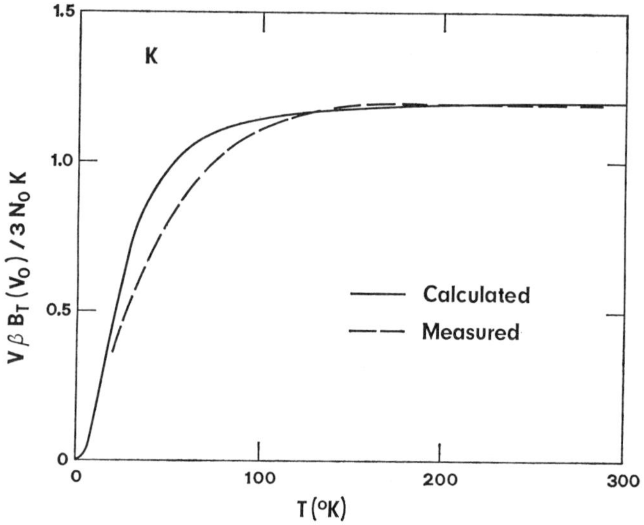

Figure 43. Comparison of theoretical (harmonic approximation) and experimental curves of $V\beta B_T/3N_0K$, evaluated at the fixed volume V_0, for K.

calculated in the harmonic approximation by evaluating (16.83). Although the anharmonic contribution to this quantity should be small for Na and K, it may not be negligible at high temperatures (see Section 31), and the slight decrease in the experimental curve for Na above 250°K might be due to anharmonicity. The calculated and measured results are in agreement within experimental error in the high-temperature region for both Na and K, but at intermediate temperatures the calculated curves are too high. Furthermore, the overall agreement between the harmonic theory and experiment cannot be substantially improved by varying the theoretical potential parameters.

Table 22. Some quantities calculated for the present pseudopotential models for Na and K

Quantity	Na	K
E_{BS} (Ry)	−0.0026	−0.0023
$\langle\gamma_{\kappa}\rangle$	1.183	1.207
$\langle\gamma_{\kappa}^2\rangle$	1.418	1.470
$\langle\xi_{\kappa}\rangle$	1.67	2.04
$d \ln \Theta_{H\infty}/d \ln V$	−1.205	−1.246
$\tilde{B}(d^2\tilde{B}/dP^2)$	−3.12	−3.17

Finally, in Table 22 we list some additional results of the present pseudopotential model calculations for Na and K. Here E_{BS} is the band-structure energy per atom, given by (34.4). The quantities $\langle \xi_\kappa \rangle$ and $\tilde{B}(d^2\tilde{B}/dP^2)$ are generally difficult to calculate, and the values listed in Table 22 will serve as useful estimates until a better lattice-dynamics calculation is carried out. In a recent publication,* the free energies for hcp and bcc Na are compared, and the procedure for finding the crystal configuration as a function of T at $P = 0$, by minimizing the free energy with respect to the configurational parameters, is also illustrated (see the discussion on p. 193).

Aluminum

We have also applied the local pseudopotential perturbation theory to lattice-dynamics calculations for Al.† Here we used two different models for the local pseudopotential, namely the Harrison model for which $w_B(q)$ is given by (26.41) in terms of the two parameters $\hat{\beta}$ and $\hat{\rho}$, and the Heine–Abarenkov model for which $w_B(q)$ is given by (26.44) in terms of the two parameters $\bar{\beta}$ and $\bar{\rho}$. A Born–Mayer repulsion $\alpha_B e^{-\gamma_B R}$ between the ions was also included. For Al this simple theory does not reproduce such an extensive amount of experimental data as it does for Na and K; hence for Al the potential parameters were determined to obtain the best overall fit to the measured phonon frequencies. In carrying out this fitting procedure, it was first found that the shape of the calculated dispersion curves is quite poor in any case where a significant Born–Mayer repulsion is included; in particular the repulsive potential acts to straighten out the curves of ω *vs* **k**. The parameter α_B was therefore set equal to zero, and for each model pseudopotential the two pseudopotential parameters were determined as follows. With one parameter as a variable, the second was adjusted so that the theoretical $\langle \omega_\kappa^2 \rangle$ agreed with the experimental value listed in Table 12, and the first was then varied to obtain good graphical fits to the measured Ω_κ for **k** along symmetry directions. The phonon frequencies were calculated from the dynamical matrix expression (27.76).

Aluminum crystallizes in the fcc lattice, and the model calculations were carried out for the volume V_0 corresponding to $T = 0$ and $P = 0$. The experimental quantities used in the calculation are as follows.

$$z = 3, \qquad V_A = 110.6a_0^3, \qquad r_s = 2.065a_0, \qquad \xi = 1.90. \qquad (34.14)$$

The pseudopotential parameters determined by fitting the calculated ω_κ to

* G. K. Straub and D. C. Wallace, *Phys. Rev.* **B3**, 1234 (1971).

† D. C. Wallace, *Phys. Rev.* **187**, 991 (1969); **B1**, 3963 (1970).

the measured Ω_κ are as follows, for the two different pseudopotential models.

$$\begin{aligned} &\text{Harrison:} && \hat{\beta} = 47.5 \text{ Ry } a_0^3, && \hat{\rho} = 0.24a_0; \\ &\text{Heine–Abarenkov:} && \bar{\beta} = 0, && \bar{\rho} = 1.117a_0. \end{aligned} \tag{34.15}$$

A rather surprising result is that the best-fit phonon-dispersion curves are nearly identical for the two pseudopotential models, the ω_κ for the two models differing by about 0.5% on the average and by 1.5% at most. The fit to experiment is quite good, as shown by Figure 44. At the temperature at which the neutron-scattering experiments were carried out, namely 80°K, the anharmonic contribution to the frequencies should be quite small, say $\lesssim 1\%$.

With regard to the model potentials determined here, it is not surprising that the Born–Mayer repulsion is negligible for Al. In comparison with Na, for which the repulsion is small, Al has much smaller ion cores and has more conduction electrons to hold the cores apart. It is interesting that the well depth $\bar{\beta}$ in the Heine–Abarenkov model is zero for the best phonon frequency fit; this special case of the Heine–Abarenkov model is an Ashcroft pseudopotential.

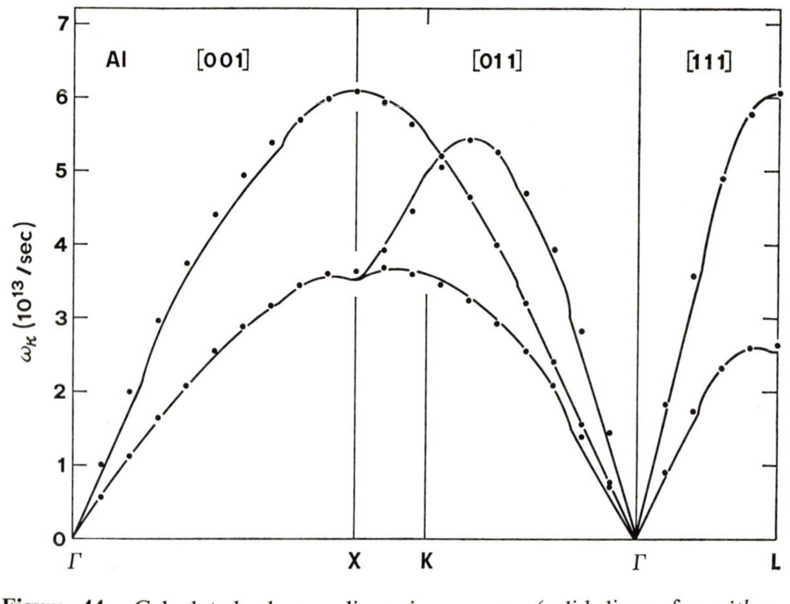

Figure 44. Calculated phonon-dispersion curves (solid lines, for *either* pseudopotential model) for fcc Al, compared with the 80°K experimental results (reference 41, Appendix 2). The Kohn anomalies have been slightly exaggerated in the theoretical curves to make them more apparent.

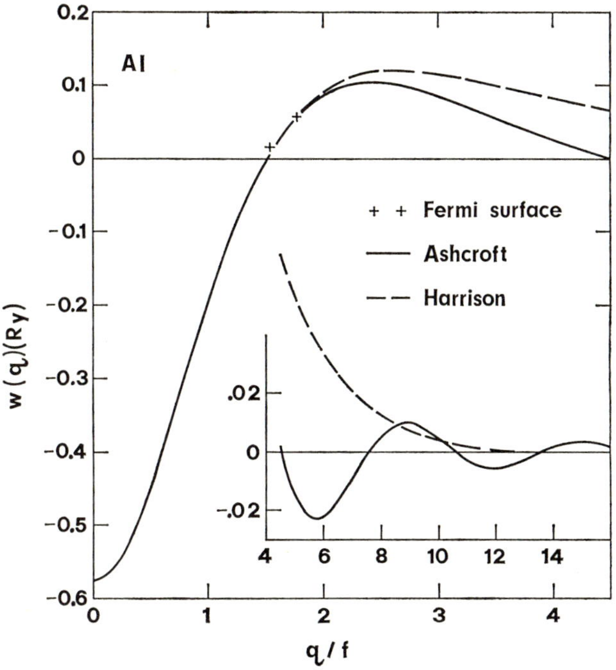

Figure 45. Screened form factors for the two pseudopotentials whose parameters were determined to give the best overall fit to the measured phonon frequencies for Al.

The screened form factors $w(q)$ for the two pseudopotential models for Al are shown in Figure 45, where they are seen to be in good agreement with the values of $|w(Q)|$, evaluated for $|\mathbf{Q}|$ at the first two inverse-lattice vectors, as determined from Fermi surface measurements.* The two model form factors are identical for $0 \leq q \leq 1.8f$, where f is the free-electron Fermi wave vector magnitude, but are quite different in the large q region. It is easy to see why the two form factors are the same in the small q region. First, according to (26.87), each model must satisfy $w(q) \to -\frac{2}{3}e_{\text{F}}$ as $q \to 0$. Further, the calculated phonon frequencies depend only on the magnitude $|w(q)|$, and the complete phonon spectrum for **k** throughout the Brillouin zone depends strongly on $|w(q)|$ for q lying in the zone, i.e., for $0 \leq q \leq 1.0f$ for Al. It is apparent from our calculations, however, that the phonon frequencies also depend on $|w(q)|$ for large q, but here they depend more or less on an average $|w(q)|$. We note that in order to calculate the ω_κ to an accuracy of 0.5% it is necessary to extend the sums over $\mathbf{Q}$, which appear in

* N. W. Ashcroft, *Phil. Mag.* **8**, 2055 (1963).

(27.72) for the dynamical matrix elements, to values of $|\mathbf{Q}| = 14f$ for the Harrison model, and to $|\mathbf{Q}| = 20f$ for the Heine–Abarenkov model. Some curves showing the convergence of ω_κ as the inverse-lattice sums are extended to larger values of $|\mathbf{Q}|$ are shown in the original reference.

As it was pointed out in Section 27, the total band-structure energy may be expressed as a sum of central potentials $v(R)$ between the ions, plus a volume-dependent term. The function $v(R)$ is written in (27.16), and the curve for the Harrison model for Al is shown in Figure 46. This curve shows that $v(R)$ is a complicated long-range function, and that this potential is small compared to the crystal binding energy. The main contributions to Φ_0/N_0 arise from the uniform electron gas energy E_{EG} and the electrostatic energy E_{ES} [see (34.3) and (34.5)], and the band-structure energy E_{BS}, which includes the central potential $v(R)$, is of little importance in determining the crystal binding and equilibrium. Hence, the positions of the shells of neighboring ions, shown also in Figure 46, are not significantly influenced by $v(R)$. It is worth noting again [see the discussion following (27.16)] that it is

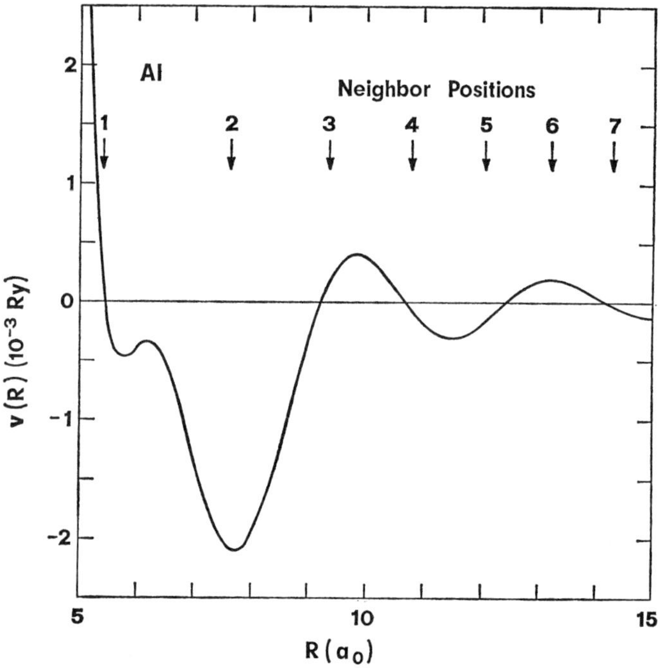

Figure 46. The effective two-body central potential between Al ions, which arises from the band-structure energy, for the Harrison pseudopotential model. This potential represents screened ion–electron–ion interactions.

Table 23. Comparison of theoretical and experimental $B_{\alpha\beta}$ and their pressure derivatives for Al. The theoretical values were calculated by the method of long waves for the Harrison model; the measured $B_{\alpha\beta}$ are at $T = 0°\text{K}$ and the measured pressure derivatives are at 77°K.

Quantity		Al	
Theory	Experiment	Theory	Experiment
$V_A\tilde{B}_{11}$	V_AB_{11} (Ry)	0.60	0.859
$V_A\tilde{B}_{12}$	V_AB_{12} (Ry)	0.32	0.466
$V_A\tilde{B}_{44}$	V_AB_{44} (Ry)	0.26	0.238
$V_A\tilde{B}$	V_AB (Ry)	0.41	0.597
$d\tilde{B}_{11}/dP$	$(\partial B_{11}^T/\partial P)_T$	5.1	6.94
$d\tilde{B}_{12}/dP$	$(\partial B_{12}^T/\partial P)_T$	3.0	3.64
$d\tilde{B}_{44}/dP$	$(\partial B_{44}^T/\partial P)_T$	2.0	2.25
$d\tilde{B}/dP$	$(\partial B_T/\partial P)_T$	3.7	4.74

inconvenient to formulate lattice-dynamics calculations in terms of $v(R)$, because of the slow convergence of the $\mathbf{q}$ integral in $v(R)$, and then the slow convergence of the lattice sums in the dynamical matrices.

The elastic stiffness coefficients $\tilde{B}_{\alpha\beta}$, as well as their pressure derivatives, were calculated by the method of long waves for the Harrison model for Al. These calculations are compared with experimental results in Table 23, where the experimental values are taken from Tables 7 and 8. Here the agreement between theory and experiment is only qualitatively respectable, and an improved theory is obviously needed. We expect that the additional contributions to the dynamical matrices, expressed in (28.35), will be important in the phonon frequencies atlong wave lengths, and perhaps sufficiently important at all wavelengths so that the pseudopotential parameters will have to be redetermined in order to maintain agreement with the measured frequencies.

The experimental Debye temperature $\Theta(V_0)$, corresponding to the lattice contribution to the heat capacity $C_V(V_0)$, is shown in Figure 47. The structure in the curve of $\Theta(V_0)$ *vs* T is due to phonon dispersion, and the initial increase of $\Theta(V_0)$ at low temperature may be interpreted as follows. Let us expand the phonon frequencies $\omega(\mathbf{k}s)$ to second order in $|\mathbf{k}|$ for small $|\mathbf{k}|$ as

$$\omega(\mathbf{k}s) = c(\hat{\mathbf{k}}s)\,|\mathbf{k}| + d(\hat{\mathbf{k}}s)\,|\mathbf{k}|^2 + \cdots. \tag{34.16}$$

The value $\Theta_{\mathrm{H}0}$ at $T = 0$ is determined by the velocities $c(\hat{\mathbf{k}}s)$. As T increases from $T = 0$, an initial *increase* of Θ_{H} shows that the quantities $d(\hat{\mathbf{k}}s)$ must be *positive* for a significant number of the low-lying (transverse) phonon

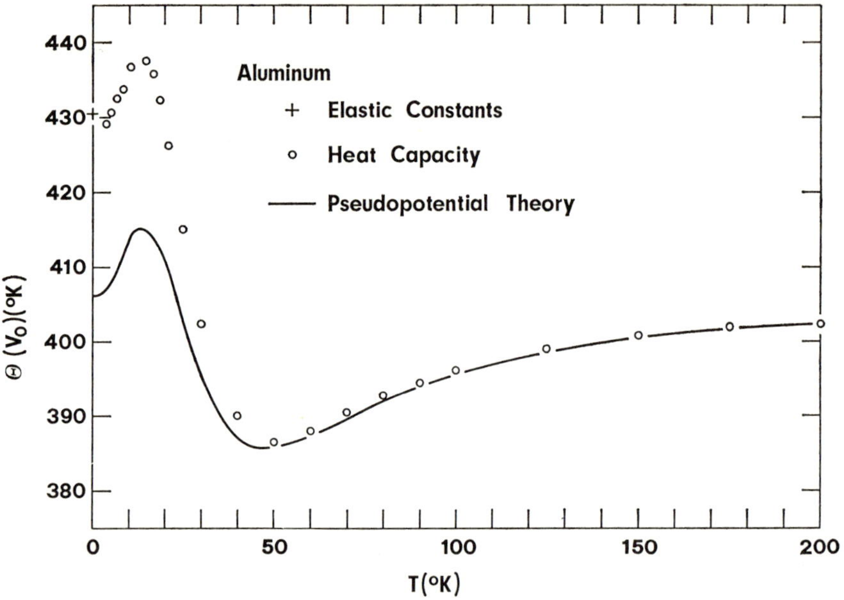

Figure 47. Comparison of the theoretical harmonic Debye temperature $\Theta_H(V_0)$ with the experimental lattice Debye temperature $\Theta(V_0)$, for Al.

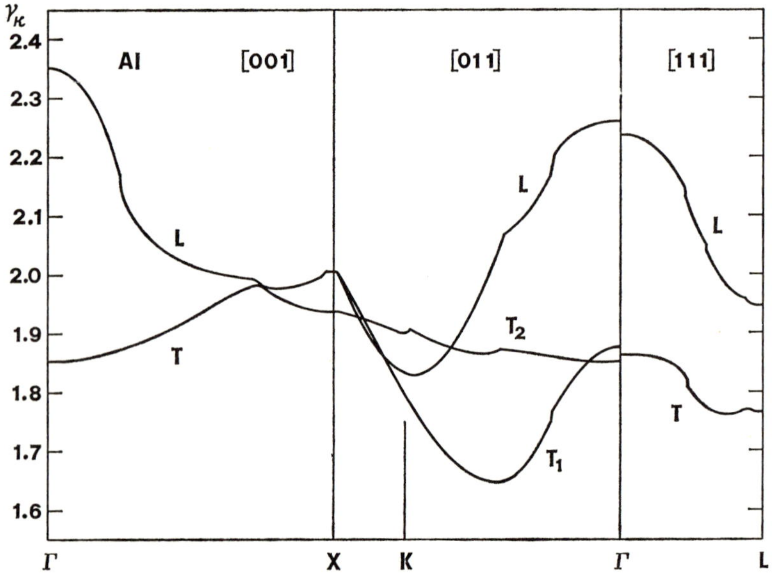

Figure 48. Calculated phonon Grüneisen parameters for the Harrison model for Al. Many Kohn anomalies are apparent in these curves (the anomaly locations are the same as for the phonon-dispersion curves, as listed in Table 25).

branches. Of course the experimental $\Theta(V_0)$ also includes the anharmonic contribution, but this should be quite small for Al. At about 200°K the experimental $\Theta(V_0)$ begins to level off toward $\Theta_{H\infty} = 408°K$, and the high-temperature anharmonic contribution to $\Theta(V_0)$ is not yet apparent. The theoretical curve of $\Theta_H(V_0)$ for the Harrison model for Al is also shown in Figure 47; this curve qualitatively reproduces the structure of the experimental $\Theta(V_0)$ curve, and is in good agreement with experiment for $T \geq 50°K$. At low temperatures the theoretical $\Theta_H(V_0)$ is too low, with $\Theta_{H0} = 406°K$ calculated from the method of long waves and $\Theta_0 = 428°K$ from the measured heat capacity.

The Grüneisen parameters γ_κ calculated for the Harrison model are shown for **k** along symmetry directions in Figure 48. Again it would be of interest to compare these calculations with direct measurements of the pressure-dependence of the phonon frequencies. The theoretical (harmonic approximation) and experimental curves of $V\beta B_T(V_0)/3N_0K$ are compared in Figure 49. The calculated curve is lower than the measured curve by about 12% for all temperatures up to 400°K, and this discrepancy is much larger than experimental error. Further, the anharmonic contribution to $V\beta B_T(V_0)/3N_0K$ is estimated to be less than 1% at 400°K, while the electronic contribution is about 0.04 (or 2%) at 400°K, so it is clear that our theoretical γ_κ are more or less uniformly too small by about 10%.

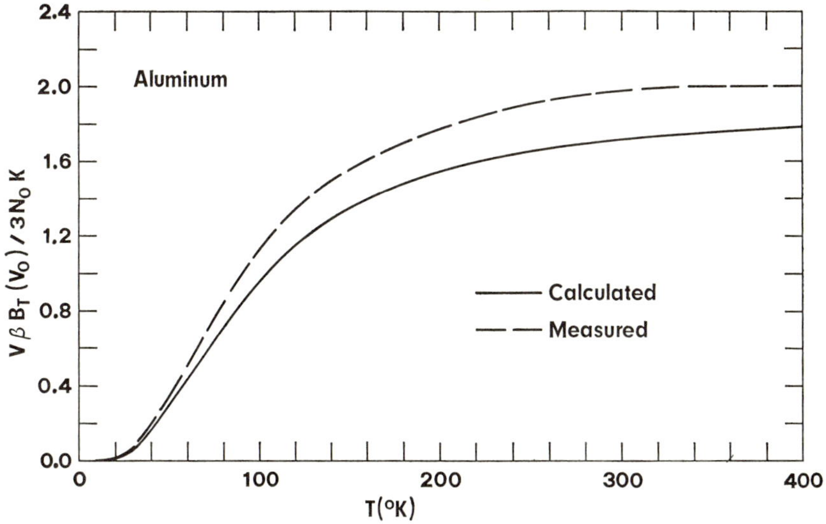

Figure 49. Comparison of the theoretical (Harrison model, harmonic approximation) and experimental curves of $V\beta B_T/3N_0K$, evaluated at the fixed volume V_0, for Al.

Finally it is useful to list some additional results of our calculations for the Harrison model for Al.

$$
\begin{aligned}
E_{\mathrm{BS}} &= -0.558 \text{ Ry}, \\
\langle \gamma_\kappa \rangle &= 1.872, \\
\langle \gamma_\kappa^2 \rangle &= 3.513, \\
\langle \xi_\kappa \rangle &= 3.33 \\
d \ln \Theta_{\mathrm{H}\infty}/d \ln V &= -1.895.
\end{aligned}
\tag{34.17}
$$

Kohn Anomalies

The Kohn anomalies were discussed briefly in Section 28. For the local pseudopotential perturbation theory, a Kohn anomaly may appear in the curve of $\omega(\mathbf{k}s)$ *vs* $\mathbf{k}$ whenever $\mathbf{k}$ is such that $|\mathbf{Q} + \mathbf{k}| = 2f$ for some inverse-lattice vector $\mathbf{Q}$. This condition depends only on the lattice structure and the number z of conduction electrons per ion. Again for the simple pseudopotential theory, the relative magnitudes of the different Kohn anomalies for a given metal may be estimated by the function $K(\mathbf{k}s)$ given by (28.20). Table 24 lists the locations of the anomalies for $\mathbf{k}$ along symmetry directions, as well as the corresponding values of $K(\mathbf{k}s)$, for the alkali metals in the bcc lattice with $z = 1$. Here the $\mathbf{k}$ vectors along symmetry lines are described as follows (see also the phonon-dispersion curves for Na and K in Figures 36 and 37, and the bcc Brillouin zone portion in Figure 55, p. 451).

$$
\begin{aligned}
&\text{Line } \Gamma\text{H:} \quad \mathbf{k} = (\sqrt{3}\pi/R_1)(0,0,\zeta), \quad 0 \le \zeta \le 1; \\
&\text{Line } \Gamma\text{N:} \quad \mathbf{k} = (\sqrt{3}\pi/R_1)(0,\zeta,\zeta), \quad 0 \le \zeta \le \tfrac{1}{2}; \\
&\text{Line } \Gamma\text{P:} \quad \mathbf{k} = (\sqrt{3}\pi/R_1)(\zeta,\zeta,\zeta), \quad 0 \le \zeta \le \tfrac{1}{2};
\end{aligned}
\tag{34.18}
$$

Table 24. Locations and estimated relative magnitudes of Kohn anomalies in the bcc alkali metals. The phonon polarizations are denoted by L for longitudinal modes and T_1 and T_2 for transverse modes, and the values of $K(s)$ are in arbitrary units.

$\mathbf{k}$ direction	ζ at anomaly	$K(L)$	$K(T_1)$	$K(T_2)$
[001]	0.2656	2.157	2	
[001]	0.7593	1.539	0	
[011]	0.1227	1.539	0	0
[011]	0.3598	0.157	2	4
[111]	0.1273	2.618	1	
[111]	0.7163	1.539	0	
[111]	0.7940	2.618	1	

where R_1 is the nearest-neighbor distance in the bcc lattice. The line PH corresponds to $\mathbf{k} = (\sqrt{3}\pi/R_1)(\frac{1}{2} - \zeta, \frac{1}{2} - \zeta, \frac{1}{2} + \zeta)$ for $0 \le \zeta \le \frac{1}{2}$; this line is equivalent to $\mathbf{k} = (\sqrt{3}\pi/R_1)(\zeta,\zeta,\zeta)$ for $\frac{1}{2} \le \zeta \le 1$, where equivalence means that the eigenvalues along the two lines are the same. Note, however, that the eigenvectors are not the same along the two lines, and this distinction is important in calculating the anomaly magnitudes $K(\mathbf{k}s)$. It is customary to measure phonon frequencies for **k** along [111] with ζ out to 1, and to plot the results along the line ΓPH, according to

$$\text{Line } \Gamma\text{PH:} \qquad \mathbf{k} = (\sqrt{3}\pi/R_1)(\zeta,\zeta,\zeta), \quad 0 \le \zeta \le 1. \tag{34.19}$$

For the fcc lattice, the **k** vectors along symmetry lines are described as follows (see also the phonon-dispersion curves for Al in Figure 44, and the fcc Brillouin zone portion in Figure 53, p. 449).

$$\begin{aligned} &\text{Line } \Gamma\text{X:} && \mathbf{k} = (\sqrt{2}\pi/R_1)(0,0,\zeta), \quad 0 \le \zeta \le 1; \\ &\text{Line } \Gamma\text{KX:} && \mathbf{k} = (\sqrt{2}\pi/R_1)(0,\zeta,\zeta), \quad 0 \le \zeta \le 1; \\ &\text{Line } \Gamma\text{L:} && \mathbf{k} = (\sqrt{2}\pi/R_1)(\zeta,\zeta,\zeta), \quad 0 \le \zeta \le \tfrac{1}{2}; \end{aligned} \tag{34.20}$$

where R_1 is the nearest-neighbor distance in the fcc lattice. Note that we are taking ΓK along [011] with $0 \le \zeta \le \frac{3}{4}$, and from K to X we again follow the custom and plot points along [011] with $\frac{3}{4} \le \zeta \le 1$; this last line segment is *not* equivalent to the line KX. A list of locations and estimated relative magnitudes of Kohn anomalies in Al for **k** along symmetry directions is given in Table 25.

Table 25. Locations and estimated relative magnitudes of Kohn anomalies in Al (fcc lattice, $z = 3$). Notation is the same as in Table 24.

k direction	ζ at anomaly	$K(L)$	$K(T_1)$	$K(T_2)$
[001]	0.2545	5.083	0	
[001]	0.7558	12.331	4	
[001]	0.9594	4.331	8	
[011]	0.2415	6.166	4	0
[011]	0.4058	5.083	0	0
[011]	0.4288	8.166	0	2
[011]	0.7358	2.166	0	8
[011]	0.9795	8.331	8	4
[111]	0.2307	7.248	4	
[111]	0.3016	5.083	0	
[111]	0.4359	7.248	4	

Table 26. Locations and estimated relative magnitudes of Kohn anomalies in the $\gamma(\mathbf{k}s)$ curves for bcc Li. The values of $K(s)/[\omega(s)]^2$ are in arbitrary units.

k direction	ζ at anomaly	$\frac{K(L)}{[\omega(L)]^2}$	$\frac{K(T_1)}{[\omega(T_1)]^2}$	$\frac{K(T_2)}{[\omega(T_2)]^2}$
[001]	0.2656	1.25	1.78	
[001]	0.7593	0.23	0	
[011]	0.1227	1.26	0	0
[011]	0.3598	0.02	7.81	1.46
[111]	0.1273	1.29	2.83	
[111]	0.7163	0.99	0	
[111]	0.7940	0.88	0.14	

It was also noted in Section 28 that for a given metal the anomalies should be more pronounced in curves of $\gamma(\mathbf{k}s)$ than in curves of $\omega(\mathbf{k}s)$, and that the relative magnitudes of the anomalies in the $\gamma(\mathbf{k}s)$ curves might be estimated by the function $K(\mathbf{k}s)[\omega(\mathbf{k}s)]^{-2}$. The positions and estimated relative magnitudes of the anomalies in the $\gamma(\mathbf{k}s)$ curves for bcc Li are listed in Table 26. We have used a Harrison model local pseudopotential to calculate

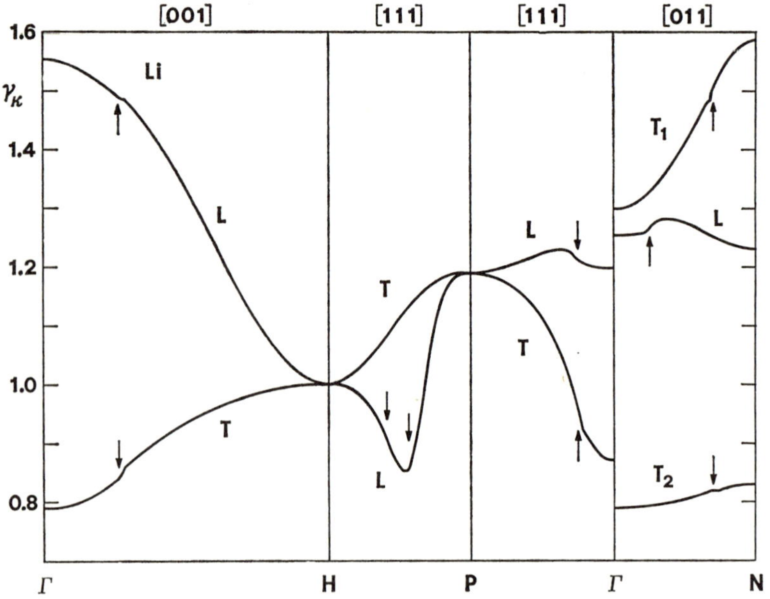

Figure 50. Calculated curves of $\gamma_K = \gamma(\mathbf{k}s)$ for bcc Li. The arrows mark the locations of expected Kohn anomalies.

the phonon frequencies and Grüneisen parameters for bcc Li,* and although the Kohn anomalies are too small to be seen (less than 0.1%) in the calculated $\omega(\mathbf{k}s)$ curves, they are quite apparent in the $\gamma(\mathbf{k}s)$ curves shown in Figure 50. Here the arrows mark the nine places where the anomalies should be large, according to Table 26, and the anomalies are observed in all but two of them, namely for the longitudinal branch with $\mathbf{k}$ along [111] and $\zeta = 0.7163$ and $\zeta = 0.7940$.

Nonadiabatic Effects

For the pseudopotential perturbation theory the total free energy arising from excitation of the conduction electrons F_e, plus the contribution due to electron–phonon interactions F_{ep}, is given by (28.52) as

$$F_e + F_{ep} = -\tfrac{1}{6}\pi^2(KT)^2 n(e_F)[1 + \Delta_W^{(0)} + \Delta_W^{(2)} + \Delta_\Gamma]. \qquad (34.21)$$

Here $n(e_F)$ is the free-electron contribution to the density of states at the Fermi energy, and $\Delta_W^{(0)}$, given by (28.42), represents the band-structure contribution. The terms $\Delta_W^{(2)}$ and Δ_Γ arise from electron–phonon interactions. Approximate low- and high-temperature expressions for $\Delta_W^{(2)}$ are given by (28.44), and from (25.86)

$$\begin{aligned} \Delta_\Gamma &= n(\mu_0)G(\mu_0,\mu_0), && \text{at low } T, \\ \Delta_\Gamma &\approx (\Theta_0/T)n(\mu_0)G(\mu_0,\mu_0), && \text{at high } T, \end{aligned} \qquad (34.22)$$

where the pseudopotential evaluation of $n(\mu_0)G(\mu_0,\mu_0)$ is given by (28.51). Each of the Δ quantities is of second order in the pseudopotential, and (34.21) is correct to second order. The Δ quantities were evaluated for the present Harrison pseudopotential models for Na, K, and Al, and the results are listed in Table 27. As it was expected, according to the discussion preceding (25.75), the term $\Delta_W^{(2)}$ is very small compared to 1, and may be neglected for temperatures up to three or four times Θ_0. The band-structure term $\Delta_W^{(0)}$ is also small, but the major electron–phonon term Δ_Γ is large, especially for Al.

In order to compare the theory with experiment, it is useful to write the total free energy (34.21) as $-\tfrac{1}{2}\hat{\Gamma}T^2$ and denote the free-electron contribution to $\hat{\Gamma}$ as Γ_{fe}:

$$\Gamma_{fe} = \tfrac{1}{3}\pi^2K^2n(e_F). \qquad (34.23)$$

The theoretical values of Γ_{fe}, and of $\hat{\Gamma}/\Gamma_{fe}$ at $T = 0$, are listed in Table 27; the difference from 1 of the ratio $\hat{\Gamma}/\Gamma_{fe}$ shows immediately the importance of

* D. C. Wallace, *Phys. Rev.* **178**, 900 (1969).

Table 27. Theoretical and experimental results for quantities related to the electronic plus electron–phonon free energy, for Na, K, and Al.

Quantity	Na	K	Al
$\Delta_W^{(0)}$	0.0023	0.0061	0.0307
$\Delta_W^{(2)}$ (low T)	0.00073	0.00046	−0.0015
$\Delta_W^{(2)}$ (high T)	0.0019 (T/Θ_0)	0.0012 (T/Θ_0)	−0.0038 (T/Θ_0)
Δ_Γ (low T)	0.25 ± 0.01	0.16 ± 0.01	0.64 ± 0.02
Γ_{fe} (10^{-4} cal/mole °K^2)	2.612	4.005	2.156
$\hat{\Gamma}/\Gamma_{\text{fe}}$ (theory, $T = 0$)	1.25	1.17	1.67
$\hat{\Gamma}/\Gamma_{\text{fe}}$ (experiment, $T = 0$)	1.26	1.24	1.50

the band-structure and electron–phonon terms. These ratios are in reasonably good agreement with the corresponding experimental results, where $\hat{\Gamma}$ was determined from low-temperature heat capacity measurements (see Table 5).

We have also calculated the volume derivatives of $\hat{\Gamma}$ at $T = 0$ for the pseudopotential models, in order to find the electronic and electron–phonon contributions to the low-temperature thermal expansion. Neglecting $\Delta_W^{(2)}$, $\hat{\Gamma}$ is

$$\hat{\Gamma} = \Gamma_{\text{fe}}[1 + \Delta_W^{(0)} + \Delta_\Gamma], \tag{34.24}$$

and the volume derivative is

$$\frac{d \ln \hat{\Gamma}}{d \ln V} = \frac{d \ln \Gamma_{\text{fe}}}{d \ln V} + \frac{\Gamma_{\text{fe}}}{\hat{\Gamma}} \frac{V d(\Delta_W^{(0)} + \Delta_\Gamma)}{dV}. \tag{34.25}$$

Now $n(e_F)$, and hence also Γ_{fe}, is proportional to $V^{2/3}$, so that $d \ln \Gamma_{\text{fe}}/d \ln V = \frac{2}{3}$. The calculated volume derivatives of $\Delta_W^{(0)}$ and Δ_Γ are listed in Table 28. Theory and experiment are in good agreement for $d \ln \hat{\Gamma}/d \ln V$

Table 28. Theoretical and experimental results for quantities related to the electronic plus electron–phonon thermal expansion at low temperature, for Na, K, and Al.

Quantity	Na	K	Al
$V\, d\Delta_W^{(0)}/dV$	−0.013	−0.027	0.074
$V\, d\Delta_\Gamma/dV$ (low T)	0.65	0.43	1.53
$d \ln \hat{\Gamma}/d \ln V$ (theory, $T = 0$)	1.18	1.01	1.63
$d \ln \hat{\Gamma}/d \ln V$ (experiment, $T = 0$)			1.8

for Al as shown in Table 28, where the experimental result is from low-temperature thermal expansion measurements (see Table 6). We are in hopes that the electronic thermal expansion will soon be measured for Na and K. It may be observed that the large value of $\Delta_\Gamma = 0.64$ for Al, and the large contributions from Δ_Γ to $d \ln \hat{\Gamma}/d \ln V$ for all three metals in Table 28, suggests that higher-order perturbation contributions of the electron–phonon interactions might be important. On the other hand, recalling the discussion at the beginning of Section 25, an improved band-structure calculation might significantly alter the various contributions to $\hat{\Gamma}$.

35. APPROXIMATIONS AND CONCLUSIONS

Born Model for the Alkali Halides

We wish to describe briefly an extremely simple model for the alkali halides; our discussion will then serve as the basis for describing improvements in the theory. Assume that each pair of ions interacts with the point-charge Coulomb potential $\varphi_{\mu\nu}(R)$ given by

$$\varphi_{\mu\nu}(R) = z_\mu z_\nu e^2/R, \tag{35.1}$$

where R is the distance between the ions. There are two kinds of ions in the crystal, and the charge z_μ is $+1$ for the anions and -1 for the cations. When the ions are all at their equilibrium positions $\mathbf{R}(\mathrm{M}\mu)$, the total static-lattice potential due to the Coulomb interactions (35.1) is simply

$$\tfrac{1}{2} \sum_{\mathrm{MN}\mu\nu}' \varphi_{\mu\nu}(|\mathbf{R}(\mathrm{M}\mu) - \mathbf{R}(\mathrm{N}\nu)|) = -N_0 \alpha_C e^2/R_1, \tag{35.2}$$

where N_0 is the number of unit cells and R_1 is the nearest-neighbor distance. The coefficient α_C depends on the lattice structure, and is 1.74756 for the NaCl lattice. Assume also that there is a short-range Born–Mayer repulsion $\varphi_\mathrm{B}(R)$ given by

$$\varphi_\mathrm{B}(R) = \alpha_\mathrm{B} e^{-\gamma_\mathrm{B} R}, \tag{35.3}$$

and that this repulsion acts only between nearest neighbors. Again for the static lattice with all ions at their equilibrium positions, and noting that there are two ions per unit cell,

$$\tfrac{1}{2} \sum_{\mathrm{MN}\mu\nu}' \varphi_\mathrm{B}(|\mathbf{R}(\mathrm{M}\mu) - \mathbf{R}(\mathrm{N}\nu)|) = N_0 \mathscr{Z} \alpha_\mathrm{B} e^{-\gamma_\mathrm{B} R_1}, \tag{35.4}$$

where $\mathscr{Z}$ is the number of nearest neighbors and is 6 for the NaCl lattice.

The total static-lattice potential for the Born model, for any value of R_1, is then

$$\Phi_0 = N_0(\mathscr{Z} \alpha_\mathrm{B} e^{-\gamma_\mathrm{B} R_1} - \alpha_C e^2/R_1). \tag{35.5}$$

The pressure $\tilde{P}$ and bulk modulus $\tilde{B}$ in the potential approximation are obtained as usual from the volume derivatives of Φ_0, and the equilibrium condition is $\tilde{P} = 0$ at the correct (measured) value of R_1. The potential approximation results evaluated at $\tilde{P} = 0$ are as follows, where we use the abbreviation

$$\hat{\gamma} = \gamma_B R_1, \tag{35.6}$$

and where R_1 is supposed to be the measured value at $T = 0$ and $P = 0$.

$$\Phi_0 = -N_0(\alpha_C e^2/R_1)(1 - \hat{\gamma}^{-1}); \tag{35.7}$$

$$\alpha_B = (\alpha_C e^2/\mathfrak{z}\hat{\gamma}R_1)e^{\hat{\gamma}}, \quad \text{equilibrium condition;} \tag{35.8}$$

$$V_0\tilde{B} = \tfrac{1}{9}N_0(\alpha_C e^2/R_1)(\hat{\gamma} - 2); \tag{35.9}$$

$$d\tilde{B}/dP = (\hat{\gamma}^2 + 3\hat{\gamma} - 12)/(3\hat{\gamma} - 6); \tag{35.10}$$

and so on for the higher-pressure derivatives of $\tilde{B}$.

We have determined the two potential parameters α_B and γ_B for three alkali halides so that Φ_0 of (35.7) equals F_0, the measured free energy at $T = 0$, and so that the equilibrium condition (35.8) is satisfied. The experimental quantities and fitted parameters are listed in Table 29. It is interesting that γ_B is nearly the same for all three alkali halides; it is given by $1/\gamma_B = 0.30\ (10^{-8}\ \text{cm})$, and this is close to (34.1) which gives the value used in the metals calculations of Section 34.

The Born model may now be tested by comparing calculated and measured values of the bulk moduli and their pressure derivatives; the results are listed in Table 30. The theory and experiment are in excellent agreement for the bulk modulus at $T = 0$. Furthermore, the differences between the calculated $d\tilde{B}/dP$ and the values of $(\partial B_S/\partial P)_T$ measured at 300°K and $P = 0$ are not unreasonably large; we note that there should be a sizable increase in

Table 29. Experimental results for R_1 and F_0 at $T = 0$ and $P = 0$, and the potential parameters required to fit these results with the Born model in the potential approximation, for the alkali halides.

Quantity	NaCl	KCl	KBr
R_1 (10^{-8} cm)	2.7982	3.1230	3.2730
F_0 (kcal/mole)	−185	−168	−161
$\hat{\gamma} = \gamma_B R_1$	9.3	10.5	10.9
α_B (10^{-9} erg)	2.824	7.441	10.203

Table 30. Comparison of theory and experiment for the bulk modulus and its pressure derivative, and for the average $\langle \omega_\kappa^2 \rangle$, for the alkali halides. The measured $(\partial B_S/\partial P)_T$ are at 300°K.

Quantity		NaCl		KCl		KBr	
Theory	Experiment	Theory	Experi-ment	Theory	Experi-ment	Theory	Experi-ment
$V_0\tilde{B}$	$V_0 B_0$ (kcal/mole)	168	168	175	173	175	178
$d\tilde{B}/dP$	$(\partial B_S/\partial P)_T$	4.68	5.27	5.09	5.35	5.23	5.38
$\tilde{B}(d^2\tilde{B}/dP^2)$		−5.53		−6.21		−6.43	
$\langle \omega_\kappa^2 \rangle$	$\langle \omega_\kappa^2 \rangle$ (10^{26} sec^{-2})	9.67	8.53	6.07	5.52	3.91	3.40

this pressure derivative when T increases from 0 to 300°K. In view of these results, we expect the calculated values of $\tilde{B}(d^2\tilde{B}/dP^2)$, listed also in Table 30, to be at least qualitatively reliable.

The next obvious test of the Born model is to calculate the average phonon frequency squared. The procedure is straightforward, starting with the general equations (10.62) and (10.63) for $\langle \omega_\kappa^2 \rangle$, and using (9.10) for the potential energy coefficients for central potentials. As it is well known, the Coulomb potential (35.1) gives no contribution to $\langle \omega_\kappa^2 \rangle$, and the result is

$$\langle \omega_\kappa^2 \rangle = (2\alpha_B \hat{\gamma}/R_1^2)(\hat{\gamma} - 2)\langle M_\mu^{-1} \rangle e^{-\hat{\gamma}}, \tag{35.11}$$

where $\langle M_\mu^{-1} \rangle$ is the average of the inverses of the two ionic masses:

$$\langle M_\mu^{-1} \rangle = \tfrac{1}{2} \sum_\mu M_\mu^{-1} = \tfrac{1}{2}(M_1^{-1} + M_2^{-1}). \tag{35.12}$$

The expression (35.11) is evaluated for the present alkali halide models, and it is compared in Table 30 with the experimental results of the high-temperature entropy analysis given in Table 12. The calculated $\langle \omega_\kappa^2 \rangle$ are too large, being larger than experiment by 13% for NaCl, 10% for KCl, and 15% for KBr. These discrepancies are much larger than the theoretical–experimental differences ($\leqslant 1\%$) for the bulk moduli, and they provide clear evidence of the inadequacy of the simple Born model. Incidentally, this procedure demonstrates the usefulness of the average $\langle \omega_\kappa^2 \rangle$, which is generally quite easy to calculate, in testing theoretical models.

The main weakness of the Born model is the neglect of polarizations of the ions when they vibrate. However, for the static lattice each ion is at a center of crystalline symmetry, and no low-order electric moments arise on the ions; this remains true if the lattice is homogeneously strained, so that the Born model should be accurate for calculating the static-lattice energy Φ_0 and its strain and volume derivatives. Therefore to the extent that the zero-point vibrational free energy can be neglected, we expect, and find, good agreement between the calculated and measured crystal energy and its volume derivatives at $T = 0$. For NaCl, KCl, and KBr we estimate the zero-point contribution to the free energy to be about 0.7%, and to the bulk modulus to be about 2%. The measured elastic constants at $T = 0$ do not obey the Cauchy relations (see Table 7), and this is presumably due to zero-point vibrational effects; such effects may be of greater relative importance in the elastic constants than in the bulk modulus.

When the ions vibrate, they polarize, and there are strong polarization interactions among the ions which must be taken into account to calculate the phonon frequencies. Shell models have been developed to try to account for the ionic polarizations, and such models can be made to fit the phonon spectrum quite well. When the spectrum is fitted, however, the shell model parameters are often found to be physically unrealistic. An alternate course of investigation, which should prove useful in improving the theory for ionic and covalent crystals, is to develop approximations for the dielectric function, in keeping with the band-structure formulation of Chapter 5. We note that this formulation includes the effects of ionic polarizations, in principle, in the electronic band-structure energy calculated as a function of the ion positions.

Approximations for Brillouin Zone Averages

We have often noted that for a given model the Brillouin zone average $\langle\omega_\kappa^2\rangle$ and its volume and strain derivatives are easily calculated. It is therefore useful to develop approximations for other Brillouin zone averages, expressed in terms of $\langle\omega_\kappa^2\rangle$, to provide estimates of the harmonic contributions to thermodynamic functions. Several similar approximations, involving both the neglect of anharmonic contributions and approximate relations among Brillouin zone averages, were tested experimentally in Section 32.

The harmonic free energy at $T = 0$ is F_{H0}, and from (18.1),

$$F_{H0} = \tfrac{1}{2} \sum_{\kappa} \hbar\omega_\kappa = \tfrac{3}{2} n N_0 \hbar \langle\omega_\kappa\rangle; \tag{35.13}$$

this may be estimated from $\langle\omega_\kappa^2\rangle$ by means of the approximation

$$\langle\omega_\kappa\rangle \approx \langle\omega_\kappa^2\rangle^{1/2}. \tag{35.14}$$

This approximation is quite good, as shown by the results listed in Table 31, for the Lennard–Jones 12,6 potential model for Ar and Kr and for the pseudopotential models for Na, K, and Al. Hence, if $\langle\omega_\kappa^2\rangle$ is determined by a high-temperature entropy analysis as in Section 31, or $\langle\Omega_\kappa^2\rangle$ by inelastic neutron scattering, a respectable estimate of F_{H0} may be obtained from

Table 31. Evaluation of several approximate relations between Brillouin zone averages, for the lattice-dynamical models of the present chapter. Each of the ratios tabulated should be approximately 1.

Quantity	Ar and Kr	Na	K	Al
$\langle\omega_\kappa^2\rangle^{1/2}/\langle\omega_\kappa\rangle$	1.044	1.070	1.071	1.047
$\dfrac{d\ln\Theta_{H\infty}/d\ln V}{d\ln\langle\omega_\kappa\rangle/d\ln V}$	1.008			
$\dfrac{\langle\gamma_\kappa\rangle\langle\omega_\kappa^2\rangle}{\langle\gamma_\kappa\omega_\kappa^2\rangle}$	0.983	0.982	0.969	0.988
$\langle\gamma_\kappa\rangle^2/\langle\gamma_\kappa^2\rangle$		0.986	0.990	0.997
$\langle\xi_\kappa\rangle_{\text{approx}}/\langle\xi_\kappa\rangle$		1.12	1.13	1.04

(35.14). Higher positive moments of the phonon frequencies may be estimated from $\langle\omega_\kappa^n\rangle \approx \langle\omega_\kappa^2\rangle^{n/2}$, with decreasing accuracy for increasing n. Note that the positive even moments $\langle\omega_\kappa^{2n}\rangle$ determine the harmonic contributions to the entropy and heat capacity in the high-temperature region, according to (19.27) and (19.28). Negative moments are not determined well from $\langle\omega_\kappa^2\rangle$; for the simple metals studied here $\langle\omega_\kappa^{-1}\rangle/\langle\omega_\kappa^2\rangle^{-1/2}$ is about 1.2–1.4, while $\langle\omega_\kappa^{-2}\rangle/\langle\omega_\kappa^2\rangle^{-1}$ is about 2–3.

The high-temperature harmonic Debye temperature appropriate for the heat capacity is defined by (19.32) as

$$K\Theta_{H\infty} = (5/3)^{1/2}\hbar\langle\omega_\kappa^2\rangle^{1/2}. \tag{35.15}$$

With the approximation (35.14), the zero-point free energy (35.13) is written in terms of $\Theta_{H\infty}$ as

$$F_{H0} \approx 1.16 n N_0 K\Theta_{H\infty}. \tag{35.16}$$

Also since $\Theta_{H\infty}$ and Θ_0 are very approximately equal, at least for most of the materials we have studied, we can write for a rough estimate $F_{H0} \sim nN_0K\Theta_0$.

The approximation (35.14) should hold equally well at any crystal volume, and it can be differentiated with respect to volume to obtain

$$\frac{1}{2}\frac{d\ln\langle\omega_\kappa^2\rangle}{d\ln V} = \frac{d\ln\Theta_{H\infty}}{d\ln V} \approx \frac{d\ln\langle\omega_\kappa\rangle}{d\ln V}. \tag{35.17}$$

This last approximation is quite well satisfied by the model for Ar and Kr, as shown in Table 31. Again the same approximation can reasonably be extended to higher volume derivatives, or to strain derivatives, and can be used to estimate zero-point vibrational contributions to the pressure and bulk modulus, or to the stresses and elastic constants, from volume or strain derivatives of $\langle\omega_\kappa^2\rangle$.

The logarithmic volume derivative of $\langle\omega_\kappa^2\rangle$ can also be expressed in the form

$$\frac{1}{2}\frac{V}{\langle\omega_\kappa^2\rangle}\frac{d\langle\omega_\kappa^2\rangle}{dV} = -\frac{\langle\gamma_\kappa\omega_\kappa^2\rangle}{\langle\omega_\kappa^2\rangle}. \tag{35.18}$$

For cubic crystals the Brillouin zone average $\langle\gamma_\kappa\omega_\kappa^2\rangle$ gives the leading harmonic contribution to the temperature-dependence of $V\beta B_T$ at high temperature, according to (19.58). This average is easily calculated for a given model from $d\langle\omega_\kappa^2\rangle/dV$, from (35.18). On the other hand, this average may be estimated from $\langle\gamma_\kappa\rangle$ and $\langle\omega_\kappa^2\rangle$ by the approximation

$$\langle\gamma_\kappa\omega_\kappa^2\rangle \approx \langle\gamma_\kappa\rangle\langle\omega_\kappa^2\rangle. \tag{35.19}$$

This approximation has been shown in Table 15 to be quite good for the experimentally determined quantities in (35.19), with the exception of Zn and Cd; it is also quite good for the theoretical models of the present chapter, as shown in Table 31. We note that (35.19) is equivalent to

$$\langle\gamma_\kappa\rangle \approx -d\ln\Theta_{H\infty}/d\ln V. \tag{35.20}$$

Let us now study approximations involving the Brillouin zone averages $\langle\gamma_\kappa^2\rangle$ and $\langle\xi_\kappa\rangle$; these averages appear in the harmonic contributions to the temperature derivatives of B_S and B_T at high temperature, according to the results (19.60) and (19.61) for cubic crystals. Consider first the approximation

$$\langle\gamma_\kappa^2\rangle \approx \langle\gamma_\kappa\rangle^2; \tag{35.21}$$

this approximation is surprisingly well satisfied by the present models, as shown in Table 31, and gives a way to estimate $\langle\gamma_\kappa^2\rangle$ from an experimentally or theoretically determined value of $\langle\gamma_\kappa\rangle$. Now the second volume derivative of $\langle\omega_\kappa^2\rangle$ may be expressed in the form

$$\frac{1}{2}\frac{V^2}{\langle\omega_\kappa^2\rangle}\frac{d^2\langle\omega_\kappa^2\rangle}{dV^2} = \frac{\langle\gamma_\kappa^2\omega_\kappa^2 + \xi_\kappa\omega_\kappa^2\rangle}{\langle\omega_\kappa^2\rangle}. \tag{35.22}$$

This equation may be approximated as

$$\frac{1}{2}\frac{V^2}{\langle\omega_\kappa^2\rangle}\frac{d^2\langle\omega_\kappa^2\rangle}{dV^2} \approx \langle\gamma_\kappa^2\rangle + \langle\xi_\kappa\rangle, \tag{35.23}$$

and serves as a convenient way to estimate $\langle \xi_\kappa \rangle$. The approximate values of $\langle \xi_\kappa \rangle$ calculated from (35.23), denoted $\langle \xi_\kappa \rangle_{\text{approx}}$, are compared with accurate values for the present models in Table 31, and the approximation is seen to be quite good.

Such approximations for Brillouin zone averages can be extended in many different ways. For example, an alternate and convenient way to estimate the volume derivatives of the harmonic zero-point free energy F_{H0} is as follows.

$$VP_{\text{H0}} = -V(dF_{\text{H0}}/dV) \approx \langle \gamma_\kappa \rangle F_{\text{H0}}, \tag{35.24}$$

$$VB_{\text{H0}} = V^2(d^2F_{\text{H0}}/dV^2) \approx \langle \xi_\kappa \rangle F_{\text{H0}}, \tag{35.25}$$

and so on. In addition, the approximations presented here should be reasonably applicable to more complicated lattices. We should like to emphasize our point of view that these approximations are useful for estimates, but should not be relied upon for accurate results.

Conclusions

With regard to the model calculations, the present need for the rare gas crystals is a more physically realistic potential, as discussed at the end of Section 33. In view of the simplicity of the local pseudopotential theory, the simple models of Section 34 give remarkably good results for the lattice-dynamic and thermodynamic properties of Na, K, and Al. Nevertheless the theoretical–experimental discrepancies for the elastic constants and Debye temperatures at $T = 0$ indicate the need for improved models, and at the same time the theory may be improved, within the local pseudopotential approximation, by using (28.35) for the pseudopotential contribution to the dynamical matrices. The situation is more complicated for ionic and covalent crystals; at this stage there is need for a reasonably good potential model, including ionic polarization effects. In view of the success of the pseudopotential theory for metals, such a model may perhaps be developed through an approximation for the dielectric function. Finally we note that it is practical at this time to undertake lattice-dynamics computations based on the band-structure formulation of Chapter 5.

It is of course desirable to test a prescribed model by comparing theoretical and experimental results for thermodynamic properties. For such a test, certain thermodynamic properties are most valuable, because of the general availability of accurate data on the one hand, and the ease of carrying out accurate calculations on the other hand. These functions are the free energy and its volume and strain derivatives at $T = 0$, the phonon-dispersion curves measured at moderately low temperature, and the heat capacity or entropy, together with $V\beta B_T$ for cubic crystals and b_{ij} for noncubic crystals, all reduced to conditions of fixed crystal configuration for temperatures up to the Debye temperature and above. Furthermore, for a first test of the theory, it is generally sufficient to calculate these quantities in the leading order of lattice-dynamics perturbation theory; this means calculating the $T = 0$ free energy and its derivatives in the potential approximation, and calculating the phonon frequencies and the functions C_V or C_η and $V\beta B_T$ or b_{ij} in the harmonic approximation. Such leading-order calculations are easily carried out to very high precision; for the models of the present chapter these quantities were calculated to an accuracy of 0.5% or better with only small amounts of computer time. Following the leading-order calculations, estimates can be made of the harmonic contributions to the free energy and its derivatives at $T = 0$, and of the anharmonic contributions to the phonon frequencies and the thermodynamic functions such as C_V and $V\beta B_T$, to see if these contributions are large enough to remove the discrepancies between theory and experiment. These procedures provide a stringent test of any model, and in fact lead to the conclusion that most presently available models are inadequate, without the need for more complicated lattice-dynamics calculations.

For more refined calculations one can proceed to the second order of lattice-dynamics perturbation theory. This order includes the harmonic contribution to the free energy and its volume and strain derivatives at all temperatures, and gives rise to the temperature-dependences and the adiabatic-isothermal differences of the elastic coefficients. It also includes the temperature-dependent phonon frequency shifts Δ_κ and half-widths Γ_κ, and the anharmonic contributions to the heat capacity and the entropy. The theoretical expressions for these quantities are characterized by sums of terms which tend to cancel, as for example, in (16.69) for C^T_{ijkl}, or more simply in (32.1) and (32.2) for the temperature derivatives of B_T and B_S, respectively, at high temperature. The anharmonic frequency shifts Δ_κ and free energy F_A are particularly difficult to calculate because of the general tendency toward cancellation between the cubic and quartic contributions. In spite of the computational difficulties, however, we recommend avoiding approximations for these quantities in any study where their effects are important. The labor required to evaluate the γ_κ and ξ_κ is quite justified when a reliably accurate result is desired. Further, because of the cancellation between cubic and quartic terms, approximations for Δ_κ and for F_A are apt to be quite unreliable and may even give the wrong sign, while accurate computations of Δ_κ and F_A are in fact practical.*

* The temperature-dependence at $P = 0$ of Δ_κ and Γ_κ for Al has been calculated by T. R. Koehler, N. S. Gillis, and D. C. Wallace, *Phys. Rev.* **B1**, 4521 (1970). An accurate evaluation of $\mathscr{A}_2$ for the Lennard–Jones 12,6 model is given by (33.26).

In comparing the results of theory and experiment for the phonon frequency shifts Δ_κ, just as in all other thermodynamic functions, the volume effects must be taken into account. In fact, from the presently available experimental information, it appears that most of the observed temperature-dependence of $\Omega_\kappa = \omega_\kappa + \Delta_\kappa$ at $P = 0$ is due to the volume-dependence of ω_κ. We can illustrate this for Al, where Ω_κ have been measured at 80 and 300°K by inelastic neutron scattering, and the corresponding averages $\langle\Omega_\kappa^2\rangle$ have been calculated at each temperature (reference 41, Appendix 2). The experimental result is

$$\frac{\langle\Omega_\kappa^2\rangle(300°\text{K}) - \langle\Omega_\kappa^2\rangle(80°\text{K})}{\langle\Omega_\kappa^2\rangle(80°\text{K})} = -0.050. \tag{35.26}$$

However, in going from 80 to 300°K the volume increases by the relative amount

$$\frac{V(300°\text{K}) - V(80°\text{K})}{V(80°\text{K})} = 0.0125. \tag{35.27}$$

When we use the theoretical values of the first and second volume derivatives of $\langle\omega_\kappa^2\rangle$ at 80°K, from the pseudopotential model of Section 34, and the volume change (35.27), we find for Al

$$\frac{\langle\omega_\kappa^2\rangle(300°\text{K}) - \langle\omega_\kappa^2\rangle(80°\text{K})}{\langle\omega_\kappa^2\rangle(80°\text{K})} = -0.047. \tag{35.28}$$

Comparison of (35.26) and (35.28) shows that the observed decrease in $\langle\Omega_\kappa^2\rangle$ is due almost entirely to the decrease in $\langle\omega_\kappa^2\rangle$ because of the volume increase, and the explicit temperature-dependence of $\langle\Omega_\kappa^2\rangle$ at fixed volume is probably smaller than the error in the experimental result (35.26). Incidentally, the apparently small temperature-dependence of $\langle\Omega_\kappa^2\rangle$ at fixed volume is in keeping with the small value of $\mathscr{A}_2$ found for Al in the high-temperature entropy analysis (see Table 12).

Let us close with a brief summary of the qualitative behavior of thermodynamic functions, including the effects of the variation of volume with temperature at $P = 0$, and the various lattice-dynamics and electronic contributions.

Very low T: The combined electronic excitation plus electron–phonon contribution can be separated experimentally from the lattice contribution. Harmonic and anharmonic lattice contributions can be separated experimentally in principle, because of their different dependences on the ion masses. Anharmonic effects are small and volume effects are negligible.

Intermediate T: Anharmonic and electron–phonon contributions are complicated, but small. Volume effects are also small. The temperature-dependences of thermodynamic functions are due mainly to the detailed values of the phonon frequencies and Grüneisen parameters; in other words, dispersion effects are important.

High T: The combined anharmonic plus electronic excitation contribution can be separated experimentally from the harmonic lattice contribution. Volume effects are important. The electron–phonon contributions to the temperature-dependent parts of thermodynamic functions approach zero as the temperature increases.

Appendix 1
COMPUTATIONAL METHODS

36. LATTICE AND INVERSE-LATTICE SUMS

Lattice Points

In order to calculate the dynamical matrices, the stresses and elastic constants, and other lattice-dynamical functions, it is necessary to carry out sums over lattice points and sums over inverse-lattice points. When the theory is properly formulated, as in the band-structure theory of Chapter 5, the required sums converge well, and are easily evaluated by computer. The general procedure for evaluating such sums is to generate the lattice points which lie in a sphere of prescribed radius, sum the desired function evaluated at each lattice point, and add a remainder to account for the contribution arising from points outside the sphere. For rapidly converging sums, the remainder is often negligible. In any case, however, it is advisable to sum over points in a sphere; summing over points in a cube for a cubic lattice, for example, generally leads to poor convergence of a sum as the size of the cube is increased. We will discuss the summing procedures in more detail for a few simple lattices.

Let $\hat{\mathbf{x}}$, $\hat{\mathbf{y}}$, and $\hat{\mathbf{z}}$ be Cartesian unit vectors, and R_1 be the nearest-neighbor distance. For the fcc lattice the lattice vectors $\mathbf{R}(\mathrm{N})$ are given by

$$\mathbf{R}(\mathrm{N}) = (R_1/\sqrt{2})(N_1\hat{\mathbf{x}} + N_2\hat{\mathbf{y}} + N_3\hat{\mathbf{z}}), \tag{36.1}$$

where N_1, N_2, and N_3 are integers which satisfy

$$N_1 + N_2 + N_3 = \text{even integer (including zero)}. \tag{36.2}$$

Table 32. The first 10 shells of neighbors for the fcc lattice.

Neighbor	Representative N_1, N_2, N_3	$R(\mathrm{N})^2/R_1^2$	Number in shell
0	000	0	1
1	011	1	12
2	002	2	6
3	112	3	24
4	022	4	12
5	013	5	24
6	222	6	8
7	123	7	48
8	004	8	6
9	033	9	12
9	114	9	24
10	024	10	24

For each set of integers N_1, N_2, N_3, there are 48 equivalent vectors corresponding to the 48 cubic point-group operations; these 48 vectors are all the same length and constitute a shell. The equivalent vectors correspond to all combinations of $\pm$ signs and permutations of the N_1, N_2, N_3. Many shells contain less than 48 *distinct* points, since some of the combinations give the same vector. For the integers 0, 0, 2, for example, there are only six lattice vectors in the shell. The first 10 shells of neighbors for the fcc lattice are listed in Table 32. The volume per unit cell or per atom for fcc is

$$V_C = V_A = R_1^3/\sqrt{2}. \tag{36.3}$$

The inverse-lattice vectors $\mathbf{Q}$ satisfy $\mathbf{Q}(\mathrm{P}) \cdot \mathbf{R}(\mathrm{N}) = 2\pi(\text{integer})$, and are given by

$$\mathbf{Q}(\mathrm{P}) = (\sqrt{2}\,\pi/R_1)(P_1\hat{\mathbf{x}} + P_2\hat{\mathbf{y}} + P_3\hat{\mathbf{z}}), \tag{36.4}$$

where P_1, P_2, and P_3 are integers which satisfy

$$P_1, P_2, P_3 \text{ are either all even or all odd.} \tag{36.5}$$

Thus the lattice inverse to fcc is bcc, with nearest-neighbor distance $\sqrt{6}\pi/R_1$.

For the bcc lattice with nearest-neighbor distance R_1, the lattice vectors are given by

$$\mathbf{R}(\mathrm{N}) = (R_1/\sqrt{3})(N_1\hat{\mathbf{x}} + N_2\hat{\mathbf{y}} + N_3\hat{\mathbf{z}}), \tag{36.6}$$

where

$$N_1, N_2, N_3 \text{ are either all even or all odd.} \tag{36.7}$$

To have the computer select integers which satisfy (36.7), we have used the equivalent condition

$$N_1 + N_2, N_2 + N_3, N_1 + N_3 \text{ are all even.} \tag{36.8}$$

Again the bcc lattice vectors transform according to the cubic point-group operations; the first 10 shells of neighbors are listed in Table 33. The unit cell volume for bcc is

$$V_C = V_A = 4R_1^3/3\sqrt{3}. \tag{36.9}$$

The inverse-lattice vectors are given by

$$\mathbf{Q}(\mathrm{P}) = (\sqrt{3}\,\pi/R_1)(P_1\hat{\mathbf{x}} + P_2\hat{\mathbf{y}} + P_3\hat{\mathbf{z}}), \tag{36.10}$$

where the integers P_1, P_2, and P_3 satisfy

$$P_1 + P_2 + P_3 = \text{even integer.} \tag{36.11}$$

Thus the lattice inverse to bcc is fcc.

The diamond lattice is a fcc lattice with a basis, i.e., with an extra atom per unit cell. The lattice vectors are $R(\mathrm{N}\mu)$, with $\mu = 0$ corresponding to the primitive lattice (fcc) vectors and $\mu = 1$ corresponding to the basis lattice vectors; with R_1 still the nearest-neighbor distance it follows that

$$\mathbf{R}(\mathrm{N}0) = (2R_1/\sqrt{3})(N_1\hat{\mathbf{x}} + N_2\hat{\mathbf{y}} + N_3\hat{\mathbf{z}}), \tag{36.12}$$

$$\mathbf{R}(\mathrm{N}1) = (2R_1/\sqrt{3})[(N_1 + \tfrac{1}{2})\hat{\mathbf{x}} + (N_2 + \tfrac{1}{2})\hat{\mathbf{y}} + (N_3 + \tfrac{1}{2})\hat{\mathbf{z}}], \tag{36.13}$$

Table 33. The first 10 shells of neighbors for the bcc lattice.

Neighbor	Representative N_1, N_2, N_3	$R(\mathrm{N})^2/R_1^2$	Number in shell
0	000	0	1
1	111	1	8
2	002	4/3	6
3	022	8/3	12
4	113	11/3	24
5	222	4	8
6	004	16/3	6
7	133	19/3	24
8	024	20/3	24
9	224	8	24
10	115	9	24
10	333	9	8

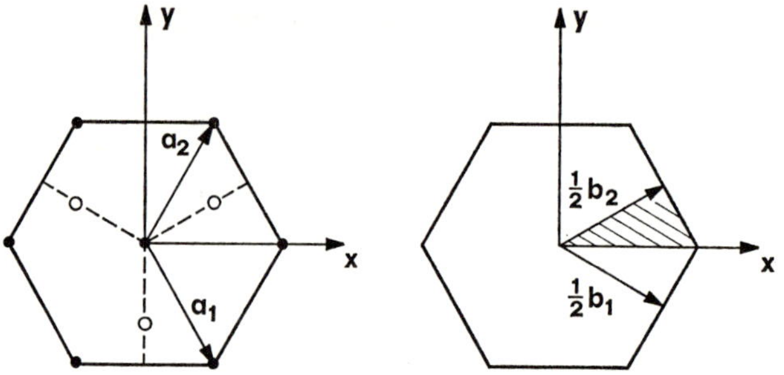

Figure 51. On the left is the basal plane of the hcp lattice, showing the primitive lattice points (dots), and the positions of the basis points (circles) which lie above and below the basal plane. On the right is the basal plane of the inverse hexagonal lattice, showing the hexagonal cross section of the first Brillouin zone, with the shaded portion denoting the $\frac{1}{24}$ portion of the zone drawn in Figure 57 (below).

where the integers N_1, N_2, and N_3 satisfy

$$N_1 + N_2 + N_3 = \text{even integer}. \tag{36.14}$$

The unit cell volume for diamond is

$$V_C = 2V_A = 16R_1^3/3\sqrt{3}. \tag{36.15}$$

The inverse lattice is, of course, the bcc lattice inverse to the primitive fcc lattice defined by the $\mathbf{R}(\mathrm{N0})$.

The hcp lattice is a simple hexagonal lattice (the primitive lattice) with one extra atom per unit cell. A convenient set of hexagonal vectors are $\mathbf{a}_1$ and $\mathbf{a}_2$, which lie in the xy plane (the basal plane) as shown in Figure 51, and $\mathbf{a}_3$, which is along the z axis. The magnitudes are $|\mathbf{a}_1| = |\mathbf{a}_2| = a$ and $|\mathbf{a}_3| = c$, and the c/a ratio is arbitrary. The lattice vectors are given by

$$\mathbf{R}(\mathrm{N0}) = N_1\mathbf{a}_1 + N_2\mathbf{a}_2 + N_3\mathbf{a}_3, \tag{36.16}$$

$$\mathbf{R}(\mathrm{N1}) = (N_1 + \tfrac{1}{3})\mathbf{a}_1 + (N_2 + \tfrac{2}{3})\mathbf{a}_2 + (N_3 + \tfrac{1}{2})\mathbf{a}_3, \tag{36.17}$$

where

$$N_1, N_2, N_3 = \text{arbitrary integers}. \tag{36.18}$$

In Cartesian coordinates, the vectors $\mathbf{a}_\alpha$ are

$$\mathbf{a}_1 = \tfrac{1}{2}a\hat{\mathbf{x}} - \frac{\sqrt{3}}{2}a\hat{\mathbf{y}}, \tag{36.19}$$

$$\mathbf{a}_2 = \tfrac{1}{2}a\hat{\mathbf{x}} + \frac{\sqrt{3}}{2}a\hat{\mathbf{y}}, \tag{36.20}$$

$$\mathbf{a}_3 = c\hat{\mathbf{z}}. \tag{36.21}$$

The Cartesian coordinates of the hcp lattice vectors may be written

$$\mathbf{R}(\mathrm{N}0) = \tfrac{1}{2}(N_1 + N_2)a\hat{\mathbf{x}} + \tfrac{\sqrt{3}}{2}(N_2 - N_1)a\hat{\mathbf{y}} + N_3c\hat{\mathbf{z}}, \tag{36.22}$$

$$\mathbf{R}(\mathrm{N}1) = \tfrac{1}{2}(N_1 + N_2 + 1)a\hat{\mathbf{x}} + \tfrac{\sqrt{3}}{2}(N_2 - N_1 + \tfrac{1}{3})a\hat{\mathbf{y}} + (N_3 + \tfrac{1}{2})c\hat{\mathbf{z}}. \tag{36.23}$$

For a given N_1, N_2, N_3 there are 24 equivalent primitive lattice vectors in a shell, these 24 being not necessarily all distinct. The vectors in a shell are given by $\pm R_x$, $\pm R_y$, $\pm R_z$ for each of the following three sets of R_x, R_y, R_z:

$$\begin{aligned} R_x &= \tfrac{1}{2}(N_2 + N_1)a, & R_y &= \tfrac{\sqrt{3}}{2}(N_2 - N_1)a, & R_z &= N_3c; \\ R_x &= \tfrac{1}{2}(2N_2 - N_1)a, & R_y &= \tfrac{\sqrt{3}}{2}N_1a, & R_z &= N_3c; \\ R_x &= \tfrac{1}{2}(2N_1 - N_2)a, & R_y &= \tfrac{\sqrt{3}}{2}N_2a, & R_z &= N_3c. \end{aligned} \tag{36.24}$$

Again for a given N_1, N_2, N_3 there are 12 equivalent basis vectors in a shell, all 12 being not necessarily distinct. These 12 vectors correspond to $\pm R_x$, R_y, $\pm R_z$ for each of the following three sets of R_x, R_y, R_z:

$$\begin{aligned} R_x &= \tfrac{1}{2}(N_2 + N_1 + 1)a, & R_y &= \tfrac{\sqrt{3}}{2}(N_2 - N_1 + \tfrac{1}{3})a, & R_z &= (N_3 + \tfrac{1}{2})c; \\ R_x &= \tfrac{1}{2}(2N_2 - N_1 + 1)a, & R_y &= \tfrac{\sqrt{3}}{2}(N_1 + \tfrac{1}{3})a, & R_z &= (N_3 + \tfrac{1}{2})c; \\ R_x &= \tfrac{1}{2}(2N_1 - N_2)a, & R_y &= \tfrac{\sqrt{3}}{2}(-N_2 - \tfrac{2}{3})a, & R_z &= (N_3 + \tfrac{1}{2})c. \end{aligned} \tag{36.25}$$

The first few shells of neighbors for the hcp lattice are listed in Table 34. Note that the ideal c/a ratio, which corresponds to closest packing of spheres in a hcp lattice and for which all 12 nearest neighbors are exactly the same distance, is $c/a = \sqrt{\tfrac{8}{3}}$. The volume per unit cell for hcp is

$$V_C = 2V_A = \sqrt{3}\,a^2c/2. \tag{36.26}$$

The lattice inverse to the simple hexagonal primitive lattice is another simple hexagonal lattice. The inverse unit cell vectors $\mathbf{b}_\beta$, defined according to (10.16) so that $\mathbf{a}_\alpha \cdot \mathbf{b}_\beta = 2\pi\delta_{\alpha\beta}$, are

$$\mathbf{b}_1 = (2\pi/a)[\hat{\mathbf{x}} - (1/\sqrt{3})\hat{\mathbf{y}}], \tag{36.27}$$

$$\mathbf{b}_2 = (2\pi/a)[\hat{\mathbf{x}} + (1/\sqrt{3})\hat{\mathbf{y}}], \tag{36.28}$$

$$\mathbf{b}_3 = (2\pi/c)\hat{\mathbf{z}}. \tag{36.29}$$

The set of inverse-lattice vectors is then given by

$$\mathbf{Q}(\mathrm{P}) = P_1\mathbf{b}_1 + P_2\mathbf{b}_2 + P_3\mathbf{b}_3, \tag{36.30}$$

Table 34. The first four shells of neighbors for the hcp lattice, with the distances $R(\mathrm{N}\mu)$ computed on the basis of the ideal $c/a = \sqrt{8/3}$. The primitive lattice points ($\mu = 0$) are given by (36.22), the basis lattice points ($\mu = 1$) are given by (36.23), and here we tabulate $M_1 = N_1 + N_2 + \mu$, $M_2 = N_2 - N_1 + \frac{1}{3}\mu$, and $M_3 = N_3 + \frac{1}{2}\mu$.

Neighbor	μ	Representative M_1, M_2, M_3	$R(\mathrm{N}\mu)^2/R_1^2$	Number of equivalent points
0	0	000	0	1
1	0	110	1	4
1	0	200	1	2
1	1	$1\frac{1}{3}\frac{1}{2}$	1	4
1	1	$0 - \frac{2}{3}\frac{1}{2}$	1	2
2	1	$2 - \frac{2}{3}\frac{1}{2}$	2	4
2	1	$0\frac{4}{3}\frac{1}{2}$	2	2
3	0	001	8/3	2
4	0	020	3	2
4	0	310	3	4
4	1	$1 - \frac{5}{3}\frac{1}{2}$	3	4
4	1	$2\frac{4}{3}\frac{1}{2}$	3	4
4	1	$3\frac{1}{3}\frac{1}{2}$	3	4

where P_1, P_2, P_3 are any integers. In Cartesian coordinates, the 24 equivalent inverse-lattice vectors which constitute a shell are $\pm Q_x$, $\pm Q_y$, $\pm Q_z$ for each of the following three sets of Q_x, Q_y, Q_z:

$$\begin{aligned} &Q_x = (2\pi/a)(P_1 + P_2), && Q_y = (2\pi/\sqrt{3}\,a)(P_2 - P_1), && Q_z = (2\pi/c)P_3; \\ &Q_x = (2\pi/a)P_1, && Q_y = (2\pi/\sqrt{3}\,a)(2P_2 + P_1), && Q_z = (2\pi/c)P_3; \\ &Q_x = (2\pi/a)P_2, && Q_y = (2\pi/\sqrt{3}\,a)(2P_1 + P_2), && Q_z = (2\pi/c)P_3. \end{aligned} \tag{36.31}$$

Symmetries of Lattice Sums

It is useful to study the symmetry properties of lattice sums of functions which depend on the lattice vector components. For a primitive lattice, let us define the sums

$$S = \sum_{\mathrm{N}}{}' f(\mathrm{N}), \tag{36.32}$$

$$S_i = \sum_{\mathrm{N}}{}' f(\mathrm{N})[R_i(\mathrm{N})/R(\mathrm{N})], \tag{36.33}$$

$$S_{ij} = \sum_{\mathrm{N}}{}' f(\mathrm{N})[R_i(\mathrm{N})R_j(\mathrm{N})/R(\mathrm{N})^2], \tag{36.34}$$

and so on, where $f(\mathrm{N})$ is a function only of the magnitude $R(\mathrm{N}) = |\mathbf{R}(\mathrm{N})|$. Since $f(\mathrm{N})$ is the same for all lattice points in a shell, the symmetry properties of these sums are found by summing the functions of lattice vector components over one shell. For fcc and bcc lattices, the following useful properties are obvious.

$$S_i = 0, \quad \text{for all } i. \tag{36.35}$$

$$S_{ij} = 0, \quad i \neq j; \; S_{ii} \text{ are equal for all } i; \; 3S_{ii} = S. \tag{36.36}$$

$$S_{ijk} = 0, \quad \text{for all } ijk. \tag{36.37}$$

$$S_{ijkl} = 0 \quad \text{unless } ijkl \text{ are equal in pairs;}$$

$$S_{iijj} \quad \text{are all equal for } i \neq j, \quad S_{iiii} \text{ are all equal;}$$

$$3S_{iiii} + 6S_{iijj} = S, \quad i \neq j. \tag{36.38}$$

The diamond lattice has two atoms per unit cell, and in calculating contributions to the dynamical matrix, for example, it is sometimes necessary to carry out sums over the primitive (fcc) lattice only, or over the basis sublattice only. Such sublattice sums are obvious extensions of the definitions (36.32)–(36.34); for a primitive lattice ($\mu = 0$) sum, for example,

$$S_{ij} = \sum_{\mathrm{N}}{}' f(\mathrm{N}0)[R_i(\mathrm{N}0)R_j(\mathrm{N}0)/R(\mathrm{N}0)^2], \tag{36.39}$$

and for the basis sublattice ($\mu = 1$) sum,

$$S_{ij} = \sum_{\mathrm{N}} f(\mathrm{N}1)[R_i(\mathrm{N}1)R_j(\mathrm{N}1)/R(\mathrm{N}1)^2]. \tag{36.40}$$

All the relations (36.35), (36.36), and (36.38) hold also for sums over the primitive lattice *or* over the basis lattice for diamond. Of course (36.37) holds for the primitive lattice sums, and for sums over the basis lattice it is

$$S_{ijk} = 0, \quad \text{for all } ijk \text{ except } xyz \text{ and permutations.} \tag{36.41}$$

The properties of lattice sums for hcp are more complicated. First of all for the primitive lattice, by considering the sum over a shell of 24 equivalent vectors, the following results are established.

$$S_i = 0, \quad \text{for all } i. \tag{36.42}$$

$$S_{ij} = 0, \quad i \neq j; \qquad S_{xx} = S_{yy} \neq S_{zz}; \qquad 2S_{xx} + S_{zz} = S. \tag{36.43}$$

$$S_{ijk} = 0, \quad \text{for all } ijk. \tag{36.44}$$

$$S_{ijkl} = 0 \quad \text{unless } ijkl \text{ are equal in pairs;}$$

$$S_{xxzz} = S_{yyzz} \neq S_{xxyy}, \qquad S_{xxxx} = S_{yyyy} \neq S_{zzzz};$$

$$2S_{xxxx} + S_{zzzz} + 4S_{xxzz} + 2S_{xxyy} = S; \tag{36.45}$$

$$S_{xxxx} = 3S_{xxyy}\,. \tag{36.46}$$

Because of these relations among lattice sums, only one second-order sum S_{ij} and one fourth-order sum S_{ijkl} are independent. For example, from (36.43),

$$S_{xx} = \tfrac{1}{2}(S - S_{zz}). \tag{36.47}$$

Also it is obvious that

$$S_{iixx} + S_{iiyy} + S_{iizz} = S_{ii}, \tag{36.48}$$

and from this relation together with (36.45) and (36.46) we can write all the fourth-order sums in terms of S, S_{zz}, and S_{zzzz}:

$$S_{xxzz} = \tfrac{1}{2}(S_{zz} - S_{zzzz}), \tag{36.49}$$

$$S_{xxyy} = \tfrac{1}{8}(S - 2S_{zz} + S_{zzzz}) = \tfrac{1}{3}S_{xxxx}. \tag{36.50}$$

All the relations (36.42)–(36.50), except (36.44), hold also for sums over the basis sublattice for hcp; for the basis sublattice sums, (36.44) is replaced by

$$S_{ijk} = 0, \quad \text{for all } ijk \text{ except } xxy \text{ and } yyy \text{ and permutations;} \qquad S_{xxy} + S_{yyy} = 0. \tag{36.51}$$

Sums and Remainders

The two steps involved in carrying out lattice sums, either for all lattice points or for a given sublattice only, are first to sum over points in a sphere, and second to add a remainder. By summing over points in a sphere we mean summing over all shells out to a largest shell. Some care is required in generating the lattice vector shells for a nonprimitive lattice; consider for example a diamond lattice. For each shell of primitive lattice (fcc) vectors, there is a sphere of sublattice vectors displaced from this shell by the basis vector $(R_1/\sqrt{3})(\hat{\mathbf{x}} + \hat{\mathbf{y}} + \hat{\mathbf{z}})$. This sphere of sublattice vectors is *not* a shell, since it is not centered on the origin, and summing over such spheres leads to poor convergence of sums as the sphere radius is increased.

To generate the lattice points in a sphere for fcc and bcc lattices, we have used the following procedure. Let L_N be the limit for the magnitude of the integers N_1, N_2, N_3; we then generate the integers in a cube

$$-L_N \leq N_1, N_2, N_3 \leq L_N, \tag{36.52}$$

keep only those in the sphere

$$N_1^2 + N_2^2 + N_3^2 \leq L_N^2, \tag{36.53}$$

and finally keep only those which satisfy the fcc condition (36.2), or the bcc condition (36.8). The numbers of lattice points so generated, for various values of L_N, are given in Table 35.

For the diamond lattice, a procedure for generating all the primitive and basis lattice vectors in a sphere is as follows. First generate all the integers

Table 35. The numbers of lattice points in spheres, not counting the primitive lattice point at the origin. For each L_N for diamond, the total number of lattice points is listed, and the number of primitive lattice points is the same as for the fcc lattice and the rest are basis points.

L_N	fcc	bcc	Diamond (total)
1			4
2	18	14	34
3	54	26	122
4	140	64	280
5	248	136	524
6	458	258	914
7	682	338	1,418
8	1,060	536	2,148
9	1,504	748	3,060
10	2,122	1,066	4,234
20	16,756	8,392	33,532
30	56,470	28,474	
40	134,008	66,952	
50	261,562	131,018	
72	781,144	390,628	

in a cube

$$-L_N \leq N_1, N_2, N_3 \leq L_N, \tag{36.54}$$

and keep only the fcc (primitive lattice) points which satisfy $N_1 + N_2 + N_3 =$ even. Then for each primitive lattice vector, generate the corresponding basis sublattice vector given by (36.13), by adding $\frac{1}{2}$ to N_1, N_2, and N_3, and finally keep all the primitive lattice points which satisfy

$$N_1^2 + N_2^2 + N_3^2 \leq L_N^2, \tag{36.55}$$

and all the basis lattice points which satisfy

$$(N_1 + \tfrac{1}{2})^2 + (N_2 + \tfrac{1}{2})^2 + (N_3 + \tfrac{1}{2})^2 \leq L_N{}^2. \tag{36.56}$$

The numbers of lattice points so generated for various values of L_N are listed in Table 35. Note that if L_N is even or odd, the outer shell generated is primitive or basis, respectively.

Another useful device in computer summing is the "split" sum technique. If one adds into a given storage register, say 10^6 times, then because of random round off after each addition the last three figures in the sum will be meaningless. To avoid this loss of accuracy, one can add into two storage registers. Suppose it is desired to sum 10^6 positive numbers, most of which are very small and whose sum is, say 1. One first establishes a cutoff α, generally by inspection, such that the sum of all numbers which are $<10^{-\alpha}$ is about 10^{-3}. One then sums all numbers $>10^{-\alpha}$ in box a, all numbers $<10^{-\alpha}$ in box b, and finally adds a and b. The large round-off error is thus accumulated in b, and then discarded when a and b are added.

We now turn to the calculation of remainders for lattice and inverse-lattice sums. The remainder is calculated by integrating the summand over all space outside a sphere, and the important point is the choice of the radius ρ of this sphere. This radius should not be the radius of the outermost shell of lattice points which was summed, but should presumably be between the radius of that shell and the next larger one. In fact if a total of T lattice points are included in the sum over shells, then a good choice for ρ is the radius of a sphere whose volume is T times the volume per lattice point. Thus for a primitive lattice, the sum of a function $g(\mathbf{R}(\mathrm{N}))$ may be written

$$\sum_{\mathrm{N}} g(\mathbf{R}(\mathrm{N})) = \sum_{\mathrm{N}=1}^{T} g(\mathbf{R}(\mathrm{N})) + V_A^{-1} \int_{\rho}^{\infty} g(\mathbf{R})\, d\mathbf{R}, \tag{36.57}$$

where the sum on the right is over the T lattice points in a prescribed sphere, V_A is the volume per atom and hence per lattice point, and the integral term is of course the remainder, with

$$\tfrac{4}{3}\pi\rho^3 = TV_A. \tag{36.58}$$

For a nonprimitive lattice we distinguish between sums over *all* lattice points, or over one or more of the sublattices. For a sum over all lattice points, the volume per lattice point is V_A and we write

$$\sum_{\mathrm{N}\mu} g(\mathbf{R}(\mathrm{N}\mu)) = \sum_{\mathrm{N}\mu=1}^{T} g(\mathbf{R}(\mathrm{N}\mu)) + V_A^{-1} \int_{\rho}^{\infty} g(\mathbf{R})\, d\mathbf{R}, \tag{36.59}$$

where ρ is given by (36.58). For a sum over the sublattice ν, the volume per lattice point is V_C and we write

$$\sum_{\mathrm{N}} g(\mathbf{R}(\mathrm{N}\nu)) = \sum_{\mathrm{N}=1}^{T} g(\mathbf{R}(\mathrm{N}\nu)) + V_C^{-1} \int_{\rho}^{\infty} g(\mathbf{R})\, d\mathbf{R}, \tag{36.60}$$

where

$$\tfrac{4}{3}\pi\rho^3 = TV_C. \tag{36.61}$$

The inverse lattice is always a primitive lattice, with the volume per inverse-lattice point being $(2\pi)^3/V_C$. Hence inverse-lattice sums are calculated like (36.57) for primitive lattice sums.

In practice we have based remainders only on the leading contribution to the summand at large R. Suppose, for example, that $g(\mathbf{R})$ goes as $\alpha R^{-6} + \beta R^{-8} + \cdots$ at large R; we calculate the remainder from the R^{-6} term as

$$V_A^{-1}\int_\rho^\infty \alpha R^{-6}\, d\mathbf{R} = 4\pi\alpha/3V_A\rho^3. \tag{36.62}$$

If $g(\mathbf{R})$ goes as $\alpha R_i R_j/R^6$ at large R, the remainder is

$$\frac{\alpha}{V_A}\int_\rho^\infty \frac{R_i R_j}{R^6}\, d\mathbf{R} = \delta_{ij}\frac{4\pi\alpha}{3V_A\rho}. \tag{36.63}$$

Suppose $g(\mathbf{R})$ goes as $\alpha R^{-6}\cos \mathbf{q}\cdot\mathbf{R}$ at large $\mathbf{R}$, where $\mathbf{q}$ is a fixed vector. By carrying out the angle integral, the remainder becomes

$$\frac{\alpha}{V_A}\int_\rho^\infty \frac{\cos \mathbf{q}\cdot\mathbf{R}}{R^6}\, d\mathbf{R} = \frac{4\pi\alpha q^3}{V_A}\int_{q\rho}^\infty \frac{\sin x}{x^5}\, dx. \tag{36.64}$$

Now the object is to integrate $x^{-5}\sin x\, dx$ by parts, so that the remaining integral contains x^{-6}, and hence develop an asymptotic series for the remainder at large ρ. Thus we write $\int x^{-5}\sin x\, dx = -x^{-5}\cos x - 5\int x^{-6}\cos x\, dx$, and the remainder (36.64) becomes

$$\frac{4\pi\alpha\cos q\rho}{q^2\rho^5 V_A} + O(\rho^{-6}). \tag{36.65}$$

Finally one should always study the convergence of lattice sums, as the number of points which are summed is increased. Remainders are particularly useful for slowly converging lattice sums, and the addition of remainders greatly improves the convergence of such sums.

37. COMPUTATION OF THERMODYNAMIC FUNCTIONS

Phonon Wave Vectors

As it was mentioned in Section 10, the phonon frequencies $\omega(\mathbf{k}s)$ are the same for all the equivalent $\mathbf{k}$ vectors which are related by point-group operations of the inverse lattice. There are 48 such operations for the primitive cubic lattice, and 24 for the primitive hexagonal lattice. This means a sum such as $\Sigma_{\mathbf{k}s} f(\omega(\mathbf{k}s))$, for any function f of the phonon frequencies, can be evaluated by a sum over $\frac{1}{48}$ of the Brillouin zone for cubic lattices, and by a sum over $\frac{1}{24}$ of the zone for hexagonal lattices. We will now describe procedures for generating the **k** vectors which lie in the representative portions of the zones. The replacement of complete zone sums with partial zone sums, with the use of weighting factors, is discussed in the following subsection. It should be recognized that quantities other than the phonon frequencies, such as elements of the phonon eigenvectors or the dynamical matrices, are not the same for each of the equivalent **k** vectors, and this has to be taken into account in summing functions of such quantities over the zone.

The construction of the first Brillouin zone for the fcc lattice is shown in Figure 52. Except for dimensions, the Brillouin zone for diamond and NaCl lattices is the same as for fcc. One of the 48 equivalent portions of the fcc zone is shown in Figure 53; in terms of the fcc nearest-neighbor distance

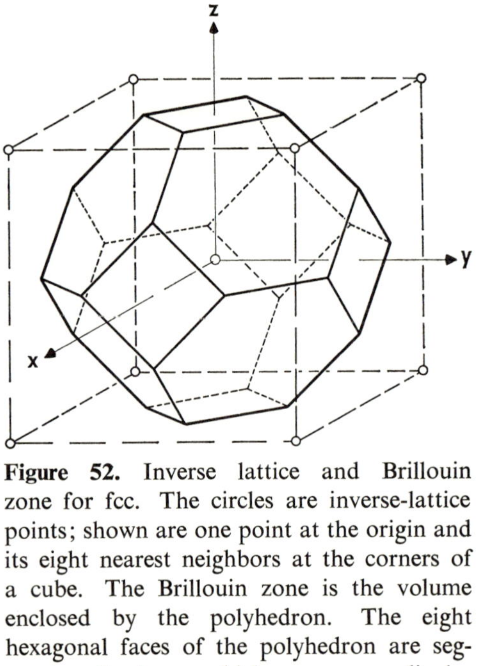

Figure 52. Inverse lattice and Brillouin zone for fcc. The circles are inverse-lattice points; shown are one point at the origin and its eight nearest neighbors at the corners of a cube. The Brillouin zone is the volume enclosed by the polyhedron. The eight hexagonal faces of the polyhedron are segments of planes which are perpendicular bisectors of lines from the origin to its nearest neighbors. The six square faces are segments of planes which are perpendicular bisectors of lines from the origin to its six second neighbors (not shown) on the coordinate axes. (From D. C. Wallace, in *Advances in Materials Research*, edited by H. Herman, Wiley-Interscience Inc., New York, 1968, Vol. 3, p. 331.)

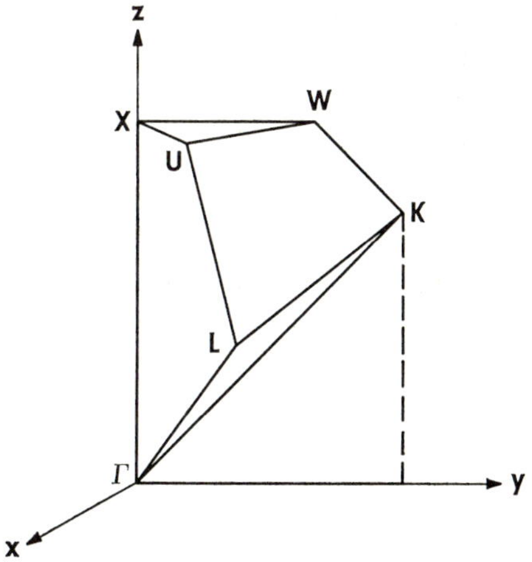

Figure 53. One of the 48 equivalent portions of the fcc Billouin zone, drawn in the same perspective as the zone of Figure 52. The point X is at the center of a square face, W is at the corner of a square face, U is at the center of a square face edge, K is at the center of a hexagonal face edge, and L is at the center of a hexagonal face. (From D. C. Wallace, in *Advances in Materials Research*, edited by H. Herman, Wiley-Interscience Inc., New York, 1968, Vol. 3, p. 331.)

R_1, the Cartesian coordinates of the symmetry points are as follows.

$$\begin{aligned} \Gamma &= (\sqrt{2}\,\pi/R_1)(0,0,0) & X &= (\sqrt{2}\,\pi/R_1)(0,0,1) \\ W &= (\sqrt{2}\,\pi/R_1)(0,\tfrac{1}{2},1) & K &= (\sqrt{2}\,\pi/R_1)(0,\tfrac{3}{4},\tfrac{3}{4}) \\ L &= (\sqrt{2}\,\pi/R_1)(\tfrac{1}{2},\tfrac{1}{2},\tfrac{1}{2}) & U &= (\sqrt{2}\,\pi/R_1)(\tfrac{1}{4},\tfrac{1}{4},1) \end{aligned} \tag{37.1}$$

The planes which form the faces of the $\frac{1}{48}$ zone shown in Figure 53 are given by the following equations.

$$\begin{aligned} &\text{Plane } \Gamma XWK\text{:} && x = 0 \\ &\text{Plane } XWU\text{:} && z = \sqrt{2}\,\pi/R_1 \\ &\text{Plane } \Gamma XUL\text{:} && x = y \\ &\text{Plane } WKLU\text{:} && x + y + z = 3\pi/\sqrt{2}\,R_1 \\ &\text{Plane } \Gamma LK\text{:} && y = z \end{aligned} \tag{37.2}$$

For a fcc crystal containing N_0 atoms there are N_0 allowed **k** vectors in the zone; for computational purposes we take a representative set of **k** vectors distributed uniformly throughout the zone. It is convenient (but not necessary) to take a simple cubic lattice of **k** vectors, defined by the integers p_1, p_2, p_3 according to

$$\mathbf{k}(p) = (\sqrt{2}\,\pi/L_pR_1)(p_1\hat{\mathbf{x}} + p_2\hat{\mathbf{y}} + p_3\hat{\mathbf{z}}), \tag{37.3}$$

where L_p is a positive integer which defines limits to be placed on the integers p_1, p_2, p_3. In order to generate the **k** vectors lying in the $\frac{1}{48}$ zone shown in Figure 53, we simply generate the integers p_1, p_2, p_3 according to the following limits.

$$0 \le p_3 \le L_p; \tag{37.4}$$

$$0 \le p_2 \le L_2, \quad \text{where} \quad L_2 = \text{minimum of } (p_3, \tfrac{3}{2}L_p - p_3); \tag{37.5}$$

$$0 \le p_1 \le L_1, \quad \text{where} \quad L_1 = \text{minimum of } (p_2, \tfrac{3}{2}L_p - p_3 - p_2). \tag{37.6}$$

The point $\mathbf{k} = 0$ is included in the conditions (37.4)–(37.6); it can be excluded by replacing (37.4) by

$$0 < p_3 \le L_p, \quad \text{excludes} \quad \mathbf{k} = 0. \tag{37.7}$$

With $\mathbf{k}(p)$ given by (37.3), and $\mathbf{R}(\mathrm{N})$ given by (36.1) for fcc, we have

$$\mathbf{k}(p) \cdot \mathbf{R}(\mathrm{N}) = (\pi/L_p)(p_1N_1 + p_2N_2 + p_3N_3). \tag{37.8}$$

The construction of the first Brillouin zone for the bcc lattice is shown in Figure 54. One of the 48 equivalent portions of the bcc zone is shown in Figure 55, and in terms of the bcc nearest-neighbor distance R_1, the Cartesian coordinates of the symmetry points are as follows.

$$\begin{aligned} \Gamma &= (\sqrt{3}\,\pi/R_1)(0,0,0) \qquad & H &= (\sqrt{3}\,\pi/R_1)(0,0,1) \\ N &= (\sqrt{3}\,\pi/R_1)(0,\tfrac{1}{2},\tfrac{1}{2}) \qquad & P &= (\sqrt{3}\,\pi/R_1)(\tfrac{1}{2},\tfrac{1}{2},\tfrac{1}{2}) \end{aligned} \tag{37.9}$$

The planes which are surfaces of the $\frac{1}{48}$ zone shown in Figure 55 are given by the following equations.

$$\begin{aligned} &\text{Plane } \Gamma HN\text{:} \quad & x &= 0 \\ &\text{Plane } HNP\text{:} \quad & y + z &= \sqrt{3}\,\pi/R_1 \\ &\text{Plane } \Gamma HP\text{:} \quad & x &= y \\ &\text{Plane } \Gamma PN\text{:} \quad & y &= z \end{aligned} \tag{37.10}$$

A primitive cubic lattice of representative **k** vectors for bcc is conveniently defined by

$$\mathbf{k}(p) = (\sqrt{3}\,\pi/L_pR_1)(p_1\hat{\mathbf{x}} + p_2\hat{\mathbf{y}} + p_3\hat{\mathbf{z}}), \tag{37.11}$$

where L_p is a positive integer. To generate the **k** vectors lying in the $\frac{1}{48}$ zone

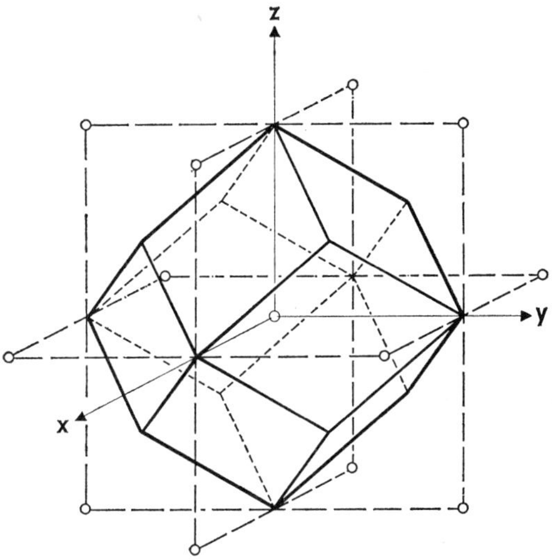

Figure 54. Inverse lattice and Brillouin zone for bcc. The circles are inverse-lattice points; shown are one point at the origin and its 12 nearest neighbors at the corners of three squares which are in the three principal planes of the coordinate system. The Brillouin zone is the volume enclosed by the polyhedron. The 12 faces of the polyhedron are segments of planes which are perpendicular bisectors of lines from the origin to its 12 nearest neighbors. (From D. C. Wallace, in *Advances in Materials Research*, edited by H. Herman, Wiley-Interscience Inc., New York, 1968, Vol. 3, p. 331.)

Figure 55. One of the 48 equivalent portions of the bcc Brillouin zone, drawn in the same perspective as the zone of Figure 54. The point H is at a face corner, N is at a face center, and P is at a face corner inequivalent to H. (From D. C. Wallace, in *Advances in Materials Research*, edited by H. Herman, Wiley-Interscience Inc., New York, 1968, Vol. 3, p. 331.)

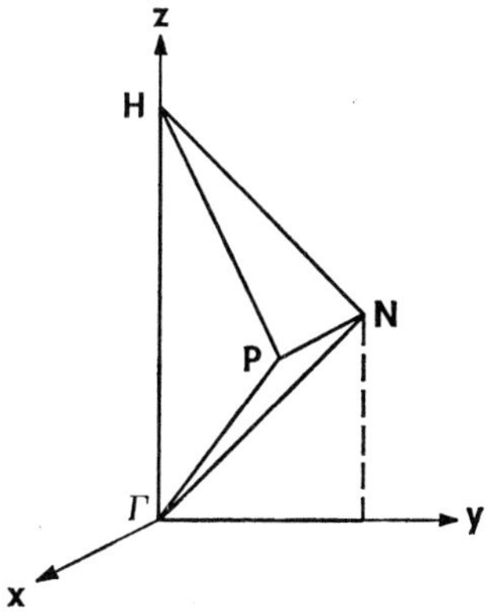

shown in Figure 55, one generates the integers p_1, p_2, p_3 according to the following limits.

$$0 \leq p_3 \leq L_p; \tag{37.12}$$

$$0 \leq p_2 \leq L_2, \quad \text{where} \quad L_2 = \text{minimum of } (p_3, L_p - p_3); \tag{37.13}$$

$$0 \leq p_1 \leq p_2. \tag{37.14}$$

To exclude the point $\mathbf{k} = 0$, (37.12) should again be replaced by the condition (37.7). With $\mathbf{k}(p)$ given by (37.11), and $\mathbf{R}(\mathrm{N})$ given by (36.6) for bcc, we again have the result (37.8) for $\mathbf{k}(p) \cdot \mathbf{R}(\mathrm{N})$.

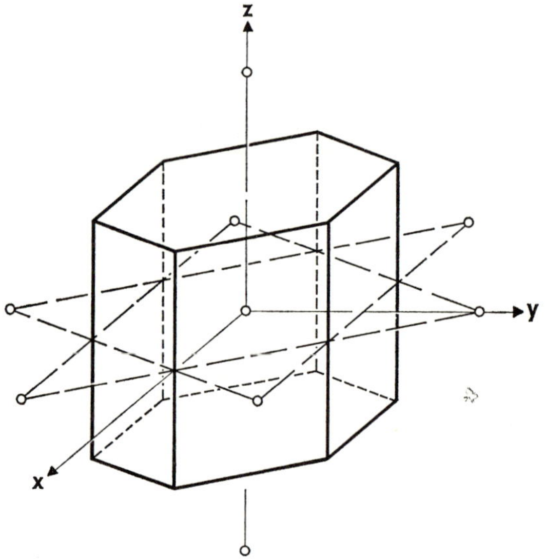

Figure 56. Inverse lattice and Brillouin zone for primitive hexagonal. The circles are inverse-lattice points; shown are one point at the origin and its eight nearest neighbors. Six of the nearest neighbors are at the corners of a regular hexagon in the xy plane, and two are on the z axis, above and below the origin. The Brillouin zone is the volume enclosed by the hexagonal polyhedron. The faces of the polyhedron are segments of planes which are perpendicular bisectors of lines from the origin to its eight nearest neighbors. (From D. C. Wallace, in *Advances in Materials Research*, edited by H. Herman, Wiley-Interscience Inc., New York, 1968, Vol. 3, p. 331.)

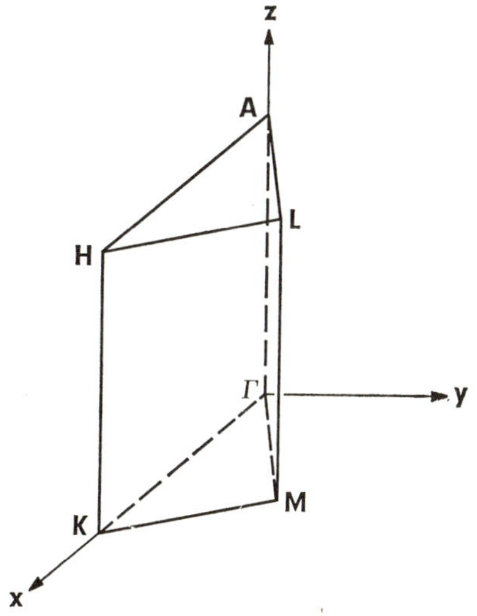

Figure 57. One of the 24 equivalent portions of the hexagonal Brillouin zone, drawn in the same perspective as the zone of Figure 56. The point A is at the center of a hexagonal face, H is at the corner of a hexagonal face, L is at the center of a hexagonal face edge, K is at the center of a rectangular face edge, and M is at the center of a rectangular face. (From D. C. Wallace, in *Advances in Materials Research*, edited by H. Herman, Wiley-Interscience Inc., New York, 1968, Vol. 3, p. 331.)

The construction of the first Brillouin zone for the primitive hexagonal lattice is shown in Figure 56; the zone for hcp is of course the same. One of the 24 equivalent portions of the hexagonal zone is shown in Figure 57, and in terms of the hexagonal lattice parameters a and c, the Cartesian coordinates of the symmetry points are as follows.

$$\begin{aligned} &\Gamma = 0 && A = (0,0,\pi/c) \\ &K = (4\pi/3a,0,0) && H = (4\pi/3a,0,\pi/c) \\ &M = (\pi/a,\pi/\sqrt{3}\,a,0) && L = (\pi/a,\pi/\sqrt{3}\,a,\pi/c) \end{aligned} \tag{37.15}$$

The planes which are surfaces of the $\frac{1}{24}$ zone shown in Figure 57 are given by the following equations.

$$\begin{aligned}
&\text{Plane } \Gamma MK: && z = 0 \\
&\text{Plane } ALH: && z = \pi/c \\
&\text{Plane } \Gamma AHK: && y = 0 \\
&\text{Plane } \Gamma ALM: && x = \sqrt{3}\, y \\
&\text{Plane } KHLM: && x + (1/\sqrt{3})y = 4\pi/3a
\end{aligned} \tag{37.16}$$

For the hexagonal lattices, we have chosen to define representative **k** vectors as points on a simple hexagonal lattice. In terms of the inverse-lattice unit cell vectors $\mathbf{b}_\beta$ given by (36.27)–(36.29), the **k** vectors are

$$\mathbf{k}(p) = (1/2L_p)(p_1\mathbf{b}_1 + p_2\mathbf{b}_2 + p_3\mathbf{b}_3), \tag{37.17}$$

where as usual p_1, p_2, p_3 are integers and L_p is a positive integer. To generate the **k** vectors lying in the $\frac{1}{24}$ zone shown in Figure 57, we simply generate the integers p_1, p_2, p_3 according to the following limits.

$$0 \leq p_3 \leq L_p; \tag{37.18}$$

$$0 \leq p_2 \leq L_p; \tag{37.19}$$

$$0 \leq p_1 \leq L_1, \quad \text{where} \quad L_1 = \text{minimum of } (p_2, 2L_p - 2p_2). \tag{37.20}$$

Note that the point $\mathbf{k} = 0$ is included in these conditions. With $\mathbf{k}(p)$ given by (37.17), and the primitive hexagonal lattice vectors $\mathbf{R}(\mathrm{N0})$ given by (36.16), and the hcp basis lattice vectors $\mathbf{R}(\mathrm{N1})$ given by (36.17), the important vector dot products are

$$\mathbf{k}(p) \cdot \mathbf{R}(\mathrm{N0}) = (\pi/L_p)(p_1N_1 + p_2N_2 + p_3N_3), \tag{37.21}$$

$$\mathbf{k}(p) \cdot \mathbf{R}(\mathrm{N1}) = (\pi/L_p)[p_1(N_1 + \tfrac{1}{3}) + p_2(N_2 + \tfrac{2}{3}) + p_3(N_3 + \tfrac{1}{2})]. \tag{37.22}$$

Brillouin Zone Sums

The lattice-dynamical expressions for thermodynamic functions are derived in Chapter 4; these expressions are sums of the form $\Sigma_\kappa f(\omega_\kappa)$, where the Σ_κ is over all phonons $\kappa = \mathbf{k}s$, and $f(\omega_\kappa)$ may depend on the temperature and other parameters as well as on the phonon frequencies ω_κ. In the old days it was customary to evaluate such sums by first constructing a phonon frequency distribution function $g(\omega)$, then transforming $\Sigma_\kappa f(\omega_\kappa)$ to $\int g(\omega)f(\omega)\,d\omega$, and evaluating the integral. We note that $g(\omega)$ has van Hove singularities, which arise from relative maxima, minima, or saddle points in the ω *vs* **k** surfaces, and these singularities cause some difficulty in obtaining accurate values of the integrals $\int g(\omega)f(\omega)\,d\omega$. In order to evaluate these integrals, it is now customary to start with a small number of measured or calculated phonon frequencies, and generate interpolated frequencies numbering in the millions so as to construct a detailed distribution function with the singularities resolved as well as possible. However, with only a modest computer, the Brillouin zone sums can be evaluated easily and accurately by direct summing over a set of representative **k** vectors lying in the zone, or a portion of it, by procedures we will outline below. Furthermore the problem due to van Hove singularities is not present in the direct summation method; in fact it is just in those regions of slowly varying $\omega(\mathbf{k}s)$ that direct summation over a few representative **k** vectors gives the most accurate results.

To make these observations more quantitative, we wish to present a theorem which is in the spirit of information theory. Consider a function $f(\omega_\kappa)$ which depends on the frequencies ω_κ, and on other parameters such as T, γ_κ, etc. Assume we have available a set of ω_κ and the other parameters required to evaluate $f(\omega_\kappa)$, either measured or calculated, and this set of values represents all our information. Then the most accurate procedure for evaluating $\Sigma_\kappa f(\omega_\kappa)$ is by direct summation over the available set of data, with the correct inclusion of weighting factors in the sum, as described below. Furthermore the generation of new values of ω_κ and of the other parameters, by interpolation procedures, introduces spurious information which reflects the interpolation formulas, and hence reduces the accuracy by which the final sum (or integral) represents the original data. Finally, with or without interpolation, the procedure of calculating $g(\omega)$ and then $\int g(\omega)f(\omega)\,d\omega$ gives a less accurate value for $\Sigma_\kappa f(\omega_\kappa)$ than does direct summation.

If the available set of data consists of ω_κ and other parameters evaluated at N different κ, then the accuracy of $\Sigma_\kappa f(\omega_\kappa)$ will be at most of order 1 part in N; the greatest accuracy may be expected for a sum in which each datum contributes more or less equally, such as in $\Sigma_\kappa \omega_\kappa$. On the other hand, the accuracy of $\Sigma_\kappa f(\omega_\kappa)$ can be considerably less than 1 part in N. For a sum such as $\Sigma_\kappa \omega_\kappa \bar{n}_\kappa$, where $\bar{n}_\kappa$ is the statistical average phonon occupation number given by (16.16), the accuracy of the sum will depend on the temperature, and will be of order 1 part in N' where N' is the number of known ω_κ which satisfy $\hbar\omega_\kappa \leqslant KT$. We are simply saying that the accuracy to which we can evaluate $\Sigma_\kappa f(\omega_\kappa)$ is governed by the number of known values of $f(\omega_\kappa)$ which contribute significantly to the sum, i.e., it is governed by the known amount of *relevant* information. Transforming the problem to $\int g(\omega)f(\omega)\,d\omega$ does not change the situation.

We now proceed to the determination of the weighting factors. For a lattice with N_0 unit cells there are N_0 allowed **k** vectors distributed uniformly over the Brillouin zone. If the allowed **k** vectors include one which lies on a zone surface, then the equivalent **k** (modulo one inverse unit cell vector) on the opposite surface should not be counted since it belongs to another zone. Alternatively, all the zone surface **k** vectors may be counted, with a weighting factor of $\frac{1}{2}$ multiplying the corresponding contributions to the Brillouin zone sum. In the same way, if the sum is carried out only over **k** vectors lying in a prescribed portion of the zone, then the weighting factor for each **k** is determined by the number of equivalent portions which share that same **k**.

Consider the Brillouin zone average

$$(3nN_0)^{-1} \sum_{\mathbf{k}s} f(\omega(\mathbf{k}s)), \tag{37.23}$$

where for a lattice with n ions per unit cell there are $3n$ values of s for each **k**. We presume $f(\omega(\mathbf{k}s))$ depends only on parameters such as $\omega(\mathbf{k}s)$ which are the same for all the equivalent **k** vectors, i.e., those **k** vectors related by point-group operations of the inverse lattice. With the weighting factors denoted by $X_\mathbf{k}$, the sum (37.23) is evaluated with a computer in the form

$$\frac{\sum_s \sum_\mathbf{k}^* X_\mathbf{k} f(\omega(\mathbf{k}s))}{3n \sum_\mathbf{k}^* X_\mathbf{k}}, \tag{37.24}$$

where $\Sigma_\mathbf{k}^*$ is over a representative set, or whatever set is convenient, of **k** vectors lying in one of the equivalent portions of the Brillouin zone. If the number of **k** vectors which are counted in the sum is $N_\mathbf{k}$, it is obvious that

$$\sum_\mathbf{k}^* X_\mathbf{k} \leq N_\mathbf{k}. \tag{37.25}$$

The $\frac{1}{48}$ zone for fcc is shown in Figure 53, and the weighting factors may be written down by inspection of that figure, as follows.

$$\begin{array}{llll}
\text{Within the volume:} & X_\mathbf{k} = 1 & & \\
\text{On any face:} & X_\mathbf{k} = \frac{1}{2} & & \\
\text{On the edge } \Gamma L: & X_\mathbf{k} = \frac{1}{6} & \text{On the edge } \Gamma X: & X_\mathbf{k} = \frac{1}{8} \\
\text{On all other edges:} & X_\mathbf{k} = \frac{1}{4} & & \\
\text{At } \Gamma: \ X_\mathbf{k} = \frac{1}{48} & & \text{At } X: & X_\mathbf{k} = \frac{1}{16} \\
\text{At } L: \ X_\mathbf{k} = \frac{1}{12} & & \text{At } U, W, K: & X_\mathbf{k} = \frac{1}{8}
\end{array} \tag{37.26}$$

A simple procedure for generating these weighting factors is to carry out the following tests, in sequence, on the integers p_1, p_2, p_3 which define the **k** vectors for fcc according to the conditions (37.4)–(37.6).

Test 1. If $p_1 = p_2 = p_3 = 0, \qquad X_{\mathbf{k}} = \frac{1}{48}.$

Test 2. If test 1 is not satisfied:

$$\text{If } p_1 = p_2 = 0, \quad p_3 \neq L_p, \quad X_{\mathbf{k}} = \tfrac{1}{8}.$$
$$\text{If } p_1 = p_2 = 0, \quad p_3 = L_p, \quad X_{\mathbf{k}} = \tfrac{1}{16}.$$

Test 3. If neither test 1 nor test 2 is satisfied:

$$\text{If } p_1 = p_2 = p_3 \neq \tfrac{1}{2}L_p, \quad X_{\mathbf{k}} = \tfrac{1}{6}.$$
$$\text{If } p_1 = p_2 = p_3 = \tfrac{1}{2}L_p, \quad X_{\mathbf{k}} = \tfrac{1}{12}.$$

Test 4. If none of tests 1, 2, or 3 is satisfied, set $X_{\mathbf{k}} = 1$ and then multiply by $\frac{1}{2}$ for each of the following five tests which is satisfied.

$$\begin{aligned} &\text{(a)} \quad p_1 + p_2 + p_3 = \tfrac{3}{2}L_p \\ &\text{(b)} \quad p_3 = L_p \\ &\text{(c)} \quad p_1 = p_2 \\ &\text{(d)} \quad p_1 = 0 \\ &\text{(e)} \quad p_2 = p_3. \end{aligned} \tag{37.27}$$

For each of several values of L_p for the fcc lattice, the number $N_{\mathbf{k}}$ of **k** vectors in $\frac{1}{48}$ of the zone, generated by the conditions (37.4)–(37.6), and the sum of the $X_{\mathbf{k}}$ for these vectors are listed in Table 36.

The $\frac{1}{48}$ zone for bcc is shown in Figure 55, and again the weighting factors

Table 36. The number $N_{\mathbf{k}}$ of **k** vectors in $\frac{1}{48}$ of the zone, and the sum of the weighting factors, for the fcc lattice. The point $\mathbf{k} = 0$ is not counted here.

L_p	$N_{\mathbf{k}}$	$\sum_{\mathbf{k}}^{*} X_{\mathbf{k}}$
4	19	5.3125
8	88	42.6458333 . . .
10	151	83.3125
12	239	143.9791666 . . .
14	355	228.6458333 . . .
16	504	341.3125
20	915	666.6458333 . . .

may be written down by inspection of that figure, as follows.

Within the volume: $X_{\mathbf{k}} = 1$

On any face: $X_{\mathbf{k}} = \frac{1}{2}$

On the edge ΓH: $X_{\mathbf{k}} = \frac{1}{8}$ On the edges ΓP, HP: $X_{\mathbf{k}} = \frac{1}{6}$

On the edges ΓN, PN, HN: $X_{\mathbf{k}} = \frac{1}{4}$

$$\begin{array}{ll} \text{At } \Gamma\text{: } X_{\mathbf{k}} = \frac{1}{48} & \text{At } H\text{: } X_{\mathbf{k}} = \frac{1}{48} \\ \text{At } P\text{: } X_{\mathbf{k}} = \frac{1}{24} & \text{At } N\text{: } X_{\mathbf{k}} = \frac{1}{8} \end{array} \tag{37.28}$$

These weighting factors may be generated by carrying out the following tests, in sequence, on the integers p_1, p_2, p_3 which define the **k** vectors for bcc according to the conditions (37.12)–(37.14).

Test 1. If $p_1 = p_2 = p_3 = 0$, $X_{\mathbf{k}} = \frac{1}{48}$.

Test 2. If test 1 is not satisfied:

If $p_1 = p_2 = 0$, $p_3 \neq L_p$, $X_{\mathbf{k}} = \frac{1}{8}$.

If $p_1 = p_2 = 0$, $p_3 = L_p$, $X_{\mathbf{k}} = \frac{1}{48}$.

Test 3. If neither test 1 nor test 2 is satisfied:

If $p_1 = p_2 = p_3 \neq \frac{1}{2}L_p$, $X_{\mathbf{k}} = \frac{1}{6}$.

If $p_1 = p_2 = p_3 = \frac{1}{2}L_p$, $X_{\mathbf{k}} = \frac{1}{24}$.

Test 4. If none of tests 1, 2, or 3 is satisfied:

If $p_1 = p_2 = L_p - p_3$, $X_{\mathbf{k}} = \frac{1}{6}$.

Test 5. If none of tests 1, 2, 3, or 4 is satisfied, set $X_{\mathbf{k}} = 1$ and then multiply by $\frac{1}{2}$ for each of the following four tests which is satisfied.

$$\begin{array}{ll} \text{(a)} & p_2 + p_3 = L_p \\ \text{(b)} & p_1 = p_2 \\ \text{(c)} & p_1 = 0 \\ \text{(d)} & p_2 = p_3. \end{array} \tag{37.29}$$

For each of several values of L_p for the bcc lattice, the number $N_{\mathbf{k}}$ of **k** vectors in $\frac{1}{48}$ of the zone and the sum of $X_{\mathbf{k}}$ for these vectors are listed in Table 37.

The $\frac{1}{24}$ zone for hexagonal is shown in Figure 57, and the **k** vectors in this zone portion are defined in terms of p_1, p_2, p_3 by (37.18)–(37.20). The corresponding weighting factors $X_{\mathbf{k}}$ may be generated by carrying out the

Table 37. The number $N_{\mathbf{k}}$ of **k** vectors in $\frac{1}{48}$ of the zone, and the sum of the weighting factors, for the bcc lattice. The point $\mathbf{k} = 0$ is not counted here.

L_p	$N_{\mathbf{k}}$	$\Sigma_{\mathbf{k}}^{*} X_{\mathbf{k}}$
4	13	2.6458333 . . .
6	29	8.9791666 . . .
8	54	21.3125
12	139	71.9791666 . . .
16	284	170.6458333 . . .
20	505	333.3125
24	818	575.9791666 . . .

following tests, in sequence.

Test 1. If $p_1 = p_2 = p_3 = 0, \quad X_{\mathbf{k}} = \frac{1}{24}$.

Test 2. If test 1 is not satisfied:

If $p_1 = p_2 = 0, \quad p_3 \neq L_p, \quad X_{\mathbf{k}} = \frac{1}{12}$.

If $p_1 = p_2 = 0, \quad p_3 = L_p, \quad X_{\mathbf{k}} = \frac{1}{24}$.

Test 3. If neither test 1 nor test 2 is satisfied:

If $p_1 = p_2 = \frac{2}{3}L_p, \quad p_3 \neq 0$ and $p_3 \neq L_p, \quad X_{\mathbf{k}} = \frac{1}{6}$.

If $p_1 = p_2 = \frac{2}{3}L_p, \quad p_3 = 0$ or $p_3 = L_p, \quad X_{\mathbf{k}} = \frac{1}{12}$.

Test 4. If none of tests 1, 2, or 3 is satisfied, set $X_{\mathbf{k}} = 1$ and then multiply by $\frac{1}{2}$ for each of the following five tests which is satisfied.

(a) $p_1 + 2p_2 = 2L_p$
(b) $p_3 = L_p$
(c) $p_3 = 0$ (37.30)
(d) $p_1 = p_2$
(e) $p_1 = 0$.

For each of several values of L_p for the hexagonal lattice, the number $N_{\mathbf{k}}$ of **k** vectors in $\frac{1}{24}$ of the zone and the sum of $X_{\mathbf{k}}$ for these vectors are listed in Table 38.

When Brillouin zone sums are to be evaluated, it is useful to study the convergence of the sums as the number of **k** vectors counted in the sums is increased, that is, as L_p is increased. We have studied the convergence of many types of zone sums for several simple lattices, based on various central potential and pseudopotential models. Our experience has been that the accuracy of a given Brillouin zone sum depends little on the lattice structure

Table 38. The number $N_{\mathbf{k}}$ of **k** vectors in $\frac{1}{24}$ of the zone, and the sum of the weighting factors, for the simple hexagonal (or hcp) lattice. The point $\mathbf{k} = 0$ is included here.

L_p	$N_{\mathbf{k}}$	$\Sigma_{\mathbf{k}}^* X_{\mathbf{k}}$
4	50	21.333 . . .
6	133	72
8	270	170.666 . . .
9	370	243

or the model potential, and depends mainly on the function summed and the number of points counted. Some convergence results for $\langle \omega_\kappa^2 \rangle$ are given in Table 39, where we list

$$\Delta\langle \omega_\kappa^2 \rangle = \frac{\text{accurate } \langle \omega_\kappa^2 \rangle - \text{zone sum of } \langle \omega_\kappa^2 \rangle}{\text{accurate } \langle \omega_\kappa^2 \rangle}, \tag{37.31}$$

and where the accurate $\langle \omega_\kappa^2 \rangle$ was calculated from a simple trace equation. In most cases the accurate result is not known, but the accuracy of a sum can be estimated from the sequence of results for increasing L_p. We note that the values of $\Delta\langle \omega_\kappa^2 \rangle$ listed in Table 39 for various L_p are approximately equal to $(3\ \Sigma_{\mathbf{k}}^* X_{\mathbf{k}})^{-1}$, as is expected.

Finally we consider Brillouin zone sums of functions which are *not* the same for each of the equivalent **k** vectors related by point operations of the

Table 39. Accuracy of Brillouin zone sums of ω_κ^2 for the fcc lattice with a Lennard–Jones 12,8 potential, presented in the form of the error $\Delta\langle \omega_\kappa^2 \rangle$ of the direct zone sum, as defined by (37.31).

L_p	$\Delta\langle \omega_\kappa^2 \rangle$
12	−0.00187
14	−0.00137
16	−0.00104
20	−0.00065

inverse lattice. An example is the function $J_{ij}(0,\mathrm{N})$ defined by (19.23) for a primitive lattice:

$$J_{ij}(0,\mathrm{N}) = N_0^{-1} \sum_{\mathbf{k}} E_{ij}(\mathbf{k})\cos[\mathbf{k} \cdot \mathbf{R}(\mathrm{N})]. \tag{37.32}$$

Consider an fcc or bcc lattice, and take a given phonon wave vector $\mathbf{k}_1$ in the representative $\frac{1}{48}$ portion of the zone. It is a simple matter to generate the 48 equivalent vectors $\mathbf{k}_\alpha$, $\alpha = 1, 2, \cdots, 48$, find the transformation properties of $E_{ij}(\mathbf{k})\cos[\mathbf{k} \cdot \mathbf{R}(\mathrm{N})]$, and average this function over α to find

$$f_{ij}(\mathbf{k}_1,\mathbf{R}(\mathrm{N})) = \tfrac{1}{48} \sum_{\alpha} E_{ij}(\mathbf{k}_\alpha)\cos[\mathbf{k}_\alpha \cdot \mathbf{R}(\mathrm{N})]. \tag{37.33}$$

Then $J_{ij}(0,\mathrm{N})$ may be calculated from the usual expression for a partial zone average:

$$J_{ij}(0,\mathrm{N}) = \frac{\sum_{\mathbf{k}}^{*} X_{\mathbf{k}} f_{ij}(\mathbf{k},\mathbf{R}(\mathrm{N}))}{\sum_{\mathbf{k}}^{*} X_{\mathbf{k}}}. \tag{37.34}$$

In most cases of practical interest the transformation properties of the summand are obvious, and it is even possible to write an analytic expression for the average over equivalent **k** vectors. It is straightforward, for example, to write $f_{ij}(\mathbf{k},\mathbf{R}(\mathrm{N}))$ for fcc and bcc lattices, and thus avoid having to evaluate the sum (37.33) for each **k** by computer. In complicated cases, the desired transformation rules can always be obtained from the transformation properties of the phonon eigenvectors, which are discussed by Maradudin and Vosko.*

Low-Temperature Integrals

The temperature-dependent parts of thermodynamic functions are derived in Section 16 as phonon sums involving the statistical average phonon occupation numbers $\bar{n}_\kappa$. Consider, for example, a sum such as $\Sigma_\kappa f(\omega_\kappa)\bar{n}_\kappa$; because of the factor $\bar{n}_\kappa$, the only terms which contribute significantly to this sum are those for which $\hbar\omega_\kappa \lesssim KT$. If the sum is carried out for various temperatures, but for a fixed number of **k** vectors in the Brillouin zone, then as T is lowered there comes a temperature at which the calculated value of the sum is meaningless, since $\hbar\omega_\kappa > KT$ for all ω_κ. An accurate value of the sum can be determined for lower and lower temperatures by choosing continually finer grids for the **k** vectors, but the value at $T = 0$ can never be calculated in principle from a Brillouin zone sum. In any case where it is desired to calculate the temperature-dependence of a thermodynamic function at low temperatures, one should use the low-temperature theory of Section 18

* A. A. Maradudin and S. H. Vosko, *Rev. Mod. Phys.* **40**, 1 (1968).

and carry out an accurate calculation of the $T = 0$ Debye temperature or its strain derivatives. It is always poor practice to try to determine Θ_0, for example, by some sort of extrapolation procedure based on Brillouin zone sums (or frequency integrals) evaluated at finite temperatures.

Suppose it is desired to calculate the angle average $\langle [C(\hat{\mathbf{k}}s)]^{-3} \rangle$ defined by (18.30). Recall that $C(\hat{\mathbf{k}}s)$ is the $T = 0$ sound velocity for branch s in direction $\hat{\mathbf{k}}$, and $(M_C/V_C)[C(\hat{\mathbf{k}}s)]^2$ for $s = 1,2,3$ are the eigenvalues of the propagation matrix $L_{ij}(\hat{\mathbf{k}})$ defined by (3.23). We presume that the propagation coefficients A_{ijkl} at $T = 0$ are available, either experimental values or the theoretical results in any desired approximation, and describe the computation of $\langle [C(\hat{\mathbf{k}}s)]^{-3} \rangle$ for the most general case of arbitrary crystal structure and arbitrary initial applied stress. Of course when the stress is isotropic pressure, the A_{ijkl} may be replaced by B_{ijkl} in the propagation matrix, and at zero stress the coefficients become C_{ijkl}.

A simple but adequate procedure for computing the angle integral is as follows. For a given positive integer L_t, generate the integers t and p according to

$$t = 0,1, \cdots, L_t - 1; \qquad p = 0,1, \cdots, 2L_t - 1. \tag{37.35}$$

For each t and p, the angles θ and φ are given by

$$\theta = \pi t/L_t, \qquad \varphi = \pi p/L_t, \tag{37.36}$$

and components of the dimensionless unit vector $\hat{\mathbf{k}}$ are

$$\hat{k}_x = \sin\theta\cos\varphi, \qquad \hat{k}_y = \sin\theta\sin\varphi, \qquad \hat{k}_z = \cos\theta. \tag{37.37}$$

Now for each t and p, compute the propagation matrix elements

$$L_{ik}(\hat{\mathbf{k}}) = \sum_{jl} A_{ijkl}\hat{k}_j\hat{k}_l, \tag{37.38}$$

diagonalize the matrix to get the three eigenvalues $(M_C/V_C)[C(\hat{\mathbf{k}}s)]^2$, and compute $C(\hat{\mathbf{k}}s)$ and $[C(\hat{\mathbf{k}}s)]^{-3}$. Finally compute the angle average from the equation

$$\langle [C(\hat{\mathbf{k}}s)]^{-3} \rangle = \frac{\sum_{tp} \sin\theta \, \frac{1}{3} \sum_{s} [C(\hat{\mathbf{k}}s)]^{-3}}{\sum_{tp} \sin\theta}. \tag{37.39}$$

There are, of course, more sophisticated procedures for carrying out such angle integrals. In our experience the accuracy of the summation procedure described by (37.39) is of order 1 part in L_t^2; more precisely, an accuracy of 0.1% is generally obtained for $L_t = 50$.

Appendix 2

EXPERIMENTAL DATA

38. TABLES OF DATA

Values of physical constants, taken from current standard references, are listed in Table 40. Of the large amount of experimental data we have collected and analyzed, some selected results are listed in Table 41 for 18 common materials. These results represent smooth curves of the data, plotted as functions of the temperature, and in some cases represent the average of several measurements which appear to be of equal reliability. The references for these data are numbered in Table 42.

Table 40. Values of physical constants, and some conversions.

$N_0 = 6.02252\ (10^{23}/\text{mole})$	Avogadro's number
$K = 1.38054\ (10^{-16}\ \text{erg}/°\text{K})$	Boltzmann constant
$N_0K = 8.31434\ (10^{7}\ \text{erg/mole}\ °\text{K})$	gas constant
$\hbar = 1.054494\ (10^{-27}\ \text{erg sec})$	Planck constant
$u = 1.66043\ (10^{-24}\ \text{g/amu})$	atomic mass unit
$m = 9.10908\ (10^{-28}\ \text{g})$	electron rest mass
$c = 2.997925\ (10^{10}\ \text{cm/sec})$	speed of light
$a_0 = \hbar^2/me^2 = 0.529167\ (10^{-8}\ \text{cm})$	Bohr radius
$1\ \text{eV} = 1.60207\ (10^{-12}\ \text{erg})$	
$1\ \text{cal} = 4.1858\ (10^{7}\ \text{erg})$	
$1\ \text{Ry} = e^2/2a_0 = 2.17971\ (10^{-11}\ \text{erg})$	
$1\ \text{kcal/mole} = 6.9502\ (10^{-14}\ \text{erg/molecule})$	
$= 3.1886\ (10^{-3}\ \text{Ry/molecule})$	
$\hbar/K = 7.63827\ (10^{-12}\ \text{sec}\ °\text{K})$	

Table 41. Selected values of experimentally determined thermodynamic functions at various temperatures and at zero pressure. The units are: T(°K), β(10^{-5} °K^{-1}), B_T(10^{10} dyne/cm^2) and C_P(cal/mole °K).

T	β	B_T	C_P	T	β	B_T	C_P	T	β	B_T	C_P
		Na				K				Pb	
20		7.41	0.935	20	5.7	3.67	2.345	20	3.30	48.7	2.630
40		7.40	2.971	40	10.6	3.65	4.494	40	5.97	48.2	4.675
60	8.2	7.36	4.223	60	14.1	3.63	5.307	60	6.80	47.7	5.360
80	11.4	7.29	4.958	80	16.1	3.60	5.685	80	7.22	47.3	5.661
100	13.7	7.21_5	5.368	100	17.4	3.56	5.890	100	7.50	46.8	5.839
120	15.5	7.12	5.623	120	18.3	3.52	6.038	120	7.70	46.3	5.939
140	16.9	7.02	5.830	140	19.0	3.47	6.157	140	7.88	45.8	6.010
160	18.0	6.92	5.983	160	19.6	3.42	6.263	160	8.02	45.3	6.070
180	18.8	6.81	6.106	180	20.0	3.36	6.353	180	8.13	44.8	6.130
200	19.5	6.69	6.213	200	20.3	3.31	6.441	200	8.24	44.3	6.184
220	19.9_5	6.59	6.322	220	20.7	3.25	6.538	220	8.34	43.8	6.235
240	20.3	6.48	6.409	240	21.1	3.19	6.644	240	8.44	43.3	6.285
260	20.5_5	6.37	6.515	260	21.6	3.14	6.754	260	8.53	42.8	6.333
280	20.7_5	6.27	6.637	280	22.3	3.08	6.893	280	8.60	42.3	6.380
300	20.9	6.17	6.781	300	23.6	3.03	7.071	300	8.67	41.8	6.422

Table 41 *(continued)*

T	β	B_T	C_P	T	β	B_T	C_P	T	β	B_T	C_P
Cu				Ag				Au			
20	0.03	142.0	0.12	20		108.7	0.40	20		180.2	0.748
40	0.69	141.8	0.91	40	1.83	108.4	2.01	40	1.90	179.7	2.692
60	1.62	141.5	2.08	60	3.13	107.9	3.43	60	2.74	178.8	3.963
80	2.48	141.0	3.11	80	3.84	107.3	4.28	80	3.21	177.9	4.669
100	3.15	140.5	3.86	100	4.29	106.6	4.82	100	3.45	176.9	5.097
120	3.61	139.9	4.37	120	4.61	106.0	5.16	120	3.64	175.8	5.351
140	3.95	139.2	4.75	140	4.85	105.3	5.39	140	3.79	174.8	5.530
160	4.19	138.5	5.04	160	5.03	104.6	5.57	160	3.89	173.7	5.657
180	4.38	137.8	5.25	180	5.18	103.9	5.70	180	3.96	172.7	5.746
200	4.53	137.1	5.40	200	5.31	103.2	5.80	200	4.02	171.7	5.812
220	4.66	136.4	5.53	220	5.43	102.4	5.88	220	4.08	170.6	5.869
240	4.77	135.6	5.63	240	5.53	101.7	5.95	240	4.12	169.6	5.916
260	4.88	134.8	5.71	260	5.62	101.0	6.02	260	4.16	168.6	5.963
280	4.97	134.0	5.78	280	5.70	100.3	6.06	280	4.20	167.6	6.005
300	5.04	133.2	5.86	300	5.76	99.6	6.08	300	4.23	166.6	6.048

Table 41 (*continued*)

T	β	B_T	C_P	T	β	B_T	C_P	T	β	B_T	C_P
Th				Zn				Cd			
20		58.1	1.106	20	0.27	66.0	0.40_5	20		52.4	1.240
40		58.0	3.355	40	2.20	65.6	1.95	40	4.79	52.0	3.180
60	2.14	58.0	4.529	60	5.20	65.1	3.25	60	6.63	51.4	4.284
80	2.44	57.9	5.132	80	6.36	64.5	4.03	80	7.47	50.9	4.920
100	2.61	57.8	5.482	100	7.08	63.9	4.57_5	100	8.01	50.4	5.284
120	2.73	57.7	5.717	120	7.59	63.3	4.98_5	120	8.38	49.9	5.519
140	2.82	57.6	5.895	140	7.96	62.8	5.26_5	140	8.64	49.4	5.682
160	2.90	57.5	6.026	160	8.21	62.2	5.46_5	160	8.82	48.9	5.798
180	2.97	57.4	6.133	180	8.38	61.7	5.62_5	180	8.96	48.4	5.886
200	3.03	57.4	6.217	200	8.52	61.1	5.73	200	9.06	47.9	5.960
240	3.16	57.2	6.362	220	8.63	60.5	5.82_5	220	9.16	47.4	6.017
280	3.27	57.1	6.483	240	8.72	60.0	5.90_5	240	9.24	46.9	6.070
320	3.38	56.9	6.591	260	8.79	59.5	5.96_5	260	9.31	46.4	6.122
360	3.47	56.7	6.699	280	8.85	58.9	6.03	280	9.38	45.9	6.171
400	3.56	56.6	6.809	300	8.91	58.4	6.07_5	300	9.45	45.4	6.224
				400	9.15	55.9	6.31				
				500	9.48	53.4	6.55				
				600	10.26	50.7	6.79				

Table 41 *(continued)*

T	β	B_T	C_P	T	β	B_T	C_P	T	β	B_T	C_P
Al				Ti				Zr			
25	0.15	79.4	0.11	25		110.1	0.153	25		97.2	0.489
50	1.14	79.2	0.91_5	50	0.33	110.0	1.135	50		97.1	2.212
75	2.46	78.9	2.07_5	75	0.90	109.7	2.39	75	0.81	97.0	3.609
100	3.66	78.4	3.11_5	100	1.33	109.4	3.43_5	100	1.17	96.7	4.460
125	4.55	77.8	3.87	125	1.61	109.1	4.15	125	1.31	96.4	4.986
150	5.16	77.1	4.41	150	1.83	108.8	4.64_5	150	1.41	96.2	5.299
175	5.62	76.4	4.83	175	2.01	108.4	5.04	175	1.48	96.0	5.525
200	6.00	75.6	5.13_5	200	2.16	108.0	5.32	200	1.53	95.7	5.691
225	6.31	74.9	5.36_5	225	2.28	107.6	5.53_5	225	1.56	95.5	5.822
250	6.57	74.2	5.55	250	2.39	107.2	5.71	250	1.59	95.2	5.908
275	6.78	73.5	5.69_5	275	2.48	106.8	5.84_5	275	1.61	95.0	5.988
300	6.96	72.8	5.82_5	300	2.55	106.4	5.98	300	1.63	94.7	6.066
350	7.23	71.3	5.98	350	2.68	105.5	6.17	350	1.67	94.2	6.22
400	7.47	69.6	6.12	400	2.79	104.7	6.30	400	1.72	93.7	6.36
450	7.68	67.8	6.27	450	2.88	103.8	6.42	450	1.78	93.2	6.49_5
500	7.92	65.8	6.42	500	2.95	103.0	6.54	500	1.85	92.6	6.63
550	8.18	63.5	6.57	550	3.01	102.2	6.66	550	1.92	92.1	6.76
600	8.49	61.0	6.72	600	3.06	101.3	6.78	600	2.01	91.5	6.88
				700	3.17	99.6	7.02	700	2.22	90.2	7.12
				800	3.28	97.9	7.26	800	2.46	88.6	7.34

Table 41 *(continued)*

T	β	B_T	C_P	T	β	B_T	C_P	T	β	B_T	C_P
NaCl				KCl				KBr			
25	0.40	26.6	0.65	25	0.50	19.7	1.297	25	1.18	17.6	2.544
50	2.93	26.5	3.62	50	3.45	19.6	5.057	50	4.98	17.5	6.788
75	5.70	26.3	6.46	75	6.05	19.5	7.793	75	7.44	17.2	9.107
100	7.58	26.1	8.34	100	7.60	19.3	9.380	100	8.78	16.9	10.305
125	8.78	25.8	9.50	125	8.58	19.0	10.305	125	9.49	16.6	10.975
150	9.60	25.5	10.28	150	9.23	18.8	10.905	150	9.95	16.3	11.410
175	10.20	25.2	10.82	175	9.70	18.5	11.300	175	10.31	16.0	11.710
200	10.68	24.9	11.20	200	10.05	18.3	11.575	200	10.61	15.7	11.915
250	11.40	24.2	11.74	250	10.63	17.7	11.970	225	10.89	15.4	12.085
300	11.95	23.5	12.10	300	11.08	17.1	12.19_5	250	11.15	15.1	12.245
350	12.38	22.8	12.34_5	350	11.50	16.4	12.33_5	275	11.43	14.8	12.382
400	12.81	21.9	12.54	400	11.88	15.8	12.45	300	11.74	14.5	12.488
450	13.24	20.9	12.73_5	450	12.25	15.2	12.61				
500	13.71	19.8	12.93	500	12.65	14.6	12.80				
550	14.18	18.7	13.12_5	550	13.08	14.0	13.00_5				
600	14.70	17.7	13.32	600	13.53	13.4	13.22_5				

Table 41 *(continued)*

T	β	B_T	C_P	T	β	B_T	C_P	T	β	B_T	C_P
	KI				Ar				Kr		
25	2.51	12.8	3.794	4	0.3	2.68	0.0432	4	1.5	3.45	0.0964
50	6.51	12.8	7.850	6	2.4	2.67	0.1659	6	5.1	3.45	0.3721
75	8.84	12.7	9.795	8	7.2	2.66	0.417	8	10.3	3.44	0.857
100	10.06	12.6	10.755	10	13.8	2.65	0.780	10	16.8	3.43	1.418
125	10.60	12.5	11.295	15	34.8	2.60	1.940	15	33.4	3.40	2.798
150	10.90	12.3	11.650	20	55.5	2.53	2.990	20	47.1	3.37	3.817
175	11.19	12.2	11.885	25	71.4	2.46	3.828	25	57.7	3.31	4.516
200	11.48	12.1	12.070	30	85.4	2.37	4.463	30	65.4	3.23	4.990
225	11.76	11.9	12.235	35	96.9	2.28	4.983	35	71.2	3.13	5.345
250	12.04	11.7	12.370	40	106.8	2.18	5.387	40	75.5	3.02	5.612
275	12.32	11.6	12.508	50	125.4	1.98	6.006	50	82.7	2.79	5.978
				60	145.0	1.75	6.528	60	90.5	2.55	6.296
				70	169.9	1.51	7.100	70	98.7	2.30	6.569
				80	203.1	1.27	7.928	80	107.3	2.05	6.824

Table 42. Reference numbers for the experimental data tabulated in Table 41. The references are listed in Section 39.

Material	β	B_T or B_S	C_P
Na	43, 44	33, 56, 57	66, 67
K	45, 46	17, 26, 45	68, 69
Cu	44	58	70
Ag	44	19	70
Au	44	19	70
Al	44, 47	20, 59	70, 71, 72
Pb	44	21	73
Th	44	60	74, 75
Zn	44	61	70, 71
Cd	44	62	70, 71, 76
Ti	44	63	70, 71
Zr	44	63	71, 77
NaCl	48, 49, 50	22, 31, 64	78, 79
KCl	48, 51, 52, 53	23, 31, 65	79, 80
KBr	48, 53	24	80
KI	53	23	80
Ar	54	25	14, 81
Kr	55	55	14, 82

39. REFERENCES FOR EXPERIMENTAL DATA

Here are the references for the experimental data presented in Chapter 7 and in Section 38.

1. D. L. Martin, *Phys. Rev.* **139,** A150 (1965).
2. D. L. Martin, *Phys. Rev.* **141,** 576 (1966).
3. L. C. Clune and B. A. Green, Jr., *Phys. Rev.* **144,** 525 (1966).
4. M. Dixon, F. E. Hoare, T. M. Holden, and D. E. Moody, *Proc. Roy. Soc. (London)* **A285,** 561 (1965).
5. B. A. Green, Jr., and H. V. Culbert, *Phys. Rev.* **137,** A1168 (1965).
6. T. A. Will and B. A. Green, Jr., *Phys. Rev.* **150,** 519 (1966).
7. J. E. Zimmerman and L. T. Crane, *Phys. Rev.* **126,** 513 (1962).
8. N. E. Phillips, *Phys. Rev.* **114,** 676 (1959).
9. B. J. C. van der Hoeven, Jr., and P. H. Keesom, *Phys. Rev.* **137,** A103 (1965).
10. P. L. Smith and N. M. Wolcott, *Suppl. Bull. Intern. Inst. Refrig.* **3,** 283 (1955).

11. D. L. Martin, *Proc. Phys. Soc. (London)* **78,** 1482 (1961).
12. N. M. Wolcott, *Phil. Mag.* **2,** 1246 (1957).
13. T. H. K. Barron, W. T. Berg, and J. A. Morrison, *Proc. Roy. Soc. (London)* **A242,** 478 (1957).
14. L. Finegold and N. E. Phillips, *Phys. Rev.* **177,** 1383 (1969).
15. J. G. Collins and G. K. White, in *Progress in Low Temperature Physics*, edited by C. J. Gorter, North-Holland Publishing Co., Amsterdam, 1964, Vol. IV, p. 450.
16. K. Andres, in *Proceedings of the Eighth International Conference on Low Temperature Physics*, edited by R. O. Davies, Butterworths, London, 1963, p. 397.
17. W. R. Marquardt and J. Trivisonno, *J. Phys. Chem. Solids* **26,** 273 (1965).
18. W. C. Overton, Jr., and J. Gaffney, *Phys. Rev.* **98,** 969 (1955).
19. J. R. Neighbours and G. A. Alers, *Phys. Rev.* **111,** 707 (1958).
20. G. N. Kamm and G. A. Alers, *J. Appl. Phys.* **35,** 327 (1964).
21. D. L. Waldorf and G. A. Alers, *J. Appl. Phys.* **33,** 3266 (1962).
22. J. T. Lewis, A. Lehoczky, and C. V. Briscoe, *Phys. Rev.* **161,** 877 (1967).
23. M. H. Norwood and C. V. Briscoe, *Phys. Rev.* **112,** 45 (1958).
24. J. K. Galt, *Phys. Rev.* **73,** 1460 (1948).
25. G. J. Keeler and D. N. Batchelder, *J. Phys. C: Solid State Phys.* **3,** 510 (1970).
26. P. A. Smith and C. S. Smith, *J. Phys. Chem. Solids* **26,** 279 (1965).
27. K. Salama and G. A. Alers, *Phys. Rev.* **161,** 673 (1967).
28. W. B. Daniels and C. S. Smith, *Phys. Rev.* **111,** 713 (1958).
29. P. S. Ho and A. L. Ruoff, *J. Appl. Phys.* **40,** 3151 (1969).
30. R. A. Miller and D. E. Schuele, *J. Phys. Chem. Solids* **30,** 589 (1969).
31. R. A. Bartels and D. E. Schuele, *J. Phys. Chem. Solids* **26,** 537 (1965).
32. P. J. Reddy and A. L. Ruoff, in *Physics of Solids at High Pressures*, edited by C. T. Tomizuka and R. M. Emrick, Academic Press, Inc., New York, 1965, p. 510.
33. W. B. Daniels, *Phys. Rev.* **119,** 1246 (1960).
34. P. W. Bridgman, *The Physics of High Pressures*, G. Bell and Sons Ltd., London, 1945, p. 160. Calibration corrections for Bridgman's data are provided by J. C. Slater, *Phys. Rev.* **57,** 744 (1940), and by C. A. Rotter and C. S. Smith, *J. Phys. Chem. Solids* **27,** 267 (1966).
35. J. A. Corll, *Office of Naval Research Technical Report No. 6*, Case-Western Reserve University, 1962.
36. A. O. Urvas, D. L. Losee, and R. O. Simmons, *J. Phys. Chem. Solids*, **28,** 2269 (1967). We have chosen the values of $(\partial B_T/\partial P)_T$ for Ar and Kr which these authors obtained by fitting the low-pressure data of J. W. Stewart, *Phys. Rev.* **97,** 578 (1955).

37. The value of $\langle \Omega_\kappa^2 \rangle$ for Na was given by D. L. Martin, reference 1; this value was based on the neutron-scattering results of A. D. B. Woods, B. N. Brockhouse, R. H. March, A. T. Stewart, and R. Bowers, *Phys. Rev.* **128,** 1112 (1962).
38. R. A. Cowley, A. D. B. Woods, and G. Dolling, *Phys. Rev.* **150,** 487 (1966).
39. E. C. Svensson, B. N. Brockhouse, and J. M. Rowe, *Phys. Rev.* **155,** 619 (1967).
40. R. M. Nicklow, G. Gilat, H. G. Smith, L. J. Raubenheimer, and M. K. Wilkinson, *Phys. Rev.* **164,** 922 (1967).
41. G. Gilat and R. M. Nicklow, *Phys. Rev.* **143,** 487 (1966). The value of $\langle \Omega_\kappa^2 \rangle$ for Al given by these authors was based on the neutron-scattering measurements of R. Stedman, L. Almqvist, and G. Nilsson, *Phys. Rev.* **162,** 549 (1967).
42. E. R. Cowley and R. A. Cowley, *Proc. Roy. Soc.* (*London*) **A292,** 209 (1966). The value of $\langle \Omega_\kappa^2 \rangle$ for KBr was based on the neutron-scattering measurements of A. D. B. Woods, B. N. Brockhouse, R. A. Cowley, and W. Cochran, *Phys. Rev.* **131,** 1025 (1963).
43. S. Siegel and S. L. Quimby, *Phys. Rev.* **54,** 76 (1938).
44. R. K. Kirby, in *American Institute of Physics Handbook*, edited by D. E. Gray, McGraw-Hill Book Co., Inc., New York, 1963, Second Ed., p. 4–64.
45. C. E. Monfort and C. A. Swenson, *J. Phys. Chem. Solids* **26,** 291 (1965). We plotted V *vs* T for K from the data of this reference and of reference 46, and then obtained β from the slope of this curve.
46. R. H. Stokes, *J. Phys. Chem. Solids* **27,** 51 (1966).
47. C. P. Abbiss, E. Huzan, and G. O. Jones, in *Proceedings of the Seventh International Conference on Low Temperature Physics*, edited by G. M. Graham and A. C. Hollis Hallett, University of Toronto Press, 1961, p. 688; D. B. Fraser and A. C. Hollis Hallett, in *Proceedings of the Seventh International Conference on Low Temperature Physics*, University of Toronto Press, 1961, p. 689.
48. P. P. M. Meincke and G. M. Graham, *Can. J. Phys.* **43,** 1853 (1965).
49. T. Rubin, H. L. Johnston, and H. W. Altman, *J. Phys. Chem.* **65,** 65 (1961).
50. F. D. Enck and J. G. Dommel, *J. Appl. Phys.* **36,** 839 (1965).
51. G. K. White, *Phil. Mag.* **6,** 1425 (1961).
52. T. Rubin, H. L. Johnston, and H. W. Altman, *J. Phys. Chem.* **66,** 948 (1962).
53. B. Yates and C. H. Panter, *Proc. Phys. Soc.* (*London*) **80,** 373 (1962).
54. O. G. Peterson, D. N. Batchelder, and R. O. Simmons, *Phys. Rev.* **150,** 703 (1966).

55. D. L. Losee and R. O. Simmons, *Phys. Rev.* **172,** 944 (1968).
56. M. E. Diederich and J. Trivisonno, *J. Phys. Chem. Solids* **27,** 637 (1966).
57. R. I. Beecroft and C. A. Swenson, *J. Phys. Chem. Solids* **18,** 329 (1961). We have used the revisions given in reference 45.
58. W. C. Overton, Jr., *J. Chem. Phys.* **37,** 2975 (1962).
59. P. M. Sutton, *Phys. Rev.* **91,** 816 (1953). These results for k_T for Al are too small by nearly 7% at 300°K, as compared to the results of reference 20. We used Sutton's data as a guide in extrapolating k_T to temperatures above 300°K, since they are the only high-temperature results we have for Al.
60. P. E. Armstrong, O. N. Carlson, and J. F. Smith, *J. Appl. Phys.* **30,** 36 (1959).
61. G. A. Alers and J. R. Neighbours, *J. Phys. Chem. Solids* **7,** 58 (1958).
62. C. W. Garland and J. Silverman, *Phys. Rev.* **119,** 1218 (1960). These results for k_S for Cd were multiplied by 1.002^6 to correct for an error in their density determination, as pointed out by J. A. Corll, reference 35.
63. E. S. Fisher and C. J. Renken, *Phys. Rev.* **135,** A482 (1964).
64. L. Hunter and S. Siegel, *Phys. Rev.* **61,** 84 (1942).
65. F. D. Enck, *Phys. Rev.* **119,** 1873 (1960). These results for k_S for KCl are about 9% too large at 300°K.
66. J. D. Filby and D. L. Martin, *Proc. Roy. Soc.* (*London*) **A276,** 187 (1963). These authors estimated C_P for bcc Na for the temperatures 3.5–30°K.
67. D. L. Martin, *Proc. Roy. Soc.* (*London*) **A254,** 433 (1960).
68. J. D. Filby and D. L. Martin, *Proc. Roy. Soc.* (*London*) **A284,** 83 (1965).
69. C. A. Krier, R. S. Craig, and W. E. Wallace, *J. Phys. Chem.* **61,** 522 (1957).
70. R. J. Corruccini and J. J. Gniewek, *National Bureau of Standards Monograph 21*, U.S. Government Printing Office, Washington, D.C., 1960.
71. R. Hultgren, R. L. Orr, P. D. Anderson, and K. K. Kelley, *Selected Values of Thermodynamic Properties of Metals and Alloys*, John Wiley & Sons, Inc., New York, 1963.
72. W. T. Berg, *Phys. Rev.* **167,** 583 (1968).
73. P. F. Meads, W. R. Forsythe, and W. F. Giauque, *J. Am. Chem. Soc.* **63,** 1902 (1941).
74. M. Griffel and R. E. Skochdopole, *J. Am. Chem. Soc.* **75,** 5250 (1953).
75. D. C. Wallace, *Phys. Rev.* **120,** 84 (1960).
76. R. S. Craig, C. A. Krier, L. W. Coffer, E. A. Bates, and W. E. Wallace, *J. Am. Chem. Soc.* **76,** 238 (1954).
77. G. B. Skinner and H. L. Johnston, *J. Am. Chem. Soc.* **73,** 4549 (1951).
78. T. H. K. Barron, A. J. Leadbetter, and J. A. Morrison, *Proc. Roy. Soc.* (*London*) **A279,** 62 (1964).

79. K. K. Kelley, *U.S. Bureau of Mines Bulletin 584*, U.S. Government Printing Office, Washington, D.C., 1960.
80. W. T. Berg and J. A. Morrison, *Proc. Roy. Soc.* (*London*) **A242,** 467 (1957).
81. P. Flubacher, A. J. Leadbetter, and J. A. Morrison, *Proc. Phys. Soc.* (*London*) **78,** 1449 (1961).
82. R. H. Beaumont, H. Chihara, and J. A. Morrison, *Proc. Phys. Soc.* (*London*) **78,** 1462 (1961).

General References

ON THERMOELASTICITY

D. C. Wallace, "Thermoelastic Theory of Stressed Crystals and Higher-Order Elastic Constants," in *Solid State Physics*, edited by H. Ehrenreich, F. Seitz, and D. Turnbull, Academic Press, Inc., New York, 1970, Vol. 25, p. 301.

ON LATTICE DYNAMICS

M. Born and K. Huang, *Dynamical Theory of Crystal Lattices*, Clarendon Press, Oxford, 1954.

G. Leibfried, "Gittertheorie der mechanischen und thermischen Eigenschaften der Kristalle," in *Handbuch der Physik*, edited by S. Flügge, Springer-Verlag, Berlin, 1955, Vol. VII/1, p. 104.

G. Leibfried and W. Ludwig, Gleichgewichtsbedingungen in der Gittertheorie, *Z. Phys.* **160,** 80 (1960).

ON ELECTRONS AND ELECTRON-PHONON INTERACTIONS

G. V. Chester, The Theory of the Interaction of Electrons with Lattice Vibrations in Metals, *Advan. Phys.* **10,** 357 (1961).

L. J. Sham and J. M. Ziman, The Electron–Phonon Interaction, in *Solid State Physics*, edited by F. Seitz and D. Turnbull, Academic Press, Inc., New York, 1963, Vol. 15, p. 221.

C. Kittel, *Quantum Theory of Solids*, John Wiley & Sons, Inc., New York, 1963.

ON PSEUDOPOTENTIALS

W. A. Harrison, *Pseudopotentials in the Theory of Metals*, W. A. Benjamin Inc., New York, 1966.

INDEX